The Origin of the Life of a Human Being

Conception and the Female According to Ancient Indian Medical and Sexological Literature

RAHUL PETER DAS

MOTILAL BANARSIDASS PUBLISHERS
PRIVATE LIMITED • DELHI

First Edition: Delhi, 2003

ISBN: 81-208-1998-5

Also available at:
MOTILAL BANARSIDASS
41 U.A. Bungalow Road, Jawahar Nagar, Delhi 110 007
8 Mahalaxmi Chamber, 22 Bhulabhai Desai Road, Mumbai 400 026
236, 9th Main III Block, Jayanagar, Bangalore 560 011
120 Royapettah High Road, Mylapore, Chennai 600 004
Sanas Plaza, 1302 Baji Rao Road, Pune 411 002
8 Camac Street, Kolkata 700 017
Ashok Rajpath, Patna 800 004
Chowk, Varanasi 221 001

Printed in India
BY JAINENDRA PRAKASH JAIN AT SHRI JAINENDRA PRESS,
A-45 NARAINA, PHASE-I, NEW DELHI 110 028
AND PUBLISHED BY NARENDRA PRAKASH JAIN FOR
MOTILAL BANARSIDASS PUBLISHERS PRIVATE LIMITED,
BUNGALOW ROAD, DELHI 110 007

Meinen Eltern

বাবা-মাকে

FOREWORD

Critical and comprehensive studies of India's medical history top the list of urgent desiderata. Far too many books appear on the subject of Indian medicine, which are published simply because the title purports to address the topic in some form or another. A perusal of such books immediately shows that little of anything new is offered, and what is presented is frequently wrong.

Thankfully, Rahul Peter Das' *The Origin of the Life of a Human Being* is not a book of this type. Not only has Das penetrated deeply into Indian medical literature, but he has also examined the Sanskrit texts of Kāmaśāstra, Indian eroticism, and other relevant literature to reveal fully and completely how the ancient Indians understood the process of conception and birth.

Never before has an authoritative study of such breadth and depth been devoted to a specific problem of ancient Indian medicine. It utilises all the tools of philological research, while critically evaluating previous scholarship on the subject. It will definitely be an important and constant source of reference for the serious scholar of Indian medicine. We are pleased to offer it to the scholarly world as a volume in the series *Indian Medical Tradition*.

KENNETH G. ZYSK DOMINIK WUJASTYK

PREFACE

This study, several years in the making, in some respects quite aptly resembles the growth of a child from an embryo. The germ, a presentation of twenty minutes' duration at an international conference held at Hamburg in 1985, grew by constant accretion and revision until it was deemed worthy of being submitted as a *Habilitationsschrift* (post-doctoral thesis) to the University of Hamburg in 1992. But even after its acceptance in 1993 the process of addition and revision continued uninterruptedly until at last I decided that if I did not set myself a fixed cut-off date after which I would stop feeding it with new information, this baby of mine would keep on growing only in my computer and never make it to the printing press and the light of day. In accordance with this decision, new material chanced upon after the end of 1996 has rarely been included, though prior to publication indexes have been added to the whole, this taking several years due to other pressing duties.

But for the generous help accorded by several persons who provided me with or assisted me in procuring relevant literature, my study would never have taken the shape it finally has. Such unselfish assistance came especially from Ronald Eric Emmerick (†), Félix Raymond Erb, Ivo Fišer, Abhijit Ghosh, Peter Khoroche, Jinadasa Liyanaratne, Gerrit Jan Meulenbeld, Karin Preisendanz, Burkhard Quessel, Arion Roşu, Karam Tej Singh Sarao, Sven Thomas, Ursula Weisser, Albrecht Wezler, Beverly Whipple, Dominik Wujastyk and Mei Yang; to them, as well as to all those unnamed, I here tender my heartfelt thanks. I am also grateful to the Wellcome Trust for a travel grant which allowed me to consult some rare works in the collections of several libraries in London in 1987. Gerrit Jan Meulenbeld and Albrecht Wezler went through drafts of this study critically, making many pertinent and helpful comments; I am particularly indebted to them. Stefan Leder and Jürgen Paul checked my translations from Arabic texts and pronounced them to be *in Ordnung*; their unselfish help is much appreciated. My thanks are also due Messrs. Motilal Banarsidass not only for agreeing to publish my study, but also for waiting patiently for several years for its completion.

In this book I also touch upon the subjects of ejaculation and orgasm. Just before its completion I came across a New Age publication* — with interesting information, but hardly describing traditional Tantrism as it claims to do — which very decidedly makes a distinction between ejaculation and orgasm of the male.** What it means by the latter, though, is intense pleas-

---

* André van Lysebeth: *Tantra. The Cult of the Feminine*. York, Maine (Samuel Weiser) 1995. (Translated from the original French version of 1992 by André van Lysebeth and Lore Dobler.)

** See especially pp.297ff. I subsequently found out that this is a theory which seems to be

ure due to prolonged sexual activity on the very brink of ejaculation, which latter is, however, restrained. I am not sure whether terminologically this too qualifies as an orgasm. On the other hand, LADAS/WHIPPLE/PERRY, pp.153-159 too refer to male orgasm, even multiple, not necessarily linked to ejaculation. It is unclear to me whether what they are referring to is the same as what van Lysebeth describes. If not, then it still seems to me not to be very relevant to my study, as it is a technique which is described as having to be actively learnt; this can thus hardly pertain to the ordinary course of events.

The matter is different in the case of the female. According to STIFTER, pp.11;156, orgasm (in the sense in which the term is usually understood, not as understood by van Lysebeth) and ejaculation in this case may not coincide, though they seem (according to the evidence of the rest of the work mentioned) often, maybe usually, to do so, provided ejaculation takes place at all.* In any case, I have refrained from going through what I have written with a view to amending references to ejaculation and orgasm of the male and, in the given context more important, the female; should I inadvertently have mixed up my terminology, I beg to be pardoned.

Though I have a number of times compared my findings with those of other traditional medical systems, I was unable to really delve into any of them. I sincerely regret this, especially in the case of Tibetan medicine, which has very close parallels to the Indian notions examined here, as even a cursory glance shows, but also differences.**

In the course of my studies I have often been extremely frustrated by some of the secondary sources I consulted, especially by the methods of research these bear witness to. Even though I have striven to avoid remarks which are polemical or *ad personam*, it may thus be that such feelings at times shine through when I discuss such works. Or it may be that, since tastes and feelings vary, especially in matters of style, some of my remarks may strike some readers as being in this vein, even though not intended to be so. I rely on the reader's indulgence in such cases, should they occur.

— Rahul Peter Das

---

common in certain circles, especially those given to New Age esoterism.

* How unknown ejaculation of women still is, is shown by a modern author who calls a vivid description of it in a Victorian work a "morass of fantasy", adding that female ejaculation is "not attested by science". See Ian Gibson: *The Erotomaniac. The Secret Life of Henry Spencer Ashbee*. London (Faber & Faber) 2001, p.191.

** See, for instance, p.87 of Nawang Dakpa: "Certain Problems of Embryology According to the Tibetan Medical Tradition", *Aspects of Classical Tibetan Medicine*. Gangtok (Sikkim Research Institute of Tibetology) 1993. Bulletin of Tibetology, Special Volume of 1993, pp.79-95.

CONTENTS

Note :The sign ☞ refers to the discussion of the following word under that heading in Appendix II (pp.511ff.)

ABBREVIATIONS*

ĀĀS	*Āyurvedābdhisāra*
AH	*Aṣṭāṅgahṛdaya* with the commentaries of Aruṇadatta and Hemādri (Bombay 1939)
Āh	Āhāravidhi
AH1	*Aṣṭāṅgahṛdaya* with the commentary of Śrīdāsa (Trivandrum 1940-1962)
AH2	*Aṣṭāṅgahṛdaya* with the commentaries of Aruṇadatta, Candranandana and Hemādri (Bambaī V.S. 2013)
AH3	*Aṣṭāṅgahṛdaya* with the commentary of Śivadāsasena (Āgarā 1942)
AH4	*Aṣṭāṅgahṛdaya* with the Kairalī commentary (Koṭṭaya 1942)
AH5	*Aṣṭāṅgahṛdaya* with the commentary of Indu (Kottayam 1956-1978)
AH6	*Aṣṭāṅgahṛdaya* with the commentary of Parameśvara (Kottayam 1950-1963)
AIGR	Altindische Grammatik
AN	*Anaṅgaraṅga*
AÑJ	*Añjananidāna*
AS	*Aṣṭāṅgasaṅgraha* with the commentary of Indu (Trichur 1913-1926)
AS1	*Aṣṭāṅgasaṅgraha* (Mumbāpura Śaka 1810)
Asa	Asamānagotrīyaśārīrādhyāya
AV	*Atharvavedasaṃhitā* of the School of Śaunaka (Berlin 1924)
AV1	*Atharvavedasaṃhitā* of the School of Śaunaka (Hoshiarpur 1990)
ĀY	*Āyurvedasaukhya*
ĀYS	*Āyurvedasūtra*
ĀYV	*Āyurvedavijñāna*
BĀ	*Bābhravyakārikā* (with the commentary of Kāmarājadīkṣita), contained in *Kāmakuñjalatā* (eighth work, with own pagination)
BH	*Bhelasaṃhitā*
BHĀ	*Bhāvaprakāśa*
BHAI	*Bhaiṣajyaratnāvalī* (Banāras 1951)
BHAI1	*Bhaiṣajyaratnāvalī* (Dillī etc. 1988)
BṚSAṂ	*Bṛhatsaṃhitā*
BYT	*Bṛhadyogataraṅgiṇī*
CA	*Carakasaṃhitā* with the commentary of Cakrapāṇidatta (Bombay 1941)
CA1	*Carakasaṃhitā* with the commentaries of Cakrapāṇidatta and Gaṅgādhara (Kalikātā Śaka 1849-1855)
CA2	*Carakasaṃhitā* with the commentaries of Cakrapāṇidatta and Jajjaṭa/Jejjaṭa (Lavapura 1941)

* For full details on the works referred to in this list please refer to the **Bibliography** on pp.594-626. Works referred to by full title or author's name in this study are of course to be found there too, and not in this list. Abbreviations have as a rule not been marked by a period sign <.>; it is assumed that there will nevertheless be no difficulty in distinguishing between works referred to in an abbreviated form and those not referred to thus.

CA3	*Carakasaṃhitā* with Bengali translation (Kalikātā Beng. era 1311)
CA4	*Carakasaṃhitā* with Hindi, Gujarati and English translations (Jamnagar 1949)
CA5	*Carakasaṃhitā* (Calcutta 1877)
CA6	*Carakasaṃhitā* with English translation (Varanasi/Delhi 1983)
CART	*Carakatattvapradīpikā*
Ci	Cikitsāsthāna/Cikitsitasthāna
CIK	*Cikitsāsāramu*
CIKS	*Cikitsāsārasaṅgrahamu*
CONCEPT	The Concept of Jaṭharāgni and Dhātwagni in Indian Medicine
DR	*Dravyaguṇa*
EIM	Encyclopaedia of Indian Medicine
FH	*Firdaus al-ḥikma fī 't-ṭibb*
fol.	folio
GAD	*Gadanigraha*
Gar	Garbhāvakrāntiśārīrādhyāya
GARBH	*Garbhopaniṣad* with French translation (Paris 1976)
GARBH1	*Garbhopaniṣad* with English translation (Amsterdam 1965-66)
HĀ	*Hārītasaṃhitā* (Calcutta n.d.)
HĀ1	*Hārītasaṃhitā* (Pondichéry 1974)
HAṬH	*Haṭhayogapradīpikā*
HDA	Handwörterbuch des deutschen Aberglaubens
Hi	Hitāhitādhyāya
In	Indriyasthāna
IOL	Catalogue of the India Office Library
Jā	Jātisūtrīyaśārīrādhyāya
Ka	Kalpa(siddhi)sthāna
KĀ	*Kāśyapasaṃhitā*
KĀD	*Kādambarasvīkaraṇakārikā* of Rājarṣi Bharata, contained in *Kāmakuñjalatā* (fourth work, with own pagination)
KĀDS	*Kādambarasvīkaraṇasūtra* of Rājarṣi Purūravas (with commentary), contained in *Kāmakuñjalatā* (third work, with own pagination)
KAL	*Kalyāṇakāraka*
KĀM	*Kāmasūtra*
KĀMS	*Kāmaśāstra* of Surūpa
KAN	*Kandarpacūḍāmaṇi*
KEL	*Kelikutūhala*
Kh	Khilasthāna
KS	*Kāṭhakasaṃhitā*
Ku	Kuṣṭhacikitsitādhyāya
MĀDH	*Mādhavanidāna*
MBH	*Mahābhārata*
ms	manuscript

mss	manuscripts
n.	note
NĀG	*Nāgarasarvasva*
NAR	*Narmakelikautukasaṃvāda* of Kavirājamukuṭadaṇḍin, contained in *Kāmakuñjalatā* (sixth work, with own pagination)
NĀṬ	*Nāṭyaśāstra*
Ni	Nidānasthāna
NYC	*Nyāyacandrikā*
PA	*Pañcasāyaka* of Jyotirīśvara, contained in *Kāmakuñjalatā* (fifth work, with own pagination)
PAU	*Paurūravasamanasijasūtra* (with the commentary of Jayakṛṣṇadīkṣita), contained in *Kāmakuñjalatā* (first work, with own pagination)
RĀM	*Rāmāyaṇa*
RAS	*Rasavaiśeṣikasūtra*
RASA	*Rasayogasāgara*
RASAR	*Rasaratnākara*
RASAT	*Rasataraṅgiṇī*
RAT	*Ratirahasya*
RATIK	*Ratikallolinī* of Dīkṣitasāmarāja, contained in *Kāmakuñjalatā* (tenth work, pp.41-69 of the last part of *Kāmakuñjalatā* comprising the ninth to twelfth works)
RATIR	*Ratiratnapradīpikā* of Prauḍhadevarāja, contained in *Kāmakuñjalatā* (ninth work, pp.1-40 of the last part of *Kāmakuñjalatā* comprising the ninth to twelfth works)
RATIŚ	*Ratiśāstra*
RATN	*Ratiśāstra* of Nāgārjuna
RATS	*Ratiśāstra* compiled by Surendranāth Bhaṭṭācārya
Śa	Śarīravicayaśārīrādhyāya
Śā	Śārīrasthāna
ŚĀR	*Śārṅgadharasaṃhitā*
ŚARĪRA	*Śarīravijñāna*
ŚĀRPA	*Śārṅgadharapaddhati*
SATA	*Saṃvarodayatantra*
ŚB	*Śatapathabrāhmaṇa*
SC	*Strīcikitsā*
Si	Siddhisthāna
SIDDH	*Siddhasāra*
SIMH	Studies on Indian Medical History
ŚIVAT	*Śivatattvaratnākara*
SMAR	*Smaratattvaprakāśikā*
SMARAD	*Smaradīpikā* of Mīnanātha, contained in *Kāmakuñjalatā* (second work, with own pagination)
ṢOḌ	*Ṣoḍaśāṅgahṛdaya*
ŚṚ	*Śṛṅgāramañjarī*
ŚṚṄG	*Śṛṅgārarasaprabandhadīpikā* of Kumāraharihara, contained in *Kāmakuñ-*

	jalatā (eleventh work, pp.70-104 of the last part of *Kāmakuñjalatā* comprising the ninth to twelfth works)
SU	*Suśrutasaṃhitā* with the commentaries of Ḍalhaṇa and Gayadāsa (Varanasi/Delhi 1980)
Sū	Sūtrasthāna
SU1	*Suśrutasaṃhitā* (Kalikātā 1835)
SU2	*Suśrutasaṃhitā* (Bombay 1945)
SU3	*Suśrutasaṃhitā* with the commentaries of Ḍalhaṇa and Cakrapāṇidatta (Kalikātā n.d.)
SU4	*Suśrutasaṃhitā* with the commentary of Cakrapāṇidatta (Agra[/Jaipur] 1939)
TA	*Tantrayuktivicāra*
TAND	*Tandulaveyāliya*
TANT	*Tantrayukti*
TS	*Taittirīyasaṃhitā* (Leipzig 1871-1872)
TS1	*Taittirīyasaṃhitā* with the commentary of Bhaṭṭabhāskaramiśra (Delhi/Varanasi/Patna/Madras 1986)
TURFAN	Sanskrithandschriften aus den Turfanfunden
Ut	Uttarasthāna/Uttaratantra
v.	verse
var	variant
vars	variants
VĀS	*Vāsiṣṭhadharmaśāstra*
Vi	Vimānasthāna
VṚK	*Vṛkṣāyurveda* of Surapāla
VS	*Vājasaneyisaṃhitā*
VŚS	*Vaidyakaśabdasindhu*
vv.	verses
YO	*Yogaratnasamuccaya*
YOG	*Yogaratnākara*
YOGAC	*Yogacintāmaṇi*

CHAPTER 1

Introduction

1.1. Of all the mysteries that this Universe confronts mankind with, the two greatest are probably those that delimit the span of life: birth and death. Two of the oldest questions mankind has striven to find an answer to are: Where do we come from?, and, Where do we go to? Intrinsically linked with these questions are two others: How have we come?, How shall we go?.

1.2. There is, however, a qualitative difference between these two pairs of questions. Life's journey begins in the Unknown, and this is also where it ends; thus the questions regarding the origin and the aim of this journey are of a similar nature. This is different in the case of the latter pair. Death is the cessation of life, it is brought about by the 'life-functions' ceasing, the actual cessation being — if we disregard border-line cases and medical definitions which rely on the use of apparatus — on the whole easily perceivable. Not so in the case of birth, which is the creation of a new life. This new life is present even before the new individual actually sallies forth alone into this world. When, however, does this life really appear? How can we know of its having appeared at the point when it actually does?

1.3. Whether we look at the matter from a religio-philosophical or a biological point of view, the problem remains. Of course these points of view are intertwined, but of the two the biological one seems more basic, for one can (at least in principle) study the problem from this angle leaving aside religio-philosophical considerations, but one cannot look for a religio-philosophical explanation unless one has at least a vague idea of what actual or supposed biological phenomena one is trying to explain. Thus to understand such an explanation and the system built around it, one will first have to grasp its premises, study these, and try to understand them.[1] The acerbic de-

[1] As regards the philosophical study of medical texts in general, the problems that may arise when these are conducted more or less purely from the point of view of the scholar of philosophy have been remarked upon in DAS,R.P. 1993A (especially pp.27ff.). Such studies often seek to superimpose what are regarded as tenets of philosophical texts onto medical ones, a typical example of such tendencies being LARSON 1987, where a few passages chosen seemingly at random (and according to unknown criteria) have been discussed with scarcely any attempt to evaluate their worth as typical examples of medical doctrines, and with, despite pp.251f., hardly any understanding of or seemingly even interest in the large variety of notions in medical texts, but with a great deal of speculation upon how doctrines from non-medical contexts have been incorporated in medical works (p.250: often not in a "precise" or "correct" manner), with no examination on whether such a-prioristic notions of 'incorporation' (or the

bate on induced abortion going on in most Western countries has made the relevancy of this painfully clear, even though the problem here is often, though not always, not so much when life as such, but rather when individually sustainable life appears.

1.4. But to understand when life appears, one obviously has to know how it is brought about. When one now turns to ancient India, one finds that the notions prevailing regarding the creation of life from a biological point of view are difficult to describe. There is a huge mass of literary evidence to show that one knew quite well that the process of conception consisted of

---

like) are in actual fact justified at all (such notions are found in many studies remarking on philosophical doctrines in Indian medical works, also in COMBA 1987 and COMBA 1990, but have recently been disputed, e.g. by FILLIOZAT,P.-S., p.111). Finally, LARSON generalises his findings based on such limited material as applicable to Indian medicine as a whole, at times even to "Indian culture generally", whatever that may be (e.g. p.255). In a vein similar to LARSON 1987, LARSON 1993 too elaborates on how Āyurveda (which, as often in such contexts, here too means basically the *Carakasaṃhitā*) appropriated the concepts of known philosophical systems and changed them, even though pp.106f. say expressly: "In other words, it is probably not possible to suggest a causal relation one way or the other, but one can argue that there was a natural affinity between the naturalistic philosophical systems and Ayurvedic medical practice which generated a mutually influential pattern of ongoing interaction", and in spite of the statement on p.111 that "it can be argued that ... both theoretical elaborations ... are remarkably alike in their fundamental postulates — fundamental postulates that probably represent a protean *Urphilosophie* of South Asia that predates as well as underlies the later more sophisticated *darśanas*".

No one will deny that there most probably is much in Indian medical texts that was taken over from philosophical systems. But that does not allow us to simply assume that anything philosophical in these texts cannot but be a loan and not a result of independent thinking. Where this may lead us is shown by the radical assumption of CHATTOPADHYAYA 1986, p.578: "Anything found in the extant *Caraka-saṃhitā* not discussing the four factors essential for medicine — specially the ideas and attitudes belonging to the contexts of ritual performance and of liberation — is to be considered as 'irrelevant' for medical science and, as a matter of fact, in flat contradiction to the genuine scientific core of the work. If we depend on this criterion, we can easily see how much of the extant *Caraka-saṃhitā* — the tedious discussions contained in it on soul, rebirth, *mokṣa* and all the theoretical correlates of such ideas — have to be scrapped for our real understanding of the tradition of ancient Indian medicine." Preconceived notions of what is or is not "original" to medical texts help us little in our endeavours to really understand them. Without attempting any evaluation of it, I may here also mention the strong tradition in India holding that philosophy is the very basis of Āyurveda (see especially ŚRĪVĀSTAVYA, pp.27-45), a notion which has also influenced WOLZ-GOTTWALD 1991 (on this work see also DAS,R.P. 1992B).

Finally, attention may be drawn to a work which seems interesting in this context, but which I unfortunately was not able to procure. It is S.D.A. Joga Rao's *Science and Indian Philosophy. Interaction between Caraka Saṃhitā and Schools of Indian Philosophy*. Visakhapatnam (Andhra University Press) 1991.

more than just the mere act of sexual intercourse. But how exactly was it explained in ancient India? Or, more specifically, by what processes in the female body was it thought of as being brought about? For it is especially this aspect of the problem which is the more difficult one, since the role of the male is by and large relatively easy to describe, in contradistinction to which describing the part the female plays is no easy task. My intention below is to study how the ancient Indians came to grips with the problem of the role of the female in conception. Since, however, no single individual can hope to cross the whole vast ocean of ancient Indian literature alone, I shall confine myself to that branch of the literature which should be expected to contain the largest amount of relevant material, namely medical[2] and sexological[3] literature, though not neglecting other sources should the need arise and the data of medical or sexological literature lead to them.[4] It must, however, be pointed out that the matter discussed in this study relates only to humans; though veterinary texts probably contain similar material of relevance, I lack the specialised knowledge to tackle such works, and must thus leave their study to others. It must be made clear too that problems relating to the time after conception, e.g. the development of the child within the body of the mother and the problem of when it was considered to be a being in its own right, shall not fall within the scope of this work; they would have bloated it excessively, and complicated it even more than it already is.[5]

1.5. According to Old and Middle Indo-Aryan texts, a child is con-

---

[2] I am of course aware of the fact that "medicine" is not a true equivalent of *āyurveda-* (see VṚK, p.1,n.2, DAS,R.P. 1987, p.6; cf. also ŚRĪVĀSTAVYA, pp.17ff.), but it serves its purpose well enough if one keeps this in mind. Nevertheless, one would do well to remember that with regard to the term "medicine" we face a problem similar to the one we come across when referring to what we call "religion" (on this latter problem see e.g. LEERTOUWER).

[3] In this case too I am using an English term which might not convey all that is meant by the term *kāmaśāstra-* or its synonyms, but it seems to be the best equivalent at our disposal.

[4] Separate studies of sources other than those this investigation is primarily concerned with clearly remain very much a desideratum, and it is to be hoped that scholars better acquainted with such sources than I am will examine them for relevant data too, thus enabling us to compare data culled from different sources. However, such studies must be really thorough and in-depth to be of real value, not like, e.g., the work discussed in §§1.8f.

[5] Let me also remark that some of the problems which led me to take up this study are also found in ancient Greek and later Graeco-Arabic medicine; I have referred to these in Appendix I to this study, though in this regard I had to rely mostly on secondary sources, as my competence in this field is minimal. There seemingly are also many parallels in East Asian, and maybe also other medical traditions, but as I am even less competent to remark on these, I can do no more than draw general attention to the matter here.

ceived when two fluids, one from the father, one from the mother, unite in the body of the latter,[6] the sex of the child thus conceived depending on the relative quantities of the two fluids: an excess of fluid from the male gives rise to a male child, an excess of fluid from the female to a female child, whereas an equilibrium results in a being which is neither one nor the other, referred to by various terms such as ☞ *napuṃsaka-* (seemingly commonest), ☞ *dviretas-*, ☞ *klība-* etc.[7] This is a notion so commonplace that I think I can

---

[6] One sporadically also comes across statements that a life-force (usually termed *jīva-* or *sattva-*) too combines with the two substances just mentioned; other statements, also sporadic, however make it seem as if the life-force were rather the result of the combination of these two substances. Cf. on this matter TIVĀRĪ 1,pp.101-107, also ROŞU 1978, pp.172ff., ROLLAND 1972, pp.131f. and 1975, pp.213f., CAILLAT, pp.50ff., COMBA 1981, p.192, WINDISCH, pp.49ff.;52ff. Since we are, however, concerned primarily with biological concepts, metaphysical factors have not been taken into account here, though for a study of the philosophical notions regarding conception they obviously are of importance. For similar reasons, metaphysical notions regarding the role of the five gross elements (☞ *mahābhūta-*) and *cetanā-* (as ☞ *dhātu-*s) in the formation of the embryo play no part in this study, though of course they too are of great importance for an investigation of philosophical nature (see, however, the discussion of *dhātu-* in Appendix II). Similarly, mother, father, ☞ *rasa-*, *ātman-*, *sattva-* and *sātmya-* as the six bases of the embryo are also not discussed even though I am aware of the fact that some of them also play a role in the formation of certain parts of the body, since they do so in the context of the development of the embryo, not of conception; I shall, however, refer to them whenever the occasion arises. (I am leaving the last three terms untranslated, as I do not wish to enter into lengthy discussions here which would have to lead wide into the realm of Indian philosophy; *rasa-*, however, will be discussed often in the course of this study). References to what medical texts have to say on these six factors are given in TIVĀRĪ 1,pp.122-136. Synoptical tables on the different parts of the foetus formed by the different factors according to CA, SU, AH, AS and the *Śivagītā* are given in COMBA 1984, pp.229-233 (*sātmya-* is not included, as it is absent in the *Śivagītā*; see p.244,n.25); some data will also be found in §5.11 below. See too ROŞU 1978, pp.172ff. (cf. on Tantric texts also HEILIJGERS-SEELEN, p.63). I may in this context also draw attention to the role of *vijñāna-* during conception last discussed by SCHMITHAUSEN, pp.37ff. (p.683 has an addendum to p.37); *vijñāna-* will be taken up again in §13.21.

[7] According to the texts, one nevertheless has the chance of changing (cf. e.g. *vivartana-* in CA,Śā 8,19) the sex of the child thus conceived by means of the Puṃsavana rite, with regard to which in medical works see especially TIVĀRĪ 1,pp.191-198 (cf. also PAI, WEZLER 1993B, p.290,n.38). BRUCKER, while remarking on the Puṃsavana, seems unaware of the fact that this rite was supposed to be able to change the sex of the child, if necessary. Thus he takes the existence of the rite to be evidence for there having been in ancient India two theories on the sex of a child, one holding that the sex is determined at birth, the other that it is determined not at birth, but later. Since BRUCKER does not take into account what seems to be one of the basic characteristics of the rite in question, namely the *change* of sex of the child, his deductions are hardly convincing. Nevertheless, in historical perspective we may perhaps indeed have here, as he holds, a fusion of two different theories on the causes leading to the specific sex of a child, but that is for further research — research that takes such factors

dispense with giving references here.[8] The fluid of the male is obviously semen, but what is that of the female? We seem, as yet, not to have any clear ideas on its character. Yet the concepts regarding this fluid are of vital importance for understanding the notions regarding conception, for it is after all one of the two basic factors on which this conception depends. Our study, concerned with fundamentals, will thus have to take the nature of this fluid as its starting point.

1.6. Once embarked upon this study, however, one soon realises that one cannot discuss the problem of this fluid in isolation, as it is intrinsically linked with the problem of the ancient Indian notions on the nature of female discharge during menstruation and coitus. We shall thus have to broaden the scope of our investigation to include these too.

1.7. A preliminary glance at medical and sexological texts shows us that what is recorded in these on the above is not at all clear; in many cases it is difficult to determine exactly what is meant, whereas in other cases various statements seem apparently contradictory or irreconcilable. On the whole, what the reader is confronted with at first sight appears to be quite puzzling.[9] But, in accordance with a fairly widespread pre-modern and also modern view prevalent in South Asia that Indian medical doctrine has from time immemorial been a basically unitary and unchanging system,[10] the dif-

---

into consideration — to decide upon. On a modern variant of the Puṃsavana ("sex reversal therapy") see BODE, pp.146f.

[8] For some references from medical literature see e.g. TIVĀRĪ, 1,pp.169-173.

[9] In this regard it must also be pointed out that the explanations of intricate points, in certain works, that native commentators (or other works) give us cannot be relied upon absolutely; there are many cases in which these explainers have obviously misunderstood what was meant, from which it follows that we can never be sure whether explanations offered by such sources are indeed correct or not. Unfortunately, modern scholars, many from India, often choose to disregard this. It may also be remarked that often matters crying for an explanation are discussed in such sources either not at all or only very perfunctorily; this may be because the commentator concerned was unaware of the problem, or because he felt unable to comment upon it, but it may also be because the matter was such common knowledge in his age and society that it simply did not light upon him that an explanation might be required.

[10] This notion of Indian medicine having from time immemorial been basically one and the same unchanging system is all the more surprising because the texts themselves expressly and often mention different and often contradictory theories and practices, as MEULENBELD 1995, pp.3f. too remarks. Cf. in this connection also CA,Vi 8,3, which states expressly that there are different schools of medicine in vogue (this does not refer to the different branches of medical science, the *aṣṭāṅga-*), and then advises the would-be physician how to go about choosing the right school to study; see too CHATTOPADHYAYA 1986, pp.573f. It may not be amiss to remember too what Gustave Liétard wrote in 1897, when the uncritical euphoria re-

ficulties have on the whole either not been noticed, or else merely glossed over — in many cases we even find attempts to circumvent difficulties by transplanting modern Western and essentially alien notions and/or perceptions such as that of the "ovum" into the Indian system, the usual reasoning being that, since ancient Indian authorities knew *everything*,[11] they naturally

---

garding ancient Indian medicine had already become quite wide-spread (LIÉTARD/CORDIER, p.241): "L'ignorance où nous sommes des remaniements qu'ont incontestablement subis les textes sanscrits et des additions qu'ils ont reçues, oblige à une grande réserve dans l'appréciation des connaissances que l'on peut attribuer à ces vieux maîtres. Il reste à faire, pour arriver à ce résultat, tout un travail d'exégèse comparative."

On the tendency to see knowledge as eternal and already fully revealed in the past in India in general see POLLOCK 1985, pp.512ff., POLLOCK 1990, pp.322ff. (with further relevant literature especially on p.344) and BUITENEN, pp.35f. (As regards BUITENEN, I may also remark — even if it does not concern the subject we are studying here — that pp.34f., in keeping with the popular notion prevailing on this matter, ascribe the term "Sanskritization" to M.N. Srinivas, though according to GUPTA it was actually used long before Srinivas by Benoy Kumar Sarkar in 1937; however, it is not clear whether Sarkar and Srinivas meant the same when they used this term. On the term "Sanskritization" cf. also ISHII, p.55.)

[11] This of course pertains to fields other than that of medicine too. In its most extreme form this notion not only regards the ancient Indians (Indo-Aryans) as unique among all peoples of the world in their knowledge and lore, but as the source from which most knowledge on this earth stems. Such convictions are a matter of belief, and usually no rational discourse is possible on this point, much as in the case of the convictions, of a different sort, referred to in n.17 below. That this problem is also perceived among Indian scholars of Āyurveda is shown by the statement of a distinguished researcher in this field: "Much labour has been expended on twisting and torturing Ayurvedic texts to squeeze modern medical meanings out of them. A keen observer among the moderns, if he were directly conversant with the Ayurvedic principles, will be amazed and shocked at the amount of intellectual dishonesty and chicanery with which the Ayurvedic texts have been treated to make them resemble the modern texts. As a result, the very base of Ayurveda has disappeared in some of the desperate attempts to sell Ayurveda to the moderns. The worst result of these attempts is that in the laborious process of wrenching modern knowledge out of ancient aphorisms, the treatment of a concept becomes so painfully tortuous and involved that the Western reader gets even more confused than he would be if he was offered a simple exposition of Ayurveda as it has existed through the centuries, leaving it to his own discretion and judgement to evaluate the entire corpus of Ayurvedic knowledge, and to accept or reject the material, in full or in part, according to his own light and requirements" (SHARMA,S., pp.209f.). Referring to the story of Procrustes, the same scholar says (p.210): "The writer who recorded this story could have little idea that the 20th century will witness a medical science, Ayurveda, being subjected to identical treatment vis-a-vis the modern science." A very famous example of this, extremely influential for subsequent developments, is SEN, in which a large part of the anatomy in Indian medical texts is rejected as spurious and only that accepted as an authentic part of the ancient tradition which is in keeping with modern Western science as it was at the turn of the twentieth century (whilst at the same time part of the terminology of ancient texts is retained, but reinterpreted).

knew about such things as the ovum too.[12] This unreflected use of "ovum" has become such a convention today that even a work which expressly says that "Probably the menstrual flow itself was regarded as possessing the power of reproduction",[13] nevertheless sticks to the use of this term, with all its implications.[14] Thus it is no surprise that very few modern Indian scholars

---

See also LESLIE,C., particularly pp.189f.;194;206. As FILLIOZAT,J. 1977 aptly remarks on p.9, those trying to enhance the glory of Āyurveda by such methods actually do it a great disservice, serving rather to discredit the whole system so much that it runs the risk of not being taken seriously any more. Loc.cit. also makes the very pertinent point that, if Āyurveda is nothing but modern medicine presented in an obfuscating manner, it retains neither any specificity nor usefulness ("Si la médecine âyourvédique n'est que la moderne obscurément exprimée, elle n'a plus de spécificité ni d'utilité"), from which it follows that one would have to ask what reason anybody should today have to study it at all.

CHATTOPADHYAYA 1995 is a strong plea for more self-criticism by Indian scholars (of nationalistic inclination) researching into ancient Indian scientific ideas; where lack of such self-criticism may lead to has been discussed in DAS,R.P. 1997. On the necessity of unbiased study of medical traditions in general (especially with regard to matters concerning reproduction) see also WEISSER 1983, p.14 (cf. also pp.50-68).

[12] The problem of such unreflected translations has also been discussed by COMBA 1981, pp.192f. In a great many instances we may quite aptly speak of the modern application of the ancient *arthavāda-* argument, on which WEZLER 1987-88, p.398 writes in another context: "Ultimately one cannot but find it amazing how convenient a label '*arthavāda*' can be for a commentator of Dharmaśāstra (and similar) texts: If for some reason or other he is not willing to take a statement at its face value, and in its literal sense, categorizing it as an *arthavāda* suffices to cope with the problem and to contend that what is meant by it is in reality something quite different — which is, however, in keeping with his own conviction." On such problems of interpretation (usually compounded by the disregard of problems such as, e.g., the possibility of diachronic evolution and synchronic diversity of the material being studied) in medical texts in general see MEULENBELD 1984A, pp.37ff., especially pp.43ff.; see also MEULENBELD 1984B, MEULENBELD 1987, ANGOT, p.15, WOLZ-GOTTWALD 1985, pp.27f., WOLZ-GOTTWALD 1991, p.21 and WOLZ-GOTTWALD/SCHMITT, p.18,n.2 (though it must be pointed out that the last three works mentioned are in other respects quite problematic; cf. in this regard also DAS,R.P. 1992B). See also ZIMMERMANN 1987, pp.162ff., though note that loc.cit. itself the translation of ☛ *dhātu-* is given as "tissue", obviously with an older meaning of "tissue" in mind (akin to that of Aristotle's *ʻomoiomerē̃*, on which see KULLMANN), which, however, leads to confusion, as the modern usage of "tissue" is different. I myself may be guilty of such transplantation of alien notions by my use of the expression "uterus" in this study, but I must admit to having been at a loss as to what other expression to use, though it must be borne in mind that "uterus" (or "womb") as used in this study to denote the receptacle of the embryo or foetus need not tally in all with what a modern physician would understand under these terms (☛ *garbhāśaya-* and its synonyms).

[13] SINGHAL/GURU, p.25,n.1.

[14] I may here also point out that the theory of the cause of the sex of the embryo given

have bothered to give us more than a few stray comments and deal with the *basic* problems mentioned above. There are of course exceptions, the most recent being, to my knowledge, TIVĀRĪ; this study, though mainly concerned with other matters, still dwells on the problems mentioned above in some detail. It is obviously influenced by a scholar known for his open-mindedness and refusal to think in dogmatic terms, namely Priyavrat Śarmā, who also wrote the foreword; nevertheless, it too is not free of dogmaticism, in the parts of especial interest to us especially apparent in its unreflected use of the word "ovum",[15] or, where the authoress has problems in this respect and cannot think of any other explanation (see e.g. 1,pp.42ff.;2,pp.12f.), "hormones", for notions of ancient Indian medical authorities, without even considering whether the use of such expressions (as also others such as "plasma", "pituitary gland" etc.) is at all justified. As in the case of many other studies published today in India, such uncritical a-prioristic assumptions, which, when confronted with the actual data, require a great deal of argumentational acrobatics on the part of the researcher, greatly depreciate the worth of the conclusions drawn. But on the whole this is nevertheless a valuable study, especially as a veritable mine of information on primary sources; we shall have occasion to refer to it often in the course of our investigation.

1.8. Turning now to other secondary sources which might be expected to have relevant information, we find that these too have, on the whole, neglected this very basic problem. Thus DATTA does not touch it, in spite of the promising title of the work, while chapter 13 of GHURYE (i.e. pp.250-292: "Exogamy And Menstruation") too is of hardly any use to us in the given con-

---

by Indian texts must stymie anyone who insists on explaining the procreatory fluid in these as referring to the ovum. TIVĀRĪ, 1,pp.84;169-173 gets around this difficulty by simply not explaining or translating the various words (e.g. *ārtava-*, *rajas-*, *rakta-* etc.; we will return to these in the next chapter) used for the procreatory fluid in such passages, but by merely repeating the expressions used in the texts and then referring to these, in the same manner as the texts, as being "more" or "less" or the like. In RAY/GUPTA/ROY, p.11 we at least find an attempt to face this problem; here the authors, speaking throughout (cf. pp.10f.) only of the "ovum", substitute "stronger" for what is actually "more" or the like in the texts ("conception of a male child occurs when the sperm is stronger than the ovum; the reverse holds good for a female child"). SINGHAL/PATTERSON, p.150 is similar. And the English translation of ĀY 4,134 speaks of an "ejaculation of *rakta* (ovum)" (a parallel to the ejaculation of semen in ĀY 4,133) of the woman before the male ejaculates during intercourse, seemingly sparing no thought on the problems caused by the notion of an ovum being ejaculated thus.

[15] Śarmā too has made use of this term elsewhere, see e.g. SHARMA,P.V. 1985, pp.357f.; 365;367. On pp.363;367 we even find *bīja-* (on this see §§2.5f. below) translated as "genes". This shows that the pervasiveness of such notions in India today has managed to influence even a great scholar of his calibre. Cf. also n.812.

text. WINTERNITZ too has no information that touches the problem we are concerned with. But this does not mean that the problem has gone totally unnoticed. As a matter of fact, the question of female secretions and discharges has been taken up in quite some detail in a recent work: O'FLAHERTY, pp.15-61 (the section entitled "Sexual Fluids"). However, one does not quite know what to do with this work. Instead of first concerning herself with basics, the authoress has tried to cover, in a *tour de force*, Indian civilisation from its very beginnings to the present day, taking in all its diverse cultures, Aryan as well as non-Aryan. Such a vast field obviously cannot be covered comprehensively or, given the fact that we know hardly anything about the basic notions involved, in anything more than a very superficial fashion. What O'FLAHERTY thus actually presents us with are, it appears, bits and pieces of information which do not create the impression of having been searched for and selected systematically or in an attempt at comprehensiveness. In doing so, she has had to depend to a very large extent on secondary sources, which seem, however, to have been used on the whole uncritically and also seem to have influenced her interpretations of the passages she cites, more or less at random, from primary texts. So all we get in the end is a hodge-podge of diverse snippets of information of very different qualities, which are then combined within the framework of comprehensive theories and explanations claiming general validity. What we do not find is the awareness that what is in fact first required are detailed and in-depth studies of specific problems in different textual layers and various contexts; it is only after such preliminary investigations that we shall be in a position to start comparing on a large scale, especially with modern data. If we try to handle everything at once and without regard to basics, then the purported geniality becomes in fact a mere embarrassment,[16] especially if we combine such attempts with smooth symbolistic and Freudian, Jungian and other psychological explanations which are — like religious dogmas — in such a context essentially a matter of belief and as such beyond proving or disproving.[17] The time has not yet come for

---

[16] Cf. in this respect the methodological remarks in DAS,R.P. 1984B, pp.319f. and DAS,R.P. 1991B, which are, *mutatis mutandis*, applicable here too. I may also add that writing a great, all-encompassing study does not absolve one from paying attention to details, cf. "Orissa (an area near Calcutta)" (O'FLAHERTY, p.37). This criticism may seem petty, but it is not so when one realises that taking the above at face value might result in beliefs and value systems of very different areas being mixed up.

[17] Using such arguments, we can 'prove' almost anything; cf. e.g. WULFF, pp.290f.: "The destructive mother imago is perhaps nowhere better illustrated than in the traditional image of Kālī. She is herself, first of all, a phallic being, the mother-with-a-penis: she stands triumphantly erect on Śiva's body, sword raised, fingers pointed, and eyes and tongue protrud-

the great genial fling, rather, we still have not passed the very preliminary stage of dogged, minute and maybe boring probing, frustrating as this may be for the individual scholar concerned.[18]

1.9. All this does not mean that this study is of no help at all as regards the matter under discussion in general. It is not without value, for it not only draws attention to subjects not generally broached, but is as a collection of material also very useful, even though neither systematic nor comprehensive, and the authoress does offer us several very interesting and at times even profound insights, which, however, one should clearly beware of taking at face value without further investigation of one's own. Compared with TIVĀRĪ, its textual base (if we regard the different varieties of texts studied) is much broader (particularly the collection of secondary sources is invaluable), though the study is much more superficial, compensating the broadness of the textual base by a shallowness of analysis.[19] Unfortunately,

---

ing. At the same time, draped with severed heads and hands, she is the bloodied image of the castrating and menstruating (thus castrated) female. Here again we find elements of repetition and overdetermination: the numerous heads and hands, for example, are both symbolic representations of the phallus and instruments in their own right of active sexuality. ... The unusual preeminence of castration anxiety in Hindu culture and the associated ambivalence toward women are responsible, argues [C.D.] Daly, for a regression to concerns derived from the anal stage of development. In this way he explains the Hindu fear of pollution and the elaborate rituals of purification"! Can one argue for or against this? Or are we not, rather, placed in a position where we either have to accept this as it is, or not? Since what is quoted here is beyond the sphere of debate, I for my part could think of no plausible argument against it, based as it is on a certain set of axioms which are regarded (they are after all axioms!) as self-evident, and on a 'model' of interpretation which, as many models of this sort tend to be, can be elastically stretched according to the interpreter's needs, much as certain theologists can, if they so wish, prove or disprove almost anything on the basis of their text(s) of reference by giving analysed words and passages fitting meanings or else having recourse to symbolistic interpretation. See also DAS,R.P. 1984A, p.241,n.15.

[18] O'Flaherty and her school are prominent in South Asian studies in the Anglo-Saxon world; thus it is important that methodical deficiencies such as the above be highlighted. Those not able to work with primary material must turn to secondary sources, and because the said school conforms to certain tenets of modern North American public discourse, its market value is considerable, which in turn affects the nature and quantity of relevant secondary sources available. Critics of the methods employed often have a non-Anglo-Saxon background, for instance GAEFFKE 1985, pp.273ff., DAS,R.P. 1991B, p.738,n.4, WEZLER 1995, p.100,n.13.

[19] I must admit that DAS,R.P. 1984A, pp.237ff. (where O'FLAHERTY has unfortunately not been utilised) shows the same tendency of spreading out (though on a much smaller scale) without really going into the required depth. However, the remarks there were meant not as a study, but only as a means of drawing attention to difficulties existing and not fully realised or else neglected, within the necessarily brief context of a review; as such, they are neither the

the material presented is, as already stated above, haphazardly collected and on the whole not taken from the texts we are here primarily concerned with — the works TIVĀRĪ analyses have hardly been utilised. As such the study, though of interest as regards the subject under discussion in general, is, in our particular case, of no great use to us even as a collection of material.

1.10. Another work which might have been expected to contain relevant information, MEYER,J.J. 1952, is also of scant use for our purposes. It has hardly any information of relevance for our special problem, even chapter 12 (pp.359-390), entitled "Woman as Child-Bearer: The Origin of Man", has only a few stray remarks on the fluid of the woman combining with semen from the male, and these too are of no great value to us. The total scope of this monograph, seeking to describe the whole role of the woman in a vast textual corpus, precludes any real in-depth study of the subject of interest to us here, and this subject was probably also considered of secondary importance by the author anyway, who seems on the whole more interested in commenting upon his findings from the viewpoint of an Indologically qualified ethnologist and social anthropologist. As such, a detailed discussion of this work and its merits and demerits becomes superfluous.

1.11. But even the scant seemingly relevant material in the two works mentioned is actually of very little relevance for the special problem we are here dealing with, namely what the medical and sexological texts have to say. Not only have MEYER,J.J. 1952 and O'FLAHERTY hardly anything gleaned from such texts, but the little they do have is based on more or less chance finds and also shows an unfamiliarity with these works, which is especially evident in the case of O'FLAHERTY.[20] Nevertheless, since these two works

---

result of in-depth study, nor should too much be expected of them (this is especially true of the remarks on female "semen" in DAS,R.P. 1984A, p.243,n.29, especially in the light of what follows below in this investigation).

[20] Cf. e.g. the remarks (O'FLAHERTY, p.36) on ☛ *rasa-* being a name for the procreatory fluid in medical texts, based on passages (CA,Śā 3,12 and SU,Sū 14,6) which cannot possibly be construed in this sense, as the authoress herself admits by uneasily referring to *rasa-* as part of the alimentary metabolic chain (on which cf. §2.4). Now there are modern theories of this sort (cf. e.g. DAS,R.P. 1992A, p.391,n.22), and maybe even ancient ones (cf. e.g. §§13.22ff. below), and one could also adduce passages (seemingly unknown to O'FLAHERTY) such as HĀ,Śā 1,p.415 (which we shall discuss below in §§11.18ff.) from medical texts, which might possibly be taken to support the authoress' contentions, but the point is that what she actually quotes does not do this. On p.37 she writes that SU,Sū 14,18 (where the appearance of semen etc. in the pubescent male is described, followed by the marks of the pubescent female) and Śā 2,38 (which describes the origin of a 'perverted' male called ☛ *āsekya-*) show that the classical medical texts "insist" that women have "seed", which she uses as a synonym of "semen" here, and not as "generative fluid". In the case of the last passage, where the 'pervert' is born

seem to deal in most detail with our field of interest in general, they can hardly be neglected when the scope of investigation is extended beyond the material we are here considering, even though for our present purposes they are only of sporadic interest.

1.12. As regards modern studies concerned primarily with Indian medicine, the only studies I know which have really taken note of the difficulties connected with the problem of female discharge or at least deemed these worthy of comment are, apart from a few cursory remarks in JOLLY, p.49 (cf. also WISE in §2.3 below), COMBA 1981, pp.191ff. and MÜLLER 1955, pp.14ff. (which is actually an update of the discussion in MÜLLER 1941-1942 and is more or less summarised in MÜLLER 1965-1966A; cf. too MÜLLER 1961B, p.84, MÜLLER 1958, pp.107-109, MÜLLER 1962, p.36, MÜLLER 1963B, pp.414-417 and MÜLLER 1965, pp.32-40). In COMBA 1981 we find a short discussion of the difficulties mentioned and the hitherto unsatisfactory manner of tackling them,

---

due to insufficient *bīja-* "seed" (this term shall be discussed in §§2.5f.) of the parents (*pitror atyalpabījatvāt*), this is hardly feasible, since 2,36 expressly refers to *ārtava-* (see on this §2.1) secreted by the woman during intercourse; cf. also §11.31. In the case of Sū 14,18, female semen *may* be meant if the word *bālānām* in the passage be of common gender; I have discussed this passage in detail in DAS,R.P. 1991A, pp.25-32;40 (with addenda in DAS,R.P. 1994, pp.287ff.) and therefore need not go into details here. However, even if female semen should indeed be meant here, the question still is whether what is then meant is a "sexual fluid" or merely the last of the "elements" (☞ *dhātu-*) mentioned in §2.4 below. I will not deny that the procreatory fluid may indeed have been taken to be "semen" by some; thus we have e.g. a reference to the procreatory fluid of women only as *śukra-* "semen" in one ms of SU (see §8.17). The point is, however, that this does not follow from what O'FLAHERTY tells us, indeed, what she writes does create the impression that she is oblivious of the very complicated issues involved here, and her work in general shows no signs of her having studied Indian theories of reproduction found in Old and Middle Indo-Aryan texts in more than rudimentary fashion. This is made even worse by obvious mistranslations; cf. e.g. what she says on p.51 on SU,Śā 2,38 mentioned above: "Suśruta recommends that a man who is incapable of erection may remedy this by drinking semen (commentator: of another man), which will expand his semen-carrying ducts (Suśruta 3. 2. 38)"! The text, however, only says that there is no doubt that this man attains an erection by drinking semen (and clearly regards this as an abnormity). Such oversights do make one liable to be wary with regard to her other translations; a particularly glaring example is given in n.795 below (in very many cases the translations seem to be but paraphrases of others' — unverified — translations anyway). In a similar vein, we learn on p.35 that "a Sanskrit term for female seed (*rati*) means 'sexual pleasure', separating the woman's erotic fluid (in the vagina) from her fertile fluid (in the breast) or her destructive fluid (blood in the veins)", which is a lot of symbolism based clearly only on *secondary* sources (O'FLAHERTY, p.41) neither discussed nor verified, and not even concerned with Sanskrit literature, but with explaining certain phenomena among speakers of Dravidian and New Indo-Aryan, even though such a meaning assigned to *rati-* cries for an explanation (cf. e.g. DAS,R.P. 1992A, pp.402ff.).

but unfortunately no attempt to go into much detail. Thus this study ultimately does not help us very much, even though it does draw attention to the problem. As regards the various studies of Müller mentioned, the problem with these, as with many of his other works, is that they are characterised by the author's well-known idiosyncrasy both of style and of method of argumentation, coupled with his proclivity to make very sweeping statements based on a rather weak knowledge of Sanskrit and linguistics;[21] moreover, what he says on the fluids in question is based basically on only a handful of references, which are not studied in any great depth. What is even worse is that Müller takes his explanations to refer to what we might call the 'original' or 'true' state of affairs, which the authors of the medical texts, however, were, according to him, not always aware of, on the contrary creating confusion and obfuscating the notions they should have adhered to.[22] Obviously, this is no way to carry out research; thus his discussion too is ultimately of little help in tackling the problem in detail, and we therefore seem to have practically no other alternative but to start our own study from scratch. This shall begin with the discussion of medical texts and then move on to sexological literature. First, however, a more detailed, though short account of the nature of the problem involved shall be given.

---

21 Such tendencies have often been remarked on by others, and they have rather unfortunately and also undeservedly served to discredit Müller and all his works in the eyes of many Indologists, despite several really valuable studies on Indian medicine from his pen.

22 MÜLLER 1941-1942, p.480 also offers some speculation on the origins of the notions he claims to have unearthed.

CHAPTER 2
The Problem

2.1. One of the chief problems regarding menstrual discharge in ancient Indian medical works is the relation of this fluid on the one hand to the fluid in the female body that is supposed to combine with semen of the male (mostly termed *śukra-*, also *retas-*[23]) to produce an embryo, on the other hand to blood in general. For all three of these fluids the texts use common names for blood such as *asr̥j-*, *śoṇita-*, *rakta-*, *rudhira-*, *lohita-*,[24] for the first two in addition *ārtava-* "belonging to the *r̥tu-*, i.e. the regularly recurring period of fecundity of a woman (considered to be the ideal time for cohabitation)"[25] and *rajas-*, originally "mist; dust; impurity",[26] and for the appear-

---

[23] *retas-* may also refer to the procreatory fluid of the female, as in ☛ *dviretas-*.

[24] Some of the words for "blood" are actually substantivised adjectives meaning "red" (blood is "the red [one]"). One could therefore raise the question whether "blood" words for the procreatory and menstrual fluids might not actually derive from adjectives meaning "red". Though I have kept an eye on this possibility, my findings are that the texts I have studied unfortunately do not offer us any means to attempt a semantic differentiation on such lines, and this is not surprising given the long history of the Indo-Aryan terminology of blood with regard to its colour, which has in several cases in the course of time clearly made the substantivisation take precedence over the adjective in the minds of the speakers, so that it is often impossible to decide, for instance, whether *raktavarṇa-* means "(of) red colour" or "(of) the colour of blood". Thus any attempts to find out the relationship between blood and the procreatory and menstrual fluids seem not to be possible along terminological lines such as these, at least as regards the material being studied here. But since I may possibly be wrong and others come to a different conclusion after a fresh evaluation of the material, I am mentioning the matter here, so that it not be lost sight of.

[25] Cf. on this COMBA 1981, pp.194f., TIVĀRĪ, 1,pp.66ff. (also for the theories on how long the *r̥tu-* lasts, on which see too JOLLY, p.50, MITRA SHASTRY, p.86, DAS,R.M., pp.185ff.; the matter will be taken up again in §5.7). Special attention may also be drawn to SLAJE 1995. The word *ārtava-* is of course also in use to denote "of the (proper) season" in general and can as such be employed in a variety of contexts, e.g. for crops, rain-clouds etc. Such occurrences have naturally been neglected in this study.

[26] *pāṃsu-*, which can be used synonymously with *rajas-* "dust", is according to MONIER-WILLIAMS, p.613c also probably found with the meaning "the menses" in the *Carakasaṃhitā*; no more information is, however, given loc.cit., and I could not at first find the word in this meaning anywhere in CA, or indeed anywhere in the 'classical' (cf. §4.1) texts. Finally, Claus Vogel of Bonn University kindly drew my attention to the entry *pāṃsú* in the smaller Petersburg Dictionary (BÖHTLINGK 1991), where on p.60c of the fourth part (in the second volume of BÖHTLINGK 1991) we find: "3) wohl *die Menses* (vgl. *rajas*) ḰARAKA 1,26."; it seems as if

ance of the first *puṣpa-*, originally "blossom, bloom" (cf. the obsolete *flowers* in English, *les fleurs* in French). The name *ārtava-* would seem to be more in keeping with the procreatory fluid than the menstrual discharge (cf. e.g. MÜLLER 1942, p.43,n.1, also MÜLLER 1958, pp.107-109), but it must be pointed out that *ṛtu-* in the above meaning is only a special application of this word; thus *ṛtu-* can mean simply "a regularly recurring period", and seems as such also to refer loosely to the period of menstrual discharge and not necessarily to the period of fecundity following this.[27] *ārtava-* could then be con-

---

Monier-Williams, continuing his habit of the first edition of his dictionary of cribbing from the larger Petersburg Dictionary (including its mistakes, as Böhtlingk does not fail to point out in his "Vorwort" to the volume of BÖHTLINGK 1991 mentioned), has, in his new edition (i.e. MONIER-WILLIAMS), cribbed in the same manner from the smaller Petersburg Dictionary, but in this case, as so often, without giving the reference found in the latter. When I tried to check on this reference myself, I understood why MONIER-WILLIAMS preferred to refer only vaguely to the *Carakasaṃhitā*, for in the list of abbreviations at the very beginning of BÖHTLINGK 1991 we find: "ḰARAKA = ḰARAKASAM̃HITÂ, Calcutta 1929 und 1877 und Hdschrr. im Besitz von ROTH (KERN und ROTH)"; obviously, especially given the vague references to manuscripts, it was too much work for Monier-Williams to check on this word on the basis of such data! I myself nearly despaired of being successful, but in the end was fortunately helped by Gabriele Zeller of the University Library at Tübingen, who kindly procured for me different hard-to-get editions of the text from the personal collection of Rudolph Roth (the co-editor of the larger Petersburg Dictionary); after much searching, I finally found the place referred to by Böhtlingk in CA5, namely in Sū 26 on p.160, where it is said of the salty taste (*lavaṇa- rasa-*): *pāṃsum upahanti*. However, the corresponding line in CA,Sū 26,43,3 reads *puṃstvam upahanti*, which refers to "manliness", i.e. the potency or procreatory power of the man, and this is the reading found in all the other editions of the text I could consult, so that the evidence of the reading of CA5 must remain doubtful until further investigations have cleared up the matter. Moreover, even if we accepted the reading of CA5 as correct, the meaning of *pāṃsu-* would still not be clear. In any case, this small example shows how necessary a critical edition of this text (as also of other major texts, SU, AH and AS at the very least) remains, and it is a crying shame that even after over a hundred and fifty years of what purports to be critical scholarship on the subject of Indian medicine no such edition(s) exist, except for SIDDH.

[27] Cf. e.g. MEYER,J.J. 1952, p.216,n.1, THIEME 1971, p.441 (on the duration of this period also TIVĀRĪ 1,pp.56f.). The period of fecundity is at times said in the medical texts to begin with the first appearance of the monthly menstrual discharge (though the days of discharge are then as a rule considered unfit for intercourse), but seems usually to refer to the following period free of discharge (as a rule taken to begin with the fourth day after the onset of menstruation), as it is only during this period that intercourse should, according to the traditional code of conduct, take place. Apart from this, the period of regularly recurring menstruation can also refer to the period of life in which a woman menstruates; in other words, saying that a woman is *ṛtumatī-* does not, in that case, mean that she has a particular *ṛtu-* at that moment, but that she is of the age when she is able to have the *ṛtu-*. This meaning of *ṛtu-* is, however, of significance primarily in a different context, namely that of the age of marriage of a girl (cf. on menstruation and this age e.g. MEYER,J.J. 1952, pp.215ff., THIEME 1971, pp.432ff., GHURYE,

nected primarily with such a usage of *ṛtu-* too, at least theoretically;[28] it is even possible that the term *ārtava-* may have been used to refer to two different fluids, each connected with a different meaning of *ṛtu-*. As regards *rajas-*, this at first glance would seem obviously to be more fitting as a name for the menstrual discharge only (cf. especially *rajasvalā-*, which denotes a woman having menses already in late Vedic texts[29]).

2.2. The fact that the nomenclature of the three fluids mentioned above seems, according to our first impression, to be hazy, may be due to metonymy or the like, but it could also be the result of certain notions pertaining to the nature and/or origin of these fluids. It is for instance theoretically possible that all three, or at least (given parallels in many other cultures of the world) blood and the menstrual discharge are in Indian medical theory (or maybe in only one or a few of several different theories) thought to be basically one and the same. Indeed, ĀY 4,68cd gives us a quotation from the lost work of Vṛddhātreya, in which a medical authority might have been asked whether the fluid called *rajas-* is indeed [the same as] blood or whether *rajas-* is without/exclusive of (i.e. different from) blood (*raktam eva rajaḥ strīṇāṃ kiṃ vā raktaṃ vinā rajaḥ*), which would, if this is indeed how the line is to be analysed,[30] show that this question must have busied at least some medical authorities. Unfortunately, the relevant part of the answer which ĀY cites says only that "the blood that arises in connection with the *ṛtu-* is authoritatively called (*procyate*) *ārtava-* by the wise",[31] which does not tell us anything with regard to the question, since it is unclear what "blood" in the answer means. We shall thus have to see below if there is any evidence for assuming that blood was held to be the same as one or both of the other fluids. But even if we should find no evidence for such an assumption in the existing medical literature, it is theoretically still possible that such a notion may have prevailed in an age older than these texts. I am, however, not in a position to comment on this without being wildly specula-

---

pp.264ff., LESLIE,I.J. 1989, pp.35;254).

[28] It must, however, be pointed out that the usage of *ṛtu-* has not yet been examined in detail. Thus our knowledge of its exact meaning in medical texts must be regarded as tentative. See also MÜLLER 1961B, pp.84;87f., MÜLLER 1960C, pp.718f. and MÜLLER 1955, pp.14ff. on this problem, and cf. also GHURYE, pp.267ff. and §5.7 below.

[29] On this term see also GHURYE, pp.267ff., and §4.1 below.

[30] One could also understand it quite differently; we shall return to this in §8.4.

[31] *ṛtusambandhajaṃ raktam ārtavaṃ procyate budhaiḥ* (4,69cd). The rest of the passage (quoted in full in n.222 below) does not contain any more information of relevance for this particular question.

tive, so I shall confine myself to discussing what the medical works we today know and which treat of the subject say.

2.3. When we turn to secondary literature to find out what has been said on the problem of the identity of menstrual discharge and the procreatory fluid, we find that not only have most scholars not only not seen any problem, but also that those that have have basically given us no more than apodictic statements. To TIVĀRĪ, 1,pp.36ff., the two are of course different, since the procreatory fluid is to her "ovum" (the proof of this equation lying, needless to say, merely in its statement). Similar statements are so common in modern Indian works on traditional Indian medicine that it seems pointless to give further examples here, though special attention may be drawn to GHĀṆEKAR 1986, p.21b, which says the same as TIVĀRĪ, but where we also find it said that *ārtava-* consists of two parts, one part, consisting of what is loc.cit. termed "Menstruation" and "Menstrual blood", being supposed not only to cleanse the womb and vagina, but also to make the vagina *sukh'saṃved'nīya* "pleasant/easy to be experienced", the womb and vagina free of obstacles for the entry of semen and the womb fit for the embryo (which looks like a description of the lubricating fluid(s) produced due to sexual stimulation), the other, also referred to by words meaning "blood", consisting of the "ovum".[32] An exception to this general tendency of explaining the procreatory fluid as the "ovum" is MURTHY 1,p.357, which says that *ārtava-* "is commonly applied to the discharge of blood from the woman's body ... known as menstruation. Ancients held the view that the menstrual blood is responsible for the production of the embryo; this view is erroneous."[33] Unfortunately, this is not substantiated by independent study or a reference to such a study or studies. Western scholars too have as a rule not given us apodictic statements on the "ovum", but what they say on the nature of the procreatory fluid is also hardly based on thorough studies. One of the earliest scholars to remark on the subject held the procreatory fluid to be identical with the menstrual fluid (see e.g. HESSLER 1887, p.142), and this opinion was shared by another early scholar who relied heavily on his work[34], namely Gustave Liétard (LIÉTARD/CORDIER, p.198). Of later scholars, COMBA 1981 holds

---

[32] Cf. the discussion on ☞ *abīja-*. For another example of what is clearly a fluid being termed "ovum" by this author see n.435 below.

[33] Just before the portion cited, MURTHY, obviously differentiating between the ancient notions as he sees them and modern notions, speaks of "the union of the śukra (male reproductive element *vis a vis* the spermatozoon contained in the semen) and ārtava (female reproductive element, *vis a vis* the ovum produced by the ovary)." Also see n.254.

[34] I.e. on HESSLER 1844-1850.

that the "blood" uniting with semen to produce an embryo is not regarded as being different from the menstrual fluid (p.192), thus also JOLLY, p.49, ZIMMER,HEN.R., pp.187f., MEYER,J.J. 1952, pp.362;364;366;368, SCHMIDT 1904, pp.204f., LESLIE,I.J. 1994, pp.67-69, ESSER, p.1644 and SUNESON, pp.111;114. WISE, p.33, seems to think the same (albeit here we find no comments on the actual problem we are concerned with), for loc.cit. "menses" is obviously used for both the menstrual and procreatory fluids. But MÜLLER 1961B, p.84, MÜLLER 1958, pp.107-109 and MÜLLER 1962, p.36 seem to hold the opposite view. According to MÜLLER 1955[35] too, Indian medical theory differentiates between a procreatory fluid called *ārtava-* and a menstrual fluid called *rajas-*, both of which are also "blood", i.e. blood in general. Unfortunately, what is said on the two fluids and blood by this author is not clear; one gains the impression that according to MÜLLER 1955 (also MÜLLER 1965-1966A) the procreatory fluid and blood are one and the same, but MÜLLER 1955, p.21 also characterises the latter as the carrier ("Träger") of the former.[36] The menstrual fluid seems according to this to be a later form of the procreatory fluid, but that too is not clear. This lack of clarity is due to Müller's confusing mode of expression, of which an example from another text, but on the same subject, may be given here: MÜLLER 1965 on p.35 most probably sees *rajas-* as being blood coming out during menstruation, but here again his statement is not quite clear; in any case, *ārtava-* is according to him clearly a procreatory fluid of women which can lead to pregnancy during the *ṛtu-*. On p.36 he writes:[37]

> "The prohibition of sexual intercourse with a woman during the time of her menstruation is continuously and without exception acknowledged in all traditions, also in the medical texts, so that a corresponding monthly bleeding can not be critically [sic] held to be a reproductory[38] substance. Otherwise it would be expected that especially that designation of blood

---

[35] The discussion of which is more or less summarised in MÜLLER 1965-1966A (cf. §1.12).

[36] This latter also seems to follow from MÜLLER 1960C, p.718: "Diese Belege beweisen, dass der Fortpflanzungsstoff *ārtava* den [sic] Abschluss eines Stoffwechsels im Inneren des Körpers ist und nicht seine Absonderung im Menstrualblut." On the same page it is also said that the negative evaluation of blood (in general), and especially of the menstruating woman, contradicts the positive evaluation of her procreatory fluid.

[37] The translation smoothes over the somewhat peculiar syntax of the original German.

[38] The word "reproductory", though a valid term, is not found in all English dictionaries, but only in larger ones; however, I find it more appropriate than "reproductive" to denote a substance leading to reproduction.

> which, as *rajas*, refers particularly to menstrual blood, would be mentioned at least once as the reproductory substance.[39] The *rajas* cannot even be considered to be a generally valid designation of blood. However, its[40] manifold appellations favoured a laxity in meaning, so that *ārtava*, by which only a notion of time is expressed, gave cause for errors in medical texts too."[41]

These confusing remarks are unfortunately typical of Müller's style already referred to in §1.12, so that his ideas ultimately remain enigmatic — at least to me.[42] In any case it is obvious that the whole matter is by no means clear.

2.4. Regarding the relation of blood to the other two fluids, a preliminary glance at medical works does create the impression that the former is something different, though, since the relation of the two latter fluids to each other has not yet been satisfactorily analysed, it is difficult to say whether it is then different from both of these or from only one. As to TIVĀRĪ, 1,pp.52-54, the authoress of course remarks only on the relation between blood and the menstrual discharge, because, since to her the procreatory fluid is simply "ovum", the question of the possible relationship of this fluid to blood does not arise. Even so, she soon runs into difficulties when confronted with statements of different medical texts running contrary to this explanation as "ovum"; some of these statements she does not comment further upon, while she circumvents one with a trick. The latter is as follows: the breaking down of food (usually *anna-* or *āhāra-*) is in Indian medical theory explained as a successive transformation into several elements (☞ *dhātu-*), namely nutrient

---

[39] Obviously, Müller means that *rajas-* is not used in this sense; he does not, however, substantiate this statement by any detailed examination.

[40] According to p.36,n.135, this refers to blood in general.

[41] "Das Verbot des Geschlechtsverkehrs mit einer Frau zur Zeit ihrer Menstruation ist fortlaufend und ausnahmslos in allen Überlieferungen anerkannt, auch in den Medizin-Texten, so dass eine entsprechende monalische [i.e. monatliche] Blutung als Fortpflanzungsstoff kritisch nicht angenommen werden kann. Es wäre sonst zu erwarten, dass gerade die Blutbezeichnung, welche sich als *rajas* besonders auf das Menstrualblut bezieht, wenigstens einmal als Fortpflanzungsstoff erwähnt würde. Das *rajas* kann nicht einmal als eine allgemeingültige Bezeichnung für das Blut gelten. Seine mehrfachen Benennungen begünstigten jedoch eine Lockerung in der Bedeutung, so dass *ārtava*, womit nur ein Zeitbegriff ausgedrückt ist, auch in Medizin-Texten zu Irrtümern Anlass gab."

[42] It is thus a great pity that HEILIJGERS-SEELEN (p.63,n.65) uncritically and without further investigation cites Müller as an authority on the nature of the female procreatory and menstrual fluids.

fluid (☞ *rasa-*), blood (*rakta-* or the like; see §2.1), flesh (☞ *māṃsa-*), fat (☞ *medas-*), bone (*asthi-*), marrow (☞ *majjan-*) and finally semen (*śukra-*) (thought to be present in females too,[43] though actually the texts mostly seem to describe male semen[44]), often also mentioning a problematic element called ☞ *āhārarasa-* too.[45] Now according to TIVĀRĪ's analysis some of the texts derive the menstrual discharge from *rasa-*, some from blood; TIVĀRĪ 1,pp.52f.[46] explains this discrepancy by simply saying that, since blood itself is derived from *rasa-*, ultimately there is no contradiction here, without even pausing to examine whether the derivation of the menstrual fluid from *rasa-* which some texts mention really does presuppose an intermediate stage of blood. On the other hand, according to COMBA 1981, p.192 blood is the direct product of "chilo" (as she translates *rasa-*), i.e. "chyle", whereas the menstrual discharge (= procreatory fluid according to COMBA 1981; see §2.3), though ultimately formed from "chyle" too, is a direct product of the penultimate element formed by the breakdown of food, namely the marrow, meaning that COMBA 1981 regards semen of the male and this fluid of the female as having the same origin, and it is in keeping with this theory that COMBA 1985 indiscriminately refers to the reproductive fluid of the female too as "semen" throughout (see especially p.921).[47] ZIMMER,HEN.R., p.185 too holds semen of the male and menstrual blood (and thus the procreatory fluid; cf. §2.3) to have the same origin.

2.5. The above has, I hope, shown that the remarks on the trustworthiness of previous studies and the help we may (or rather: may not) expect from them were justified, and that we really have no other choice but to take up the problem afresh. But before proceeding onwards now, I must draw attention to a term which I have not mentioned in §2.1, but which according to

---

[43] We shall return to this later in the course of this study.

[44] This latter problem will be taken up again in §8.9.

[45] We shall discuss *āhārarasa-* in more detail in §§7.29ff. below.

[46] With some problematic translations; cf. e.g. the explanation of Cakrapāṇidatta's commentary on SU,Sū 14,7, which incorporates statements (both in the Hindi and in the English texts) Cakrapāṇidatta himself does not make (see the Sanskrit text on p.53,n.1), since the latter remarks on the *difference* between "*ārtava*--blood" (*ārtavaśoṇita-*) and "*dhātu*--blood" (*dhātuśoṇita-*), which TIVĀRĪ takes to refer to a *derivation* of the former from the latter, maybe based on a misinterpretation of *vyāvartayati* "makes separate" in the sentence *tuśabdaḥ pūrvasmāt saumyāgneyād raktād vyāvartayati*: "The word *tu* ('however') makes [☞ *āgneya-ārtava-*] separate (i.e. distinguishes it) from the former (this clearly refers to the mention and description in the commentary on 14,4f.), [at the same time both] ☞ *saumya-* and *āgneya-*, blood."

[47] Cf. on this also §6.15.

TIVĀRĪ, 1,pp.39f. is used in the majority of cases to denote what she calls "*strī-bīj*" and "ovum", namely *bīja-* "seed, germ". Actually, when we examine the passages from medical texts the authoress has adduced loc.cit. as proof for her statement, we find that, even if we were to grant that the texts presuppose an "ovum", we cannot simply say that the quoted passages (or even most of them) clearly refer to this "ovum". In fact, one in very many cases[48] gains the impression that *bīja-* in medical texts seems to refer to the "germ" created due to intercourse (i.e. the combined state of the substances which form an embryo) just prior to its formation, but after combination; TIVĀRĪ, 1,pp.173ff. several times presupposes this meaning too, even using the term "zygote" on pp.176f.! Sometimes, however, it seems as if *bīja-* referred not to the embryo as a whole, but to the germinal form of individual organs, either in the germinal form of the embryo as such, or in the procreatory fluids from which it is formed; thus for instance Cakrapāṇidatta in his commentary on CA,Śā 4,9 speaks of the state of "being in the form of a *bīja-*" (*bījarūpatā-*) of the limbs/organs (*aṅgānām*) of the embryo, though it is unclear whether this refers to the embryo from the very beginning, or at a stage when rudimentary limbs or organs appear.[49] One could at times also consider taking *bīja-* to mean that part of each of the fluids combining which is procreatory,[50] though, if I be permitted to anticipate this part of my study here, this is a very problematic proposition with regard to what the bulk of the material which has been studied seems to presuppose.[51] As regards the latter meaning as applied to the procreatory fluid of the male, it does tempt one to translate *bīja-* in such a context as "sperm(atozoa)" and differentiate it from "semen", and to attempt the same, *mutatis mutandis*, in the case of the fluid of the female.[52] However, one should be very careful in such cases, for I have found no clear evidence that Indian medical texts speculate on the existence of something like spermatozoa[53] and their hypothetic counterparts

---

48 This also includes cases TIVĀRĪ, who quotes only some relevant passages, does not refer to loc.cit., though note also op.cit., 1,pp.173ff.

49 One cannot fail to notice that in this regard there are several points of contact with the use of *bīja-* in connection with the Buddhist concept of *ālayavijñāna-*, of which we now have a detailed study (SCHMITHAUSEN, with supplementary data in MUROJI). See on this also §13.21.

50 Cf. e.g. *subīja-* in KAL 23,18 (which we will discuss in more detail in §7.7 below).

51 References will be found in §14.2.

52 E.g. in the term ☛ *abīja-*.

53 In the West Antoni van Leeuwenhoek and Johan Ham published their discovery of them (with the help of the microscope) in 1677.

in the fluid of the female.[54]

---

[54] As it is, though one sporadically comes across passages which *might* be taken to state that the procreatory fluids have different components, none of these passages are conclusive in this regard (cf. n.51). Moreover, even if there should have, in medical texts, been the notion of a procreatory fluid consisting of different components only one or some of which are actually procreatory, that still would not allow us, in the case of the male procreatory fluid, to simply equate this procreatory portion with spermatozoa or the like, for the deduction that this very particular form was presupposed in ancient Indian medicine (even if according to this the procreatory fluid should have been taken to be composed of various components, for which assumption we have, I must repeat, no hard and fast evidence) can hardly be convincingly based simply on vague speculations, but requires solid evidence. The dangers of ignoring this are made evident e.g. by JAINI, p.179, where he translates from a Jaina text (*Yuktiprabodha-svopajñavṛtti*) as follows: "... when a man is overcome by sexual passion and engages in sexual activity, he kills 900,000 minute beings [i.e., the sperm cells in the ejaculate]." At first glance the explanation given by the author in square brackets seems very plausible. However, the same text then says that in the vagina of a woman too from 100,000 to 300,000 minute beings are formed, and that the intercourse of man and woman kills these beings, obviously due to the friction of intercourse (cf. JAINI, p.190,n.36). And before this, the text also tells us that the beings in the woman's vagina are formed due to her menstrual flow (pp.166;178); this is a reference to similar statements found in much older texts and quoted on pp.35;143. These beings are clearly *not* germinal human beings. At the places mentioned it is also said or implied that the beings generated also die (e.g. p.178: "As regards the menstrual flow in the vagina of a female and the birth and destruction of many lower forms of life therein ..."). In the face of this, the parallelism of man and woman and their mention together loc.cit. would seem rather to point to a generation of other living beings within the penis of the man due to the flow of semen, which is something quite different from what JAINI presupposes, all the more so since p.178 mentions cases parallel to the generation due to the menstrual flow, namely the birth and death of worms in the stomachs of males, and also refers to the flow of blood, phlegm and pus in males and their medication, clearly implying the death of creatures due to the latter, which of course means that these creatures are such as are generated due to the flows mentioned. From this too it seems to follow that the parallel flow of semen has some similar effect. This deduction is made even more probable by pp.35;143;166, which tell us that similar beings are generated on other parts of a woman's body too (e.g. in her armpits and navel, between the breasts), and by p.143 saying expressly that "men also [are] subject to the same faults"; from this too it seems to follow that no potential human beings and hence no sperm cells are meant on p.179. (It should, however, be pointed out that according to p.143 women generate more living beings than men, whereas according to p.179 the reverse seems to be true.) But here one may object that the passage from the *Yuktiprabodhasvopajñavṛtti* translated on p.179 of JAINI seems in actual fact a quotation from some unknown source (but is this sure?), and as such might actually be referring to something different from what the person quoting it thinks it says. This possibility cannot be fully discounted, but in the face of the other evidence, is it really a probability? In any case, we see that there is no *clear* evidence for assuming a reference to sperm cells. Now all this is just a small example, but it shows quite well how complicated the matter actually may be, and how we should avoid jumping to conclusions fitting pre-conceived notions, even though these may seem to be quite plausible

2.6. But TIVĀRĪ is not totally mistaken, for *bīja-* does seem also to refer to the procreatory fluid of both male and female (though it seems to refer to that of the former more often). There are many examples of this usage in medical texts,[55] but I have not tried to collect them systematically and thus refrain from giving any here, as such a list cannot but be haphazard and arbitrary.[56] In many cases more than one interpretation of the word *bīja-* can be considered, and in some passages we actually seem to have different meanings side by side.[57] One often feels quite confused by such pas-

---

at first sight.

[55] Cf. also the discussions on ☞ *dviretas-* and ☞ *dhātu-*.

[56] I must, however, draw attention to RAT 5,4, where *phalabīja-* is quite clearly used to refer to semen of the male (on the verse containing this expression see also SCHMIDT 1927, p.186). Here semen seems to be perceived as the "seed" from which the "fruit" (☞ *phala-*), i.e. the embryo arises.

[57] Cf. for instance CA,Śā 2,12ff. (which presuppose the theory — on this see §1.5 — of the sex of the foetus depending on the relative quantities of the two procreatory fluids involved); according to Śā 2,12 a daughter is produced through excess "blood", a son through excess semen, whereas if the *bīja-* is divided into two, a daughter and a son are born according to which *bīja-* is in excess in each part:

raktena kanyām adhikena putraṃ śukreṇa tena dvividhīkṛtena
bījena kanyāṃ ca sutaṃ ca sūte yathāsvabījānyatarādhikena.

Śā 2,13 tells us that if in a woman the *bīja-* having an excess of semen divides into two (or maybe: if the *bīja-* divides into two parts having an excess of semen), she produces twin sons, twin daughters, however, if *bīja-* with "blood" in excess divides (or maybe: if the *bīja-* divides into two parts having an excess of "blood"):

śukrādhikaṃ dvaidham upaiti bījaṃ yasyāḥ sutau sā sahitau prasūte
raktādhikaṃ vā yadi bhedam eti dvidhā sute sā sahite prasūte.

Here *bīja-* on the one hand (Śā 2,12: *yathāsvabījānyatarādhikena*) seems to refer to the respective procreatory fluid, on the other (2,12: *dvividhīkṛtena bījena*, 2,13: *dvaidham upaiti bījaṃ*) to something following the combination of semen and "blood", though one could in the latter cases also consider taking it to refer to a single procreatory fluid; the verses would then be telling us that each of these fluids divides and only then combines with the respective part of the other. This, however, creates difficulties. Śā 2,14 tells us that semen and *ārtava-* may also be divided into many parts by wind (*bhinatti ... śukrārtavaṃ vāyur atipravṛddhaḥ*) in the womb, giving rise to a corresponding number of embryos, due to *karman-* and not due to individual action. This could refer to a division of each of the fluids separately, but the compound *śukrārtava-* might also refer to both fluids together, in combination. Interestingly, the latter is exactly how TIVĀRĪ, 1,p.174 has analysed this passage. The whole is complicated by the fact that, of the parallels in other passages (listed in TIVĀRĪ 1,p.173) some tend to favour the former, some the latter explanation.

In CA,Śā 2,18, a ☞ *dviretas-* is said to arise *bījāt samāṃśād upataptabījāt* "from the *bīja-* whose parts are equal, from the heated *bīja-*". According to CA3, p.501, *bīja-* refers to *śukra-* and *śoṇita-*, which are here either of equal quantity, or else heated due to several

sages, and in some cases it seems that the mediaeval native commentators of medical texts fared no better, often tending to heighten the confusion.[58] It is thus not surprising that Gaṅgādhar Gopāl Guṇe in his Prastāvanā to KAL uses *bīja-* in a single sentence to denote successively the "seed/germ" of the male, the "seed/germ" of the female, and the elemental "seed/germ" created by the union of these two.[59] But interesting as all this is, I have not pursued the matter further. I do not deny that it may be important for our understanding of some aspects of ancient Indian notions regarding conception and reproduction, but my preliminary investigations convinced me that the problem of the usage of *bīja-* would not directly touch the basic problem concerning us here, but involve at most only a certain elaboration of it. Since, however, the complexity of the usage of this word would have involved an extensive study in addition to the one we are now embarking upon, I did not feel up to the task of undertaking this too. Thus I content myself with drawing attention to the problem here, in the hope that it may be taken up by somebody else in future.[60] I shall, however, have to remark on it now and again when the word *bīja-* crops up during the following discussion.

---

factors. Cakrapāṇidatta in his commentary takes the first *bīja-* to refer to *śukra-* and *śoṇita-* too, but explains the second *bīja-* differently, saying that it refers to that part of the heated *bīja-* which is capable of generating *bīja-* (*upataptabījād iti upataptabījajanakabījabhāgāt*). This is clearly problematic; on similar problems see Excursus I on pp.25-29.

[58] The difficulties these different meanings of *bīja-* occasion can be best exemplified by means of a discussion of CA,Ci 30,8;34f. and its parallels AS,Ut 38,pp.289b;292a and AH,Ut 33,28;39f. (these are also quoted by TIVĀRĪ 1,p.40). Since the discussion of these commentaries is quite lengthy, it has been given separately in Excursus I on pp.25-29.

[59] *śukr (puṃ'bīj) va ārtav (strībīj) ke saṃyog'se bīj dhātu ban'tā hai* (p.13); *dhātu* (a loanword from Sanskrit, i.e. ☛ *dhātu-*) here seems to be used in the sense of "primary element" in general.

[60] Such a study of *bīja-* will also have to tackle problematic passages such as AH,Śā 1,2, where the fine (*sūkṣma-*) gross elements (☛ *mahābhūta-*), through which the foetus/embryo grows within the mother after it has been formed through the union of semen and *ārtava-*, are called *bījātmaka-*, or AS,Śā 2,p.286b, where the wind in the body causing semen to sally forth to combine with *ārtava-*, and followed by the other gross elements, is called *bījātman-*. It is not possible for me to discuss such problems here; cf., however, for preliminary information the various native commentaries on these passages, as well as WINDISCH, pp.49f. On the problem caused by the use of *bīja-*, *garbha-* and *retas-* in non-medical texts, and their seeming synonymy at times, attention may be drawn to HARA 1994, pp.50ff.

EXCURSUS I

(in place of footnote 58)

CA,Ci 30,8 lists both corrupted (*praduṣṭa-*) *ārtava-* and a corruption of the *bīja-* (*bījadoṣa-*) among the factors responsible for disorders of the female genital tract. Cakrapāṇidatta in his commentary says:

> *bījadoṣād ity ārtavagatabījaduṣṭeḥ. sā paścād vaktavyā*
>
> "'*bījadoṣāt*': due to corruption of the *bīja-* contained in/relating (i.e. leading) to/having become (= being?) (or maybe, but less likely: come forth from) *ārtava-*. It is to be mentioned later."

With the latter statement he obviously (and correctly) refers to Ci 30,34f. We have here, as in 31f., the description of a disorder in a female,[61] brought about while she is still within her mother, through the receptacle (*āśaya-*) being corrupted by wind in the embryo (*garbha-*) due to a corruption of the *bīja-*.[62] Cakrapāṇidatta here equates *bīja-* and *ārtava-*.[63] But it is unclear whether the *ārtava-* of the mother or of the embryo/foetus is meant. Jejjaṭa says simply:

> *bījam atra janakam*
>
> "*bīja-* here is [that which is] generating."

He then quotes CA,Śā 3,17 (on which see also p.27 below) in a variant or paraphrased form,[64] which says that that becomes abnormal in which "the *bīja-*-part (*bījabhāga-*) is heated".

As regards *āśaya-*, Cakrapāṇidatta says, no doubt rightly, that what is meant is the uterus of the embryo/foetus,[65] drawing attention to CA,Śā 4,30.[66] Referring to the theory that certain parts of the embryo come from the mother only,[67] CA,Śā 4,30 says that individual parts of the embryo are unnatural according to whether the morbific entities (☞ *doṣa-*) in the *bīja-* or *bīja-*-part (*bījabhāga-*) relating to them are corrupted.[68] After this 4,30 gives

---

[61] In the case of 34f. the affected woman hates men, which probably implies frigidity, is flat-chested and a *ṣaṇḍhī-*, i.e. a female ☞ *ṣaṇḍha-*.

[62] *bījadoṣāt tu garbhasthamārutopahatāśayā* (30,34cd).

[63] *bījadoṣād iti sāmānyavacanenāpy ārtavarūpabījadoṣa eva jñeyaḥ.*

[64] *uktaṃ ca yasmin bījabhāga upatapto bhavati tasmin vikṛtir abhinirvartate.*

[65] Thus also Gaṅgādhara in CA1, p.3465.

[66] From which he quotes a part relating to the formation of a ☞ *vandhyā-* "barren woman".

[67] Cf. on this e.g. TIVĀRĪ 1,pp.125f., as well as §5.11 and n.6; the uterus too comes from the mother according to AS,Śā 5,p.316a.

[68] *yasya yasya hy avayavasya bīje bījabhāge vā doṣāḥ prakopam āpadyante taṃ tam avayav-*

examples, among them also "*bīja*-part for the uterus (☞ *garbhāśaya-*)" (*garbhāśayabījabhāga-*) and "portion of the *bīja*--part for the uterus" (*garbhāśayabījabhāgāvayava-*) in the "blood" (*śoṇita-*) (the latter obviously that of the mother). One could now take *bīja-* alone to refer to the "germ" of the individual member, *bījabhāga-*, however, to that part of the "germ", from which the whole embryo develops, relating to the individual member (of this embryo), i.e. *bījabhāga-* is "part of the *bīja-*", *bīja-* referring to the first stage of the embryo as a whole. On the other hand, one could also analyse *bījabhāga-* as "part which is the *bīja-*", in which case *bīja-* in *bījabhāga-* would refer to the "germ" of the individual member too. Though the CA text does not, at least at first sight, seem to speak for this second interpretation, it should not be lost sight of, especially since *garbhāśayabījabhāga-* could also be analysed as "part of the uterus-*bīja-*" or "part which is the uterus-*bīja-*", and *garbhāśayabījabhāgāvayava-* as "portion of the part of the uterus-*bīja-*" or "portion of the part which is the uterus-*bīja-*"; in these cases *bīja-* would obviously refer to the "germ" from which the individual part arises. Cf. in this connection Cakrapāṇidatta's commentary on SU,Sū 15,19 (SU3: 15,18), in which he says that all the ☞ *dhātu-*s are present in *bīja*--form in (male) semen and (female) "blood" (in the fluids singly, or in their combined form?).[69]

Cakrapāṇidatta in his commentary explains *bīja-* as referring to the whole germinal part, *bījabhāga-*, however, to a part of the *bīja-* of the individual member,[70] which creates the impression that *bīja-* as the "whole germinal part" refers to the whole individual member and not to the embryo as a whole, though that is not clear.[71] He further gives two alternative explanations of *garbhāśayabījabhāga-*, namely as that *bīja*--part[72] giving the uterus, or else that part (scil. of the "blood" of the mother) giving the uterus and the *bīja-*, i.e. *ārtava-*.[73] He also says that the embryo/foetus has *ārtava-*

---

aṃ vikṛtir āviśati.

[69] SU3, p.270, SU4, p.117: *śukraśoṇite bījarūpatayā sarvadhātūnāṃ vidyamānatvāt.*

[70] *bīja iti kṛtsna evārambhake bīje. bījabhāge vety avayavabījasyaikadeśe.*

[71] Gaṅgādhara in CA1, p.1974 is vague, for he gives the explanation *avayavārambhakabījabhāga-* "*bīja*--part beginning the [body-]part", but does not tell us whether this relates to *bīja-* or *bījabhāga-* of the text, or to both.

[72] "Part of the *bīja-*" or "part which is the *bīja-*"? In the first case *bīja-* would refer most probably to the embryo, in the second to the individual member.

[73] *garbhāśayajanako bījabhāgo garbhāśayabījabhāgaḥ. kiṃvā garbhāśayasya tathā bījasya cārtavarūpasya janako bhāgaḥ.*

since its very inception,[74] which makes it seem likely that in his commentary on Ci 30,34f. he has the *ārtava-* of the embryo/foetus in mind.

Commenting on CA,Śā 4,14 (≈ AS,Śā 2,p.289a), which too mentions the *garbhāśayabījabhāga-* along with other things making a female (*tathāyuktāni cāparāṇi strīkarāṇi*), Cakrapāṇidatta says that the latter refer to the "*yonibījabhāga-* etc."; on the translation of *yonibījabhāga-*[75] cf. the problems with regard to *garbhāśayabījabhāga-* referred to in the discussion of CA,Śā 4,30 above. Since he then says that such things happen in a *bīja-* in which *rajas-* is in excess,[76] it seems that to Cakrapāṇidatta *bīja-* in Śā 4,14 is the "germ" formed by the combination of the two procreatory fluids, i.e. the embryo as a whole, from which it would follow that *bījabhāga-* is, if *bīja-* here really means the same, part of the "germ" which is this embryo; unfortunately, we have no surety that *bīja-* means the same in both cases. However, in his commentary on CA,Śā 3,17, which deals with parts coming from the father or "fathers"[77] (i.e. parents, "the two fathers" being an ancient expression for these), Cakrapāṇidatta says that *bīja-* refers to semen and *śoṇita-*, and that "*bīja*-part" refers to that part of the *bīja-* bringing about the body members, adding that the *bīja-* of humans is the combination of the parts of the *bīja-* relating to the various members of the body.[78] The text of CA,Śā 3,17 says that those body members (of an embryo) become abnormal whose *bījabhāga-* in the *bīja-* becomes heated.[79] The parallel AS,Śā 2,p.294f. refers to heat in the *bīja-* or part of the *bīja-*.[80]

It may also be noted that CA,Śā 4,31 deals with the individual parts of the embryo coming from the father,[81] but a discussion of this would lead

---

[74] On this see §5.16.

[75] ☞ *yoni-* probably referring to the vagina, as the uterus is already referred to as the ☞ *garbhāśaya-*.

[76] *rajo'dhika eva bīje bhavantīti jñeyam.*

[77] *pitṛ-* in *pitṛsadṛśarūpa-* of the text could refer to either father or fathers.

[78] *bīja iti śukraśoṇite. bījasyāṅgapratyaṅganirvartako bhāgo bījabhāgaḥ. ... evaṃ manyate. manuṣyabījaṃ hi pratyaṅgabījabhāgasamudāyātmakaṃ svasadṛśapratyaṅgasamudāyarūpapuruṣajanakam.*

[79] *yasya yasya hy aṅgāvayavasya bīje bījabhāga upatapto bhavati tasya tasyāṅgāvayavasya vikṛtir upajāyate.* It is possible that Cakrapāṇidatta had passages such as this in mind when explaining *bīja-* in CA,Śā 2,18 (see n.57).

[80] *yasya yasyāṅgāvayavasya bīje bījāṃśe vopataptir bhavati tasya tasyāvayavasya vikṛtir upajāyate.*

[81] On these cf. e.g. TIVĀRĪ 1,pp.127f.; see also n.67 above.

us too far off the track. The above is problematic enough as it is, and this is compounded by the fact that, in spite of what Cakrapāṇidatta says, we cannot be sure that all the CA passages quoted are to be connected. Matters are not made easier when we consider other similar passages. AH,Ut 33,28 is similar to CA,Ci 30,8, whereas AH,Ut 33,39f. is similar to CA,Ci 30,34f., though it uses ☞ *yoni-* in the place of *āśaya-*. HILGENBERG/KIRFEL, pp.674f. simply translate *bīja-* with "Samen" (i.e. "seed"), which is begging the question, unless it should refer to semen (as the German word may), which would show that they have not understood what is meant. Aruṇadatta does not say anything relevant, but Śivadāsasena (in AH3) has some comments to make. On AH,Ut 33,28 (29 in AH3) he says:

> *bījadoṣeṇeti strīśarīrārambhakabīje ya ārtavabhāgo vidyate taddoṣeṇety arthaḥ. ārtavarūpabījadoṣād eva vandhyāyāḥ sambhavāt*
>
> "'*bījadoṣeṇa*': the meaning is: 'through corruption of that which is present as the part which is/part which pertains to *ārtava-* in the *bīja-* beginning the body of the woman'. Because of the production of a barren woman (☞ *vandhyā-*) due to [this] very corruption of *bīja-* in the form of *ārtava-*."

And on AH,Ut 33,39f. (40f. in AH3) he says:

> *etad evāha bījadoṣata iti. yady api śukraśoṇitasamudāya eva bījaṃ tathāpy atrārtavarūpabījadoṣa eva jñeyaḥ. taddoṣasyaiva garbhāśayopaghātakatvāt*
>
> "This very [disorder] he says is '*bījadoṣataḥ*' (i.e. from a corruption of the *bīja-*). Even though actually the combination of [male] semen and *śoṇita-* is the *bīja-*, nevertheless here a corruption of the *bīja-* in the form of *ārtava-* is to be known. Because of that very corruption's being damaging for the uterus (☞ *garbhāśaya-*)."

In his commentary on AS,Ut 38,p.289b (and the identical verse AH,Ut 33,28), Indu says that *bīja-* here refers to the *bīja-* which is the germinal form of the ☞ *yoni-*, whilst the *Kairalī* commentary adds (on the AH passage) that it has the nature of the ether.[82] Indu seems to have nothing to say on AS,Ut 38,p.292a = AH,Ut 33,39f. What the *Kairalī* says on the latter might, however, equate *bīja-* with *ārtava-*, though it might also be referring to a corruption of *ārtava-* due to a corruption of the *bīja-*, as it is unclear whether *bīja-*

---

[82] Indu on the AS passage: *bījadoṣaḥ striyo janmakāle yonyārambhakasya bījasyopahatatvam*, on the AH passage: *bījadoṣeṇa striyā janmakāle yonyārambhakasya bījasyopahatatvena*; *Kairalī*: *bījadoṣeṇa yonyārambhakasyākāśātmakasya bījasyopahatatvena*.

doṣata ārtavaduṣṭyā in the commentary means "due to a corruption of the *bīja-*, [i.e.] through corruption of *ārtava-*" or "through corruption of *ārtava-* due to corruption of the *bīja-*".[83] In his commentary on AS,Śā 2,p.289a, Indu explains *garbhāśayabījabhāga-* of the text commented upon as that *bījabhāga-* which is the cause of the ☞ *garbhāśaya-* (*garbhāśayakāraṇa-*), and remarks on *tathāyuktāś cāpare strīkarā bhāvāḥ*[84] of the text that this refers to the *bīja-bhāga-* which is capable (*samartha-*) of creating the specific characteristics of a female; this does not tell us much on what exactly *bījabhāga-* means to Indu.

SU,Ut 38,6 mentions *bījadoṣa-* (along with *praduṣṭa- ārtava-* in 5) as a cause of genital disorders in females, but does not follow this up, and neither does Ḍalhaṇa comment upon this. AS,Śā 2,p.294a first refers to the *bīja-* or part of the *bīja-* of the members deriving from the mother being corrupted,[85] then, however, refers to the *bīja-* called "blood" being corrupted;[86] but this refers to the "blood" of the embryo, the *bīja-* which gives this being situated in the womb of the mother (similarly, the *śuklākhya- bīja-* "*bīja-* called *śukla-*[87]" of the embryo is mentioned on p.294b). A further discussion would lead us too far astray from our actual problem, but even so I trust that the various remarks excerpted above have given a good example of the difficulties connected with the term *bīja-*.[88]

---

[83] *garbhe garbhādhānasamaye bījadoṣata ārtavaduṣṭyā strī jāyamānaputrī nṛdveṣiṇī ... syāt.*

[84] Which parallels *tathāyuktāni cāparāṇi strīkarāṇi* of CA,Śā 4,14 discussed above.

[85] *garbhasya yasya yasya mātṛjasyāvayavasya bīje bījāṃśe vā doṣaḥ prakopam āpadyate ...* .

[86] *yadā hy asyāḥ śoṇitākhye bīje garbhāśayasthe nirvartakaṃ bījaṃ praduṣyati ...* .

[87] I.e. *śukra-*.

[88] On *bīja-* and *bījabhāga-* cf. also NYC's commentary in n.2007 and §11.31.

CHAPTER 3

The Bower Manuscript

3.1. One of the oldest medical texts known to us[89] and containing information on our problem (the texts in the sixteenth chapter of the *Suvarṇa-(pra)bhāsottamasūtra*,[90] in LÜDERS, pp.585-591, and in TURFAN, 1,p.288 and 3,pp.47-54;167 are of no use to us in this respect) are contained in the *Bower Manuscript* (which includes the *Nāvanītaka*), and they may be as old as the beginning of the Christian era or even older. In them we find *ārtava-* used (3,p.183,48) in expressions referring to women with "standing" (blocked?) and spoilt *ārtava-* (*sthitārtavā-* and *naṣṭārtavā-* respectively), which would make it seem as if the menstrual fluid were meant here, but this is not certain. Nothing is said about the origin of this *ārtava-* or the reasons for or symptoms of the disorders. 1,p.8,114 refers to women who discharge spoilt *rajas-* (*rajo duṣṭaṃ sṛjanti*), which could be the same as the spoilt *ārtava-* mentioned above, but need not be so. We have no further information in this case either. *rajonigraha-* "withholding of *rajas-*" in 2,p.47,424 could be the same as the blocked *ārtava-* mentioned above, but here again we cannot be certain, nor is any further information available. An unhealthy discharge seems to be called *asadrajas-* in 2,p.28,36, 2,p.33,142, 2,p.40,285 and 2,p.43,341; however, *asad-* means both "wrong" or "bad" as well as "unreal". *asṛgdara-* (on the symptoms of this — provided, of course, that the same disease is meant — see §5.14 below) is mentioned in 1,p.4,41, but from this too we learn nothing, as the causes mentioned loc.cit. below need not necessarily apply here. Thus all this does not help us much in determining the nature and eventual relationship of the procreatory and menstrual fluids to each other and/or blood in general.

---

[89] The texts in the *Bower Manuscript* are clearly quite early because of the antiquity of the manuscript, which cannot be later than the first few centuries of the Christian era, approximately the sixth century at the utmost. Whether these texts are older than the 'classical' texts (to be discussed in the next chapter) *in their present form*, is a matter of controversy which I am leaving untouched here.

[90] This chapter is called *Vyādhipraśamanaparivarta*. In the edition used it is to be found on pp.174-182, the medical portion of it being on pp.177-180.

CHAPTER 4

The 'Classical' Medical Texts

4.1. Let us therefore move on to the four so-called 'classical' medical texts, which are regarded as authoritative in India. These are: CA, SU, AS and AH.[91] Before beginning with their discussion, I must, however, point out that finding relevant information in these voluminous works (as well as in the other texts discussed in this study) is not always easy. Not only were they written as a rule by specialists for specialists, thus leaving much that was common knowledge to such specialists (though unfortunately not to modern scholars) unmentioned, but one is also handicapped by a general lack of adequate resources for their study (e.g. indexes etc.). So it is possible that I may have overlooked something relevant, though I have done my best to leave out no scrap of information that may somehow or other be connected with

---

[91] The exact dates of these four works are uncertain and need not be discussed here in detail. CA and SU have each been redacted at least once, and it is known that several of their portions are later additions of the redactors. The oldest versions now known are those commented upon by Cakrapāṇidatta (CA and SU, only parts of the commentary on the latter is today extant) and Gayadāsa (SU, only parts of the commentary being extant) (Cakrapāṇidatta and Gayadāsa lived in approx. the 11th century A.D.; cf. MEULENBELD 1974, pp.398f.;400-402, on Cakrapāṇidatta cf. also DATTA CAUDHURĪ, pp.26ff. and SUBHAKTHA, pp.54f.), and we do not know exactly how old these versions are; we do have a portion of Jejjaṭa's commentary on CA too (Jejjaṭa may have lived in about the seventh century; cf. MEULENBELD 1974, pp.406-408), but this is fragmentary, and moreover preserved only in a version later revised. Nevertheless, portions of CA and SU, especially those here concerning us most, seem to be about a thousand years — if not more — older than Cakrapāṇidatta and Gayadāsa. As regards the other two texts, they may or may not be by the same author; one nowadays tends to assume the former, and many hold that AS gradually evolved out of AH, though this is still not proved beyond the shadow of a doubt (cf. too AH,Ut 40,80, which, however, could also be a later addition), though it may be noted that to Aruṇadatta AS and AH were definitely composed by the same author; cf. his commentary on AH,Śā 1,8 (*ata evāyam eva tantrakāro 'nyathā saṅgrahe jagāda* ...). In any case, whichever one of these is older must have been composed around 600 A.D. Both seem to have obtained their present shapes by the ninth or tenth century at the latest. Regarding all other questions concerning the date and authorship of these as well as the other medical texts I shall discuss below, I request the reader to consult the voluminous work on the history of Indian medical literature Gerrit Jan Meulenbeld is preparing and which will, it is to be hoped, be published soon; for preliminary information see MEULENBELD 1974, pp.403-406;423-425;431f. and HĀLADĀRA (on the problem of AH and AS see also pp.(4)-(15) of the introduction of AH3).

our problem. I have, however, neglected the word *rajasvalā-* (cf. §2.1),[92] as this always refers to a woman having menses,[93] but offers no other useful information.

4.2. I am aware of the fact that, since these texts are not homogeneous in origin, it is methodically not quite correct to treat them as if they were. Thus I shall do my best to keep not only the discussions in different texts, but also those of different theories enumerated in the same text separate. However, it will be seen that certain obscure points in one text or in one part of one text can be elucidated with the help of the information in another text or another part of the same text. In such cases one must of course reckon with a potential methodical unsoundness, though if on the other hand one should find that, leaving such cases aside, the theories extracted from each of the texts mentioned or from different parts of these are not in conflict with each other, then the potential for error is greatly reduced. But all these are theoretical considerations; how our study ultimately turns out will of course depend on the data we are able to extract from the individual texts.

[92] E.g. CA,Ci 30,165, SU,Ni 12,7, Ci 24,114;121, Ut 61,5, to name only a few occurrences.

[93] Very often combined with the injunction (so common in Indian literature) against having intercourse with her then.

CHAPTER 5
The Carakasaṃhitā

5.1. The data from CA are quite scant in comparison with those of the other texts and also leave the impression that the topics interesting us were not deemed worthy of especial consideration by the author(s). *rajas-* is used to denote the fluid uniting with semen to form the embryo in Śā 2,34. In Śā 4,7 "old" (*purāṇa-*) *rajas-* is said to leave the body and be replaced by new, this condition being then that of the *ṛtu-* (on this see §2.1), in which conception takes place due to intercourse.[94] These uses of *rajas-* could point to the menstrual flow being taken to be the flow of the fluid which unites with semen, after this fluid has lost its power of fecundity, i.e. grown "old". But the text does not say so expressly, and there is a possibility that menstruation may have been taken to be a general cleansing of the body and womb quite unconnected with the procreatory fluid, and that "old" procreatory fluid was held to leave the body by some process other than menstruation or be otherwise absorbed by the body, especially since in the same passage we find the procreatory fluid mentioned and being called not *rajas-*, but *ārtava-* (cf. §5.4).

5.2. In Śā 2,3 and Si 2,16 we find *rajaḥkṣaya-*, "*kṣaya-* of *rajas-*", mentioned. The term *kṣaya-* means "loss, destruction, waste, diminution, decline",[95] so that its exact meaning here is not clear; nevertheless, it seems probable (even though we cannot be fully sure) that in Śā 2,3 menstruation is meant, for the text says that intercourse to create a child should take place when this *rajaḥkṣaya-* is over (*rajaḥkṣayānte*). But Si 2,16, where *rajaḥkṣaya-* is a disorder brought about by wind, is more problematic. In this passage we also find *retaḥkṣaya-*, where *retas-* seems to refer to semen, as it does in Ci 28,22 where we find, in a passage which lists disorders caused by wind again, *garbhaśukrarajonāśa-* "loss/destruction of the embryo/foetus, semen and

---

94 *gate purāṇe rajasi nave cāvasthite śuddhasnātāṃ striyam avyāpannayoniśoṇita-garbhāśayām ṛtumatīm ācakṣmahe. tayā saha tathābhūtayā yadā pumān avyāpannabījo miśrībhāvaṃ gacchati ...*
"When the old *rajas-* is gone and the new is present, we regard the woman, pure [after the first three days of the *ṛtu-*] and [ritually] bathed, with uncorrupted ☛ *yoni-* (here the vagina?), blood (on this term see §5.4 and §5.8) (or: *yoni-*-blood, i.e. blood of or from the *yoni-*?) and womb (☛ *garbhāśaya-*), as being characterised by the *ṛtu-*. If the man, with uncorrupted *bīja-* (see §2.6), commingles with her, the one who is so ... ."

95 MEULENBELD 1974, pp.458f. prefers only "decline" for *kṣaya-*.

rajas-".[96] In these two passages *rajas-* could be used parallelly to "semen", in which case it could refer to the procreatory fluid. On the other hand, what could be meant is the non-visibility of a discharge that one usually can see (this would also apply to semen), in which case the menstrual fluid could be meant. Since a woman who cannot menstruate as a rule does not, unless she be pregnant already and does not menstruate because of this, possess the general capability of conception either, we would in each case have a reference to infertility, whatever be the meaning of *rajas-* here. Thus *rajas-* in Śā 2,3 and in Si 2,16 (as well as in Ci 28,22) need not mean the same.

5.3. Be that as it may, the use of *kṣaya-* to describe the flow of *rajas-* could, however, be of importance, as it could point to a parallel, regarding the female, existing to the notion prevalent in almost all cultures that seminal discharge in connections other than procreation is a "waste" or "loss"[97].[98] This latter is, however, only speculation unless it is substantiated further. It may be pointed out that according to HĀ 3,48,p.361 *dhātukṣaya-*, i.e. "*kṣaya-* of the element", makes a woman "crow-barren"[99] or induces other forms of childlessness; ☞ *dhātu-* (which is used in a variety of meanings in medical texts)[100] here seems to refer to the female procreatory fluid (see also p.555). In this passage *kṣaya-* appears to be more than a mere discharge, though this is not sure. In the same text, however, we find another use of *dhātukṣaya-* (in HĀ,Śā 1,p.413): *mṛtyur dhātukṣayāt* "death [results]

[96] It must, however, be pointed out that *kṣaya-* and *nāśa-* may not refer to exactly the same; cf. e.g. AS,Sū 28,p.213, where we find *śukrārtavastanyanāśa-* and *rajaḥkṣaya-* mentioned side by side, the problem being compounded by the fact that at this stage of our investigation we do not know whether *ārtava-* and *rajas-* here refer to the same or different substances. It is in this context, however, interesting to note that two different expressions for "semen" occur in this very same passage, nearly side by side: *śukra-*, already quoted above, and *retas-* (in *retodoṣa-*); we similarly find *retaḥkṣaya-* and *śukrasaṅga-* nearly side by side in CA,Si 2,16, one of the passages we are discussing here.

[97] On the belief in the extremely harmful effects of semen loss (spermatorrhoea, spermatic leakage) in Indian as well as Western cultures see especially BOTTÉRO.

[98] One could in this connection point to SU,Sū 14,6 (*raktaṃ rajaḥsaṃjñaṃ ... yāti pañcāśataḥ kṣayam*) and Śā 3,11 (*asṛk ... yāti pañcāśataḥ kṣayam*), where it is said that at the age of fifty the "blood" (SU,Sū 14,6: "blood called *rajas-*") flowing monthly attains *kṣaya-*. But here *kṣaya-* does seem rather to mean "cessation" (since the menopause is being described), even though we cannot at this stage of our investigation fully discard the possibility that the word *kṣaya-* may refer to some unknown theory on the origin of the menopause, in which it could quite well mean "waste" etc.

[99] *kākavandhyā-* (see on this p.580). On ☞ *vandhyā-* see Appendix II.

[100] See *dhātu-* in Appendix II.

from *kṣaya-* of the elements [upholding the body]"; here *dhātu-* is probably used in the manner of §2.4 above for the elements from the transformation of food, which would be in keeping with a great many similar passages in other texts which speak of death or illnesses arising from the *kṣaya-* of these elements.[101] The factors leading to such *kṣaya-* are various, and it is not quite clear in each such case in what manner this *kṣaya-* is supposed to take place, even though obviously some sort of loss or diminution is meant;[102] this also holds true for CA,Sū 17,63ff. (see also Ci 30,184), where the effects of the *kṣaya-* of different substances are described. I may also point out that these CA passages, in contradistinction to e.g. SU,Sū 15,12 (which will be discussed later in §6.10), do not mention *ārtava-*, *rajas-* or the like in this connection. In any case, they do not help us much in determining beyond the shadow of a doubt what *kṣaya-* in §5.2 actually signifies, so we shall have to stick to the tentative deductions given there.

5.4. All in all, however, we do get an inkling of the notions regarding the relation between menstrual discharge and the procreatory fluid from the above. Unfortunately, this seems to be more or less all the information available on this topic, other statements not being very helpful. The procreatory fluid (uniting with semen from the male) is called *ārtava-* in Śā 2,14, and 4,7 and one of the common words for "blood" (see §2.1) in all other clear cases I could find,[103] without our being able to gain any more relevant information from this apart from the fact that the nomenclature used could suggest that the procreatory fluid was viewed as being blood or maybe like blood (which are two different notions), though this need not necessarily be the case, for there could be other reasons — reasons which to us may not be obvious and not having anything to do with its colour — for calling this fluid "blood". The appearance of the menstrual flow is called *puṣpa-* in Śā 8,5;[104] in Śā 8,23-26 discharges which appear during pregnancy are called

101 Cf. MEULENBELD 1974, pp.458f., also n.211 below.

102 As it is, *kṣaya-* is an extremely common term in medical texts to denote the loss or diminution or the like of various individual substances of the body too (not just of the elements mentioned), individually or together.

103 Vi 8,95, Śā 2,6f.;12f.;16;34, Śā 3,3;8;11, Śā 4,5;30, Śā 8,3;6;17, Ci 11,77 (*śoṇita-* as a cause of urinary disorder in Sū 19,4,1, however, refers to blood in general, in spite of being directly preceded by *śukra-*; cf. Ci 26,43f.). Śā 2,12;18 state the well-known theory (cf. §1.5) of the predominance of the female procreatory fluid giving rise to a girl, of male semen to a boy, whereas an equilibrium results in a being which is neither, and referred to by CA,Śā 2,18 not as ☞ *napuṃsaka-*, but as ☞ *dviretas-*:

... *strīpuṃsaliṅgī bhavati dviretāḥ*

"... a *dviretas-* characterised by the signs of [both] female and male develops".

104 *Not* in In 2,4 (quoted by GHĀṆEKAR 1986, p.81a), where *puṣpa-* has to do with the

puṣpa-, without our being able to say what they are composed of, apart from the fact that, since *puṣpa-* commonly denotes the menstrual flow, one would assume that they must be like this. In Ci 28,10f. *ārtava-* is mentioned in connection with wind, which latter "hurls (out)" (i.e. expels) (*sṛjati*) semen (*śukra-*), urine, faeces, *ārtava-* and the embryo/foetus; what exactly *ārtava-* is is unclear, as it might be a periodic discharge like urine and faeces, or some other discharge linked with the embryo/foetus or parallel to semen. In Ci 28,230 we find mentioned what seems to be an excessive flow (*ativartana-*) of *rajas-*; the meaning of *ativartana-* is, however, unclear, since *ati√vṛt* "to pass beyond" is also used to mean "transgress, violate", and as such *ativartana-* might also refer to a delayed flow or a flow at the wrong time, both of which are "transgressions" or "violations" of the usual course of events too. The same applies for *ativṛtti-* in the parallel passage AH,Ni 16,45. The reading of the last passage accepted by AH1 and also Śrīdāsa, as well as that of AS,Ni 16,p.82a, is, however, *ativṛddhi-* "excessive growth". But whatever be the correct reading or meaning, *rajas-* in any of these cases could refer to the menstrual fluid, but we cannot be sure on this point. CA,Ci 13,32, however, tells us that women put *rajas-*[105] into the food of men, seemingly to make these enamoured of them; Ci 23,233 is similar. Here *rajas-* surely refers to the menstrual fluid, though that does not necessarily also apply to other passages in which it occurs.

5.5. In Śā 2,8f. it is said that a disorder of the wind may result in a blocking of "blood" (no details are given), leading to the signs of a false pregnancy (here too no details are given). The text says that this "blood" may, due to several causes, ultimately flow out of the body and be seen. It is, however, not possible to determine the exact nature of this "blood" (here called *asṛj-* throughout), for since what we have here is an irregular flow, it could, even though one can see it (just as one can see menstrual blood), refer to the procreatory fluid, in which case this would then, if we follow the possibility mentioned in §5.1 on some as yet unknown manner of leaving the body or being absorbed into it, maybe leave the body in a manner not its usual one. One could also consider the possibility of blood in general being referred to here; since, however, the passage speaks of the "blood" being trapped in the body (Cakrapāṇidatta's commentary refers explicitly to the ☞ *garbhāśaya-* or womb[106]), i.e. seemingly hindered from leaving this as it

---

signs of impending death and has nothing to do with any discharge or the female specifically.

105 Along with stool, urine, nails, hair etc. On this passage and its parallels cf. also DAS, R.P. ?, n.12.

106 *garbhāśaye hi vivardhamānaṃ rudhiraṃ prabhāvād garbhaliṅgāni kānicid darśayatīty*

normally does, that would speak for this blood, if indeed it should be meant here, having a special function as procreatory or menstrual fluid. Or it could simply be the menstrual fluid, hindered from egressing in its usual manner.[107] In Sū 13,47, where *rudhira-* is mentioned (along with *retas-*, here probably "semen") as being "dried up" (*saṃśuṣka-*), *rudhira-* seems to refer to the menstrual flow, but that is not certain either.[108] In Ni 4,7 we find "blood" (*śoṇita-*) mentioned in a list also containing semen (*śukra-*); due to the mention of semen "blood" could refer to the corresponding fluid of the female, but whether that would be procreatory or menstrual is quite unclear. Moreover, it is also not clear whether this list, which otherwise lists only substances common to both males and females, really refers to the special "blood" of females and not to blood in general, since semen too was thought to be common to both males and females (cf. §2.4). Unclear too is Ci 18,105 referring to disorders (*doṣa-*) and "drying up" of semen and "blood", or to "drying up" of semen, "blood" and the morbific entities (*doṣa-*) (*śukrāsṛgdoṣaśoṣeṣu*). Though it looks as if "blood" here might refer to a fluid of the female, we cannot be fully certain that this is so;[109] if it does refer to a fluid of women, then we are still in the dark as to what exactly is meant. In Ci 30,

---

arthaḥ.

107 I may add regarding CA,Śā 2,9c that this line, which CA gives as *dṛṣṭvāsṛg ekaṃ* (var: *evaṃ*) *na ca garbhasaṃjñaṃ*, is also found in the form *dṛṣṭvāsṛg eke na ca garbham ajñāḥ* (e.g. in CA3, p.500), while CA1 (p.1899) gives it as *dṛṣṭvāsṛg ekaṃ na ca garbham ajñāḥ* (var: *garbhasaṃjñaṃ*).

108 The text has *saṃśuṣkaretorudhirā[ḥ]* "whose *retas-* and *rudhira-* are dried up". The expression is here masculine (plural), but may, according to the rules of Sanskrit grammar, subsume the feminine too, if both should be meant. It is unclear whether only males or males and females are meant; in the former case (cf. *narāḥ* in 49) *rudhira-* would obviously refer to "blood" literally, i.e. to blood in general. Otherwise, it could be that a state is meant in which a discharge usually visible is not seen, which would speak for *rudhira-* referring to the menstrual discharge. However, the expression may also be an oblique way of saying "sterile", in which case *rudhira-* might rather refer to the procreatory fluid, which may or may not be (according to our study so far) different from the menstrual fluid. And there is even the possibility of *rudhira-* referring to the blood (in general) of both men and women. Cf. in this connection SU,Ci 31,17, which mentions people with "dried up" semen and blood (*śuṣkaretoraktā[ḥ]*) in a similar list of persons for whom a certain medicine is beneficial, subsequently labelling all those enumerated as *narāḥ*, which might mean "men" or (when both men and women are mentioned) "persons" (in this case the masculine form subsumes the feminine). Clearly, here too "blood" would refer to blood in general if men alone were meant; if women too were meant, then it could, however, refer to some fluid of theirs, though what this would be is unclear.

109 Cf. the discussion in n.108.

where disorders of the female generative organs are described, *ārtava-*, *rajas-* and the "blood" words are used throughout the chapter, seemingly without any system.[110] Ci 30,8 refers to polluted (*praduṣṭa-*) *ārtava-* being one of the causes of these disorders, but the exact nature of *ārtava-* remains unclear, especially when we consider 30,11-14 and a variant to 30,36,[111] where various diseased discharges are called *ārtava-* too. Here *ārtava-* is probably not blood in general, but that is all we can say with some certainty. It could refer to (diseased) menstrual discharge, but need not. Other passages in Ci 30 will be discussed in §5.8 and §5.10 below.

5.6. In Si 9,62 we find it said that a woman is to be treated by ☞ *uttaravasti-* during the *ārtavakāla-* "time of *ārtava-*", for during this time the *garbhāsanā-* (probably the uterus, i.e. ☞ *garbhāśaya-*, as Cakrapāṇidatta in his commentary also opines, though he says too that others take it to be the ☞ *yoni-*) is *apāvṛta-* "laid open; uncovered" and thus easily accepts what is injected[112] (similarly AS,Sū 28,p.223a, AH,Sū 19,77f.)[113]. Here *ārtavakāla-* seems to refer obliquely to the time of the *ṛtu-*,[114] i.e. the period of fecundity, not of menstruation (cf. §2.1), for, though CA does not say so expressly, we know from other texts (see TIVĀRĪ, 1,pp.68f. for references) that the ☞ *yoni-*[115] of a woman is at times other than this supposed to close (scil. contract?) and not to permit any semen to enter it. Should this be so, then we could deduce from this that here *ārtava-* is the procreatory fluid, for it would seem to be that which is capable of combining with semen when the uterus is "open". Cakrapāṇidatta's commentary on the CA passage (that of Jejjaṭa is unfortunately missing), however, operates with "uncovered" in a different sense, saying that *apāvṛta-* means "from which the covering/obstruction in the form of accumulated *rajas-* is gone";[116] since it is hardly feasible to assume that Cakrapāṇidatta means some sort of covering (due to

---

[110] Neither does there seem to be any differentiation on the basis of impurity (according to whether a red fluid issues within or outside the 'usual' time of menstruation), as one finds in non-medical texts according to e.g. LESLIE,I.J. 1989, pp.283f. (see also LESLIE,I.J. 1994, pp.74f.). On this latter point cf. also n.124.

[111] *asaṃvṛtamukhī sārtiḥ saphenārtavavāhinī* (see on this n.137).

[112] *garbhāsanā sukhaṃ snehaṃ tadādatte hy apāvṛtā.*

[113] *strīṇām ārtavakāle tu yonir gṛhṇāty apāvṛteḥ* (var: *apāvṛtā*).

[114] This would be in accord with the basic meaning of *ārtava-*: "belonging to the *ṛtu-*".

[115] Here most probably the uterus (scil. its orifice or cervix?) is meant; we shall return to this problem in §8.19.

[116] *apāvṛtety apagatasañcitarajorūpāvaraṇā.*

dried fluid residues or the like) over the opening of the uterus into the vagina, as this would presuppose some steady trickle of fluid into the vagina and accumulation there, and as there seem to be no parallels of any sort for such a theory of regular covering of the opening of the uterus in Indian medical literature, this would in effect mean a uterus no more filled by useless old fluid obstructing or hindering other processes, i.e. an empty uterus.[117] Now does the "open" of the text not seem a bit strained if it should indeed refer to an empty uterus? The explanations given in the commentaries on AH and AS of *apāvṛti-* (obviously "the lying open; uncoveredness") or the variant *apāvṛta-* (thus Indu, maybe also Candranandana) are: "due to the obstruction/concealment/covering [being] gone" (Aruṇadatta: *apāvṛter apāvaraṇāt kāraṇāt*), "due to unconcealedness/openness" (Hemādri: *apāvṛter vivṛtatvāt*; Śrīdāsa: *apāvṛter vivṛtatvād dhetor*), "due to the uncovering of the channel (☞ *srotas-*)" (Parameśvara: *apāvṛteḥ srotovivaraṇāt*),[118] "the ☞ *yoni-* is *apāvṛta-*, [i.e.] its mouth is unconcealed/open" (Indu on AH and AS: *yonir apāvṛtā vivṛtamukhā bhavati*) and "because at that time the ☞ *yoni-* does not offer the *śukra-* [of the male] an obstruction (*antar*)" (Candranandana on AH2,Sū 19,66: *yasmāt tasmin kāle yoniḥ sā śukraṃ nāntaḥ prayacchati*); all of these are in keeping with the fact that these texts contain the theory of the periodic closure and opening of the *yoni-*. Why has Cakrapāṇidatta not explained the CA passage in a similar fashion? One does gain the impression that this is because of the non-mention of this theory in CA; the theory would, however, fit in nicely with Si 9,62. But even if this theory should not be presupposed here, Cakrapāṇidatta's statement that the uterus is empty would still preclude *ārtavakāla-* meaning the time of menstruation, for that would already be over. Thus even in this case, i.e. even if CA should not presuppose the theory of closure of the uterus at times other than the period of fecundity, *ārtava-* would probably refer to the procreatory fluid. I must admit, though, that this identification, however we may ultimately arrive at it, does not actually help us much in determining the relationship of this fluid to the menstrual fluid.[119]

---

[117] What exactly this fluid presupposed by Cakrapāṇidatta could be is discussed in n.119.

[118] Only the relevant words in the cited portions of the commentaries of Aruṇadatta, Hemādri, Śrīdāsa and Parameśvara have here been translated.

[119] As regards *rajas-* in Cakrapāṇidatta's commentary, how exactly are we to understand this? To my reasoning a *garbhāsanā-* "from which the covering/obstruction in the form of accumulated *rajas-* is gone" makes sense if we take *rajas-* to refer to the menstrual fluid, which leaves the body periodically. If we now assume that *ārtava-* must, on etymological grounds, be present during the *ārtavakāla-*, it could follow from this that we have here, according to the

5.7. It may, however, be pointed out that the theory of the closure of the ☞ *yoni-* is, in the context of the theory that male semen and female procreatory fluid combine to form an embryo, a good explanation for the problem of why then conception does not invariably occur each time sexual intercourse takes place, provided we assume that the procreatory fluid is, like semen, *always* capable of combining. This, however, could be in conflict with the possibility considered in §5.1, namely that the procreatory fluid at some time loses its power of fecundity and has to be renewed. But it is possible that this loss of fecundity is independent of the duration of the *ṛtu-*[120]. In this context I may point to SU,Śā 2,31, which tells us that intercourse during the first few days of the *ṛtu-*, i.e. during menstruation (cf. n.27), cannot lead to conception, because the semen (or embryo)[121] is washed out by the flow of menstrual discharge,[122] just as an object capable of floating follows the current of a stream and does not swim against this; from this we could conclude that conception could occur if only the semen succeeded in reaching the uterus or the embryo could succeed in staying within this, and we do indeed learn from this passage, as well as from ĀY 4,107f., AS,Śā 1,p.275b, BH,Śā 5,5 and KĀ,Śā,Jā 5, that, even though the fluid being discharged by the woman may and often does hinder it, conception may yet occur due to intercourse during menstruation, resulting, however, in a child which is abnormal, seemingly because a part of the fluids from which it is formed is washed away (cf. too TIVĀRĪ 1,pp.64-66).[123] This bears testimony to know-

---

commentary, two fluids, namely the menstrual fluid (here *rajas-*) and the procreatory fluid (here *ārtava-*). However, this need not be, for Cakrapāṇidatta's *rajas-* and CA's *ārtava-* could refer to the same fluid, but in different stages, namely old and new respectively, and this naturally reminds us of the possibility considered in §5.1. It would also follow then that this fluid is not always present in the uterus (for if it were, this could not be empty). But all this would apply to Cakrapāṇidatta's comments, which do not necessarily explain the CA passage. correctly; in fact, if the CA passage is to be understood in the way we have just considered possible, then they most probably indeed do not, unless Cakrapāṇidatta is hinting at the same in an oblique manner, in which case, however, the mention of *rajas-* is puzzling.

[120] On *ṛtu-* cf. n.25 and n.27.

[121] On the word *bīja-* used see §§2.5f.

[122] A similar theory of the Greek physician Galen is mentioned by NICKEL, pp.37,n.3;57f.

[123] The SU passage states:
tatra prathame divasa ṛtumatyāṃ maithunagamanam anāyuṣyaṃ puṃsāṃ bhavati. yaś ca tatrādhīyate garbhaḥ sa prasavamāno vimucyate (var: *vipadyate*). *dvitīye 'py evaṃ sūtikāgṛhe vā. tṛtīye 'py evam asampūrṇāṅgo 'lpāyur vā bhavati. caturthe tu sampūrṇāṅgo dīrghāyuś ca bhavati. na ca pravartamāne rakte bījaṃ praviṣṭaṃ guṇakaraṃ bhavati. yathā nadyāṃ pratisrotaḥ plāvidrav-*

---

yaṃ prakṣiptaṃ pratinivartate nordhvaṃ gacchati tadvad etad draṣṭavyam.

Cf. on this SU passage BHĀ 1,3,12-14:

pravahatsalile kṣiptaṃ dravyaṃ gacchaty adho yathā
tathā vahati rakte tu kṣiptaṃ vīryam adho vrajet
āyuḥkṣayabhayāt bhartā prathame divase striyam
dvitīye 'pi dine ratyai tyajed ṛtumatīṃ tathā
tatra yaś cāhito garbho jāyamāno na jīvati
āhito yas tṛtīye 'hni svalpāyur vikalāṅgakaḥ,

as well as RATIŚ, p.76:

ṛtau śoṇitaṃ nāryā (to be read as *nārīyā*?) *vegena vahate dhruvam*
bījaṃ puṃsaḥ saṅgame tu na cāntaḥ praviśet kvacit
prathame 'hni mahādevi yadi nārī patiṃ vrajet
acirāt mriyate (sic) *tasyāḥ patiputrau na saṃśayaḥ*
dvitīye 'hni phalaṃ tadvat patisaṅge tu jāyate
sūtikāgāra evāsau suto nūnaṃ vinaśyati
tṛtīye 'hni kṛte saṅge bhavaty alpāyur arbhakaḥ
jāyate vikalāṅgaś ca suśruteneritaṃ (printed: *śuśru°*) *tathā,*

RATS, p.69:

ṛtumatī yadā nārī tridinaṃ munipuṅgava
vegena vahate gātre śoṇitaṃ nātra saṃśayaḥ
gamane tatkāle caiva apūrṇo jāyate sutaḥ
akāle maraṇaṃ tasya vikalāṅgo 'thavā bhavet,

and RATS, pp.72f.:

ṛtumatī[ṃ] prathame 'hni gacched yo madanāturaḥ
svalpāyuḥ sa bhaved brahman akāle mriyate sutam (sic)
dvitīye ramaṇīṃ gacched yo naraḥ kāmatāpitaḥ
gatāsuḥ santatis tasya jaṭhare nātra saṃśayaḥ
tṛtīye 'hni mahābhāga ramaṇīṃ yāti yo naraḥ
alpāyus tasya putraḥ syāt svayaṃ cirarogī bhavet.

The ĀY passage tells us:

aśuddhe pravartite 'sre (var: *pravartamāne rakte ca*) *yadi gacchati mūḍhadhīḥ* (107cd)
doṣayukto syād bahulo na bījaṃ garbhadhārakam
yadi garbhaḥ prajāyeta vyaṅgonmatto 'lpajīvitaḥ (var: *vyaṅgo matto 'lpajīvanaḥ*)
bhavet tasmān na gaccheta na paśyet tu kadācana (108).

The AS passage remarks at the same time upon the effects of improper age of the parents, unhealthy procreatory fluids, and intercourse during the first three days after the onset of menstruation:

ūnapañcaviṃśativarṣeṇonaṣoḍaśavarṣāyām āhito garbhaḥ kukṣistha eva vināśam āpnuyād alpāyur balārogyavibhavo vā syād vikalendriyo vā. ṛtāv ādye ca rātritraye tathā duṣṭe śuklārtava iti.

The BH passage states:

tryahaṃ tu khalu purāṇarudhiraṃ parivarjayet. kiṃ ca tat purāṇam. yad āditas tryahaṃ parisravad ṛtukāle tat purāṇam. tasmin tryahe garbhopakramaṇe (ms:

ledge of the actual fact that intercourse during the time of menstruation may in some cases indeed lead to conception (RAY/GUPTA/ROY, p.11 and GHĀṆEKAR 1986, p.36a too remark upon this); from Dharmaśāstra texts too it follows that a woman was thought of as being able to conceive if she had intercourse during menstruation.[124] RATIŚ, p.59 actually says that the *ṛtu*- lasts for sixteen days [after the appearance of the menstrual blood], and that conception may take place on any of these days, the first four days, however, not being fit for intercourse (as also the eleventh and thirteenth days).[125] All this need not, however, necessarily be at variance with §5.1, for it is possible to conceive of new fecund procreatory fluid present at the same time as the old fluid is discharged, and from this one could deduce that conception is in principle always possible, but the fluid involved at the same

---

°krameṇa) na tiṣṭhati. avasthito vā nāyuṣi samartho bhavati... .

The KĀ passage says:

> *rajasvalāyāś cet prathame 'hani garbha āpadyate taṃ vātagarbham ācakṣate*
> *viphalaṃ vātapuṣpam ivodbhidānām. dvitīye 'hani cet sraṃsate cyavate vā.*
> *tṛtīye 'hani sūtikāsane mriyate na vā dīrghāyur bhavati hīnāṅgaś ca jāyate*
> (the continuation of the text is given in n.127).

Note that *vātapuṣpa*- "wind-flower" in the last passage obviously refers to a flower which does not give a fruit, from which it seems to follow that all flowers were thought of as being in principle capable of producing fruits; this would be further confirmation that the ancient Indians did not know of the sexual mode of reproduction of plants (cf. e.g. VṚK, pp.246;249ff., DAS,R.P. 1997, p.214; more confirmation will be given in n.1151).

124 See e.g. GHURYE, pp.272;278;282 (COMBA 1991, p.196 too draws attention to the "testi giuridici" in this regard, but unfortunately does not give any details). On similar notions in Balinese medical texts (which are heavily influenced by Indian theories) cf. WECK, p.105. Of especial interest is SLAJE 1995, who points out that the actual flow of menstrual blood does not necessarily correspond to the normative flow as seemingly described in various texts, and concludes that, at least in Vedic times, intercourse was prescribed even during the period of the flow of blood (cf. on this point also n.110), but proscribed for the first few days, maybe for fear of producing female offspring. This interesting conclusion deserves to be investigated further, though there seems to be little material in medical texts to facilitate such an investigation.

125 *ṛtuḥ svābhāvikaḥ strīṇāṃ rātrayaḥ ṣoḍaśa smṛtāḥ*
yasmin kasmin dine tv eṣāṃ gamane garbhasambhavaḥ
eṣām ādyāś catasras tu ninditaikādaśī ca vā
trayodaśī ca śeṣās tu praśastā (printed: *prasastā*) *daśa rātrayaḥ*.

Note the use of the masculine/neuter form *eṣām* instead of the feminine *āsām* "of these", which we would have expected in connection with the feminine plural *rātrayaḥ* "nights", as well as the singular *ninditā* after the plural *ādyāś catasras tu* (or do we have a case of secondary contraction of an original, and correct, hiatus *ninditā ekādaśī* — this then, however, standing against the usual metre — followed by the insertion of either an unoriginal *ca* or an unoriginal *vā*, whose standing together here is indeed most peculiar?).

time constantly being replenished. On the whole, though, we do not yet have enough material to deal in detail with the problem of whether conception is possible at times other than the *ṛtu-*.[126] Nevertheless, it could be that there were theories (maybe not accepted by all) in vogue in this regard; note that As,Śā 1,p.281b and KĀ,Śā,Jā 5 seem to speak expressly of children conceived *after* the time of the *ṛtu-*.[127] What is even more significant is that

---

[126] In this context a statement of Kāśyapa quoted by NYC in the commentary on SU,Śā 2,40 (fol.21v) seems of interest, namely that a man with phlegmatic semen having intercourse with a woman without *rajas-* (*arajaskā*) during her fecund period produces a sexual 'pervert' called *kumbhila-*:

> *arajaskāṃ yadā nārīṃ śleṣmaretā vrajed* (ms: *vṛjed*) *ṛtau*
> *anyāsaktibhavā* (ms: *anyāśakti°*) *prīti[r] jāyate kumbhilas tadā.*

Ḍalhaṇa in his commentary on loc.cit. also quotes this verse, expressly saying that it is a verse of Kāśyapa quoted by Gayadāsa. Pāda c is read by him as *anyasaktā* (var: *anyāsaktā*) *bhavet prītir*. Āḍhamalla in his commentary on ŚĀR 1,7,171f. too quotes the above verse (with *śleṣmalas tu vrajed ṛtau* in pāda b, *anyāsaktā bhavet prītir* in pāda c and *kumbhikas* for *kumbhilas*), saying that it is a verse of Gayadāsa. The verse could be taken to refer to a woman either too young or too old or for some other reason not able to menstruate, but even then still having a period of fecundity, i.e. having procreatory fluid. This is, however, highly problematic. Even more problematic would, however, be the case if we took this passage to refer to a woman not menstruating at the exact moment of intercourse. One could take *rajas-* to refer to the procreatory fluid, but then how could the woman have a *ṛtu-*? Actually, what seems to be meant here is not an absence of some fluid, but of passion or desire, which is another meaning of *rajas-*; this also fits the next line of the quotation, which speaks of love felt for someone else.

[127] The passages seem to imply that conception is possible even after the *ṛtu-*; one could also hold that what is meant is that in some cases the *ṛtu-* lasts longer than usual, but that is not the apparent impression one gains. The latter is true especially of the KĀ passage, which tells us that a child conceived during this time is weak and unhealthy like corn grown at the wrong time (continuing the text given in n.123):

> *ata ūrdhvam ṛtur dvādaśāhaṃ brāhmaṇīnām ekādaśāhaṃ kṣatriyāṇāṃ daśāhaṃ vaiśyānāṃ navarātram itarāsām. ṛtur bījakālam avekṣata ity āhur maharṣayaḥ. ata ūrdhvam akālajam āhuḥ. akālajaṃ hīnaṃ durbalam asthiram adṛḍham apīnabhaṅguraṃ dhānyam iva bhavati.*

Since corn can in many parts of India basically be made to grow at almost any time, but grows well only at some, it would follow from this simile that a child can be conceived at any time, but is normal and healthy only if conceived during the right time, namely the *ṛtu-*. (As regards the link between caste and the duration of the *ṛtu-* mentioned here, similar statements are found in some late medical texts — including BHĀ 1,3,2 — and sexological works). The AS passage too says:

> *ataḥ paraṃ* (i.e. after the actual *ṛtu-*; Indu: *ato 'nantaraṃ trayodaśyādiṣu rātriṣu*) *tūttarottaram evāyurādīnāṃ hrāsaḥ.*

TIVĀRĪ, 1,p.69 too notes these passages, but does not comment on the fact that, if indeed a conception *after* the *ṛtu-* should be meant, this collides with her assumptions that *ṛtu-* refers

AS,Śā 1,p.280b, where four different theories on the duration of the *ṛtu-* are given, says that according to some the *ṛtu-* may last a month (*śuddha-yonigarbhāśayārtavāyā māsam api kecit*); according to my understanding of this statement, it can mean only that the *ṛtu-* may last, according to this theory, from one menstrual cycle to the next.[128] Note in this connection too the theory[129] which states expressly that the *ṛtu-* is, according to some authorities, not even associated with the appearance of *ārtava-*, in this context obviously menstrual blood. All this could mean that in CA,Si 9,62 mentioned in §5.6 we could maybe have a theory not necessarily in keeping with other statements of this text regarding the procreatory fluid. But since unfortunately no more data on this point seem to be available from CA, we must content ourselves with merely drawing attention to this problem here without being able to offer any solution.

5.8. Let us therefore turn our attention to other passages in CA. *rajas-* in Ci 30,17 quite probably refers to menstrual flow (the woman is without *rajas-* — *arajaskā-* — and discharges only some uncommon fluid), but in 25f. and 207-209 its exact meaning is difficult to ascertain, as also that of the apparent synonyms *ārtava-* and *rajas-* in 26. The exact meaning of "blood", often used in Ci 30, also remains unclear. The difficulties are to a large extent caused by the fact that the chapter elaborates on diseased discharges of various sorts, as well as on irregular menstrual discharges, without our being able to differentiate the two clearly or even know whether they were to be differentiated at all (see especially 30,207ff.). However, 222 mentions the *raktamārga-* "passage/channel of blood", obviously meaning the vagina, which means that here at least *rakta-* denotes the menstrual discharge. And in 30,225f., which describes the healthy — red — menstrual discharge, *ārtava-*, the expression used, quite obviously denotes this discharge.[130] In Śā 4,16, one of the signs of conception is given as the non-appearance of *ārtava-*, which leads one to see the menstrual discharge in this here too. But all that is hardly much by way of information. Similarly, Ci 28,143 and Si 4,11, where

not only to the period of fecundity, but also that the theory behind all this was in actual fact more or less the same as that of modern Western medicine. GHĀṆEKAR 1986, pp.76a;79b, however, draws attention to theories stating that conception is always possible.

[128] The other three theories relate to a period of twelve and sixteen nights after the onset of menstruation, and to the *ṛtu-* not being connected to menstruation at all (see below). All four theories are mentioned by Indu in his commentary on AH,Śā 1,20 too.

[129] In SU,Śā 3,6 and AS,Śā 1,p.280b, but not CA; see also TIVĀRĪ, 1,pp.66f., GHĀṆEKAR 1986, pp.75ff., and n.128 and especially n.183 below.

[130] This passage will be taken up again in §7.7, where the Sanskrit text is given in n.375.

we find the mention of medicines guaranteeing progeny[131] and removing disorders of semen and *rajas-*, are also not very informative: *rajas-* could be taken to refer to the procreatory fluid, but since it is possible that disorders hindering conception in the female were diagnosed on the basis of the appearance of the menstrual fluid, this is not certain. Cf. in this connection also Ci 30,225f., just mentioned, where the healthy *menstrual* fluid is described, and Ci 30,337, which states that the characteristics of healthy semen (of the male; cf. 30,133) and *rajas-* have been described in this chapter (Ci 30), as such obviously referring, as regards the latter, to 30,225f. The same difficulties pertain to Ci 26,173 and 30,67, where semen and *asṛj-* "blood" are mentioned. Ci 29,120, where disorders of semen and *asṛj-* are mentioned too, is also unclear, for not only do we have here the same problems as in the case of Ci 28,143 and Si 4,11, but we are not even sure whether *asṛj-* refers to special female "blood" or to blood in general. As regards pure *śoṇita-* in Śā 4,7, here too it is unclear whether this refers to the procreatory or the menstrual fluid. The meaning of *rajas-* in a variant to Śā 4,14 is also unclear: this passage states that certain parts of a human are visible already in the embryo, while others appear after birth, among these latter being "the becoming distinct" (*vyaktībhāva-*), which Cakrapāṇidatta explains as referring to the appearance of semen and *rajas-* (*vyaktībhāvaḥ śukrarajasor āvirbhāvaḥ*), which is also what the variant in effect states (*śukrarajovyaktībhāva-* for *vyaktībhāva-*), without elaborating further on this. As regards the woman called *anārtavā-* "without *ārtava-*" in Si 12,19,1, it is unclear what *ārtava-* here refers to, though it must be something special to women.

5.9. Thus, apart from the clue given by the use of *rajas-* mentioned in §5.1, we gain practically no information on the menstrual and procreatory fluids from the aforesaid. However, Ni 3,13f. is of interest. This passage describes the origin and symptoms of *śoṇitagulma-* "unusual growth [in the abdominal region] due to blood". This is, the text tells us, to be found in women only, because only they have a flow of *ārtava-* into the uterus (*garbhakoṣṭha-*, a synonym of ☞ *garbhāśaya-*). The reason for its appearance is that this *ārtava-* is hindered from flowing out by wind entering the ☞ *yonimukha-* and blocking (*uparuṇaddhi*) the *ārtava-*, which leads to its collecting and increasing month by month and distending the abdomen.[132] (Here it is for

131 Expressly mentioned in the former, but only in a variant to the latter passage.

132 13: *śoṇitagulmas tu khalu striyā eva bhavati na puruṣasya garbhakoṣṭhārtavāgamanavaiśeṣyāt. pāratantryād avaiśāradyāt satatam upacārānurodhād vā vegān udīrṇān uparundhatyā āmagarbhe vāpy acirapatite 'thavāpy aciraprajātāyā ṛtau vā vātaprakopaṇāny āsevamānāyāḥ kṣipraṃ vātaḥ prakopam āpad-*

now immaterial whether the problematic term ☞ *yonimukha-* refers to the uterus, the orifice of the uterus or — in this particular case rather unlikely — the external orifice of the vagina, as the result is the same in each case.)[133] This then leads to all the signs of pregnancy appearing (including the presence of milk in the breasts — the import of this will become apparent in §§6.17ff. further below)[134]. Ci 5,18f. tells us simply (without mentioning the blockage of a fluid) that the *gulma-* called *raktabhava-* and *raudhira-*, i.e. "originating from blood" appears in women due to certain emotions, actions or disorders. We have seen in §5.5 that Śā 2,8f. describes a similar disorder, though CA does not seem to consider this to be the same as *śoṇitagulma-* (on this see also TIVĀRĪ, 1,pp.401ff.); in any case, we have no further description of it loc.cit. But be that as it may, the description of Ni 3,13f. seems to fit the menstrual fluid best, so that we seem to be able to deduce that this menstrual fluid collects in the uterus and is discharged each month.[135] But if its

---

yate. 14: *sa prakupito yonimukham anupraviśyārtavam uparuṇaddhi. māsi māsi tad ārtavam* (or: *tadārtavam*) *uparudhyamānaṃ kukṣim abhivardhayati.* ...

The reading *tad ārtavam* in 14 makes sense here, as we would have a reference to the *ārtava-* already mentioned in the line just preceding: "month by month this *ārtava-* being blocked distends the abdomen". On the other hand, *tadārtavam* too would make sense: "month by month the *ārtava-* being blocked then distends the abdomen". The same meaning could, however, also be conveyed by *tad ārtavam*, since *tad* can also have a meaning similar to that of *tadā*. See n.194 for a similar case.

133 It may, however, be pointed out here that sexual intercourse (not necessarily urination, as only the vagina, not the vulva as such could be meant) would be impossible if the external orifice of the vagina were blocked. This could lead us to deduce that ☞ *yonimukha-* must mean "mouth of the uterus" here. However, there is always the possibility that the theory of *śoṇitagulma-* does not take this problem into account, especially since several medical texts proscribe sexual intercourse during pregnancy (see TIVĀRĪ, 1,pp.233f.), though this very proscription shows that the possibility must have been reckoned with (cf. the similar case of intercourse outside the period of fecundity referred to in n.1984), and indeed, some texts forbid only excessive intercourse (cf. too n.1368). See in this context also CA,Ci 30,35f. mentioned in §5.10. Note also that the fact that the wind is supposed to "enter" the *yonimukha-* in Ni 3,13f. could be taken to mean that the latter expression refers to the uterus, as this is a cavity or hollow that can easily be referred to as being "entered"; however, an opening too could be thought of as being "entered" in the sense that something which passes *through* the opening (or, as would be the case here, blocks it) would have to go *into* the opening first.

134 CA,Śā 6,23, which Cakrapāṇidatta in his commentary quotes in connection with the formation of milk, also describes the formation of milk in the pregnant woman, but from ☞ *rasa-*. We shall discuss this later, in §10.6, as it is relevant primarily in a different context.

135 In this connection attention may also be drawn to §5.6. If in the passages we are here examining a periodic opening and closing of the uterus should have been presupposed, then the blockage of the egress of fluid becomes relevant only when the uterus should normally be

egress is hindered, at the mouth of either the uterus or the vagina,[136] then it follows that this would hold true for the procreatory fluid too, unless we assume that this leaves the body by some other means, as was hypothetically considered in §5.1 (though it is clear that this raises the very pertinent question of what other means there could be), or is otherwise absorbed. As the text seems to presuppose only one fluid being blocked, one could then deduce that this means that the procreatory and menstrual fluids are one and the same.

5.10. It is, however, obvious that this non-mention of another fluid by no means gives us a watertight case. Moreover, there just may be a chance that the release of procreatory fluid (different from menstrual fluid) has the entry of male semen into the womb as a prerequisite; but this semen can obviously not enter if the opening of the womb is blocked, which would mean that no procreatory fluid can be released, so that the question of its discharge becomes irrelevant, thus leaving us with only the menstrual fluid. I admit that all this may appear far-fetched to some, but there are enough examples to show that such seeming far-fetchedness is in actual fact often not so, thus warning us rather to err on the side of caution than to be overly confident. From this it follows that we still have basically only §5.1 on which to base our deductions, which means that we shall have to wait for more evidence before coming to any definite conclusion. I may here also draw attention to CA,Ci 30,35f., which mentions a disease in which the wind does something, described by the verb *viṣṭambhayet*, to the "mouth" (*mukha-*) of the uterus (☞ *garbhāśaya-*) and of the *yoni-* (cf. ☞ *yonimukha-*), resulting, among other disorders, in a diseased discharge;[137] this discharge is called *ārtava-*

---

open, since at other times it is closed anyway.

[136] On the latter possibility see, however, n.133.

[137] This passage describes the disorder termed *mahāyoni-* "great (i.e. excessively dilated) *yoni-*":

> *viṣamaṃ duḥkhaśayyāyāṃ maithunāt kupito 'nilaḥ*
> *garbhāśayasya yonyāś ca mukhaṃ viṣṭambhayet striyāḥ* (35)
> *asaṃvṛtamukhī sārtī rūkṣaphenāsravāhinī* (var: *sārtiḥ saphenārtavavāhinī*)
> *māṃsotsannā mahāyoniḥ parvavaṅkṣaṇaśūlinī* (36).

The parallel AH,Ut 33,40 = AS,Ut 38,p.292a (which does not mention any discharge) speaks of the wind doing something which is termed *viṣṭabhya* (i.e. a form of the same verb as *viṣṭambhayet*) to the "mouth" (*āsya-*) of the *yoni-* and the uterus (*garbhakoṣṭha-*), mentioned separately:

> *duṣṭo viṣṭabhya yonyāsyaṃ garbhakoṣṭhaṃ ca mārutaḥ.*

This parallel passage, and SU,Ut 38,19 (which has no other information), also say that the *yoni-* is *vivṛta-* "open", whereas the CA passage characterises the *yoni-* as *asaṃvṛtamukha-*

in a variant (see n.137). One is inclined to take *vi√stambh* in the common meaning "stop, restrain" here and see in this passage a disorder similar to the ones discussed in §5.9, but in that case how should a discharge come out of the body? Most probably the verb here is to be taken in its older meaning "hold asunder/apart",[138] though HILGENBERG/KIRFEL, p.675, translating the AH parallel (see n.137), prefer the verb *versteifen* "stiffen" (obviously in an open position). But all this gives us no clue as to what the discharge is composed of.

5.11. As regards the relation of the procreatory and menstrual fluids to blood in general, the information on this too is meagre. That blood in general was somehow connected especially with females is borne out by passages in which blood as well as the liver (*yakṛt-*) and spleen (*plīhan-*), which are in medical texts generally held to be the "seat" of or otherwise specially connected with blood,[139] are given to the child by the mother,[140] whereas its semen is given to it by the father,[141] in which connection it needs to be remembered that semen is generally held to be present throughout the whole

---

"whose mouth is uncovered". What is meant seems to be a dilated vagina which can be looked into from without, in which case *mukha-* in *asaṃvṛtamukha-* of the CA passage would refer to the opening to the exterior. The AH and AS passages also say that the *yoni-* is *srasta-*, which can mean "hanging down", but also "loosened". According to both meanings, what could be meant is a flaccid vagina, according to the former one could also think of a vaginal prolapse. In both cases we would have a dilated, visible vagina. One could, however, also see in *srasta-* a reference to a prolapse of the uterus (cf. *yonibhraṃśa-* in the discussion on ☛ *yonimukha-*); in that case *mukha-* in *asaṃvṛtamukha-* in the CA passage could refer to the uterine orifice, as that would then be "uncovered" and visible. But be that as it may, it does not alter the fact that *mukha-* in v.35 of the CA passage, in a genitival relationship with both *garbhāśaya-* and *yoni-* (*garbhāśayasya yonyāś ca mukham*), seems to refer to both the openings of the uterus into the vagina and of the vagina to the exterior. This being so, it is difficult to determine exactly what the "mouth" (*āsya-*) of the AH and AS passages actually refers to; it could be the orifice of the uterus (thus also HILGENBERG/KIRFEL, p.675), but it could equally well be the external opening of the vagina.

[138] Cakrapāṇidatta too glosses *viṣṭambhayet* as *vistārayet*, from *vi√stṝ* "expand, spread". Śivadāsasena in his commentary on the AH parallel (see n.137) (vv.41f. in AH3) glosses *viṣṭabhya* with *ākramya* "having approached/attacked".

[139] See e.g. CA,Ni 2,4, Vi 5,8, Ci 4,10, SU,Sū 21,16, Śā 4,10;25, Śā 7,6, Śā 9,12, AH,Ni 3,4, AS,Śā 5,p.319a;p.320a, Ni 3,p.24a, to mention only a few instances from the 'classical' texts. See also §7.10 and n.420.

[140] E.g. Śā 3,6. Thus also SU,Śā 3,33, AS,Śā 5,p.316a. See too TIVĀRĪ, 1,pp.125f., SUNESON, p.116, as well as n.6.

[141] E.g. Śā 3,7. Thus also SU,Śā 3,33, AS,Śā 5,p.316a. See too TIVĀRĪ, 1,pp.127f., SUNESON, p.116, as well as n.6.

body of both males and females in Indian medical theory.[142] But this is hardly much by way of information. Though the origin of blood in general (as part of the chain of elements resulting from the breakdown of food; see §2.4) is commented upon, that of the special female "blood" is nowhere mentioned, except possibly in Ci 15,17,[143] to be discussed in §§5.12f. Even in the list describing the quantities of different substances present in the human body to be found in Śā 7,15, only blood in general (*śoṇita-*) is mentioned, but — in contradistinction to similar lists in other texts[144] — no special female fluid or "blood" (cf. also CA,Vi 5,8 with SU,Śā 9,12). True, some editions also list a variant of Śā 7,15 mentioning the quantity of *ārtava-*, but this seems secondary (Cakrapāṇidatta too does not comment upon or mention it) and as such most probably an attempt to bring CA in line with other texts.

5.12. Ci 15,16 enumerates the substances or elements derived due to the breakdown of food in succession as given in §2.4 above. 15,17 then mentions further products of individual members of this chain:[145]

> *rasāt stanyaṃ tato raktam asṛjaḥ kaṇḍarāḥ sirāḥ*
> *māṃsād vasā tvacaḥ ṣaṭ ca medasaḥ snāyusambhavaḥ*[146]
> "From ☞ *rasa-* milk, from it/further blood (*rakta-*), from blood (*asṛj-*) large cords (☞ *kaṇḍarā-*), vessels (☞ *sirā-*), from flesh tissue-fat/muscle-fat (☞ *vasā-*) and the six skins (☞ *tvac-*), from fat (☞ *medas-*) the occurrence of cords (☞ *snāyu-*)[147]."[148]

asṛj- quite obviously refers to blood in general. But what does *rakta-* refer to? Is it too blood in general, which would be in keeping with §2.4? But it is known quite well, and shall also be discussed in detail below, that there are theories which set the special female "blood" (whatever this may in each particular case be) too in relation to the elements of the chain of the breakdown of food. Could we — especially since the verse mentions milk, which is a sub-

142 Cf. on this also the discussion in DAS,R.P. 1984A, pp.237ff.

143 Cf. on this also DAS,R.P. 1984A, p.239.

144 AH,Śā 3,80f., AS,Śā 5,p.326, BYT 2,199. But KĀ,Śā,Śa, pp.51f. and BH,Śā 7,6 too do not mention the special fluid of the female, whereas SU does not have such a list. Cf. also KIRFEL 1954, pp.353f., and see too §6.11.

145 The text does not elaborate further on this verse, so that all we have to go on is the statement of the latter.

146 Var: *snāyusandhayaḥ*.

147 According to the variant mentioned in n.146, the translation is: "from fat cords and junctures (☞ *sandhi-*)".

148 A puzzling and most probably spurious verse mentioned by PUROHIT and completing the statement of CA,Ci 15,17 will be discussed in §11.8.

stance special to females — assume a similar theory here? Does *rakta-* therefore refer to this female "blood"? But if it does, what does *tatas* mean? Does it mean "further"? This would mean that *rakta-* here is derived from *rasa-*. On the other hand, *tatas* might mean "from it". What is "it" then: *rasa-* or milk? And if milk should indeed be meant, would it not follow then that this would mean that milk is in principle *always* present in the body of a woman capable of conceiving? Of course all this depends on the assumption that *rakta-* here refers to the special female "blood"; I must, however, stress that this is merely a possibility we are here considering.[149]

5.13. However, the line *rasāt stanyaṃ tato raktam* has a variant which runs:

rasāt stanyaṃ striyā raktam[150] ...

"From ☞ *rasa-* milk, the blood of a/the woman ... ."

This obviously refers to the special "blood" of women (whose exact nature here is still, however, to be determined). It is a much smoother reading, not as ambiguous as that of §5.12, which does lead one to the impression that the latter, the *lectio difficilior*, could be the original reading. This impression is strengthened by the reading of Cakrapāṇidatta in DR,Āh 22 (= the passage above) and his commentary on SU,Sū 14,10, where we find *tathā* "likewise" in the place of *tato*, and it looks as if this is also the reading he comments upon in his commentary on CA,Ci 15,17;[151] *tathā* is of course much nearer to *tato* (i.e. *tatas*) than *striyā[ḥ]*. But even if the latter should be original, our problems are still not at an end, for it could be that, according to the construction of the verse, *stanya-* and *rakta-* are not separate substances, but the same, standing in apposition (cf. also §5.9). And to top it all, the

[149] There is also the theoretical possibility of this verse telling us that blood in general derives from milk. This would, however, be so much at variance with the general theory mentioned in §2.4 that it seems highly improbable, especially since it would not account for blood in the male.

[150] This is also the reading of CA1, p.2921.

[151] Cakrapāṇidatta's commentary on CA,Ci 15.17:

rasāt stanyaṃ prasādajaṃ. tathā raktam api rajaḥsaṃjñaṃ rasād eva prasādabhāgajam

"From *rasa-* milk is born of the pure part (☞ *prasāda-*). Likewise *rakta-* too, called *rajas-* (cf. also Cakrapāṇidatta's terminology in n.119), is born of the part which is the pure part [of *rasa-*] from *rasa-* itself."

Ḍalhaṇa in his commentary on SU,Sū 14,10 quite obviously paraphrases CA,Ci 15,17, taking both milk and *ārtava-* (= *rajas-*) to be derived from *rasa-*, but he unfortunately does not give us the exact reading of the CA verse he presumes.

mode of transformation is unclear in any case.[152] Thus here too we get some tantalising bits of information, but in the end not really much to build up a case with. Our difficulties are compounded by Gayadāsa, who in his commentary on SU,Ni 1,25-29 quotes CA,Ci 15,17 not only with the variant of n.146, but also with pāda a running:

rasād raktaṃ tathā stanyam,

which says more or less the same as *rasāt stanyaṃ tathā raktam*. The trouble is, Ci 15,16, which describes the chain of ☞ *dhātu*-s as in §2.4, begins with:

rasād raktaṃ tato māṃsaṃ ...,[153]

so that we must also reckon with the possibility of a contamination between 15,16 and 15,17 having taken place with regard to Gayadāsa's reading. Gayadāsa himself, while commenting on the verse he quotes, seems to take *rakta*- in *rasād raktaṃ tathā stanyam* to stand for *ārtava*-, saying too that both this and milk are *abhinna*- from blood.[154] Now *abhinna*- can mean "not different (from)"; is Gayadāsa here saying that blood in general and *ārtava*-, as well as milk, are actually the same? The rest of the statement in n.154 must, however, serve to caution us, for here Gayadāsa also says that the "skins" (☞ *tvac*-) are *abhinna*- from ☞ *rasa*-, blood and flesh; quite obviously Gayadāsa is here referring to the chain of elements as given in §2.4, in this case particularly its three successive parts *rasa*-, blood and flesh, from which latter, according to the verse Gayadāsa quotes, the "skins" *develop*. In other words, *abhinna*- here seems rather to mean "not separated"[155], as the substances mentioned are all linked together in a chain, and in this regard *ārtava*- (for *rakta*- of the verse) and milk, since they are also derived from *rasa*- just like blood, are also *abhinna*- from the latter in Gayadāsa's opinion.

5.14. At any rate, Ci 30,207-209 make it more or less certain that blood in general (here *rakta*-) and *rajas*- (as it is called here, see §5.8) are different, as the origin of *asṛgdara*-, here = *pradara*- (a name applied to various diseased or irregular discharges), is said to be a combination in the uterus (☞ *garbhāśaya*-) of *rajas*- with excessive blood, due to excessive wind, which forces the latter into the vessels[156] (☞ *sirā*-) carrying *rajas*- to the

---

[152] Cakrapāṇidatta's explanation in n.151 need not be that which CA actually presupposes.

[153] I.e. blood arises from *rasa*-, flesh from blood, etc.; the full text is given in n.408.

[154] *tatra stanyārtavayo raktābhinnatvāt tvacāṃ ca rasaraktamāṃsair abhinnatvāt tatrasthavāyoḥ saiva kriyā rogāś ca.*

[155] Which is nearer to its original meaning "uncut; not split".

[156] A plural number is mentioned, which leads to a problem to be discussed below in §6.25.

uterus.[157] But what *rajas-* actually is remains unclear. Attention may also be drawn to Si 12,16,1, where in a list of diseases or disorders we find °*kukṣiśūlamūtrakṛcchrāsṛgrajovisarpapravāhikā*°; is, since *visarpa-* is a known name of a skin-disease and *mūtrakṛcchra-* a known urinary disorder, *asṛgrajas-* another name of *asṛgdara-*? For one could hold that *asṛgrajas-* means "*asṛj-* and *rajas-*", i.e. maybe "*rajas-* [which flows] together with *asṛj-* (scil. blood)", which would in effect state the same as Ci 30,207-209. But the problem with this explanation is that no parallels for such terminology seem to be available. Is thus *visarpa-* to be construed with the preceding, in which case it would probably refer to something like "diffusion" (in the sense of "issuing"?) of *asṛj-* and *rajas-*? But what that should refer to I do not know, as I know of no parallels in this case either. Nevertheless, irrespective of the correct explanation of Si 12,16,1, the mention of *asṛj-* "blood" and *rajas-* side by side would seem to show that they are here two different substances, as Ci 30,207-209 have already made probable. Ci 30,204cd, however, states that the description of *pradara-* in the verses following is an amplification on what was stated before, and this refers, as also the commentator Cakrapāṇidatta notes, probably to Ci 30,16:

> *raktapittakarair nāryā raktaṃ pittena dūṣitam*
> *atipravartate yonyāṃ labdhe garbhe 'pi sāsṛjā*
>
> "If, through [effects] causing blood-bile (☞ *raktapitta-*) the blood (*rakta-*) of a woman is made impure by the bile, it flows copiously into the *yoni-*[158]. Even if an embryo is [already] conceived (or: it flows copiously. Even if an embryo is [already] conceived in the *yoni-*), she (i.e. the woman? or the — feminine — *yoni-*, as in 9-15?) is with (= characterised by?) blood (*asṛj-*)."[159]

---

[157] *garbhāśayagatāḥ sirāḥ | rajovahāḥ.*

[158] On the problems connected with this expression see the discussion on ☞ *yoni(mukha)-* in Appendix II. Here "uterus" could fit, but note the differentiation between *yoni-* and uterus in the next verse (see n.159).

[159] Verse 17 reads:

> *yonigarbhāśayasthaṃ cet pittaṃ sandūṣayed asṛk*
> *sārajaskā matā kārśyavaivarṇyajananī bhṛśam*
>
> "If the bile situated within the *yoni-* and the uterus (☞*garbhāśaya-*) makes the blood impure, she (the woman?, the *yoni-*?) is held to be without *rajas-* (*arajaska-*) (or: [then] that [disorder] is held to be *arajaskā-*), to be a producer, in an excessive manner, of thinness and being off-colour (this probably refers to the secretion; cf. also §7.5)."

What is "blood" here? It is corrupted in the *yoni-* and the uterus, so it does seem to be not

It can be debated whether *sāsṛjā* might not, on the lines of the alternative translation of *sārajaskā matā* in n.159, be taken to mean "[then] that [disorder] is *asṛjā-*". This is indeed how TIVĀRĪ 2,pp.20f. analyses the verse (see n.161), also taking the variant *sāprajā* (see below) to similarly refer to a disorder called *aprajā-*.[160] According to Cakrapāṇidatta's commentary, the disorder mentioned in CA,Ci 30,16 (which he terms *raktayoni-*[161]) results in the embryo being washed out, and a variant of *sāsṛjā*, namely *sāprajā* ("she is without offspring") seems to say the same. According to Jejjaṭa the verse is to be read with the latter reading, and also with *yonyā labdhe* instead of *yonyāṃ labdhe*, which would say either that the blood flows from (and not into) the ☞ *yoni-* (*yonyāḥ* as ablative), or else that the embryo is conceived by

---

blood in general, but the special "blood" of females. The question then is what the blood which is corrupted in v.16 refers to; is it the same fluid, or blood in general? As it is, vv.16 and 17 need not necessarily be referring to the same, as they describe disorders with different symptoms, seemingly treated together because both are caused by an impurity of bile. The whole is complicated even more by Jejjaṭa, who in his commentary reads *saṃśodhayed asṛk* in the place of *sandūṣayed asṛk*; should this mean "purifies the blood" (it is difficult to conceive of another meaning), then it is obvious that we have a great problem here, which I for my part am unable to solve.

160 On such problems of nomenclature see ☞ *yonivyāpad-* in Appendix II.

161 Of the other 'classical' texts, AH,Ut 33,43 = AS,Ut 38,p.292b, AS,Śā 1,p.276a do mention the name *raktayoni-*, but do not really describe the disease mentioned, simply stating laconically that it arises due to excessive flow of blood. It is questionable whether SU mentions the disease at all, though see Ut 38,7;12 (on which cf. also TIVĀRĪ 2,pp.21f., as well as pp.24f.). Cakrapāṇidatta in his commentary on CA,Ci 30,9-15;204-224 says that another name for *raktayoni-* is *asṛgdara-*, and this is supported by AS,Śā 1,p.276a, which equates *asṛgdara-*, *pradara-* and *raktayoni-*:

> *atiprasaṅgenānṛtāv ṛtau vā tad evāsṛgdaraṃ pradaraṃ vyāpadaṃ ca raktayonisaṃjñāṃ labhate.*

On this cf. also the description of *asṛgdara-* in *Nāḍījñānadarpaṇa* 251:

> *tad evātiprasaṅgena pravṛttam anṛtāv api*
> *asṛgdaraṃ vijānīyād ato 'nyad raktalakṣaṇam.*

But there seem to be terminological problems in the classification of diseases as *raktayoni-*, *asṛgdara-* and *pradara-* (this latter also mentioned by Cakrapāṇidatta in his commentary on CA,Ci 30,9-15); see on this the extremely detailed discussion in TIVĀRĪ 2,pp.178ff. (pp.22f. too, where we also find what TIVĀRĪ takes to be *asṛjā-* — see above — explained as the feminine of *asṛ-* + *ja-* "born/produced of blood", this hypothetical *asṛ-* supposed to be the same as *asṛj-*, which is, however, hardly feasible, to say the least). Interesting in this connection is also ĀAS 5750, which is introduced by the statement that the common symptoms of *pradara-* will be discussed (*atha pradarasya sāmānyaṃ lakṣaṇam āha*), but says that everything is *asṛgdara-* (*asṛgdaraṃ bhavet sarvaṃ sāṅgamardaṃ savedanam*); is *asṛgdara-* here an adjectival qualification? Intriguingly, the line on *asṛgdara-* cited is the same as SU,Śā 2,19ab with some slight changes, but in a rather different context.

the *yoni-* (*yonyā*, instrumental, or *yonyāḥ*, genitive used instrumentally with the passive participle), which would ultimately be the same as the alternative translation above. But this hardly changes anything (since the blood flowing into the *yoni-* is obviously taken to flow out too), as also not the variants he mentions for *labdhe garbhe 'pi sāprajā*, namely *labdhe bīje 'pi sāsṛjā*[162] and *sāsṛjā pittajā yonir* "that is the *yoni*-[-disease?] born from bile through/with blood".[163] In other words, the verse in any case tells us that "blood" flows and washes out the embryo, reminding us of what was said in §5.7. Thus all these variants do not basically alter matters. But a problem remains, namely that the disease ☞ *raktapitta-* is, as far as I know, taken to result from blood in general; so could this verse be referring to this fluid? As it is, it says merely that blood flows even if a woman conceives and hence becomes pregnant (only the last variant cited does not refer to conception), but it does not say what this blood is, i.e. it may be blood in general and not the special "blood" of women. On the other hand, what flows may indeed be the latter, since the description of *pradara-* above tells us that this is made to flow by excess blood (in general) in the body.[164] But the verse actually simply says that the factors causing the disorder are those causing *raktapitta-* (too?), without saying that the disorder here is also to be subsumed under *raktapitta-*. Moreover, the very next verse might be referring to bile causing a disorder of the special female "blood" (see n.159). What complicates the matter further is not only that 30,16 may not actually be connected with 30,207-209,[165] but also that Aruṇadatta and Parameśvara in their commentaries on AH,Sū

---

[162] On *bīja-* here in the place of *garbha-* cf. §§2.5f.

[163] Some more for us immaterial variants are given by the editor in brackets; I could not ascertain on what basis.

[164] Note too that CA,Sū 24,11f. lists *pradara-* as one of the disorders caused by blood, and that CA,Sū 28,11 and BHĀ 1,7,7,61-63 say the same of *asṛgdara-*. According to KĀ,Sū 27,62f., *pradara-* is caused by blood. Interestingly, SU,Ni 1,37 does not mention blood, but says that *asṛgdara-* is caused by bile corrupting the *apāna-* wind; a variant even has *raktamūtratā* in the stead of *syād asṛgdaraḥ*, i.e. does not mention *asṛgdara-* at all. The editor of AH and TIVĀRĪ 2,p.180 take CA,Ci 28,230, AH,Ni 16,45 and AS,Ni 16,p.82a (= AH,Ni 16,45var) (which have already been discussed in §5.4 and refer to *atipravartana-*, *ativṛtti-* and *ativṛddhi-* of *rajas-* respectively due to the same cause) as referring to *asṛgdara-* too, but this equation seems quite arbitrary; it rather looks as if we had different aetiologies and/or disorders here, especially if we consider what CA says on *asṛgdara-/pradara-* in the other two passages cited above. See also n.161.

[165] Cakrapāṇidatta in his commentary on 204-224 says that this whole passage has been held by previous commentators to be spurious (*pūrvaṭīkākṛtām asammatam*), but that he explains it nonetheless, as it is to be found in many manuscripts.

19,77f. tell us that ☞ *raktapitta-* making its appearance by way of the ☞ *yonimukha-* is called *asṛgdara-*.[166] We must, however, also bear in mind the possibility that Aruṇadatta may be wrong, having simply misunderstood passages similar to CA,Ci 30,16 quoted above. On the other hand, we have also seen just a little above that according to CA,Ci 30,207-209 corrupted blood enters the vessels of *rajas-*, so that one could hold that in the case of *raktayoni-* too something similar happens, blood in general thus being discharged instead of or mixed with the special fluid of women, especially if *asṛgdara-/pradara-* and *raktayoni-* should ultimately be the same (see n.161). But be that as it may, the flow of blood, if it should indeed occur, is something exceptional. And if the fluid flowing from the *yoni-* should not be blood in general, but (one of) the special fluid(s) of women, or if it should consist of this fluid mixed with blood, then, since CA,Ci 30.16 expressly mentions a flow during pregnancy, it follows that it is present in the body during pregnancy too, though it does not as a rule flow then. We can not, however, say definitely what exactly it is in this case, for not only the menstrual fluid, but the procreatory fluid too could be meant by this irregular flow.

5.15. To sum up: according to the information in CA, the menstrual discharge could be the procreatory fluid which has lost its power of fecundity (see §5.1), but we have no watertight case to uphold this statement. From the terminology employed and the descriptions given we can deduce that both fluids were thought to be some kind of "blood", which could be taken to mean that they were thought to be blood-like (we have no evidence so far to assume that they were thought to be identical with blood in general, rather evidence to the contrary), though note in this context the caveat in §5.4. Their exact relationship to blood as well as their origins, however, remain obscure. A purely female "blood" quite probably is derived ultimately from ☞ *rasa-* (cf. §§5.12f.), as blood in general too is, but what the former "blood" thus derived actually is, and how it is ultimately thus derived, are unclear. As to the procreatory fluid, there is a possibility that this is basically always capable of combining with semen from the male to form an embryo (see §5.7).

5.16. The reason for this — on the whole — meagreness of information could be that CA is even more male-oriented than other medical texts[167]

---

[166] *raktapittaṃ yonimukhenāgacchat strīṇāṃ asṛgdaram* (Parameśvara: *asṛgdara*) *ucyate*.

[167] Indeed, according to CA,Ci 30,5f. the reason for considering female disorders is given as the inability of women to beget children if they have such disorders (cf. on this subject also SIDDH 29,1), i.e. not any basic concern for the well-being of women as such. On the male-oriented character of Indian medical texts cf. DAS,R.P. 1991A, pp.31f. Such orientation holds good for Greek medicine too (cf. WEISSER 1983, pp.88f., though DEAN-JONES 1994 takes pains to point out that this is a post-Hippocratic development, in the Hippocratic system men and

(this impression is substantiated by further reading) and seems to have given little thought to the special characteristics of females. Thus we find for instance a detailed description of the process by which male semen is collected from all over the body and made to combine with the corresponding female fluid to create an embryo,[168] but we have nothing on what the female

---

women being regarded as more or less different beings), and was indeed a characteristic of Western medicine until only a few centuries ago. The reason for the latter was, if we follow LAQUEUR 1989 and 1990, the perception of women as imperfect men, who were as such in no need of special separate treatment because what applied to men automatically applied to them too (cf. also BYNUM, pp.186f. and 212,n.97, as well as BOS, p.224; but note that CRAWFORD, pp.110ff. warns against the unqualified blanket application of the one-sex model in Western thought and draws attention to the fact that the respective organs are, though anatomically indeed considered similar, greatly different in size, though whether such differences were actually explained by different economies of fluid in the male and female bodies, as she maintains, remains a matter for further research to decide upon). The belief that men and women are basically the same in *all* respects, and that not only are the male genitals present in the female, but the female genitals present in the male too, is indeed still to be found in modern South Asia (cf. e.g. CHÖPEL, pp.56;196, as well as DAS,R.P. 1992A, p.413 on male "menstruation"). In any case, the usual reference to males only must always be considered when drawing conclusions relating to humans in general from Indian medical texts, the problem being compounded by the fact that such texts seem as a rule to concern themselves with rich (*āḍhya-*, AH,Sū 1,29) and noble patients and not with the general populace (cf. on this e.g. DAS,R.P. 1989, cols.920f., ROŞU 1986A, p.593, ANGOT, p.20,n.13; see also ZIMMERMANN 1989A, p.40).

[168] Śā 4,7, Ci 2,4,46-49, Ci 15,33-35 (as regards the latter passage, however, note that Cakrapāṇidatta does not comment on Ci 15,22-35, which thus might not be authentic; this problem of authenticity is discussed in n.420). The *śukra-* is collected from all over the body due to the exertion and heat of passion and coitus and then ejaculated (due to the pressure of the agitated wind). This mode of seminal ejaculation in males is presupposed by or expressly stated often in the texts I have seen, right down to ĀY (e.g. 4,86), so that I can refrain from giving further citations. An Arabian source for Indian medicine too tells us of the role of wind during intercourse. SIGGEL 1950 unfortunately takes *al-hawāʾ* "Luft" of the text to be a mistake for *al-hawā* (from *hawan*) "Liebe, Leidenschaft, Zuneigung" or its plural *al-ahwāʾ* (p.52 = 1146), which has led him to turn the wind referred to as one of the three things excited as a prerequisite to intercourse into "Liebessehnen" (p.21 = 1115); this is all the more puzzling because SIGGEL 1941-1942, p.259 not only has the translation "Luft", but even says expressly (p.259,n.11): "Gemeint ist sicherlich der Wind, der in der indischen Medizin als treibende Kraft eine Rolle spielt".

On the absolute quantities of male and female fluids capable of combining see also ĀY 4,129f. and BHĀ 1,3,40(4f.); cf. also Ḍalhaṇa's comments on SU,Śā 3,5. ĀY 4,128, quoting from the *Yogapradīpa*, tells us that during the time of creation of the embryo 23 parts of *rajas-* and 14 parts of *śukra-* arise in the female and male respectively:

trayoviṃśarajobhāgāḥ śukrabhāgāś caturdaśa
garbhasañjanane kāle puṃstriyoḥ sambhavanti hi;

this must mean that the female and the male are in a position to beget a child when these

counterpart does. As regards the mention of semen to the exclusion of any fluid only women have, it must also be pointed out that this is ultimately always justifiable from the point of view of ancient Indian medicine, since semen was thought to be present in women too (cf. e.g. §5.11),[169] even though it was clearly often seen only in its function as a fluid of the male.[170] So, apart from the male-orientedness of medical theory, the notion of semen being present in all humans too must have been a powerful factor leading to what to us moderns is an asymmetry with regard to females, for which CA has many examples. Ci 15,21, for instance, tells us that, according to some (*kecid*), it takes six days, but according to the author(s) of CA probably an indefinite period of time,[171] for semen to be ultimately formed in the normal course of events, but we have no similar statement on the female fluid. The list Śā 7,15 mentioned in §5.11 contains semen (*śukra-*), but not only no female procreatory (or menstrual) fluid, but also no milk — in

---

quantities of the fluids are in their bodies, not, as the editor surmises, that these quantities combine to create "male and female offsprings respectively". More material is given in §6.11. (Note also the theory — e.g. in SU,Sū 15,9 — that, if a male possesses too little semen, he may ejaculate semen mixed with blood.)

[169] One example for this is probably the list of substances derived by the embryo/foetus from mother and father (cf. §5.11 and n.6); semen comes from the father, blood from the mother, this blood not corresponding to semen, but being blood in general — no procreatory (or menstrual) fluid of women is found in such a context. But if semen has been seen here as being present in all humans, male or female, then there is actually no asymmetry as regards the two sexes in this list, since nothing special only to males would be mentioned. One could of course also hold that the non-mention of any special fluid of the female might point to blood and this fluid having been regarded as being one and the same at the time of formulation of the list, but that would be nothing but speculation.

[170] This latter problem has been taken up again in §8.9.

[171] The text does not say so directly, but says that this transformation is *cakravat* "like a wheel". As the commentator Cakrapāṇidatta rightly explains, this refers also to an incessant transformation (thus also Gaṅgādhara in CA1, p.2923; see also ŚĀSTRĪ, pp.4f.), but this incessant transformation would also be presupposed by all other theories, as the period of time obviously refers separately to each new bit of food that is digested. Thus the simile of the wheel must ultimately refer primarily to the time actually taken for the transformation of each bit of food as it is digested, and Cakrapāṇidatta, probably rightly, says that this refers to the varying speed of rotation of a wheel (he mentions an irrigation wheel), which varies according to the force (which is not constant) applied; this force is similar to the digestive fire (☛ *agni-*) in the body, which is at times strong, at times weak. (I may also add that, in his commentary on SU,Sū 14,16 — SU3, p.241, SU4, p.106 —, Cakrapāṇidatta presupposes a reading of CA,Ci 15,21 which is different from the one in our edition of the text, but more or less the same as ĀY 4,359f. According to this reading, CA mentions not only the opinion regarding six days, but also one which assumes seven days as the time needed for the transformation.)

contradistinction to several other similar lists (see n.144; KĀ and BH, however, agree with CA). Of course one could infer from this that both the female procreatory and/or menstrual fluid(s) as well as milk are, contrary to semen, in CA thought not to be present in the body at all times, but what we have seen above seems to speak against any such assumption, at least as regards the procreatory fluid; indeed, Cakrapāṇidatta in his commentary on Śā 4,30[172] states expressly that *ārtava-* is, just like teeth, facial hair etc., actually present in the embryo from the very beginning,[173] even though it does not appear in the child at once.[174] Now Cakrapāṇidatta may not necessarily be correct in his explanation of CA, but his comments do show that a tradition to this effect did exist, as is also evinced by KĀ,Śā,Jā 4, which says that "blood" and semen (*śoṇitaśukre*) are present in the child right from the beginning, and Ḍalhaṇa's commentary on SU,Sū 14,18 saying that this passage, which states that semen, though first perceivable only at puberty, is present in the body from the very beginning, applies to milk (*stanya-*) and *ārtava-* too, and *rajas-* and milk are in fact mentioned in such a context in AS,Śā 1,p.277b[175].[176] But even if it be the case that the procreatory or the menstrual fluid was not considered in the case of CA,Śā 7,15 to be always present in the body, why should such a fluid be treated differently from stool (*purīṣa-*) and urine (*mūtra-*), which are also mentioned in the passage? True, these latter are more or less daily excrements, but the point is that they are also voided periodically, and such periodical voiding seems, as we have seen, to have been presupposed for the procreatory fluid too, even if not necessarily by way of menstruation. Thus there seems to be no reason for this general neglect of the fluids of women only, unless we do assume a certain disinterest for women in general, which would incidentally fit well with the reason given by CA,Ci 30,5f., and mentioned in n.167, for discussing female diseases, namely not the well-being of women, but their inability to beget children if they have such disorders.

---

[172] On which and similar passages see Excursus I at the end of Chapter 2 above.

[173] More data on the notion of preformation in Indian medical texts is given in §13.18 below.

[174] *ārtavaṃ ca yady api dvādaśavarṣād ūrdhvaṃ vyajyate tathāpy ārtavotpattir garbhakāla* (var: *kanyotpattisamakālam*) *eva bhavati. yena satām evārtavadantaśmaśruprabhṛtīnāṃ kāle vyaktir bhavati tenārtavārambhakasyāpi bījasya garbhakāle pradoṣa upapannaḥ* (var: *utpannaḥ*).

[175] Cf. in this connection also n.298 for a possible natural explanation of this statement as regards milk.

[176] I may add that passages such as these have caused TIVĀRĪ, 1,pp.42-44 such headache that the only explanation she has been able to think of is that they refer to hormones!

5.17. Thus, all in all, it seems that CA presents us with a mode of thinking in which one was of the opinion that women had "blood" (maybe a blood-like fluid[177]), different from blood in general, which united with male semen to produce an embryo and which somehow was — probably, though this assumption is still very tentative — connected with menstruation; however, one either did not give much thought to this fluid at all, or else (and this seems more likely) did not deem it necessary to say much about it apart from in connection with its role in procreation.

[177] Note §5.4 though.

CHAPTER 6

The Suśrutasaṃhitā (1): Two Fluids or One?

6.1. In SU, which has much more information, the fluid corresponding to or uniting with male semen to form the embryo is called *ārtava-* in several places,[178] but we also find words meaning "blood" (*śoṇita-*, *lohita-*).[179] Sū 3,15 refers to the chapter Śā 2 by means of the word *rajaḥśuddhi-* "purification of *rajas-*", whereas the colophon of Śā 2 as well as Śā 2,1 have *śukra-śoṇitaśuddhi-* "purification of [male] *śukra-* and [female] *śoṇita-*", which is in actual fact what the chapter contains. The menstrual flow is called *rakta-* in Śā 5,10 (*rakta-* flows out of the vagina[180]) and Śā 2,31 (cf. §5.7); *puṣpa-* in Ci 37,125 also refers to it.[181] *ārtava-* in Śā 10,16 clearly denotes the menstrual fluid (visible for the first time after childbirth),[182] and *ārtava-* in Śā 3,6 too refers to this fluid,[183] as also does *rajas-* in a variant to Sū 10,5 in

---

[178] E.g. Sū 15,5,[3], Śā 2,5, Śā 3,3-5. See also Śā 2,48 (we shall return to this passage in §8.6). Śā 3,5 has, like CA,Śā 2,12;18 (cf. n.103), the theory of the predominance of the female procreatory fluid giving rise to a girl, of male semen to a boy, and of their equilibrium resulting in a being which is neither, and here called ☞ *napuṃsaka-*.

[179] E.g. Sū 24,5, Ni 5,28, Śā 4,4;63, Śā 5,3 and Śā 10,57, and Śā 1,16 (var: *ārtava-*) respectively. As regards *rakta-* in SU,Ci 31,17, it is not clear whether this refers to blood in general or to the special fluid of women; cf. n.108. *śoṇita-* in Śā 10,16 too could be taken to refer to blood in general as a morbific entity (☞ *doṣa-*), in accordance with Ḍalhaṇa's interpretation of *doṣa-* in the same passage in his commentary (*doṣaśabdo raktavacanaḥ*); it is, however, not clear what exactly *doṣa-* means both in Ḍalhaṇa's commentary and in the passage commented upon (we shall discuss the SU passage and this statement in §7.3 below). On blood as a morbific entity see the discussion on ☞ *doṣa-*.

[180] Here termed ☞ *srotas-*.

[181] The text speaks of an "excess" of *puṣpa-* (*puṣpodreka-*), which Ḍalhaṇa, however, explains as a discharge at a time not the normal one (*akālapravṛtti-*).

[182] This is clear from the context:

anena vidhinādhyardhamāsam upasaṃskṛtā vimuktāhārācārā vigatasūtikābhidhānā syāt. punar ārtavadarśanād ity eke.

[183] See on this also §5.7. The passage states:

ṛtus tu dvādaśarātraṃ bhavati dṛṣṭārtavaḥ. adṛṣṭārtavāpy astīty eke bhāṣante

"The *ṛtu-*, in connection with which *ārtava-* is seen, is, however, [of the duration] of twelve days. 'One (i.e. a woman), in connection with whom *ārtava-* is not seen, is also [in her period of fecundity],' say some."

The reading *adṛṣṭārtavāpy* is found in most editions. NYC, fols.28v-29r, states: *adṛṣṭārtavatā-*

SU1, p.30, where *rajas-* appears together with excretions (wind, faeces, urine) to be voided.[184] A breastless woman, a *ṣaṇḍī-* (i.e. a female ☞ *ṣaṇḍ(h)a-*), is said to be without *ārtava-* in Ut 38,18 (*anārtavastanā*[185]); this could refer to the procreatory fluid, but it seems rather that, like breastlessness, it refers to a condition which can usually be perceived, i.e. lack of the menstrual fluid. SU1, p.44 (in a variant of SU,Sū 14,18) mentions *ārtava-* and could be telling us that this appears in the female during puberty, which would make it seem as if the menstrual fluid were meant here; see on this difficult variant, however, DAS,R.P. 1991A, p.56,n.86. The term *ārtava-* is used in Ni 7,11 to describe something women put into the food of men; here too the menstrual fluid seems to be meant, like *rajas-* in the parallel passage CA,Ci 13,32 referred to in §5.4. SU,Ni 2,17 tells us that pile-like growths (*arśas-*)[186] in the ☞ *yoni-* damage (*upaghnanti*) the *yoni-* and the *ārtava-*. Now in the same passage similar growths affecting the penis are said to destroy the penis and damage the "manliness".[187] The commentaries of Ḍalhaṇa and Gayadāsa take the latter to refer to semen; this could mean that, parallelly, *ārtava-* here refers to the procreatory fluid. But "manliness" may simply refer to potency, in which case this argument would not apply. The procreatory fluid could nevertheless be meant, but we see that we cannot state this with certainty. The parallels AH,Ut 33,9-11, AS,Ut 38,p.285 are of no help, as they say more or less the same.[188] Matters are not made easier by passages like

---

py āgamaprasiddhā, but since in the manuscript the text commented upon is not given, we cannot deduce from this that the latter had *adṛṣṭārtavatā* ("the being *adṛṣṭārtava-*") for *adṛṣṭārtavā*, for the former could merely be explaining the latter. Some editions of the text (similar to AS,Śā 1,p.280b and ĀY 4,111 too; see also the variants of the latter) read *adṛṣṭārtavo 'py* ... (e.g. SU2) ("'There is also one [*ṛtu-*], in connection with which *ārtava-* is not seen,' say some"). In any case, it seems clear that the following vv.7f., describing the appearance of a woman in her fecund period, refer to such a woman when no *ārtava-* is seen (as both Ḍalhaṇa and NYC also assume), for what would be the point of detailing these signs if the menstruation shows the onset of the fecund period anyway?

184 *vātamūtrapurīṣarajasāṃ pravṛttyapravṛttī* instead of *vātamūtrapurīṣāṇāṃ pravṛttim apravṛttiṃ* of SU.

185 BYT 137,11 reproduces this as *anārtavā stanī*, which is clearly a mistake for *anārtavāstanī*. I may add that in the description of the *aṇḍinī-* in 11cd too there appears to be a mistake: *atikāmagṛhītā* is probably wrong for *atikāyagṛhītā* (cf. e.g. TIVĀRĪ 2,pp.61f.).

186 It looks as if condylomata were meant.

187 *śepho vināśayanti upaghnanti ca puṃstvam.*

188 The relevant line states: *te 'rśāṃsy upekṣayā ghnanti meḍhrapuṃstvaṃ bhagārtavam* (thus AH; AS and AH, var: *meḍhrapuṃstvabhagārtavam*, AH, var: *meḍhraṃ puṃso bhagārtavam*). *puṃstva-* is explained by the *Kairalī* commentary (on the AH passage) as being either

ŚĀR 1,5,46 and BYT 2,62,[189] in which *vīrya-* and *bīja-*[190] respectively, seemingly referring to semen, are mentioned together with *pauruṣa-* "manliness" as being in the testicles; Āḍhamalla explains *pauruṣa-* as referring to potency. But as this problem is only peripherally of concern to our subject matter, we may be excused for not pursuing it further. Turning therefore to other SU passages dealing with *ārtava-*, we find that according to Ci 1,110, an ☞ *uttaravasti-* is to be applied to cure a disorder of the *ārtava-* (*ārtavadoṣa-*), but it is unclear what *ārtava-* refers to here. In Ni 1,19 it is said that wind causes disorders by "dragging" (scil. overpowering/injuring) (*karṣati*) faeces, urine, semen, the embryo and *ārtava-*; the mention of semen and the embryo could point to *ārtava-* being a procreatory fluid here, but faeces and urine might speak for its being something voided as well, so that what *ārtava-* is here is unclear. Lastly, in Ka 3,4f. we find the poisonous *ārtava-*, and in Ka 8,85;88 the poisonous *rajas-* (the two are obviously synonyms) of spiders mentioned, but it is quite unclear what is meant.

6.2. *puṣpa-* in SU,Śā 3,12 presents some difficulty:

puṣpakāle śucis tasmād apatyārthī striyaṃ vrajet

"Therefore the one wishing progeny should go, pure, to the woman [for sexual intercourse] at the time of the *puṣpa-*."

Is the image here maybe that mentioned in §5.6, according to which the ☞ *yoni-* of a woman closes like a day-lotus at the onset of night when the period of fecundity is over? Then the period of fecundity would be the "bloom" of the woman when the *yoni-* is open, *puṣpa-* thus being the equivalent of *ṛtu-* (on which see §2.1). Or is "time of the *puṣpa-*" equivalent to "time commencing with the appearance of the *puṣpa-*"? We would then have a reference to menstruation here, which is more in keeping with the usual meaning of the term in such contexts. It must, however, also be pointed out that just before 3,12 it is stated in 3,11 that the fecundity of a woman

---

semen or potency according to the quotation from it given in the AS edition, loc.cit. (*meḍhre cen meḍhram. puṃstvaṃ strīgamanasāmarthyaṃ śuklaṃ vā*), though not according to AH4, p.428, which seems corrupt (*meḍhre cen meḍhrapuṃstvaṃ strīgamanasāmarthyaṃ vā*). *ārtava-* of the AH passage is explained by the *Kairalī* as "*rajas-* qualified for *ṛtu-*-hood" according to AH4 (*bhage ced bhagam ṛtutvayogyaṃ rajaś ca hanti*), but as "*ārtava-* (this seemingly an adjective: of/belonging to the *ṛtu-*) blood qualified for *ṛtu-*-hood" according to the edition of AS (*bhage ced bhagam ṛtutvayogyam ārtavaṃ raktaṃ ca ghnanti*); on *ṛtu-* see §2.1. This does look as if to the *Kairalī ārtava-* here referred to the procreatory fluid, but the statement is not precise enough to make this certain; moreover, what a commentary says is not necessarily what the text commented upon actually presupposes.

[189] Quoted in n.229.

[190] Cf. on this also §§2.5f.

lasts from the age of twelve to the age of fifty; thus *puṣpakāla-* could mean "time [during the life-span of a woman] in which the *puṣpa-* can be [regularly] seen" (i.e. "time in which a woman is fertile"). In this case too *puṣpa-* would denote menstruation, as usual.[191] I may, however, also point to a puzzling passage in a rather late sexological compilation (though the verses it has compiled need not necessarily be late too), in which having intercourse with a woman when the "lotus" (*kamala-*) is complete (*pūrṇa-*) (scil. in full bloom?) is recommended, since when the lotus is incomplete (*apūrṇa-*) the progeny does not live long and is unhealthy.[192] The context speaks for the "lotus" being within the woman's body, but what is meant is nevertheless not clear to me. The mention of the "lotus" could, however, point to Tantric notions, found e.g. among certain groups in Bengal today, in which the womb and the female genitals may be referred to as "flower" or "lotus" (a few examples are given in DAS,R.P. 1992A, pp.401;416), or in which one of the plexuses presumed to be in the body plays a role during menstruation.[193] But if we should in truth have a Tantric notion in the sexological work, then could we really consider SU,Śā 3,12 to refer to something similar? Since Tantric notions are as a rule not found in the 'classical' medical texts (though there are sporadic cases of agreement), this does not seem very probable; nevertheless, it cannot be categorically rejected. But in any case, what the SU passage cannot be saying is that the man should have intercourse with a menstruating woman, as that would involve going against a very strict injunction (in non-Tantric contexts), which may in actual fact have been disregarded (cf. §5.7), but surely not in such a way that disregarding it was not only recommended, but even said to be ideal for fathering a child.

---

[191] An interesting case of *puṣpa-* "menstruation" is found in a verse quoted by PUROHIT, p.117 (attributed there to Vātsyāyana; I was unable to find it in KĀM or the commentary of Yaśodhara): if the *puṣpa-* of a girl over twelve years of age does not issue to the exterior, then it remains hidden within like the flowers of certain *Artocarpus* or *Ficus* species:

dvādaśābde vyatīte tu yadi puṣpaṃ bahir na hi
antaḥ puṣpaṃ bhavaty eva panasodumbarādivat.

What is interesting here is the express comparison with a flower.

[192] The passage is found in RATS, p.70:

gacchen nārīṃ mahābhāga pūrṇaṃ (printed: *pūrṇas*) *tu kamalaṃ yadā*
apūrṇe kamale brahman na gacched ramaṇīṃ sudhīḥ
apūrṇe kamale brahman yo gacchen ramaṇīṃ naraḥ
tatsutaś ca sutā vāpi satataṃ (printed *sadataṃ*) *mriyate dhruvam*
apūrṇe kamale brahman ramaṇīṃ yadi gacchati
tasya putro vā kanyā [vā?] cirarogī bhaved dhruvam.

[193] Cf. in this regard also n.1549.

6.3. SU,Śā 3,10 states:
māsenopacitaṃ kāle dhamanībhyāṃ tad ārtavam/tadārtavam
īṣatkṛṣṇaṃ vivarṇaṃ ca vāyur yonimukhaṃ nayet
"In time/At the [appropriate] time the wind [in the body] leads this (or: the) (*tad*) *ārtava-*/the wind [in the body] then (*tad/tadā*) leads the *ārtava-*,[194] [whose colour is] slightly dark and *vivarṇa-*[195], increased/accumulated in a month, through/from two ducts (☞ *dhamanī-*), to the ☞ *yonimukha-*."

This verse is obviously to be connected with Śā 3,6 mentioned in §6.1; indeed, it has to be, for the fact that the colour of the *ārtava-* is mentioned shows that it is something which can be perceived, and that would have to be the menstrual fluid. But what does the rest of the verse mean? That depends in no small degree on how *dhamanībhyāṃ* "through/from two ducts" is to be construed; I have tried to bring out the ambiguity of the original in my translation above. The term ☞ *yonimukha-* too is problematic. Does it refer to the external vaginal orifice? Then, since the ancient Indians of course knew how uterus and vagina were connected (i.e. that they were not connected by a pair of ducts, but directly through the uterine orifice; cf. on this also the discussion on ☞ *yonimukha-*), *dhamanībhyāṃ* could be connected with *upacitaṃ*, i.e. the collection takes place in some place by means of or from the ducts, which open out into this place of collection, which would then seem to imply that, prior to the discharge by the agency of wind, through the vagina, the fluid has to pass through the uterus, which would thus be connected with

[194] We have here a problem similar to that mentioned in n.132. If we read *tad ārtavam* in pāda b, then we could translate with "this *ārtava-*", which would refer to a fluid mentioned before. The verses immediately preceding 3,10, however, do not refer to the fluid; 3,7f. give us the signs of a woman in her fecund period (see on this n.183), whereas 3,9 describes the closure of the ☞ *yoni-* (on this cf. §5.6), so that the reference would have to be to 3,6 (on which see §6.1 and n.183), which is so far removed from 3,10 that this creates a difficulty, though similar cases are not unknown. But *tad ārtavam* might also simply mean "the *ārtava-*", though it has to be examined whether this article-like use of *tad* is indeed common in SU. On the other hand, *tadārtavam* would not create any difficulties; "then" would refer to the time after the closure (which follows on the fecund period), when a new period of menstruation and subsequent fecundity begins. But the same meaning could also be obtained by reading *tad ārtavam*, since *tad* can approach *tadā* in meaning too. This would rather speak for the analysis as *tadārtavam*, of as *tad ārtavam* with *tad* meaning "then" or the like. Nevertheless, the difficulties regarding the other analysis are not of such a nature that they can serve to disqualify it altogether. In the version of this verse given in BHĀ 1,3,205 *tad ārtavam* is probably correct (see n.395), but there the context is not the same, so that we cannot simply make a similar inference for the SU passage.

[195] This term will be discussed below; see §7.8.

the place of collection in some way. Or else the ducts connect the uterus with the place of collection. On the other hand, the uterus itself — then obviously joined with the ducts connected with the collection — might be the place of collection, the discharge then flowing directly into and from this. It is of course theoretically possible that the fluid was thought to be discharged directly into the vagina from some part of the body through some openings in its wall not connected with the uterus, but this does not seem probable, as I know of no evidence which would lead us to assume that a notion of such openings for the menstrual fluid existing in the vaginal wall was prevalent in ancient India; moreover, there seems also no evidence to assume that there was a theory stating that menstruation involved a discharge directly out of the vagina by bypassing the uterus. But be that as it may, it is not clear where in the body the collection is supposed to take place: it could be in the uterus, but it could also be in some other place, the uterus serving only as the place into which the discharge after collection takes place due to wind, after which the whole flows out via the vagina.

6.4. One could, however, also consider combining *dhamanībhyāṃ* with *vāyur yonimukhaṃ nayet*; in that case, the wind would be forcing the discharge out through or from the ducts. That would, however, mean that these ducts would have to open into the uterus, since we have seen above that we have no evidence for assuming that a notion of such ducts opening directly into the vagina was prevalent. In other words, though the discharge would ultimately exit via the vaginal orifice (☞ *yonimukha-*), the process of discharge would not be a direct one. Moreover, the place of collection could not be the uterus itself, as the fluid would have to be collected before being led through or from the ducts.

6.5. If *yonimukha-* refers not to the external vaginal orifice, but to the uterine orifice, then basically nothing is altered as regards the deductions in §6.3f., for the path of the discharge would be the same.

6.6. But *yonimukha-* might also refer to the uterus. That would mean that, depending on how one construes the above verse with respect to *dhamanībhyāṃ*, the fluid either collects through or from the ducts at some place in the body and is then discharged, due to the wind, into the uterus, or else that the fluid collects in some part of the body and is then discharged, due to the wind, through or from the ducts — which must in that case be connected directly with it — into the uterus. Ultimately, this discharge must of course come out via the vaginal orifice.

6.7. To sum up: Śā 3,10 tells us that the menstrual fluid collects, in the course of a month, somewhere in the body and is then discharged to the exterior via the vagina. The place of collection could be the uterus, but it could also be some other part of the body, the discharge then first reaching the uterus before it reaches the vagina. I may add that v.11 tells us that this fluid, here referred to as *asṛj-*, is present in a woman only between the ages of twelve and fifty, after which it attains *kṣaya-*; here *kṣaya-* (on which see also §§5.2f.) obviously refers to a "loss" in the sense of "vanishing" or the like, for what is meant is clearly the menopause.

6.8. We have discussed the two ducts mentioned in SU,Śā 3,10. Such ducts are also mentioned in Śā 9,7, where it is stated:

> *adhogamās tu vātamūtrapurīṣaśukrārtavādīny adho vahanti. ... śukravahe dve śukraprādurbhāvāya. dve visargāya. te eva raktam abhivahato*[196] *visṛjataś ca nārīṇām ārtavasaṃjñam.*
>
> "Those [ducts[197]] going downwards, however, carry wind, urine, faeces, semen, *ārtava-* etc. downwards. ... Two carrying semen are for the appearance of semen, two for discharge. These same two (variant, on which see n.196: These same ones) convey and discharge the blood of women termed *ārtava-*."

This passage creates a few problems of interpretation, to which we shall, however, return later (see §6.12), since they do not touch what is of utmost importance to us right now, namely that here semen and *ārtava-* are clearly mentioned on a par, and that the total number of ducts actually enumerated in the passage tallies with their number mentioned in the same (viz. thirty) only if we once count the two pairs of ducts connected with semen and once the two pairs connected with *ārtava-*.[198] In other words, we here have a reference once to the male, once to the female, which means not only that the

---

[196] Var: *tā eva raktam abhivahanti*. This would have made sense if there had also been *visṛjanti* in the place of *visṛjataḥ* (thus referring in the plural to the four ducts and not in the dual to the two pairs of ducts), but as it is the variant seems to be merely a mistake.

[197] ☛ *dhamanī-*; the whole chapter discusses these ducts.

[198] Cakrapāṇidatta, commenting on SU,Sū 15,12, says that the *ārtava-*-carrying and the *śukra-*-carrying channels (☛ *srotas-*) are the same:

> *ārtavasrotaś ca śukravaham eva. yad ucyate śārīre te eva raktam abhivahato nārīṇām ārtavasaṃjñam* (i.e. SU,Śā 9,7) *iti*.

But what does this actually mean? Does Cakrapāṇidatta mean to say that males and females have similar channels, but with different functions, or does he mean that in the female *both* fluids are carried by the *same* channels (which in the male carry only one fluid)? The latter interpretation of course would lead us to the problem of female semen, which we shall discuss later below. (Note that here ☛ *srotas-*, which I have translated as "channel", is obviously used in the sense of ☛ *dhamanī-* by Cakrapāṇidatta, since he expressly refers to SU,Śā 9,7.) Attention may also be drawn to the in many respects similar remarks of NYC in Excursus III on pp.101-105 at the end of this chapter.

semen here is seen in its specifically male-related role (cf. §2.4), but that *ārtava-*, since it is a parallel to this semen, must refer to the procreatory fluid.[199] But this also means that both the procreatory fluid (in 9,7) and the menstrual fluid (in 3,10) are connected with ducts, which are — since no other ducts are mentioned in 9,7 — in all probability identical,[200] and this then must lead to the conclusion that the procreatory and menstrual fluids are one and the same, i.e. that there is only one procreatory-menstrual fluid. Now it may of course be that Śā 3,10 and Śā 9,7 actually speak of different ducts, but since the procreatory fluid must also, somehow or other, enter the uterus, this does not seem very probable. Moreover, the conclusion that there is only one procreatory-menstrual fluid is very well in keeping with the tentative deduction made from CA,Śā 4,7 (cf. §5.1 and §5.15), and it also explains why the procreatory fluid too is labelled with words meaning "blood".

6.9. The other data of SU not only do not speak against this deduction, but in several cases specifically support it. Thus, Śā 2,12 says that *ārtava-* can be purified (also by means of an ☞ *uttaravasti-*), and in the following verses (Śā 2,13-16) we find the mention of abnormal states of the *ārtava-*. On the basis especially of SU,Śā 2,5, where diseased *ārtava-* is said to be ☞ *abīja-*,[201] of the fact that a discussion of semen precedes that of *ārtava-*, and of 2,24, which mentions both *śukra-* and *ārtava-* as parts of the male and the female [involved in reproduction] respectively,[202] this fluid is here a counterpart of male semen, combining with this in procreation, and thus definitely the procreatory fluid, irrespective of whether it issues forth to the exterior or not.[203] The diseased states[204] consist of clotting or foul smell and pus-like character, which obviously refers to a discharge, and it seems — for the text

---

[199] Trying to explain *ārtava-* in relation to one pair of ducts (of the two pairs in this passage) as the "ovum", in relation to the other pair as the menstrual fluid, as TIVĀRĪ, 1,p.14 does, simply because it seems "more reasonable" (*adhik yuktisaṃgat*), is merely a means of twisting a statement found not to be in keeping with a pre-conceived theory, so as to make it be in line with this theory.

[200] We are for the present leaving aside the question of which of the two pairs of ducts in Śā 9,7 corresponds to that of Śā 3,10.

[201] Since the fluid is called *abīja-*, it must in any case be procreatory.

[202] *evaṃ aduṣṭaśukraḥ śuddhārtavā ca.*

[203] We could not say the same with regard to the various fluids mentioned by CA and discussed in §5.8, because there we had no statements similar to those of SU on the procreatory nature of the fluid in question.

[204] These are here enumerated, in contradistinction to SU,Ut 38,31 and Ci 37,125, which merely mention diseased *ārtava-* and *śoṇita-* respectively.

does not say anything about a discharge at a time other than the usual (in contrast to e.g. 2,18) — that this discharge is the menstrual discharge, which would point to our having one and the same fluid here. That this deduction is justified follows from 2,17, where the characteristics of pure and healthy *ārtava-* are enumerated, chief among these being a certain *red* colour,[205] which must refer to menstrual blood. And this more or less follows from Sū 15,5,[3] too, which tells us that *ārtava-*, which is capable of creating an embryo, has the characteristics of blood,[206] which seems at the very least to refer to the colour of the fluid.[207] This very fluid, we are told in Śā 2,18, may flow in too large a quantity[208] or at an irregular time,[209] and with an appearance different from that described in Śā 2,17; the disorder is then called *asṛgdara-*, which is described in Śā 2,18-21. From Śā 2,17-21 it is quite clear that what is described is an excessive flow of *ārtava-*, with no blood in general mentioned in this connection; this disorder is to be treated like *raktapitta-*, but is obviously not the same as this latter. Cf. in this context §5.14, especially n.164.

6.10. Śā 2,21-23 refer to *ārtava-* which is lost or disappears (*naśyati*) due to obstruction of its passage-way (*āvṛtamārgatvāt*). This is according to the text to be treated in the same way as the "lost/disappeared/wasted" *rakta-* (*kṣīṇa-* ... *rakta-*, and *naṣṭarakta-*) mentioned earlier, and this cannot but refer (as Ḍalhaṇa too notes in his commentary) to Sū 15,12, where the *kṣaya-* (on this see term §§5.2f.) of *ārtava-* is described,[210] as part of a description of the effects of the *kṣaya-* of different bodily substances[211]

---

[205] This passage will be discussed in more detail below; see §§7.4ff.

[206] *raktalakṣaṇam ārtavaṃ garbhakṛc ca*; this line too will be taken up again later, in §7.10.

[207] Such a clear statement on the colour of the procreatory fluid (cf. in this context also §5.4) is not to be found in CA, which is why in §5.9 we broached the possibility of there perhaps having been a notion that the procreatory and menstrual fluids leave the body by different means.

[208] *atiprasaṅgena*, on which Ḍalhaṇa says: *atiprācuryena*. One could, however, also consider translating *atiprasaṅgena* as "due to too much intercourse", but to my knowledge excessive intercourse does not seem to have been considered to be the cause of *asṛgdara-*.

[209] *anṛtau* "at a time not the *ṛtu-*"; on *ṛtu-* see §2.1.

[210] This *kṣaya-* seems to be an actual diminution (as, in a different context, in §6.7 above), whereas what Śā 2,21 refers to seems to be a mere seeming loss (as Ḍalhaṇa too observes), as in this latter case we have something impeding the outflow. See also TIVĀRĪ, 2,pp.173-175.

[211] According to Sū 15,12, *ārtavakṣaya-* occasions, among various other things, pain of the

(see also TIVĀRĪ, 2,pp.152f.;169).[212] Since it is clear from preceding passages in the same chapter (see §6.9) that Sū 15,12 refers to the procreatory fluid,[213] whereas Śā 2,21-23 seem to refer to a fluid which is hindered from appearing, i.e. the menstrual fluid, and since Sū 15,12 tells us that *ārtava-kṣaya-* prevents an appearance at the correct time (*yathocitakālādarśana-*), it follows from this too that we actually have only one, namely the procreatory-menstrual fluid. And it is this same fluid which must, flowing in a diseased state or at an abnormal time, be meant in Ut 38, where such discharges are discussed and referred to as *ārtava-*, *rajas-* and "blood", in the same manner — we can now say — as in CA,Ci 30 (see above, §§5.5ff.). Note in this context also Ut 38,9f., where it is stated that a ☞ *yoni-* which is *udāvarta-* (*udāvartā*) emits (*muñcati*) *rajas-* only with difficulty and in a frothy state, whereas one which is *vipluta-* (*viplutā*) is to be understood as being one whose *ārtava-* is spoilt (*naṣṭa-*)[214] and which is barren (*vandhya-*[215]).[216]

6.11. So basically what we have seen so far from SU tends to corroborate what we tentatively deduced from CA (see §5.15), namely that women were thought to have a procreatory-menstrual fluid which accumulates in the body and then comes out as the menstrual discharge if no fertilisation takes place.[217] We could also deduce from the foregoing that the procreatory-

---

☞ *yoni-*, here most probably "vagina".

[212] Cf. for similar descriptions e.g. CA,Sū 17,63ff. (discussed in §5.3 above), AH,Sū 11,15f., AS,Sū 19,p.142b (all without mention of *ārtava-* or the like). Note also SU,Śā 6,38 on *kṣaya-* due to damage to the ☞ *marman-*: *tato* (i.e. *hate [marmaṇi]*) *dhātukṣayāj jantur vedanābhiś ca naśyati*. See also MEULENBELD 1974, pp.458f. SU,Ci 35,5 mentions the loss/destruction/disappearance (*nāśa-*) of *śukra-*, *ārtava-* and milk (*stanya-*), without our being able to ascertain what *ārtava-* here stands for.

[213] This then obviously holds good for 15,16 too, where the effects of an abnormal increase of *ārtava-* are described, along with those of milk (*stanya-*) and of the embryo/foetus (*garbha-*).

[214] This passage has caused TIVĀRĪ 2,p.57 no little headache; in the end she decides not to explain *ārtava-* as "ovum" (cf. §2.4), but as the menstrual fluid.

[215] On *vandhya-* cf. also ☞ *vandhyā-* in Appendix II.

[216] That *udāvartā* and *viplutā* refer to the *yoni-* is made very probable by Ut 38,8, in which the *yoni-* is expressly mentioned; this would especially be the case if one followed the reading of SU,Ut 38,6f. in the form found in ĀĀS and cited in n.2004. However, one could also consider whether they could apply to the woman afflicted. It must of course also be pointed out that commonly they are today taken to be names of the disorder, but that seems most unlikely here. Cf. on this problem of nomenclature the discussion on ☞ *yonivyāpad-* in Appendix II. (On the diseases *viplutā-* and *udāvartā-* see also TIVĀRĪ 1,pp.58;26f. and 2,pp.34-36 respectively, on the latter also p.566 below.)

[217] Ḍalhaṇa, quoting (like NYC, fol.18r) a verse from an unidentified source, also says in

menstrual fluid seems to have been conceived of as always being present in some quantity in the body (cf. on this also §5.16). According to the AH, AS and BYT passages mentioned in n.144, this quantity is fixed at four *añjali*-s. But this does not necessarily mean that exactly this quantity is *always* present in the body, for the measures of, for instance, semen and milk are given too, and since these are not discharged periodically, but sporadically, their measures in actual fact cannot be conceived of as being constant at some fixed time, as one could perhaps opine is the case of the procreatory-menstrual fluid. Indeed, CA,Śā 7,15 (which does not mention the procreatory-menstrual fluid; see §5.11) tells us that the measures mentioned in it are ideal, they being liable to increase or decrease and [as such then only] inferable/estimable.[218] Similarly, the KĀ passage mentioned in n.144 says that it gives ideal measures,[219] and according to AH, AS and BYT the measures refer to those in a body which is in a state of equilibrium (*samadhātu*-). It is, however, not clear what all this means exactly with regard to substances which are voided periodically. ĀY 4,129f. contains the interesting statement (ascribed to Gopura, known as the author of a lost work) that the quantity of *rajas*- in the body is sixteen *prasṛti*-s[220] (which would, according to the general conversion, be equal to eight, and not four *añjali*-s); the text further says that only one *prasṛti*- is found in the body of the woman on the fourth day of her period.[221] Even more interesting is the citation from Vṛddhātreya's work (ĀY 4,69-72) telling us that four *añjali*-s of the blood called *ārtava*- leave the body through the vagina after a month, one *prastha*- (of this measure?, or of the whole fluid, voided or unvoided?) being left, which the wise call *rajas*-! This latter measure seems, however, to be an ideal

---

his commentary on SU,Śā 2,25 that menstruation is the discharge of "old" or "decayed" *śoṇita*-, through which the woman becomes "pure/clean" (*[nārī] śuddhā jīrṇaśoṇitāpagamane*). The verse quoted runs:

nave ṛtau (var: *tanau*) *ca sañjāte vigate jīrṇaśoṇite*
nārī bhavati saṃśuddhā puṃsā saṃsṛjyate tadā.

218 *tat paraṃ pramāṇam abhijñeyam. tac ca vṛddhihrāsayogi tarkyam eva* (on *tarkya*- see 7,14 with Cakrapāṇidatta's commentary).

219 *etat pramāṇam uddiṣṭam utkṛṣṭaṃ sarvam eva tu* (p.52).

220 ĀY 4,129ab says:

śukrapramāṇaṃ prasṛtī rajaḥprasṛtiṣoḍaśaḥ
"The measure of semen is a *prasṛti*-, [which is] one-sixteenth of the *prasṛti*-s of *rajas*-."

In the edition of ĀY this line is wrongly printed as:

śukrapramāṇaṃ prasṛtiḥ rajaḥ prasṛti ṣoḍaśaḥ.

221 We shall return to this passage later, in §8.4.

measure, for the text then goes on to tell us that a larger amount may also be left over.[222] (The text has some more interesting information, but this is not the proper place to discuss it.)[223] Cf. also n.168.

6.12. The actual mode of collection of the procreatory-menstrual fluid within the body is also not clear; the fluid could have been thought to collect steadily somewhere in the body and then sort of overflow — after passing a certain limit — once a month, or else to be let out in full. According to our discussion of Śā 3,10 above (§§6.3ff.), this place of collection could be the uterus, or else the fluid could collect elsewhere in the body and then flow out once a month via the uterus (in which it would thus not be steadily collected).[224] The only thing that we could say with some certainty was that the ducts mentioned in §§6.3ff. play some role in the process. On the other hand SU,Śā 9,7 (see §6.8), mentions *two* pairs of ducts for semen as well as for the procreatory-menstrual fluid. One of these pairs is for the "appearance" (*prādurbhāva-*) of the respective fluid. But this "appearance" does not refer to a discharge, as the other pair of ducts is responsible for this. Now we have seen above (cf. n.168) that semen is present all over the body and has first to be collected prior to its ejaculation; this makes it very probable that the "appearance" refers to a "coming out" in the sense that the semen spread all over the body is concentrated somewhere through the ducts (the parallel passage AS,Śā 6,p.330b speaks of two ducts which "carry" semen [in men] and *ārtava-* in women)[225]. Since the other pair serves to ejaculate

---

[222] 69cd (where the fluid is identified as *ārtava-*) has already been quoted in n.31. 69cd-72 state:

ṛtusambandhajaṃ raktam ārtavaṃ procyate budhaiḥ (69cd)
strīṇāṃ garbhāśaye vāyur yad raktaṃ labhate śanaiḥ
catvāro 'ñjalayo māsāc cyavate yonimārgataḥ (70)
srutāvaśeṣaṃ tatprasthaṃ raja ity ucyate budhaiḥ
srutāvaśeṣam athavā yadi tiṣṭhati vādhikam (71)
tadā śukrasya saṃsargi jāyate malavarjitam
yathā malinakhaṇḍasya malaṃ ca harate balāt
kṣīraṃ kṣāras tathā vidur garbha ārtavaśuddhikṛt (72).

The above is badly edited in the printed edition. Thus the text as given has *catvāro jalayo māsācyavate* (which is gibberish) in 70 (translated as: "it flows ... for 4 days every month"!), whereas obviously only the dot of an *ṃ* (for the *ñ* given above) has been left out. Moreover, it has *tadāśu śukrasya* instead of *tadā śukrasya* in 72, even though this is patently against the metre and the latter reading is expressly mentioned as a variant. Also, it has *saṃsargī* instead of the grammatically correct *saṃsargi*.

[223] We shall return to this information in §8.16 below.

[224] WISE, p.33 seems to presuppose the latter, but without any relevant references. Note too the use of "menses" for both the procreatory and the menstrual fluids (cf. §2.3 above).

[225] *dve śuklaṃ* (i.e. *śukraṃ*) *vahato dve ca muñcataḥ. te eva nārīṇām ārtavam*. As regards

the semen, it follows that this pair must then arise from the place (or two places?) of collection.[226] Then, however, the same most probably holds true, parallelly, for the procreatory-menstrual fluid too. But we have seen above that this fluid must in any case pass through the uterus, and since it is highly improbable that a pair of ducts was assumed to exist between uterus and vagina (cf. §6.3), therefore we can deduce firstly that in the theory presupposed here the uterus must be connected with one pair of ducts, and that, secondly, since there are two pairs of ducts and no pair of ducts connects uterus and vagina, it cannot itself be the place of collection of the fluid. In other words, the procreatory-menstrual fluid is collected not in the uterus, but somewhere else.[227] Where this is, is, however, unclear, and this holds true, *mutatis mutandis*, for the male too, for nowhere is it said that the place of origin of semen is the testes.[228] Indeed, in a medical diagram from Nepal, described in more detail in §6.14 below, we find a receptacle of semen in the thoracic region; not only is this not connected in any way with the testes, but these latter are also not even shown in the drawing! Of course the special link between the testicles and semen was known to Indian medical authorities,[229] which could lead one to assume that the pair of

---

the whole passage as given on p.330, cf. also the variants in the quotation in Aruṇadatta's commentary on AH,Śā 3,39f; see also Śrīdāsa's commentary on loc.cit. in AH.

[226] One may of course also hold that the first pair leads only to a place of storage, the second pair then leading to a discharge independently of this. But then what happens to the semen in the place of storage? Obviously, this must ultimately reach the place of discharge too, so that then we would have to postulate yet another connection between the place of storage and the place of discharge, so that this line of thought does not seem correct.

[227] True, one could postulate that one of the pairs of ducts mentioned empties directly into the other pair, which would mean that there is no intermediate place of collection, from which it would follow that ultimately the collection does take place in the uterus. But then what would be the point of postulating *two* pairs of ducts and not only one pair? Two pairs of ducts only make sense if one pair leads to some receptacle and one pair away from this receptacle.

[228] Cf. on this also MÜLLER 1955, pp.10f.

[229] See e.g. SU,Śā 9,12, AS,Śā 6,p.331b and also DAS,R.P. 1984A, p.238. ŚĀR 1,5,46 says that the testicles are the receptacles of vessels (*śirā-* = ☞ *sirā-*) carrying semen:

vīryavāhiśirādhārau vṛṣaṇau pauruṣāvahau.

BYT 2,62 is similar:

medaḥkaphāsramāṃsānāṃ prasādād vṛṣaṇau matau
bījavāhiśirādhārau viśeṣāt pauruṣāvahau.

On *vīrya-*, *pauruṣa-* and *bīja-* cf. §6.1. We find a similar statement in BHĀ 1,3,87f.:

vṛṣaṇau bhavataḥ sārāt kaphāsṛṅmāṃsamedasām (var: *kaphāsṛgbhyāṃ ca medasām*)

ejaculatory ducts are in fact the seminal ducts, but as what should we then identify the other pair of ducts which must then lead to them?

6.13. As it is, CA,Śā 2,18f. mention a receptacle of semen (*śukrāśaya-*) in the male, which neither CA nor Cakrapāṇidatta equate with the testicles, and which moreover appears in the singular. CA,Ci 26,72 too mentions a *śukrāśaya-*, but gives no further information on this, whereas according to CA,Śā 2,19 this receptacle has an opening (*dvāra-*); however, neither the *śukrāśaya-* nor its opening are further elaborated upon.[230] This must nevertheless remind one of SU,Śā 4,22,[231] which tells us that the semen of a man[232]

---

vīryavāhiśirādhārau tau matau pauruṣāvahau.

This passage is also found in a corrupt form at the right of the head of the figure on the cover of SIMH and COMBA 1991 mentioned in §6.14. See also §6.19. But I must also point out that SU,Ci 7,36 (on which see also n.308 below) and AS,Ci 13,p.213a mention a structure or structures carrying semen *and* ☞ *srotas*-s coming from the testes, damage to the former causing "*klība*--hood" (*klaibya-*), but of the latter ☞ *dhvajabhaṅga-*; *dhvajabhaṅga-* is commonly taken to refer to impotence, i.e. lack of the faculty of erection, which, whatever the other expression (i.e. *klaibya-*) may mean, is not the same as sterility and as such not necessarily to be connected with semen. Interesting is also CA,Ci 30,188, which tells us that incurable *klaibya-* is according to "some" (*kecid*) said to arise due to amputation of the penis and removal of the testes:

kecit klaibye tv asādhye dve dhvajabhaṅgakṣayodbhave
vadanti śephasaś chedād vṛṣaṇotpāṭanena ca.

Whatever be the meaning of *klaibya-*, this seems to imply that this opinion was not necessarily the opinion of all, including the author of these lines. Note also that in this passage one sort of incurable *klaibya-* is said to arise from *dhvajabhaṅga-*. Statements such as these lead to the very complicated, but nonetheless highly interesting, subject of the ancient Indian terminology for sexual dysfunctions and their causes, which is obviously of importance as regards the subject of the functions attributed to the testes. The complexity of the matter hinders a discussion in this footnote, but this discussion will be found in Appendix II under ☞ *klība-*, where there are also further references; also cf. *vātikaṣaṇḍhaka-* under ☞ *ṣaṇḍha-*.

230 CA,Śā 2,18f. run:

bījāt samāṃśād upataptabījāt strīpuṃsaliṅgī bhavati dviretāḥ
śukrāśayaṃ garbhagatasya hatvā karoti vāyuḥ pavanendriyatvam (18)
śukrāśayadvāravighaṭṭanena saṃskāravāhaṃ kurute 'nilaś ca
mandālpabījāv abalāv aharṣau klībau ca hetur vikṛtidvayasya (19).

Variants: *gatvā* for *hatvā* in 18, *hi karoti vāyuḥ* for *kurute 'nilaś ca* in 19. Cf. on these verses also ☞ *dviretas-* and ☞ *klība-* in Appendix II.

231 *dvyaṅgule dakṣiṇe pārśve bastidvārasya cāpy adhaḥ*
mūtrasrotaḥpathāc chukraṃ puruṣasya pravartate.

232 The verse says expressly that it refers to a man's semen (*śukraṃ puruṣasya*; see n.231). That the body of a man is here meant follows also from the fact that 4,22 is sandwiched between two passages which expressly refer to males too. Śā 4,21 speaks of *śukra-* in the body of males (*śarīreṣu tathā śukraṃ nṛṇāṃ vidyād bhiṣagvaraḥ*); one could here regard "males"

leaves his body by means of the channel (☞ *srotas-*) of urine, two fingers to the right of the opening of the urinary bladder and below it; this verse is also found in BHĀ 1,3,201 and Āḍhamalla's commentary on ŚĀR 1,5,46, without an ascription. As an editorial footnote to SU,Śā 4,22 rightly points out (the work quoted there[233] even considering amending the text of SU), the mention of only one ☞ *srotas-* here seems in conflict with SU,Śā 9,7, according to which two ☞ *dhamanī-*s are responsible for the discharge of semen (cf. §6.12). The matter is, however, complicated by the fact that we do not know whether *srotas-* and *dhamanī-* here are synonyms (see on this §6.18), or whether the latter may not have been taken to discharge into the former. AS,Ci 13,p.211a and AH,Ci 11,42 too mention a *śukrāśaya-* of the male (which in the case of a certain urinary disorder is to be cleansed by eating strengthening food and indulging in intercourse with women) but unfortunately have no more information on this; as the word is part of a compound (*śukrāśaya-śodhanārtham* and *śukrāśayaviśuddhaye* respectively), we do not even know whether it is singular, dual or plural. SU,Ut 39,88 = MĀDH 2,54[234] also mentions a *śukrasthāna-*, but this, parallel as it is to *rasa-*, blood, flesh, fat, bones and marrow, seems to refer merely to any site of *śukra-* in general (i.e. the *śukra-* in the body)[235] and not to any specific organ (as expressly stated by the commentary *Madhukośa* too[236]); similar is the case as regards *śukra-*

---

as subsuming females too, but this is impossible as regards 4,23, where the ejaculation of semen of the person having intercourse with women (*strīṣu vyāvacchataḥ*) is described.

[233] Namely SEN (see 1,p.77).

[234] As well as a variant to the additional verse found after MĀDH 2,54 (see MEULENBELD 1974, p.186,n.226).

[235] Cf. also n.168.

[236] *śukraṃ ca tatsthānaṃ ceti śūkrasthānam. na tu śukrasya sthānaṃ śukrasthānam* (this latter line is missing in some mss). *śukrasya sarvadehagatvena niyatasthānāsambhavād iti kārtikaḥ.*

(On Kārtikakuṇḍa see MEULENBELD 1974, pp.394f.) It is incidentally interesting to see how the modern Hindi commentary of MĀDH tries to twist the quite clear statement of the *Madhukośa* so as to show that there are two sorts of *śukra-* (more information on this theory will be given in n.812), one all over the body (this bringing out the signs of maleness: *yah śarīr meṃ pauruṣ cihnoṃ ko prakaṭ kar'tā hai*), and the other in the testes and the prostate, etc.; for "prostate" the neologism *pauruṣ'granthi* (i.e. *pauruṣagranthi-* "manliness-gland"; cf. on *granthi-* n.256) is given, seemingly as if it were an old medical term! True, CA,Vi 5,8, which states that the *mūla-*s of the ☞ *srotas-*s of semen are the testes and the penis (on this see §6.18; cf. also n.229) is quoted, but that does not tell us anything about the *sthāna-* of semen. Incidentally, the explanation just given conveniently neglects the mention of the penis (rather embarrassing for the theory, actually, if Āyurveda presupposes modern science in everything!). As it is,

sthāna- in Ca,Ci 3,82. Bhā 1,3,179(p.57) expressly refers to the "place of *śukra-*" being the whole body, a part of the *rasa-* of food going thither to unite with semen.[237] Ḍalhaṇa in his commentary on Su,Ut 39,88, however, also mentions the "being concealed/separated by the ☞ *kalā-*" of semen;[238] according to Su,Śā 4,20, though, the *kalā-* containing semen is spread all over the body.[239] But in the case of these passages (Su,Ut 39,88 and those following) semen seems to be seen as being present in humans generally, i.e. in both males and females, and not only in males,[240] so that considerations quite different from those applying to the passages mentioned before them might be involved. A further problem is created by As,Śā 5,p.319b[241] and a passage from an unidentified source quoted in Aruṇadatta's commentary on Ah,Śā 3,9f., which tell us that the ☞ *kalā-* of semen just mentioned causes the latter to revolve/move,[242] and permeates the whole body,[243] but is at the same time situated two fingers to the right of the opening of the urinary bladder and below it, attached to the path of urine,[244] which is nearly the same as the description of the path by which semen leaves the body in Su,Śā 4,22 (the passage of Aruṇadatta even incorporating the exact wording of 4,22ab), and leads to the same problems and seeming discrepancies as regards the number of ☞ *srotas-*s mentioned (one). Moreover, the passages just mentioned seem to refer to humans in general, like the other passages relating to the *kalā-* containing semen mentioned above, whereas Su,Śā 4,22 expressly refers only to the male; could the passage of As and that quoted by Aruṇadatta therefore be an attempt to build a bridge between the other passages relating to the *kalā-* and Su,Śā 4,22? But if they are, then it is obvious

---

given the wealth of often contradictory data in §§6.12f. and §§6.18f., one must wonder at the notions on methodology which allow a single statement chosen seemingly at random to form the basis of an argument with such far-reaching consequences.

[237] *śukrasya sthānaṃ sakalaṃ śarīraṃ gatvā.*

[238] *sarvaśarīrasthasyāpi śukrasya kalāntaritatvena tvacāṃ krameṇa marmaprāpteḥ.*

[239] *yā sarvaprāṇināṃ sarvaśarīravyāpinī.*

[240] Cf. in this regard also §2.4; further matter will be given in §8.17.

[241] Quoted also by Bhā 1,3,201, right after Su,Śā 4,22, given above, has been quoted, as well as by Ḍalhaṇa in his commentary on Su,Śā 4,22. Śrīdāsa in his commentary on Ah,Śā 3,9f. conflates this with Indu's paraphrase of it.

[242] Aruṇadatta's passage: *vartayati*, As: *pravartyati*; the latter could also mean "pours forth" (transitive).

[243] Aruṇadatta's passage: *śarīraṃ vyāpya sakalam*, As: *sakalaśarīravyāpinī.*

[244] Aruṇadatta's passage: *mūtramārgam upāśritā*, As: *mūtramārgam āśritā.*

that they lead to problems which I for one am at a loss to solve.

6.14. All in all, the matter is so unclear that I personally consider it prudent to desist from speculation on the nature of the receptacle of semen and not to try, for instance, to identify it apologetically with the testicles or, failing these, maybe with organs such as the prostate or the glandulae vesiculosae, at least until some sort of hard and fast evidence turns up (in which context I have even neglected the fact of the receptacle of semen being referred to in the singular). But the connection with the urinary tract mentioned above[245] does bring to mind CA,Ci 15,34f.,[246] which tell us that semen (evolved from the marrow) is collected from all over the body (which it permeates) due to sexual excitement (cf. n.168). The heat of intercourse then makes this, "set free from [its] place" (*sthānavicyuta-*), exit from the body after it has first collected in the urinary bladder.[247] It is not clear whether *sthāna-* "place" here refers to the body as a whole, as the place where semen is diffused, or to a place of storage into which the semen flows from all over the body due to sexual excitement. In the first case one could hold the bladder to be an equivalent of the *śukrāśaya-* mentioned above, whereas in the latter case the bladder would be something the semen flows into after it has left the place of storage. CA,Ci 2,4,46-49, though similar in some respects, are actually no help in this connection, for they say merely that semen in the body removes itself (*pracyavate*) from its place (*sthāna-*) when man and woman unite, and is ejaculated. In any case, CA,Ci 15,34f. are not in keeping with the description of SU,Śā 4,22. What makes the whole even more problematic is that semen here is first described as part of the chain of alimentary transformation in humans as such, but then suddenly seems to be seen purely in the function of male semen; this is a familiar problem in Indian medical texts.[248] Or are we to deduce that the semen of women too is ejaculated during intercourse in this way? That would lead to a most difficult problem, which will be taken up in detail in chapter 8. I may add that this problem would also apply to CA,Ci 2,4,46-49 if we regarded this passage in isolation; it is, however. part of the chapter on *vājīkaraṇa-* or aph-

[245] Cf. also the remark of Gayadāsa in his commentary on SU,Ni 3,19, where *pauruṣa-* "manliness" is connected with the urinary bladder; according to Gayadāsa *pauruṣa-* refers to the tube (☛ *nāḍī-*) carrying semen (*pauruṣaṃ śukravahanāḍī*). Ḍalhaṇa, however, explains *pauruṣa-* as *meḍhra-* "penis", which raises the question of whether Gayadāsa's "tube" might be referring to the same or to an internal structure.

[246] In a passage which may not be an authentic part of CA; cf. n.420 and §7.17.

[247] The full text and translation have been given, in another context, in n.491.

[248] We shall have occasion to remark on this problem again in §8.9.

rodisiacs — for the *male* only — and as such most probably applies to males only, which increases the probability that this is also true of CA,Ci 15,34f. Unfortunately we cannot be quite sure on this latter count. Turning again to the definitely male seminal receptacle, we find that a recently discovered Nepalese diagram (probably late 18th century) of a male, of (pseudo-?)medical character (but quite different from Tantric and similar diagrams of the human body, showing the *cakra*-s, *nāḍī*-s etc., one often finds), has a receptacle of *śukla-* = *śukra-* (*śuklasthāna-*) situated just above where we would seek the heart (☞ *hṛdaya-*; on this see the footnote below), a *rasasthāna-* or receptacle of ☞ *rasa-*, however, opposite it on the other side of the body.[249] Several particulars of this diagram seem to be different from what Indian medical texts in general presuppose, and thus the location of the receptacle of semen in this diagram need not coincide with that presupposed by older 'classical' works, but it nevertheless shows quite clearly, perhaps more clearly than textual descriptions alone could, that we should not take what seems obvious to us for granted when researching on other cultures. This consideration must then also similarly apply to all our investigations into the nature of other organs connected with the ejaculation of semen, and this must include the passages (ducts or the like) through which this takes place. But the same must, as the deliberations in §6.12 have shown, then also apply in the case of the ducts, last mentioned in §6.12, leading to the uterus in women; it is obvious that, in the light of the problems detailed above with regard to the receptacle of semen, we face problems here too. And if anyone should consider taking these ejaculatory ducts in the female to be the Fallopian tubes

---

[249] I worked with quite unclear reproductions of the original, which is kept in the library of the Wellcome Institute for the History of Medicine in London (accession number 347674). The diagram is reproduced on the covers of SIMH and COMBA 1991. It is most interesting because it has receptacles for various substances distributed throughout the abdominal-thoracic cavity (including one for *indriya-*); unfortunately, their legends are not always legible, and probably unclear in the original too. Under the (round) *śuklasthāna-* are two receptacles, a conch-shaped one a little to its left and a pear-shaped one right underneath. The first seems to be labelled *phokto*, reminding one of the Nepali *phokso* "lung", and could even be this very same word with the Nāgarī *s* damaged. In the second I could make out *kta* followed by what looked like the upper part of an *s* (as in the ligature *sthā* in the *śuklasthāna-* above), followed by an unidentifiable letter; I suspect that what we have here is a ☞ *raktasthāna-* "receptacle of blood". Beneath the conch-like receptacle is an irregular receptacle with what appears to be *aujasthāna*, i.e. *ojaḥsthāna-* "receptacle of ☞ *ojas-*", while touching it on the left and a little below is a pear-shaped organ whose legend *could* be *hṛdaya*, though the writing is so indistinct that this is basically conjecture. I may add that there are several text passages on the sides of the diagram. Those I attempted to identify were all found in BHĀ; they are all hopelessly corrupt, and show that the scribe can have had only a rudimentary knowledge of Sanskrit at most.

(whose openings are incidentally extremely minute and very hard to perceive even for a trained eye) and hence the ovaries to correspond to the *śukrāśaya-* (singular!) of males mentioned above, then the difficulty is even greater than in the case of the testes, for we have no evidence at all that the ovaries were takjen to be (since the ovum was unknown) connected with procreation; indeed, none of the texts seems to even mention them.[250] Again, Śrīdāsa in his commentary on AHı,Śā 3,67 (p.125) quotes a passage[251] which states that women too have a *śukrasthāna-*, and that this is the ☞ *yonimukha-*. Whatever *yonimukha-* be here,[252] this complicates matters even further, for it becomes difficult to conceive of ejaculatory ducts in the female leading away from this (whether it be the uterus or its or the vagina's orifice) and discharging semen somewhere. All this has not, however, deterred some scholars from simply projecting modern 'loan-translations'[253] such as ☞ *ḍimba-* or ☞ *aṇḍa-* "egg", which are neologistically used for "ovum"[254,255] into Old Indo-Aryan medical works, which then serves as a convenient means for bringing in *modern* terms such as *ḍimbagranthi-*[256] "ovary" or *ḍimbanālī-/ḍimbapraṇālī-* "Fallopian tube". This is not to say that there may not be a possibility that it is the Fallopian tubes that might have been meant by the ducts (or the like) leading to the uterus, but the point is that we have no concrete evidence to back this assumption, so that statements assuming such an identity are but speculation.[257]

---

250 A discussion of modern attempts to show nevertheless that the ovaries were presupposed by ancient Indian medical authorities will be found in Excursus II on pp.97-100 below.

251 Discussed in some detail in §11.12.

252 We shall take up the matter in §11.12.

253 I.e. newly formed equivalents or calques, deriving from the vocabulary of Sanskrit, of terms found in other modern languages (usually English).

254 MURTHY 1,p.357 also uses such modern terminology: "The spermatazoa [sic] (śukrāṇu) produced by the testes (vṛṣaṇa) of the man and ovum (aṇḍāṇu) produced by the ovaries of the woman are actually the male and female seed respectively, the union of the spermatozoon and the ovum gives rise to the formation of the embryo." Since MURTHY has, just before these lines, quite clearly stated that the ancient Indians regarded the menstrual fluid as being (as a procreatory fluid) responsible for the formation of the embryo (see p.17 above), his remarks cited here seem to be an explanation of what actually happens (in accordance with present-day knowledge), though the terms *śukrāṇu-* and *aṇḍāṇu-* might generate confusion.

255 Cf. in this context the use of Arabic *bayḍa* "egg; testicle" referred to on p.574.

256 *granthi-* "gland" is another modern 'loan-translation' often projected into old medical texts.

257 A case in point are the statements on the *aṇḍādhāra-*, on which see the discussion on

6.15. Let us, however, return to the problem of the collection of the procreatory-menstrual fluid in the body. In this context I would also like to draw attention to the theory in Tibetan medicine that the female's procreatory fluid (which, like in Indian medicine, is also the menstrual fluid) is collected gradually in a separate organ called the *bsam se'u*, from which it flows via two tubular structures into the uterus periodically to come out as the menstrual discharge, these being forced to open by the pressure of the fluid within them. This theory can be most conveniently read about in MEYER,F. 1983, pp.109ff. Since Tibetan medicine has been heavily influenced by Indian medicine, this does look like a confirmation of what we have deduced above. But I may also point out that MEYER,F. 1983, pp.155f. is of the opinion that the *bsam se'u* is, at least as regards its name, a borrowing from Chinese medicine, though the function of the similarly named organ in Chinese medicine is, according to him, quite different. Note also that according to most authors the *bsam se'u* is (also) a receptacle of semen,[258] in which case it would correspond to the *śukrāśaya-* mentioned in §§6.13f. MEYER,F. 1983, pp.111;155, however, disputes this, saying that there is no evidence that semen is mentioned in this context, but he himself states (p.155) that the *bsam se'u* serves as the receptacle for what he terms "suc" (i.e. "juice"), the last of the elements formed in the same manner as in Indian medicine (see §2.4) in the chain of digestion, this "juice" (Tibetan *khu-ba*) then being transformed (pp.132f.) to semen and what the author terms menstrual blood.[259] Whether semen and "menstrual blood" indeed have, as the author supposes, the same origin,[260] is immaterial here, for what I want to stress is that his own statements make it likely that *bsam se'u* is a name given to the receptacle of the procreatory fluid in both male and female. In this regard I would like to draw attention to the wealth of material on this organ (referring to *both* male and female) in DBANG 'DUS, pp.664-667. We

☛ *aṇḍa-* in Appendix II.

[258] Thus also in WALSH, p.1242. MÜLLER 1955, p.10,n.28, referring to WALSH, points out that according to the drawings reproduced there the *bsam se'u* has *no* connection with the testes, and also not with the urinary bladder; cf. too MÜLLER 1941-1942, p.468 (continuation of n.6 on p.467).

[259] The chart on p.132 is drawn a bit awkwardly, and could be taken to mean that the "fluid of vitality" (Tibetan *mdangs*), which (like ☛ *ojas-*) is the subtlest part of the "juice" left over after the separation of the procreatory fluid, is in the *bsam-s'eu*, but this is clearly not in accord with p.155.

[260] Note in this connection also what COMBA 1981 and 1985 and ZIMMER,HEN.R., p.185 state with regard to Indian medicine; see §2.4.

thus see that Tibetan medicine seems to offer us parallel evidence proving the validity of our deductions made above. Nevertheless, the fact that other medical systems have also influenced Tibetan medicine must make us be cautious in using such evidence, and this caution must be applied in the case of any Indian influence too, for Tibetan medicine has not only been heavily influenced by Indian medicine, but also shows many similarities with non-medical Indian Tantric theories (in which connection I am leaving aside the difficult question of where such theories came from in the first place). This note of caution applies to the drawing from Nepal mentioned in §6.14 too, as Nepal has been heavily influenced not only by Indian, but also by Tibetan culture. On the other hand, no amount of caution can do away with the fact that we do seem to have a remarkable correspondence between what we have seen in §§6.12ff. and the theories of a medical system which is known to have been influenced by Indian medicine. What makes the matter even more intriguing is that WADDELL posits that the Tibetan anatomical drawings WALSH has described actually give us Indian conceptions; MÜLLER 1955, p.10,n.28 too writes of "bei deutlicher altindischer Grundlage" in this connection, and MÜLLER 1960C, p.720 says that an East Asian influence is improbable..[261]

6.16. In any case, that the procreatory-menstrual fluid is most probably collected in some other place within the body and not in the uterus, this, however, being connected with the place of collection, need not necessarily mean that the ducts mentioned in Śā 3,10 are connected with the uterus, for one of the two pairs of ducts mentioned in Śā 9,7 could equally well be meant, since the net result in each case is the same (cf. the discussion in §§6.3ff. and §6.12). In other words, *dhamanībhyām* in 3,10 might be connected either with *upacitaṃ* or with *vāyur yonimukhaṃ nayet*. The reason I am stressing this here becomes obvious when one turns to AH,Śā 1,22cd-23ab = AS,Śā 1,p.280b, which has nearly the same wording as SU,Śā 3,10, but with some noteworthy differences. The verse runs:

> *māsenopacitaṃ raktaṃ dhamanībhyām ṛtau punaḥ*
> *īṣatkṛṣṇaṃ vigandhaṃ ca vāyur yonimukhān nudet*[262]
>
> "Again, during the *ṛtu-*[263], the wind [in the body] thrusts the [menstrual] blood, [whose colour is] slightly dark and *vigandha-*, accumulated/increased in a month, through/from two

---

[261] I lack the specialised knowledge to comment upon the validity of this assertion; cf. also MÜLLER 1925, pp.550f. and FLEET.

[262] It is found in this form also in CIKS, p.41 and YO 21,39.

[263] On this see §2.1.

ducts, [out] from/via the ☞ *yonimukha-*."

Leaving aside the problem of *vigandha-* to be taken up later (in §7.9), we see that the main difference, causing a difficulty, is *yonimukhān nudet* in the place of *yonimukhaṃ nayet*, for if we connect this with *dhamanībhyam* (as Aruṇadatta does in his commentary on the AH passage),[264] we could analyse this as stating that the ducts lead away from the *yonimukha-*, which would then have to be the uterus, to the exterior (or at least the vagina), and this seems, in the light of what we have seen so far (cf. §§6.12ff.), highly improbable. But another analysis too is possible, namely that the ducts lead *to* the *yonimukha-*, i.e. the wind leads the blood out of the body [first] through the ducts (*dhamanībhyām ... nudet*) and [subsequently] via or from the *yonimukha-* (*yonimukhān nudet*), and this would be in keeping with our deductions above. Indu actually could be thinking so, for in his commentary on the AH passage he says that the wind brings the "blood" collected in a month to the *garbhasthāna-* "place of the embryo/foetus" (i.e. uterus) (cf. ☞ *garbhāśaya-*) through the ducts. Unfortunately, we have no surety on how he really understands the whole verse.[265] No problem arises, however, if we connect *dhamanībhyām* with *upacitaṃ*, as in §6.3, as this would say that the blood is collected by means of the ducts, not discharged through them.[266]

6.17. A different problem is posed by SU,Śā 4,24, which states that in

---

[264] *kena nudet. dhamanībhyāṃ srotoviśeṣābhyām.*

[265] The printed text runs:

vāyuḥ strīṇāṃ māsenopacitaṃ raktaṃ sarvaśarīrād garbhasthānaṃ dhamanībhyām ānīya yoni(mukhāt prerayet)

"The wind, having brought the blood of women, collected in a month, from the whole body to the place of the embryo/foetus through two ducts, (pushes [it out] from) the *yoni(mukha)-*."

The part within round brackets is printed thus in the edition and could be either an editorial conjecture or else show that this part is missing in one or more mss. Since no more information is given (at least I could not find any), we thus cannot be sure that this part actually has Indu's own words, or, to put it differently, that Indu is actually commenting upon the very same reading *yonimukhān nudet* we have cited, the whole being complicated by the prose word order of the contents of the verse (the *anvaya-*, seemingly appended by the editor), which reads *yonimukhaṃ* instead of *yonimukhān* (... *vāyuḥ dhamanībhyāṃ yonimukhaṃ nudet*). But in any case it does seem as if to Indu *yonimukha-* (whatever its case ending actually be) and *garbhasthāna-* were identical.

[266] I may here also add that ĀY 4,76 quotes the above verse with the line:

īṣadvigandhi kṛṣṇaṃ ca vāyur yonimukhaṃ nayet.

This seems to be a contamination of the different readings given here and in §6.3. Śrīdāsa in his commentary on AHI,Śā 1,23cd-24ab (AH,Śā 1,22cd-23ab) lists a variant which has *īṣaduṣṇaṃ* "slightly warm" in the place of *īṣatkṛṣṇaṃ*.

pregnant women the apertures or channels (☞ *srotas*-s) carrying *ārtava*- are blocked by the foetus, and that therefore this *ārtava*-, unable to flow downwards, is not seen during pregnancy. Rather, it flows upwards and, being collected again/subsequently (*aparam*), forms the ☞ *aparā*- or afterbirth; flowing still further upwards, it makes the breasts swell.[267] Since the *aparā*- is within the womb, it would follow from this that the procreatory-menstrual fluid starts collecting within the womb and then, when this is full, in parts of the body lying above this and — this is an obvious inference — connected with this by some means. The deduction from this is that the procreatory-menstrual fluid enters the womb even during a pregnancy,[268] but cannot leave it. Now during the process of birth there is also a flow of blood, especially in the stage after the child has been delivered and the other contents of the uterus come out; the blood mingles with the other fluid contents of the uterus and colours them red, often making the whole quite a messy affair, as anyone who has witnessed a birth can confirm. Since the procreatory-menstrual fluid too is "blood", one can quite well imagine a connection having being made between it and the mentioned discharge. If thus the text is really telling us that the procreatory-menstrual fluid collects in the uterus after conception, then from Śā 5,10, where the so-called "external" ☞ *srotas*-s (apertures leading to the exterior) of the body are enumerated (one of these carrying "blood"; cf. §6.1 with n.180), it would seem that in Śā 4,24 we must have a reference to a blockage of the orifice leading to the exterior; but since the opening of the vagina itself can hardly have been taken to be blocked,[269] it would seem that the opening of the uterus into the vagina is meant, loosely taken to be a blockage of the vagina (and in any case being a blockage of the path to the exterior). However, the SU passage refers to a plural number of ☞ *srotas*-s. The commentary to be found in NYC, fol.37v also refers to a plural number of apertures of tubes (☞ *nāḍī*-), and not only may *nāḍī*- and ☞ *srotas*- at times denote the same structures, but the last part of the commentary also does indeed seem to equate the two terms, since it

---

[267] The paraphrased passage states in full:

gṛhītagarbhāṇām ārtavavahānāṃ srotasāṃ vartmāny avarudhyante garbheṇa. tasmād gṛhītagarbhāṇām ārtavaṃ na dṛśyate. tatas tad adhaḥ pratihatam ūrdhvam āgatam aparaṃ copacīyamānam aparety abhidhīyate. śeṣaṃ cordhvataram āgataṃ payodharāv abhipratipadyate. tasmād garbhiṇyaḥ pīnonnatapayodharā bhavanti.

BHĀ 1,3,206 has more or less the same.

[268] On the entry of the procreatory-menstrual fluid into the uterus at times other than the *ṛtu*- see §5.7 and §5.14 too.

[269] On this point cf. also n.133.

uses *nāḍī-* and *srotas-* in the same context. However, from the wording of the commentary it seems that the *nāḍī-*s referred to here are not those generally leading the fluid out of the womb, but rather structures through which the blocked blood now flows upwards backwards, and therefore do not seem to be the same as the structures we are here concerned with.[270] But even if the commentary seems no help in this matter, we still have the statement of the SU passage itself. Now one could say that, since this also speaks of "women" and not of a single woman, the plural of *srotas-* is due to this plural number of women, only a single *srotas-* in each woman being meant. But why is *garbha-* "embryo/foetus" in the singular then? Moreover, the parallel passage AS,Śā 2,p.288a[271] speaks only of a single "woman".[272] One may of course opine that this latter passage may presuppose something else than the SU passage, or simply be an incorrect rendering of it, but all in all we do have reason not to simply disregard the plural.

6.18. If we take the plural seriously, we must conclude that the "external" ☞ *srotas-*, i.e. "aperture", mentioned above cannot be the (only?) *srotas-* meant. Now ☞ *srotas-* may at times subsume various hollow and tubular structures within the body, including both ☞ *sirā-*s "vessels" and ☞ *dhamanī-*s "ducts";[273] but if what is meant here is a blockage of the egress of the procreatory-menstrual fluid, then the "ducts" of §6.12 seem hardly to be meant, for these carry the procreatory-menstrual fluid *into* the uterus, which means that their blockage would not allow the fluid to form the *aparā-* within the latter. But what other *srotas-*s could then be meant here? ☞ *srotas-* is used in a number of different meanings in medical texts, which does not make the whole any easier. If we, however, confine ourselves to SU, we find that, apart from Śā 5,10 mentioned in §6.17, the *srotas-*s of the body are also described in Śā 9,12. Here we have a description not of the "external", but of the "internal" *srotas-*s, which we may most aptly call "channels". These do not

---

270 The text, translation and discussion of this commentary is given as Excursus III at the end of this chapter (pp.101-105), as the whole is too long to fit into a footnote.

271 Cf. on this also the discussion on ☞ *aparā-* below.

272 *tasyāś ca rajovāhināṃ srotasāṃ vartmāny uparudhyante garbheṇa. tasmāt tataḥ param ārtavaṃ na dṛśyate. tatas tad adhaḥ pratihatam aparam aparaṃ copacīyamānam aparety āhuḥ. jarāyur ity anye. sthite rakte romarājiḥ prādurbhavati. jarāyuśeṣaṃ cordhvam asṛk pratipadyate. tasmāt pīnakapolapayodharatā kṛṣṇauṣṭhacūcukatvaṃ ca. stanāśrayam eva ca kaphoparañjitaṃ stanyatām upagataṃ prasūtāyāḥ punar āhārarasenāpyāyate.*

On the last line cf. §7.37 and §9.6.

273 See on this the discussion on ☞ *srotas-* in Appendix II to this study.

open to the exterior, but carry substances within the body (cf. also 9,13). These *srotas*-s all have a *mūla*- "base, basis", which may consist of one or more organs. *mūla*- is explained in 9,13 as *kha*-, "cavity, hollow";[274] Cakrapāṇidatta in his commentary on CA,Vi 5,8 says: *mūlam iti prabhavasthānam* "'*mūla*-' is the place of origin", which can, however, hardly mean that not only the *srotas*-s, but the substances connected with the *mūla*-s too *originate* in them (as the correspondences given in the texts make quite clear). As it is, the relation of the substances mentioned to the *mūla*-s of the *srotas*-s of these substances is unclear. Of the channels mentioned in 9,12, two carry *ārtava*-, their *mūla*- being the womb (☞ *garbhāśaya*-) and the ducts (*dhamanī*-) carrying *ārtava*-;[275] piercing these channels leads to infertility, incapability to stand intercourse and loss/destruction (*nāśa*-) of *ārtava*-. As regards the "ducts", these are most probably the same as those in §6.12. Whether the channels are linked with each of the womb and the ducts directly, or whether this link is indirect with regard to either womb of ducts, is not clear; should the link be indirect, then the channels may either be linked only with the ducts (which in their turn are linked with the womb) or only with the womb (which in its turn is linked with the ducts). Clearly only in the case of the channels having openings into the womb would their blockage through something within the womb be possible.

6.19. Since, however, the channels are wholly internal, all this cannot be taken as proof that a theory of the sort discussed in §5.1 and §5.9 did in effect exist, namely, that the procreatory fluid was thought to leave the body by some means other than the genital opening, especially since we now know that this fluid was actually thought to be the same as the menstrual fluid, and this was definitely not thought to leave the body by any means other than the vagina. But that is all we can say as regards the link of the channels carrying the procreatory-menstrual fluid with the womb (and the ducts carrying this fluid), for the nature of this link is nowhere stated. I may, however, point out here that if we should surmise that the channels (i.e. a plural number inside a single woman) blocked lead out of the womb into the body, then it would follow from this that there is some sort of circulation of procreatory-menstrual fluid within the body via the womb. Not only have we as yet found no evidence for this, but we have also to consider the fact that such a theory seems not to stand in any connection with a periodic voiding of this fluid; such a voiding is, however, expressly presupposed by the texts given in n.267

---

[274] Which Ḍalhaṇa glosses further with *chidra*- "opening".

[275] *tayor mūlaṃ garbhāśaya ārtavavāhinyaś ca dhamanyaḥ*.

and n.272.[276] But quite apart from this, I must also confess that, though I can come to terms with the picture of an embryo or foetus even in the early stages (i.e. still quite small) blocking the uterine orifice to the vagina,[277] I find it very difficult to imagine it blocking several openings within the uterus, unless these be concentrated in the same area in which the opening into the vagina is present. One could point out that the foetal envelopes are much larger that the foetus itself and that thus these might have been thought to fill up the whole uterus and block all openings connected with it, but against this one must hold that these envelopes are to my knowledge nowhere stated to be formed at the very beginning; this must lead one to be very wary of seeing in them the prime cause of blocking. Moreover, if ☞ *aparā-* here should subsume these envelopes (cf. Appendix II), then ultimately the text would be saying that it forms itself through its own action, which would be absurd. But be all that as it may, we would be stuck with the problem of some ☞ *srotas-*(s) leading away from the womb (part of their *mūla-*) in any case. Actually, the whole concept of "internal" *srotas-*s (channels) carrying *ārtava-* as given in SU,Śā 9,12 seems to be unique to this passage. I know of no similar statement in other texts (and neither does TIVĀRĪ, 1,p.12 mention any); not even AS,Śā 6,p.331, which is a parallel to SU,Śā 9,12, mentions such *srotas-*s carrying *ārtava-* or the like, even though in the same text we have a parallel (AS,Śā 2,p.288a) to SU,Śā 4,24. It is also interesting to note that according to SU,Śā 9,12 (as well as CA,Vi 5,8 and AS,Śā 6,p.331b) the *mūla-* (see §6.18) of the channels carrying semen consists of several organs (not the same as the *mūla-* of the channels carrying *ārtava-*), among which, however, the ducts (☞ *dhamanī-*) carrying semen are *not* listed, i.e., in this case, if semen be taken to mean the semen of the male, we have a dissimilarity between the male and the female, since the corresponding ducts (*dhamanī-*) in the female are, as we have seen above, part of the *mūla-* of the corresponding channels (☞ *srotas-*). Could we take all this to mean that the notion of channels carrying *ārtava-* as given in SU,Śā 9,12 is a later

---

[276] Inasmuch as the non-appearance of the fluid due to the blockage caused by the pregnancy is expressly remarked upon (*ārtavaṃ na dṛśyate*).

[277] Of course this orifice would have to be closed even before that, right from the time when semen and procreatory-menstrual fluid combine (and have not yet had the chance to create anything perceivable), as otherwise the contents of the uterus would flow out at once. This matter has been discussed in §§8.19f. But one might assume that this closure pertains only to a limited period of time, namely to as long as the embryo/foetus is still too small to block the opening itself. Or it may well be that the theory of blockage due to the embryo/foetus simply did not take the problem of keeping the fluids within the womb before that into account.

accretion and not an original part of the general theory of channels, which rather presupposes channels carrying *śukra-* in both male and female, since *śukra-* (semen) was thought to be present in the latter too (see §2.4)? True, the mention of *śukra-* and *ārtava-* one after the other in the SU passage does create the impression that first the male, then the female fluid is being described, but on the other hand the text does not say expressly that *śukra-* pertains to the male only. The mention of the *muṣkau* as part of the *mūla-* of these channels in the AS passage (Śā 6,p.331b) seems, however, to speak against this, for, though the dual of *muṣka-* can refer to the testicles as well as the female pudenda,[278] its usage in the latter meaning appears to be confined to Vedic texts. Nevertheless, a residue of uncertainty remains, as one cannot fully count out the possibility of this latter usage of the term having sporadically survived in later times. But the mention of the testicles (*vr̥ṣaṇau*, a dual) in the CA (Vi 5,8) and SU (Śā 9,12) passages and of the penis (*śephas-*) in the CA passage does rather speak against the female too being meant, especially since the CA passage — in contradistinction to the others — does not mention any other organs.[279] Thus we cannot simply dismiss the channels carrying *ārtava-* in SU in the above manner, even though they are unique to the passage mentioned. But whether the notion pertaining to them be original or not, these channels are internal and do not lead to the exterior. If we indeed have to assume some sort of circulation of procreatory-menstrual fluid (and I cannot think of any other deduction if we consider these channels to lead away from the womb, either directly or indirectly via the *dhamanī-*s), then we have a closed system in which the amount of fluid stays constant regardless of whether the flow is blocked in some place, or, if new fluid constantly enters the closed system, steadily goes on accumulating, whether there be a blockage within the system or not. The theory of an ac-

---

[278] JAMISON 1996, p.276,n.134 suggests that this may refer to the labia, which makes sense, if we assume that this refers to the fleshy, rounded external parts of the pubic region on both sides of the genital opening, which are a pronounced characteristic of many women.

[279] SU,Śā 9,12 and AS,Śā 6,p.331b also subsume the breasts under the *mūla-*. The expression used (*stana-*, here as the dual *stanau*) is according to our dictionaries used only for the breast(s) of the female, but it is in medical texts sometimes used to describe the two halves of the chest in general, i.e. also of males (e.g. in CA,Śā 7,11, SU,Sū 35,12) (as it obviously is here too), the clearest examples of this I know being SU,Śā 9,5 (*narasya stanābhyām*) and AS,Śā 6,p.330a (*stanāśrite ... narasya*) (see n.299). HOERNLE 1907, pp.14f. is of the opinion that *stana-* in similar cases is used "in the wider sense of breast-piece, rib-piece; it is synonymous with *pārśva*, *pārśvaka* of Charaka and Suśruta, and denotes the ribs"; he also translates the dual with "the two rib-pieces". The ☛ *marman-*s called *stanamūla-* and *stanarohita-* are of course also nothing particular to women. Cf. too DAS,R.P. 1984A, p.238.

cumulation due to blockage only makes sense if without blockage there is some egress to the exterior outside this closed system. This being so, the ☞ *srotas*-s having part of their *mūla*- in the uterus can hardly be meant in the context of SU,Śā 4,24 (and AS,Śā 2,p.288a), and the difficulties caused by the plural of *srotas*- remain; the problem is not resolved.

6.20. Regarding openings leading out of the uterus, one may also consider SU,Ni 8,3, which says that due to various causes the embryo/foetus may drop down from its moorings like a ripe fruit from its stalk, leave the uterus and fall down into the visceral cavity, causing various disorders; it thus aggravates the wind, which forces it out together with blood.[280] This statement, repeated in AS,Śā 4,p.310b, and part of the aetiology of *mūḍhagarbha*-, is highly problematic, and I know of no other contexts in Indian medical texts which presuppose that the embryo can leave the uterus and enter the general visceral cavity, even though it is difficult to presume that the ancient Indians did not observe cases of extra-uterine gravidity. The parallel passages on *mūḍhagarbha*- in other texts (see e.g. TIVĀRĪ 1,p.506) do not seem to presuppose anything similar and are thus of no great help. TIVĀRĪ 1,pp.510-513 too is aware of the problem and tries to solve it by explaining *ati√kram* (in *atikramya* "having transgressed [the uterus]") as meaning "irritation or stimulation of uterus due to hyper-activity of extremities of fetus caused by detachment of it's [sic] bonds" (p.512). This, however, requires such semantic contortions that it hardly seems feasible. As it is, there seems to be evidence in non-medical texts for (ancient?) theories not presupposing a clear knowledge of the uterus; cf. in this regard also n.12. Thus MBH 12,318,23ff.[281] tell us that the embryo is formed in the same place in which digestion occurs, and ask why it is nevertheless not digested too; they tell us too that the embryo finally issues out in the same manner as urine and faeces, involuntarily.[282] Similarly, MBH 12,695* (after 12,245,10) says

---

[280] ...*garbhaśātanaprabhṛtibhir viśeṣair bandhanān mucyate garbhaḥ phalam iva vṛntabandhanād abhighātaviśeṣaiḥ. sa vimuktabandhano garbhāśayam atikramya yakṛtplīhāntravivarair avasraṃsamānaḥ koṣṭhasaṅkṣobham āpādayati. tasyā jaṭharasaṅkṣobhād vāyur apāno mūḍhaḥ pārśvabastiśīrṣodarayoniśūlānāhamūtrasaṅgānām anyatamam āpādya garbhaṃ cyāvayati* (var: *vyāpādayati*) *taruṇaṃ śoṇitasrāveṇa*.

[281] I am indebted to MÜLLER 1941-1942, p.467 for this and the following reference. The MBH references are also given by MÜLLER 1965, p.39,n.152 in connection with his statement on p.152 that SU,Śā 5,39 presupposes that the embryo lies in the visceral cavity unprotected by a uterus, even though the text says explicitly that it lies within the *garbhaśayyā*- = ☞ *garbhāśaya*- (*pittapakvāśayayor madhye garbhaśayyā yatra garbhas tiṣṭhati*); this is because Müller idiosyncratically insists that the receptacle of the embryo or foetus cannot refer to the uterus (see the discussion on ☞ *garbhāśaya*- in Appendix II).

[282] 12,318,23-26:

that the embryo is not digested like food in the stomach; obviously, this is regarded as something wondrous. However, this follows on a statement that the embryo lies in a place where it is enclosed by great heat (clearly the digestive "fire"),[283] and this makes us wonder whether the statement on the non-digestion of the embryo must really mean that the embryo lies in exactly the same place as food; it could perhaps just mean that it is susceptible to the influence of digestive heat in its vicinity, a statement which would not preclude its lying within the uterus — what would then be wondrous would be that the heat, even though in the vicinity of the uterus, does not influence what is in the latter. In the light of this line of reasoning, MBH 12,318,23ff. too could presuppose something similar, for even the statement that the embryo is "in that very stomach" (*tasminn evodare*) in which ingested food and drink are digested (318,24) need not necessarily mean that the very same place within the stomach is meant, but might just refer to the belly as such. As it is, one variant of 12,695* seems to reason in a similar manner, for it says clearly that the individual reaches "the place of the embryo" (and not "the state of the embryo"),[284] which could be a reference to the womb. Thus the passages cited do not constitute proof beyond the shadow of a doubt for a theory holding that the embryo/foetus is situated within the digestive tract. Real proof seems, however, to be available in *Garuḍapurāṇasāroddhāra* 6,9-11, which tell us that the embryo lies in a pit together with faeces and urine, where it is tormented by the worms residing there and by the food the mother eats,[285] and in 6,20, where the embryo laments that it

---

saṅgatyā jaṭhare nyastaṃ retobindum acetanam
kena yatnena jīvantaṃ garbhaṃ tvam iha paśyasi
annapānāni jīryante yatra bhakṣāś ca bhakṣitāḥ
tasminn evodare garbhaḥ kiṃ nānnam iva jīryate
garbhamūtrapurīṣāṇāṃ svabhāvaniyatā gatiḥ
dhāraṇe vā visarge vā na kartur vidyate vaśaḥ
sravanti hy udarād garbhā jāyamānās tathāpare
āgamena sahānyeṣāṃ vināśa upapadyate.

The variants given in the edition do not substantially alter what is said here.

[283] The whole verse runs:
mahoṣmāntargataś cāpi garbhatvaṃ samupeyivān
daśa māsān vasan kukṣau naiṣo 'nnam iva jīryate.

[284] I.e. *garbhasthānam upeyivān* instead of *garbhatvaṃ samupeyivān* in the text of n.283 (the other variants given are of no great import).

[285] *mātur jagdhānnapānādyair edhaddhātur asammate*
śete viṇmūtrayor garte sa jantur jantusambhave (9)
kṛmibhiḥ kṣatasarvāṅgaḥ saukumāryāt pratikṣaṇam

has fallen into a pit of faeces and urine and is being tormented by the digestive fire, and asks how it can come out of this place.[286] This would seem to prove the existence of a theory holding that embryo and food (and the digestive products of this latter) are in the same place,[287] from which it would follow that those adhering to this theory were unaware of the actual nature or even existence of the uterus (cf. on this n.12); note in this connection also the many descriptions of the impurity of the womb in Indian literature,[288] which might be connected with such a belief. Things are, however, not that easy, for *Garuḍapurāṇasāroddhāra* 6,11 states that the embryo is enclosed by the *ulba-*, and this by *tantra-*s (probably string-like or sinewy organs); according to ABEGG, p.93,n.7 *ulba-* here refers not to a covering membrane, but to the uterus. Should this indeed be so, then we would clearly have a problem with regard to our deductions above, so that ultimately the evidence from this source too would lead to the same inconclusiveness as that from MBH. Unfortunately, the exact meaning of the word ☞ *ulba-* has to my knowledge not yet been examined in detail, so that the matter remains unclear. ABEGG, p.93 moreover follows a reading in which not *tantra-*s, but *antra-*s ("intestines") are mentioned, which complicates the matter still further. We thus do not seem to be able to come to a definitive conclusion from the evidence available. Nevertheless, there are well-known Indian legends of women becoming pregnant by ingesting semen of males through their mouths; cf. in this regard also the passage from the *Milindapañha* translated by WINDISCH, pp.21f., which states expressly that such a mode of conception is actually possible. This could be taken to be evidence that there does indeed seem to have been a belief holding that the embryo is in the same place as food. Even though the passage from SU (and AS) quoted above in this paragraph would not seem to presuppose such a belief too, the fact that it speaks of the

---

mūrchām āpnoty urukleśas tatrasthaiḥ (printed: *tatratyaiḥ*) *kṣudhitair muhuḥ* (10)

kaṭutīkṣṇoṣṇalavaṇarūkṣāmlādibhir ulbaṇaiḥ
mātṛbhuktair upaspṛṣṭaḥ sarvāṅgotthitavedanaḥ
ulbeṇa saṃvṛtas tasmin tantraiś ca bahir āvṛtaḥ (11).

6,15 speaks of a brother/creature sharing the same belly (*sodara-*; the translation "womb" for *udara-* should obviously be avoided here) and living in faeces. The embryo, which is unable to remain at one and the same place because of the winds of birth, resembles this creature (a worm?; cf. *kṛmi-* in 6,10): *naikatrāste sūtivātair viṣṭhābhūr iva sodaraḥ*.

[286] *viṇmūtrakūpe patito dagdho 'ha[ṃ] jaṭharāgninā*
icchānvito vivasituṃ kadā niryāsyate bahiḥ.

[287] For parallels to the *Garuḍapurāṇasāroddhāra* passages see ABEGG, pp.93-95.

[288] For some references see e.g. HARA 1980, p.155,n.54.

embryo leaving the uterus could be taken to show that it has been influenced by or is a remnant of it; note too the notions on "foetuses gone astray" in Dhammapāla's commentary on *Therīgāthā* 224f., on which see KLOPPENBURG, p.164. But I fear that I can say nothing more on this subject, and thus on the problem of the statements of the medical texts, or on the problem of how the embryo/foetus finally leaves the body (seemingly via the vaginal channel) if it is in the visceral cavity, and of what "blood" in this context is. What is, however, noteworthy here is that if the embryo/foetus is taken to leave the uterus, the presumption might be that it does so via passages which lead out of this, and that of course assumes relevancy in connection with a possible flow of procreatory-menstrual fluid out of the uterus. But in this case too the flow would be internal, and thus the same arguments as in §6.19 would apply against such possible passages being meant in SU,Śā 4,24 (and AS,Śā 2,p.288a). One could, however, object that what is presupposed is a flow out of the uterus into the visceral cavity and subsequently out via the vagina, but I really wonder if this would not constitute an over-interpretation of SU,Ni 8,3 and AS,Śā 4,p.310b, and an impermissible generalisation at that; moreover, such a deduction seems to go so much against the theory of the flow of fluid out of the uterus to the exterior via the vagina that I for one would be extremely wary in this regard.

6.21. We have, however, also already discussed several CA passages pertaining to the procreatory-menstrual fluid's egress being blocked. True, the blocking agent is not the foetus, but the results of the blockage are similar, namely the appearance of (in these cases false) signs of pregnancy. Of these passages, Śā 2,8f. (see §5.5) merely says that the fluid is somehow hindered from flowing without giving any further details. But Ni 3,13f. (see §5.9) tell us that wind enters the ☞ *yonimukha-* and hinders the fluid from flowing (see also n.135). If *yonimukha-* refers to the orifice of the uterus or of the vagina, then the hindrance would of course be in the form of a blockage of this orifice. If it, however, refers to the uterus, then what is meant could still be a blockage of the orifice of the uterus leading to the vagina, but, if this orifice should *not* be meant, then it could equally well be something akin to what SU,Śā 4,24 might presuppose, whatever that may actually be.

6.22. Before discussing this possibility, let us regard similar passages in SU. We have in Ut 42,13-15 a description of the *gulma-* of women which is produced by "blood" (*asṛgbhava-*)[289] more or less (but not fully) on the

[289] Apart from *asṛj-* and *rakta-* (the latter also in 42,8), "blood" here is also referred to by the kenning *kṣataja-* "wound-produced".

lines of CA,Ci 5,18f. (see §5.9), i.e. without mention of the blockage of the fluid. But SU,Ut 42,15 seems to state that the abdomen does not grow,[290] even though the other typical signs of pregnancy appear. Now since exactly the reverse, namely that the abdomen is distended, is stated in CA,Ni 3,14,[291] AH,Ni 11,51[292] and AS,Ni 11,p.59b,[293] one could deduce from this that the SU passage actually presupposes not a blockage hindering the fluid from flowing out of the uterus, but rather one hindering the former from flowing *into* the latter, in which case the ducts carrying the fluid into the uterus would be blocked. But would this not seem not to be in keeping with the descriptions of *raktagulma-* in CA, AS and AH (cf. TIVĀRĪ, 1,pp.407-411)? According to Ḍalhaṇa's commentary on the SU passage, some authorities take the two *na*s of the passage (see n.290) not to refer to *spandate* and *udaram eti vṛddhim* respectively, but to actually be a double negation of *spandate*, i.e. *na spandate na* = *spandate*, so that the line in n.290 would mean "palpitates; the abdomen distends" and be in keeping with what the other texts just mentioned presuppose. On the other hand, AH,Ni 11,55, AS,Ni 11,p.59b state that it is actually the *gulma-*, but not the *kukṣi-* itself, which grows.[294] What does this mean? The commentaries give us no explanation, even though these very same texts have, just a few lines previously, said expressly that the *kukṣi-* distends. One could hold that what is meant is that the *gulma-* forces the *kukṣi-* to distend, the latter not growing of its own, but then one must ask what is supposed to happen in the case of a pregnancy. Could *kukṣi-* here — and in contrast to its use in n.291, n.292 and n.293 — mean not "abdominal cavity" in general, but "uterus", as it also may? In that case the statements mentioned need not preclude a distension of the abdomen (which the texts quoted in fact expressly mention; see above), but could refer to a distension of this through the pressure of fluid building up *outside* the uterus, in which case these texts too (and thus maybe CA also) could presuppose a hindrance of the ingress of fluid into the latter. This would tally with the fact that none of the texts tells us that the blood collects

[290] *na spandate nodaram eti vṛddhiṃ* "does not palpitate; the abdomen does not distend". In ĀĀS 2810, this line is quoted as *yaḥ spandate piṇḍata eva nāṅgaiḥ*, with the following lines quite different from the SU text.

[291] *tad ārtavam uparudhyamānaṃ kukṣim abhivardhayati.*

[292] *kukṣiṃ karoti.*

[293] *kukṣim abhinirvartayati.*

[294] AH: *na cāsyā vardhate kukṣir gulma eva tu vardhate*, AS: *gulma eva vardhate na kukṣiḥ*. The tables given by TIVĀRĪ, 1,pp.415f. refer, as regards this particular point, to these passages, not to the KĀ passage quoted on pp.414f.

in the uterus; a collection within the uterus is mentioned only by Cakrapāṇidatta's commentary on CA,Śā 2,8f. (see n.106), but it should be noted that this CA passage refers only to a false pregnancy and *not* to a *gulma-*, so that we may have a different cause presupposed here, and, moreover, what a commentator says is of course not necessarily what the text commented upon presupposes. Nevertheless, a collection of fluid outside the uterus is only a possibility I am broaching and not an explicit statement of the texts; for all we know, they may actually presuppose the exact opposite, i.e. a hindrance of the egress of fluid from the uterus (though in that case the statements of AH and AS just referred to remain unexplained, which may, however, very well be due simply to our imperfect knowledge of the matter). Anyway, KĀ,Kh 9,58f. state that the foetus grows rapidly, whereas the *gulma-* grows very slowly; KĀ also does not tell us expressly that the abdomen distends (cf. TIVĀRĪ, 1,pp.407-411). From this we could deduce[295] that the SU passage, if it does indeed state that the abdomen does not grow, may actually mean that the growth is so minimal as not to be readily perceptible, or else adheres to a tradition different from that of CA, AS and AH in that it really does not presuppose any growth of the abdomen. If the latter should be true, then could we also deduce from this that the *gulma-* according to SU actually does not grow at all? Whether there be no or only a minimal growth of the *gulma-*, it could be argued that this means that no further, or only very little "blood" enters the uterus. But since all the other symptoms of pregnancy appear, including those brought about by "blood" pent up in other parts of the body, one could be tempted to infer that, for SU at least, *raktagulma-* results not when the egress from, but the ingress of blood into the uterus is stopped, so that this, and hence the abdomen, does not distend. This would be even more the case if *kukṣi-* should refer not to the abodmen, but to the uterus.[296] We have seen above that we could explain (albeit very speculatively) the other texts as being in line with this deduction; but even then there could still be a difference of traditions with regard to the question of the growth of the *gulma-*, and this difference would of course clearly be there if there

[295] Though the KĀ passage poses a problem, which we shall discuss in n.1059.

[296] In this connection one might think of looking at SU,Ut 42,4, referring to *gulma-* in general, which says that the term refers to movable or immovable knots (tumours?) between heart and bladder (i.e. obviously: in the region between these):

hṛdbastyor antare granthiḥ sañcārī yadi vācalaḥ
cayāpacayavān vṛttaḥ sa gulma iti kīrtitaḥ.

One could be tempted to take this to refer to an extra-uterine knot, but, since the uterus too is situated in the region mentioned, one cannot rule out the possibility of a knot within the uterus being meant (this knot being felt through the uterus).

should be different traditions between SU and the other texts with regard to the blockage. Nevertheless, one may ask whether, if the explanation of the nature of blockage given above should turn out to be correct, the building up of fluid in *raktagulma-* and pregnancy differs only in that in one case the ingress, in the other, however, the egress of fluid is stopped. But all this is very unsure, based on surmises which may turn out to be nothing more than speculation. We shall thus probably do better to shelve the matter of the *gulma-* and the distension or non-distension of the abdomen due to it until we have collected more material on it (see §§10.24ff.). Let us concentrate here on the matter of pregnancy.

6.23. Ḍalhaṇa in his commentary on SU,Sū 14,18 says that some authorities insert the following in this passage (14,18):

> *rajasi copacīyamāne śanaiḥ śanaiḥ stanagarbhāśayayonyabhivṛddhir bhavati*
>
> "and whilst *rajas-* is accumulated, there gradually is growth of the breasts, the uterus (☞ *garbhāśaya-*) and the vagina/vulva".

This reading is also found in SU1, p.44, TIVĀRĪ, 1,p.54 and GHĀṆEKAR 1977A, p.82a. The context of the passage makes it quite clear that this is part of a description of puberty in females,[297] when, as the passage itself tells us, the uterus has still to mature. This means that, as far as the accumulation of the procreatory-menstrual fluid is concerned, this passage cannot be seen in the same light as those mentioned in §6.17 and §§6.21f., but rather serves as proof that the notion that the procreatory-menstrual fluid gradually accumulates in the body as a whole did actually exist. But in another respect it fits in very well with these passages, for in it too we find the same connection as in many of them, viz. the connection between procreatory-menstrual fluid and breasts and/or milk.[298]

6.24. This connection brings to mind the statement in SU,Śā 9,5 that two ducts (☞ *dhamanī-*) leading upwards from the navel carry milk from/for the breasts in women, semen in men.[299] Here what corresponds to semen

---

[297] I have discussed SU,Sū 14,18 quite extensively in DAS,R.P. 1991A, pp.25-32;40, adding some information in DAS,R.P. 1994, pp.287ff., and thus refrain from giving details here. Some supplementary remarks, which have no direct bearing on our special problem, though, are to be found in Excursus IV (pp.105-107) at the end of this chapter, as they are too extensive for a footnote.

[298] Pregnant women very often discharge a milky fluid from their breasts long before they have delivered, a circumstance which cannot have escaped the notice of the ancient Indians. Many young girls too discharge a milky fluid from their breasts at puberty (some even right after birth). This makes us ask about possible notions in ancient India regarding milk always being present in women; this problem will be taken up later on pp.309ff. (cf. also §5.12).

[299] SU,Śā 9,5: *dve stanyaṃ striyā vahataḥ stanasaṃśrite. te eva śukraṃ narasya stanābhyām*

is milk, and not the procreatory-menstrual fluid as in 9,7;12, which serves as further proof that some sort of connection was thought to exist between the latter two fluids. Whether milk and semen too were in any way thought to be connected more closely than in their (here obviously parallel) functions as fluids of the female and the male respectively, is a matter that calls for further investigation. According to Priyavrat Śarmā, the similarities between milk and semen are only functional and confined to certain common aspects, and no more (see TIVĀRĪ, 1,pp.(9)f.), but this is stated in a rather apodictic manner, which does simplify matters too much. However, this problem is too complicated to handle here; on it and on probably different theories on the origin of milk see e.g. DAS,R.P. 1984A, pp.237ff., also O'FLAHERTY, pp.15-61 (cf. §1.8).[300] Note too that the probable connection between milk and the procreatory-menstrual fluid also brings to mind CA,Ci 15,17 discussed in §§5.12f. as well as Ḍalhaṇa's commentary on SU,Sū 14,18 mentioned in §5.16. We cannot, however, say much more on this at the present stage of our investigation, and shall therefore shelve the matter for the present. In any case, it does not seem to help us as regards the problem of the plural in Śā 4,24 (§§6.17ff.).

6.25. But I must point out that so far we have mostly been considering the possibility that the foetus hinders the procreatory-menstrual fluid from flowing out of the body from the uterus to a large extent because we surmised that, because the ☞ *aparā-* is situated within the womb, what forms it would have to be within the womb too (see §6.17). But the placenta is not without its connection with the uterine wall. What if, therefore, there might have been a theory in existence surmising that it grew out of this wall? In that case the collection of procreatory-menstrual fluid (usually leaving the body via the uterus) need not take place within the uterus at all, but could be surmised to take place within the body outside the uterus (see also §6.22), sort of pushing the placenta into the latter due to pressure on the uterine wall from outside.[301] Then we might equate the channels (☞ *srotas-*) in

---

abhivahataḥ. The parallel passage AS,Śā 6,p.330b says: *dve ca stanāśrite nāryāḥ stanyaṃ narasya śuklam*.

300 In this connection I may also be permitted to draw attention to SU,Ni 10,19-23 (also found, without the source being named, in a variant form in the *Kairalī* commentary on AH,Ut 1,18; BHĀ 1,4,9 is an excerpt in variant form, with no source named either), which says that, just as semen of males is present throughout the body and may be made to appear due to sight, thought, sound, touch etc. in connection with a woman, so too the milk of a woman can appear, just like semen, when she touches, sees, thinks about or takes up her child. On the appearance of milk described in this passage see also pp.309ff. Cf. too §7.40.

301 There is also another possibility in line with WEISSER 1983, p.187, whence we learn of

SU,Śā 4,24 with the ducts (☞ *dhamanī-*) leading to the uterus (cf. §6.18), which are hindered from discharging into the uterus. The problem is, however, that it seems that only two ducts open into the uterus. Now the dual has indeed been replaced by the plural in later Indo-Aryan, and we have sporadic cases of the plural seemingly standing for the dual in Vedic too, but on the other hand the 'classical' medical texts do not seem to substitute the plural for the dual, so that here too the plural in SU,Śā 4,24 remains problematic. Of course we could argue that the passage actually means the two *pairs* of ducts of Śā 9,7, which are probably connected somehow (see §6.12), even though only one pair seems to open into the uterus. We might even consider arguing that the plural refers to channels in a variant theory (i.e. not two only as in Śā 9,12) which have as their *mūla-* these ducts and the uterus (cf. §6.18), but from the discussion in §6.19 it appears highly doubtful whether channels such as these should be considered in this connection at all. The other arguments regarding the plural too seem somewhat Procrustean. Should we therefore see in Śā 4,24 (and its parallel AS,Śā 2,p.288a) a reference to a theory in which not two, but several ducts (or channels or something similar) carrying procreatory-menstrual fluid open into the uterus? What comes to mind at once is the plural number of ☞ *nāḍī-*s Gayadāsa mentions in NYC (see §6.17). One is indeed inclined to connect these with the *srotas-*s of the SU passage when one considers that the version of SU,Śā 3,10 (see §6.3) as given by BHĀ 1,3,205 has the plural *dhamanībhyaḥ* in the place of the dual *dhamanībhyām* of the SU passage (see n.395), i.e. has a plural number of *dhamanī-*s carrying *ārtava-* to the uterus. Though the relationship of the reading of BHĀ to that of SU is unclear, it could serve as evidence that there was a theory current, according to which *ārtava-* entered the uterus through a plural number of tubular structures. The age of this postulated theory remains unknown, but it nevertheless brings to mind the ☞ *sirā-*s "vessels" (plural!) carrying *rajas-* to the uterus mentioned in §5.14.[302] Since *srotas-* may also subsume *sirā-*, we could be tempted to link SU,Śā 4,24 and CA,Ci 30,207. As the latter says expressly that the "vessels"

---

a Galenic theory further developed by Arab physicians that the procreatory fluids of male and female first combine to form an inner lining of the uterus, from which the embryo/foetus is suspended. One could hold that this lining is the actual cause for blocking the ingress of fluid into the uterus, the growth of the structures within the latter, and of the embryo, taking place through its medium since it is in contact with the fluid it hinders from entering the uterus. The trouble with this surmise is that not only do we not have any evidence to support it in the Indian texts, but that SU,Śā 4,24 also expressly says that the foetus (*garbha-*) is the cause for the accumulation of fluid (see §6.17). Nevertheless, we should not lose sight of this possibility.

302 CA,Ci 30,207: *garbhāśayagatāḥ sirāḥ* | *rajovahāḥ*.

reach the uterus, it would then follow that this would hold true for the "channels" of the former too. However, there are so many "maybe"s connected with all this that it cannot serve as evidence leading to definitive results regarding the interpretation of SU,Śā 4,24. But be that as it may, even if the problem of the plural in SU,Śā 4,24 is still not satisfactorily solved, we must nevertheless hold on to the fact that, if not SU,Śā 4,24 too, then at least CA,Ci 30,207 (and also BHĀ 1,3,205), as well as the NYC on SU,Śā 4,24 could stand as evidence for a theory according to which the procreatory-menstrual fluid reaches the uterus not by two ducts or the like, but by a plural number of passages (called vessels in CA,Ci 30,207). We shall, however, not pursue this matter further now, but wait until we have more material from other texts (cf. especially §§10.22ff.). Apart from the problem of the plural, however, postulating a blockage of openings leading *to* the uterus presents less difficulties than the notion of openings leading from the uterus being blocked, even when we consider the problem of the size of the foetus as detailed in §6.19. At first fluid would of course be hindered from flowing out of the uterus, and we could even consider the embryo or foetus in this regard (in spite of n.277); what is, however, relevant in this connection is that the size of the embryo or foetus at this stage is not problematic, as no other openings in the womb need to be blocked to hinder the egress of its contents. The fluid can thus continue entering and accumulate[303] until the foetus is big enough to fill out the uterus completely, at this stage stopping the ingress of further fluid. Whether this explanation, which on the face of it seems plausible, is in actual fact correct cannot yet be said due to lack of material; the problem shall thus be shelved for the present, to be returned to later. I may, however, add here that Indu in his commentary on AS,Śā 2,p.288a (see §6.17) explains the *srotas*-s carrying *rajas*- as pertaining to such as carry impure "blood" out of the body during menstruation,[304] but this can hardly refer to channels linking uterus and vagina by some path different from that of the uterine orifice.[305] Does Indu therefore mean a singular *srotas*- (the opening of the uterus/vagina to the exterior), using the plural only because it stands in the passage he comments upon, or does he use the plural in the sense of its referring to women, not to the single woman in the

303 We have seen in §6.17 that the theory of procreatory-menstrual fluid being within the uterus would tally with the actual presence of blood within this, though in this regard we must of course be cautious and not simply uncritically transpose modern notions into the ancient texts.

304 *yaiḥ srotobhir malabhūtaṃ raktaṃ ṛtukāle bahir vahati.*

305 Cf. §6.3; the link between uterus and vagina is too obvious to have been overlooked.

passage commented upon? If he should, however, refer to channels carrying impure "blood" to the uterus for subsequent excretion, then we could have a confirmation for what we have tentatively deduced above.

EXCURSUS II

(in place of footnote 250)

Several Indian scholars have tried to bring the ovaries into the scheme of ancient Indian medical theory as devised by them, but none has been able to cite any passage from an ancient text where these are actually mentioned clearly as such. For instance, PUROHIT, pp.58;117;165;203 (see also the diagram facing p.115) adduces SU,Śā 5,41 as proof that ovaries are described in Indian medical texts. This passage says:

puṃsāṃ peśyaḥ purastād yāḥ proktā lakṣaṇamuṣkajāḥ (var: °*muṣkayoḥ*)
strīṇām āvr̥tya (var: *āśritya*) *tiṣṭhanti phalam antargataṃ hi tāḥ.*

This refers to the muscles (☞ *peśī-*) in the female body. Males have, says 5,37, one muscle in the penis (*meḍhra-*) and two attached to the testicles (*vr̥ṣaṇa-*, dual); according to the above verse, in females these three muscles enclose (*āvr̥tya tiṣṭhanti*) or are attached to (*āśritya tiṣṭhanti*) the ☞ *phala-* (singular!) situated within (them?, or the body?). I do not see how this could be taken to refer to ovaries, and it would be interesting to know how the author reconciled the fact that there are only two ovaries, whereas three muscles are mentioned (and only one *phala-*), especially in the face of the fact that other passages expressly mention two different sets of three muscles linked with the womb, but in a different context (see on this §8.19). I may also point out that Ḍalhaṇa is quite definite that *phala-* here refers to the uterus (*phalaṃ garbhāśayaḥ*); of course he need not necessarily be right, but can we simply neglect his statement? Ḍalhaṇa incidentally also says that Gayadāsa does not comment upon this verse (which according to Ḍalhaṇa gives us the theory of a different school anyway), sticking instead to the theory that women have, in the above case, three muscles less than men; unfortunately NYC, fol.50v has a lacuna in the commentary from Śā 5,38 to about Śā 6,8 (no pages are missing). BHĀ 1,3,147 also refers to this theory of Gayadāsa.

As regards the number three, other modern Indian scholars have of course seen the difficulty overlooked by PUROHIT, and have ingeniously hit upon the explanation that the three muscles are situated in the clitoris and the ovaries (cf. on this TIVĀRĪ, 1,p.24)! In the lengthy discussion explaining the verse as referring to the clitoris and the labia majora, TIVĀRĪ, 1,pp.24-27 reasons that the *phala-* ("uterus" to her too) would after all be situated between these (pp.25;27)[306]. This really is sharp reasoning! But in spite of equating *phala-* with the uterus, TIVĀRĪ, 1,pp.15f.;24-27 nevertheless takes the term to refer to the ovaries too,[307] as, according to her, in SU,Ci 7,38 (loc.cit.: 7,20).[308] Actually, the probability that *phala-* there refers to some-

---

306 The latter states: "testes and their counterpart ovaries do not contain muscles, the fibres present are microscopic structure [sic], thus with *muṣka* scrotum should be considered and analogous to the scrotum female [sic] possess labia majora, and behind these uterus is situated". On "scrotum" here cf. the discussion on ☞ *aṇḍa-* in Appendix II.

307 1,p.27: "it may be concluded that word '*phala*' in dual number indicates ovaries and in singular number the uterus".

308 *[marmāṇi:]*
sevanī śukraharaṇī srotasī phalayor gudam
mūtrasekaṃ mūtravahaṃ yonir bastis tathāṣṭamaḥ.

DAS,R.P. 1990, p.55 considered *śukraharaṇī* to be a dual qualifying *srotasī*, but that does not seem feasible because *srotasī* and *phalayoḥ* belong together as one *marman-*, and taking *śukraharaṇī* to belong to *srotasī* too would give us seven *marman*-s and not the eight the verse mentions. That does not, however, do away with the problem of what *śukraharaṇī* stands for. It clearly refers to the same as *śukravaha-* in Ci 7,38 (see the text in n.309). But the latter is part of a compound and in the middle of this, so that it is unclear whether it is to be taken as being in the singular, the dual or the plural. According to TIVĀRĪ, 1,p.15, *śukraharaṇī* should be the dual *śukraharaṇe* as there are two ☞ *srotas*-s for *śukra-* (cf. SU,Śā 9,12); on p.16 she identifies *śukravaha-* as "spermatic cords", clearly as a pair. The problem is of course that neither SU,Ci 7,38 not SU,Ci 7,36 give us any clear statements on this, so that it remains possible that what we have here is a reference to a singular structure called *śukravaha-* and *śukraharaṇī-*, the latter then a feminine. What this structure would then refer to is something I am unable to comment upon. One could hold it to be something that collects semen from other organs or structures and carries it to the penis, but at present that seems to be pure speculation. In any case, we are once again reminded of the problem of the place of storage of semen in the body discussed in §§6.13f. (see also n.229). If, however, *śukraharaṇī* should indeed be a dual, then the problems discussed in DAS,R.P. 1990, p.55 remain, even though the word does not seem to qualify *srotasī*, as was assumed. A possibility not considered there is whether *śukraharaṇī* might not be a mistake for the dual *śukrahariṇī*. What causes difficulties for this assumption is not only that *śukraharaṇī* does not seem to have any variant readings, but also that, since *harin-* "carrying (away)" is extremely rare (is it attested at all?), we would rather have expected *śukrahāriṇī*. One could, of course, simply assume that the latter is the original reading, but for the present no independent substantiation seems to be available for this assumption, which requires even more tampering with the text. Thus the problem remains. Moreover, there is the added difficulty that it is unclear whether in Ci 7,36 we are to connect *srotas-* only with *mūtravaha-*, or also with *śukravaha-*. The parallel 7,38 would seem to speak against this. However, what if *srotasī* is actually to be construed *both* with *śukraharaṇī* and with *phalayoḥ*? Then *śukravaha-* in 7,36 could indeed also be construed with *srotas-*, and, moreover, *śukraharaṇī* would have to be dual. But that would ultimately not alter the nature the problem of our still having different structures for semen and connected with the testes.

The singular *mūtravahaṃ* too, obviously the same as the compounded, and therefore

thing in the female body is remote, since *srotasī phalayoḥ* "channels (☞ *srotas-*, dual) of the *phala-*s (dual)" in SU,Ci 7,38 clearly has as its equivalent *muṣkasrotas-* in 7,36.[309] TIVĀRĪ 1,p.15 too holds both to denote the same, but sees a discrepancy between the dual *srotasī phalayoḥ* and what to her is the singular *muṣkasrotas-*. But in what other form than the singular could the latter appear as part of a compound? I thus do not see the relevance of this remark. Anyway, the expression in SU,Ci 7,36 can refer only to the male, not only because *muṣka-* means "testicle",[310] but also because an injury here is expressly said to cause ☞ *dhvajabhaṅga-*,[311] and this is something only males can have; moreover, the parallels AH,Ci 11,63 and AS,Ci 13,p.213a refer to the male too.[312] Even TIVĀRĪ 1,p.15 cannot but take 7,36 to refer to the male. But this does not hold her back from taking *phala-* to refer to the ovaries (loc.cit. above), her reasoning being that the uterus is not mentioned as a ☞ *marman-* in passages listing these, and that we know that damage to the uterus does not produce the results usually linked with damage to the *marman-*s, whereas damage to the ovaries "can produce at least permanent disability" (p.27). This is even sharper reasoning! Actually, not only are the other organs given above, except for *guda-* and *basti-*, not mentioned in *marman-*-lists either, even though 7,37 expressly calls them *marman-*s, but we seem moreover to have here a class of *marman-*s as a rule not mentioned in our texts generally, namely *srotoja-* (according to 37); cf. on this also CA,Si 9,4 in Excursus XI at the end of chapter 7. Quite apart from this we have the methodological error of simply projecting that which from our modern point of view seems fitting into an old text, without even bothering to find out the point of view of the latter; note particularly that the ovaries enter the picture *not* because of any hints the text may give us, but merely because TIVĀRĪ

---

in any case singular, *mūtravaha-* in 7,36, is problematic; cf. e.g. Śā 9,12, where the number of channels (☞ *srotas-*) of *mūtra-* (and also *śukra-*) is given as two. One could hold that the singular actually stands for a dual number, but we may also point out that anyone with experience in reading manuscripts knows how easily *ṃ* and superscribed *e* can be confused; could *mūtravahaṃ* therefore be a mistake for the dual *mūtravahe*? Ḍalhaṇa is unfortunately silent on this point.

309 *mūtravahaśukravahamuṣkasrotomūtraprasekasevanīyonigudabastīn aṣṭau pariharet.*

310 TIVĀRĪ, 1,p.16 writes: "*muṣkasrota* [sic] or *phalasrotas* (testies) [sic]", but since *muṣka-* itself would already give this meaning, this does not seem feasible. Rather, the text must refer to *srotas-*s connected with the testes.

311 And nothing else: *muṣkasrotaüpaghātād dhvajabhaṅgaḥ.*

312 This effectively precludes any speculation that we might have a continuation of the Vedic usage of the dual of *muṣka-* to refer to the female pudenda, on which latter cf. p.86.

feels that they should, since they are, in her opinion, analogous to the testes (pp.15f.;25;27). Such striving to bring in organs of the female even when only those of the male are mentioned[313] does create the impression that ancient Indian medical texts too were watched over by a committee of female equal rights advocates!

It is interesting to see that, in 'identifying' the ducts (☞ *dhamanī-*) of SU,Śā 9,7 which we have discussed several times in this chapter,[314] TIVĀRĪ, 1,p.14 has not even considered the Fallopian tubes (or even the seminal ducts with regard to the male) even though she is definite that the texts mention the ovaries; maybe this is because a ☞ *dhamanī-* to her is an "artery".[315] Speaking of Fallopian tubes, I must clarify here that I do not mean to say that ancient Indian medical authorities cannot have had any knowledge about the ovaries or the Fallopian tubes, but only that we do not seem to have any clear references to them, especially in the contexts of interest to us. Whether they indeed knew about them is a problem connected with the question of their knowledge of anatomy based on actual dissection.[316] In any case, should the ancient Indians have known about these organs, then they do not seem to have seen their functions in the same way we do today; one could of course argue that the two ducts leading the procreatory-menstrual fluid into the uterus could be the Fallopian tubes, but even that would not correspond with their true function. However, we could hold that, apart from the problem of function, this identification remains a possibility, but, since any substantiating evidence seems lacking, nothing can be said about it without being wildly speculative; in any case, the openings of the Fallopian tubes into the uterus are difficult to perceive even when one is trained to do so, these openings and they themselves being, moreover, visible only through dissection, and thus I personally would not dare to indulge in such speculation unless more solid evidence should turn up.[317]

---

[313] Cf. on this also §5.16, especially n.167.

[314] For instance in §6.8 and §§6.12f.

[315] The Fallopian tubes are considered in quite another context in 1,pp.18;20-25;27; cf. also the discussion on ☞ *gavīnikā-* in Appendix II.

[316] For a preliminary overview of this problem see ZYSK 1986, whose deductions, which probably require more debate to prove of disprove them, I am unfortunately unable to comment upon (though cf. too n.1698 below).

[317] Matters may even be more complicated and not allow a simple "yes/no" alternative, since there may have been theories in vogue allowing for more than two ducts or the like opening into the uterus; on this subject see §§6.17ff. and §6.25 in this chapter (it will be taken up several more times in the course of this study).

Excursus III

(in place of footnote 270)

In his commentary on SU,Śā 4,24[318] Gayadāsa says (NYC, fol.37v): *prakarṣeṇa pravartate. sarvāṅgeṣu punar apakarṣaṃ ca vadanti* (ms: *ca* | *vadanti* |) *mūtrasrotasy eva* (ms: *mūtraśrotrasyaiva*) *śukrapravṛttadvāreṇa. ārtavam* (ms: °*dvāreṇa ārtavam*) *api strīṇāṃ pravartate. tasya ca sarvadaiva vidyamānatve 'pi yena hetunā garbhiṇyā ārtavaṃ na dṛśyate taṃ darśayann āha. gṛhītagarbhāṇām ityādi. adhaḥ pratihatam ityādi. pravahat salilam* (ms: °*hac chali*°) *iva nāḍīnirodha* (ms: °*dhe*) *upacīyate copari. upary upacayenāpareti* (ms: °*nāpaceti*) *nirucyate* (ms: °*ucyato*[319]). *yoṣitaś cainām amalām ity āhuḥ. śāstrakāro* (ms: *śāstrarā*) *jarāyusaṃjñaṃ* (ms: *jarāyasaṃ*°) *mūḍhagarbhakarmaṇi* (ms: °*karmaṇi* |) *yad āha carakaḥ. jarāyu ca dehoparigataṃ garbhaśarīrāvaguṇṭhakaṃ kṣīrasantānikāsadṛśam. tathāmalākhyaṃ* (ms: *tathāmālā*°) *cobhayam evaitad* (ms: *evautat* | or *evai* | *tat* |) *ārtavajaṃ rasavahanāḍī ca garbhanābhināḍīsambaddhā* (ms: *garbhānā*°). *tad uktaṃ bhoje.*

raktā[j] jarāyur bhavati nāḍī caiva rasātmikā
sā nāḍī garbham āpnoti tayā garbhasya vartanam.

śeṣam aṅkasthānopacitād anyan (ms: *anyāt*) *nāḍīmukhaiḥ pratyāvṛt[t]yā āvṛttaṃ cord[h]vata[ra]m āgataṃ payodharāv abhipratipadyate* (ms: °*dharau pra*°, *abhi* added on the margin). *payodharayoḥ śukravahasrotomūlatvāc* (ms: °*vahaśroto*°) *chukravahārtavavahayoś ca srotasor* (ms: *śrota*°) *abhedāt. ārtavasya raktatvāt.*

"[Semen of the male] issues intensely/through pushing forth (or: through drawing out). And again, with respect to all the limbs, they relate the drawing forth through the aperture of the issuing (*pravṛtta-*)[320] of *śukra-* (or: aperture through which *śukra-* is issued) right in/into the ☞ *srotas-* of urine. The *ārtava-* of women too issues. Showing the cause due to which the *ārtava-* of a pregnant [woman] is not seen in spite of its *always* being present,[321] [the author] says: *gṛhītagarbhāṇām* etc. (see

[318] The text of which is given in n.267.

[319] A *daṇḍa* sign seems to have amalgamated with the last letter to turn *e* to *o*.

[320] Cf. AIGR II,2,§436.

[321] Literally: "Due to which cause the *ārtava-* ..., showing that".

n.267). 'Impeded below' etc.: like flowing water in [the case of] blockage of the tube (viz. pipe) [through which it flows], [it] too (*ca*) accumulates above. Through the accumulation above the ☞ *aparā-* [is formed]: thus [it] is explained. And women call it '*amalā-*', which the maker (i.e. author) of the [well-known medical] treatise, Caraka, mentions as that whose appellation is *jarāyu-*, in [the passages on] the action [to be performed] in the case of an obstructed/dead foetus (*mūḍhagarbha-*).[322] And the *jarāyu-* is the covering/envelope of the body of the foetus, surfacial to the body [of the foetus] (or: risen upwards in the body [of the woman]), resembling the cream of milk (i.e. formed from the fluid like the skin on milk). And so *both* these, [*aparā-* and *jarāyu-*,] called *amalā-*,[323] are produced from *ārtava-* — and [also] the tube (☞ *nāḍī-*) carrying ☞ *rasa*, connected with the navel-tube of the foetus. This is said in [the work of] Bhoja[324]:

> From blood the *jarāyu-* arises, also indeed the tube having the nature of *rasa-*. This tube (scil. umbilical cord) reaches the foetus; through it the sustenance of the embryo [takes place].

The rest, different (i.e. not changed in the above manner) from that collected in the place of the lap (i.e. probably: the uterus) and reversed [in its flow] through [its] turning back through the mouths (scil. apertures) of tubes (☞ *nāḍī-*),[325] attains, [having] arrived [in a] higher [region], the breasts, because of the breasts being the *mūla-* of the ☞ *srotas-*s carrying *śukra-* and because of the identicalness of those carrying *śukra-* and those carrying *ārtava-*. Because of *ārtava-* being blood."

The last line is puzzling. One could understand it as explaining that *ārtava-* in the SU passage and *rakta-* in the passage by Bhoja quoted refer to the same substance by different names. One could, however, also see in it a reference to blood, milk and the procreatory-menstrual fluid being *abhinna-*, as Gayadāsa says in another part of his commentary (see §5.13, which also

---

[322] Obviously a reference to CA,Śā 8,31: *tasya garbhaśalyasya jarāyuprapātanaṃ karma ...* .

[323] Are *aparā-* and *jarāyu-* thus the same according to Gayadāsa, or actually different, though both called *amalā-*?

[324] On him cf. MEULENBELD 1974, pp.418f.

[325] Because of the instrumental it does not seem possible to translate *nāḍīmukhaiḥ* as "from the mouths of tubes".

discusses what *abhinna-* means). But one could see in it too an introduction to the commentary on SU,Śā 4,25[326]; in that case, however, it becomes problematic (this will be discussed in §11.32).

As regards the first part of the above, NYC has just commented upon Śā 4,22, which tells us that in men semen issues into the urinary tract (see §§6.13f.), and then upon Śā 4,23, which details the process by which semen in males is collected from all over the body and then ejaculated (cf. n.168). This does decidedly speak against reading this part, as the punctuation of the ms might lead one to presume, as:

> *prakarṣeṇa pravartate. sarvāṅgeṣu punar apakarṣaṃ ca vadanti. mūtrasrotasy eva śukrapravṛttadvāreṇārtavam api strīṇāṃ pravartate*
>
> "[Semen of the male] issues intensely/through pushing forth (or: through drawing out). And again, they relate the drawing forth with respect to all the limbs. The *ārtava-* of women too issues through the aperture of the issuing of *śukra-* (or: aperture through which *śukra-* is issued) right in/into the ☞ *srotas-* of urine."

Should one nevertheless analyse the passage in this way, then it would state either that *ārtava-* and the "semen" of women (we will discuss this problematic fluid in detail in chapter 8) both issue through the same opening, through or into the urinary channel,[327] or else, and more probably, that the aperture through which *ārtava-* issues corresponds to that through which semen in the male issues through or into the urinary channel. In both cases the issuing into the urinary channel is problematic; it could mean that what NYC calls the urinary channel of the female is seen analogous to that of the male, i.e. is a channel for both urine and the procreatory-menstrual fluid (and eventually female semen), from which it would follow either that it must refer not to the actual urethral channel, but to that part of the female genitals through which both urine and procreatory-menstrual fluid issue, i.e. the vulva (via which both the urethral channel and the vagina open), or that the urethral and vaginal channels are regarded as one and the same, which, since the medical texts do expressly remark on the difference between the two, seems not very probable, though we cannot reject the possibility out of hand. The statement of NYC could, alternatively, mean that *ārtava-* (and eventually female semen) do indeed flow into the urethral channel proper, in which case *ārtava-* in the context of intercourse cannot be the same as the

---

[326] Which begins: *garbhasya yakṛtplīhānau śoṇitajau ...*, and is discussed in §7.10.

[327] On this possibility cf. also §8.5.

procreatory-menstrual fluid, since this flows into the uterus to form the embryo, though in the context of menstruation it could be this fluid, provided Gayadāsa assumes that the uterus is linked to the urethral passage, regarded as being identical with the vaginal passage. But the *ārtava-* is mentioned, in the context in which the problematic line we are discussing occurs, as a parallel to male semen, so that it does seem as if conception, and not menstruation, is the issue here, though after that the subject changes to the latter. This means that it is not very probable that a flow of procreatory-menstrual fluid into the urethral passage is presupposed here, which would mean that only the possibility of *ārtava-* not referring to the procreatory-menstrual fluid seems feasible. However, since NYC in the rest of this passage of the commentary clearly means the procreatory-menstrual fluid when it uses *ārtava-*, and since we have no independent evidence to assume that it here refers to another fluid by the same name too, the latter alternative does not seem viable, which means that "urinary channel" would pertain to the external opening of both urinary and reproductive tracts in the body of the woman, namely the vulva.[328] But the context makes it very improbable that the beginning of the passage of NYC's commentary quoted above is to be read in this manner anyway.

With reference to the line before the last,[329] in the description of the "internal" channels of SU,Śā 9,12 (discussed in §§6.18f.), the ☞ *srotas*-s of *śukra-* do indeed have the breasts (and testicles) as their *mūla*-s.[330] However, the *mūla*-s of the *srotas*-s of *śukra-* and *ārtava-* are different, which makes it difficult to follow NYC's argument based on their identicalness. On the other hand, this argument would apply to the ☞ *dhamanīs*-s of *śukra-* and *ārtava-* mentioned in SU,Śā 9,7 and discussed in §6.8. It does seem as if NYC were here combining this passage as well as SU,Śā 9,5, which links the *dhamanī*-s carrying *śukra-* with the breasts (this is taken up again in §6.24), with SU,Śā 9,12 to explain why *ārtava-* is able to reach the breasts. Though the argument here has several similarities with that of Cakrapāṇidatta given in n.198, there are differences, the most important being that in the case of NYC we do not simply have a case of *srotas-* and *dhamanī-* being used as synonyms to describe the same structures; rather, the fact that the two terms can

---

[328] We shall come across similar cases in §§12.16f.; from these it follows that such terminology here would not be something impossible.

[329] I.e. to the line *payodharayoḥ śukravahasrotomūlatvāc chukravahārtavavahayoś ca srotasor abhedāt* "because of the breasts being the *mūla-* of the ☞ *srotas*-s carrying *śukra-* and because of the identicalness of those carrying *śukra-* and those carrying *ārtava-*".

[330] On *mūla-* see §6.18, on the breasts n.279.

be used synonymously seems to have been used to combine the descriptions of different sets of structures, since the *dhamanī*-s in SU,Śā 9,5;7 and the *srotas*-s in SU,Śā 9,12 are not the same according to SU itself (see especially 9,13[331]).

EXCURSUS IV

(in place of footnote 297)

As additional information to my detailed discussion of SU,Sū 14,18 in DAS,R.P. 1991A, pp.25-32;40 (with some additions in DAS,R.P. 1994, pp.287ff.), I would like to add that GHĀNEKAR 1977A (not utilised in DAS,R.P. 1991A) incorporates the additional line mentioned in §6.23 in the text he comments upon, with the following interpunctuation:

> ... *evaṃ bālānām api vayaḥpariṇāmāc chukraprādurbhāvo bhavati, romarājyādayaś ca viśeṣāḥ; nārīṇāṃ rajasi copacīyamāne śanaiḥ śanaiḥ stanagarbhāśayayonyabhivṛddhir bhavati*
> "... thus the appearance of the semen of children/boys[332] too occurs due to the ripeness of age — and the special characteristics, the *romarājī*-[333] etc. And whilst *rajas*- is accumulated, there gradually is growth of the breasts, the uterus and the vagina/vulva of women."

In doing so, he says basically the same as DAS,R.P. 1991A, pp.28f., in that he does not connect *romarājyādayaś ca* with *nārīṇāṃ*, the difference, however, being that he construes *viśeṣāḥ* with *romarājyādayaś ca* and not with *nārīṇāṃ*, thus making the *romarājī*- a characteristic of all children or boys only. On the face of it, this makes excellent sense and is moreover a very smooth way of analysing the above. However, the fact that the text commented upon not only by Ḍalhaṇa, but also by Cakrapāṇidatta (see SU3, pp.244f.), ends with *nārīṇām* or *strīṇām*,[334] and that Cakrapāṇidatta does *not* mention the additional reading mentioned by Ḍalhaṇa, shows that we actually cannot separate *viśeṣāḥ* from *nārīṇām*, at least not without better evidence. Indeed, it rather seems as if *rajasi ... bhavati* is a later accretion, serving (at least this is my suspicion) to explain *romarājyādayaś ca viśeṣā nārīṇām*,[335] probably by tak-

---

[331] Also referred to in the discussion on ☞ *srotas*- in Appendix II.

[332] On the alternative translations see DAS,R.P. 1991A, loc.cit.

[333] On this too see loc.cit.

[334] This is the word Cakrapāṇidatta comments upon in SU3, though in SU4, p.107 the word commented upon is *nārīṇām* too, as in the case of Ḍalhaṇa.

[335] Which is the traditional — and most probably wrong — way of construing this portion

ing the whole from *romarājyādayaś ca* to *bhavati* to be one sentence (as TIVĀRĪ, 1,p.54 does). Note in this connection that Ḍalhaṇa states expressly that the part beginning with *rajasi copacīyamāne* is only recited by "some" (*kecid*), the terminology showing — as is usual in the technical language of commentators — that he does not agree with their view, as is also evinced by the fact that he does not comment upon this additional portion. As it is, I have not found any evidence in the 'classical' texts that would lead us to assume that the growth of the reproductive tract and the breasts is the result of the accumulation of procreatory-menstrual fluid within the body at puberty; rather, the accumulation due to its non-expulsion from the body leads to growth of the breasts and enlargement of the uterus during pregnancy or disorders leading to the symptoms of a pregnancy. Of course one cannot deduce from this that a theory on the procreatory-menstrual fluid's accumulation leading to the growth of the primary sexual characteristics cannot by any means have existed, but on the evidence sifted so far it does rather seem as if some old commentator has wrongly taken such growth and enlargement *in general* to be due to accumulation as in the case of pregnancy. See also DAS,R.P. 1991A, p.56,n.90a.

As it is, the parallel passage AS,Śā 1,p.277b. makes it very probable that SU,Sū 14,18 really does refer to a growth of hair of the male, for this AS passsage says:

> *yathā ca bālasya jīrṇasya vā taroḥ puṣpaphalaṃ puṣpaphalasya vā tadvidhasya gandho nopalabhyate tathaiva nonaṣoḍaśātītasaptativarṣasya puṃsaḥ śukraromādayo yoṣitaś conadvādaśātītapañcāśadvarṣāyā raja[ḥ]stanyādaya iti*
>
> "And as in an immature or old tree flowers and fruits are not obtained, or the smell of flowers and fruits of such a one (i.e. of such a tree)/or the smell of such (i.e. immature or old) flowers and fruits, thus indeed no semen, body-hair etc. of a man below sixteen and above seventy years [of age], and *rajas*-, milk etc. of a woman below twelve and above fifty years [of age]."

It is clear that *roman*- "body-hair" is problematic here, for this does not disappear with old age; I therefore suspect that in the case of *roman*- the nonappearance of certain hairs before puberty is drawn attention to,[336] or else the AS passage has been influenced by the reading of SU Ḍalhaṇa quotes, or a similar reading. But whatever the origin of the reading of AS be, the asso-

---

of the above passage in the reading of Ḍalhaṇa and Cakrapāṇidatta.

[336] The disappearance of hair after the age of seventy remains a puzzle, though.

ciation of *roman-* with the male here does make it very likely that *romarājī-* in the SU passage too was originally associated with the male, as I assumed.

I would also like to draw attention to Āḍhamalla's commentary on ŚĀR 1,5,16, which latter states that milk and *rajas-* appear and depart at the [appropriate] time in women;[337] Āḍhamalla says:

> *tathā ca romarājyādayaś cāviśeṣā nārīṇāṃ kālāntareṇābhivyaktiṃ gacchantīti sambandhaḥ. atrādiśabdāt stanyārtavādayo gṛhyante*
>
> "And likewise [the statement (obviously an alternative reading of SU,Sū 14,18)]: 'and the *romarājī-* etc., the non-special characteristics[338] of women, attain manifestation after some time', is a(n) association/connection (i.e. a kindred statement which is to be connected). Here milk, *ārtava-* etc. are obtained by means of the word 'etc.'."

The translation "non-special characteristics" is, however, conditional upon *cāviśeṣā[ḥ]* indeed being the correct reading and not just a mistake for *ca viśeṣā[ḥ]*, the usual reading. See also pp.309ff.

On the *romarājī-* see also n.1059.

---

[337] *stanyaṃ rajaś ca nārīṇāṃ kāle bhavati gacchati.*

[338] I.e. characteristics not different from those of pubescent males, which would refer e.g. to pubic hair.

CHAPTER 7

The Suśrutasaṃhitā (2): The Problem of Blood and Metabolism

7.1. We have seen in the last chapter that the procreatory and the menstrual fluids are actually one and the same, namely the procreatory-menstrual fluid. What does SU tell us about the relation of this fluid to blood? From the data of CA we have tentatively deduced that the procreatory-menstrual fluid and blood were both thought ultimately to be derived from *rasa-*, but not to be identical (cf. §5.15). We shall now examine whether the data of SU too allow such a deduction or speak against this.

7.2. The first passage we shall examine is Ni 9,26-28, which tells us that in women who have had a miscarriage/abortion or abnormal delivery (*apaprajātā-*), or who have had a normal delivery (*prajātā-*) without all the "blood" (*asṛj-*) leaving the body (*kāyād aniḥsṛtam*), a disease called "*vidradhi-* due to blood" (*raktavidradhi-* and *raktaja- vidradhi-*), also called *makkalla-*, arises in the abdomen/womb (*kukṣi-*). We shall not go into the details of the symptoms etc. of this disease or the meaning of the term *vidradhi-*. What interests us here is the fact that of the several *vidradhi-*s listed in Ni 9,6, one is caused by blood — in both males and females, since what is said applies to the patient in general. This makes it obvious that "blood" in 9,6 is blood in general. After describing all these types of *vidradhi-* as they appear externally, the text states that there are also internal types of *vidradhi-* (9,15),[339] and that these have the same symptoms as the external types (9,18), from which one could deduce that this also refers to the *vidradhi-* of blood mentioned in 9,6. Then would, if the internal *vidradhi-*s should have the symptoms of the external *vidradhi-*s due to it, this mean too that this blood is also a factor causing internal *vidradhi-*s? True, 9,16 mentions only the morbific entities (*doṣāḥ*), but does this mean that no other substances could be meant? Of course all this refers only to possibilities, which may in actual fact be remote, but are, in view of the vague statements of the text, on the other hand not improbable either. But if SU should indeed mean to say that blood in general too causes internal *vidradhi-*, then what about the *vidradhi-* due to "blood" only women can have mentioned in 26-28? Is this then only a special form of the *vidradhi-*, caused by blood in general, both men and women can have internally, or is it something else, a type only women can have? This is indeed a problem, and not an easy one.[340] The connection with pregnancy

---

[339] *ābhyantarān atas ūrdhvaṃ vidradhīn paricakṣate.*

[340] For details on this see TIVĀRĪ, 1,pp.606-608, who quotes various medical authorities.

makes one think spontaneously of the procreatory-menstrual fluid, and this is indeed what some authorities take "blood" in 26-28 to be (cf. loc.cit. in n.340); this would be unproblematic, as this fluid would of course be expected to be in the uterus. But if blood in general should be meant, then we end up with the conclusion that somehow blood is released into the uterus here (cf. in this connection also §5.14), or that the procreatory-menstrual fluid and blood are one and the same.[341] If the last should indeed be meant, then we would of course come to a conclusion diametrically opposed to what we have tentatively deduced from CA. The disease *makkalla-* is, however, described in SU,Śā 10,22 as arising in a woman after delivery whose body is emaciated/dried (*rūkṣa-*) when unpurified (*aviśodhita-*)[342] "blood" (*rakta-*) is extensively blocked (*atisaṃruddha-*) by wind which has entered its place (*taddeśagata-*) (which cannot but refer to the uterus) and as a result causes "knots" (*granthi-*) in various organs below the navel. From this it follows that the disease as described here is occasioned by the "blood" leaving the uterus and entering other parts of the body, the lack of other fluids in these (this follows from *rūkṣa-*) seemingly facilitating this. All this would of course speak for the procreatory-menstrual fluid being meant by "blood" here. But can we be sure that the *makkalla-* described here[343] is indeed the *vidradhi-* called *makkalla-* in Ni 9,26-28? Ḍalhaṇa and NYC obviously think so, saying that the cause of both is the same "blood" of birth.[344] Unfortu-

---

As regards the authoress' own view, to her "the blood accumulated in uterus is not different from general blood of body" (p.607), which is in keeping with her opinion that what we now know to be the procreatory-menstrual fluid is actually the "ovum" (see §1.7) and as such naturally not a fluid like blood.

341 TIVĀRĪ, 1,pp.607f. has, since to her there is no procreatory-menstrual fluid, but only the ovum and menstrual blood (which is ultimately the same as blood in general), not seen this problem. According to her, blood in general (cf. n.340), unexcreted after delivery, causes a uterine abscess, which is this type of *vidradhi-* (GHĀṆEKAR 1977B, p.54a uses the term "endometritis"). Of course this explanation simply transplants modern Western medical notions on menstrual blood and female diseases into the text uncritically, but to the authoress this is no problem (cf. §1.7).

342 A variant has "dried up/out" (*avaśoṣita-*), but that does not fit here.

343 On this *makkalla-* see also TIVĀRĪ 1,pp.602f.

344 NYC says:

> *prajātāyāḥ prajananaśoṇitasañjanitaśūlaṃ makkallam. tadraktapākajanito vidradhir api makkallaḥ kārye kā[raṇo]pacārāt* (5th last folio,r).

Ḍalhaṇa opines:

> *prajātāyāḥ prajananaśoṇitasañjanitaśūlo makkallaḥ. avaruddharaktapākajanito vidradhir api makkallaḥ kārye kāraṇopacārāt.*

nately, both do not elaborate on what this "blood" is. True, in the commentary on SU,Śā 10,57, which speaks of the flow of "blood" (*rakta-*, *śoṇita-*) in connection with the embryo/foetus aborting, NYC tells us that due to the abortion the mouths of the *srotas*-s of *ārtava-* are uncovered (which must obviously be seen in connection with SU,Śā 4,24 in §§6.17ff.) and that hence *ārtava-* flows,[345] but that does not necessarily mean that this explanation could be applied to Śā 10,22 too, and moreover we do not know whether it is also the view of SU in the first place. Now SU,Śā 10,57 does mention *śoṇita-* as the fluid uniting with semen (see n.179), but this is in a different context and thus not necessarily relevant here. So ultimately all this hardly helps us to understand SU,Ni 9,26-28.[346]

7.3. A similar problem is found in SU,Śā 10,16, which tells us that after giving birth a woman is to be treated so as to remove wind, and then, if some *doṣa-* remains, to be treated further *āduṣṭaśoṇitāt*, i.e. until the corrupted "blood" (*śoṇita-*) has flowed out and uncorrupted "blood" has appeared.[347] As we have already seen in n.179, one could hold that Ḍalhaṇa

---

In their commentaries on Ni 9,27 both Gayadāsa and Ḍalhaṇa say:

> *makkallasaṃjñitaṃ makkallajanitaṃ kāraṇe kāryopacārāt. ghṛtam āyur iti yathā.*

[345] 3rd last folio:

> *garbhasraṃsanenārtavavahasrotomukhavivṛter* (ms: *garbhaśraṃs°* and *°khavavṛter*) *ārtavadarśanam.*

This view of Gayadāsa is referred to in Ḍalhaṇa's commentary too.

[346] As regards *makkalla-* in general, a problem is also caused by Ḍalhaṇa saying in his commentary on SU,Sū 33,13 that this can also occur in a woman who has not yet delivered (*aprajātā-*). On this problem, as well as on what might be an attempt to solve it by Āḍhamalla, who says in his commentary on ŚĀR 1,7,181 that there are two types of *makkalla-*, one arising during pregnancy, one after delivery, see TIVĀRĪ 1,pp.529;531f.

[347]

> *atha sūtikāṃ balātailābhyaktāṃ vātaharauṣadhaniṣkvāthenopacaret. saśeṣadoṣāṃ tu tadahaḥ pippalīpippalīmūlahastipippalīcitrakaśṛṅgaveracūrṇaṃ* (a variant adds: *snehavimiśraṃ*) *guḍodakenoṣṇena pāyayet. evaṃ dvirātraṃ trirātraṃ vā kuryād āduṣṭaśoṇitāt.*

As regards the correct interpretation of *āduṣṭaśoṇitāt*, we must also bear in mind that the following line obviously speaks of treatment when the *śoṇita-* is pure:

> *viśuddhe tato vidārigandhādisiddhāṃ snehayavāgūṃ kṣīrayavāgūṃ vā pāyayet trirātram.*

(Ḍalhaṇa too on this: *viśuddhe śoṇite*). Since this is obviously the stage after the treatment which is to last *āduṣṭaśoṇitāt*, one could think of taking this latter to mean "as long as the blood is corrupt". This would, however, be very difficult to combine with the usual meaning of *ā* in combination with the ablative, and as such does not seem feasible. "Until (*ā*) the corrupted blood [appears] (*duṣṭaśoṇitāt*)" is on the other hand possible, but implies that the flow ap-

explains *doṣa-* here as referring to blood;[348] this would obviously have to be connected with the corrupted "blood" (*śoṇita-*) mentioned after this, which too would thus appear to be blood in general. One could in this connection then point to *ārtava-*, which clearly denotes the procreatory-menstrual fluid (see n.182), mentioned in the same passage; the use of another term (*ārtava-*, not "blood") here too could be taken to speak for "blood" being a fluid different from the procreatory-menstrual fluid. However, all this is based not on the SU passage itself, but only on a certain interpretation of Ḍalhaṇa's interpretation of it. But let us for the sake of argument grant that *doṣa-* indeed means "morbific entity" here. Though it does look as if this referred to the "blood" following, we actually have no proof that it must needs do so. Then, however, the wind-removing treatment mentioned just before the line containing *doṣa-* could speak for the latter referring to wind (which is after all a *doṣa-*). On the other hand, the fact that *doṣa-* can refer to different corrupted substances (see n.179) might speak for it referring here not to a morbific entity, but to a special corrupted substance, namely the procreatory-menstrual fluid, and if this is indeed so, then Ḍalhaṇa's commentary might actually be referring to this. The parallel passages could support each of these theses.[349] One could even consider a more general meaning for *do-*

---

pears only as a result of the treatment. Does the text really presuppose this? If so, then would we not have to deduce from this that until this blood appears there is no flow of any other blood? But in actual fact it can easily be perceived that a flow *is* present more or less continously for some time after delivery; this is such a commonplace that it would be peculiar if ancient Indian medical authorities too had not observed it. So would such a flow have to be differentiated from the flow referred to by *āduṣṭaśoṇitāt*? This is problematic; thus this interpretation does not seem feasible either. But one could also consider the meaning "until [all] the corrupted blood [appears]", which would mean that the passage actually refers to the ceasing of the corrupted flow present throughout; this need not mean that the whole flow ceases, but merely that it changes from a corrupted to pure flow. (It is very tempting too to consider "as long as the corrupt blood [is present]", which would ultimately mean the same, but *ā* can hardly be taken to apply to the duration of a stretch of time, as we have already seen above.)

A very elegant way of solving our problem would be to take *āduṣṭaśoṇitat* to be a combination not of *ā* and *duṣṭaśoṇitāt*, but of *ā* and *aduṣṭaśoṇitāt*; it would then mean "until uncorrupted blood [appears]", which would of course fit this passage excellently and refer to the flow changing its character, signalling that it has become pure. However, I feel a bit uneasy about this solution, because it presupposes a combination (*ā* + *aduṣṭaśoṇitāt*) which is ambiguous, since, on the strength of the many similar combinations, one would at first spontaneously analyse the combination as *ā* + *duṣṭaśoṇitāt*. In any case, both "until [all] the corrupted blood [appears]" and "until uncorrupted blood [appears]" ultimately boil down to the same.

[348] This is followed by TIVĀRĪ 1,pp.575;578.

[349] See on these parallels TIVĀRĪ 1,pp.573-581; nearly all of them (including the SU pas-

ṣa-.[350] All in all, the passage is not clear,[351] but we nevertheless see that there is no need to needlessly complicate matters by referring to blood in general, though of course, as often in such cases, one can never be fully sure that this might not, after all, be meant. What we have to go on ultimately are different grades of probabilities, and it must be admitted that the probability that blood in general is meant here is much smaller than the probability that something else is. This is all the more so when one considers what Ḍalhaṇa has to say on SU,Ut 42,7f. Here different types of *gulma-* (cf. §6.22) are enumerated, after which SU says (8ab):

> *puruṣāṇāṃ tathā strīṇāṃ jñeyo raktena cāparaḥ*
> "[The *gulma-* is that] of males. And, likewise, another one of females is to be known as through blood".

This is traditionally taken to mean that the types mentioned before this may affect males (which would subsume females), whereas there is another type which females may have; this would be in accord with Ut 42,13-15 discussed in §6.22. However, Ḍalhaṇa takes the statement to mean that both males and females may have a *gulma-* derived from blood in general, whereas only females may have one derived from *ārtava-* blood.[352] Obviously, Ḍalhaṇa takes the fluids to be different somehow (which would fit our deductions so far), but more than that we cannot say yet. What complicates matters, though, is not only that what we have given here is the opinion of Ḍalhaṇa and not necessarily of SU itself, but also the fact that NYC does not comment upon SU,Śā 10,14-16, the ms not seeming to have any lacuna. This may not mean anything, but it might also mean that this portion of the text (on which

---

sage; see n.347) speak of taking measures to counteract the effects of wind. The measures mentioned, especially the tight girdling of the abdomen, seem to serve to free the uterus of wind (not necessarily to the exterior), so that the wind cannot create disorders in this due to a lack of opportunity (*anavakāśatvāt*, as some texts say explicitly). On the other hand, the texts also speak of corrupted "blood". As it is, we might have different theories here, since some of the parallels speak of a corruption of wind to be avoided, others of a corruption due to wind to be avoided.

350 Thus GHĀṆEKAR 1986, p.269 obviously takes *doṣa-* to refer to impurities and residues in the uterus in general, not only lochia, but also parts of the afterbirth.

351 Which explains the helplessness of some translators having to cope with it. Thus BHISHAGRATNA 2,p.222 says: "If there still be any abnormality in the condition of the Doshas (the discharge of vitiated blood *i.e.*, lochia) ...", whilst HESSLER 1844-1850 2,p.41 has: "Mulieri autem, reliquis humorum vitiis tentatae ...".

352 *raktenārtavaraktena. cakārād dhāturaktenāpi gulmo jāyate. sa ca puṃsāṃ strīṇāṃ ca bhavati. tasya ca pṛthag upādānaṃ na kṛtaṃ paittikagulmasamānalakṣaṇatayā tatraivāvarodhāt.* SHARMA,P.V. 1982, p.38 has also drawn attention to Ḍalhaṇa's statement.

Ḍalhaṇa comments) was not part of the text Gayadāsa commented upon.[353]

7.4. Unfortunately SU gives us no other information on this problem, so that we shall probably never know what Ni 9,26-28 actually mean to say. But fortunately other passages leave no doubt that on the whole (though that does not exclude the possibility of there being variant theories) at least the possibility that blood and the procreatory-menstrual fluid are one and the same is remote. Thus Śā 2,5 states that *ārtava-* can, in combination with — amongst other substances — blood (*śoṇita-*), become ☞ *abīja-*, i.e. probably infertile or sterile; as such this is a parallel to 2,4, which states that blood may corrupt semen in males. According to Ci 39,27, blood (*rudhira-*) issues with/like semen and with *rajas-* in certain disorders.[354] Śā 9,12 states that, of the ☞ *srotas-*s of the body, some carry blood (*rakta-*), some *ārtava-*. Blood and *ārtava-* (here obviously seen primarily in its function as the menstrual discharge) have different colours and properties too: according to Sū 14,22, healthy blood (*śoṇita-*; see 14,21) is of the colour of the *indragopaka-* ("rain-mite" according to LIENHARD 1978[355]), not coagulated (*asaṃhata-*) and *avivarṇa-*,[356] whereas *ārtava-* has, according to Śā 2,17 (already mentioned in

---

353 Another problematic passage with "blood" and *doṣa-* is CA,Ci 28,15, which lists the excessive flow of *doṣāsṛj-* (*doṣāsṛksravaṇād ati*) as one of the causes for disorders of wind. The context makes it seem that what is referred to here is something having to do with patients in general, and that would exclude any special fluid of women, i.e. "blood" seems to refer to blood in general. But the expression remains problematic nonetheless. It is unclear how Cakrapāṇidatta analyses it, but he does seem to see a copulative compound here ("flow of *doṣa-* and blood"), since he says that *doṣa-* subsumes faeces too (*doṣāsṛgityādau doṣaśabdena purīṣam api gṛhyate*). What seems to me, however, much more probable is that what is meant here is the excessive flow of corrupted (*doṣa-*) blood, i.e. excessive blood-letting. This would ultimately entail an adjectival usage of *doṣa-*. The problem is that I do not know of other examples of such usage. True, an example seems at first glance to be present in CA,Ci 14,177:

pravṛttam ādāv arśobhyo yo nigṛhṇāty abuddhimān
śoṇitaṃ doṣam anilaṃ tad rogāñ janayed bahūn.

A comparison with other editions shows, however, that there is a printing mistake in the relevant part of this verse, the correct form of which is *śoṇitaṃ doṣamalinaṃ* "blood tainted through an impurity", and not *śoṇitaṃ doṣam anilaṃ* (which is difficult to analyse meaningfully).

354 *rudhiraṃ śukravac cāpi sarajaskaṃ pravartate.*

355 I would, however, beg to disagree with the opinion of LIENHARD 1978, pp.183f. that *indra-* in words such as *indradhanus-* and its synonyms (commonly translated as "Indra's bow") denoting "rainbow" must mean "rain" and not "Indra", as Indra is not characterised by a bow. Whatever other arguments may speak for such a meaning of *indra-*, the one mentioned can not, as Indra *is* characterised by a bow, albeit in Vedic texts (see e.g. SRINIVASAN, p.547b).

356 *indragopakapratīkāśam asaṃhatam avivarṇaṃ ca prakṛtisthaṃ jānīyāt.* (The meaning

§6.9), the colour of hare's blood[357] or "lac-juice" (*lākṣārasa-*) (whatever that may be; see §7.7) and does not discolour clothes.[358] (Sū 21,17, by contrast, tells us simply that blood in general — *śoṇita-* — is red in colour: *raktaṃ ca varṇataḥ.*)

7.5. Some particulars of the different characteristics of blood and the procreatory-menstrual fluid as given above must, however, be remarked upon. According to Ḍalhaṇa's commentary on SU,Sū 14,22, *avivarṇa-*, a qualification of blood, means either that blood is "variously/differently coloured" ("slightly variously/differently coloured, though of the colour of an *indragopaka-*"),[359] i.e. that its colour is like that of an *indragopaka-*, but still varied

---

of *avivarṇa-* will be taken up in §7.5.)

[357] MÜLLER 1965, p.34,n.128 apodictically takes "hare" to refer not to the animal, but to the moon, without telling us what "the moon's blood" is in that case supposed to mean!

[358] The latter is usually taken to mean that *ārtava-* can be washed out of clothes, cf. the parallels (which also mention hare's blood and "lac-juice") AH,Śā 1,18 (*dhautaṃ yac ca virajyate*) and AS,Śā 1,p.277b (*dhautaṃ ca virajyamānam*; on this verse see also n.370). CA is silent on this point. The whole verse in SU runs:

śaśāsṛkpratimaṃ yat tu yad vā lākṣārasopamam
tad ārtavaṃ praśaṃsanti yad vāso na virañjayet.

(Āḍhamalla, commenting on ŚĀR 1,6,12, quotes this with *yac ca* for *yat tu*, whereas *Nāḍījñānadarpaṇa* 250 is identical with the SU verse apart from a misprinted *lakṣārasopamam*; BYT 134,7f. and BHĀ 1,3,206 also have the variant, and BYT 147,30 has *śaśāsrapra°* for *śaśāsṛkpra°*.) Commenting on the verse, Ḍalhaṇa states that the last part means that clothing is not coloured by it (*yad ārtavaṃ vāsaḥsthitaṃ śuṣkam udakaprakṣālitaṃ vāso na virañjayed rāgopetaṃ vastraṃ na kuryāt tat praśaṃsanti*). Not only BHĀ 1,3,206, but also NYC (fol.16v) say the same, the latter in the process quoting a verse of Hiraṇyākṣa (an authority quoted in many works — including CA — and commentaries) to the same effect:

śaśāsṛg ityādi. apsu dhautaṃ yac ca virajyate. pānīyadhautaṃ virāgaṃ bhavati. atroktaṃ hiraṇyākṣeṇa.

indragopakasaṅkāśaṃ snigdhaṃ ca madhugandhi ca
apicchalam atiktaṃ (ms: *atī°*) *ca yad vāso na virañjayet*
adoṣaliṅgasaṃsṛṣṭaṃ tad rajaḥ śuddham ādiśet.

yenoktaṃ vāsaḥ prakṣālitaṃ vigatarāgaṃ syāt tad ārtavaṃ śuddham.

According to this, healthy *rajas-* is also like the *indragopaka-*, unctuous/viscous, smelling like honey, not slimy/slippery, and not bitter. Śrīdāsa in his commentary on AH,Śā 1,23f. quotes this verse as follows:

indragopakasaṅkāśaṃ snigdhaṃ yad api gandhi ca
apicchilam aśītaṃ ca yac ca vāso na rañjayet
adoṣaliṅgasaṃyuktaṃ tad rajaḥ śuddham ucyate,

according to which the *rajas-* simply smells (the smell of honey is not mentioned) and is not not bitter, but not cold.

[359] *avivarṇam iti. indragopakavarṇam apiṣad vividhavarṇam. etena padmālaktakaguñjā-*

enough to accommodate other hues too, or else it means that it cannot be washed out of clothes,[360] i.e. is "not discoloured" (discolour = change colour, in this context obviously in the sense of: lose colour). The former seems to be an attempt to combine SU,Sū 14,22 and CA,Sū 24,22, which latter names other hues of red too (note that Ḍalhaṇa too mentions *padma-*, *alaktaka-* and *guñjāphala-*),[361] but does not have *avivarṇa-* or anything similar; Cakrapāṇidatta in his commentary on the SU passage (see SU3, p.247, SU4, p.108) actually quotes CA,Sū 24,22.[362] The second explanation of Ḍalhaṇa could be taken to be an oblique reference to SU,Śā 2,17 and an attempt to differentiate blood from *ārtava-*. The difficulty herein lies in the fact that CA and AH are not only silent on such a property of blood, but CA even says the opposite; CA,Si 6,80 states that a dry white dress, stained with blood and then washed with lukewarm water, shows a corruption of the blood by bile if it is *vivarṇa-*,[363] healthy blood if it is *śuddha-* "pure".[364] Since bile is often

---

phalavarṇam ity uktam.

360 *athavā vastrādilagnaṃ sat prakṣālyamānam api na vivarṇatāṃ yātīty avivarṇam.*

361 The colours mentioned by CA are: that of molten gold, the *indragopa-*, the (obviously red) lotus or water-lily, lac, and the berry of *Abrus precatorius* Linn. The whole passage runs:
tapanīyendragopābhaṃ padmālaktakasannibham
guñjāphalasavarṇaṃ ca viśuddhaṃ viddhi śoṇitam.
This is reformulated in prose in AS,Sū 36,p.251a.

362 Incidentally, according to Cakrapāṇidatta the different colours of blood described in these SU and CA passages depend on the constitution (*prakṛti-*) of the individual, i.e. on whether one or two of the morbific entities (wind, bile and phlegm) in the body predominate(s), or on whether they are in a state of equilibrium. This explanation (which Ḍalhaṇa does not repeat) looks to me like a later apologetic attempt to come to grips with the problem caused by different colours of blood being mentioned (and especially by the word *avivarṇa-*); as far as I can see, the four 'classical' medical texts at least have no similar statement. (On *prakṛti-* see e.g. CA,Vi 6,12ff., SU,Śā 4,62ff., AH,Śā 3,82ff. and AS,Śā 8,p.344aff.; see also MEULENBELD 1974, p.478., as well as LIENHARD 1959, RAI, pp.58f.;171f.;275, ROȘU 1978, pp.116-119, HILGENBERG, and — with reference to women only — RAI, pp.59;172f.;275f.;309, ŚR,Introduction, p.28, KĀMS, p.20,n.8, KEL 6,40ff., SCHMIDT 1922, pp.173-175. This classification is also applied to elephants; cf. e.g. ZIMMER,HEI., pp.93-99, HILGENBERG, p.161,n.5.)

363 Cf. also *vaivarṇya-* in CA,Ci 30,17 quoted in n.159.

364 The CA passage (also to be found in AH,Ka 3,34) runs:
śuklaṃ vā bhāvitaṃ vastram āvānaṃ koṣṇavāriṇā
prakṣālitaṃ vivarṇaṃ syāt pitte śuddhaṃ tu śoṇite (var: *pittaṃ śuddhaṃ tu śoṇitam*).
A variant of Aruṇadatta's commentary on AH,Śā 5,75 quotes this verse with *śuddhaṃ* for *śuklaṃ*. (Aruṇadatta's commentary is also quoted by Śrīdāsa in his commentary on AH1,Śā 5,74.; this quotation seems to read *pittaṃ śuddhaṃ tu śoṇite* in pāda d of the CA passage, which is quoted here too.) More or less the same is said in prose in AS,Ka 3,p.333ab:

characterised as a powerful colouring agent by the texts and in CA,Ni 2,24 incurable "blood-bile" (☞ *raktapitta-*) is expressly said to colour clothing,[365] it would seem to follow that the *vivarṇa-* cloth in Si 6,80 refers to a coloured cloth, so that the "pure" cloth must consequently be a cloth which is not coloured by the blood (thus also Aruṇadatta in n.367 below). Cakrapāṇidatta, commenting on Ni 2,24, is, since to him blood *in general* colours clothes, irritated by the fact that "blood-bile" and blood seem to have the same characteristic according to Ni 2,24, and considers emending the text so that blood-bile does not colour clothing![366] Other authorities seem to have had similar problems, since AH,Śā 5,75 (a parallel to CA,Ni 2,24) has two diametrically opposed readings regarding the colouring of clothing.[367] To

---

ubhayatra ca jīvaraktapittayor jñānārthaṃ rakte śuklaṃ picu plotaṃ vā kṣipet. tasminn āvāne koṣṇāmbukṣālitaśuddhe jīvaraktam. vivarṇe tu raktapittaṃ vidyāt.

(On the term *jīvarakta-* cf. §7.12.)

[365] CA,Ni 2,24cd:

raktapittam asādhyaṃ tad vāsaso rañjanaṃ ca yat

"That blood-bile [which has the characteristics related in the aforegoing line], and which is a colourer of a garment, is incurable."

[366] *vāsaso rañjanaṃ ca yad iti prakṣālitam api sad vāso rañjayati na kṣālanenāpaitīti boddhavyam. anyathā sarvam eva raktaṃ vāsaso rañjanaṃ bhavatīti* (I have been unable to identify this quotation) *vacanam anarthakaṃ syāt. kiṃvārañjanam iti pāṭhaḥ. tena raktasya vāsaso rañjakasyārañjanam eva vikṛtiḥ* "'*vāsaso rañjanaṃ ca yat*': even though it is washed, it colours a garment; 'does not go away through washing' is to be understood. Otherwise the statement '*all* blood is a colourer of a garment' would be meaningless[, since blood-bile is a form of blood]. Or the reading is '*arañjanam* (i.e.: non-colouring/a non-colourer)'. Through this, *non-colouring* is a changed condition of blood, [which is actually] a colourer of a garment."

According to the latter, *vāsaso rañjanaṃ* of the text in n.365 would thus actually be *vāsaso 'rañjanaṃ*.

[367] The text of AH,Śā 5,74-76ab presupposed by Aruṇadatta (and also by the editors of AH and AH1) runs (*nihanti* is to be supplied from 5,71):

raktapittaṃ bhṛśaṃ raktaṃ kṛṣṇam indradhanuṣprabham
tāmrahāridraharitaṃ rūpaṃ raktaṃ pradarśayet.
romakūpapraviṣṭaṃ kaṇṭhāsyahṛdaye sajat
vāsaso 'rañjanaṃ pūti vegavac cāti bhūri ca.
vṛddhaṃ pāṇḍujvaracchardikāsaśophātisāriṇam.

This means that clothing is not coloured: *raktapittaṃ ... vāsaso 'rañjanaṃ* (5,75). Aruṇadatta's commentary:

vāsaso 'rañjanam iti vāsasaḥ arañjanam (printed correctly thus) *ity akārapraśleṣo 'tra bodhyaḥ* (var: *akāreṇa praśleṣo boddhavyaḥ*). *vastrasya hi jīva-*

complicate matters still further, AS,Sū 36,p.251a says of blood: *dhautaṃ ca virajyamānam* "and [is], [when] washed, *virajyamāna-*". The commentator Indu takes this to mean that it colours garments etc. and cannot be washed out;[368] according to the usual medio-passival meaning of *vi√rañj*, however, we would rather translate the AS passage as "and [is], [when] washed, dis-

---

> *śoṇitaṃ śuddhaśoṇitarūpatvān na rañjakam. atas tena raktaṃ vastraṃ samanantaraṃ* (var: *vastram anantaraṃ*) *prakṣālitaṃ śuddhaṃ sampadyate, na rāgayuktam. tathā hy asmin vyādhau raktapittābhyāṃ parasparam avibhāgāpannatvād* (var: *avibhāgopapannatvād*) *vastrasya rañjanaṃ prāptam. raktena hi pittasahitena vastraṃ rajyate. na ca tad* (var: *naitad*) *dhautaṃ śuddhaṃ sampadyate. tathā coktam* (here follows the verse given in n.364).

The gist of this commentary (cited also by Śrīdāsa in his commentary on AH1,Śā 5,73-75) is that healthy blood does not colour cloth, whereas ☞ *raktapitta-* "blood-bile" does; this latter is, however, diametrically opposed to the reading of AH Aruṇadatta himself presupposes and comments upon here, even though this clearly refers only to "blood-bile"! (HILGENBERG/KIRFEL, p.201, following Aruṇadatta, say that blood-bile colours clothes; MEULENBELD 1974 p.326,n.28, following the AH text given above, says that it does not.) However, commenting on 5,75a, where it is stated that contaminated blood issues out of the pores of the body (*romakūpapravisṛtaṃ*), Aruṇadatta says that this issued blood is colourless (*vināpi varṇai romakūpebhyaḥ samastebhyaḥ prakarṣeṇa visṛtaṃ*) (HILGENBERG/KIRFEL, p.201 follow Aruṇadatta in their translation: "kommt sie [sogar farblos] aus den Hautporen"). It could be that Aruṇadatta connects *arañjana-* "not colouring".in 75c with this supposed colourlessness, in which case clothing would obviously not be coloured. This would then be a very special form of "blood-bile", which the text of AH itself does not seem to presuppose, for 74ff. merely state that *raktapitta-* consists of blood having different colours, issuing from the pores of the body, adhering to throat, mouth and heart, not colouring clothes, stinking etc. Now there may be different forms of *raktapitta-* meant here, but the text does not say so expressly, since it merely lists, consecutively, different symptoms, and it definitely does not say that the non-colouring of clothing is to be connected with the issuing out of the pores, though of course one could forcibly construe the verse thus if one so wished. Śrīdāsa in his commentary on AH1,Śā 5,73-75 quotes Hāṭaka who, seemingly commenting on *vāsaso 'rañjanaṃ* in the AH verse above, says that this refers to the non-coloration of cloth by pure blood (*vāsaso 'rañjanaṃ vastrasya hi jīvaśoṇitaṃ śuddhaśoṇitarūpatvān na rañjanakṛt*, which is a repetition of Aruṇadatta's words), for which I fail, as in the case of Aruṇadatta, to see any justification in the text. According to the editor's note loc.cit. in AH, Candranandana explains 75c as *vastrarāgakṛt-*, i.e. "colouring clothing", which means that he reads the text as *vāsaso rañjanaṃ*, i.e. the opposite of Aruṇadatta's reading. This reading is also given in AH5,Śā 5,74 (Indu's commentary seems unfortunately to be defective here, but because of the *api* "even" in his commentary it does look as if he too read this way: *vāsaso rañjanaṃ dhautasyāpi X X*). The editor of AS gives the reading of the same verse as *vāsaso 'rañjanaṃ* in AS,Śā 11,p.359b and Indu's commentary (p.360a), but it could just as well be *vāsaso rañjanaṃ* for all we know, as Indu does not explain it. Indeed, AS1,Śā 11,p.247 does read *vāsaso rañjanaṃ*!

368 *dhautaṃ vastrādi rañjayati.*

coloured (i.e. its colour goes away)",[369] especially since we have *virajyate* and *virajyamāna-* in the AH and AS passages mentioned in n.358 (referring to the procreatory-menstrual fluid and not to blood in general) — the AS extract quoted there (Śā 1,p.277b)[370] even having exactly the same wording as the AS extract on blood just quoted above! —, where *vi√rañj* most probably means "be discoloured", as the commentators too (including Indu himself!)[371] also say. Of course one could here point to SU,Śā 2,17 quoted in §7.4, where we find *vi√rañj* used to refer to colouring and not to a loss of colour, but here we have not a medio-passival form, but a causative with a nasal infix (i.e. "discolour" and not "be discoloured"), and what is, according to the verse, not discoloured here is not the fluid, but the garment to which it attaches, and this seems to mean that this does not lose its original colour by becoming red, i.e. the colour of the fluid is not fast, and that would be in keeping with the parallels given in n.358.

7.6. One may, however, hold against this that SU,Śā 2,17 could be telling us the exact opposite, i.e. the discolouring of the garment refers not to the original colour of this, but to the colour it has attained through contact with the fluid, which would mean that, since the garment is not discoloured, it does not lose *this* colour, i.e. retains the colour of the red fluid. Then the SU passage would not be in keeping with those in n.358. Indeed, we actually find a variant to SU,Śā 2,17d which states exactly this, namely, that the *fluid* is not discoloured in water,[372] i.e. that the colour due to it is fast and cannot be washed out. Now this other reading may in fact be of secondary origin and thus spurious, but if it is, it does seem to show that there must have been some who analysed the passage in the alternative manner we have done and then reworded it so as to leave no doubt as to this analysis. On the other hand, *na* "not" in this reading may simply be a mistake for *ca* "and"; note the line *apsu dhautaṃ yac ca virajyate* of NYC given in n.358, which may show that NYC actually comments (with the explanation *pānīyadhautaṃ virāgaṃ bhavati*) upon a line like the one in n.372, but with *ca* for *na*. However, since

---

[369] Which would conform to what Aruṇadatta presupposes in n.367.

[370] The whole passage runs: *ārtavaṃ punaḥ śaśarudhiralākṣārasopamaṃ dhautaṃ ca virajyamānaṃ śuddham āhuḥ*. This is given a nonsensical translation by MÜLLER 1963B, p.416,n.35: "Das *ārtava* ist ferner aus dem Blute (des Hasen im Monde) als vortrefflicher Saft, lautergewaschen, entfärbt und rein angesagt."

[371] *dhautaṃ sad vastrād virajyamānam apasarac ca.*

[372] *yad vāpsu na virajyate* instead of *yad vāso na virañjayet*. The verse as given in BYT 134,7f. reads nearly the same: *yad apsu na virajyate*; similarly, YOG 2,Strīrogādhikāra 8 (p.397) has *yac cāpsu na virajyate*.

NYC does not expressly say that this is part of the text commented upon, this is not sure. Thus we cannot make any clear statement one way or the other and still have a problem, namely that the SU passage (mentioning the procreatory-menstrual fluid) is either in keeping with those in n.358 and in that case says the same as AS,Sū 36,p.251a, which, however, refers to blood and not to the procreatory-menstrual fluid, or else is not in keeping with the passages in n.358, in which case it does not say the same as AS,Sū 36,p.251a either. As regards the former alternative, one would be inclined to assume that the fact that a certain property of the procreatory-menstrual fluid (its not colouring cloth) is expressly mentioned in SU implies that blood has the opposite property. And this is indeed confirmed by SU,Ci 34,14, which tells us that blood in general (*jīvaśoṇita-*; on this see §7.12) colours cloth or cotton-wool (*picu-*) when it is pure, the opposite obviously being true in the case of ☛ *raktapitta-*.[373] This passage is thus not only similar to Cakrapāṇidatta's commentary on CA,Ni 2,24 in §7.5, but also in keeping with Ḍalhaṇa's commentary on SU,Sū 14,22 (see §7.5), and one could assume that Ḍalhaṇa had it in mind when commenting on Sū 14,22, were it not for the fact that in his commentary on Ci 34,14 he says the exact opposite, namely that an absence of colour in the cloth or cotton-wool held into the blood, i.e. a retention of the white colour, speaks for pure blood, a discolouring for *raktapitta-*![374] But be that as it may, if we could be certain that SU ascribes dif-

---

[373] The colouring due to *jīvaśoṇita-* is fast even in hot water. Further, a dog eats *jīvaśoṇita-*, but not *raktapitta-*:

> *jīvaśoṇitaraktapittayoś ca jijñāsārthaṃ tasmin picuṃ plotaṃ vā kṣipet. yady uṣṇodakaprakṣālitam api vastraṃ rañjayati taj jīvaśoṇitam avagantavyam. sabhaktaṃ ca śune dadyāc chaktusammiśraṃ vā. sa yady upabhuñjīta taj jīvaśoṇitam avagantavyam. anyathā raktapittam iti.*

According to CA,Si 6,79, AH,Ka 3,33 and AS,Ka 3,p.333b, what a dog or crow (CA: bird) eats is *jīva-* (scil. blood), what it does not is *pitta-* (obviously a similarly shortened form). Note also what KAL 22,75cd says on what it calls *jīvaśoṇita-* in 75b:

> *yac coṣṇodakadhautam apy atitarāṃ naivāpasaṃsajyate*
> *śvāpad bhakṣayatīha* (printed: *svāpabhdakṣa°*) *śoṇitam idaṃ cānyatra pittānvitam*
>
> "And that which, even though washed profusely in hot water, is not at all detached/removed, [which] a beast of prey (viz. dog?) eats, this is in this case blood, and otherwise connected with bile."

The eating by dogs (and crows) is referred to in CA,Vi 4,7 too; what is eaten is *dhārilohita-* (on this expression cf. §7.12), what is not eaten is *raktapitta-* (*śvakākabhakṣaṇād dhārilohitam abhakṣaṇāl lohitapittam ity anumātavyam*).

[374] Note that Ḍalhaṇa here refers to the white cloth or cotton-wool coloured by *raktapitta-* as *vivarṇa-* (cf. the discussion in §7.5):

ferent properties related to washing to blood and the procreatory-menstrual fluid, then from this it would follow that Śā 2,17 does indeed tell us that the latter does not colour cloth. But we have no guarantee that the different parts of the text are indeed to be explained thus with reference to each other — we may after all simply have contradictory theories in the same text. Similarly, the fact that AS ascribes the same property to both fluids may also not mean much, for the author of AS might simply have made a mistake, but it also may be that there were actually different notions in vogue on the properties of blood and the procreatory-menstrual fluid, and that the author is simply referring to two different traditions. In any case, it does strike one as peculiar that Indu explains exactly the same phrase in two quite opposite manners in two different passages.

7.7. That there were indeed different notions in vogue regarding the properties of the two fluids is made very probable by several incompatibilities in their descriptions. Thus the colours ascribed to blood by CA,Sū 24,22 (see n.361) are nearly all ascribed to the procreatory-menstrual fluid too in CA,Ci 30,226[375] (whereas we have seen that in SU the description of the colour is different in the respective passages). Then too, Hiraṇyākṣa states (see n.358) that the procreatory-menstrual fluid has — like *blood* in CA,Sū 24,22, SU,Sū 14,22 (see n.356), AH,Sū 27,1[376] and AS,Sū 36,p.251a[377] —

---

tasminn iti vānte 'dhaḥpravṛtte dhātudrave picuṃ karpāsatūlakaṃ (var: *picuṃ arkatūlakaṃ*) *plotaṃ vastraṃ śuklaṃ kṣipet. tad yadi śuklam uṣṇodakaprakṣālitaṃ`rāgarahitaṃ bhavet tadā jīvaśoṇitaṃ vivarṇe ca raktapittam iti.*

[375] Note that some pādas of Sū 24,22 also appear verbatim in Ci 30,226:
guñjāphalasavarṇaṃ ca padmālaktakasannibham (var: *ca yad vālakta°*)
indragopakasaṅkāśam ārtavaṃ śuddham ādiśet.

Before this, Ci 30,225 states:
māsān niṣpicchadāhārti pañcarātrānubandhi ca
naivātibahu nātyalpam ārtavaṃ śuddham ādiśet.

[376] 27,1f. run:
madhuraṃ lavaṇaṃ kiñcid aśītoṣṇam asaṃhatam
padmendragopahemāviśaśalohitalohitam.
lohitaṃ prabhavaḥ śuddhaṃ (var: *pravadec chuddhaṃ*) *tanos tenaiva ca sthitiḥ ... .*

[377] Also quoted in part by Hemādri in his commentary on the introduction of AH,Sū 27. The AS says:

tatrāmbu śārīram āhārasārabhūtaṃ rasākhyam avikṛtam avikṛtena tejasā rañjitam indragopākāraṃ ca śaśaśoṇitaguñjāphalālaktakapadmasuvarṇavarṇaṃ dhautaṃ ca virajyamānaṃ madhuram īṣallavaṇaṃ snigdham asaṃhatam aśītoṣṇaṃ guru pittaikacayaprakopopaśamaṃ saumyāgneyaṃ prakṛtyā raktam āhuḥ. tathā doṣam. dūṣyam iti kecit. ubhayātmakam anye.

the colour of the *indragopaka-*. Again, according to the AH and AS passages just mentioned, blood in general is like hare's blood, whereas we have seen in §7.4 that this is according not only to SU,Śā 2,17, but also according to the parallels of this passage in AH and AS (see n.358) a characteristic of the procreatory-menstrual fluid. Lastly, there is the possibility that the colour of *lākṣārasa-* "lac-juice" as a characteristic of the procreatory-menstrual fluid in SU,Śā 2,17[378] may refer to the same colour as in CA,Sū 24,22, where we find the colour of lac (*alaktaka-*) mentioned — but as a characteristic of blood.[379] The exact meaning of *lākṣārasa-* is, however, not known, though it does not seem to be the same as *lākṣātaila-*, as some dictionaries suppose.[380] But we see nevertheless that all in all we do have evidence of several conflicting theories existing. And in KAL 23,18 we actually find a description of *ārtava-* which is called *subīja-* (either "whose *bīja-* is good" — i.e. having good *bīja-*; would this refer to a part of the *ārtava-*?[381] — or "through which the *bīja-* is good" or else "[which is] good *bīja-*"; cf. §§2.5f.), of which it is said: *prakṣālitaṃ na ca virajyati* (printed *virajyata*).[382] The Hindi translation takes this to mean that a stained cloth will lose its colour when it is

---

Hemādri has only: *tatrāmbu ... avikṛtena tejasā rañjitaṃ raktam āhuḥ*. Cf. also n.361. As regards the discussion at the end of this passage (*tathā doṣam* etc.), this is a very clear statement that there was among medical authorities a controversy on the classification of blood, whether as a morbific entity (cf. too the discussion on ☞ *doṣa-* in Appendix II), whether as a substance (an element, i.e. ☞ *dhātu-*?) capable of being corrupted (☞ *dūṣya-*), or whether as both (*ubhayātmaka-*). MEULENBELD 1991, pp.98;102 too draws attention to this AS passage.

378 See §7.4, also n.358 for parallels.

379 I may mention in this context that the explanation of BHĀ 1,3,206 (= SU,Śā 2,17; see note 97) as given by the author of BHĀ himself explains the different colours of the procreatory-menstrual fluid as being due to different constitutions (*ārtavasya varṇadvayābhidhānam. vātādiprakṛtibhedena varṇabhedāt*), whereby he repeats the argument of Cakrapāṇidatta (who also refers to CA,Sū 24,22), which, however, pertains to blood in general (see n.362)! Needless to say, this explanation of BHĀ does not seem to be substantiated by the 'classical' texts either. The texts are of course full of descriptions of different coloured diseased discharges of the female due to the agency of, among other substances, the morbific entities, but that is something quite different.

380 Cf. e.g. SIDDH 5,119, where *lākṣārasa-* is *part* of a *taila-* called *lākṣādika-*.

381 Cf. on this problem especially §§11.15ff.

382 The whole verse runs:
śuddhārtavaṃ maṇiśilādravahaṃsapādi-
paṅkopamaṃ śaśaśarirajaraktavac ca
lākṣārasapratimam uj[j]valakuṅkumābhaṃ
prakṣālitaṃ na ca virajyati (printed: °*yata*) *tat subījam*.

washed,[383] but the fact that the *ārtava-* itself is the subject of the verb, and the meaning of this verb (discussed in §7.5), show that this is probably not what is meant. I would translate: "and [which], washed, will not be discoloured"[384] (i.e. its colour is fast), which is the exact opposite of what the passages quoted in n.358 and probably also SU,Śā 2,17 tell us. Moreover, the KAL passage too mentions the colour of *lākṣārasa-* as belonging to the procreatory-menstrual fluid, along with the colour of hare's blood and of several other substances. What is more, whatever the KAL passage may actually be saying, it seems to be the exact opposite of BṚSAṂ 77,19, which uses the same terminology, but *without* the negation.[385] Nevertheless, though this obviously adds to the confusion in this regard, that fortunately does not matter much, for what the actual properties ascribed to the two fluids in question are is only of secondary importance to us compared with the fact that there obviously was some confusion in this regard and in the differentiation of the two (see also §7.12). But however much confusion there be with regard to different texts and even to different parts of the same text, it does

---

[383] *jo raj ... vastr par lage hue ko dhone par chūṭ jāveṃ, kapʳṛe ko na raṃge.*

[384] Actually: "does not discolour (intransitive)". The printed form *virajyata*, though metrically sound, is very problematic. It is formally the second person plural of the imperative of the verb conjugated according to the fourth verbal class. The trouble is that, if one translates it as such, one can only obtain sense if one assumes that *vi√rañj* is not used medio-passively here, but actively to mean "discolour" ("and [which], washed, you shall not discolour"), even though it does not follow the usual nasal conjugation of the transitive, which would be very odd, quite apart from the extremely peculiar mode of expression. It is thus justified to assume that something is wrong here. One could consider whether *virajyata* could stand here *metri causa* for *virajyeta* or *virajyate*. The latter would pose no problems: "and [which], washed, is not discoloured" (i.e. its colour is fast); the former would, however, make good sense too: "and [which], if washed, is not discoloured" (on this way of translating the optative cf. VṚK, pp.508f.). But this seems highly unlikely too; one could have maybe a medial for an active form, or possibly a singular for a plural, or the like, but hardly a form so different from what it would stand for. As it is, the form *virajyata* actually looks like an injunctive (third person singular medium), and the function of the injunctive, namely to merely mention an occurrence without specifically fixing it in time (cf. especially HOFFMANN 1967, pp.266ff.), fits well with the fact that this characteristic of good *ārtava-* is always valid. The trouble is that the injunctive is (except in certain special cases, e.g. with the prohibitive *mā*) more or less a Vedic form; we can therefore hardly postulate an injunctive here. Thus the emendation proposed seems the most likely reading, especially since it too is metrically correct.

[385] This verse, 78,19 in other editions, runs:
śaśaśoṇitasaṅkāśaṃ lākṣārasasannikāśam athavā yat
prakṣālitaṃ virajyati yac cāsṛk tad bhavec chuddham.
Note that the colour of the fluid is here said to resemble that either of hare's blood or of *lākṣārasa-*.

appear as if SU itself at least mostly[386] differentiated not only between the procreatory-menstrual fluid and blood as such, but also with regard to their properties.

7.8. This brings me back to the problem of *avivarṇa-* in the description of blood in SU,Sū 14,22. *avivarṇa-* literally means "not *vivarṇa-*", and *vivarṇa-* means either "whose colour is off/away" (i.e. "off-colour" or "colourless"; cf. *vivarṇatā-* in Ḍalhaṇa's commentary in n.360) or "whose colour is different", from which it would follow that *avivarṇa-* means either "not off-colour" or "not colourless" or "not different-coloured".[387] But if that is so, then the first explanation of this word given by Ḍalhaṇa in §7.5 (see n.359) seems to be quite impossible, for this explanation appears to presuppose a meaning "different-coloured" for *avivarṇa-*,[388] i.e. the exact opposite of what has been considered above. The second one, namely that the fluid (blood) has a fast colour (i.e. does not lose its colour and become off-colour/colourless when washed) may, after all we have seen in §§7.5ff., be possible, but it does bother one a great deal that the passage Ḍalhaṇa comments upon (see n.356) mentions neither clothing nor washing and/or water, and that saying that blood does not become off-colour, colourless or different-coloured in certain circumstances is very different from saying that it *is* so. But what could *avivarṇa-* mean then? Now we actually have already found *vivarṇa-* in a description of what we by now know is the procreatory-menstrual fluid, namely in SU,Śā 3,10 (see §6.3). Might it not be that Sū 14,22 is actually telling us that blood does not have this property, namely being *vivarṇa-*, that the procreatory-menstrual fluid has in Śā 3,10? But if this should indeed be so, then it would follow that the text is telling us that blood is different form the procreatory-menstrual fluid in that the latter may be not only "slightly dark" (*īṣatkṛṣṇa-*), but also *vivarṇa-*, i.e. either of a colour different from this, or else colourless, whereas the former is not *vivarṇa-* in this sense, i.e. always has one and the same colour (namely red), which does not change. Ḍalhaṇa's first explanation of *vivarṇa-* in SU,Sū 14,22 could actually be referring to exactly this, for if we look at the wording given in n.359 closely, we see that *avivarṇa-* need not necessarily be explained by *vividhavarṇa-*, but that *indra-*

386 I have used "mostly" because of the problem of the fastness of the colour drawn attention to in §7.6.

387 Examples for the different uses of *vivarṇa-* and *avivarṇa-* discussed here can be adduced from other contexts in the relevant texts. Since the matter is too lengthy to fit into a footnote, it has been discussed in Excursus V on pp.204-207.

388 Here in the sense of *vividhavarṇa-* "variously/differently coloured", qualified by *īṣat* "slightly" so as not to lose the connection with the colour of the *indragopaka-*.

gopakavarṇam apīṣad vividhavarṇam could be a further qualification of *avivarṇa-*: the blood, though *avivarṇa-*, i.e. having only the colour of an *indragopaka-*, nevertheless has slight variations of colour, which brings in the other shades of red Ḍalhaṇa mentions. In this case *avivarṇa-* would by implication mean, according to Ḍalhaṇa, "not different-coloured", as it would refer to only the colour of the *indragopaka-* (basically). Of course this is just a possibility (though it would do away with the problem of the meaning of *avivarṇa-* which arises otherwise), and in need of further substantiation. However, irrespective of what Ḍalhaṇa actually means, by comparing SU,Śā 3,10 and SU,Sū 14,22 we do seem to have arrived at a possible differentiation between blood (unchangeable in colour) and the procreatory-menstrual fluid (changeable in colour), at least according to this text. There is no denying that it is, at least at this stage, anything but certain that we are on the right track, but it is also clear that these deliberations do actually do away with some of the problems mentioned above. But it is obvious that if this should indeed be what the text says, then we have a great problem on our hands as regards the nature of the procreatory-menstrual fluid, for this does not seem to fit what we would spontaneously associate with the visible menstrual fluid. However, we do not, at the present stage of our investigation, have enough material to tackle this, so that I see no alternative to setting it aside for the time being; we shall return to it in §§8.3ff.

7.9. It must also be pointed out that this problem concerning the colour of the procreatory-menstrual fluid does not necessarily apply to ancient Indian medicine as a whole. Thus, though SU2 has *vivarṇa-* like SU in Śā 3,10, SU1, p.321 reads *vigandha-*, which latter is also to be found in GHĀṆEKAR 1986, p.80a and TIVĀRĪ, 1,p.57. Both Ḍalhaṇa and NYC are silent on this point in their commentaries on Śā 3,10, so that we do not know what the reading they presuppose here is, but in his commentary on Sū 15,16 Ḍalhaṇa quotes Śā 3,10 with the reading *vigandha-* in the place of *vivarṇa-*, connecting this *vigandha-* with *daurgandhya-* "bad smell" in Sū 15,16. However, Sū 15,16 refers to a condition in which the body as a whole has *too much* procreatory-menstrual fluid (cf. also n.213), leading to acute bodily pain, excessive flow and bad smell,[389] the latter two not explained further, but probably referring to the fluid itself. But since Śā 3,10 refers to menstruation in the normal course of events, I do not quite see how this can be combined with Sū 15,16, even though according to our dictionaries *vigandha-* may indeed mean "smelling bad".[390] If, however, such a connection is not given, then we have no

---

389 *ārtavam aṅgamardam atipravṛttiṃ daurgandhyaṃ ca.*

390 To complicate matters still further, Ḍalhaṇa also mentions a variant for *daurgandhya-*

need to try to make *vigandha-* in Śā 3,10 fit in with Sū 15,16. And indeed, Aruṇadatta, commenting on *vigandha-* in AH,Śā 1,22cd-23ab = AS,Śā 1,p.280b (already quoted in §6.16), explains this as meaning *visragandharahita-* "without a musty smell", which is seconded by Śrīdāsa in his commentary on AH1,Śā 1,23cd-24ab,[391] while JOLLY, p.49 writes that the menstrual fluid is "geruchlos", which is indeed — and not surprisingly, since *vi* can also mean "without" — also given as a meaning of *vigandha-* by our dictionaries.[392] Cakrapāṇidatta, commenting on *vigandha-* in CA,Si 3,11, where the expression occurs as a characteristic of an animal bladder fit for use with a syringe (especially for a clyster)[393], says: "that whose putrid smell is gone is *vigandha-*",[394] though actually simply "without smell" would fit just as well here. Whatever the exact meaning of *vigandha-* be, however, we see quite clearly that we cannot be sure that *vivarṇa-* as a characteristic of the procreatory-menstrual fluid is clearly presupposed by SU beyond the shadow of a doubt, even though BHĀ 1,3,205, which has very nearly the same wording as SU,Śā 3,10 given in §6.3,[395] reads *vivarṇa-* too, though here the *ārtava-* is described as "slightly *vivarṇa-*" and also as "dark". Still, the fact remains that some tradition at least obviously thought that *ārtava-* could be *vivarṇa-* too.

7.10. This tradition too shows that blood and the procreatory-menstrual fluid were thought to be different (cf. also n.391). But one might point

---

in Sū 15,16, namely *daurbalya-* "weakness".

391 *nāpi pittena tathābhūtena yuktaṃ yena visragandhi* (Aruṇadatta)/*visragandhaḥ* (Śrīdāsa) *syāt*. This statement that the musty smell comes through the (here absent) association with bile could be taken to be an oblique reference to blood, which — according to the above then obviously in contrast to the procreatory-menstrual fluid — was held to be very closely connected with bile (cf. e.g. §7.16); indeed, SU,Sū 14,9, should it indeed refer to blood in general (cf. §§7.11f.), would characterise this as musty. That would further substantiate the correctness of the deduction that blood and the procreatory-menstrual fluid were considered to be different.

392 MÜLLER 1942, p.43,n.1, however, translates *vigandha-* in the passage just quoted as "übelriechend"; maybe he has been influenced by Ḍalhaṇa.

393 On this instrument, *vasti-* or *basti-*, see also MEULENBELD 1974, p.481.

394 *vigataḥ pūtigandho yasya sa vigandhaḥ*.

395 The version of BHĀ runs:
māsenopacitaṃ kāle dhamanībhyas tad ārtavam
īṣadvivarṇaṃ kṛṣṇaṃ ca vāyur yonimukhaṃ (var: *yoner mukhaṃ*) *nayet*.
Here the analysis *tad ārtavam* is most probably correct, as v.1,3,204 immediately preceding (= SU,Sū 14,6 in the variant form beginning *rasād eva rajaḥ strīṇāṃ* ...; see §7.11) mentions the fluid, and *tad* would thus be taking this up (cf. n.194). On the plural number of ☛ *dhamanī*-s mentioned here see §6.25.

to SU,Śā 4,25, which tells us that certain parts of the body derive from blood: liver (*yakṛt-*) and spleen (*plīhan-*)[396] are formed of blood, the ☞ *phupphusa-* of the foam of blood, and the *uṇḍuka-*[397] of the impurity (☞ *kiṭṭa-*) of blood.[398] Now vv.4,26ff., referring to the formation of various other parts of the body, in which blood in general plays a part, do make it very probable that 4,25 too refers to blood in general, as the intimate relationship between blood and liver and spleen (see §5.11) too would suggest. The parallel passages also clearly or seemingly speak of blood in general.[399] But 4,24 immediately before 4,25 refers to the blockage of procreatory-menstrual fluid and the formation of the ☞ *aparā-* (cf. §6.17), so that the context makes it appear as if 4,25 too referred to the embryo/foetus, in which the organs mentioned are formed; one may thus be inclined to see in "blood" here (4,25) a reference to the procreatory-menstrual fluid too. This would make both the fluids ultimately the same. But actually that is no convincing argument for 4,25 referring to the procreatory-menstrual fluid as forming the organs mentioned, since transformations within the body of the embryo/foetus could be taken to develop blood, out of which (and not out of procreatory-menstrual fluid within the womb) the mentioned organs are then formed. Indeed, since these organs seem to belong to a stage of development after that in which, in the preliminary form of the embryo arising from semen and procreatory-menstrual fluid, the necessary basic substances for further development have already been formed, there does not seem to be any need to refer to the procreatory-menstrual fluid any more with regard to them. This also applies to the parallel KĀ,Śā,Gar 5,[400] the same argu-

---

396 On these two organs see too n.139.

397 On the *uṇḍuka-* see §7.28.

398 This passage (on which cf. also p.103 and §11.32) states:
garbhasya yakṛtplīhānau śoṇitajau. śoṇitaphenaprabhavaḥ phupphusaḥ. śoṇitakiṭṭaprabhava uṇḍukaḥ.
AS,Śā 5,p.320a is similar.

399 This is quite clear in the case of AS,Śā 5,p.320a and the passage both Aruṇadatta and Śrīdāsa quote in their commentaries on AH,Śā 3,12 (also quoted in ĀY 4,232). ĀY 4,233f. (attributed to Pārāśara) too seem to presuppose the same; at least I can find nothing to connect them with the procreatory-menstrual fluid. See also Āḍhamalla's commentary on ŚĀR 1,5,43f.

400 This passage speaks of "its" (*tasya*) heart arising from blood, from the heart the *yakṛt-*, from this the *plīhan-*, and from this the *phupphusa-*, and it seems pretty obvious that "its" (*tasya*) must refer to the embryo mentioned at the beginning of the chapter:
śoṇitād dhṛdayaṃ tasya jāyate hṛdayād yakṛt
yakṛto jāyate plīhā plīhnaḥ phupphusam ucyate.

ments being valid as in the case of SU,Śā 4,25.[401] Thus SU,Śā 4,25 cannot be used as an indication for a theory equating blood in general and the procreatory-menstrual fluid.[402] Nevertheless, we find both fluids being termed "blood" in probably all medical texts, SU being, as we have already seen, no exception. Of course the menstrual discharge must — obviously — look like blood even to those who do not regard it as being blood, but one would not expect people of the calibre of the authors of our medical texts to have overlooked the difficulties caused by calling two to them different substances by the same name in a matter as fundamental as this. And in truth, we do find the term "blood", as used for the procreatory-menstrual fluid, qualified several times, e.g. in SU,Śā 9,7[403], Sū 14,6[404] and Sū 14,7[405]. Cf. also Sū 15,5,[3]:

> *raktalakṣaṇam ārtavaṃ garbhakṛc ca*
> "*ārtava-* is that whose characteristics are [those of] blood and which is embryo-making",[406]

---

401 The procreatory-menstrual fluid could, however, be meant in KĀ,Śā,Asa, p.46:
prāṇas tu bījadhātuṃ hi vibhajaty asthisaṅkhyayā (var: *asthisaṃsthayā*)
praviṣṭamātraṃ bījaṃ hi raktena pariveṣṭayet
śukrād asthy asthito māṃsam ubhābhyāṃ snāyavaḥ smṛtāḥ
sarvendriyāṇi garbhasya sarvāṅgāvayavās tathā
tṛtīye māsi yugapan nirvartante yathākramam.
Unfortunately, the whole of the chapter preceding this is lost, so that the exact context of these lines is unclear. But *bījarūpo dhātuḥ* (var: *bījadhātuḥ*) in CA,Śā 4,7 (see on this ☛ *dhātu-* in Appendix II) makes it highly probable that *bījadhātu-* here refers to semen of the male; *bīja-* (on which see §§2.5f.) most probably too refers to this, especially since this *bīja-* is said to be surrounded by *rakta-* as soon as it has entered (*praviṣṭamātram*), which does look like a reference to an entrance into the womb. These lines would then be saying that this semen, divided by the *prāṇa-* according to the number of bones in the body [to be formed], is surrounded by "blood", then giving rise to bones and so on, until in the third month all the organs of the foetus appear simultaneously. But this would ultimately only be telling us that semen and the procreatory-menstrual fluid ("blood") combine to form the beginnings of the embryo, which then develops a dynamism of its own, and this is something so basic that the passages cited with regard to the formation of liver and spleen must presuppose it too, though not necessarily in the form of the theory stated in the passage quoted here.

402 Whether SU,Śā 4,25 and its parallels might nevertheless point to older theories having existed equating blood and the procreatory-menstrual fluid (theories of which they might then be remnants), is a matter one can only speculate upon (cf. also n.169).

403 *raktam ... ārtavasaṃjñam* "blood ... called *ārtava-* (cf. §6.8).

404 *raktaṃ rajaḥsaṃjñam* "blood called *rajas-*".

405 *ārtavaṃ śoṇitam* "blood which is *ārtava-*, i.e. belonging to the *ṛtu-*"; on *ṛtu-* see §2.1.

406 Ḍalhaṇa: *raktalakṣaṇam ārtavam iti raktasya yāni lakṣaṇāni tāny asya sambhavantīti*

which is incidentally also another proof of the oneness of the procreatory and menstrual fluids (cf. §6.9). The adjective *ārtava-* in Sū 14,7 could perhaps make one surmise that this is the remnant of an older theory in which there was no differentiation between blood and the procreatory-menstrual fluid (cf. §2.2 and §2.4), the menstrual fluid being merely blood that flows at a certain time. This would, however, be highly conjectural, and even if it should be true, then the very fact that this special flow is qualified would still make it seem that simply "blood" as a name for the procreatory-menstrual fluid is *here* regarded as being essentially an abbreviation or a metonym.[407]

7.11. But if this fluid and blood are different, what are their respective origins according to SU? Sū 14,10 tells us that blood comes from *rasa-*, and thus forms part of the chain of the breakdown of food which is mentioned in §2.4.[408] But before that there is a description of the origins of both blood and the "blood" of women (i.e. the procreatory-menstrual fluid). After Sū 14,3, in which *rasa-*, whose seat is the heart (*hṛdaya-*), but which permeates the whole body, is called[409] the essence (*sāra-*) of fully transformed (*samyakpariṇata-*) food (*āhāra-*),[410] we find (the variants are, when not other-

---

svabhāvakathanam. Ḍalhaṇa also mentions a variant of the SU passage which runs:

raktalakṣaṇam ārtavaṃ yathoktam akṛtrimaṃ garbhakṛc ca.

The additional words of this are, however, irrelevant for our discussion. SU4 (p.113) has the reading:

yathoktaraktalakṣaṇam ārtavaṃ garbhakṛc ca.

Cakrapāṇidatta says (also SU3, p.261): *yathoktaraktalakṣaṇam ity ārtavarūparaktasya yallakṣaṇam uktaṃ* ..., i.e. does not refer *rakta-* to blood in general.

[407] Cf. in this context also Ḍalhaṇa's commentary on Ut 42,8, where *rakta-* is differentiated along the lines of *ārtavarakta-* and *dhāturakta-*, the latter quite obviously blood in general.

[408] The full wording of 14,10 describes the chain as in §2.4:

rasād raktaṃ tato māṃsaṃ māṃsān medaḥ prajāyate
medaso 'sthi tato majjā majjñaḥ śukraṃ tu jāyate.

BHĀ 1,3,189 too has this verse, but with *śukrasya sambhavaḥ* for *śukraṃ tu jāyate*. This variant is also mentioned by Ḍalhaṇa in his commentary on SU,Sū 14,10, another variant he mentions being *pravartate* for *prajāyate*; *pravartate* is indeed the reading Cakrapāṇidatta presupposes in his commentary on 14,10. The similar verse CA,Ci 15,16 (= AH,Śā 3,62cd-63ab) runs:

rasād raktaṃ tato māṃsaṃ māṃsān medas tato 'sthi ca
asthno majjā tataḥ śukraṃ śukrād garbhaḥ prasādajaḥ (AH: *prajāyate*).

A variant of AH in pāda b reads: *māṃsān medo 'sthi medasaḥ*. A version found in KAL has several differing readings; it is given in n.1095. Śivadāsasena in his commentary on DR,Āh 21 mentions and discusses the variant *garbhaprasādajam* (for *garbhaḥ prasādajaḥ*).

[409] As in Sū 46,528 (referred to later again in §7.37).

[410] Cf. CA,Sū 28,4: *rasa-* is the ☞ *prasāda-*, i.e. pure portion of food. I may in this connexion mention the following definition of food (*āhāra-*): *āharanti tasmād rasam ity āhāraḥ* "from

wise mentioned, given in SU):

4. *sa khalv āpyo raso yakṛtplīhānau prāpya rāgam upaiti.*

5. *ślokau cātra bhavataḥ* (var:[411] *bhavataś cātra*).

rañjitās tejasā tv āpaḥ śarīrasthena dehinām
avyāpannāḥ prasannena raktam ity abhidhīyate.[412]

6. *rasād eva striyā raktaṃ rajaḥsaṃjñaṃ pravartate* (var:[413] *prapadyate*)
tad varṣād dvādaśād ūrdhvaṃ (var:[414] *ūrdhve*) *yāti pañcāśataḥ kṣayam*

(var: *rasād eva rajaḥ strīṇāṃ māsi māsi tryahaṃ sravet*
tad varṣād dvādaśād ūrdhvaṃ yāvat pañcāśataṃ samāḥ).[415]

7. *ārtavaṃ śoṇitaṃ tv* (var:[416] *ārtavaśoṇitaṃ tv*, var:[417] *ārtavaṃ śoṇitaṃ ca*) *āgneyam agnīṣomīyatvād* (var:[418] *agnisomīyatvād*) *garbhasya.*

8. *pāñcabhautikaṃ tv apare jīvaraktam āhur ācāryāḥ.*

9. *visratā dravatā rāgaḥ spandanaṃ laghutā tathā*[419]
bhūmyādīnāṃ guṇā hy ete dṛśyante cātra śoṇite

(var of cd: *pṛthivyādiguṇās tv ete dṛśyante śoṇite yataḥ*).

---

this (i.e. food) they (i.e. living beings) get/take (*āharanti*) *rasa-*, thus *āhāra-*" (*Kāśikā* on Pāṇini's *Aṣṭādhyāyī* 3,3,19); on this passage cf. also WEZLER 1990, p.133.

[411] In SU1 (p.43), SU3 and GHĀṆEKAR 1977A.

[412] The verse as given in ĀY 4,62 reads *saṃsiddhās tejasā hy* for *rañjitās tejasā tv* in pāda a. (As for the singular *abhidhīyate* with the plural nominative *āpaḥ*, cf. SPEIJER, §19.I on p.14.)

[413] In Śivadāsasena's commentary on DR,Āh 22.

[414] ŚARĪRA 2,6.

[415] BHĀ 1,3,204 has a hybrid version:
rasād eva rajaḥ strīṇāṃ māsi māsi tryahaṃ sravet
tad varṣād dvādaśād ūrdhvaṃ yāti pañcāśataḥ kṣayam.

[416] Mentioned by Ḍalhaṇa and presupposed by Cakrapāṇidatta (cf. also SU3, p.234 and SU4, p.102), also in ŚARĪRA 2,7.

[417] Mentioned by Cakrapāṇidatta; see SU3, p.234. SU4, p.102 suppresses the *ca*, but the context makes it clear that it is to be read; see n.441.

[418] ŚARĪRA 2,7.

[419] SU3, p.234 and Āḍhamalla in his commentary on ŚĀR 1,6,10f. read *syandanaṃ* for *spandanaṃ*, and SU lists a variant *tanutā* for *laghutā*.

4. "This watery *rasa-* (i.e. *rasa-* of watery nature or connection), you should know, becomes coloured on attaining liver and spleen.[420]

[420] On the close connection of blood with liver and spleen see n.139 and §7.10 above. On the process of colouring (most probably by means of ☞ *pitta-* "bile", as Cakrapāṇidatta in his commentary also presupposes) cf. SU,Sū 21,10 as well as the references given by MEULENBELD 1974, p.474 (*pitta*). I must here, however, draw attention to the fact that according to ŚĀR 1,6,9cd-10ab, the colouring of ☞ *rasa-* by bile, which gives rise to blood, takes place in the heart (☞ *hṛdaya-*) — liver and spleen are not mentioned, in contradistinction to ŚĀR 1,5,31, which tells us that it is bile in the liver which colours *rasa-* to form blood. Āḍhamalla in his commentary on 1,6,9f. says that *rasa-* is at first only lightly coloured, being fully coloured to give blood only after a week; this seems to imply that only the initial colouring takes place in the heart (this is also Kāśirāma's explanation), the final colouring in the liver (and spleen; cf. his commentary on 1,5,31). This could explain away the discrepancy, though whether this deduction from the commentaries would indeed explain the problematic text of ŚĀR 1,6,9cd-10ab correctly is a matter of debate. Note in this context also AS,Sū 19,p.145a and AH,Sū 11,37 (also BHĀ 1,3,184), from which it follows that ☞ *ojas-* (which may also stand for *rasa-* and is also said to be slightly reddish-yellow) is, though situated in the heart, spread all over the body (AH and BHĀ say so explicitly: *hṛdayastham api vyāpi*); Āḍhamalla in his commentary on ŚĀR 1,6,9cd-10ab too expressly refers to the heart as the place of *ojas-*. AS,Sū 20,p.148a and AH,Sū 12,13 clearly state that *rasa-* is coloured by bile present in the receptacle of undigested food (*āmāśaya-*); liver and spleen are not mentioned. Now one may take this to mean that, though the bile is stored here, the actual colouring takes place elsewhere, but since the parallel passage SU,Sū 21,10 not only states like SU,Sū 14,4 that *rasa-* is coloured in the liver and the spleen, but also that the colouring bile is located in these organs, it does look as if AS and AH are saying that the colouring too takes place in the *āmāśaya-* (cf. in this context also §7.34). BHĀ 1,3,123(6), however, says that the bile situated in the receptacle of fire (*agnyāśaya-*) between the receptacles of undigested and digested food (*āmāśaya-* and *pakvāśaya-* respectively) acts through the receptacles of bile situated elsewhere by performing certain functions, such as for instance colouring *rasa-* in liver and spleen (according to BHĀ 1,3,139 *rasa-* is coloured and becomes blood in the liver); this looks very much like an attempt to bring about a synthesis of different views, though of course it would be difficult to prove clearly that it is indeed so. (On the abode of bile see also §7.35, with further references.) As it is, one of the causes for the confusion we see here may be the fact that there seems to have been a theory according to which ingested food and drink first entered the *heart* and then the *āmāśaya-*, which would obviously also be of relevance for *rasa-*; this theory has been discussed in Appendix II under ☞ *hṛdaya-*. I may in this context also draw attention to RANADE/PARAMJAPE (a work of the sort mentioned in §1.7), p.91, where the awareness of problems similar to those discussed above may have led to the statement (offered with not a shred of supportive evidence) that "... according to our opinion, a major and important site of dhātwagni is the yakrut (liver). It has been proved that the major portion of āhārarasa first enters yakrut, before going to the heart. The purpose of the same is obvious. Samśraya of śūkṣma pacana [sic] (metabolic digestion) takes place in liver".

ĀĀS 6933 gives us the same verse as ŚĀR 1,6,9cd-10ab (quoted above), but with an important difference: instead of *hṛdayaṃ* "heart" it has *grahaṇī[ṃ]*! This is then followed by an explanation similar to that which we have deduced Āḍhamalla might be giving, namely that

5. "And there exist two ślokas to this effect (var: And there exist two to this effect):

> Water, however, coloured (var in n.412: Water, indeed, thoroughly prepared[, i.e. cooked]), [but] not destroyed/spoiled through the pure fiery energy situated in the body of corporeal beings, is called '*rakta-*'.

6. "The blood of a woman, called *rajas-*, is from *rasa-* itself. It arises (var: commences/takes form) subsequent to the twelfth year, attains *kṣaya-* (cf. §§5.2f.) from/at fifty. (Var: The *rajas-* of women is from *rasa-* itself. It flows monthly for three days subsequent to the twelfth year, up to [the age of] fifty years.)[421]

---

first *rasa-* goes to the heart and then to liver and spleen, the colouring thus taking place in two stages. The discrepancy between the explanation referring to the heart and the verse referring to the ☛ *grahaṇī-* is not commented upon, so that one may consider whether the mention of the *grahaṇī-* might not be a secondary emendation.

The colouring of *rasa-* by bile to give blood is also mentioned (without a specific place being named) in CA,Ci 15,28 (see n.491). However, Cakrapāṇidatta does not comment on Ci 15,22-35, which thus might not be authentic. Aruṇadatta, though, in his commentary on AH,Śā 3,62f. does quote this passage (with the exception of 35ab) as a part of CA, but Aruṇadatta is younger than Cakrapāṇidatta (cf. MEULENBELD 1974, pp.389-391;400-402); anyway, Aruṇadatta probably quotes from another version of the text (cf. §7.17), and as it is much more probable that our texts have expanded in the course of time due to accretion than that they have been trimmed, Cakrapāṇidatta does appear to be more trustworthy in this regard than Aruṇadatta. Note too that even Aruṇadatta (and, seemingly following him, also Śrīdāsa in his commentary on AH1,Śā 3,62f.; 62 is wrongly printed loc.cit. as 61) leaves out a line of the text we have today. Also note ĀY 4,56-59, which is a quotation of CA,Ci 15,23-26; the quotation is prefaced *not* by *atha carakāt* or the like, but by *athaupadhenavāc cikitsāsārasaṅgrahāt* (on Aupadhenava and his work, also *aupadhenava-*, cf. e.g. SU,Sū 1,3 and 4,9).

[421] Cf. also §6.7. A similar verse is found in AH,Śā 1,7 and YO 21,37:

māsi māsi rajaḥ strīṇāṃ rasajaṃ sravati tryaham
vatsarād dvādaśād ūrdhvaṃ yāti pañcāśataḥ kṣayam
"Subsequent to the twelfth year the *rasa-*-arisen *rajas-* of women flows monthly for three days [and] attains *kṣaya-* from/at fifty."

In his commentary on AH1,Śā 1,7, Śrīdāsa mentions a variant *ajasraṃ* "plentiful" for *rasajaṃ* "*rasa-*-arisen", but says that this variant cannot be correct. In BYT 2,7cd we find only: *māsi māsi rajaḥ strīṇāṃ tryahaṃ sravati yonitaḥ*. *Nāḍījñānadarpaṇa* 248 says:

dvādaśād vatsarād ūrdhvam āpañcāśat samāḥ striyaḥ
māsi māsi bhagadvārā prakṛtyaivārtavaṃ sravet
"Subsequent to the twelfth year until fifty years [of age], by nature the *ārtava-* of a woman is wont to flow monthly through the portal of the vagina."

7. "The blood belonging to the *ṛtu-* (cf. §2.1) is, however, (var: The blood belonging to the *ṛtu-* is, however,/That belonging to the *ṛtu-* and blood are, however,) (another var: That belonging to the *ṛtu-* and blood are) fiery because of the [both] fiery and somic nature[422] of the embryo.[423]

8. "But other authorities say that *jīvarakta-* is composed of the five [gross] elements [earth, water, fire, wind and ether],[424]

9. "for (var: since) these qualities of earth etc., [namely] mustiness, fluidity, colour, pulsation (var in n.419: flowing) [and] likewise lightness (var in n.419: thinness), are indeed (*ca* = *eva*) perceived here (var: are perceived[425]) in blood."

7.12. Let us first turn our attention to 14,8f. The theory given here (and found also in Āḍhamalla's commentary on ŚĀR 1,6,9-11) seems to be one that the author himself does not adhere to, since in Indian scholarly texts saying that "others" say something is a usual way of distancing oneself from the view propounded. Nevertheless, this passage shows that there were different theories on the nature of *jīvarakta-*; note too that the classificatory system presupposed in 8f.[426] seems to be different from that of 7[427]. The trouble is, we do not know for sure what *jīvarakta-* is. Ḍalhaṇa tells us that it refers to *rakta-* similar to life or the life-force (*jīva-*), as blood is seen only in a living (*jīvat-*) body, not in a corpse;[428] from this it follows that according to

---

[422] *agnīṣomīyatva-*. On the problem of what "fiery" and "somic" actually refer to, see the discussion on ☛ *āgneya-* and ☛ *saumya-* in Appendix II.

[423] Sū 14,3, in answer to the question of whether ☛ *rasa-* is ☛ *saumya-* "somic" or *taijasa-* "of the nature of fiery energy (☛ *tejas-*)", clearly a synonym of ☛ *āgneya-* here, has just stated that it is *saumya-*. The embryo is both "somic" and "fiery" (see also e.g. Śā 3,4); so, because semen is "somic" (Śā 3,3), the "fiery" part must be the procreatory "blood" (called "fiery" in Śā 3,3 too), even though this comes from "somic" *rasa-* (cf. on *rasa-* also ☛ *ojas-* in Appendix II).

[424] I.e. of the ☛ *mahābhūta-*s.

[425] *tu* here is an untranslatable gap-filler.

[426] Based on the gross elements, on which see ☛ *mahābhūta-* in Appendix II.

[427] Based on the theory of "fiery" and "somic", on which see ☛ *āgneya-* in Appendix II.

[428] *jīvaraktam iti jīvatulyaṃ raktam. kutaḥ. jīvaccharīre raktadarśanān mṛtaśarīre cādarśanāt.* According to most interpreters, SU,Sū 14,44 would expressly call blood *jīva-* "life(-force)":

tasmād yatnena saṃrakṣyaṃ raktaṃ jīva iti sthitiḥ

Actually this is ambiguous and could mean either:

Ḍalhaṇa *jīvarakta-* here refers to blood in general. In SU,Ci 34,13f. (cf. §7.6) *jīvaśoṇita-* (*śoṇita-* = *rakta-*) too quite clearly refers to blood,[429] and this is also the case as regards *jīvarakta-* in ŚĀR 1,5,9 = ĀY 4,242,[430] *jīvarakta-* in CIK, p.5 and *jīvāsra-* in BYT 2,51. Similarly, *jīvarakta-* in Cakrapāṇidatta's

---

> "'therefore blood is to be carefully protected in the/a living [creature] (*jīve*)': thus [is] the maxim",

or:

> "therefore blood is to be carefully protected; 'blood is (the) life(-force)' is the maxim",

according as to whether one takes *jīva* to stand for the locative *jīve* or the nominative *jīvaḥ*. A similar problem crops up in the parallel ĀY 4,78:

> *pañcatvam upapannasya dehe raktaṃ na dṛśyate* (var: *vidyate*)
> *tasmād raktaṃ jīva iti pūrvācāryair niveditam.*

(Healthy) blood is, however, clearly called *jīva-* in CA,Si 6,78:

> *atitīkṣṇaṃ mṛdau koṣṭhe laghudoṣasya bheṣajam*
> *doṣān hṛtvā vinirmathya jīvaṃ harati śoṇitam,*

and in AH,Ka 3,32:

> *atiyogāc ca bhaiṣajyaṃ jīvaṃ harati śoṇitam*

(according to the rest of the latter passage, this is technically called *jīvādāna-* "taking away, i.e. loss, of *jīva-*, and Cakrapāṇidatta similarly explains the CA passage as referring to *jīvādāna-*). This *jīva-* blood is subsequently differentiated, as healthy blood, from ☞ *raktapitta-* "blood-bile" in CA,Si 6,79f. and AH,Ka 3,33. A nominative *jīva-* "life" is improbable (it would have to be a neuter noun anyway, and, though "life" can indeed sporadically be neuter, as a rule it is masculine), but an accusative noun could be considered. However, what seems more probable is that *jīva-* is used here as an adjective meaning "causing to live, vivifying", and that is exactly how Cakrapāṇidatta, commenting on the CA passage, explains it (*jīvam iti jīvanahetudhāturūpaṃ śoṇitam*) (cf. also CA,Si 7,82). Aruṇadatta and Indu in their commentaries on the AH passage, however, explain it as referring to "blood called *jīva-*" (*jīvākhyaṃ śoṇitam*) (Hemādri: *jīvaśoṇitaṃ ojaḥ*; cf. on this ☞ *ojas-* in Appendix II). Similar are the commentaries on AH,Sū 18,26 (Aruṇadatta and Parameśvara: *jīvākhyasya śoṇitasya*; Śrīdāsa in AH1, Indu on AH5,Sū 18,24 and Candranandana on AH2,Sū 18,17: *jīvākhyasya raktasya*) and Indu's commentary on AS,Sū 27,p.204a (*jīvākhyasya raktasya*), both these passages referring to *jīvaśoṇita-* "blood which is *jīva-*", which is ambiguous, as it could refer to *jīva-* as a noun or as an adjective. Commenting on *jīvādāna-* in CA,Sū 15,13 and 20,14, CART explains *jīva-* as *jīvanahetudhāturūpaśoṇita-* and *jīvanahetudhātusvarūpaśoṇita-* respectively (which is more or less the same as Cakrapāṇidatta's remark, quoted above, in his commentary on CA,Si 6,78), and then in the former case quotes AH,Ka 3,32, which, as we have seen above, contains a definition of *jīvādāna-*: taking away (*harati*) of blood which is *jīva-* (KAL 22,76 simply calls it the coming forth of blood: *jīvādānam asṛkpravṛttiḥ*). Cakrapāṇidatta in his commentary on *jīvādāna-* in CA,Si 10,39;41 too glosses *jīva-* as *jīvaśoṇita-*. *jīvādāna-* is also mentioned in a context similar to those discussed here in SU,Ci 34,3, and seems to refer to the vomiting and spitting out of blood (*śoṇita-*) mentioned in 34,12; cf. also the remarks which follow below on *jīvaśoṇita-* in 34,13f.

429 Ḍalhaṇa's commentary says: *viśuddhaṃ śoṇitam* "pure blood".

430 Āḍhamalla: *jīvatulyaṃ raktam*; cf. Ḍalhaṇa's words in n.428.

commentary on SU,Sū 15,21f. (20f. in SU3) can hardly but refer to blood in general, as it is characterised as being situated in the heart (*hṛdaya-stha-*).[431] According to Aruṇadatta in n.367 and Cakrapāṇidatta on CA,Si 6,79f. too, *jīvaśoṇita-* is blood in general; in his commentary on CA,Vi 4,7,[432] where *dhārilohita-* "supporting/upholding blood" is differentiated from *lohitapitta-* (= ☞ *raktapitta-*),[433] Cakrapāṇidatta glosses this with *jīva-śoṇita-* (see also his commentary on CA,Si 6,78-84), and *jīvarakta-* is actually found in a similar context in AS,Ka 3,p.333. In his commentary on SU,Sū 14,21, which describes blood in various diseased states, Ḍalhaṇa quotes a variant which says: *jīvaśoṇitam anyatropadekṣyāmaḥ* "we shall give instructions about *jīvaśoṇita-* elsewhere"; here too *jīvaśoṇita-* is obviously healthy blood. SU,Sū 15,5,1 says *raktaṃ ... jīvayati* "blood makes/keeps alive", which too would speak for *jīvarakta-* being blood in general. BHĀ 1,3,140 says that blood, present all over the body, is the receptacle (*ādhāra-*) of life. See further the passages quoted in n.428.[434] GHĀṆEKAR 1977A, p.78b, however, states in his notes on SU,Sū 14,8f. (loc.cit.: 9f.) that *jīvarakta-* here could refer either to blood in general or to *ārtava-*, the latter option being mentioned by several commentators.[435] I do not know to which commentators this refers, but Ḍalhaṇa in his commentary on SU,Sū 15,5,[3] (cited in §7.10) says that *ārtava-* too performs the same function as blood, this function being called *jīvana-* "vivifying, giving life";[436] Śrīdāsa in his commentary on AH1,Śā 1,7 seems to say expressly that some authorities equate *jīvarakta-* with *rajas-*,[437] but this equation is not quite clear, as there is also the possibility that *jīvarakta-* might actually be referring to blood in general (though in that case blood and *rajas-* would in all probability not be equated in the sense of being the same fluid) — this problem will be discussed later in n.884.

---

[431] On this passage cf. also ☞ *raktasthāna-* in Appendix II.

[432] Quoted already in n.373.

[433] Just as *jīva-* blood is differentiated from *raktapitta-* in CA,Si 6,79f. and AH,Ka 3,33 referred to in n.428.

[434] Cf. also *jīvitaśoṇita-* on p.532, and *jīvarasa-* in n.614.

[435] *is'kā arth jīvan'poṣak'śoṇit yā dhātuśoṇit hotā hai. dūs'rā arth jīvanotpādak śoṇit yā ārtav'śoṇit bhī hotā hai. kuch ṭīkākār sūtr nau kā arth ārtav'śoṇit par kar'te haiṃ.* Note how here it is obvious that a fluid which may be mixed up with blood and which the author himself calls *jīvanotpādak śoṇit* "life-producing blood" can hardly refer to something like the "ovum". It is typical that the author does not even see this problem, or else chooses to ignore it, as it cannot fit in with his fixed notion on the "ovum" (cf. §1.7 and §2.3).

[436] *cakārād raktakarmāpi jīvanasaṃjñaṃ karoti.*

[437] *jīvaraktam raja iti kecid āhuḥ.*

In ĀĀS 6938 *ārtava-* might be called *jīva-*, but it is unclear whether the latter really refers to *ārtava-*, or to itself alone.[438] That there was indeed some confusion on what "*jīva-* blood" actually refers to is also evident from the variant to 14,7 mentioned by Cakrapāṇidatta (see n.417), that both *ārtava-* and blood are "fiery";[439] Cakrapāṇidatta, who personally holds *jīvaśoṇita-* (= *jīvarakta-*) here to be blood in general, remarks on this that those who hold this view at the same time opine that the following *agnīṣomīyatvād garbhasya*, undoubtedly problematic if applied to blood in general too,[440] holds good only for *ārtava-*:

> *ye tv ārtavaṃ śoṇitaṃ ceti*[441] *paṭhitvārtavaṃ dhātuśoṇitaṃ cāgneyam icchanty abhedaṃ ca tayor vadanti tanmate 'gnīṣomīyatvād iti hetur na dhātuśoṇite saṅgatas tantrāntaravirodhaś ca. dhātūpadhātvoś caikarūpatā svatantraparatantrāsammateti doṣaḥ* (SU3, pp.234f., SU4,p.102)
>
> "In the opinion of those, however, who, reciting '*ārtavaṃ śoṇitaṃ ca*', regard *ārtava-* and the *dhātu-* blood as fiery and affirm their non-difference (*abheda-*),[442] the reason '*agnīṣomīyatvād*' is not apposite with regard to the *dhātu-* blood, and a contradiction of other systems/works.[443] And the uniformity of ☞ *dhātu-* and ☞ *upadhātu-* is without the consent of [their?, or our?] own system/work and other systems/works: thus is the fault [in taking *agnīṣomīyatvād* to refer to the *dhātu-* blood too]."

Indeed, in his commentary on 14,4f., after first quoting a verse of Bhoja[444]

---

438 *garbhāśaye gataṃ śukram ārtavaṃ jīvasaṃjñakam*
prakṛtiḥ sarvavikārā ca tat sarvaṃ garbhasaṃjñakam.
The *ca* is against the metre and probably to be deleted.

439 On the properties of blood see also pp.522f.

440 As that would imply a sameness of both fluids (one a ☞ *dhātu-*, one an ☞ *upadhātu-* according to Cakrapāṇidatta).

441 SU4 has *śoṇitam iti*. But leaving aside *ca* "and" spoils the whole argument, as it gives us no reason for taking *ārtava-* and *śoṇita-* as different substances, rather letting the former be an adjectival qualification of the latter ("*ārtava-* blood").

442 It is clear that this non-difference (*abheda-*) pertains to both the fluids being "fiery" and does not mean that they are the same. Cf. also n.447.

443 This seems to refer to systems or works different from those of the authorities holding the views referred to, and as such might subsume the system(s) or work(s) preferred by Cakrapāṇidatta himself.

444 The same as that mentioned in n.1770.

which says that blood is "hot" only when it is inflamed (*vidagdhaṃ coṣṇam*),[445] Cakrapāṇidatta expressly says that blood is "fiery" only when it becomes hot due to the effect of corrupted bile; healthy blood is both "fiery" and "somic".[446] Obviously referring to this, he also says in his commentary on 14,7 that *tu* "however" is used in 14,7 to differentiate the procreatory-menstrual fluid from blood, which has been described (by him) as "fiery" and "somic" before. Ḍalhaṇa in his commentary on SU,Sū 14,6 remarks on the "non-difference" (*abheda-*) between blood in general and the "blood" of females, in that both originate from ☞ *rasa-*.[447] In his commentary on 14,7, in which he takes *ārtavaṃ śoṇitaṃ tu* (which is clearly what he reads)[448] to refer to *ārtavaṃ śoṇitaṃ ca*, i.e. *ārtava- and* blood (and not to *ārtava-* blood),[449] he then adds that both, though originating from "somic" *rasa-*, are "fiery". As regards the variant *ārtavaśoṇitaṃ tu* (see n.416), he quotes an analysis explaining this as "*ārtava-* blood", i.e. "blood belonging to the *ṛtu-*",[450] saying that according to those reciting this variant *ārtava-* is "fiery"; according to SU the authorities Ḍalhaṇa refers to (*ācāryāḥ manyante*) hold blood to be *anuṣṇaśīta-*, whereas according to SU3 blood is not held to be [of the same sort as] *ārtava-* (*ārtavatayā na manyante*) because it is *anuṣṇaśīta-*.[451] The dis-

---

445 In the Bhoja verse as quoted by Āḍhamalla, however, it is not said that blood is hot when inflamed, but that its inflammation is like that of bile (cf. n.1770).

446 Cf. the discussion on ☞ *āgneya-* in Appendix II.

447 *raktārtavayor abhedaṃ darśayann āha rasād evetyādi*. This does not mean that Ḍalhaṇa, who all through his commentary of SU (also on that of the nearly adjacent Sū 14,10) otherwise regards them as different substances, here nevertheless presupposes that both fluids are one and the same. The differentiation is also clear from the commentary on 14,7 quoted in n.451, where he speaks explicitly of a "non-difference in character" (*svabhāvābheda-*). Cf. also *abheda-* in Cakrapāṇidatta's commentary quoted above (see n.442), as well as Gayadāsa's use of *abhinna-* in §5.13.

448 *tuśabdo 'tra bhede*.

449 I can think of no reason for this except maybe the *abheda-* he has postulated in his commentary on 14,6 (see above), for he writes in his commentary on 14,7 (see n.451): *ārtavaśoṇitayoḥ svabhāvābhedād ārtavasyāgneyatva ukte śoṇitasyāgneyatvam uktam eveti*.

450 *ārtavaṃ ca tac choṇitaṃ cety ārtavaśoṇitam*; on *ṛtu-* see §2.1.

451 The commentary in the reading of SU runs:

raktārtavayoḥ saumyarasasambhūtayor api svabhāvabhedaṃ darśayann āha ārtavam ityādi. tuśabdo 'tra bhede. tena rasāt saumyāj jātam apy ārtavaṃ śoṇitaṃ cāgneyam. kutaḥ kāraṇād ity āha agnīṣomīyatvād garbhasyeti. ārtavam āgneyaṃ. śukraṃ saumyaṃ. tābhyām agnīṣomīyo garbho bhavati. nanu yady ārtavam api saumyaṃ bhavati tadā saumyārabdhaḥ saumya eva garbho bhavati. na caivam. agnīṣomīyatvād garbhasyety atrārtavasyāgneyatvam ucyate. ārta-

cussion on ☞ *āgneya-* in Appendix II shows that "hot" and "cold" are not necessarily synonymous with "fiery" and "somic" respectively; here, however, they seem to be synonyms, because in the variant theory Ḍalhaṇa quotes "fiery" *ārtava-* is contrasted to blood which is "not hot or cold", and thus obviously neither "somic" (= "cold") nor "fiery" (= "hot"). Exactly this latter seems, however, to be presupposed by Gayadāsa in his commentary on SU,Ni 2,13, for he says that blood is "fiery" like bile, the "fiery" substance par excellence,[452] which would be in keeping with the variant statement *ārtavaṃ śoṇitaṃ ca* of SU,Sū 14,7 noted above (see n.417), but lead to the same problems as discussed above when connected with the following statement regarding the "fiery" and "somic" nature of the embryo. Much of what the commentators say incidentally seems to show the same sort of confusion regarding the properties of blood and the procreatory-menstrual fluid we saw in §§7.4ff.[453] But I must stress that this pertains to the *properties* of the fluids;

> *vaśoṇitayoḥ svabhāvābhedād ārtavasyāgneyatva ukte śoṇitasyāpy āgneyatvam uktam eveti. ārtavaśoṇitaṃ tv āgneyam ity anye paṭhanti. ārtavaṃ ca tac choṇitaṃ cety ārtavaśoṇitaṃ. tuśabdo 'trāvadhāraṇe. tenārtavasyāgneyatvam avadhārayatīti vyākhyānayanti. raktaṃ punar anuṣṇaśītam evam ācāryā manyante.*

SU3, p.233 has *raktārtavayoḥ saumyarasasambhūtayor api* etc. until *agniṣomīyo garbho bhavati* as in SU, and then runs:

> *nanu yady agniṣomīyatvād garbhasyety atrārtavaśoṇitayoḥ svabhāvābheda* (printed °*daḥ*) *etad ārtavasyāgneyatva ukte śoṇitasyāpy āgneyatvam uktam eveti. ārtavaśoṇitaṃ tv āgneyam ity anye paṭhanti. ārtavaṃ ca tac choṇitaṃ ceti tad ārtavaśoṇitam. tuśabdo 'trāvadhāraṇe. tenārtavasyāgneyatvam avadhārayeti vyākhyānayanti. raktaṃ punar anuṣṇaśītam evārtavatayā na manyante.*

In the reading of SU3 a part of the text clearly has been left out. The reading of SU tells us that if *ārtava-* were "somic", the embryo too would be so (obviously because then both fluids forming it would be "somic"). But since it is both "fiery" and "somic", it follows that the fluid of the female is "fiery", and that blood in general must be "fiery" too because both *ārtava-* and blood are the same in this respect. This is in keeping with Ḍalhaṇa's other remarks here. By contrast, the text as given in SU3 says that if the embryo were both "fiery" and "somic", blood too would be "fiery" because *ārtava-* would be "fiery" and both fluids have the same characteristic. This clearly makes no sense.

From the above it is also obvious that HOERNLE 1897, p.90,n.222 is wrong in referring to "Ḍallaṇa, who reads *ārthavaṁ* [sic] *çoṇitañ=ca*", as *ārtavaṃ śoṇitaṃ ca* is merely Ḍalhaṇa's explanation of *ārtavaṃ śoṇitaṃ tu*. As regards the statement loc.cit. that according to Ḍalhaṇa the reason for both fluids being "fiery" (loc.cit.: "hot") is "the heat of the ordinary blood being inevitably involved in that of the menstrual blood", there is nothing in the commentary that can support this — indeed, it would be most peculiar if there were something.

[452] *āgneyatve raktapittasya caikatvāt.*

[453] HOERNLE 1897, p.90,n.222 too remarks on the difference in this respect between the

we cannot deduce from this that some authorities thought that they were one and the same. This does not mean that such a theory may not have at some time been current (though not necessarily to be found in our medical texts), but what we have seen so far does not allow us to remark on this possibility one way or the other (cf. also §§7.2f. and §7.10). But in any case, if SU,Sū 14,8f. should refer to blood in general, then, as it seems to give an alternative opinion, it would make sense if Sū 14,7, while speaking of the procreatory-menstrual fluid, would at the same time also refer to this blood. If on the other hand 8f. should refer to the latter fluid, then that would speak more for blood in general not being mentioned in 7, though it would not preclude it. An answer to the problem does not seem obtainable.

7.13. Let us therefore turn our attention to the rest of the passage. We find here difficulties of the sort caused by CA,Ci 15,17 discussed in §§5.12f. above. Though there is no doubt here that the SU passage states not only the origin of the special female "blood", but also that it is derived from ☞ *rasa-*, the exact mode of origination is unclear, as is also not only the mode of origination of blood in general, but even the nature of the substance from which this originates, for a different terminology is used in its case. The first difficulty we encounter is the qualification of *rasa-* in 4: *āpya-* "watery". What does this mean? In AS,Sū 18,p.132a we find: *rasaḥ khalv āpyaḥ prāg avyaktaś ca* "*rasa-*, you should know, is watery and at first undeveloped". This *rasa-*, however, refers not to the element *rasa-* as part of the chain of the breakdown of food, but to *rasa-* "taste" in general, in an undefined state before splitting up into the six well-known individual tastes (sweet, sour etc.).[454] CA,Sū 26,9 too speaks of the undeveloped state (*avyaktībhāva-*), but of all the *rasa-*s (note the plural!), and says that water is their matrix (*yoni-*).[455] CA,Si 10,14 also calls water *rasayoni-* "the matrix of *rasa-*". SU,Sū 42,3 mentions *āpyo rasaḥ* "watery *rasa-*", and CA,Śā 1,27 a *rasa-* intimately connected with water. In these two latter passages what is meant is "taste" in general (note the singular!) as connected intimately with the gross element

---

commentaries of Cakrapāṇidatta and Ḍalhaṇa just discussed. Cf. on loc.cit. also n.451.

[454] See on ☞ *rasa-* "taste" Appendix II. See also Śivadāsasena's commentary on DR,Dhānyavarga 1, on the *rasa-* which is *āpya-* or *avyakta-*, as well as Cakrapāṇidatta's commentary and CART on CA,Sū 1,64.

[455] The latter statement precedes the former:

> *teṣāṃ ṣaṇṇāṃ rasānāṃ yonir udakam. ... avyaktībhāvas* (var: *avyaktabhāvas*) *tu khalu rasānāṃ prakṛtau bhavati.*

Cakrapāṇidatta says on the latter:

> *prakṛtau kāraṇe jala ity arthaḥ. avyaktatvaṃ ca rasasāmānyamātropalabdhir madhurādiviśeṣaśūnyā. sā ca jale bhavati.*

(☞ *mahābhūta*-) water in contradistinction to the other gross elements.[456] In the same manner, *rasa*- is called *āpya*- in CA,Śā 7,16, SU,Śā 1,19, AH,Śā 3,4, BH,Śā 2,12 and KĀ,Śā,Gar 4 and *āmbhasa*- (which means the same) in AS,Śā 5,p.316a. This refers to *rasa*- "taste" being, along with the tongue and fluids like *rasa*- (mentioned explicitly by CA and AS[457]), blood (mentioned by CA, KĀ and AS, whereas SU and AH speak only of fluids, and BH mentions only *rasa*- "taste" and the tongue), bile, phlegm, urine, sweat etc., intimately connected with water as part of a theory setting different parts of the body in relation to one of the gross elements. CA,Sū 26,11 tells us that all substances which abound in (among other things) the quality or qualities (☞ *guṇa*-) of *rasa*- "taste" (*rasaguṇabahula*-) are *āpya*- (just as loc.cit. substances abounding in the quality of *gandha*- "smell" are related to earth, of *sparśa*- "touch" to wind etc.).[458]

7.14. However all these various passages mentioned in §7.13 may or may not be connected,[459] they all seem to be no help in explaining SU,Sū

---

456 This theory linking sound, touch/tangibility, shape, taste/flavour and smell with the five gross elements as their ☞ *guṇa*-s or "qualities" (or "attributes") is known so well from general philosophical works that I can dispense with references to other texts. Cf. on the medical passages just mentioned also Śivadāsasena, loc.cit. in n.454.

457 Indu in his commentary on the latter says expressly that one *rasa*- refers to taste, the other to the ☞ *dhātu*- *rasa*-. In his commentary on AH,Śā 3,4 where only *rasa*- "taste" is mentioned, he tries to bring in the second *rasa*- by stating that this too is implied.

458 See on the theory linking different substances of the body to different ☞ *mahābhūta*-s or gross elements PREISENDANZ 2,pp.424-441, also EIM, pp.138-141. However, EIM is wrong in taking (p.139) *rasa*- in this theory to refer to "vital sap", i.e. the element *rasa*- of the chain of substances derived from food. Since *rasa*- is, in the passages cited, mentioned parallelly to sound, touch, sight and smell, it is obvious that it must refer to taste. According to CA,Śā 5,5 all the wetness (*kleda*-) in the body is the gross element water; similarly, BH,Śā 5,12 says: *yad dravaṃ snigdhaṃ mṛdu vā tad audakam* "that [in the body] which is fluid/flowing, viscous/unctuous or soft, is watery", just as CA,Sū 26,11 (just mentioned) too tells us that substances abounding in the attribute(s) of flowing (*drava...guṇabahula*-) are *āpya*- (on this latter passage see also SHARMA,P.V. 1985, pp.212f.). SU,Sū 46,422 says that all *rasa*-s have the nature/character of water (*toyātmakāḥ sarvarasāś ca dṛṣṭāḥ*); this probably refers to the six tastes sweet, sour etc., but there is also the chance that all substances capable of being drunk are meant. Interestingly, v.34 of KIRFEL 1954, p.338 does not mention *rasa*-, but only saliva, urine, semen, marrow and blood as being those substances of the body connected with water; note also that these five substances are called the ☞ *guṇa*-s of water in the same way as in v.33 five substances of the body are listed as the *guṇa*-s of earth, whereas the five *guṇa*-s of fire, wind — of this we have two different series of five — and ether listed in vv.35-38 on p.339 are not substances, but feelings and actions of the body. On the connection between water and blood see also PREISENDANZ 2,pp.438f.

459 Note in this context also CA,Sū 1,64:

14,4, since it is difficult to see in 14,4 any theory in which we could take "watery" ☞ *rasa-* to stand in the sense of "taste". Indeed, 14,3 states expressly that food, of which *rasa-* (here obviously not "taste", but the ☞ *dhātu-*) is said to be the essence (cf. §7.11), is made up not only of all the "tastes", but also of all the five ☞ *mahābhūta-*s or gross elements,[460] so that it does strain our credulity to postulate that this very essence of food is now supposed to be intimately connected with merely one of those gross elements in accordance with the passages mentioned in §7.13, especially since other, non-liquid substances of the body too are derived from it — substances which are as a rule *not* intimately connected with the gross element water, in contradistinction to liquid substances *per se* (see n.458). One may get over this by postulating that what is meant is that, even though all the other gross elements too are present in it, water predominates in *rasa-* "nutrient fluid" (this is after all a liquid), but would it then still be the essence of food made up of *all* the gross elements? Moreover, what holds good for *rasa-* should hold good for the liquid blood too; why is this not called *āpya-* here? After all, a Vedic theory did connect blood expressly with water.[461] As it is, the theory of the nature of blood given in 14,8f. (and also to be found, I may add, in Āḍhamalla's commentary on ŚĀR 1,6,10,[462] though ŚĀR itself does not have this theory) shows that the connection with *all* the gross elements was felt by some authorities to extend from food to blood too, i.e. to the substance derived from *rasa-*, itself derived from food, which means that the link between food and blood, namely *rasa-*, should logically be — at least according to these author-

---

rasanārtho rasas tasya dravyam āpaḥ kṣitis tathā
"*rasa-* is for tasting; its substance (☞ *dravya-*) is water, further earth."
Cakrapāṇidatta explains this as referring to the fact that earth obtains its *rasa-* from the *rasa-* of water in it (*kṣitis tv apām eva rasena nityānuṣaktena rasavatīty ucyate. yato nityaḥ kṣitijala-sambandhaḥ*) (cf. too SHARMA,P.V. 1985, p.51); see also CART on this passage. Here *dravya-* seems to refer to the gross elements (☞ *mahābhūta-*), probably as concrete manifestations; cf. on this ☞ *dravya-* in Appendix II.

SU,Sū 36,13 (in other editions Sū 36 and Sū 37 are interchanged) tells us that the *rasa-* of water is undeveloped (see §7.13), being, however, developed due to the *rasa-* of earth:
avyaktaḥ kila toyasya raso niścayaniścitaḥ
rasaḥ sa eva cāvyakto vyakto bhūmirasād bhavet.

460 *pāñcabhautikasya ... ṣaḍrasasya* (var: *ṣaḍrasopetasya*) ... *āhārasya ... sāraḥ ... rasaḥ*. Cf. also Sū 46,526 (*pañcabhūtātmake dehe hy āhāraḥ pāñcabhautikaḥ*) and the statement of Aruṇadatta — pertaining to *all* the ☞ *dhātu*-s! — in his commentary on AH,Śā 3,60 quoted in n.506.

461 See e.g. n.690 and n.1031; cf. also n.537.

462 *tad raktaṃ sarvaśarīrasthitaṃ bhavati pāñcabhautikatvāt.*

ities — connected with all the gross elements too. In other words, this as well as the differentiation of *rasa-* "taste" from our *rasa-* "nutrient fluid" in 14,3 would make it appear most likely that the theory of the build-up of the body connected with the elements of the breakdown of food is a theory different from that connecting different parts of the body with different gross elements,[463] from which it would then follow that *āpya-* in 14,4 probably means something else than *āpya-* in the other passages cited.[464] In this con-

---

[463] As regards the theory (on which see also §7.22), formulated by Vāryovida as part of a discussion on the origins of man and his diseases and given in CA,Sū 25,13, to the effect that all living beings originate from *rasa-* and that water is "characterised by *rasa-*", what is meant is unclear:

> *rasajāni tu bhūtāni vyādhayaś ca pṛthagvidhāḥ*
> *āpo hi rasavatyas tāḥ smṛtā nirvṛttihetavaḥ.*

Does Vāryovida perhaps mean that all the tastes (*rasa-*) are contained in water? Note that the number of *rasa-*s is not clear from the verse, as the word appears only in compounded forms. Cakrapāṇidatta explains *nirvṛttihetu-* as a characteristic of *ap-* by saying that this means that water is the origin of man and disease because it is the cause of *rasa-*, which causes these, or that water is the cause of the origination of the *rasa-*s, or that it is the origin of origination (i.e. birth?) (or: growth?) because it is *rasavat-*:

> *rasajānītyādau smṛtā nirvṛttihetava iti vyādhipuruṣayoḥ. etena vyādhipuruṣajanakarasakāraṇatvenāpaḥ kāraṇakāraṇatayā puruṣavikārayoḥ kāraṇaṃ bhavanti. kiṃvāpo nirvṛttihetava iti rasānām. kiṃvā yasmād rasavatya āpas tasmāt tā nirvṛttihetava iti yojanā.*

CART explains *bhūta-* as referring to the gross elements, which are derived from *rasa-*, thus ultimately making man derive from *rasa-* too. *nirvṛtti-* is explained as referring to both diseases and man, through which water is the cause of the origination of man and diseases because it is that whose "employment/action is included in/subordinate to *rasa-*" (a very tentative translation of *avāntaravyāpāra-*; the meaning of *avāntara-* which — if I am right — seems to be presupposed here is usually not noted by older dictionaries):

> *bhūtānītyādi pṛthivyādīni. etena teṣāṃ rasajatvena_tanmelakasya puruṣasyāpi rasajatvam iti bhāvaḥ. nirvṛttihetava iti vyādhipuruṣayor iti śeṣaḥ. etena rasāvāntaravyāpārā āpaḥ puruṣavyādhinirvṛttihetava ity arthaḥ.*

Gaṅgādhara (CA1, p.855) says that water is the cause of the origination of *rasa-* of the nature of the five gross elements, and goes on to say that from the waters the *rasa-*s (i.e. the tastes) originate:

> *hi yasmād āpo rasavatyas tā evāpaḥ pāñcabhautikarasasya nirvṛttihetavaḥ. tat pañcabhūtārabdhe 'dbhya eva madhurādayo nirvartante.*

All these commentaries beg the problematic question of what *rasavat-* actually means, because formally "characterised by *rasa-*" does *not* mean "giving birth to/origin of *rasa-*" (WEZLER 1993A, p.120 unfortunately too does not explain what is meant by "das Wasser ist durch *rasa* charakterisiert"). And then too there is the problem of whether *rasa-* here does in actual fact refer to the tastes, as Gaṅgādhara says explicitly and Cakrapāṇidatta seems to assume (because of the plural *rasānām* in his commentary). See on this §7.22.

[464] I may in this connection be permitted to draw attention to the use of ☞ *dhātu-* to refer

nection it is most interesting to note that SU,Sū 45,6 differentiates between water in general and the gross element water, the former obtaining its *rasa-* "taste" (text: *salilarasa-*) according to which of the gross elements predominates in the ground in which it is situated; thus the gross element water (which is sweet according to some philosophical systems like e.g. Vaiśeṣika) makes water sweet (cf. also MEULENBELD 1987, p.6). A predominance of the ether (*ākāśa-*) gives water "whose taste is undifferentiated" (*avyaktarasa-*), since the ether itself is undifferentiated (*avyaktaṃ hy ākāśam*). AS,Sū 6,p.32b and KAL 5,4;7 are similar. It is thus only consequent that according to DR,Pānīyavarga 3 water falling from heaven (*khāmbu-*) is characterised as *avyaktarasa-*, Śivadāsasena's commentary drawing express attention to SU,Sū 45,6 in this regard.[465] SU,Ka 3,23 says the same of water from heaven (*antarikṣāt*), CA,Sū 27,200 of water from rain, hail and snow, and AS,Sū 6,p.32a and AH,Sū 5,1 of the heavenly water of the Ganges. These different usages of "water" show how problematic this term can be,[466] and should caution a-

---

to the gross elements (☛ *mahābhūta-*). It may be that some terminological confusion has arisen due to the fact that *dhātu-* in medical terminology is most common as a term not for these gross elements, but for the elements of the chain of transformation of food, though the gross elements too are referred to by this term at times in medical texts.

Note too that according to the discussion above wind would be a *dhātu-* not only as a ☛ *doṣa-*, but also as a gross element, which might possibly be a source of confusion. In this connection cf. also CA,Sū 12,5f.: 5 lists certain factors responsible for the excitement (*prakopa-*; cf. on this term also MEULENBELD 1974, p.464) of wind, and then says that this is so because such factors are responsible for the increase (*vṛddhi-*; cf. MEULENBELD 1974, loc.cit.) of the *dhātu-*s; 6 says the same with respect to the abatement (*praśama-*) of wind and the *dhātu-*s. Clearly wind is here one of the *dhātu-*s, as factors pertaining to these are the reason for what happens to it, but what exactly *dhātu-* is is unclear. In Sū 12,13 the three morbific entities are enumerated by name, and one could thus argue justifiedly that they are the *dhātu-*s mentioned in 5f. Cakrapāṇidatta in his commentary on CA,Si 12,41 draws express attention to Sū 12,5, telling us that though wind is primarily meant here the *dhātu-*s "*rasa-* etc." are also meant; this seems to mean that to him wind is not a *dhātu-*, which is problematic:

> *tatra vātasyeti vaktavye yad ayaṃ samānaśabdaṃ dhātūnām iti karoti, tena yathā vāyos tathā rasādīnām api samānaguṇābhyāso vṛddhikāraṇam iti gamyate.*

K.R. Srikanta Murthy (Moorthy) too seems to see a problem here, though what he says on the subject is unclear; in his remarks on TA he says (on p.XIII after the edition of the text) with reference to CA,Sū 12,5 that, though *dhātu-* refers to wind here, it is not used exclusively for wind, "but holds good even in other places", the example for the latter given being *dhātu-vṛddhi-*, which is enigmatic.

[465] Cf. also SU,Sū 45,3: *pānīyam āntarikṣam anirdeśyarasam.*

[466] In CA,Sū 25,40 we find both *udaka-* and *jala-* (both terms for "water") mentioned as substances with specific functions (*udakam āśvāsakarāṇām*, var: *udakam āpyāyanakarāṇām*,

gainst an uncritical explanation of SU,Sū 14,4, for instance linking it with the theory of the gross elements. Such problems may be exemplified by a reference to CA,Sū 26,39f. 26,39 says that heavenly water is not only of undifferentiated taste (*avyaktarasa-*), but also ☞ *saumya-*;[467] *saumya-* "somic" here would appear to be a reference to the classification along the lines of "fiery" (☞ *āgneya-*) and "somic". Cakrapāṇidatta explains *saumya-* as connected with the divinity ☞ *soma-*, i.e. the moon: *saumyāḥ somadevatākāḥ* "*saumya*-s are [those] whose divinity is *soma-*".[468] The text goes on to say that the water, once it has descended from above, combines with the five gross elements (☞ *mahābhūta-*), satiates/refreshens (*abhiprīṇayanti*) all material beings and things, and is transformed into the six tastes (*rasa-*). 26,40 then tells us that the sweet (*madhura-*) taste is due to the prevalence of the quality/qualities (☞ *guṇa-*) of *soma-*. The other tastes are all formed due to combinations of two gross elements, and it is interesting to see that the gross element water too is mentioned in this context, namely in its combination with fire to give the salty (*lavaṇa-*) taste. The word used here is *salila-*, a common term for water.[469] *soma-* of course could be taken to mean "water" too, and Cakrapāṇidatta does indeed explain it thus (as *ap-*),[470] saying that this same

---

and *jalaṃ stambhanīyānām*); Cakrapāṇidatta takes both to be the same substance (the text is reproduced nearly verbatim by CART):

> *tena karmadvaye 'pi jalasyānanyasādhāraṇatāṃ darśayati. militvā guṇapāṭhe hy ubhayakarmakartṛṣv eva prādhānyaṃ syāt*

(see on this also SHARMA,P.V. 1985, p.202). Before that, however, Cakrapāṇidatta says that *jala-* refers to a gross element, obviously as a concrete manifestation (this is reproduced verbatim by CART):

> *dravyāṇīti mahābhūtāni. yathā jalaṃ stambhanānāṃ vāyuḥ prāṇasaṃjñāpradānahetūnām ityādi.*

Indeed, *jala-* is mentioned together with wind (*vāyu-*), fire (*agni-*) and earth (*mṛd-*), which lends credibility to this view, though note that ether, the fifth gross element, is absent here, as in CA,Sū 5,6 (see p.526). *udaka-*, on the other hand, is mentioned after food (*anna-*) and before several foodstuffs and beverages.

[467] It is also "cold": *saumyāḥ khalv āpo 'ntarikṣaprabhavā prakṛtiśītāḥ laghvaś cāvyaktarasāś ca.*

[468] Gaṅgādhara (CA1, p.943) mentions the moon explicitly: *somaś candramās tasyemā iti saumyāḥ*. One could in this regard also point to passages such as SU,Sū 45,59f., where milk milked in the morning is called, among other things, "cold" (*śīta-*) due to the *soma-*-quality (☞ *guṇa-*) of the night. Since right after this (60f.) milk milked in the afternoon is said to be that of animals heated by the sun (*divākarābhitapta-*), one tends to consider seeing a reference to the moon in *soma-* here. But of course it must be stressed that, even if this deduction should be correct, it cannot simply be generalised without further examination.

[469] In 26,41 we find, beside *salila-*, also *udaka-*.

[470] Later commentators have had difficulties with the characterisation of *madhura-*

water also forms the salty taste, but the terminological differentiation, coupled with the use of *saumya-* in 26,39 in a sense obviously *not* referring to the gross element water must make us be wary of unquestioningly following Cakrapāṇidatta, for we have to ask whether this text might not perhaps be a conscious attempt (or continue a tradition derived from such an attempt) to harmonise two ancient and divergent theories, one referring to the pair "fiery" and "somic" (39), the other to the gross elements (40). This is not to say that *soma-* cannot refer to the gross element water here, but it could also refer to something else as yet unclear, so that we cannot simply postulate that it does refer to the gross element water without examining the other implications of its use; cf. in this context also the statement that water and earth are *saumya-* in n.470, where *saumya-* can hardly be connected with the gross element water as *soma-*.[471]

---

seeming to be due only to a prevalence of water, and have brought in the gross element earth too; cf. CART, the commentary of Gaṅgādhara (CA1, p.943f) and the editor's note in the edition of CA, loc.cit., from which we gather that there is even a variant mentioning earth too (*pṛthvīsomaguṇātirekān madhuro rasaḥ*). CART says that *soma-* might refer to the deity water (*jaladevatā-*), but alternatively also to water and earth, both being *saumya-*; the latter explanation is followed by Gaṅgādhara. On the problem of the element(s) in the *madhura-* taste see also MEULENBELD 1987, p.6,n.14.

[471] The "water" in the body is expressly called *soma-* in certain later texts (e.g. BHĀ, YOG) in connection with the disease *somaroga-*, which is said to be due to a loss/diminution (*kṣaya-*) of this "water", and consist of a discharge of clear, pure, cool, smell-less, pain-less and white liquid through the urethra of *women only* (for details see TIVĀRĪ 2, pp.253ff.). In YO 15,328ff., however, the patients described are very clearly (and expressly) *males*. Be that as it may, what is important for us here is the terminology of the disease and its description, which are not very different in YO. The disease *somaroga-* is mentioned also by ŚĀR 1,7,63, but not specified. Āḍhamalla in his commentary gives its symptoms and makes some other remarks on it, but, like ŚĀR, does not tell us whether the disease afflicts only males, only females, or both (TIVĀRĪ 2, p.257 has overlooked this point; but then, the YO passage is unknown to TIVĀRĪ); it is only Kāśirāma who opines that the disease is found only in women. The disease *somaroga-* is also mentioned in ĀĀS 5804ff. (pp.851ff.). 5804f. tell us that the water in the body gets agitated, leaves its places and flows to the bladder and then out; one of the causes for this seems to be given as having too much recourse to women (*strīṇām atiprasaṅgena*), from which it would follows that the disease being described here is one men can have (too?). However, one could also construe *strīṇām* with *āpaḥ sarvaśarīrasthāḥ kṣubhyanti prasravanti hi* in 5805ab, which would then tell us that it is the water in the body of women which gets agitated and flows out due to too much intercourse (*prasaṅga-*); the problem is that expressions referring to intercourse with women as a cause of various disorders are so common in medical texts that one is spontaneously inclined to take this meaning for *strīṇām atiprasaṅga-* here. In 5805c we then have a feminine *tasyās tāḥ* "those (scil. the waters) of her"; but this is in the singular, and could thus also be a later addition. Later, 5809 says that the disease is one only women have, in this context also calling the water of the body *soma-* (the disease is due

7.15. Could we see a similar attempt at synthesis in the use of *āpya-rasa-*, *ap-* and *rasa-* in SU,Sū 14,4-6? Of course *rasa-* here does not, at least at first glance, lend itself to an interpretation similar to that of the CA passage, as it clearly does not refer to taste (i.e. something not a fluid) and thus to something completely different. But a similar development should at least be considered, even if only to ultimately disprove its probability. An attempt at harmonising different theories must, however, have as a precondition the perception of the existence of differing notions. Such perception is indeed to be found in the passage quoted and translated in §7.11, for with regard to the argument brought forward in §7.14 regarding the connection of blood with all the gross elements based on SU,Sū 14,8, we cannot overlook the fact that the wording of 14,8, namely that "other authorities" (*apare ... ācāryāḥ*) hold the view that blood is made up of all the gross elements, shows not only that we have a divergent view here, but also that the author of 14,8 himself most probably did not agree with this view (cf. §7.12), and this is not surprising, since, quite apart from the characterisations of blood in the framework of the system of "fiery" and "somic" classification,[472] it is in conflict with the theory that blood, along with other body fluids, is connected with the gross element water (see §7.13). Now we do not know whether the individual parts of Sū 14,3-9 are homogeneous or heterogeneous in origin. Let us therefore examine what each of the statements causing us difficulties may mean if we for the present do not try to harmonise them with each other. 14,3 tells us that *rasa-* is "somic" (☞ *saumya-*), but this refers to a pair of opposing characteristics of substances ("somic" and "fiery") in a different system of coordinates and as such not connected — at least in the same way — with the theory of the gross elements as set forth in §7.13 regarding the connection of different constituents of the body to different gross elements. True, bile, the "fiery" substance par excellence, is connected with the gross element fire, and the "somic" element par excellence, phlegm,[473] with water

to a loss of this latter). Note too that YO 15,333, part of the description of *somaroga-*, is ascribed to Aurabhra, which is extremely problematic, as the latter seems to be a very ancient authority. Coupled with the problematic description of eight types of *masūrikā-* (this classification is found only in texts which are not early) ascribed to the same authority (here called U-rabhra) by the commentary of Indu on AS,Ut 36,p.269b and the *Kairalī* commentary on AH,Ut 31,8, this raises grave questions as to the bona fides of this ascription.

472 See ☞ *āgneya-* in Appendix II and §7.12.

473 On the connection between *agni-/soma-* and bile/phlegm cf. MEULENBELD 1974, p.516 (add: KĀ,Kh 1,53f.). AH,Śā 3,90 and 3,96 go so far as to *equate* bile and fire (*pittaṃ vahnir vahnijaṃ vā yad asmāt pittodriktas ...*) and phlegm and *soma-* (*śleṣmā somaḥ śleṣmalas tena saumyo ...*) respectively; see also *Harivaṃśa* 30,45.

in CA,Śā 7,16, but in the very same passage bile is — logically, since it is fluid — also connected directly with water, and is thus the only substance in this passage to have two affiliations. This does rather look like an attempt to do justice to two different theoretical considerations.[474] Even more confusingly, AS,Sū 20,p.147b tells us that in bile situated between the receptacles of undigested and digested food (see on this also §7.35), even though it is of the nature of all the five gross elements, the quality/qualities of ☞ *soma-* are diminished/suppressed due to a predominance of fiery energy (☞ *tejas-*);[475] here *soma-* seems to refer to the gross element water. As regards SU,Sū 14,4, one could opine that "watery" refers to the gross element water, in which case this passage would conform to the theory that fluids in the body are intimately connected with this gross element, though it is peculiar that *rasa-* in the passages mentioned in §7.13 clearly refers to *rasa-* "taste", *rasa-* "nutrient fluid" not being mentioned (except by CA,Śā 7,16 and AS,Śā 5,p.316a), though blood and other body fluids are. On the other hand, "watery" may merely be a characterisation of the state of *rasa-*, i.e. its fluidity, and nothing else. Even more problematic is "water" in 14,5. Could this refer to the gross element? Then this "water" and *rasa-* can hardly be synonyms, as *rasa-* would, if it indeed be connected intimately with this gross element, merely be subsumed under this term. Or is "water" something else? One could think of, e.g., the fluid part of ingested food (separated out through digestion),[476] but something quite different may be meant.[477] In any case, the use of "water" here raises the question of whether there might not have been a theory in which "water" occupied the place of *rasa-* in the theory of the breakdown of food. Then *rasa-* and "water" may ultimately mean the same, i.e. be true synonyms,[478] but it could also be that, though they occupy the same place in different theories, they are not quite identical. That 14,6, commenting on the origin of the procreatory-menstrual fluid, speaks of its originating from "*rasa-* itself" (*rasād eva*), thus creating the impression that the same substance is meant as in 14,5 ("water"), is irrelevant as long as we do not know

---

[474] Cf. on this p.525 in Appendix II.

[475] *tatra yad āmāśayapakvāśayamadhyastham pañcamahābhūtātmakatve 'pi tejoguṇotkarṣāt kṣapitasomaguṇam ... .*

[476] Cf. e.g. AS,Sū 19,p.138b, which tells us that urine carries off *annakleda-* "the moisture of food", as well as Ḍalhaṇa's commentary on SU,Sū 15,5,2 to the effect that urine is the moisture of food. See also Excursus VIII on pp.213-215 below.

[477] Cf. on this also §7.43.

[478] Cf. in this connection also *rasodbhava-* as a kenning for the pearl, in which *rasa-* seems to refer to water (see LÜDERS, p.179).

whether 14,5 and 14,6 are homogeneous in origin. But I must also draw attention to the discussion[479] by several ancient authorities on the actual number of *rasa*-s (the term seemingly not always used in the same sense) to be found in CA,Sū 26,8f. (see also Cakrapāṇidatta's commentary and CART), and which is resolved by Ātreya Punarvasu stating that there are six *rasa*-s (the tastes mentioned in §7.13), all originating from water. Most interesting to us is the opening statement of Bhadrakāpya, who states that there is only one *rasa*-, perceivable by means of the tongue (*jihvāvaiṣayika*-), and identical with water (*udakād ananyaḥ*). This seems to be a very ancient theory which had fallen into disrepute at the time of the composition of CA.[480] This expressly stated identicalness of *rasa*- and water is highly intriguing, but unfortunately *rasa*- does not, since it is expressly said to be tasteable, seem to refer to the *rasa*- we are here concerned with, but is rather part of the theory of tastes, so that in the end our difficulties remain. To complicate matters further, GARBH 2 tells us that blood, flesh etc. derive from *rasa*- which is qualified as *ṣaḍvidha*- "six-fold"; such a qualification must of course make us think at once of *rasa*- "taste". Even though GARBH is not a true medical text, we cannot therefore neglect the possibility of the different meanings of *rasa*- having been mixed up in medical texts too, so that, in spite of the fact that *rasa*- in SU,Sū 14,4 on the face of it seems not to refer to "taste" (cf. also n.959), the possibility cannot be dismissed out of hand. In this regard KĀ,Kh 9,43 is highly intriguing, for it speaks of a plural number of *rasa*-s being the cause of the increase of the ☞ *dhātu*-s,[481] and from the following verses it is clear that these *rasa*-s are the tastes.[482]

7.16. Some of these difficulties have been seen by Ḍalhaṇa and Cakra-

---

479 On the nature of such a discussion (*kathā*-) in general see WEZLER 1993A, especially pp.118ff.

480 BH,Sū 12,2ff., where similar theories are discussed, in 12,2 only says that some maintain that there is only one *rasa*-, without mentioning the characterisations of CA. As it is, the theory of one *rasa*-, also characterised as *avyakta*- (cf. §7.13), is complicated further by a passage in a very ancient medical text found in Turkestan, in which ten *rasa*-s are mentioned, one of which is *avyakta*-, and one *vyakta*-. Cf. on this LÜDERS, pp.587ff.

481 The meaning of this term being unclear; cf. on this Appendix II.

482 KĀ,Kh 9,43-45ab:
dohadaṃ yat karotīti śṛṇu tatrāpi kāraṇam
ya eva hi rasāḥ prāyo dhātūnāṃ vṛddhihetavaḥ (43)
teṣām evābhilāṣaḥ syād yonisādharmyatattvataḥ
vātapittānvitaṃ raktaṃ cīyamānaṃ vikāravat (44)
kaṭvamlalavaṇādīnāṃ rasānāṃ gṛddhim āvahet (45ab).
On *dohada*- cf. also n.1594.

pāṇidatta too. Now SU,Sū 15,8 tells us that a loss/diminution (*kṣaya-*[483]) of the three morbific entities in the body is to be combatted by "[substances] increasing the[ir] own matrix" or (alternative translation) "[substances] increasing that whose matrix is the[ir] own [matrix] (i.e. which has a similar matrix)" (*svayonivardhanāni*); similarly SU,Sū 15,10 says that the *kṣaya-* of the seven ☞ *dhātu-*s *rasa-* to semen is to be combatted by "the employment of substances increasing the[ir] own matrix" or "the employment of substances increasing that whose matrix is the[ir] own" (*svayonivardhanadravyopayogaḥ*).[484] What this appears to mean is that a substance will increase another substance with the same matrix; the two substances may be identical, or else they may have the same properties, deriving from their mutual origin.[485] (See on this also §7.19.) In his commentary on SU,Sū 15,8, which incorporates a commentary on Sū 15,9f. too (SU3, p.263, SU4, p.114), Cakrapāṇidatta says on the matrices of the *dhātu-*s: "Because of/Through belonging to/consisting of the five gross elements, *rasa-* is watery, blood belongs to/consists of fiery energy (☞ *tejas-*) (here obviously = "fire"; cf. too §7.19) and water, flesh is earthy (i.e. belongs to earth), ...",[486] without explaining this further. The first part of the statement (i.e. that there is a connection with the gross elements) obviously applies to all the *dhātu-*s (and their impurities, ☞ *mala-*s), as Cakrapāṇidatta refers to these in the line previous to that in n.486.[487] Why, however, does it follow from this that only *rasa-* is "watery"? If Cakrapāṇidatta is referring to the theory of the affiliation of fluids of the body mentioned in §7.13, then should not blood too be "watery"? Of course this theory also leaves scope for combinations of the gross elements, and so saying that blood is connected with water and fire (which does away nicely with the problem concerning blood mentioned in §7.14) could be taken to mean that blood is basically connected with water, but also has a strong percentage of fire. However, it could also be that Cakrapāṇidatta is using a trick here. Sū 14,5 has told us that blood is formed through *tejas-*. Now this probably presupposes a connection with bile (cf. n.420), irrespective of whether bile and the "fire" in the body are the same substance or linked (cf. ☞ *agni-*), and

---

[483] On this cf. §§5.2f.

[484] The theory is summarised again in Sū 15,29.

[485] Since I lack specialised knowledge in Indian philosphy, I am not in a position to comment on the examination of this passage by COMBA 1990 to find out whether it has connections with Vaiśeṣika concepts.

[486] *tatra pañcabhūtātmakatvena rasa āpyo raktaṃ tejojalātmakaṃ māṃsaṃ pārthivaṃ ...* .

[487] *evaṃ dhātumalakṣaye 'pi svayonivardhanaṃ vyākhyeyam.*

whether *tejas-* is "fiery energy" (i.e. maybe a part of "fire") or a synonym of "fire". But regardless of what SU presupposes, Cakrapāṇidatta in his commentary on 14,4 (SU3: 14,5) expressly mentions bile in this connection.[488] But bile is also characterised as "fiery" (☞ *āgneya-*), as a rule, however, in the system of coordinates classifying according to "somic" and "fiery" (see §7.15).[489] Could Cakrapāṇidatta have used this to build a bridge, in his commentary on 15,8ff., to the theory of the gross elements, having in mind also the wide-spread theory that held bile to be a ☞ *mala-* of blood, or otherwise closely connected with blood (for instance as its homologue),[490] which itself was held to be "fiery" or both "somic" and "fiery" (see §7.12)? In this connection it is very interesting to note that in the same passage Cakrapāṇidatta gives the affiliation of urine to water and fire by using a word for "fire" (and not "fiery energy"), namely *anala-* (*mūtraṃ jalānalātmakaṃ*), i.e. using the terminology common with respect to the gross elements, but refers to the procreatory-menstrual fluid by the same word used in Sū 14,7, namely "fiery" (Cakrapāṇidatta: *ārtavam āgneyaṃ*); but in 14,7 we definitely have a reference to the system of "somic" and "fiery" and (at least primarily) not to that of the gross elements. Thus the possibility that Cakrapāṇidatta is mixing up two separate sets of classification does not seem to be any help as regards the problem of *rasa-* that is our concern here; indeed, it only serves to add to the confusion.

7.17. One could, however, point to CA,Ci 15,28-33, where the transformation of food is brought into relation with the influence of the gross elements. In answer to the question of a disciple on how the transformation of each of the seven ☞ *dhātu-*s of the chain of metabolic change into the next *dhātu-* actually occurs (15,22-27; cf. also ĀY 4,56-62), the teacher explains in these verses how the gross elements produce changes in the elements of the chain of transformation to form the following elements.[491] But here the

---

[488] *yakṛtplīhānāv ity anena tatrastharañjakapittasambandhaṃ* (thus SU3; SU4: *tatra rañja°*) *darśayati*.

[489] Cf. in this regard also *taijasa-* in 14,3 (referred to already in n.423), which is, however, clearly not to be connected with *tejas-* in 14,5.

[490] Cf. also ☞ *raktapitta-* in Appendix II.

[491] The full text of 28-35 runs.

tejo rasānāṃ sarveṣāṃ manujānāṃ yad ucyate
pittoṣmaṇaḥ sa rāgeṇa raso raktatvam ṛcchati (28)
vāyvambutejasā raktam ūṣmaṇā cābhisaṃyutam
sthiratāṃ prāpya māṃsaṃ syāt svoṣmaṇā pakvam eva tat

(variant of this verse:
śoṇitaṃ svāgninā pakvaṃ vāyunā ca ghanīkṛtam

gross elements act as agents bringing about change, and the text does not derive any of the elements of the chain of alimentary metabolism from them. Thus this passage (which seems to voice a variant theory on the evolution of the *dhātu*-s anyway) can not be brought into relation with Cakrapāṇidatta's

---

tad eva māṃsaṃ jānīyāt sthiraṃ bhavati dehinām) (29)
svatejo'mbuguṇasnigdhodriktaṃ medo 'bhijāyate
pṛthivyagnyanilādīnāṃ saṅghātaḥ svoṣmaṇā kṛtaḥ (var: *saṅghātaḥ śleṣmaṇāvṛtaḥ*) (30)
kharatvaṃ prakaroty asya jāyate 'sthi tato nṛṇām
karoti tatra sauṣiryam asthnāṃ madhye samīraṇaḥ (31)
medasas tāni pūryante sneho majjā tataḥ smṛtaḥ
tasmān majjñas tu yaḥ snehaḥ śukraṃ sañjāyate tataḥ (32)
vāyvākāśādibhir bhāvaiḥ sauṣiryaṃ jāyate 'sthiṣu
tena sravati tac chukraṃ navāt kumbhād ivodakam (33)
srotobhiḥ syandate dehāt samantāc chukravāhibhiḥ (var: *sametaṃ śukravāhi°*)
harṣeṇodīritaṃ vegāt saṅkalpāc ca manobhavāt (34)
vilīnaṃ ghṛtavyd vyāyāmoṣmaṇā sthānavicyutam
bastau sambhṛtya niryāti sthalān nimnād ivodakam (35)

"That *rasa*- which is called the fiery energy of all the *rasa*-s (the tastes?) of humans attains the state of [being] blood through the colour(ing) of the heat of bile. And [when] blood is joined with heat through the fiery energy of wind and water, it is, attaining firmness, flesh (cf. VṚK, pp.508f. on the optative). Cooked indeed through its own heat, that becomes fat, (var: Blood is cooked through its own fire and made dense by wind. That indeed one should know as flesh. [The flesh] of corporeal beings is firm. Fat is produced,) oleaginous and augmented through its own fiery energy and the qualities of water. The combination of earth, fire, wind etc., effected by its own fiery energy (var: enveloped by phlegm), produces its hardness; from that the bone of men is produced. Thereupon the wind effects porosity within the bones. They are filled from/by fat; the oleaginous substance is then known as marrow (☞ *majjan*-). Semen, however, comes forth from the oleaginous substance that is produced from that marrow. Through the substances (*bhāva*-) wind, ether etc. porosity is produced in the bones. Through that/Because of that/Thus that semen flows like water from a new pitcher. Agitated/Made to move through [sexual] pleasure due to excitement, due to resolve/desire and due to love, it streams from the entire body (var: streams, come together from the body) through ☞ *srotas*-s conveying semen. Molten like clarified butter through the heat of exertion, set free from [its] place, it exits, after having collected in the bladder, like water from a low place (cf. on this §§6.13f.)."

sthalān nimnād "from a low place" is problematic; is a sloping surface meant? CA1, p.2926 and CA3, p.361 read *sthalān nimnam*, i.e. "downwards from a [high] place". CA6 has *sthalaṃ nimnaṃ* "towards a low place". Another possibility would be to read *sthalān nimnān* "towards low places".

statements on the characteristics of different constituents of the body as regards their affiliation with the gross elements. Hence, even though here the gross elements are mentioned in connection with the elements in the chain of the transformation of food, this passage is of no real relevance for the above. Moreover, CA,Ci 15,22-35 are not even commented upon by Cakrapāṇidatta (cf. n.420), which does create the impression that these verses are actually not original to CA,[492] though if ĀY (loc.cit.) is correct in ascribing them to the ancient authority Aupadhenava, then they could be quite ancient nevertheless (which would even be in keeping with their formulation of a theory on the evolution of the *dhātu*-s seemingly at variance with the one later generally accepted). Note also the plural and the singular of *rasa*- in 28a and 28d respectively; it looks as if the first refers to the tastes, the latter to *rasa*- as we are here considering it. This reminds us of the terminological confusion with regard to this term I have drawn attention to in §§7.13-15. In this regard attention may be drawn to the commentaries of Aruṇadatta and Śrīdāsa on AH,Śā 3,62f., where the passage discussed above is also cited and attributed to CA, but with so many differences that it seems that they quote another version of it (cf. also n.420); in the line corresponding to 15,28 of the version of CA Aruṇadatta has *rasānāṃ sarveṣām ambujānāṃ* "of all the *rasa*-s originating from water" for CA's *rasānāṃ sarveṣāṃ manujānāṃ* "of all the *rasa*-s of humans ", whereas Śrīdāsa has *rasānāṃ sarveṣām annajānāṃ* "of all the *rasa*-s originating from food". In the light of our discussion above, these readings do seem to speak for the identification of these *rasa*-s as the tastes.

7.18. Ḍalhaṇa's commentary on SU,Sū 15,10 too is interesting. Before discussing it, I must, however, remark on the puzzle regarding digestion as described in Indian medical texts. The 'standard' medical theory in the 'classical' works seems to be that food is first broken down by the central, abdominal fire. But there are moreover also five digestive "fires" (☞ *agni*-) or "heats" (*ūṣman*-) in the body, each connected with one of the gross elements (☞ *mahābhūta*-, also simply *bhūta*-), respectively transforming the constituent of ingested food connected with that particular element, and these transformed constituents (earthy, watery etc.) then replenish the corresponding constituents of the body respectively (earthy, watery etc.). And then there are also seven "fires" or "heats" responsible for the transformation of each of the

[492] Gaṅgādhara in his commentary too hints at this when he says that the reading is that of "some" (*atra kecid paṭhanti*) (CA1, p.2924), whilst a footnote on the same page, probably stemming from the editors, is even more explicit in that it states that it is to be debated whether this reading has been accepted by Cakrapāṇidatta or not (*ataḥ pāṭho 'yaṃ cakrapāṇisammato na veti cintanīyam*).

elements of the chain of breakdown of food (☞ *rasa-*, blood etc.). Nowhere, however, do we find a clear description of how these processes are actually supposed to correlate to one another. Though RANADE/PARAMJAPE maintain (without adducing any evidence) that "The action of bhūtāgni takes place probably after the action of jaṭharāgni and dhātwagni on food" (p.92), one does gain the impression that, when the breakdown according to the gross elements is mentioned at all in the context of that according to the seven elements, which is mostly not the case, the transformation due to the five fires is held to take place before that of the seven fires[493] — and CA,Ci 15,14f., AS,Śā 6,pp.333f. and AH,Śā 3,60f. actually seem to say so.[494] But it is then rather difficult to understand how food which is broken down into its five constituent parts can also not only have an essence or essential part[495] which pertains to it as a whole (would we not rather have expected five different essences or essential parts?), but which moreover is then also further broken down into different elemental parts which are themselves linked with different (and as a rule more than one) gross elements.[496] Are all the five constituents broken down at first reunited prior to their sub-

---

[493] Cf. e.g. CONCEPT, pp.27f., EIM, pp.58f. DASH 1971, p.74 voices a similar opinion.

[494] At least the order of mention of the various forms of digestion seem to point to such an order of the forms of digestion being presupposed. One could, however, also argue that the order is irrelevant, what is meant simply being two different processes, without any further information on how these correlate. Thus we have no definitive proof in this regard.

[495] Which we have also found mentioned in SU,Sū 14,3 (see §7.11).

[496] DASH 1971, p.74 tries to get over this difficulty thus: "The foregoing description of *bhūtāgnipāka* resembles the description of auto digestion, comparable to anaerolic [sic] reactions. This step would seem to be necessary, as the food consumed are [sic] foreign to the body i.e. *vijātīya* and unless they are suitably processed they may not be converted as organism-specific i.e. *sajātīya* substances. This can be illustrated with the example of starches, fats and proteins of the food which by the process of digestion are rendered fit to be re-synthesised as organism specific carbohydrate, fat and proteins." The substances mentioned, we are told, are broken down, transformed, and then *rebuilt* (e.g. fats are broken down to fatty acids and glycerols and then rebuilt to lipids). When thus p.73 says that the digestion due to the central fire (*jāṭharāgni-*) "renders food substances fit for being assimilated into and built up as parts of corresponding *bhūta* class of substances present in the *dhātus* after the same has been subjected to the action of *dhātvagnis*", what is obviously meant is that the gross elements thus transformed then serve to be transformed into the "elements", seemingly not as separate entities (glycogen alone, lipids alone etc.), but in varying combinations. This is a most ingenious theory to explain away difficulties the author obviously perceived, but is unfortunately not substantiated by statements of the texts themselves (though clearly that will not deter apologists from 'finding' such statements). (On the nature of DASH 1971 cf. also p.585 in the discussion on ☞ *srotas-* in Appendix II.)

sequent transformation? Or are the different gross elements even reunited in various ways by the "fires" of the seven elements of the chain of breakdown of food to give different elements of this chain?[497] The texts do not say anything on the lines of either of these notions.[498] Moreover, CA,Ci 15,7f., AH,Sū 12,10f. and AS,Sū 20,p.147b tell us that it is the central fire which forms the essential part of food (together with the corresponding impurity), no other fires being mentioned as part of the process.[499] Since digestion by the central fire seems in any case to be the first stage of digestion, does therefore the essence or essential part of food, serving as the basis for digestion by the seven fires of the elements of the chain of breakdown of food and formed prior to the digestion through the five fires of the gross elements, somehow retain its character even after the digestion by the five fires? Or are we to deduce that digestion according to the gross elements follows upon that of the seven elements? In both cases there then seems to be no point in digestion according to the gross elements; moreover, the latter alternative would be in conflict with the statements of CA,Ci 15,14f., AH,Śā 3,60f. and AS,Śā 6,pp.333f. seemingly to the contrary.[500] Moreover, it is also hard to imagine a process of digestion in which the breakdown of food proceeds parallelly on each of the lines described, since that would mean that part of the food is treated differently from the rest. And what indeed is the actual relationship of the process of digestion by the abdominal fire to that by other fires, since even the few examples given above have thrown up problems on the demarcations?[501] Also, the texts themselves, when they refer to digestion in passages not primarily dealing with its theoretical foundations, seem to mean as a rule digestion in which the series of seven "ele-

---

[497] Cf. also n.496.

[498] One could opine with CONCEPT, p.28 that "Really every cell of the body, representative of tissue, is **Pāncabhautic** in nature but when it is stated that a particular constituent of the body is made of a particular **Bhūta**, it is only to indicate the predominance", but that is no help, as food as such would still be composed of all the gross elements and digested accordingly, part by part.

[499] See also Excursus VIII on pp.213-215 below and the parallels mentioned in §7.27, as well as Cakrapāṇidatta's commentary on CA,Ci 15,13 quoted in n.509.

[500] As regards the essential part (*sāra-*) of food, according to these AS and AH passages this seems to be formed (together with the corresponding impurity) between the digestions due to the fires of the gross elements and the fires of the seven elements of the chain of breakdown of food, whereas in CA it is according to Ci 15,7f. formed by the *central* fire (see above).

[501] Cf. too the data of EIM, pp.108f. Modern apologetic explanations of the function of the *jāṭharāgni-* are found e.g. in DASH 1971, pp.66ff. and CONCEPT, pp.6ff.

ments" (☞ *dhātu-*) is formed from food, without any mention of a breakdown according to the gross elements. This too is hard to reconcile with the theory of replenishment by the products of the transformation through the first five fires. True, CA,Ci 15,28-33 (which could be spurious) does bring the gross elements into the process of digestion forming the elements *rasa-* etc., but there the gross elements are something acting on the process from outside it (see §7.17), i.e. we would, if we chose to see it in the light of the splitting up of food into the gross elements, have to say that the gross elements formed from food subsequently help in transforming the products of this very same food formed quite independently of them, which is extremely problematic. One could also opine that, since the texts tell us that the fires of the gross elements set free the "qualities" (☞ *guṇa-*) of the respective gross elements of the food, which then augment the respective "qualities" in the body,[502] the fires of the "elements" subsequently act on that part of the food which is not these "qualities". This, however, raises the very pertinent question of what this part could be, since it is hard to conceive of "qualities" of food or the body existing without something to adhere to, as the commentators expressly say too,[503] and is complicated by the fact that the "elements" (*dhātu-*) of course also have "qualities". So do the fires of the gross elements and of the *dhātu-*s somehow overlap? But if so, how? ZYSK 1988, p.126 rather helplessly states that the five gross (loc.cit.: "basic") elements "cooperated together to produce seven elementary constituents [= *dhātu-*s] from digested food", but that is obviously no great help. Other modern scholars writing on the theory of digestion in Indian medical texts do not offer us a clear picture either, and when they do propound a theory, it seems more often than not preconceived, and, moreover, based on only a sampling of material.[504] The mediaeval commentators too offer no clear picture; indeed, many of the problems referred to above are mirrored in the remarks of Cakrapāṇidatta,[505] and the commentaries of Aruṇadatta and Śrīdāsa too bear ample testimony to the fact that the whole matter is quite problematic.[506] See also Śivadāsasena's

---

[502] Cf. e.g. SU,Sū 46,526, CA,Ci 15,13f., AH,Śā 3,59f., AS,Śā 6,p.333b.

[503] See Excursus VI at the end of this chapter, and n.506.

[504] Some modern Āyurvedic practitioners, obviously trying to bring concepts of modern Western medicine into the picture, have hit upon an ingenious explanation of digestion through the agency of 'micro-organisms' (called ☞ *kṛmi-* in this theory), passing this theory off as ancient. Cf. BODE, p.154,n.32.

[505] Since the discussion of this commentary is too lengthy for a footnote, it has been shifted to Excursus VI on pp.207—210 at the end of this chapter.

[506] In his commentary on AH,Śā 3,60 = CA,Ci 15,14 Aruṇadatta says that the ☞ *guṇa-*s

(which appear in this verse as well as in AH,Śā 3,59 = CA,Ci 15,13), referred to as being "cooked", are "cooked" together with their receptacles, the gross elements (☛ *mahābhūta-*), as a *guṇa-* cannot exist without a receptacle:

> *guṇānāṃ ca nirādhārāṇām asambhavāt sahādhāreṇa pṛthivyādinā mahābhūtena saha pakvā ete mahābhūtaguṇās tathāvidhān eva mahābhutaguṇān puṣṇantīti sāmarthyalabhyo 'rthaḥ.*

(See on this also the statement of Ḍalhaṇa in n.514.) Then he says that, since the ☛ *dhātu*-s too are made up of the gross elements, they too would be "cooked" again by the five fires of the gross elements when their turn came, giving (together with the abdominal fire) eighteen fires in all:

> *nanu pārthivādyūṣmabhiḥ pakvasya punar dhātūṣmabhiḥ pākakāle dhātūnām api pāñcabhautikatvāt tatrāpi pārthivādyūṣmabhāvaḥ. taiś ca pārthivādyūṣmabhiḥ punaḥ pāka ity evam aṣṭādaśa prāpnuvanti.*

(From this it also follows that Aruṇadatta postulates a consecutive, and not simultaneous, action of the two sets of fires, as is also evident from AH,Śā 3,61f. and Aruṇadatta's commentary on these). But there are in actual fact not eighteen fires, as the five fires of the gross elements are transposed and are then designated as the heats of the elements, just as water in the body is, when it is transposed, called serous fluid (☛ *lasīkā-*) etc. (cf. on this comparison §§10.10ff.):

> *ta eva pañcoṣmāṇaḥ pārthivādayaḥ sthānāntaraprāptā dhātūṣmāṇa iti vyapadeśam āsādayanti. yathodakaṃ sthānāntaragataṃ lasīkādivyapadeśaṃ labhate.*

Unfortunately, Aruṇadatta does not tell us how this transposition tallies with the fact that, of the total number of thirteen fires he himself also mentions, there are seven fires of the *dhātu*-s, but only five of the gross elements (the remaining one is the central fire).

He then quotes "others" to the effect that there are thirteen fires, further seven hundred fires in seven hundred vessels (☛ *sirā-*, whose total number the medical texts usually place at seven hundred; cf. MEULENBELD 1974, p.514), five hundred in five hundred muscles (☛ *peśī-*, whose number is five hundred only according to some authorities; cf. MEULENBELD 1974, p.477) of flesh:

> *anye tv evam āhuḥ. ete trayodaśāgnayaḥ. tathā saptasu sirāśateṣu saptāgniśatāni pañcasu māṃsapeśīśateṣu ca pañcāgniśatāni.*

Then he says that thus ☛ *rasa-*, the essence of food composed of the five gross elements, is produced, and that through the *guṇa*-s of the gross elements in it (scil. the *rasa-*) the *guṇa*-s in the body, which depend on this *rasa-*, are in turn augmented:

> *evaṃ pañcamahābhūtātmakād āhārād yathāsvam agninā pakvād rasa āhārasāraḥ. tatrasthair mahābhūtaguṇair dehagā mahābhūtaguṇā rasadhātvāśritā yathāsvam abhivardhante.*

This does not seem to be in agreement with the text, according to which the augmentation of the *guṇa*-s depends on the products of the breakdown of food through the individual products of the five fires of the gross elements (cf. 3,59f.). Nevertheless, this attempt at an explanation shows that Aruṇadatta does seem to have seen a problem, it being in truth difficult to understand how *rasa-* (or any other of the ☛ *dhātu*-s) could be something left over after the *guṇas*-s have been distilled out by the five fires, since the *dhātu*-s too of course contain *guṇa*-s. In his commentary on 3,62 (see n.668) Aruṇadatta tells us that *rasa-* is transformed to blood not only by its own fire (situated in the receptacle of the *dhātu- rasa-*: *rasadhātvāśayasthena*

commentary on DR,Āh 20 (= CA,Ci 15,15). All this does create the impression that we actually have two different theories on digestion, one saying that food splits up into its constituents according to the gross elements of which it is composed, the other opining that the transformation is of the sort presupposed by the theory of the chain composed of *rasa-*, blood, flesh etc., and that in our 'classical' texts the two theories are merely placed side by side without really enmeshing.[507] It is in this regard highly intriguing that, as far

---

vahninā), but with the help of the fires of the gross elements; similarly blood [etc.]. And in his commentary on 3,70 he tells us that the essence (*sāra-*) (seemingly = *rasa-*) and waste matter (☞ *kiṭṭa-*) of food are formed by the abdominal fire (*annaṃ sārakiṭṭatayā yat pariṇamati tad etaj* — var *eva* — *jāṭharāgneḥ karma*), which of course makes us wonder why the fires of the *dhātu-*s cannot take over from here without the fires of the gross elements entering the scene, and also clashes with AH,Śā 3,61f., according to which *sāra-* and ☞ *kiṭṭa-* originate only after the "cooking" of the fires of the gross elements (3,59f.). (Could all this also have been influenced by the equation of *guṇa-* and *dhātu-* discussed on p.209 below?)

Aruṇadatta's commentary is quoted by Śrīdāsa in his commentary on AH1,Śā 3,60f., 62 (wrongly printed as 61) and 70, who then says in his commentary on 3,62f. that if the fires of the gross elements cook all the food and the abdominal fire (acting before them) does so too, one could say that there is no need to postulate the fires of the gross elements, from which it follows that the abdominal fire forms the basis for the other fires, ruling over them like a king, on which he then quotes a *sauśruta-* verse, i.e. verse of (the school of?) Suśruta, not found in SU (this is also quoted, in prose form and without an ascription, in Cakrapāṇidatta's commentary on CA,Ci 15,13 discussed in Excursus VI at the end of this chapter):

> *yathā jāṭhareṇaivāgnināhārasya pāka ārabdhe pratiniyataphalābhinirvṛttiśaktitvāt tadāśrayāṇām eva bhṛtyasthānīyānāṃ pañcānāṃ bhūtāgninām pākaḥ. nanu ca yady ete bhūtāgnayaḥ pañcāhāradravyāṇi pacanti na ca tadvyatirekeṇāpara āhāro 'pi vidyate yaj jāṭharaḥ pacati. atha samudāyaṃ pacati jāṭhara iti bhūtāgnikalpanā vyarthā. atrocyate. samudāyāvayavapākasampādakatvāt sa mūlo jāṭharo rājavat sarveṣām anugrāhaka ity uktam. tathā ca sauśrutaṃ vacaḥ.*
>
> *jāṭhareṇāgninā pūrvaṃ kṛte saṃhatibhedena*
> *paścād bhūtāgnayaḥ pañca svaṃ svaṃ dravyaṃ pacanti hi.*
>
> *iti.*

Later he says that the fires of the gross elements become the fires of the *dhātu-*s, being called differently not just because of their transposition, but also their different action, just as a person is able to do different actions (i.e. have different occupations) and is accordingly designated differently, in which context he again gives the example of water, which may become sweat and ☞ *lasīkā-*:

> *sthānaviśeṣāt kāryaviśeṣāc ca bhūtāgnaya eva dhātvagnisamākhyāṃ labhante. yathaiko bahūnāṃ karmaṇāṃ karaṇasamartho bhavati. yad yat karma kurute tasya tasya karmaṇaḥ sadṛśaṃ tannāmaviśeṣaṃ prāpnoti pācakādivat. yathā codakasya daśāñjalayaḥ. tad eva romakūpebhyo niṣpatat svedaśabdaṃ labhate. tad eva vraṇaprāptaṃ lasīkāśabdam. tadvad agnayo 'pīty anavadyam.*

This creates the same difficulties as the similar explanation of Aruṇadatta.

[507] Indu in his commentary on AH,Śā 3,62, in which digestion through the seven fires of

as I can see, SU does *not* refer to the seven fires of the elements (*dhātu-*), even though a breakdown of food along the lines of these elements is also referred to in this text (thus also in Sū 14,3ff., the starting point of our discussion in §§7.11ff.); in fact, SU does not even seem to have an explicit reference to the five fires of the gross elements, though Sū 46,526, which says that food is split up into five components in line with the gross elements could (but need not) presuppose these. As it is, the total number of fires in the body seems to have been maintained to be different by different authorities (cf. e.g. DASH 1971, p.28), which does appear to show that we are on the right track in presuming that there might have been different theories on digestion current, but subsequently harmonised. I may in this context also point out that *dhātu-* is a term used not only for the elements in the chain *rasa-*, blood etc., but also for the gross elements (see ☞ *dhātu-* in Appendix II). That there were in any case different theories on digestion current is evinced e.g. by AS,Śā 6,p.332b, where some authorities are quoted as holding the divergent theory that the internal fire (☞ *agni-*) has no connection with bile (☞ *pitta-*),[508] contrary to the theory which has gained ascendancy, as well as by the controversies connected with the concept of *vipāka-* "(product of) digestive alteration; post-digestive taste" (connected especially with changes of *rasa-* "taste"), on which see especially MEULENBELD 1987, pp.7-11 (particularly p.7,n.18), MEULENBELD 1974, pp.499-501, DASH 1971, pp.68f., SHARMA,P.V. 1982, pp.94f. and also Cakrapāṇidatta's commentary on CA,Ci 15,9-11.[509] The disagreements here are on different matters, but show that

---

the *dhātu-*s is described, says that this digestion follows after the digestion through the central fire (*pradhānāgni-*), and makes no mention of the digestion through the five fires of the gross elements, even though this has been described in 3,60, obviously as a stage before the digestion through the seven fires.

[508] Cf. the text given in n.1731.

[509] See also CONCEPT, pp.11ff., DASH 1971, pp.55ff.;66ff., EIM, pp.33;216-218; cf. too COMBA 1990, p.31. EIM seems to have misunderstood the meaning of *vipāka-*, since this is equated on pp.33;216 with *niṣṭhāpāka-* (of which it is actually taken to be the result); the author further says that this is the culminating stage of digestion after the preliminary *prapāka-* or *avasthāpāka-* stage (pp.33;216) (cf. on *prapāka-* also n.631 and below), but just prior to the separation of the elements ☞ *rasa-* etc. (p.216). As it is, the processes of *avasthāpāka-* and *niṣṭhāpāka-* seem not to be linked, but are similar in many respects; according to MEULENBELD 1987, loc.cit. *vipāka-* is associated only with the latter. Whether this indeed be so or not, this shows that the matter deserves more extensive and careful investigation, not apodictic statements of the sort given by EIM or by DASH 1971, pp.55f. (according to the latter, *avasthāpāka-* has two phases, *prapāka-* and *vipāka-*). In this connection we must also differentiate between commentaries and the texts commented upon, something the Indian authors mentioned above usually do not do. But even these commentaries are problematic; thus Cak-

our impression on two different theories on the breaking down of food into its constituent elements maybe having existed is not to be dismissed out of hand, especially when we consider MEULENBELD 1987,p.10: "One of my conjectures is that the theory of *vipāka* is secondary and has been modelled upon that of *avasthāpāka*." See also Excursus VII and Excursus VIII at the end of this chapter; it is even possible that more theories on digestion were in vogue (cf. n.681).[510]

7.19. Further studies[511] may show that this subjective impression that different theories might have been combined is not justified in actual fact, but some such development would help us understand the rather puzzling commentary of Ḍalhaṇa on SU,Sū 15,10, the starting point of our discussion in §7.18. Referring to the theory on the counteraction of the *kṣaya-* of substances as detailed in §7.16 above, Ḍalhaṇa says that what is meant is that either substances of the same sort are to be ingested (blood to increase blood, flesh to increase flesh etc.), or else substances having the same quality (☞ *guṇa-*). The example given for the latter is: blood is replenished by substances (☞ *dravya-*) originating from fiery energy (*taijasa-*), or else having an

---

rapāṇidatta's commentary on CA,Ci 15,9-11 refers to *prapāka-* in a manner which could show that to him it is an equivalent of *avasthāpāka-*, as EIM opines, but also that it is only the first stage of the latter, in accordance with the opinion of DASH 1971 (the CA verses do *not* mention *avasthāpāka-*, but only *prapāka-*). Neither is it clear whether *aṇupāka-*, which Cakrapāṇidatta seems to use to refer to the process he calls *avasthāpāka-*, refers to *prapāka-* too (note that the CA verses commented upon do *not* use the term *aṇupāka-* either; neither is *sthūlapāka-* found in Ci 15,6-8, though Cakrapāṇidatta uses this term in referring to these verses). It is also to be noted that Cakrapāṇidatta remarks expressly on the differences between *avasthāpāka-* and *niṣṭhāpāka-* in his commentary (the latter term too is *not* used by CA here), and seems to imply that they follow upon each other, though this is not quite clear. In his commentary he mentions *vipāka- only* when referring to CA,Sū 26,58, where the term is used, but *not* otherwise, and it is unclear what its relation to the other *pāka-*s mentioned here is, for though he compares it with *avasthāpāka-*, he does not clearly equate or link the two. (Cakrapāṇidatta also refers to *vipāka-* in his commentary on CA,Ci 15,13: *tatra jāṭharāgniḥ sarvān evāhārarasamalavipākān pacati bhautikās tv agnayaḥ svān svān gunāñ janayanti*, but since he says nothing on this and the context also gives no hint, it is unclear what context he has in mind when mentioning this *vipāka-*.)

[510] In this connection we have not even considered Yogic-Tantric theories on digestion of the sort mentioned e.g. in n.1086.

[511] Which will have to start more or less from scratch, as it seems that apart from MÜLLER 1960A, pp.31f., where attention has briefly been drawn to the problem of the two different systems of fires (five and seven by number), the problem has not yet been deemed worthy of in-depth study.

excess of the *guṇa-* of fiery energy (*tejoguṇabhūyiṣṭhadravya-*).[512] It is obvious that *guṇa-* here is not used in the sense of the series of qualities such as heavy and light etc. referred to on p.544, since in such lists ☞ *tejas-* is not listed as a *guṇa-*; i.e. *tejoguṇa-*, if it should mean "*guṇa-* which is *tejas-*", which is debatable, cannot be used in this sense. Is *guṇa-* maybe used for the attributes of the five gross elements (☞ *mahābhūta-*),[513] an impression that is strengthened by a reference to *guṇa-* by Ḍalhaṇa in exactly this sense in his commentary on the relevant passage Sū 46,526?[514] Then *tejoguṇa-* would probably be "*guṇa-* of *tejas-*", in which case *tejas-* would signify the gross element fire, as it does for instance in AS,Sū 18,p.132b (see also §7.16). In any case, Ḍalhaṇa does seem to be seeking a link with the gross element fire. This is confirmed by the passage from an unknown source quoted in his commentary on Sū 15,10,[515] which links each of the alimentary elements with

---

[512] Cf. also the parallels mentioned in Excursus X at the end of this chapter.

[513] Though not necessarily to signify only the attributes mentioned in n.456.

[514] Commenting on *vipakvaḥ* (scil. *āhāraḥ pāñcabhautikaḥ*) *pañcadhā samyag guṇān svān abhivardhayet* ("Fully cooked/digested [food composed of the five gross elements] serves to increase, in five ways, at the same time/in the manner befitting [each], the own qualities [of each gross element]"), he says:

> *guṇān iti guṇaśabdenātra guṇinaḥ pṛthivyādaya ucyante. tena pārthivāḥ pārthivān abhivardhayanti. evaṃ śeṣeṣv api. anye tu guṇān iti gandhādīn. svān iti pañca guṇino dravyāṇi pārthivādīni. etena gandhādīn ātmanaś ca pārthivādīni dravyāṇy abhivardhayed ity arthaḥ*
>
> "'*guṇān*': through the word *guṇa-* here the [substances] earth etc., [which are] characterised by *guṇa-*s, are meant (cf. p.209 in Excursus VI). Through this [explanation it follows that] [substances] of earthy nature increase [substances] of earthy nature. Thus also as regards the rest. But others [say of] '*guṇān*': smell etc. 'Own': the five substances characterised by *guṇa-*s, [namely] [substances] of earthy nature etc. Through this [explanation] the meaning is: 'serves to increase smell etc. and the substances of earthy nature etc. of the individual' (i.e. *guṇān* in this case is not qualified by *svān*, but the latter is an independent substantivised adjective)."

Whether Ḍalhaṇa interprets the passage he comments upon correctly is beside the point here. On this passage attention may also be drawn to *guṇa-* in the meaning "derived matter", referred to in n.522.

[515] The relevant lines are:

> *yady api pañcabhūtānāṃ vācyaḥ pāko dvidhā punaḥ*
> *tathāpy apāṃ pradhānatvād rasaḥ saumyo 'bhidhīyate.*
> *atiriktā guṇā rakte vahner māṃse tu pārthivāḥ*
> *medasy ambubhuvor asthni pṛthivyanilatejasām.*
> *majjñi śukre ca somasya* (var: *saumyasya*) *mūtre 'mbuśikhinor guṇāḥ*
> *bhuvo viśy ārtave tv agneḥ prasvedastanyayor apām.*

the gross elements by giving the *guṇa*-(s) of the latter present in the former, and which links blood with fire, though not water and fire as in Cakrapāṇidatta's commentary in §7.16. Cf. in this connection also the characterisation of blood by Ḍalhaṇa as "fiery" (though in a different system of reference) in §7.12. Noteworthy too in the passage in n.515 is not only the use of *soma-* for what obviously is the gross element water, but also the statement that *rasa-* is called *saumya-* because the gross element water predominates in it (cf. in this regard §7.14). In the light of §7.13 this makes sense, as *rasa-* is after all a fluid in the body; however, we have also seen loc.cit. that it is *rasa-* "taste" which is as a rule connected with the gross element water. Is this therefore an apologetic attempt to connect *rasa-* "nutrient fluid" with water in the same way as *rasa-* "taste"?[516] What is also interesting is *dvidhā* "twofold/in two ways". Could this be a reference to the theory SU,Sū 40,10-12 cite approvingly and refer to as traditional (*āgame hi*) and which is the only one found in SIDDH (1,27), according to which there are two kinds of digestive transformation of taste, namely to sweet (*madhura-*) and pungent (*kaṭu-*)? The former transformation is heavy, the latter light; they are occasioned by the heaviness and lightness (i.e. *guṇa*-s, as SU expressly says) respectively of the gross elements in the substances being digested, heaviness being, as SU tells us, a characteristic of water and earth, lightness of the others.[517] In this theory we find the gross elements linked with digestion, but in a manner quite dif-

---

iti dhātumaleṣūktā guṇāḥ prādhānyataḥ sthitāḥ
prāyeṇa bhūguṇā garbhe stokā hy anyaguṇā api (var: *stokastokādyanugrahaḥ*).
"Even though, moreover, the maturation/digestion (*pāka-* "cooking") of the five gross elements is to be called twofold/explained in two ways, *rasa-* is nevertheless called *saumya-* because of the predominance of water. The predominating (this must be what *atirikta-* means here; 'surplus' or the like would make no sense) *guṇa*-s in blood (*rakta-*) are those of fire, in flesh (☛ *māṃsa-*), however, those belonging to earth, in fat (☛ *medas-*) of water and earth, in bone (*asthi-*) of earth, wind and *tejas-* (here obviously: fire). In marrow (☛ *majjan-*) and semen (*śukra-*) the *guṇa*-s are of *soma-* (obviously: water) (var: of that belonging to *soma-*), in urine of water and fire, of earth in faeces, in *ārtava-*, however, of fire, in sweat and milk of water. Thus the *guṇa*-s in ☛ *dhātu*-s and ☛ *mala*-s are enumerated as settled/established/fixed according to [their] predominance. Generally/Mostly the *guṇa*-s of earth are in the embryo, other *guṇa*-s of course slightly too."

(I am unable to translate the variant *stokastokādyanugrahaḥ* — or is it *stokas tokādyanugrahaḥ*? — adequately.)

516 See in this connection also the discussion in n.959.

517 See also EIM, p.217, MEULENBELD 1987, p.7,n.18, SHARMA,P.V. 1985, pp.227ff., DASH 1971, pp.68f. and Cakrapāṇidatta on CA,Sū 26,63.

ferent from that of §7.18; note too that this theory on digestive alteration is not found in other 'classical' medical texts (cf. *vipāka-* in §7.18). Or does the text in n.515 perhaps refer to the two-fold process of digestion that we have mentioned in §7.18, pertaining once to the gross elements and once to the *dhātu*-s? We have after all seen in the commentaries referred to in §7.18 that the "cooking" relating to the *dhātu*-s was taken to subsume a "cooking" of the gross elements too,[518] which thus would logically be "cooked" twice. On the other hand, CA,Ci 15,19cd = AH,Śā 3,64cd = DR,Āh 24 refers to the breaking up of the individual *dhātu*-s into a pure part (☞ *prasāda-*) and an impurity (☞ *kiṭṭa-*) by saying that they, in these forms, proceed twofold/in two ways from their "cooking",[519] which is a repetition of the same terminology in CA,Ci 15,15 = DR,Āh 20.[520] Could this breaking up be meant by *dvidhā* above? But I do not see how this could be combined with the gross elements.

7.20. All in all, both Cakrapāṇidatta's (§7.16) and Ḍalhaṇa's (§7.19) comments are not only not particularly enlightening as regards our problems relating to *rasa-*, water and watery *rasa-* in SU,Sū 14,4-6 (cf. §7.15), but problematic in themselves too. That they create problems has also been perceived by others, e.g. MITRA, p.56: "Blood (*rakta*) belongs to *Ap dhātu* but the commentators add *tejas* to it ..." (referring to the passages of Cakrapāṇidatta and Ḍalhaṇa just discussed).[521] What makes the matter more difficult

518 Cf. Excursus VI (referred to in n.505), p.210, and n.506.

519 *prasādakiṭṭe* (CAI, p.2922 and AH: *prasādakiṭṭau*) *dhātūnāṃ pākād evaṃ dvidharcchataḥ* (CAI and a variant of CA: *evāvigarhataḥ*). CA is wrongly printed as *evaṃvidharcchataḥ*, even though Cakrapāṇidatta presupposes *evaṃ dvi°* and moreover an instrumental *evaṃvidhā* would be highly archaic and thus most improbable here. DR has two mistakes: *prasādakiṭṭa* for *prasādakiṭṭe*, and *dvidharcutaḥ* for *dvidharcchataḥ* (even the commentary of Śivadāsasena is printed as *ṛcuto bhavata ity arthaḥ*!).

520 The texts of CA and DR are identical:

saptabhir dehadhātāro dhātavo dvividhaṃ punaḥ
yathāsvam agnibhiḥ pākaṃ yānti kiṭṭaprasādavat (CA, var: *°dataḥ*).

AH,Śā 3,61 is different, though not in the relevant respect:

kiṭṭaṃ sāraś ca tat (var: *sāras tathā*) *pakvam annaṃ sambhavati dvidhā*
tatrāccham kiṭṭam annasya mūtraṃ vidyād ghanaṃ śakṛt.

521 But MITRA also creates new terminological problems unnecessarily. Thus pp.293f. explains SU,Śā 2,33 as: "A coordination of the four factors: menstrual blood (*ṛtu*), healt[h]y womb (*kṣetra*), nutrient liquid, i.e. *rasa* of digested food (*ambu*)[,] healthy semen (*bīja*) and proper observance of the rules if [sic] necessary for the conception and development of a healthy child." This statement, a nearly verbatim reproduction of BHISHAGRATNA 2,p.129, is problematic. Actually, what the text states is that four factors combine to form the embryo, just as [in agriculture] the [proper] season (*ṛtu-*), field (*kṣetra-*), water (*ambu-*), and seed (*bīja-*) combine to form a sprout. The time of *rajas-* (not simply *rajas-*), womb, *rasa-*, and *ārtava-*

is the possibility that the discussion in these commentaries, much of it having to do with the term ☞ *guṇa-*, may in fact be concerned, wholly or in part, with this term in a meaning not presupposed in the discussion, namely as perhaps in CA,Śā 6,16f., in which passage we also find ☞ *dravya-* used in the sense of ☞ *dhātu-* to denote the elements *rasa-* to semen.[522] These prob-

---

and semen (*bījaṃ strīpuṃsayor ārtavaśukre*) respectively are the *parallels* to these as regards human conception according to Ḍalhaṇa (*not* SU, which does not enumerate the four factors). (Cf. also p.520 in the discussion on ☞ *abīja-* in Appendix II.) ☞ *rasa-* is here obviously seen in its function as a nourisher of the embryo; we shall see more on this below (e.g. in §§10.5f.).

522 16: *pariṇamatas* (var: *pariṇāmatas*) *tv āhārasya guṇāḥ śarīraguṇabhāvam āpadyante yathāsvam aviruddhāḥ. viruddhāś ca vihanyur vihatāś ca virodhibhiḥ śarīram*. 17: *śarīraguṇāḥ* (var: *śarīradhātavaḥ*) *punar dvividhāḥ saṅgraheṇa malabhūtāḥ prasādabhūtāś ca. tatra malabhūtās te ye śarīrasyābādhakarāḥ syuḥ. tad yathā śarīracchidreṣūpadehāḥ pṛthagjanmāno bahirmukhāḥ paripakvāś* (var in Cakrapāṇidatta's commentary: *bahirmukhā aparipakvāś*) *ca dhātavaḥ prakupitāś ca vātapittaśleṣmāṇaḥ. ye cānye 'pi kecic charīre tiṣṭhato bhāvāḥ śarīrasyopaghātāyopapadyante sarvāṃs tān male* (var: *malākhyān*) *sañcakṣmahe. itarāṃs tu prasāde* (var: *prasādākhyān*). *gurvādīṃś ca dravāntān guṇabhedena. rasādīṃś ca śukrāntān dravyabhedena*

"The *guṇa*-s of food being transformed (var: of food due to transformation/in the end), however, attain the state of (i.e. become) *guṇa*-s of the body (*śarīraguṇa-*), respectively not adverse [with regard to the corresponding *guṇa-* of the body]. And [should they be] adverse, [they] afflict (on the optative cf. VṚK, pp.508f.) [the *guṇa*-s of the body], and the[se] afflicted [*guṇa*-s] together with the contradictory [*guṇa*-s of food] [in their turn afflict] the body. The *guṇa*-s (var: *dhātu*-s) of the body, again, are in sum of two kinds: [those] become ☞ *mala*-s and [those] become ☞ *prasāda*-s. With regard to that, [those] become *mala*-s are those which cause the body trouble. As for instance excretions, of different (scil.: various) origins, directed outwards (i.e. coming to the exterior), in the openings of the body, and fully (scil.: overly?) ripe (var: not fully ripe) *dhātu*-s, and excited wind, bile and phlegm. And whatever other substances abiding in the body occur/are fit for (i.e. occasion) damage to the body, all those too we enumerate under [the heading of] (var: as [substances] called) *mala-*. The others, however, under [the heading of] (var: as [substances] called) *prasāda-*. And those [qualities(?) enumerated in the list of CA,Śā 6,10] beginning with 'heavy' and ending with 'liquid' by means of (i.e. in/as) the division '*guṇa-*' (or: as special *guṇa*-s). And those [substances in the list] beginning with *rasa-* and ending with semen (i.e. the *dhātu*-s of the chain of alimentary transformation) by means of (i.e. in/as) the division '*dravya-*' (or: as special *dravya*-s)."

☞ *dravya-* here is used in a sense very similar to one we shall meet with in §§10.10f. As regards ☞ *guṇa-*, this term is in 6,17 obviously not applied only to qualities, but also to substances. Now if in 6,16 *guṇa-* should denote only a quality, then we would end up with a difficulty inasmuch as *śarīraguṇa-* there would be something different from *śarīraguṇa-* in 6,17.

lems become even greater when we consider that our texts also have a threefold classification based on ☞ *saumya-* ("somic"), ☞ *āgneya-* ("fiery") and *vāyavya-* ("windy", i.e. "of the nature of/related to wind"), e.g. in AS,Sū 23,p.170b and Śā 7,pp.341b-342a and CA,Ni 1,4; the latter passage, which classifies diseases along these lines, makes it seem as if the former two terms stood for "phlegmatic" and "bilious" respectively, and that is exactly how Cakrapāṇidatta explains them,[523] though this need not necessarily apply to the other passages mentioned.[524] However, AS,Ut 44,p.362a expressly links the bites of spiders which are "fiery" (*āgneya-*) with corruptions of bile, of those which are "somic" (*saumya-*) with corruptions of phlegm, and of those which are "windy" (*vāyavya-*) with corruptions of wind. AS,Ut 43,p.351, referring to *kīṭa*-s[525], is similar, whereas the parallel AH,Ut 37,2f. does not specifically combine wind etc. with "windy" etc., and further lists only *vāyavya-* and *āgneya- kīṭa*-s, but none that are *saumya-*, having *kaphādhika-* "having an excess of phlegm" instead; the *Kairalī* commentary, however, says that the latter

---

Nevertheless, Cakrapāṇidatta in his commentary takes *śarīraguṇa-* in 6,16 to refer to qualities, but then seems to shift his explanation to encompass substances, as the example he adduces demonstrates:

> *āhārasya kaṭhino bhāgo māṃsāsthyādikaṭhinabhāgapoṣako bhavati. dravāṃśas tu śoṇitādirūpo bhavatītyādi*
>
> "'The hard part of food is/becomes the nourisher/augmenter of the hard part of/which is flesh, bone etc.; the liquid portion, however, is/becomes (i.e. takes/assumes) the form [of] blood' etc."

This is reminiscent of the equation of *guṇa-* and ☞ *dhātu-* discussed in Excursus VI at the end of this chapter. (In his commentary on 6,17 he says nothing of relevance, even though he clearly reads *śarīraguṇāḥ* and not *śarīradhātavaḥ* as in the variant, the propagator of which obviously saw a difficulty with regard to the term *guṇa-*.) We should in this regard also bear in mind, however, that according to a theory predating that of classical Vaiśeṣika *guṇa-* may refer to derived matter, which includes, if we take the ☞ *mahābhūta*-s as an example, things such as colour, taste, etc., and what are in later times "qualities", such as heavy, light etc., which are thus understood as some sort of derived, but independently existing substance; see in this regard PREISENDANZ 2,p.525 ("Hierbei ist jedoch zu beachten, dass die 'Eigenschaften' in diesem Rahmen noch nicht gegenüber den Elementen als ihren Substraten einen untergeordneten ontologischen Status haben, sondern als Produkte der Elemente in eigenem Recht existieren"). Cakrapāṇidatta's seeming shift in meaning as regards the expression *guṇa-* also bears examination in this light.

523 *āgneyāḥ paittikāḥ saumyāḥ kaphajāḥ*.

524 SU, Sū 21,8 tells us that the three morbific entities uphold the body in the same manner as moon (*soma-*), sun and wind uphold the world (cf. also Cakrapāṇidatta's commentary). This is, however, strictly speaking only a comparison; as such no special link is necessarily implied.

525 On this term see MEULENBELD 1974, p.455.

stands for *saumya-*. In CA,Ci 3,38f. we find the three substances wind (*vāyu-*), fiery energy (☞ *tejas-*) and ☞ *soma-* mentioned, the latter linked with *śīta-* "cold";[526] in 3,82 wind (*vāyu-*), fire (☞ *agni-*) and *soma-* are mentioned. The substances *agni-*, *soma-* and *vāyu-* (clearly in the form of a triad) are also mentioned in SU,Śā 4,3, where they are, along with other substances (*sattva-*, *rajas-* and *tamas-*, the five *indriya-*s, and the *bhūtātman-*) collectively called the *prāṇa-*s;[527] this very same list is also found in BHĀ 1,3,320, where the substances mentioned vivify/nourish (*sañjīvayanti*) the embryo/foetus, but with the addition of earth (*mahī-*) and ether (*nabhas-*), thus making *agni-*, *soma-* (here then obviously "water") and *vāyu-* members of the pentad of gross elements (☞ *mahābhūta-*)! SU,Sū 42,5 links the morbific entity wind with itself, i.e. wind,[528] and calls the morbific entities bile and phlegm "fiery" and "somic" respectively;[529] it is noteworthy that Ḍalhaṇa in his commentary explains "somic" as referring to a derivation from *soma-* without telling us what this means,[530] whereas in his commentary on Śā 4,3 he explains *soma-* as being the presiding deity (*adhidaivatvam āpannaḥ*) of the mind presiding as the power (*śakti-*) of all wet parts of the body and the sense of taste. Cakrapāṇidatta in his commentary glosses *saumya-* with *jalātmaka-* "of the nature of water", which brings us back to the problems referred to in §7.19. On the other hand, the movements of wind, sun (*arka-*) and *soma-* (here probably "moon") of the macrocosm are compared with those of the morbific entities wind, bile and phlegm of the microcosm in CA,Ci 28,246; in this connection cf. also SU,Sū 21,8 (see n.524), CA,Sū 6,5, and Ḍalhaṇa's commentary on SU,Sū 6,14 mentioning the triad of *soma-*, sun and wind, where *soma-* seems to mean the moon.[531] I may also draw attention to CA,Sū 26,43, where the six tastes (*rasa-*) of Sū 26,39f., already discussed in §7.14 as examples of syncretising tendencies, are discussed according to their properties. It is most interesting to note that the functions of the sweet (*madhura-*) taste have many similarities with those of phlegm, those of the sour (*amla-*) taste with those of bile, whilst those of the pungent (*kaṭuka-*), bitter (*tikta-*) and astringent (*kaṣāya-*) tastes have much similarity with

---

[526] Cf. on this ☞ *āgneya-* in Appendix II.

[527] In his commentary on SU,Sū 15,21f. (15,20f. in SU3) Cakrapāṇidatta says that the *prāṇa-*s are *agni-*, *soma-* and *vāyu-*, *sattva-*, the five *indriya-*s, *bhūtātman-* and *manas-*.

[528] *vāyor ātmaivātmā.*

[529] Cf. in this respect also KIRFEL 1951, pp.13f.

[530] *saumya iti somād utpadyate.*

[531] Cf. p.526 of the discussion on ☞ *āgneya-* in Appendix II.

those of wind, the sixth taste, *lavaṇa-* "salty", however, being difficult to classify. Turning back to the classification along the lines of "windy", "fiery" and "somic", we find that SU,Śā 6,16 classes the ☞ *marman*-s along these lines,[532] the latter two expressly linked with fire (*agni-*: Ḍalhaṇa: *tejas-* "fiery energy") and the unexplained *soma-* respectively, which latter is said to be characterised by *śaitya-* "coldness"; in this context MBH 12,16,11 and 14,12,3 are highly interesting, for these passages mention three ☞ *guṇa*-s of the body, namely cold (*śīta-*), hot (*uṣṇa-*) and wind (*vāyu-*).[533] In Śā 6,35 it is said that *soma-*, wind and fiery energy (*tejas-*), together with *rajas-*, *sattva-*, *tamas-* and the *bhūtātman-*,[534] are situated in the *marman*-s. Note too in this context the theory of triadic division along the lines of food (*anna-*), fiery energy (*tejas-*) and water (*ap-*) in the Upaniṣads,[535] and also found mentioned later together with the theory of the three morbific entities (see SHARMA,P.V. 1972, pp.16f.).[536] Such terminological and other similarities of different triadic theories shows how great the chances for terminological and categorial confusion must have been, and all our efforts at systematic reconstruction will be doomed to failure from the start if we do not keep this in mind.[537] As a case in point, note that SU,Śā 6,16, just cited above as

---

[532] Cf. also AS,Śā 7,pp.341b-342a mentioned above.

[533] On the problem of hot and cold, and *āgneya-* and *saumya-* respectively, see the discussion on ☞ *āgneya-* in Appendix II.

[534] Cf. Śā 4,3 mentioned above.

[535] Cf. RUBEN 1971, pp.84ff. (also p.95), HANEFELD, pp.178;180.

[536] On this theory cf. also §7.43.

[537] I would like to draw attention to just a few further ancient theories which could be a possible source of confusion. Thus we find semen (*retas-*) and water intimately linked in *Aitareyopaniṣad* 1,4 and 2,4 (this is a link deriving one from the other, and not the sort of general identification one finds for instance in *Śvetāśvataropaniṣad* 4,2, where the *ātman-* is identified with fire, sun, wind, moon, semen (*śukra-*), *brahman-*, water and *prajāpati-*). Is this connected with the characterisation of semen as "somic" (cf. also §7.19)? See in this connection, however, also SU,Sū 5,28, which asks water to make semen grow/thrive (*retas tv āpyāyayantv āpaḥ*); this is part of a passage (19-32) in which various parts of the body are linked with various deities (both personages and natural substances and phenomena), including ☞ *agni-* and ☞ *soma-* (linked with the tongue and the breath called *vyāna-* respectively). (Water is here, needless to say, in any case clearly not a gross element.) *Bṛhadāraṇyakopaniṣad* 3,2,13 says that when a person dies, his semen (*retas-*) dissolves in water. The passage further says that this happens to blood (in general) (*lohita-*) too; this connection between blood and water reminds us of the problematic discussion on this subject in §7.14. *Aitareyopaniṣad* 4,2 tells us that during intercourse a man pours his semen (*retas-*), assembled as ☞ *tejas-* "fiery energy" from all the limbs, into the woman. This not only seems to presuppose something akin to the

mentioning a triadic division of *marman*-s, says that the pain connected with the "fiery" and "windy" *marman*-s is according to others rather linked with the five gross elements (☞ *mahābhūta*-).[538] And the *Madhukośa* commentary on MĀDH 2,45-47 mentions *saumyāgneyadoṣakoṭidvayamelāpa*- "the union of the two aspects of the morbific entities, (i.e.) their *soma*-like and fiery aspect",[539] where the triadic division of the morbific entities is combined with the division of "somic" and "fiery". In this light it is small wonder that the commentators, who strive to explain their texts as homogeneous and homogenous parts of a single system, are often more a hindrance than a help.[540]

7.21. We shall thus have to look for other relevant information to help us understand SU,Sū 14,4-6 (quoted in §7.11) properly. Unfortunately, what commentators and translators have so far said on it is of very little use in this respect. Thus Ḍalhaṇa states in his commentary on Sū 14,4:

āpyo 'pi raso rāgam upaiti lohitatvaṃ gacchati
"Even though [of] watery [nature], the *rasa*- becomes coloured, [that is:] turns red."

This does not at first glance appear to be a great help, though if, as we have seen above (§7.12 and §7.19), blood be taken to be affiliated with fire, we would at least have an explanation for the characterisation of *khalu* of the text as meaning *api* "even though" (*khaluśabdo 'pyarthaḥ*). "Water" in 5 Ḍalhaṇa explains as being *rasa*-.[541] Cakrapāṇidatta (SU3, p.232, SU4, p.101) says more or less the same as Ḍalhaṇa on 4:

āpyo 'pi rāgam upaiti

---

mode of collection of semen we find in medical texts (see n.168), but also connects semen with a substance which stands, as can be seen in §10.19, in antagonism to water; cf. on *tejas*- also §7.16 and §7.19. The connection of semen with *tejas*- is, however, found in medical texts too; see p.533 in Appendix II (where Cakrapāṇidatta's commentary on CA,Ci 15,3 equating *tejas*- and semen is also mentioned). Does this not seem to contradict the connection between semen and water we have just seen too? On the other hand, we also have the very old notion of fire and/or *tejas*- hidden in water; see on this JACOB, pp.74-77 and *Śvetāśvataropaniṣad* 6,15, as well as OLDENBERG, pp.112ff.

[538] While listing the effects of *marmans*- singly or in pairs (the combination of *saumya*- and *vāyavya*- is, however, absent), it says:

rujākarāṇy agnivāyuguṇabhūyiṣṭhāni. viśeṣataś ca (var: *viśeṣeṇa*) *tau rujākarau. pāñcabhautikiṃ ca rujām āhur eke.*

[539] Translation by MEULENBELD 1974, p.128.

[540] It may also be pointed out that in the light of the discussion above the comments of ZIMMERMANN 1993, pp.322f. on the origins of the *tridoṣa*--doctrine are much too simplistic, since they pay no attention to the complexity of the relevant material.

[541] *āpo 'tra rasaḥ.*

"Even though [of] watery [nature], [it] becomes coloured."
He also explains "water" in 5 as referring to water in the form of *rasa-*.[542] As regards his use of *api* here, we have already seen in §7.16 that Cakrapāṇidatta too connects blood with fire (see §7.16) — and water.[543] (As detailed, in the discussion on ☞ *āgneya-* in Appendix II and in §7.12, he also says that blood is both "fiery" and "somic".) He further remarks that in those cases in which something, though derived from *rasa-*, does not take up the "somic" (☞ *saumya-*) property of ☞ *rasa-* too, the words "water" (*ap-*) and "watery" (*āpya-*) are not used, but merely "*rasa-*".[544] As an example for this he then cites SU,Sū 14,6 which says that the procreatory-menstrual fluid ("fiery" according to 14,7) derives from *rasa-*. But though all this shows that to Cakrapāṇidatta *āpya- rasa-* in 4, *ap-* in 5 and *rasa-* in 6 are one and the same, it does not solve the problems we have been discussing so far. The translations I consulted were no help either. BHISHAGRATNA 1,p.107 substitutes "Saumya" for "watery (*āpya-*) in his translation of 4, and "Rasa" for "water" in that of 5.[545] GHĀṆEKAR 1977A, p.77a substitutes *ann'ras*[546] for "water" in 5 (loc.cit.: 6) too.[547] Both these authors give no explanation for their substitutions; did they not see any problem here? HESSLER 1844-1850 1,p.28 on the other hand has at least seen that the passage is problematic, for he has used "humores" in 5, "chylus" in 6;[548] true, "humores" is not "water"

---

[542] *āpa iti rasarūpā āpaḥ.*

[543] Indeed, in his commentary on 14,5 he expressly remarks on the colour showing that blood becomes affiliated with fire through the transformation, even though it is of the nature of all five gross elements (☞ *bhūta-*):

> *tathā pāñcabhautikatvābhidhāne 'pi śoṇite rāga evāgnyātmakatāgamako 'bhidhātavyaḥ.*

[544] *tathā yatra rasasya kāraṇatve 'pi taddharmasaumyatvānugatir na bhavati tatrāpa iti āpya iti* (SU4: *iti ca*) *bhāṣāṃ parityajya rasasaṃjñaiva* (SU4: *rasasyaiva*) *kāraṇatvaṃ vadati.*

[545] The translation of 4-6 is: "The Rasa, though a Saumya fluid, obtains its characteristic pigment (Rágam) in its passage through the spleen and liver. ... The Rasa or the lymph chyle, coloured through the effect of the healthy normal dyeing heat of the body, obtains the name of blood. The Rasa is transformed in to the catamenial flow in women which commences at the age of twelve and ceases at fifty."

[546] = ☞ *annarasa-*; cf. on this term also §7.37.

[547] *āpya-* in 4 (loc.cit.: 5) is translated as *jal'rūp* "having the form of/being water". The translation of 4-5 runs: *vah jal'rūp ras yakṛt· aur plīhā meṃ prāpt hokar lāl ho jātā hai. manuṣyoṃ ke śarīr meṃ rah'ne vāle viśuddh tej se lāl huā yahī prasād'rūp ann'ras rakt kah'lātā hai.*

[548] The translation of 4-6: "Chylus revera aquosus est; hepar et lien assecutus ruborem adipiscitur. ... Colorati et vi praediti humores in hominum corporibus, non interfecti a puro,

(cf. "aquosus" in 4) and thus requires an explanation (which is lacking), but nevertheless some attempt at differentiation has been made here. HOERNLE 1897, p.90 too has differentiated, for he translates 14,4 as "Now this chyle, though originally a watery fluid, assumes a red colour", but uses "watery fluid" for *ap-* in 14,5 (and "chyle" for *rasa-* in 14,6). However, this too does not appear to help us much.

7.22. We are thus left with the problem drawn attention to in §7.15, which ultimately boils down to the following: do "watery" ☞ *rasa-* in SU,Sū 14,4, "water" in 5 and *rasa-* in 6 denote the same or not? Now both "watery" *rasa-* in 4 and "water" in 5 are said to be coloured and to turn to blood. This does make it seem that actually one and the same fluid is here meant. But even though according to the order of the passages the same fluid seems to be meant, can we be sure that each of these passages, taken by itself, does indeed refer to the same fluid? We might have an attempt at conscious harmonising of different theories here, but it is equally possible that actually all we have here is different theories with different terminologies placed side by side (cf. also §§7.14f.). According to AS,Sū 36,p.251a however, we would have the same fluid in both cases, for this parallel passage states that blood is obtained by the colouring of *rasa-* through fiery energy (☞ *tejas-*), this *rasa-* expressly being said to be both bodily "water" as well as the essence of food (*āhārasāra-*).[549] This need not, however, hold good for SU, especially since it is known that AS is not only younger than SU, but also tries (like AH) to create a synthesis between different texts and theories current at the time of its composition (cf. §9.2). Therefore we cannot rely on AS as regards the nature of "watery" *rasa-* and "water" in SU. Moreover, CA,Ci 22,16 states that the body arises from *rasa-*, and this from water;[550] since a *kṣaya-*[551] of

---

sanguinis vocantur. Ex chylo etiam feminae sanguis, menstrua appellatus, procedit, qui ab anno aetatis decimo prodit, et quinquagesimo cessat."

[549] *tatrāmbu śārīram āhārasārabhūtaṃ rasākhyam* etc. (see n.377) (cf. also Śā 5,p.326: *āhārasārasya rasākhyasya*). Aruṇadatta cites this passage in his commentary on the introductory lines of AH,Sū 27, and *one* ms of the commentary has a variant *āhārarasasārabhūtaṃ* for *āhārasārabhūtaṃ*. This variant does look like a scribal error, but even if it is not, it would not be of relevance for us, as we would merely have here a theory seeing in ☞ *āhārarasa-* (which we shall discuss in detail in §§7.29ff.) a link between food and ☞ *rasa-* and to all purposes a substance different from *rasa-* (cf. also n.668), which is definitely not the theory of SU, as we shall see below (§§7.29ff.).

[550] *deho rasajo 'mbubhavo rasaś ca*; Cakrapāṇidatta glosses *ambubhava-* with *āpya-*, which reminds us of the terminology of SU,Sū 14,4.

[551] On this term see §§5.2f.

this *rasa-* causes thirst,[552] "taste" cannot be meant here, which means that this *rasa-* is some other *rasa-* (maybe even the element *rasa-* of the chain of metabolic transformation?), which could be substantiated by the fact that our texts often stress the vital importance of *rasa-* (whose nature is, however, not always clear) for upholding the body.[553] But this *rasa-* is then not the same as, but a product of water! However, what is this water? The text does not tell us anything, and so we cannot find an answer to our problem from this passage either. BH,Śā 4,33 too has a similar statement saying that man originates from *rasa-*, but does not mention "water(y)" in this context,[554] so that this is no help either. Actually, CA,Ci 22,16 might be referring to an old[555] and seemingly discredited (but obviously nevertheless influential) theory which Vāryovida enunciates[556] in CA,Sū 25,12f., namely that all beings derive from *rasa-* (which is not the same as saying that they depend on this), and that water is *rasavat-* "characterised by (scil. filled with?) *rasa-*";[557] it thus appears only consequent that in his commentary on Ci 22,15 Cakrapāṇidatta says that due to the *kṣaya-* of *rasa-* a *kṣaya-* of water takes place, which then leads to thirst.[558] But can we be sure that Vāryovida is not actually referring to the tastes (cf. n.463)? Note that Cakrapāṇidatta in his commentary on CA,Sū 25,13 uses the plural of *rasa-* (see n.463), i.e. seems to have taken *rasa-* in exactly this sense, though unfortunately he does not specify what he means. We see that we are here entering a veritable maze which, far from leading to the solution of our problem, leads us ever and ever anon to fresh ones.

7.23. Let us therefore try another approach. In this respect AS,Śā 6,p.333b is most interesting, for it tells us expressly that *rasa-* is *accha-*, i.e. "transparent" or "clear".[559] This clearness might explain why *rasa-* is called

---

552 *tasya kṣayāc ca tṛṣyed dhi.*

553 See also EIM, pp.172f. and the discussion on ☞ *rasa-* in Appendix II.

554 *rasodbhavaḥ pumāṃs tasmād raso jīvanam ucyate.*

555 Cf. in this context also the theory discussed by FALK 1986, pp.83f.

556 In the course of a *kathā-* (on this expression cf. n.479).

557 CA,Sū 25,13:
rasajāni tu bhūtāni vyādhayaś ca pṛthagvidhāḥ
āpo hi rasavatyas tāḥ smṛtāḥ nirvṛttihetavaḥ.
Could this be a reflex of an ancient theory taking everything to be formed from water (cf. RUBEN 1971, pp.94f.)? On Vāryovida's theory cf. also n.463 and p.557 in the discussion on ☞ *dhātu-* in Appendix II.

558 Cf. in this connection also §§10.17ff.

559 *evaṃ ca pakvād āhārād vividhauṣadhagarbhād iva snehād acchaḥ sārabhūto rasākhyaḥ*

"watery". Indeed, in the light of the AS passage the commentaries of Ḍalhaṇa and Cakrapāṇidatta mentioned in §7.21 make sense.[560] However, what AS states is not necessarily that which SU presupposes (see also §7.22). And one may also ask why a scholar of the calibre of BHISHAGRATNA did not then explain the passage in this obvious fashion (cf. §7.21). Indeed, I have been unable to find a characterisation of *rasa-* as "clear" in SU.[561] On the other hand, I may of course simply have overlooked such a characterisation. Moreover, what is the use otherwise of telling us in SU,Sū 14,4 that *rasa-* is coloured to give blood? But even if *āpya-* in 14,4 should be a reference to the colour of *rasa-*, that does not solve the problem of 14,5, where not *rasa-*, but only water is mentioned. "Water" and "watery" are not the same. It may be that 14,5 too refers to a clear fluid, but that still does not tell us why this is not called *rasa-*, but water. As mentioned in §7.22, it could be that we have different theories placed side by side here, one deriving blood from *rasa-*, the other from water, *rasa-* and water being either different substances altogether, or else named according to differing terminologies in different theories. Note also the ancient theory stating that *rasa-* and water are identical given in §7.15, even though *rasa-* in this theory does *not* seem to refer to the nutrient fluid. But whatever *rasa-* and water may originally have signified, here, in our SU passage, they seem to have the same function as regards blood, so that it does seem safe to see in them, *at the level of the text of SU* (i.e. maybe secondarily) and irrespective of whether "watery" is a reference to the colour of the fluid, the same substance. True, this assumption does not solve the problem of the actual meaning of "watery" in 14,4 and "water" in 14,5, but we are fortunately not directly concerned with this aspect of the problem, but only with the problem of the derivations of blood and the procreatory-menstrual fluid.

7.24. But even if 4 and 5 should refer to the same fluid, that does not solve our actual problem, for is this fluid then the same as *rasa-* in 6? If the two are the same, then "water(y)" would refer simply to a certain state or characteristic of this *rasa-*. If they are not, then we would have a case similar to that of *śoṇita-*, and *śoṇita-* qualified by *ārtava-*. Does SU offer us any data which might help us in solving this problem?

---

kiṭṭākhyaś ca malo 'bhinivartate.

560 I may also point to the passage from the late CIK quoted in n.614 which tells us that the *rasa-* derived from food is at first a clear, sour and watery (*āpa- kāñjika-*) fluid, which then turns to blood.

561 EIM, p.173 states that *rasa-* is "transparent", but does not give any references for this characterisation.

7.25. It could be. According to Ut 40,6, diarrhoea (*atīsāra-*, also *atisāra-*) is caused by the "element of water" (*apāṃ dhātuḥ*). The actual wording of Ut 40,6 is very interesting:

> *saṃśamyāpāṃ dhātur antaḥ kṛśānuṃ*
> *varcomiśro mārutena praṇunnaḥ*
> *vṛddho 'tīvādhaḥ saraty eṣa yasmād*
> *vyādhiṃ ghoraṃ taṃ tv atīsāram āhuḥ*
>
> "Because the watery element, after appeasing the internal fire, mixed with the excremental matter, pushed forth by the wind and having increased, flows downwards to an extreme degree, this dreadful illness is called ['flowing too much/through', i.e.] diarrhoea."[562]

This cause of diarrhoea is also presupposed (at times in a variant form) by MĀDH 3,4 = BYT 63,1 = BHĀ 2,2,5 = YOG 1,Atisāranidāna,5 (p.254) = ĀĀS, p.102. Now according to the aetiology of *atīsāra-* given by CA,Ci 19,5-7 urine, sweat, bile and phlegm are forced into the faecal receptacle (*purīṣāśaya-*) and there liquefy the faeces.[563] In this theory too the substances

---

[562] I quote the translation of MEULENBELD 1974, p.211,n.19, with an insertion of my own.

[563] See also MEULENBELD 1974, pp.211,n.19;212,n.20. The wording of the relevant portions of CA,Ci 19,5 is:

> *vāyuḥ prakopam āpadyate. paktā copahanyate. sa vāyuḥ kupito 'gnāv upahate mūtrasvedau purīṣāśayam upahṛtya tābhyāṃ purīṣaṃ dravīkṛtyātisārāya prakalpate*
>
> "The wind gets into an excited state and the maturing (fire) is damaged. When the fire has been damaged, the excited wind brings the urine and sweat to the faecal receptacle, liquefies the faecal matter by means of these two (substances) and (thus) makes things ready for (the appearance of) diarrhoea."

The relevant portion of 19,6 states:

> *pittaṃ prakopam āpadyate. tat prakupitaṃ dravatvād ūṣmāṇam upahatya purīṣāśayavisṛtam auṣṇyād dravatvāt saratvāc ca bhittvā purīṣam atisārāya prakalpate*
>
> "The bile gets into an excited state. This excited (bile) damages the heat (of the body, i.e. the fire), by its liquidity, spreads to the faecal receptacle, loosens the faecal matter by its hotness, liquidity and ability to give rise to the flowing out (of impure matter) and (thus) makes things ready for (the appearance of) diarrhoea."

In 19,7 the relevant portion runs:

> *śleṣmā prakopam āpadyate. sa svabhāvād gurumadhuraśītasnigdhaḥ srasto 'gnim upahatya saumyasvabhāvāt purīṣāśayam upahatyopakledya purīṣam atisārāya kalpate*
>
> "Phlegm gets into an excited state. Being heavy, sweet, cold and oleaginous

enumerated damage the digestive "fire" (except for urine and sweat, these being propelled by wind, which is the damaging agent), but it is expressly stated that they do so *before* entering the faecal receptacle to liquefy the faeces, whereas the SU passage only says that the "watery element" (actually: "element of water") is "pushed forth" mixed with stool, and prior to leaving the body, no mention being made of whether the damaging of the fire takes place in this same receptacle or before it. Now this could be taken to mean that the "element of water" passes through the receptacle of the "fire" first, then enters the faecal receptacle, and is then pushed out of this latter, subsequently leaving the body. Then SU and CA would say the same.

7.26. But urine, sweat, bile and phlegm of CA and the "element of water" of SU are no synonyms. Even if we take the "element of water" to refer to the gross element (☞ *mahābhūta-*) water, we run into difficulties, not only because there are more liquid substances than those just mentioned connected with this gross element in the body, but also because bile is usually connected with fire (though see also §7.15). True, AH,Ni 8,3f. and AS,Ni 8,p.46a (see also MEULENBELD 1974, p.211,n.19) have the same aetiology as CA, but with the "element (of) water" (*abdhātu-*) instead of urine etc. being pushed by the wind (bile and phlegm are not mentioned), which reminds us of the SU passage. Since these texts are, however, of a later date and moreover known for their syncretising tendencies (see §7.22), this does not mean much. Further, they — like SU — do not explain the term they use. As to commentators (see MEULENBELD 1974, pp.197;212,n.21.), these explain "element (of) water" by merely listing various liquid substances in the body, not only not agreeing amongst themselves, but also mentioning variant explanations. This does not help us much in comparing the aetiologies of SU and CA. However, we learn from CA,Śā 7,15 that water in the body (seemingly *not* the gross element water, whatever it may actually be), if it be in excess, adheres to faeces leaving the body, as well as to urine, blood and other elements of the body; the same is said with regard to faeces by BH,Śā 7,6, while KĀ,Śā,Śa, p.51 tells us that due to water stool issues in diarrhoea, and urine too.[564] If the CA passages mentioned in §7.25 may be connected with these

---

> on account of its own nature and having become lax, it damages the fire, damages the faecal receptacle because of its *soma*-like nature, moistens the faeces and (thus) makes things ready for (the appearance of) diarrhoea."

The translations are those of MEULENBELD 1974, loc.cit.

[564] The CA passage:

> *daśodakasyāñjalayaḥ śarīre svenāñjalipramāṇena. yat tu pracyavamānaṃ purīṣam anubadhnāty atiyogena tathā mūtraṃ rudhiram anyāṃś ca śarīradhātūn.*

BH says:

passages, this would mean that the former ultimately presuppose water as the primary cause of diarrhoea too. Of course we cannot be sure that such a linkage is actually permissible, but it does strike one that CA,Śā 7,16, right after 7,15, connects bile with water (though there the gross element is meant) and that bile is mentioned among the passages of §7.25 too. In this respect it is highly significant that KAL 9,83 says that diarrhoea is caused by inflamed/corrupted (*vidagdha-*) *bile*, which, together with blood, phlegm and wind, enters the receptacle of impurities (*malāśaya-*), damages the digestive "fire", and liquefies the impurity there (probably: faeces); there is no mention of water.[565] And in CA,Vi 2,7 we find it stated that in a person imbibing of too much liquid,[566] the three morbific entities become corrupted in the receptacle of undigested food (*āmāśaya-*) and then cause various diseases, diarrhoea being caused by bile (but *not* by wind or phlegm). So ultimately there might indeed be agreement between CA and SU in this matter.

7.27. But SU and CA could nevertheless not have the same theory. The descriptions of the latter create the impression that the liquid substances mentioned come from other parts of the body, whereas the SU passage gives us no hints as to where the "element of water" comes from. Could it come from within the digestive tract itself? According to SU,Sū 21,10 (see also 21,6), AH,Sū 12,10f., AS,Sū 20,pp.147f. and BHĀ 1,3,123(6), the bile which is situated (BHĀ: in the *agnyāśaya-*, the receptacle of fire) between the receptacles of undigested and digested food (*āmāśaya-* and *pakvāśaya-* respectively)[567] digests ingested food and drink,[568] separating out (AH, AS) their essence (*sāra-*) and waste matter (☞ *kiṭṭa-*), or (SU, BHĀ) their ☞ *doṣa*-s,[569]

---

daśodakasyāñjalayaḥ śarīre pracyavamānaṃ purīṣam anubadhnāty atiyoge.

KĀ states:

tenodakena ... atīsāre purīṣaṃ ca tato mūtraṃ pravartate.

565 Also, no other cause is mentioned, though subsequently the symptoms of diarrhoea due (secondarily, it seems) to wind, bile and phlegm, alone and all together, are given. 9,83 runs in the original:

pittaṃ vidagdham asṛjā kaphamārutābhyāṃ
yuktaṃ malāśayagataṃ śamitodarāgnim
kṣipraṃ malaṃ visṛjati dravatām upetam
taṃ vyādhim āhur atisāram iti pravīṇāḥ.

566 *yo ... dravais tṛptim āpadyate bhūyas.*

567 Cf. §7.35.

568 On the connection between bile and the digestive fire see the discussion on ☞ *agni-* in Appendix II.

569 On this problematic statement cf. n.681.

rasa-, urine and faeces.[570] According to SU,Śā 9,7 too, the *rasa-* of food and drink (*annapānarasa-*) (seemingly with urine, faeces and sweat) arises in the receptacle of bile (*pittāśaya-*).[571] The wording of SU,Ut 40,6 is such (see §7.25) that it could be saying that the "element of water" mixed with faeces actually leaves not merely some faecal receptacle, but the abode of the digestive "fire". This would fit the passages just mentioned, as these seem to be saying that impurities are formed in this abode of the "fire". What could then be meant is that the internal digestive "fire", weakened by factors which the text lists, is not able to (further?) digest the "watery element" (whatever this be) properly. Then *atīsāra-* would not be too different from the closely allied disease of the ☞ *grahaṇī-* (*grahaṇīroga-*),[572] which seems to refer to severe,

---

570 Note, however, the difficulty occasioned by AH,Ci 9,1, according to which *atīsāra-* is mostly connected with the *āmāśaya-* after it has damaged the digestive fire (*atīsāro hi bhūyiṣṭhaṃ bhavaty āmāśayānvayaḥ | hatvāgnim*), which Aruṇadatta explains as meaning that it originates in the *āmāśaya-* (*āmāśayo 'nvayo janma bhūr utpattir yasya sa āmāśayānvayaḥ*); this is echoed by AS,Ci 11,p.191a (*prāyeṇātisāro hy agnim upahatya prāg āmādhiṣṭhāno bhavati*) and Indu's commentary on the AH passage. Does the wording of the AH and AS passages not make it seem as if the digestive fire were here situated *before* the *āmāśaya-*? Of course the absolutives used (*hatvā* and *upahatya*) need not refer to an action before the action of the finite verb of the respective sentence, but could theoretically refer to one taking place at the same time. Then, however, *prāk* "before" (in *upahatya prāk*) becomes problematic. Moreover, even if the actions were to take place at the same time, that would place the fire not *after*, but probably *in* the *āmāśaya-*; cf. on this also ☞ *grahaṇī-* in Appendix II.

571 The substances mentioned are, according to this passage, transported through ☞ *dhamanī-*s:

> *tās tu pittāśayam abhiprapannās tatrastham evānnapānarasaṃ vipakvam auṣṇyād vivecayantyo 'bhivahantyaḥ śarīraṃ tarpayanti. arpayanti cordhvagānāṃ tiryaggāṇāṃ* (var: *cordhvagatānāṃ tiryaggatānāṃ*) *ca. rasasthānaṃ cābhipūrayanti. mūtrapurīṣasvedāṃś ca vivecayanti. āmapakvāśayāntare ca tridhā* (var: *trividhā*) *jāyante ... .*

It is not to be overlooked that this passage and the other passages just quoted create the impression that *rasa-* is formed before it reaches the receptacle of "cooked" food. In this context I would also like to draw attention to one of the theories on the cause of fever (*jvara-*), according to which the corrupted morbific entities combined enter the receptacle of undigested food, then follow the course of the food and throw the internal fire out of the "place of cooking" (*paktisthāna-*), in the process also corrupting, according to most of the texts, *rasa-* (for references see MEULENBELD 1974, pp.87f.;155,n.14). Now we cannot be quite sure what the "place of cooking" mentioned here is, but it must be before the receptacle of "cooked" food (*pakvāśaya-*), or at the very least the same as this. In any case, from this aetiology it would follow that *rasa-* if formed before or at least in the *pakvāśaya-*. Cf. also CA,Ci 3,129ff., the reading of 130b (*paktiṃ sthānān nirasya ca* or *paktisthānān nirasya ca*), however, not being clear.

572 Cf. on the differentiation between the two diseases especially MEULENBELD 1974, p.197.

probably chronic, diarrhoea, in which the fire in the *grahaṇī-*, not being able to digest food properly, releases this in an undigested or badly digested state.[573] Indeed, AS,Ci 12,p.202a tells us that when this disorder is caused by corrupted phlegm, the "element (of) water" in the body puts the fire (literally: ☞ *tejas-* "fiery energy") out;[574] Indu says that the "element (of) water" refers to that part of the body which is water,[575] an explanation as puzzling as what is explained. But the "element (of) water" does constitute a parallel to what we have seen above in §§7.25ff. with respect to diarrhoea. If the two disorders are indeed similar, the difference might lie in the severity of the disorder (diarrhoea being the milder form), or else in the stage at which indigestion sets in.[576] According to BHĀ 1,3,172f. the ☞ *grahaṇī-* is actually the place where the impurities of food are situated after separation from the food,[577] i.e. these must have been formed in (or before) it, which would tally with our discussion on SU,Sū 21,10 and Śā 9,7. Regarding impurities situated in the *grahaṇī-*, cf. CA,Ci 15,56f., which tells us that the *grahaṇī-*, the seat (*adhiṣṭhāna-*) of the (digestive) fire situated above the navel, is so called because of its seizing (*grahaṇa-*) food, taking it up in a raw state and releasing it "cooked".[578] Also, Gayadāsa in his commentary on SU,Ni 7,6,

---

573 See SU,Ut 40,169-172, as well as MEULENBELD 1974, pp.219-229.

574 *abdhātur hi pravṛddhas tejo nirvāpayati.*

575 *abdhātur dehe jalabhāgaḥ.*

576 Though Cakrapāṇidatta in his commentary on CA,Ci 15,41 says that the similarity is only superficial, the two diseases having different causes and treatments.

577 See Excursus XI at the end of this chapter; this impurity is then further divided, first urine separating out, and then the remaining part, namely faeces, being sent to and stored in the *malāśaya-* "receptacle of impurities".

578 *agnyadhiṣṭhānam annasya grahaṇād grahaṇī matā*
nābher upary agnibalenopaṣṭabdhopabṛṃhitā (56)
apakvaṃ dhārayaty annaṃ pakvaṃ sṛjati pārśvataḥ
durbalāgnibalā duṣṭā tv āmam eva vimuñcati (57).

The printed edition wrongly has *uparyahy* for *upary*, probably a hybrid due to the variant *nābher upari sā hy agnibalopastambhabṛṃhitā* of 56cd. 57c too has a variant: *durbalāgnibalād duṣṭā*. The *Madhukośa* commentary on MĀDH 4,2f. quotes the above with the reading of the variant of 56cd and *cāpy adhaḥ* in the place of *pārśvataḥ* in 57b. Parts of the above are to be found in YOG 1,Grahaṇīnidāna,3f. (p.279):

nābher upary agnibalenopastabdhopabṛṃhitā
ṣaṣṭhī kalā pittadharā yā purvaṃ samudāhṛtā (3)
pakvāmāśayamadhyasthā grahaṇī sā prakīrtitā
apakvaṃ jārayaty annaṃ pakvaṃ sṛjati pārśvataḥ (4).

CA,Ci 15,56 is also quoted by Āḍhamalla in his commentary on ŚĀR 1,5,8, with the second

while explaining *koṣṭha-* as the receptacle of [the result of] "cooking" of food (*āhārapākādhāra-*),[579] equates this with the *grahaṇī-*, saying either that this is the receptacle of the constituents (scil. obviously: of food) *rasa-*, *doṣa-*, urine and faeces, or else that it is the receptacle for the separation of *rasa-*, *doṣa-*, urine and faeces (*rasadoṣamūtrapurīṣavibhāgāśraya-*),[580] in which regard it resembles SU,Sū 21,10 and BHĀ 1,3,123(6), which have been discussed above in this paragraph, but in which the *grahaṇī-* is not mentioned.

7.28. But even if faeces should not be in the abode of the digestive "fire" itself, that would not change much, for then the faeces would probably be stored or at least formed in the receptacle of digested food, the *pakvāśaya-*. Now there seems to be some confusion on where faeces are stored in the body. The passages of the 'classical' texts listing the "receptacles" (*āśaya-*) in the body do not mention any faecal receptacle.[581] The AH and AS passages giving the aetiology of *atīsāra-* (see §7.26) too do not mention a faecal receptacle like CA, but merely the visceral cavity after/alongside/near to faeces (*anuśakṛtkoṣṭha-*); in AS,Sū 19,p.144a[582] the receptacles of faeces and urine are mentioned,[583] but no other receptacles, so that we cannot make comparisons. On the other hand, CA, which has no separate list of receptacles, but does enumerate a few (not all) of these in Śā 7,10, mentions, along with the *pakvāśaya-*, a receptacle of faeces (*purīṣādhāra-*); a receptacle of faeces (but no other receptacles, so that we cannot compare) is also mentioned in the CA passages referred to in §7.25. CA,Śā 3,6 lists *both* a *purīṣādhāna-* "receptacle of faeces" and a *pakvāśaya-* as separate entities which the embryo receives from its mother,[584] and in Sū 20,8 too *both* are enumerated in the same list of organs serving as receptacles of wind;[585] with regard to this, it is interesting that BYT 2,51 lists a receptacle of wind (*pavanāśa-*

---

line, however, running:

nābher upari sa hy agnibalopastambhavāhi ca.

[579] Cf. p.589 in Appendix II (☞ *srotas-*).

[580] The problem of these alternative translations is discussed in n.681.

[581] See SU,Śā 5,6;8 (also Ci 2,12f.), AS,Śā 5,p.320a, AH,Śā 3,10f. On the *āśaya-*s in general see too MEULENBELD 1974, pp.457f., EIM, p.29, RASA, Upodghāta,pp.161ff., PUROHIT, pp.6f., ŚĀR 1,5,7-10, ĀY 4,240cd-243cd, BHĀ 1,3,215-218, ĀĀS 6870-6873.

[582] On this passage see also §7.30.

[583] *viṇmūtrayoḥ sāśrayayos tatra tatropadekṣyate ... .*

[584] Cakrapāṇidatta nevertheless explains *purīṣādhāna-* as *pakvāśaya-*.

[585] In this case too Cakrapāṇidatta simply equates the *purīṣādhāna-* with the *pakvāśaya-*, and the same is done by CART (see on this also SHARMA,P.V. 1985, p.169).

ya-[586] lying beneath the *pakvāśaya-*) and a receptacle of urine (*mūtrāśaya-* = *vasti-*), but none of faeces or impurities. Here too there thus could be a difference between SU (which does not mention any faecal receptacle in connection with diarrhoea) and CA. SU,Śā 5,5 and 5,7, however, mention a visceral organ called *uṇḍuka-*,[587] of which Ḍalhaṇa in his commentary on 5,8 says that it is a "receptacle of divided/separate(d) impure matter" or "divided/separate(d) receptacle of impure matter" (*vibhaktamalādhāra-*)[588] situated in a part (inside?, or outside?) of the *pakvāśaya-* (*tasyaikadeśe*), which is divided/different (*bhinna-*) from it.[589] However, SU itself, which mentions it again in Ci 2,13, says nothing on the seat of this organ, though Śā 4,16f. tell us rather puzzlingly first that the fifth of the seven ☞ *kalā-*s, called *purīṣadharā-* "carrying faeces", separates/divides the impurities and is situated in(/at?) the *pakvāśaya-*, but then that this *kalā-* (now called *maladharā-* "carrying impurities"), situated close to other visceral organs, separates/divides the impurities in the *uṇḍuka-*;[590] this has occasioned the commentators Gayadāsa (NYC, fol.36) and Ḍalhaṇa difficulties.[591] AS,Śā 5,p.319a as

---

586 Wrongly printed as *pavanāśalayo (nigadito)*

587 Also *unduka-*, *undaka-*; see also MEULENBELD 1974, p.457f. In connection with his remarks on this organ not having been identified so far, MÜLLER 1965, p.32,n.117 says that it cannot refer to the stomach for the simple reason that the ancient Indians did not know of the stomach at all; he also says that this can be proved, but gives no such proof. Gananath Sen (on whom see n.11) identified the *uṇḍuka-* as the caecum (SEN 2,p.212).

588 Cf. Śā 4,16f. below.

589 Another name of this organ is, according to Ḍalhaṇa's commentary on Śā 4,16f. (also on 4,25), *poṭṭalaka-*. Ḍalhaṇa also says that the *guda-* ("anus; rectum") is situated near it (*uṇḍukagrahaṇena sānnidhyād gudo gṛhyate*). KATRE 1957, p.243 too cites Ḍalhaṇa and translates *poṭṭalaka-* as "the stomach or abdomen", obviously unaware of the issues involved.

590 *pañcamī purīṣadharā nāma yāntaḥkoṣṭhe malam abhivibhajate pakvāśayasthā.*
yakṛtsamantāt koṣṭhaṃ ca tathāntrāṇi samāśritā
uṇḍukasthaṃ (var: *unduka°*) *vibhajate malaṃ maladharā kalā.*

591 In spite of his extensive commentary Ḍalhaṇa ultimately says only that faeces are situated in the *pakvāśaya-*, and impurities separated/divided in the *uṇḍuka-/unduka-*, which is not much by way of explanation, a complication being added by his statement that the separation/division of impurities refers to a separation/division into urine and faeces:

malaṃ vibhajate mūtrapurīṣarūpatayā vibhāgaṃ karoti.

Ḍalhaṇa quotes and also paraphrases NYC, the latter not being much different, though shorter. I refrain from giving the reading of the ms of NYC here, as it seems corrupt; parts given by Ḍalhaṇa are missing, and other parts require text-critical editing, which would have been too time-consuming for me and taken me too far off the track. I may, however, draw attention to the last part of the commentary of NYC, from which it follows that there are several receptacles of faeces (*purīṣādhāra-*), the chief among these being the *uṇḍuka-* (here

well as a passage from an unknown source quoted by Aruṇadatta in his commentary on AH,Śā 3,9f. echo SU, but say that the *kalā-* adheres to both *āmāśaya-* and *pakvāśaya-*,[592] which throws up even more problems regarding the location of the *uṇḍuka-*,[593] for BHĀ 1,3,214 tells us that there is a *kalā-* situated between *āmāśaya-* and *pakvāśaya-*,[594] this, however, being termed the ☞ *grahaṇī-*![595] The whole passage 1,3,214-216 listing the receptacles in

---

undaka-/unduka-):

undukastham vibhajate nirasyati vibhaktaṃ karoti śarīrāt. undakasya purīṣādhāreṣv antreṣu prādhānyāt. ata eva carake undukaḥ purīṣāntraśabdenoditaḥ. malaṃ purīṣam. ata eva purīṣadhareti.

In his commentary on SU,Śā 4,25 (fol.38r) Gayadāsa calls the *uṇḍaka-* (sic) a/the receptacle of faeces: *purīṣādhānam uṇḍakaḥ.*

592 The AS passage states:

pañcamī purīṣadharā nāma. sā hy atrāmapakvāśayāśritā koṣṭhāntar uṇḍukasthaṃ malaṃ vibhajati.

The passage quoted by Aruṇadatta states:

pañcamy āsāṃ (var: *anyā*) *viḍādhārā sāmapakvāśayāśrayā* (var: *sā ca pakvāśayāśrayā*) *undukasthaṃ* (var: *tatraikasthaṃ*) *vibhajate malaṃ ... .*

593 Which Indu in his commentary laconically calls the receptacle of faeces (*uṇḍukaṃ purīṣādhāraḥ*).

594 According to e.g. SU,Śā 6,25, AS,Śā 7,p.338a and AH,Śā 4,12f., the navel is situated between these two *āśaya-*s.

595 *āmāśayād adhaḥ pakvāśayād ūrdhvaṃ tu yā kalā*
grahaṇī nāmikā saiva kathitaḥ pācakāśayaḥ.

1,3,215ab (≈ ŚĀR 1,5,8ab, ĀY 4,241ab and ĀĀS 6870cd) mentions a receptacle of fire (*agnyāśaya-*) at the navel, no *grahaṇī-* being mentioned in this connection: *ūrdhvam agnyāśayo nābher madhyabhāge vyavasthitaḥ* (the ŚĀR, ĀY and ĀĀS passages have *vāmabhāge* in the place of *madhyabhāge*, i.e. place the receptacle to the left of the navel); Āḍhamalla in his commentary on ŚĀR 1,5,8, however, equates the two (see n.578), and in his commentary on 1,5,30 says that the *agnyāśaya-* is situated between the *āmāśaya-* and the *pakvāśaya-* (*pakvāmāśayamadhye 'gnyāśayam iti vācyam*), which would fit the ☞ *grahaṇī-* (cf. also SU,Ci 2,12f., which list the receptacles of undigested food, fire and digested food in this order — as well as the *uṇḍuka-*, and BYT 2,51, which places the *pakvāśaya-* below the receptacle of fire, on which latter see below). Note that BHĀ 1,3,214 is not found in ŚĀR or ĀY, i.e. 214 and 215 do not seem to be by origin related. Could the juxtaposing of BHĀ be an attempt to accommodate passages such as BH,Śā 4,11f. (found also in ĀY 4,11f., but attributed to Bhāluki, on whom cf. MEULENBELD 1974, p.417), according to which the fire (called *jāṭhara-* in 4,14) is situated in the *sūryamaṇḍala-*, which is within the *somamaṇḍala-*, which itself is in the navel (*nābhimadhye-*)? As it is, ŚĀR 1,5,8ab = ĀY 4,241ab = ĀĀS 6870cd, locating the *agnyāśaya-* on the left of the body, agrees with e.g. CA,Si 3,24c on the *grahaṇī-*: *vāmāśraye hi grahaṇīgude ca* (var: *vāmāśrayo 'gnir grahaṇī gudaṃ ca*), on which Jejjaṭa expressly remarks that this refers to a location both above the navel and on the left, which is similar to Cakrapāṇidatta's statement in his commentary on *pārśvataḥ* in CA,Ci 15,57b (cf. n.578 above) and also similar

the body does not, however, mention any separate receptacle of faeces, but expressly calls the *pakvāśaya-* the receptacle of impurities (*malāśaya-*). In BH,Śā 7,4 and ŚĀR 1,5,7-10 = ĀY 4,240cd-243cd = ĀĀS 6870-6873 there is no mention of a *pakvāśaya-*, a *purīṣādhāna-* or *malāśaya-*[596] being listed instead, which Āḍhamalla in his commentary equates with the *pakvāśaya-* (note that ŚĀR 1,6,8 = ĀĀS 6932 says that *mala-* is stored in the *pakvāśaya-*), though he also states that this is different from the *uṇḍuka-*, without saying what the latter actually is, except that in CA *purīṣāntra-* is allegedly used in the sense of *uṇḍuka-*;[597] this could be a reference to the *purīṣādhāra-* in CA,Śā 7,10 (see above), as Ḍalhaṇa expressly equates the *purīṣādhāra-* (var: *purīṣāntra-*) in CA with the *uṇḍuka-* in his commentary on SU,Śā 4,16f., as also does NYC (see n.591).[598] I may also point out that the description of the *malāśaya-* in the ŚĀR passage in n.581 (and above) tallies with that of the *pakvāśaya-* (equated there with the *malāśaya-*) of the BHĀ passage cited there. SU,Ci 23,6 mentions an *āmāśaya-*, with a *pakvāśaya-* below this and a receptacle of faeces (*varcaḥsthāna-*) below that.[599] KĀ,Śā,Gar 4, mentions a *purīṣadhāraṇa-* but no *pakvāśaya-*, whereas the opposite (a *pakvāśaya-*, but no receptacle of faeces) holds good for Śā,Gar 7 and Śā,Śa, p.50, and also SU,Śā 9,12, CA,Vi 5,8 and AS,Śā 6,p.331b, where one of the *mūla*-s[600] of the ☞ *srotas*-s carrying faeces is the *pakvāśaya-* and not a *purīṣāśaya-* or something similar (but what of CA,Śā 7,10 and CA,Sū 20,8, mentioned above on p.176, then?). Conversely, KĀ,Kh 9,16ab[601] = KĀ,Śā,Gar 9ab mentions both a *pakvāśaya-* and a receptacle (*āśaya-*) of faeces (*viṣ-*);[602] both these

---

to BYT 2,51 on the receptacle of fire (*nābher vāmata ūrdhvam īṣad analasya*); Jejjaṭa also draws attention to CA,Śā 8,6:

> *vāmāśraye 'gnir grahaṇīgude ceti nirdeśād evaiṣāṃ vāmapārśve sthitatvaṃ bījā[vakrāntau śārīre] 'bhihitam. vāme pārśve pittaṃ tad asyāḥ prapīḍitaṃ vidahati raktaśukle* (Śā 8,6) *iti*.

[596] Exact references for this: ŚĀR 1,5,9, ĀY 4,242, ĀĀS 6871.

[597] *carake tu purīṣāntraśabdenoṇḍukaḥ pratipādita iti*.

[598] Could similar considerations have led Gaṅgādhara to say in his commentary on CA,Śā 7,10 (CA1, p.2041) that *kloman-* (on this see p.213 in Excursus VIII below) in this passage actually refers to two organs, one of these being the *uṇḍuka-*?

[599] *doṣāḥ śvayathum ūrdhvaṃ hi kurvanty āmāśayasthitāḥ*
pakvāśayasthā madhye ca varcaḥsthānagatās tv adhaḥ.

[600] Cf. §6.18.

[601] The full wording is given in n.1042.

[602] Apart from, amongst other receptacles, also a receptacle of urine and a very puzzling receptacle of worms (*krimi-* = ☞ *kṛmi-*).

receptacles (the latter called *purīṣādhāna-*) are also mentioned by Jejjaṭa in his commentary on CA,Ci 30,292f. (see on this §7.35). As it is, the whole question of the receptacle of faeces seems to have caused the commentators some headache.[603] But even if the *uṇḍuka-* should be the faecal receptacle otherwise missing in SU, that still need not alter much, for a faecal receptacle may (though it need not) refer to a place where faeces are merely *stored*, not necessarily *formed*, and it goes without saying that an action involving stored faeces (which have so to say already been taken out of the direct process of digestion) may not necessarily be of the same sort as an action involving faeces being formed; according to BHĀ 1,3,172f.[604] the impurity in the ☞ *grahaṇī-* is further divided, first urine separating out, and then the remaining part, namely faeces, being sent to and stored in the *malāśaya-*.[605] (This is quite apart from the consideration that different parts of SU may have different theories anyway; cf. §4.2.) Thus, despite the problem posed by the receptacle of faeces, SU,Ut 40,6 could still be telling us that diarrhoea is basically caused by an inability to digest the "watery element" properly.

7.29. Now I do not mean to say that the theory of SU must needs be

---

[603] Thus Aruṇadatta in his commentary on AH,Śā 3,10f. simply explains the *pakvāśaya-* "receptacle of digested food" as the receptacle of food which has become faeces (*purīṣatāṃ prāpta-*) (thus also Śrīdāsa in his commentary on AH1, loc.cit.). This reminds one of the passages in §7.27 on the digestive "fire" between *āmāśaya-* and *pakvāśaya-*, from which it could follow that faeces are indeed formed before and thus enter the *pakvāśaya-*. But this does not necessarily mean that they are also stored there. Cakrapāṇidatta however, commenting on CA,Ci 19,5, where the *purīṣāśaya-* is mentioned in connection with diarrhoea, equates this with the *pakvāśaya-*. Commenting on Ci 15,11 he states that the origination of wind from food "which has attained the *pakvāśaya-* and is being parched by fire" (CA: *pakvāśayaṃ tu prāptasya śoṣyamāṇasya vahninā*) refers to food "which has entered the *pakvāśaya-* as an impurity" (*pakvāśayaṃ tu prāptasya malarūpatayā pakvāśayaṃ gatasya*), there being no *real* digestion in this organ (*yataś cādhogamane samyag vahnivyāpāro nāsty ataḥ pacyamānasyeti padaṃ parityajya śoṣyamāṇasyeti kṛtam*) (see also EIM, pp.32f.), though Indu, commenting on the parallels AS,Śā 6,p.333a and AH,Śā 3.58, and Aruṇadatta and Śrīdāsa, commenting on AH,Śā 3,58, mention no impurity, the latter two even stating expressly that there is indeed renewed digestion (after that in the *āmāśaya-*) in the *pakvāśaya-* (note too their comments on the *pakvāśaya-* in Excursus VII at the end of this chapter). But whether there be real digestion or not, the commentators' statements (not necessarily those of the texts commented upon) imply that faeces at the very least enter or are formed in the *pakvāśaya-*, whether they be stored there or not. Note also Aruṇadatta's statement that the digestive fire is situated within the *āmāśaya-*, a statement which we also find in CA,Ci 13,27 (*nihanty āmāśaye vahniṃ ...*); cf. on this also the discussion on ☞ *grahaṇī-* in Appendix II.

[604] On this problematic passage see Excursus XI at the end of this chapter.

[605] The matter of this footnote has been transferred to Excursus VII at the end of this chapter (pp.211-213), as it is too lengthy.

different from that of CA, but the discussion above, even though a part may ultimately turn out to be mere hair-splitting, has shown that there is at least a possibility that it could be so. And this possibility should be enough to make us be very careful in our analysis of the SU passage and not resort to the very tempting means of trying to explain it with the help of other texts, but to first and foremost analyse it entirely by itself, which of course also means taking into account the possible theory on *atīsāra-* suggested in §7.27 without regard to whether it fits in with what other texts say. Our first deduction based on this theory would be that, if the "element of water" is something undergoing digestion, it must be something in or derived from the ingested food. In this context the word ☞ *dhātu-* could be important. Now faeces (here: *varcas-*) are as a rule said to be the ☞ *mala-* of food; since they are mentioned here as still being mixed with the "element of water", this would mean that we have here an early stage in the process of digestion, where the faeces have just been formed, in this case, because of impaired digestion, not being separated further, but remaining with the "element of water". We know that in the chain of the transformation of food consisting of "elements" (*dhātu-*) there is a substance which according to common current opinion seems to occupy a sort of intermediate position between the actual chain of seven "elements" (beginning with ☞ *rasa-*) and food: it is the essential part of food, usually called ☞ *āhārarasa-*, but often merely *rasa-*. This is often referred to as a *dhātu-* too, and is said to consist of the pure part of food.[606] The very name *rasa-* "juice; liquid" suggests that it is liquid.[607] Moreover, is it not strange that in the list of the impurities (☞ *mala-*) of the different substances in the chain in the 'classical' texts, each of the seven elements beginning with *rasa-* (usually, though not always, with the exception of semen, of which it is then stated that it cannot be broken down further) has one impurity, whereas food has two, namely faeces and urine, while *āhārarasa-* does not figure in the list? True, according to MEULENBELD 1974, p.488, CA,Sū 28,4 mentions the impurity of *āhārarasa-* too, but this is actually not so, for the text states that a pure portion of food called *rasa-* and waste matter (*kiṭṭaṃ ca malākhyam*) are derived from food, then saying that *āhārarasa-* serves to nourish or satiate the "elements" *rasa-*, blood etc.,[608] the impurity to nourish or satiate various substances which are enumerated and

[606] Its own pure part we often find described as being *rasa-*, that of *rasa-* blood and so on; on this cf. especially §§7.38f.

[607] Note in this context SU,Sū 21,10 (see §7.27) on the place of separation of faeces and *rasa-* from food.

[608] This will be taken up in §§7.37ff. The relevant text is given in n.658.

obviously also ☞ *mala*-s, but not individually linked with any other individual substances.[609] EIM, p.59 states that ☞ *annarasa*- (i.e. seemingly *āhārarasa*-) has as its impurities urine and faeces, but fails to mention any passage substantiating this claim; I could not find any such passage. However, since the texts list two impurities of food, but none of *āhārarasa*-, one could get the impression that *āhārarasa*- is, as it is the essential part of food, subsumed under this. That would mean that its impurity would be looked upon as that of food too, and this impurity would in that case probably be urine, since faeces are the "grossest" part of food and as such liable to be filtered out first. SU,Ut 40,6 would then tell us that diarrhoea is in fact nothing but indigestion in which food, after breaking down into *āhārarasa*- and the impurity faeces, is not broken down further because the *āhārarasa*- does not separate from the faeces, but follows them down past or through the place where it should have been digested.[610] If this were the case, then it could explain why certain severe types of diarrhoea are caused by a corruption of only the *dhātu*-s blood etc. (CA,Ci 19,9; see also MEULENBELD 1974, p.214,n.36), or are accompanied by a discharge of only these *dhātu*-s (Aruṇadatta on AH,Śā 5,82), and not *rasa*- too. I must, however, also point out that the data in n.609 does rather create the impression that there was a theory that *āhārarasa*- has a parallel impurity whose relation to one or both of the impurities of food, faeces and urine, is of the same sort as that of *āhārarasa*- to *rasa*-. Though it does not look as if this theory, if it existed, gained general acceptance, we must still reckon with difficulties caused by its existence. Its possible existence, however, does not invalidate our deductions based on SU, but does add a further factor of uncertainty to them.

7.30. Let us, however, pursue our line of reasoning based on ☞ *āhāra*-

---

609 I.e., we find no reference to the theory, found for instance in CA,Ci 15,18f., that urine and faeces are the impurities of food, phlegm that of *rasa*-, bile that of blood, etc., which dominates later medical works. Indeed, several of the passages referred to during the discussion on diarrhoea (§§7.25ff.) speak of an undifferentiated "impurity" formed from food, and CA,Sū 28,4 actually tells us that this impurity nourishes or satiates, amongst the impurities mentioned, faeces and urine too. Interesting in this context is also AS,Śā 6,p.333b, which echoes CA,Sū 28,4 by referring to an (unspecified) impurity (☞ *kiṭṭa*-, ☞ *mala*-) of food, formed along with *rasa*- which is the essence (*sāra*-) of food, both of these acting in the same manner as in the CA passage. AS,Śā 6,p.334b (the actual wording is to be found in n.1112) then tells us, however, (like e.g. AH,Śā 3,61), that the clear impurity of food is urine, the dense impurity faeces, without any explanation on what the other undifferentiated impurity mentioned earlier is in that case.

610 Could *grahaṇīroga*- (see §7.27) then be explained as indigestion of the stage even before the separation of *āhārarasa*-?

rasa- further and look for additional material on it. In AÑJ 41 we actually find ☞ *rasa-* standing where SU,Ut 40,6 has *apāṃ dhātu-*,[611] and, since *rasa-* can also stand for *āhārarasa-*, this could serve as confirmation for the above. Similarly, Cakrapāṇidatta's commentary on CA,Ci 22,16 states that *rasa-* arises from *āhārarasa-*,[612] whereas CA itself derives *rasa-* from water.[613] Note also CIK, p.5, which tells us that the *rasa-* of ingested food first of all is a watery *kāñjika-*, which in this context probably means "clear sour liquid" (cf. also §7.23), which then turns to blood and is called *prāṇarasa-* "life-*rasa-*" and *jīvarasa-* "life-*rasa-*" or "vivifying *rasa-*".[614] Then there is also the rather puzzling passage KĀ,Śā,Śa, pp.51f. (already referred to in §7.26), which, while listing the quantities of different substances in the body, also gives us a description of one of these, namely "water". Though like other similar passages (see §5.11 and n.144) this one too lists water and *rasa-* (here called *paktirasa-*) as separate substances (their measures being nine and ten *añjali-*s respectively),[615] it also tells us that

---

611 *vṛddhaḥ saviḍ rasaḥ kṣiptavahnir vātātisāritaḥ*
atisāraḥ sa ṣoḍhā taiḥ pṛthak sarvaiḥ śucāmataḥ.
"*rasa-*, increased, combined with faeces, thrown into the [digestive] fire [and] caused to pass through (obviously: without being changed subsequently) by wind, is *atisāra-*. It is sixfold through [each of] them (i.e. wind, bile and phlegm; cf. 42) separately [and] combined, through grief [and] through rawness/immature matter (cf. on this latter MEULENBELD 1974, pp.196-199)."

Note the compound *kṣiptavahni-*, obviously the same as *vahnikṣipta-*; cf. on this DAS,R.P. 1990, pp.49-53.

612 See Excursus IX at the end of this chapter.

613 See n.550.

614 Cf. also §7.12. The whole passage runs:
bhuktānnasya rasaḥ pūrvam āpaṃ bhavati kāñjikam
paramāṇudravo hy eṣa bhaved āpomayo nṛṇām
prāṇādhmātaḥ śuddhim eti sahoṣmā raktatāṃ vrajet
sa hi prāṇaraso jñeyas sañcaraty api sarvadā
tanmūlaṃ jīvanaṃ jñeyaṃ tasmāj jīvarasas smṛtaḥ.

According to the copy I made at the India Office Library, the CIK verse is printed with *āpomaye* instead of *āpomayo*, the reading above being my correction (it also reads *tammūlaṃ* instead of *tanmūlaṃ*, but this is probably due to a scribal convention). The text is printed in the Telugu script.

AS,Sū 36,p.255a says that blood too is *prāṇa-*: *prāṇaḥ prāṇabhṛtāṃ raktaṃ tatkṣayāt kṣīyate 'nalaḥ.*

On *kāñjika-* see MEULENBELD 1974, p.454.

615 This listing, with separate measures, of water and *rasa-* is one of the reasons for the *Madhukośa* commentary on MĀDH 49,25cd to say unequivocally (for the context cf. n.965 and

> "Through this water the elements (☞ *dhātu-*) blood etc. are nourished/satiated."[616]

Now this may say more or less the same as the CA and BH passages quoted in n.564, which refer to the function of water with respect to other substances when this water in the body is in excess. On the other hand, the terminology is almost identical with that used to describe the "nourishing/satiating" of the seven elements of the chain of alimentary breakdown through *(āhāra)rasa-*, a matter which we shall discuss in detail in §§7.37ff. But *(āhāra)rasa-* is in general said to function in this way with respect to *all* the seven elements (including the first, *rasa-*); however, there are also some puzzling statements[617] which say that it does so only with regard to the elements excluding *rasa-*, i.e. blood etc. Is the KĀ passage actually linked with statements of this latter sort? But then this KĀ passage could be saying exactly the same of water as is stated of *(āhāra)rasa-*, at least according to the statements just mentioned, and this, along with the fact that the role of water in diarrhoea is expressly mentioned in this passage, does make one wonder whether different theories relating to water and *rasa-* might not have been mixed up in it,[618] or whether water here does not actually stand for *āhārarasa-* (different from the *paktirasa-* expressly mentioned). Unfortunately, we are not offered a clear statement in KĀ, so that we should not put too much stress on what the passage says. But we still have the AÑJ passage linking *rasa-* and diarrhoea. If thus our explanation of what SU might be presupposing is correct, then it also follows that the terminology used made this process incomprehensible to most of those following after the author(s) of SU, though I must

---

§10.12) that water and *rasa-* are different (*udakāc ca raso bhinnaḥ*).

[616] *tenodakena puṣyanti dhātavo lohitādayaḥ.*

The whole passage runs:

> *śoṇitāñjalayo 'ṣṭau tu nava paktirasasya tu*
> *daśaivāñjalayaḥ proktā udakasya tvagāśrayāḥ*
> *tenodakena puṣyanti dhātavo lohitādayaḥ*
> *atīsāre purīṣaṃ ca tato mūtraṃ pravartate*
> *vraṇe lasīkā pūyaṃ ca picchā cātaḥ pravartate*
> *bhavanti tasmin duṣṭe ca dadrukaṇḍūvicarcikāḥ*
> *tvagāmayāḥ kilāsāni pāmā keśavadhas tathā*
> *tad agnimārutodviddhaṃ [kū]pakaiḥ sveda ucyate.*

As regards *tvagāśraya-* in connection with water, we shall return to this in §10.12.

[617] These we shall discuss in n.663, and Excursus IX and Excursus X at the end of this chapter.

[618] Cf. in this respect also CA,Vi 2,7 in §7.26, which tells us that ingested liquid matter ultimately causes diarrhoea, but by means of agitating bile, and not directly.

point out that CA,Vi 1,25,4 and AS,Sū 10,p.89b tell us that, if one eats again before the food of the previous meal has been digested, the not metabolised (*apariṇata-*) *rasa-* derived from the latter (*pūrvasyāhārasya*) mixes with that of the new food (CA calls the new *rasa- āhārarasa-*; Cakrapāṇidatta is unsure what this means) and agitates the morbific entities, which of course brings to mind the role of the morbific entities in diarrhoea as stated by CA (see §7.25). This line of thought could be confirmed by Indu in his commentary on AS,Sū 19,pp.143f.: the AS passage tells us that *rasa-*s (plural) may corrupt the morbific entities, which then, as well as the seven ☞ *dhātu-*s (themselves corrupted by the morbific entities), may corrupt the impurities (☞ *mala-*) of the body, amongst them faeces, which latter Indu explains as leading to diarrhoea.[619] The problem here is, however, the plural of *rasa-*; the commentaries take this to refer to the tastes, which would be something different from what we have seen above. Nevertheless, we see that according to Indu the corrupted morbific entities can ultimately cause diarrhoea in a context in which *rasa-* plays a role, even though this term is problematic. But even if all this might offer us a means of combining dissimilar aetiologies, we must of course first examine whether such a harmony may not in actual fact be secondary;[620] note also that there is a difference between opining that *rasa-* takes a part in diarrhoeal discharge and that *rasa-* agitates the morbific entities, which then bring about diarrhoea.

7.31. But such questions need not bother us too much right now, as we are here concerned mainly with what SU says. Now a confirmation for the proposed identification of water and *āhārarasa-* in §7.29 could be contained in SU,Śā 9,12, where we have a list of substances borne by the internal ☞ *srotas-*s in the body (cf. §6.18). These substances are enumerated in the following order: breath, food, water, *rasa-*, blood, flesh, fat, urine, faeces, semen, *ārtava-*. Except for the first member (*prāṇa-* "breath"), we have here the chain of digestive breakdown from food to fat in the usual order, followed by substances that leave the body, first those common to males and females, and then (so it seems) those peculiar to males and females respectively; semen (*śukra-*) seems here to be seen more as something paralleling *ārtava-* than as the seventh and last element of the chain of alimentary

---

[619] *viśi* (i.e. *viṣi*) *purīṣe mūtre ca tayoś cāśraye sthitā doṣā yat kurvanti tat tatra mūtrāghātātisārādāv upadekṣyate*. Aruṇadatta, Hemādri, Śrīdāsa, Parameśvara and Candranandana (see AH2,Sū 11,39) in their commentaries on the parallel passage AH,Sū 11,35f. have no such statement, and neither does Indu.

[620] Provided of course that there really is, in the first place, a difference in the aetiologies as they present themselves.

breakdown. The only disturbing substance seems to be *udaka-* "water". But "water" seems to fit wonderfully into the chain if it is actually to be taken to mean the same as *āhārarasa-*. Note in this context also Sū 14,16:

> *sa śabdārcirjalasantānavad aṇunā viśeṣeṇānudhāvaty evaṃ śarīraṃ kevalam*
>
> "Like a continuous flow of sound, light and/or water, it (scil. *rasa-*) thus follows [the course of] (i.e. permeates) the whole body in a minute manner/fine process(?; *aṇunā viśeṣeṇa*)."[621]

7.32. If the above inferences are correct, then we have shown that ☞ *āhārarasa-* is also called "water". But if we now apply this to Sū 14,4-6 in §7.11, we end up with a problem: blood in general would then be a product of *āhārarasa-*. Would this not be a theory at variance with the usual one in which blood is a product of ☞ *rasa-*? Moreover, is the procreatory-menstrual fluid then a product of this *āhārarasa-* (then called merely *rasa-* in 6) too? If so, where does *rasa-* enter the whole picture? Or is *rasa-* in 6 not *āhārarasa-*? Whichever way we decide, we end up with problems because of *āhārarasa-*. Are we then on a wild-goose chase with regard to *āhārarasa-*? Is the "element of water" referred to in SU,Ut 40,6 after all the gross element water, even though §7.26 seems to speak against this possibility? Then we could have here a reference not to the process of digestion in which the seven elements *rasa-* to semen are formed, but to the other process in which the constituents of food are broken down according to which gross element they belong to (cf. §§7.18f.). But where do faeces fit into this theory? These are usually connected with the theory of the elements *rasa-* to semen.[622] Moreover, "fire" (☞ *agni-*) without any further qualification (except for "internal", "maturing" or the like), as in the passages on diarrhoea quoted in §7.25 above, as a rule refers to the central "fire" (*jāṭharāgni-* or the like); but that which is digested by this is the ingested food as such, after some preliminary preparation. In other words, if this fire be indeed meant, then "water" can hardly be something broken down already, but must be something awaiting breaking down or still in the first stage(s) of being broken down. The gross element water (or rather, that in the food which belongs to this), however, if

---

[621] I may here point out too, in connection with the above discussion on *atīsāra-*, that Ḍalhaṇa states in his commentary on Sū 14,16 that some authorities maintain that *rasa-* flowing *jalasantānavat* ("like a continuous flow of water") is a special characteristic of those with a weak digestion, in those with a strong digestion the flow being like sound, in those with a middling digestion like light. Cakrapāṇidatta too in effect says the same, though sound and light are interchanged in his explanation.

[622] Though cf. also the remarks on the end of digestion on p.212 in the Excursus VII at the end of this chapter.

it should indeed be meant, must then already be the result of advanced breaking down which divides food into its constituent parts. As such, it is extremely doubtful whether it can be meant here. But what is the "element of water" then?

7.33. In this respect SU,Sū 21,12 is of great interest, for it tells us, in a discussion prefaced by the remark that now the receptacles of phlegm shall be discussed,[623] that not food as such, but food liquefied through "watery qualities" and disintegrated, is digested in the receptacle of undigested food (*āmāśaya-*).[624] This liquefaction is traditionally ascribed to phlegm in the *āmāśaya-*, and indeed, 21,14 tells us that phlegm present in the *āmāśaya-* is characterised by "watery qualities" (*audaka- guṇa-*), supporting (*anugrahaṃ karoti*) the other phlegms in the body by means of its "action of water" (*udakakarman-*); cf. also SU,Sū 15,4, which mentions the "action of water" of phlegm.[625] AH,Sū 12,16f. and AS,Sū 20,p.148b tell us expressly that phlegm in the *āmāśaya-* wets the food,[626] though the "action of water" is mentioned in connection with the phlegm in the chest (in 12,15 in AH). Of this latter phlegm EIM, p.112 states that "this aspect [of phlegm] represents the storehouse of the dosha: for, on it depends the nature and efficiency of the other aspects of kapha [= phlegm]. Textual descriptions suggest that the body-water is mostly this aspect of kapha." Could phlegm thus be the "element of water"? This would indeed be in accord with the opinion of ROY, pp.154f., where we also find: "Both *ap* and *kapha* signify the fluid-matrix in which all the operations of life are possible" (p.155). But it is the "element of water" that finally comes out of the body in diarrhoea according to SU (see §7.25). Now if the SU passage really tells us that diarrhoea is due to incomplete digestion, what happens to the ingested substances that are not properly digested, if only phlegm and stool exit? For does phlegm not merely act as an additive to produce a reaction here? But be that as it may, EIM does not mention any specific passages supporting the claim made, and as for the secondary literature referred to by ROY, this either does not mention ☞ *kapha-* in this context, or else *speculates* on its character in a general matter and without actual substantiation based on medical texts. Moreover, the lists of liquids in the body mentioned in §5.11 (also n.144) mention *both* "water" and phlegm — as well as *rasa-*. As it is, "water" seems to be used in so

---

623 *ata ūrdhvaṃ śleṣmasthānāny anuvyākhyāsyāmaḥ.*

624 *āmāśayaḥ ... caturvidhasyāhārasyādhāraḥ. sa ca tatraudakair guṇair āhāraḥ praklinno bhinnasaṅghātaḥ sukhajaro bhavati.*

625 *°śleṣmā pañcadhā pravibhakta udakakarmaṇānugrahaṃ karoti.*

626 Cf. also the commentaries of Āḍhamalla and Kāśirāma on ŚĀR 1,5,34f.

many different meanings and contexts in medical texts that it is premature (one might even say: foolhardy) to make sweeping statements such as the above at this stage of our research.

7.34. I would also like to draw attention to SU,Sū 21,13, where it is stated that sweet, cold phlegm arises in the receptacle of undigested food because of "sweetness, slipperiness and being characterised by wetness".[627] Ḍalhaṇa in his commentary states that the latter refers to the characteristics of ingested food (*āhārasyeti śeṣaḥ*), which is not easy to understand and is not stated by Cakrapāṇidatta. What is also interesting is that Ḍalhaṇa says that the verse refers not to the appearance of phlegm, but to its excitation, as phlegm is derived from *rasa-*.[628] And in truth, phlegm is the ☞ *mala-* of *rasa-*. Ḍalhaṇa's difficulty is understandable, for if phlegm serves to prepare the food, of which *rasa-* is a product, for digestion, should it not already be there before this digestion and not be formed due to this? Actually, it could be both, for if *rasa-* is the first product of the digestion of food, its ☞ *mala-* may be separated in the *āmāśaya-* and then stored somewhere else, a part of it re-entering the *āmāśaya-* to help in digestion. Of course, this would be in direct contradiction to the theory of *rasa-* being formed *after* food has been processed (or at least while this processing is taking place) by the digestive "fire" present *below* the *āmāśaya-*,[629] but this cannot be helped, for Sū 21,13 expressly tells us that phlegm does arise in the *āmāśaya-* (*āmāśaye sambhavati*).[630] One could now also say that, actually, phlegm is produced independently of *rasa-*, but that would then be at variance with the theory that the former is the ☞ *mala-* of the latter, as SU itself maintains (e.g. in Sū 46,529). I may also draw attention to CA,Ci 15,9-11, where it is said that phlegm, bile and wind *udīryate*, the first right after ingestion, the second while the food is being pushed out of an unspecified *āśaya-* (Cakrapāṇidatta: *āmāśaya-*), and the third in the *pakvāśaya-*. Now *udīryate* means "rises up" or "issues forth", and as such it may mean "arises" or "is produced", which in this case would fit in with the SU passage just mentioned. However, it may also mean "is roused" or "is excited"; could Ḍalhaṇa thus be obliquely referring to the CA passage, taking it in the latter meaning? What complicates matters further is

---

627 *mādhuryāt picchilatvāc ca prakleditvāt tathaiva ca*
āmāśaye sambhavati śleṣmā madhuraśītalaḥ.

628 *sambhavatīti prakupyati. na punar abhūtaprādurbhāveṇa. kaphasya rasadhātuta evotpannatvāt.*

629 Cf. §7.27, also n.603, and Excursus VII at the end of this chapter.

630 See also n.420 on the colouring of *rasa-* in the *āmāśaya-*, from which it follows that it must have been formed in or before this organ according to this theory.

that this passage refers to a transformation of ingested food on the basis of *rasa-* "taste",[631] which does not serve to lessen the difficulties touched upon in §7.18 (see also n.681). Cf. also CA,Ni 1,20;23;26, where it is stated that agitated wind, bile and phlegm, having entered the *āmāśaya-*,[632] mix with heat, *follow* the *dhātu-* named *rasa-*,[633] block the ☞ *srotas-*s carrying *rasa-* and sweat,[634] damage the digestive fire and force the heat out of the "place of digestion" (*paktisthāna-*), to cause fever. What is the "place of digestion" — the *pakvāśaya-*? According to the commentary of Cakrapāṇidatta, it is the abode of the digestive fire.[635]

7.35. It is thus understandable that Ḍalhaṇa has difficulties, which would have been even greater had he noted the discrepancy between SU,Śā 5,8, where *both* the *āmāśaya-* and the receptacle of phlegm (*śleṣmāśaya-*) are listed, and SU,Sū 21,6, where only the *āmāśaya-* is said to be the abode of phlegm[636] — in contrast to bile, whose own receptacle (*pittāśaya-*), mentioned in Śā 5,8, is also mentioned in Sū 21,6, and to wind, which though in Sū 21,6 not said to be present in a special receptacle (*vātāśaya-*) as in Śā 5,8, is, in contrast with phlegm, not connected with any of the other receptacles mentioned in Śā 5,8, but with organs (which thus could just possibly be its "receptacles") not mentioned there, namely *śroṇi-* and *guda-*.[637] Bile is according to Sū 21,6 (as in 21,10) situated between *āmāśaya-* and *pakvāśaya-* (thus also e.g. KAL 3,50, whereas according to SIDDH 1,17-18 it is situated in the *pakvāśaya-*).[638] Matters get even more complicated when we consider not only CA,Ni 1,20;23;26 in §7.34, but also e.g., CA,Sū 20,8, where different abodes (*sthāna-*) of wind, bile and phlegm are enumerated,[639] it being said

---

631 *prapāka-*; cf. on this especially EIM, pp.32f.;58f.;168, as well as *vipāka-* in §7.18, particularly n.509.

632 Whether bile enters the *āmāśaya-* or is already there is not clear; cf. also Cakrapāṇidatta's commentary on 1,26 and 1,20 in the discussion on ☞ *grahaṇī-* in Appendix II.

633 *āhārapariṇāmadhātuṃ rasanāmānam anvavetya.*

634 Cf. on these §6.18.

635 *paktisthānāt pācakāgnisthānat* (on 1,20).

636 *āmāśayaḥ śleṣmaṇaḥ [sthānam].*

637 The former organ probably refers to the pelvic region in general (cf. DAS,R.P. 1991A, p.54,n.49) (which would apply also to Ḍalhaṇa's gloss *kaṭi-*), whereas the latter refers to the anus/rectum.

638 Cf. also n.603.

639 Cf. also CA,Ci 15,9-11 above in §7.34.

that the special abode of *bile* is the *āmāśaya-*, of wind the *pakvāśaya-*, and of phlegm the chest (*uras-*) (but according to 20,19 the chief abodes are probably the chest *and* the *āmāśaya-*)[640]; 20,8 incidentally also mentions other abodes of the morbific entities, among them, as regards wind, a receptacle of faeces (*purīṣādhāna-*) beside the *pakvāśaya-* just cited (cf. on this §7.28). Jejjaṭa in his commentary on CA,Ci 30,292f. also expressly states that the *āmāśaya-* is the principal abode of bile;[641] on this location of bile in the *āmāśaya-* cf. also n.570, n.603, and ☞ *grahaṇī-* in Appendix II.[642] Cakrapāṇidatta in his commentary on CA,Sū 20,8 remarks that bile is situated in the lower, phlegm in the upper part of the *āmāśaya-* — he has obviously seen a problem here. Jejjaṭa (loc.cit.) also lists both a *purīṣādhāna-* and a *pakvāśaya-* as receptacles of wind, though the latter is according to him the chief of such abodes.[643] AS,Sū 20,p.146b and AH,Sū 12,1-3 say that of the several receptacles of wind, the chief is the *pakvāśaya-* or *pakvādhāna-*, whereas of the several receptacles of bile (which include the *āmāśaya-*, but *not* the *pakvāśaya-*), the chief is the navel (*nābhi-*; cf. also n.594), and of phlegm (one of whose receptacles is also the *āmāśaya-*) the chest (*uras-*). Note also that, whereas according to CA,Sū 20,8 ☞ *rasa-* is one of the receptacles of bile, according to AS,Sū 20,p.146b *rasa-* is one of the receptacles of phlegm. And AH,Sū 12,2 tells us that *rasa-* is one of the abodes of bile, whereas AH,Sū 12,3 tells us that *rasa-* is one of the abodes of phlegm! Finally, KĀ,Kh 9,16 = KĀ,Śā,Gar 9 in its list of receptacles lists both the *āmāśaya-* and a receptacle of phlegm (*kaphāśaya-*) (as well as a *pakvāśaya-* and a receptacle of wind, but none of bile; see also §7.28). We are, however, happily not so much concerned with analysing theories of digestion or of receptacles as

---

[640] See on this also SHARMA,P.V. 1985, p.177.

[641] Five abodes are mentioned in all:
svedo raso lasīkā rudhiram āmāśayaś ceti pittasthānāni. atrāpy āmāśayo viśeṣeṇa pittasthānam.
This seems like a literal quote from CA,Sū 20,8, the only difference being *ca* and *tatrāpy* there for Jejjaṭa's *ceti* and *atrāpy* respectively.

[642] On the receptacle of bile see also §7.15, §7.27, §7.41, n.420 and n.1025, on that of phlegm §7.33, n.420 and n.1025. For further references to the five abodes and the corresponding kinds of phlegm, bile and wind respectively, see MEULENBELD 1974, pp.452;474;497 (on wind see also §7.28 above).

[643] This too seems like a nearly literal quote from CA,Sū 20,8, which runs:
bastiḥ purīṣādhānaṃ kaṭiḥ sakthinī pādāv asthīni pakvāśayaś ca vātasthānāni. tatrāpi pakvāśayo viśeṣeṇa vātasthānaṃ.
The only difference is that Jejjaṭa has *kaṭī* for *kaṭiḥ*, simply *ca* for *pakvāśayaś ca*, and *atrāpi* for *tatrāpi*.

such,[644] but only inasmuch as they are of relevance to our problems. In this respect the origin of phlegm is not so important, what interests us more is what it does to food in the *āmāśaya-*. Now it seems to have been overlooked so far that SU,Sū 21,12 does *not* mention phlegm as the liquefying agent of food; this is only something the later texts AH and AS maintain, while CA is silent on the point. Literally, the SU passage tells us only that "watery qualities" cause the liquefaction (not even necessarily the disintegration) and this could mean that the "watery qualities" are primarily part of the matter to be digested itself (from which phlegm subsequently arises), which would also explain Ḍalhaṇa's remark (see §7.34) on food being meant in SU,Sū 21,13.[645] This may very well be over-interpretation or hair-splitting, but it nevertheless is a possibility if we take the text literally. If, however, the ingested food should indeed have such "watery qualities", should we, if we consider whether phlegm may be the "element of water", not consider this even more in the case of *rasa-*, its homologue[646] and origin (phlegm is the ☞ *mala-* of *rasa-*; see §7.34), which is left over after the solid part (faeces) has been separated out from food?[647]

7.36. This must of course remind us of the AÑJ passage quoted in §7.30. Ultimately, we seem to end up again with *rasa-*. In this respect it is significant that CA,Ci 16,9 describes certain forms of diarrhoea in which the ☞ *dhātu-*s blood etc. (*śoṇitādi-*) are corrupted. Why is the *dhātu- rasa-* not mentioned here? This would make sense if it is *rasa-*, already formed from food, which forms the basic matter of diarrhoea. Of course, this need not necessarily be presupposed here, especially when we consider the aetiology of diarrhoea of CA (see §§7.25f.), but it does give one food for thought, especially when we also consider the KĀ passage mentioned in §7.30. Thus we see that here too we are forced — be it rightly or wrongly — to consider *rasa-*, and this is even more so the case when we consider §7.34 too,[648] where theories on *rasa-* being formed already in the *āmāśaya-* were discussed; *rasa-*

---

644 "Happily", since we would otherwise also have to consider the theory of food entering the heart and maybe also being digested there (see ☞ *hṛdaya-* in Appendix II).

645 The connection between "water" or substances closely connected with it and the *āmāśaya-* may explain Gayadāsa's remark in his commentary on SU,Ni 1,23 that the *kloman-* is situated in or on the *āmāśaya-*. On this matter see the discussion in Excursus VIII on pp.213-215 at the end of this chapter.

646 AH,Sū 11,8: *raso 'pi śleṣmavat* (cf. also MEULENBELD 1991, p.105,n.38).

647 This separation would, if we follow the line of reasoning developed here, probably take place in the *āmāśaya-*.

648 Also n.647.

already formed prior to its first contact with the digestive "fire" would fit the "element of water" in SU,Ut 40,6 even better, even though if such a theory were postulated here it would be at variance with SU,Sū 21,6;10 (see §7.27). But what about ☞ *āhārarasa-*, on which we chanced due to the discussion of SU,Ut 40,6 in §7.29? Even if our deduction from the latter passage should be wrong, the nature of *āhārarasa-* and its place in the chain of metabolism nevertheless remains a problem. And until we have come to some conclusion in this regard, we cannot arrive at a satisfactory conclusion regarding the origins of blood and the procreatory-menstrual fluid based on SU,Sū 14,4ff.

7.37. In this context SU,Sū 14,11 is of interest: the *rasa-* of food and drink (*annapānarasa-*) is the satiating or nourishing agent of the elements *rasa-* to semen,[649] a statement that is echoed by Sū 46,528, which, however, uses only the term *rasa-*.[650] This *rasa-* of food (*anna-*) and drink (*pāna-*) could be *āhārarasa-*, since it is called *annarasa-* in 14,19, where it is stated to be non-satiating or not nourishing in the aged,[651] and since *anna-* is a synonym of *āhāra-*. In Sū 15,33 too, *rasa-* acts as a satiating or nourishing agent,[652] and this *rasa-* obviously corresponds to the *annarasa-* mentioned in 15,32[653] and 15,34[654], especially since *rasa-* appears in 15,32 too,[655] and *rasadhātu-* in 15,33.[656] Moreover, CA,Sū 28,4 (cf. §7.29) too states that the elements *rasa-* etc. derived from the metabolism of food are nourished[657]

---

649 *tatraiteṣāṃ dhātūnām annapānarasaḥ prīṇayitā*. Cakrapāṇidatta (see SU3, p.239 or SU4, p.104) reads *sarvadhātūnām* in the place of *dhātūnām*, and SU3 and GHĀṆEKAR 1977A moreover have *tatraiṣāṃ* for *tatraiteṣāṃ*. *All* the *dhātu*-s (mentioned in 14,10) are here obviously meant; cf. also HESSLER 1844-1850 1,p.29 ("Horum corporis elementorum exhilarator est chylus ex cibo et potu"), HOERNLE 1897, p.91 ("Thus chyle, being itself the essence of food and drink, nourishes these (other six) *dhātu* or 'constituent parts of the body'") and BHISHAGRATNA 1,p.108 ("The Rasa which is originated from the digested or assimilated food and drink pre-eminently strengthens all the fundamental principles of the body").

650 *viṇmūtram āhāramalaḥ sāraḥ prāgīrito rasaḥ*
sa tu vyānena vikṣiptaḥ sarvān dhātūn pratarpayet.

651 *sa evānnaraso vṛddhānāṃ ... aprīṇano bhavati*.

652 The passage states that due to some causes *rasa-* is hindered from fulfilling this function: ... *upaśoṣito rasadhātuḥ śarīram ananukrāman na prīṇāti*.

653 *āma evānnaraso madhurataraś ca śarīram ananukrāmann atisnehān medo janayati*. Further on we read: *āvṛtamārgatvād eva śeṣā dhātavo nāpyāyante*.

654 *annarasaḥ śarīram anukrāman samān dhātūn upacinoti*.

655 *rasanimittam eva sthaulyaṃ kārśyaṃ ca*.

656 ... *upaśoṣito rasadhātuḥ śarīram ananukrāmann alpatvān na prīṇāti*.

657 Not "produced", as the translation to CA4, p.561 opines.

out of *āhārarasa-*, which is loc.cit. also called the "pure portion" (☞ *prasāda-*) of food,[658] just as in SU,Sū 46,528 *rasa-* is called the essence (*sāra-*) of food (see n.250), making it seem quite probable that in SU,Sū 46,528 too *āhārarasa-* is meant. This, however, could then hold true for Sū 14,3ff. too, since Sū 46,528 merely repeats what Sū 14,3 has already stated, namely that *rasa-* is the essence of food (see §7.11). And indeed, AS,Śā 6,p.334a does appear to say that *āhārarasa-* nourishes or satiates all the seven elements,[659] also using the term *annarasa-* subsequently (p.335a), as well as in Ci 10,p.182b, where *annarasa-* is said to nourish and develop the body. And AS,Śā 2,p.288a tells us that the milk of a woman who has delivered is "increased/augmented" (*āpyāyate*) by her *āhārarasa-* (see n.272). But what "nourishing" or "satiating" or the like[660] actually is and how it is accomplished remain unclear; the 'classical' texts (except AS; cf. n.666) do not seem to comment upon this. But we get an idea especially from AS,Śā 6,pp.334f. and the commentaries of Cakrapāṇidatta on SU,Sū 14,10f., Aruṇadatta on AH,Śā 3,62f.;68f., Śrīdāsa on AH1,Śā 3,62f.;65;67;68f. (3,62 is wrongly numbered as 3,61), Indu on AH5,Śā 3,54, Śivadāsasena on DR,Āh 21f., and Cakrapāṇidatta on CA,Sū 28,4 and Ci 15,16. Their relevant portions are too lengthy and complicated to treat here in full, and I therefore refer the reader to the discussion of ZIMMERMANN 1987, pp.162ff. (also for references to other relevant studies), and SHARMA,P.V. 1985, pp.253-255; see also ŚRĪVĀSTAVYA, pp.181ff., PREISENDANZ 2,pp.433ff., GHĀṆEKAR 1977A, pp.78f., and n.923 below.[661]

7.38. The gist of the arguments of the *commentators* cited is, however, clear. It is that *rasa-* in the body is of two sorts, one kind being *sthāyin-*

---

658 *tatrāhāraprasādākhyo rasaḥ kiṭṭaṃ ca malākhyam abhinirvartate* (var: *tatrāhāraḥ prasādākhyaṃ rasaṃ kiṭṭaṃ ca malākhyam abhinirvartayati*). *kiṭṭāt svedamūtrapurīṣavātapittaśleṣmāṇaḥ karṇākṣināsikāsyalomakūpaprajananamalāḥ keśaśmaśrulomanakhādayaś cāvayavāḥ puṣyanti. puṣyanti tv āhārarasād rasarudhiramāṃsamedo'sthimajjaśukraujāṃsi pañcendriyadravyāṇi dhātuprasādasaṃjñakāni śarīrasandhibandhapicchādayaś cāvayavāḥ*.

659 *tatrāhārarasо vyānavikṣipto yathāsvaṃ saptasu dhātvagniṣu kramāt pacyamānaḥ svātmabhāvapracyutisamanantaram eva prāptaraktādidhātusaṃjñakaḥ kālavad askhalitabalapramāṇo deham ūrjayitvā dhātūn dhātumalāṃś ca puṣṇāti*. That all the seven elements are meant follows from the mention of the seven digestive fires through which the *āhārarasa-* has to pass (*saptasu dhātvagniṣu kramāt pacyamānaḥ*). On the other hand, the text also states that this *āhārarasa-* takes up the names of the ☞ *dhātu-*s blood etc. in succession (*prāptaraktādidhātusaṃjñaka-*); since *rasa-* is not mentioned, it would seem to follow from this that *rasa-* and *āhārarasa-* are not differentiated, and that would be in keeping with what is said in §7.38.

660 Another possible English equivalent is "invigoration", which e.g. HARA 1987 prefers.

661 For an interesting modern variant cf. p.9 of the *Prastāvnā* of KAL, where Gaṅgādhar Gopāl Guṇe also brings in the three morbific entities as active agents in this process.

"constant", i.e. already there (from a previous transformation of food), and to be nourished or satiated (*poṣya-*), whereas the other is formed anew from the food last ingested and acts as a nourishing or satiating agent (*poṣaka-*).[662] But both sorts are essentially the same, as they are both *rasa-*. The subsequent elements (blood, flesh etc.) too are, in the same way, *sthāyin-* in the body and as such *poṣya-*, and in their case too the *rasa-* newly formed is *poṣaka-*. How this *rasa-* acts is, however, controversial, i.e. whether the *rasa-* as such nourishes or satiates each element (here too different modes being possible), or whether it is itself transformed, each product of the transformation then nourishing or satiating the corresponding element.[663] But what-

---

[662] This terminology is not found in all the ancient commentaries cited. In the modern secondary literature one also finds the *rasa-* called *poṣaka-* referred to as *asthāyin-* too (see e.g. EIM, p.54, DASH 1971, pp.83-85;97;117). Is this really a term used by the relevant texts, or is it rather a new formation? The evidence seems to suggest the latter (though I cannot state this with surety); at least the relevant passages I know of contrast only *poṣaka-* to both *poṣya-* and *sthāyin-*.

[663] Ḍalhaṇa in his commentary on SU,Sū 14,11 says that the elements blood etc. are nourished by *annapānarasa-*:

> *eteṣāṃ dhātūnāṃ prāguktānāṃ raktādīnām annapānarasa annapānasya sāraḥ prīṇayitā tarpayitety arthaḥ.*

This creates the impression that he has misunderstood the text, which obviously refers to *all* the elements (cf. n.649). But AS,Śā 6,p.334a and the commentaries mentioned in §7.37 make it appear as if he might be, albeit very succinctly, referring to the theory (just mentioned above) in which the nourishing *rasa-* changes to blood, blood to flesh and so on (these elements then nourishing the elements already there), though the explanation's wording does cause difficulties in this respect. Cf. in this context the statement of Hiraṇyākṣa quoted in Śrīdāsa's commentary on AHI,Śā 3,67:

> *dhātūnāṃ śoṇitādīnāṃ sa mūlaṃ vardhano rasaḥ.*

Unfortunately, Ḍalhaṇa's statement is not explained further, so that we do not know for sure what he actually means to say. We might thus consider whether he may have some other theory on the nourishing or satiating role of *rasa-* in mind, which maybe assumes only that *rasa-* changes to the following ☞ *dhātu*-s in succession and does not subscribe to the theory of its being of two kinds (*poṣya-* and *poṣaka-*), one kind nourishing or satiating the other. In other words, Ḍalhaṇa might subscribe to a theory approaching that maybe mentioned in Śivadāsasena's commentary on DR,Āh 21 (see p.217 below), where, however, the *poṣya- rasa-* functions as nourisher or satiator in this way, independently of the *poṣaka- rasa-*, which functions differently. This supposition regarding Ḍalhaṇa is, however, problematic, for in his commentary on SU,Sū 46,528 he shows familiarity with one of the theories pertaining to the nourishing or satiating of *all* the elements by *rasa-*:

> *sa rasaḥ. ... sa ca kulyākedāranyāyena sarvān dhātūn pratarpayati.*

And in truth, it does seem difficult to believe that a scholar of the calibre of Ḍalhaṇa would so misrepresent the exact wording of the text he is commenting upon. Nevertheless, we cannot simply neglect the possibility that Ḍalhaṇa may have misinterpreted the text of SU, especially

ever be the different views prevailing in this matter,[664] there seems to be agreement on the point that there is a continuous process of change on the one hand and replenishment on the other, the *sthāyin-* elements undergoing a change to the next part of the elemental chain,[665] while at the same time continously being replenished by *rasa-*. Now it is unclear whether what later texts, commentators and secondary studies say is actually presupposed by the 'classical' texts,[666] but it does seem as if they, irrespective of details, at any rate subscribe fundamentally to the theory of *rasa-* being both a nourisher or satiator of *all* the elements of the metabolic chain (cf. the examples in §7.37) and the first of these elements,[667] and the evidence suggests that CA and

---

since Indu in his commentary on AS,Śā 6,p.334a too says the same as Ḍalhaṇa:

> *āhārasya ... yo raso bhavati sa ... dhātūn raktādīn dhātumalāṃś ca kaphādīn puṣṇāti,*

whereas the text he comments upon seems rather to mention all the seven elements (cf. n.659), merely saying that the *āhārarasa-* mentioned changes to blood etc. in succession, which is something quite different, being indeed in keeping with one of the theories on the nourishing or satiating of the elements the commentators mentioned in §7.37 too discuss, as has just been mentioned above. (See in this connection also the discussion on KĀ,Śā,Śa, pp.51f. in §7.30, which might be important in this context.) One could also consider whether Ḍalhaṇa's statement might not have to be seen in connection with a probable variant list of *dhātu-*s (see on this §10.29); since, however, in Ḍalhaṇa's times medical scholars as a rule (and Ḍalhaṇa seems to have been no exception, especially when one considers his efforts to explain away discrepancies as detailed in §§10.9f.) hardly seem to have deviated from what had come to be accepted as the 'standard' list of *dhātu-*s, such a proposition is quite problematic.

664 I may here mention that the views ascribed to CA (Ci 15) and SU (Sū 14) by EIM, pp.56;173, are actually to be found in the commentaries on these.

665 Each serving thus to increase or nourish the following element. Cf. also SU,Sū 15,5,1.

666 We simply do not know what CA, SU, and AH have to say on the actual mode of nourishment or satiation. As regards AS, we have seen in §7.37 that this text does give details, but the theory or theories supposed may not be exactly the same as those presupposed by the other texts — we must not forget that AS is younger than CA and SU at any rate, and may present us with theories which are actually later developments. On the other hand, the theories given by AS and the various commentators, even though they do not agree in all points, do not seem to leave much room for still other theories, so that one is rather inclined to ascribe them to the other 'classical' texts too. But one must nevertheless be cautious, for what is 'obvious' or 'common-sense' to us may not have been so to medical authorities in olden days.

667 It may also be pointed out that in his commentary on *parasparopasaṃstambdhā dhātusnehaparamparā* in CA,Ci 15,20, Cakrapāṇidatta remarks that blood also may strengthen/increase *rasa-*, flesh blood etc.:

> *samprati poṣyeṇāpi raktena rasasyāpyāyanaṃ kriyate. tathā māṃsenāpi raktasyetyādi parasparaṃ dhātūnām utpādakatvam āha parasparetyādi. dhātu-*

SU basically agree on this (though I cannot present irrefutable proof), i.e. that *rasa-* has two functions, corresponding to those of the two sorts later technically termed *poṣaka-* and *poṣya-* respectively, though it should be noted not only that Aruṇadatta does not follow this terminology,[668] but also

---

snehaparamparā parasparopasaṃstambdhety anyo'nyam upasnehena saṃstabdhatvāt santarpiteti yāvat.

This theory is repeated by Śivadāsasena in his commentary on DR,Āh 25 (= CA,Ci 15,20).

Cf. in this context also CA,Sū 21,4, where it is stated that in an obese person, it is only fat which is increased, not the other elements:

tasya hy atimātramedasvino (var: *°traṃ meda°*) *meda evopacīyate na tathetare dhātavaḥ. tasmād asyāyuṣo hrāsaḥ.*

According to SU,Sū 15,32, the fat generated blocks the passages by which the other elements are "increased":

āvṛtamārgatvād eva śeṣā dhātavo nāpyāyante.

How this is supposed to happen (maybe, but not necessarily, also according to the authors of the CA and SU passages) is explained by Aruṇadatta in his commentary on AH,Śā 3,62f.:

medasa āhārapūrṇasya (with variants, see loc.cit.) *sa āhāraraso medastvaṃ prāptaḥ punaḥ pratisrotonyāyena māṃsasya viśeṣavatīṃ vṛddhiṃ karoti. anantaraṃ raktasya. anantaraṃ rasasya,*

i.e. the *āhārarasa-*, after "increasing" the members of the chain up to fat and, at least according to Aruṇadatta, turning into the latter (*medastvaṃ prāptaḥ*), goes on performing the same function, but in the reverse order, i.e. fat to *rasa-* again. The other elements (bone to semen) do not "increase" because their ☛ *srotas*-s are blocked by fat:

asthyādayaś cāsthyādisrotasāṃ medasāvṛtatvān na tathopacīyante.

This is repeated, with slightly different words, by Śrīdāsa in his commentary on this passage. See also Ḍalhaṇa's commentary on SU,Sū 15,32, as well as AS,Sū 24,pp.179f.

668 Indeed, Aruṇadatta's commentary could create the impression that he subscribed to the theory, which seems to be followed by several modern scholars, that *āhārarasa-* is actually a sort of link between food and *rasa-*, the latter being derived from it (cf. also n.674). However, the commentary on 62f. is in this point not so clear that it precludes our giving Aruṇadatta the benefit of the doubt, for in some cases he could be playing the *advocatus diaboli*, while in others his wording could just possibly refer to what we have termed "nourishing" or "satiating", and not to a direct derivation. In his commentary on 69, though, we do find a seemingly unequivocal statement relating to the element *rasa-* being the pure part (☛ *prasāda-*) of *āhārarasa-*: *rasadhātur āhārarasaprasādākhyaḥ*. (What, one may ask, is the impurity in this case?; cf. §7.29.) But there is also a variant reading of this line, namely *rasadhātur āhāraprasādākhyaḥ*, according to which *rasa-* is derived — as usual — from food, so that this line cannot serve as unassailable proof.

The commentary on Śā 3,62ab, however, would seem to offer us a means of deciding this question, for here Aruṇadatta tells us that *rasa-* freshly formed from food goes to the heart, and is then expelled from this to unite with *rasa-* and then be turned to the essence called blood:

tato yaḥ sāraḥ prasādākhyaḥ sa pūrvaṃ hṛdayaṃ yāti. tato hṛdayāntaḥsthena vyānena vikṣiptas tatsthābhiḥ sirābhir daśabhir mūlabhūtābhiḥ kṛtsnaṃ śarī-

that there are several problems in this regard which bear further investigation.[669] It must, however, be made very clear that the terms *sthāyin-*, *poṣya-* and *poṣaka-* are not terms of the 'classical' texts themselves (this includes the AS passage cited in §7.37), even though much modern secondary literature would have us believe this. Some modern works also are of the opinion that *sthāyin-* characterises an *unchanging* form of the respective ☞ *dhātu-*. This is not only not borne out by the texts themselves, but even the commentaries do not — at least to my knowledge — seem to presuppose this. As it is, all we can be sure of with regard to the 'classical' texts is that according to these too *rasa-* seems to nourish or satiate all the *dhātu-*s including *rasa-*, so that *rasa-* in them is of two sorts with different functions (which need not, however, necessarily mean that one excludes the other).

7.39. A perusal of the texts also suggests that what is later called *poṣaka- rasa-* is *āhārarasa-*. This is maybe a way of differentiating the two types of *rasa-*, but it cannot be seen on a par with the use of "blood", and of "blood" characterised by *ārtava-*[670] to signify the procreatory-menstrual fluid, since *rasa-* and *āhārarasa-* are *not* two different substances. However, the terminological differentiation of the two types of *rasa-* does not seem to have been systematic; thus cf., for instance, CA,Sū 28,4 mentioned in §7.29 and quoted in n.658, which calls the pure portion (☞ *prasāda-*) of food *rasa-*, but then says that *āhararasa-* nourishes or satiates all the elements *rasa-* to semen (and other substances), this *āhārarasa-*, however, clearly being the same as the *rasa-* first mentioned, as it is parallel to ☞ *kiṭṭa-* in the passage. We have also already seen in §7.37 that SU seems to use *rasa-* and ☞ *annarasa-* indiscriminately. On the other hand, Aruṇadatta, commenting on AH,Śā 3,62f., quotes the ancient authority Nagnajit as saying that the ☞ *dhātu- rasa-* increases (*vivardhate*) through *āhārarasa-*, and that the *dhātu-*s beginning with

---

ram uhyamāno rasadhātunā sahaikatāṃ prāpto rasadhātvāśayasthena vahninā bhūtāgnisahāyena pacyamāno raktadhātusāratayā pariṇamati.

(On this passage see also n.506; on the theory of *rasa-* entering the heart see the discussion on ☞ *hṛdaya-* in Appendix II.) Though Aruṇadatta does not say so explicitly, the first *rasa-* mentioned here does seem to be *āhārarasa-*, which, since it is said to *mingle* with the other *rasa-*, does *not* seem to act as a direct derivatory link between it and food. The matter is clinched by the commentary on AH,Śā 3,67, which says that *āhārarasa-* mingles with the ☞ *dhātu- rasa-* to form *rasa-*, blood etc.:

āhāraraso rasadhātunā sahaikatām āpanno rasaraktādibhāvena pariṇamati yāvac chukraprasādo garbhatvenety uktam.

See also n.549.

[669] Cf. §7.39 and especially Excursus X at the end of this chapter.

[670] Which qualification is often simply left out, for which we by now have ample evidence.

rasa- then act similarly with regard to the *dhātu-*s following them in the known chain of *dhātu-*s.[671] Things are complicated by passages such as SU,Ni 7,6 and AH,Ut 15,21, which speak simply of *annasāra-* "essence of food".[672] And KAL 20,2 calls the *rasa-* which is the first part in its enumeration of the chain of alimentary transformation *āhṛtasānnapānarasa-* "extracted *rasa-* from drink together with food".[673] As it is, the differentiation between *āhāra-* and *anna-* in the passage (KIRFEL 1954, p.339) given in n.748 below, where the former must denote ingested food (undifferentiated), the latter, however, since it and water are formed from the former, only the solid part of this, shows that we must also reckon with the possibility that *āhārarasa-* and ☞ *annarasa-* may by origin not be synonyms, which brings up the problem of "water" too, since in that case *annarasa-* would refer only to the *rasa-* derived from the solid part of food. I have not, however, found any evidence in the medical texts examined that such distinctions were made (though of course I might have overlooked relevant information); still, the matter must be kept in mind for a historical study of the development of the notions of *āhārarasa-*, *annarasa-* and *rasa-* in the context of digestion. But at any rate, we can be sure that *āhārarasa-* is *not* a link in the chain of alimentary transformation between food and *rasa-*;[674] the two *rasa-*s are actually the same, but the latter is produced before the former in the sense that it is already present in the body, whereas the former is formed anew from food.[675] On the other hand, the latter *rasa-* is also produced after the for-

---

[671] On Nagnajit see e.g. KĀ,Upodghāta, pp.99f., HĀLADĀRA, p.74. The passage quoted by Aruṇadatta runs:

tatrāhārarasāt pūrvaṃ rasadhātur vivardhate
raktadhātū rasāc caiva raktākhyān māṃsam eva ca
medodhātus tato māṃsād asthyākhyo medasas tathā
majjadhātus tato 'sthyākhyān majjñaḥ śukrākhya eva ca
śukrāt paraṃ ca garbhasya sambhavaḥ parikīrtitaḥ.

A variant has *raktaṃ dhāturasāc caiva* for *raktadhātū rasāc caiva.*

[672] Explained by Ḍalhaṇa and Gayadāsa and, according to the editor of AH, Candranandana as *rasa-*, by Śivadāsasena (AH3) as *annarasa-*. Cf. also n.549.

[673] The plural of *annarasa-* (*annarasāṃś ca*) in CA,Ci 23,25, however, refers to the "tastes" in food and as such has nothing to do with all this. (A variant has *anurasāṃś ca* for *annarasāṃś ca*, but this is against the metre.)

[674] This does not preclude the possibility of such a notion having existed already in ancient times. Thus for instance the passage in n.671 Aruṇadatta attributes to Nagnajit, in which *āhārarasa-* increases *rasa-*, this blood, this flesh, etc., could maybe be based on such a theory. But even if not, it and similar passages might very well have contributed to the development of such a theory in later times.

[675] Cakrapāṇidatta in his commentary on SU,Sū 14,11 says expressly that there are not

mer in the sense that it is nourished or satiated by this former, i.e. depends on this.[676]

7.40. Now the procreatory-menstrual fluid is not, like the *dhātu*-s *rasa*-, blood etc., an integral part of the chain of the breakdown of food. Our discussion above has, however, shown us too that *āhārarasa*-, though it is also *rasa*-, itself basically functions outside this chain proper. Should we not therefore consider linking the procreatory-menstrual fluid with *āhārarasa*-? This too would be in accordance with SU,Sū 14,6 (see §7.11). Indeed, SU,Ni 10,21 would seem to confirm such a derivation, for here it is stated expressly that milk is derived from *āhārarasa*-,[677] and we have seen in §§6.23f. that milk and the procreatory-menstrual fluid seem to be either the same or at any rate very closely linked. On the other hand, Ni 10,19-23 also speak of a close parallelism between milk and semen, even likening the coming out of the former from the body to the ejaculation of semen.[678] Now semen is the last *dhātu*- to be derived from *rasa*-, but even as such, one could still say that it is somehow ultimately derived from *āhārarasa*-, even if only because it is nourished (or the like) by it, whether or not the other *rasa*- be in this context seen primarily in its function different from that of *āhārarasa*- or primarily as a substance which is actually the same as this. Now of course the statement that milk is derived from *āhārarasa*- may not in actual fact be linked in any way with a (possible) theory linking semen and milk (cf. on this subject §6.24). Nevertheless, the theoretical possibility this has led us to take into account, namely that one could theoretically still link semen with *āhārarasa*- even if it should not be directly derived from this, does make us consider whether some similar link might not have been postulated for the procreatory-menstrual fluid, i.e. that this may have been linked to, but not derived from *āhārarasa*-. But such a connection between this fluid and *āhārarasa*- must of course remain tenuous, as one cannot at this stage be sure that the special female "blood" was not actually somehow or other thought to be derived from the *rasa*- of the chain of *dhātu*-s proper in an as yet unexplained manner. One must also not overlook the fact that *āhārarasa*- is, if

---

two *rasa*-s, but only one. The discussion of his commentary and what it leads to will be found in Excursus IX on pp.215-217 at the end of this chapter.

[676] In this context I would also like to draw attention to Cakrapāṇidatta's commentary on CA,Ci 8,39. However, since the discussion of this is too long for a footnote, it has been shifted to Excursus X at the end of this chapter (pp.218-220).

[677] *āhārarasayonitvād* (var: *āhārarasavīryatvād*) *evaṃ stanyam api striyāḥ* ... . On Ni 10,21-23 cf. also n.1077.

[678] Cf. n.300.

our reasoning in §7.37 above is sound, in SU,Sū 14 clearly mentioned in 14,11 at the earliest, i.e. *after* 14,6, and that 14,6 contains an emphasizing *eva* after *rasād*, which is stated to be the origin of the procreatory-menstrual fluid. And to complicate matters even further, SU,Ni 10,18 also remarks on the origin of milk, deriving it, however, not from *āhārarasa-* (as 10,21, above, does), but merely from *rasa-*, calling it the "pure part"[679] of *rasa-*.[680] (Ḍalhaṇa in his commentary on SU,Sū 15,12 too says that milk is derived from *rasa-*.) One may in this context also compare CA,Ci 15,17, discussed in §§5.12f., in which milk is derived from *rasa-*.

7.41. On the other hand, due to the fact that *rasa-* and *āhārarasa-* are basically the same, we must also reckon with the possibility of *āhārarasa-* simply being termed *rasa-*. In this light SU,Ni 10,18 and 10,21 in §7.40 need not contradict each other, provided of course that this section of the text is homogeneous. Cf. in this context also SU,Sū 21,10, where we find it stated that bile, situated between the *āmāśaya-* and the *pakvāśaya-*, breaks down (*vivecayati*) ingested food and drink into the ☛ *doṣa*-s, *rasa-*, urine and faeces (cf. §§7.27f.),[681] but of which Ḍalhaṇa mentions a variant which enumerates

---

[679] Cf. also n.1100.

[680] *rasaprasādo ... kṛtsnadehāt stanau prāptaḥ stanyam ity abhidhīyate.*

[681] SU,Sū 21,10 says:

viśeṣeṇa pakvāmāśayamadhyasthaṃ pittaṃ caturvidham annapānaṃ pacati vivecayati ca doṣarasamūtrapurīṣāṇi.

BHĀ 1,3,123(6) (also referred to in §7.27) has a similar statement:

pācakaṃ pittam āmapakvāśayamadhyasthaṃ ṣaḍvidham āhāraṃ bhojyaṃ bhakṣyaṃ carvyaṃ lehyaṃ cūṣyaṃ peyaṃ pacati doṣarasamūtrapurīṣāṇi pṛthakkaroti.

And Gayadāsa in his commentary on SU,Ni 7,6 (mentioned in §7.27 too) says:

koṣṭhaśabdenāhārapākādhāro rasadoṣamūtrapurīṣavibhāgāśrayo grahaṇyabhidhāna ucyate.

☛ *doṣa-* could refer to the morbific entities wind, bile and phlegm, and that would imply that these are won from food together with the elements *rasa-* etc., which is a most problematic proposition. One could of course consider translating *doṣarasa-* as "*rasa-* of the *doṣa-*(s)", but what that should mean I am frankly at a loss to say, and I think it unlikely that some such thing should be meant here, especially since Gayadāsa in his commentary on SU,Śā 4,16f. in NYC (fol.36r), quotes the SU passage with *rasadoṣamūtra°* in the place of *doṣarasamūtra°* (Ḍalhaṇa in his commentary on loc.cit. last quotes NYC as having only *rasamūtra°*), and in his commentary on Ni 7,6 cited above also mentions *rasadoṣamūtra°*. Now we may hold that Gayadāsa's remark in the latter case, as well as what BHĀ says, cannot be cited as evidence for the SU passage itself, as they could very well themselves have been influenced by SU,Sū 21,10 and thus not be independent evidence for *doṣa-* appearing in such a context. But neither can we simply dismiss *doṣa-* in the SU passage as a mistake, not only because of the reading of NYC, but also since Ḍalhaṇa's commentary on Sū 21,10 too has this reading, although Ḍal-

☞ *annarasa-*, urine and faeces (*anna-* = *ahāra-*)[682] (along with a problematic variant enumerating wind, urine and faeces,[683] i.e. the three chief excrements of the body often mentioned together[684]). This being so, CA,Ci 15,17, which we have discussed in §§5.12f. (see also §7.13), could perhaps refer to a derivation of the procreatory-menstrual fluid from *āhārarasa-* too, even though it only mentions *rasa-*. But on the other hand, this need not necessarily be so. Unfortunately, I know of no other medical passages which could help us to arrive at a definite conclusion with regard to this verse. As it is, the derivation of the other substances mentioned as coming from blood, flesh and fat, the other ☞ *dhātu-*s mentioned in the verse, is (except for that of ☞ *vasā-*) also most peculiar, and the verse as a whole seems to have no

---

haṇa does not comment upon it. However, if *doṣa-* should be original here, then that would mean that we might have yet another theory of digestion over and above those discussed in §7.18 and Excursus VII below. On the other hand, there obviously was some confusion as regards the reading of this line; not only do we have the variants Ḍalhaṇa himself mentions (see below), but Cakrapāṇidatta in his commentary also seems to read only *rasa* in the place of *doṣarasa*:

> *agneḥ śarīrādahanaṃ tathānnapacanaṃ rasamūtrādivivecanādi ca kāryaṃ ...*

— though only according to SU4, p.160, for this passage is missing in SU3, p.368. But note in this context also SU,Sū 40,10, where the word *doṣa-* crops up during a discussion of *vipāka-* (on this cf. §7.18), and seemingly not in the meaning "morbific entity":

> *iha sarvadravāṇy abhyavahṛtāni samyaṅmithyāvipakvāni guṇaṃ doṣaṃ vā janayanti.*

I must also point out, though, that the morbific entities did crop up in our discussion on changes of *rasa-* "taste" during digestion in §7.34. To compound our difficulties, *rasa-* "taste" in connection with digestion must willy-nilly also bring up the whole problem of the different meanings of ☞ *rasa-* discussed in §§7.13ff.

(On the terms *bhojya-* and *bhakṣya-* in BHĀ 1,3,123(6) cited above, and on the classification of ingested food in general, particular attention may be drawn to YAGI, a study of the terminology used in the classification of food. Unfortunately, the medical literature has been only very sporadically utilised, so that a detailed study of this remains very much a desideratum. Interestingly, *aśita-* and *khādita-* hardly seem to figure in the post-Vedic non-medical terminology analysed by YAGI, whereas we find them several times in medical texts; cf. YAGI, p.385, also e.g. CA,Vi 2,15;17 and SU,Śā 4,19, where food is divided into *aśita-*, *khādita-*, *pīta-* and *līḍha-*. Ḍalhaṇa, commenting on the latter passage, explains *aśita-* and *khādita-* as being derived from verbal roots used in the senses of *bhojane* and *bhakṣaṇe* respectively.)

682 *annarasamūtrapurīṣāṇi.*

683 *vātamūtrapurīṣāṇi.*

684 Cf. in this context §§7.34f. on the different statements regarding the abode of the wind; could the variant mentioned by Ḍalhaṇa point to a confusion existing between wind as a morbific entity and wind as an impurity? Cf. in this regard n.464 on wind as both gross element and morbific entity.

parallel in the rest of medical literature. Thus the CA passage too is as problematic as SU,Sū 14,6, for which we ultimately end up with a *non liquet*.

7.42. It must also be pointed out that, if the "watery *rasa*-" and "water" of SU,Sū 14,4f. indeed are *rasa*- as part of the chain of breakdown of ingested food proper (cf. §§7.23f.), then, though this fits in better with the general prevailing theory on this chain, it does not fit in with what we have deduced out of SU,Ut 40,6 in §7.29 and Śā 9,12 in §7.31. regarding the "element of water" and "water" respectively. Ut 40,6, however, poses no special problem: we would only have to substitute *rasa*- for *āhārarasa*- (i.e. the *poṣya*- for the *poṣaka*-), or else postulate that the two *rasa*-s are here seen merely in their character of being *rasa*-, i.e. as being intrinsically the same, not according to their functions (cf. in this context especially AÑJ 41, §7.30 and §7.36); the process of diarrhoea postulated would remain the same.[685] What is, however, problematic is that urine as the impurity of *āhārarasa*- does not fit into the picture anymore. However, my argument regarding urine was based on the principle of symmetry, and it is quite probable that in actual fact the chain of metabolism has no such symmetry in this particular case. Nevertheless, if SU,Sū 14,6 should indeed refer to *āhararasa*- as being that out of which the procreatory-menstrual fluid is formed (a possibility, not a certainty), then it would follow that even after nourishing or satiating the seven ☞ *dhātu*-s it would have to break down further to form this fluid, so that it could maybe have an impurity (which the *poṣya*- *rasa*- would not have, at least not in this form) of its own too. But would not such an impurity be near the end of the chain of metabolism, and thus hardly on a par with faeces? Moreover, since urine and faeces are both excretions, a postulated link between the two, even though one is liquid and the other solid, irrespective of any notions of symmetry, should not surprise us. However, the problem may be shelved here, as it is secondary to our actual discussion, the more so since there is also the problem of the other, not further specified impurity discussed in n.609. Could this be the missing impurity of *āhārarasa*-? If so, then in the theory postulating it, it would seem to have the same function as *āhārarasa*-; if, however, this latter is actually not different from *rasa*-, then something similar might hold true with respect to this impurity and urine (and/or faeces?) too.[686]

---

[685] See also §10.16.

[686] The remarks on *mūtra*- here apply of course to a possible older theory of its origin, not the one which has gained general acceptance in our texts and according to which food has two ☞ *mala*-s, faeces and urine. On this latter see the discussion in Excursus XI on pp.220-221 at the end of this chapter.

7.43. SU,Śā 9,12 presents even greater difficulties. If the "element of water" of Ut 40,6 is not *āhārarasa-*, then the argument brought forward in §7.31, based on Śā 9,12, equating water and this would seem to fall flat. Moreover, CA,Vi 5,7f. list the internal ☞ *srotas*-s not in the order of the SU passage (see §7.31), but in the following: breath, water, food, *rasa-* to marrow, semen, urine, faeces, sweat; AS,Śā 6,p.331 repeats this order, whereas AH,Śā 3,42 has the order breath, "elements", "impurities", water and food. In BYT 2,137f. we find the order: breath, food, *rasa-*, blood, water, flesh, *ārtava-*, semen (*ṛetas-*), fat, urine, ☞ *mala-* "impurity" (faeces?), etc. These different orders of enumeration too could speak against water being the same as *āhārarasa-*; rather, we could here have a parallel to e.g. SU,Sū 14,11 (see §7.37), where both food and water are mentioned. However, "water" may also be something quite different and as yet unclear to us (see in this context also the remarks in §7.23). In any case, cf. in this connection also CA,Śā 7,15, where water and the "element from the transformation of food"[687] (the latter being expressly called *rasa-* by the text) are differentiated. However, even though water and *āhārarasa-* may not have been equated, SU,Sū 14,4f. does seem to equate *rasa-* and water, and we have seen in §7.22 that this holds good for other passages too; we shall probably have to leave it at this, at least for the present, and profess our inability to go into any more details of this problem here, for a really thorough study of this whole complex of problems cannot confine itself to medical literature alone. Thus I may draw attention to *Chāndogyopaniṣad* 6,5, which, while adhering to the Upaniṣadic triadic theory[688] mentioned in §7.20, distinguishes between ingested food and water, saying too that the thickest *dhātu-*[689] of the former is faeces, of the latter, however, urine.[690] And there are also Upaniṣadic passages which say

[687] *āhārapariṇāmadhātu-*.

[688] Which lists food, fiery energy and water.

[689] The use of ☞ *dhātu-* here to describe not only faeces and urine, but also flesh, blood, bones, marrow and other substances (see the text in n.690), shows that at the time this Vedic text was composed the term was not yet used in the technical sense we have in medical texts. Cf. also EDHOLM, pp.103f. It is quite probable that the use of *dhātu-* to describe the impurities of the chain of alimentary breakdown in medical texts too is a remnant of this older usage.

[690] The relevant portion, *Chāndogyopaniṣad* 6,5,1-3, states:

1: *annam aśitaṃ tredhā vidhīyate tasya yaḥ sthaviṣṭho dhātus tat purīṣaṃ bhavati yo madhyamas tan māṁsaṃ yo 'ṇiṣṭhas tan manaḥ.*

2: *āpaḥ pītās tredhā vidhīyante tāsāṃ yaḥ sthaviṣṭho dhātus tan mūtraṃ bhavati yo madhyamas tal lohitaṃ yo 'ṇiṣṭhaḥ sa prāṇaḥ.*

3: *tejo 'śitaṃ tredhā vidhīyate tasya yaḥ sthaviṣṭho dhātus tad asthi bhavati yo madhyamaḥ sa majjā yo 'ṇiṣṭhaḥ sā vāk.*

that water is food, which in several cases looks as if the latter subsumed the former, but seems not always to be so.[691] Needless to say, this shows once again how problematic "water" is (and is also of relevance for the discussion on the impurity urine in §§7.41f.), and that we cannot cope with this problem solely on the basis of the material we are here studying.[692]

7.44. Leaving aside the problem of *rasa-* and water,[693] we are now, all in all, able to arrive at a picture of the theory regarding blood and the procreatory-menstrual fluid in their relation to each other according to SU. The two are different (though there is some uncertainty with regard to their respective properties).[694] But both are derived from the same substance, namely *rasa-*, though it is unclear whether in each case the same *rasa-* is meant (in the case of the procreatory-menstrual fluid it could be *āhārarasa-*); also, blood is part of the regular chain of the breakdown of food, whereas we do not know (at least not at this stage) how the procreatory-menstrual fluid is produced.

EXCURSUS V

(in place of footnote 387)

In §7.8 we saw that *vivarṇa-* means either "whose colour is off/away" (i.e. "off-colour" or "colourless") or "whose colour is different", *avivarṇa-* correspondingly meaning either "whose colour is not off/away" (i.e. "not off-colour" or "not colourless") or "whose colour is not different" (i.e. "not different-coloured"). As regards the first meaning of *(a)vivarṇa-*, attention may be drawn to Aruṇadatta's remark on AH,Ni 12,26 that *vivarṇa-* means *aniścitavarṇa-* "of un(as)certain(ed) colour";[695] this refers to the colour of the enlargened abdomen (*udara-*), which is said in the text to be reddish or *vivarṇa-* and with blue-black or yellowish lines.[696] Obviously, this does not mean that *aniścitavarṇa-* is to be taken to mean "no fixed colour" in the sense of "*any* colour", but rather that the colour is, though no more that of the skin, not easily discernible as any specific colour, i.e. most probably pale, for

---

[691] For examples see e.g. VISHVA BANDHU, p.16f. (nos. 5;6;15;16;22;24;26). Cf. also the equation of ☞ *soma-* and *anna-* mentioned on p.525 in Appendix II in the discussion on ☞ *āgneya-*.

[692] See also n.1031, as well as HARIPRAPANNAJĪ, pp.1f.

[693] To which, in the context of medical literature, we shall return in §§10.11ff.

[694] Cf. in this context also the difficulties discussed in §§7.2f.

[695] Thus also Śrīdāsa on AH1, loc.cit.

[696] *aruṇābhaṃ vivarṇaṃ vā nīlahāridrarājimat.*

otherwise the whole statement becomes meaningless. Cf. in this context that Ḍalhaṇa in his commentary on SU,Ka 1,21 explains *vivarṇa-* referring to a person's face (*vivarṇavaktra-*) as *malinavarṇa-* "whose colour is grey/dirty", which is the same as saying "pale" in the case of a person of dark complexion. Similarly, the explanation of Cakrapāṇidatta and the alternative (see below) explanation of Ḍalhaṇa of *vivarṇa-* in SU,Sū 12,16 as *pāṇḍuvarṇa-* "(of) a pale colour" refers to skin scorched by improper cauterisation and thus, it seems, having lost its pigmentation. And in SU,Śā 6,26 and AS,Śā 7,p.338b, a person dying due to loss of blood is actually said to be both *pāṇḍu-* "pale" and *vivarṇa-*, as well as *hīnarūpa-* "(one) whose outward appearance has left (i.e. is poorly)".[697]

With regard to the second meaning of *(a)vivarṇa-*, namely "whose colour is (not) different", cf. in this context also *vivarṇa-* in §7.5 (in CA,Si 6,80) referring to cloth stained by "blood-bile"; the cloth now has a different colour from what it previously had. SU,Ut 22,16 refers to discharge from the nose which is clear (*accha-*), like water (*salilaprakāśa-*) and *avivarṇa-*; this latter can only mean "not discoloured" in the sense of "not with a colour other than clear".[698] According to BH,Vi 4,4 illness can be ascertained from the *vivarṇatā-* "*vivarṇa-*-hood" of faeces and urine, which most probably refers to their having a colour not the usual one in a healthy person; since this is a general statement, and since we often find that some particular colour or the other (which includes paleness, though rarely in the case of faeces) shows an illness in the medical texts, we cannot operate with "colourless" or "off-colour" in the sense of "having lost its colour" here, since in that case we would not have a general statement applying to all changes of colour any more. Similarly, *vivarṇa-* in CA,Ci 23,121, where water unfit for use is said to be of vile smell, dirty and *vivarṇa-*,[699] a loss of colour can hardly be meant, and this is most probably also the case in AS,Ut 3,p.36b = AH,Ut 3,24, which speaks of the *vivarṇa-*-ness (*vaivarṇya-*) of stool.[700]

There are, however, several cases in which both "off-colour" (which may take on the shade of "colourless") and "with a different colour" make sense, often ultimately leading to the same analysis. Thus in the commentaries of Hemādri and Parameśvara on AH,Sū 5,3, *avivarṇa-* as a characteristic

---

[697] SU also has a variant *hīnadeha-* "whose body is poorly".

[698] Ḍalhaṇa in his commentary explains this as *aviśiṣṭavarṇa-* "without a distinct/distinguished/particular colour".

[699] *durgandhaṃ sakaluṣaṃ vivarṇaṃ ca.*

[700] *vivarṇa-* as a characterisation of semen in CA,Ci 30,139 might be similar, but this is not definite.

of heavenly Ganges water is explained as *varṇāntararahita-* (var: *varṇānyatva-rahita-*) "without a different colour" and *pūrvavarṇam eva* "of the very same colour as before" respectively, but it could just as well refer to water not without the characteristic colour of heavenly Ganges water. Indu says that *vivarṇa-* referring to clothing in AS,Śā 12,p.363b means *viśiṣṭavarṇa-*; this most probably means "whose colour is distinguished/distinct/particular" in the sense of "whose colour is different (from what it should be)", i.e. "discoloured". This would fit the context well, for *vivarṇa-* must refer to something undesirable, since other attributes of the clothing are: worn out (*jīrṇa-*), soiled (*malina-*) and torn (*chinna-*). But one could also make a case for *vivarṇa-* referring primarily to a loss, and not a change of colour; ultimately in this case both would amount to the same. Ḍalhaṇa, while commenting on SU,Sū 45,64, where milk which is spoilt is described as being *vivarṇa-*, explains this as meaning "having obtained a colour other than white",[701] which obviously refers to a different colour, though this different colour might also be taken to be colourlessness, since this is a different colour from white; the same arguments apply *mutatis mutandis* to *vivarṇa-* and *vaivarṇya-* "*vivarṇa*--hood" said of diseased urine in CA,Si 9,47, AH,Sū 11,22 and AS,Sū 19,p.140a, and in CA,Sū 17,71 respectively.[702] In SU,Ni 10,25 *avivarṇa-* describes pure/unspoilt (*prasanna-*) milk; Ḍalhaṇa explains this as meaning milk "coloured slightly yellow etc." (*īṣatpītādivarṇa-*), whereas Gayadāsa says that the milk is "coloured naturally, slightly yellow etc." (*īṣatpītādiprakṛtivarṇa-*) — here too the above considerations apply. And in his commentary on SU,Sū 12,16 Ḍalhaṇa explains *vivarṇa-* referring to one of the types of improper cauterisation (*agnikarman-*), namely *pluṣṭa-* "scorched", as meaning *prākṛtavarṇād anyavarṇam* "(of) a colour different from the natural colour", but also gives an alternative explanation of it as *pāṇḍuvarṇa-* (see above in this Excursus). Both sorts of explanation can also be considered in explaining *vivarṇa-* in SU,Ka 8,87 and AS,Ut 44,p.363b, where the colour of a bite (i.e. obviously: of the region bitten) is described. SU,Ut 12,36 speaks of destroying the *vaivarṇya-* "*vivarṇa*--hood" of/due to *śukra-*, a white discolouration of the iris;[703] the white colour could be explained as a new colour or as a loss of colour, the meaning of *vaivarṇya-* being in accord. One could also consider a similar explanation for *vivarṇa-* as a characteristic of an intumescence

---

701 *śuklād varṇāntaram āpannam.*

702 Here we have particular cases of urine being *vivarṇa-*, where being different coloured or colourless would equally fit, and not a generalised case as in BH,Vi 4,4 above, which does not leave us with this choice.

703 Ḍalhaṇa explains this as referring to bringing about blackness.

(*śopha-*) in AH,Sū 29,3 and AS,Sū 38,p.261b, but in the AH passage this is also characterised as inflamed red (*rāgin-*), so that the commentaries are probably correct in explaining *vivarṇa-* as referring to a colour not that of skin.

EXCURSUS VI

(in place of footnote 505)

Cakrapāṇidatta in his commentary on CA,Ci 15,13 first gives us two explanations for the statement of 15,13 that each of the fires relating to the five gross elements[704] "cooks" the *āhāraguṇa*-s belonging to it.[705] In the first case he obviously takes *āhāraguṇa-* to mean "quality (☞ *guṇa-*) of food". This may mean, he says, that the fires situated in/peculiar to (*vyavasthita-*) the "substances" (☞ *dravya-*) belonging to earth etc. obtain their strength through the abdominal fire,[706] after which, though actually the "substance" (*dravya-*) of the food is in each case "cooked" by the corresponding fire of the gross element, a corresponding "substance" is formed by this and a "quality" (*guṇa-*) peculiar to that gross element is produced in the "substance".[707] In his commentary on 15,14 he remarks that augmentation/nourishment of a "quality" is impossible without a simultaneous augmentation/nourishment of the "substance" [harbouring this quality].[708]

I may add that since in his commentary on Ci 15,13 Cakrapāṇidatta mentions smell etc. (originating from the gross elements) in the context of the *guṇa*-s,[709] and even says that the *guṇa*-s in 15,13 are not different from

---

704 Not called ☞ *mahābhūta-* or ☞ *bhūta-* in the text, but there is no doubt that these are meant.

705 *bhaumāpyāgneyavāyavyāḥ pañcoṣmāṇaḥ sanābhasāḥ*
pañcāhāraguṇān svān svān pārthivādīn pacanti hi.

706 *pārthivādidravyavyavasthitā jāṭharāgnisandhukṣitabalā[ḥ].*

707 *yady api bhūtāgninā pārthivādidravyaṃ pacyate tathāpi pārthivādidravyāṇāṃ pākenaitad eva jananaṃ yad viśiṣṭaguṇayuktatvam. tena pākena janyamāne 'pi dravye guṇā eva janyanta ity abhiprāyeṇa pārthivādīn āhāraguṇāñ janayantīty ucyate.*

708 *dravyapoṣaṇaṃ tu guṇapoṣaṇena labhyate guṇapoṣaṇaṃ dravyapoṣaṇam antarā kartum aśakyatvāt.* Cf. on this also CA,Sū 1,50 (with the commentaries of Cakrapāṇidatta and CART):

samavāyo 'pṛthagbhāvo bhūmyādīnāṃ guṇair mataḥ
sa nityo yatra hi dravyaṃ na tatrāniyato guṇaḥ.

In this regard cf. also HALBFASS, p.299 (under "*samavāya*" and its derivatives) and pp.97ff.;146, KANNU, p.244 ("Samavāya").

709 *bhaumādyaḥ pañcoṣmāṇaḥ pārthivādidravyavyavasthitā jāṭharāgnisandhukṣitabalā antarīyaṃ dravyaṃ pacantaḥ svān svān pārthivādīn pūrvapārthivagandhatvādyavilakṣaṇān guṇān*

those mentioned earlier,[710] he is here obviously also thinking of Ci 15,12, which says that through the smells etc. (*gandhādi-*) — not called *guṇa*-s here! — which are in it food augments/nourishes the smells etc. and the corresponding sense organs in the body,[711] i.e. to Cakrapāṇidatta *guṇa-* is here used in the sense we have met in §7.13, namely to denote the five qualities or attributes sound, touch/tangibility, shape, taste/flavour and smell with the five gross elements. Note also that *vyavasthita-*[712] in Cakrapāṇidatta's commentary as cited in n.709 and translated above in this Excursus is not clear; it may mean "peculiar to", but it may also mean "situated in". In the latter case we would have a statement that the five fires linked with the gross elements are situated in the "substances" (*dravya-*), obviously then the corresponding "substances" within the body, though one could also take it to refer to the "substances" in the food, which would be puzzling.[713]

Be that as it may, the outcome of Cakrapāṇidatta's first explanation of CA,Ci 15,13 is that the production of the "qualities" (☞ *guṇa-*) is mentioned here (CA,Ci 15,13), not of the "substances" (☞ *dravya-*).[714]

Note in this context that in connection with the *cetanā-* (= *brahman-* = *ātman-*, etc.) taking up what are clearly the gross elements (the term ☞ *mahābhūta-* is not used here), CA,Śā 4,8 says:

> *sa guṇopādānakāle 'ntarikṣaṃ pūrvataram anyebhyo guṇebhyo upādatte.*

Since according to this the ether (*antarikṣa-*) is said to be taken up before the other *guṇa*-s, this looks as if here the gross elements were called *guṇa*-s.[715] However, in the same passage we also find them referred to as ☞ *dhātu*-s with *guṇa*-s.[716] Cakrapāṇidatta's commentary, to which attention has also been drawn by DASGUPTA 2,p.314,n.1, takes *guṇa-* in the first instance to refer to the gross elements (☞ *bhūta-*) characterised by *guṇa*-s, by

---

nirvartayanti.

[710] *pūrvapārthivagandhatvādyavilakṣaṇa-.*

[711] *dehe prīṇāti gandhādīn ghrāṇādīnīndriyāṇi ca* (var: *prāṇādīn indriyāṇi ca*).

[712] In *pārthivādidravyavyavasthitā[ḥ].*

[713] Cf. also ŚRĪVĀSTAVYA, p.167, who says that the fires of the gross elements can digest the food only after the abdominal fire has strengthened them, using the term *āhār'gat* to describe these fires, which we can translate as "contained in the food [itself]", or as "having reached the food".

[714] *anena guṇajananam evāgninocyate na dravyajananam.*

[715] On CA,Śā 4,8 see also the discussion on p.556.

[716] *tataḥ krameṇa vyaktataraguṇān dhātūn vāyvādikāṃś caturaḥ.*

the carrying over of identicalness with respect to the *guṇa-* and that which has this *guṇa-* (*guṇin-*).[717] This fits his statement on CA,Ci 15,14 given above, and is also similar to Ḍalhaṇa's commentary on SU,Sū 46,526 (in §7.19, especially n.514), BHĀ 1,3,170, and Indu's remark on AS,Śā 6,p.333b: *sarvatra guṇaśabdo 'vayavavācī*; this latter brings to mind *guṇa-* "derived matter" referred to in n.522, but which does not seem to be considered by Cakrapāṇidatta in the context we are discussing here. Indu's remark actually anticipates the express statement of the AS in the following paragraph (also p.333b) that the word *guṇa-* is a synonym of ☞ *dhātu-* referring to both the pure and impure parts of food (see n.1906),[718] which, however, does not fit at all with the statement in the preceding passage speaking of the five *guṇa*-s of the gross elements,[719] and looks rather like an ad hoc solution by the author of AS to a problem similar to the one discussed in §7.18.

The other explanation of *āhāraguṇa-* in Ci 15,13 given by Cakrapāṇidatta is that this is to be understood as "food and ☞ *guṇa*-s", "food" subsuming "substance" (☞ *dravya-*), the latter "having the form of the substratum of food" or "having the form of that whose receptacle is food" (alternative translations of *āhārādhikaraṇarūpa-*).[720] This means that, though the abdominal fire "cooks" (or here maybe better: "brings to maturity") all the transformations (*vipāka-*)[721] of ☞ *āhārarasa-* and "impurities" (☞ *mala-*),[722] the fires of the gross elements bring forth their respective "qualities" (*guṇa-*);[723] in this context he quotes a line from an unknown source telling us that the abdominal fire breaks up the combination [of all the substances], after which the fires of the gross elements (☞ *bhūta-*) cook their respective substances.[724] Śrī-

---

[717] *guṇaśabdena guṇaguṇinor abhedopacārād guṇavanti bhūtāny ucyante. kiṃvā guṇo 'pradhānam. pradhānaṃ cātmā. tadvyatiriktāni ca bhūtāni guṇāḥ. ... anyebhyo guṇebhya iti vāyvādibhyaḥ.*

[718] *tābhyāṃ ca sāramalābhyāṃ tadātmakānām eva śarīraguṇānāṃ* (ASI,Śā 6,p.231: *śārīraguṇānāṃ*) *dhātvākhyānāṃ yathāsvaṃ srotāṃsi pāramparyeṇavicchinnasantānam āpūryante. ... tataś ca dhātvākhyāḥ prasādamalāḥ svaṃ svam eva mānam anuvartante.*

[719] 333b: *tataś caivaṃ viklinna āhāre pañca pañcātmakā mahābhūtāgnayo vāyunā vyastān yathāsvaṃ pañcaiva bhūtaguṇān āhārasthān pacanti.*

[720] *kiṃvāhārāś ca guṇāś ceti vigrahād āhāraśabdenāhārādhikaraṇarūpaṃ dravyam api gṛhyate.*

[721] On this term see p.157.

[722] *sarvān evāhārarasamalavipākān pacati.*

[723] *bhautikās tv agnayaḥ svān svān guṇāñ janayanti.*

[724] *jāṭhareṇāgninā pūrvaṃ kṛte saṅghātabhede paścād bhūtāgnayaḥ pañca svaṃ svaṃ dravyaṃ pacanti.*

dāsa in his commentary on AHI,Śā 3,62f. also quotes this in verse form and calls it a *sauśruta-* statement (see n.506).

After this discussion Cakrapāṇidatta finally remarks on the relationship of the two sets of fires (of the gross elements and the ☞ *dhātu-*s) to each other, saying that this function of the fires of the gross elements also relates to the *dhātu-*s, as these are composed of the five gross elements, and that the function of the fires of the gross elements and the *dhātu-*s is in accordance with the abdominal fire, which seems to mean that the explanation of CA,Ci 15,13 one chooses from the two given above will apply to both sets of fires.[725] This shows that Cakrapāṇidatta must have seen some sort of problem here, but of course this laconic statement on the action of the sets of fires overlapping presents a great problem itself, which is not lessened by the statement in CA,Sū 28,4 that the "substances" (*dravya-*) of the sense organs (*pañcendriyadravyāṇi*), i.e. the gross elements,[726] are not only called the ☞ *prasāda-* ("pure portion") of the *dhātu-*s, but are also either equated with the latter or else are along with them (the seven elements *rasa-* to semen),[727] together with ☞ *ojas-* augmented/nourished by ☞ *āhārarasa-*,[728] (which leads to a discussion taken up in §§7.37ff.) — what sort of digestive process is presupposed here?

In this context I may also mention SU,Sū 46,526, which causes an additional difficulty inasmuch as in it the augmentation of the *guṇa-*s of the body as a result of digestion by the five fires of the gross elements is detailed right after a passage (46,514-525) which lists twenty *guṇa-*s in the sense common in medical texts (see ☞ *guṇa-* in Appendix II), which is not the same as that in §7.13 referred to above on p.208.

---

[725] *ayaṃ ca bhūtāgnivyāpāro dhātuṣv apy asti yato dhātuṣv api pañcabhūtāni santi. tatrāpi dhātvagnivyāpāro bhūtāgnivyāpāraś ca jāṭharāgnikrameṇaivokto jñeyaḥ.*

[726] See CA,Sū 8,9: *pañcendriyadravyāṇi khaṃ vāyur jyotir āpo bhūr iti*, and Sū 30,27: *dravyāṇāṃ pṛthivyādīnām*, though, as the discussion above has shown, these "substances" seem to be concrete manifestations and not just abstract categories, as is also stated by CA,Sū 26,9: *pañcamahābhūtavikārās tv āśrayāḥ .... teṣv āśrayeṣu dravyasaṃjñakeṣu guṇā[ḥ]*; cf. also CA,Sū 1,64.

[727] Depending on whether *rasarudhiramāṃsamedo'sthimajjaśukraujas-* and *pañcendriya-dravya-* refer to the same, or are parallel.

[728] *puṣyanti tv āhārarasād rasarudhiramāṃsamedo'sthimajjaśukraujāṃsi pañcendriyadravyāṇi dhātuprasādasaṃjñakāni śarīrasandhibandhapicchādayaś cāvayavāḥ.*

EXCURSUS VII

(in place of footnote 605)

The equation of the receptacles of digested food and faeces considered in §7.28 could be held to disregard the fact that according to the predominant theory of digestion the faeces are, along with urine, the ☞ *mala-* of food, and separated as such near the beginning of the whole transformation as a waste product. This means that, if the impurity faeces should be stored in the receptacle of digested food (*pakvāśaya-*), and if food undergoing digestion should pass through this organ on its way and go on being transformed even after that, then this food would keep on coming into contact with the impurity and thus become ever more impure. In this respect the postulation of another organ (i.e. the *uṇḍuka-*) as a receptacle of faeces makes sense, as one may then say that faeces are created either in or just before, but not stored in the *pakvāśaya-*, especially since it is difficult to conceive of faeces being called "digested food", which is after all what the receptacle is named after (though we may here also ask what exactly "digested food" is). Viewed in this light, the separate receptacle of faeces of CA referred to in §7.28 makes sense too, whether or not it be the same as the *uṇḍuka-*. We must, however, be honest enough to point out that in CA,Ci 19,8,[729] which describes diarrhoea due to the concerted action of all the three morbific entities, the *pakvāśaya-* is mentioned where in the other types of diarrhoea we found (see n.563) the *purīṣāśaya-*, though there is a difference, for the text says that in this type of diarrhoea the morbific entities injure the digestive fire "even more" (*bhūya evāgnim upahatya*), then enter the *pakvāśaya-* and cause diarrhoea; the actual effect on faeces is, in contrast to the other types of diarrhoea, not mentioned. I do not know whether these distinctions are important and point to a significant difference. This is, however, a problem special to CA, for *pakvāśaya-* could indeed refer to the faecal receptacle in many other cases.

However, apart from the problem of contamination through stored impurities, in such cases *pakva-* "cooked" too causes a problem, for the "cooking" of food is clearly not over, as the "fires" (☞ *agni-*) of the seven elements seem to begin their work only when or after the food has reached the *pakvāśaya-*. Now one may counter that these "fires" seem not to be localised in the digestive tract itself, so that what they act on is no more the food as such, but metabolised parts of it; indeed, this is the common terminology used especially in modern apologetic works on Indian medicine. Whether, however, the texts themselves make such a distinction may well be debated, as they

---

729 Cf. also MEULENBELD 1974, p.210,n.3.

seem to refer to "cooking" throughout. But even if it should be shown that they do not consider food or parts of this no more within the alimentary tract to be food, we still have to reckon with theories assuming further "cooking" within the *pakvāśaya-* itself, and these clearly refer to food. Thus Aruṇadatta in his commentary on AH,Śā 3,57f. states expressly that food (*anna-*) that enters the *pakvāśaya-* is further digested there.[730] This is repeated by Śrīdāsa in his commentary on the same passage. What sort of digestive process is presumed here, and is this process in accord with other presupposed digestive processes? For it cannot be overlooked that a further digestion of food defined by its nomenclature as already digested raises the question of whether a theory or theories seeing the end of digestion as already attained in the *pakvāśaya-* may not have been amplified to make room for yet further digestion, in which connection the problem of different sets of fires mentioned in §7.18 also springs to mind. The matter bears looking into.

With regard to *pakva-* we further have the problem of whether impurities too were regarded as "cooked" food; this question cannot but crop up if the *pakvāśaya-* serves also as the receptacle of impurities, and becomes even more pressing if these impurities should have been taken to be not only formed, but to be stored there too. In this regard it is clear that if impurities were taken to *enter* the *pakvāśaya-*, the chances that they were presumed to be stored there are greater than if they were taken to be *formed* there (could Aruṇadatta's statement just quoted be understood in this latter manner?), for in that case one could at least argue that formation does not automatically imply storage. This matter is just as clearly connected with the problem of the formation of the essential part (*sāra-*) of food and the corresponding impurity/impurities, which on its part is linked with the question about the role the various sorts of digestive fires play. This is not the place for a detailed study of the problem, so that all I can do is draw attention to the data in §7.18 and §7.27 (cf. also §7.34), which are of course anything but exhaustive.[731] Thus, whichever way we look at the subject, we are faced with problems, which are compounded by SU,Ni 3,20, which calls the *bladder* (*mūtrāśaya-*) a receptacle of impurities (*malādhāra-*), but different from the *pakvāśaya-*, to which according to 3,21f. it is connected by urine-carrying

---

[730] *yataḥ pakvāśayasthasyāgninā pāko 'nupapannaḥ. dūravyavadhāno hy agnir āmāśayasthaḥ. tasmāt pakvāśayasthasyānnasya pakvatvam anupapannam (var: na pakvatvam upapannam). atrocyate. pakvāśaye 'py asti pākaḥ. tantrakāravacanāt.*

[731] Note also that AS,Śā 6,p.330b, the parallel of SU,Śā 9,7 cited in §7.27, mentions the *pakvāśaya-* where the latter has the *pittāśaya-* between *āmāśaya-* and *pakvāśaya-*.

tubes[732] whose mouths one, however, cannot see because they are so fine.[733]

Incidentally, EIM, p.59 and DASH 1971, p.74 opine that the digestion due to the fires of the gross elements takes place in the *adhaāmāśaya-* "lower *āmāśaya-*". This seems to be a modern theory whose origins I do not know. EIM refers to CA,Ci 15,13 in support of it, but neither here nor in Cakrapāṇidatta's commentary is there anything relevant. Cf. on this also the discussion on ☞ *grahaṇī-* in Appendix II.

EXCURSUS VIII

(in place of footnote 645)

In §§7.33ff. attention was drawn to the links presupposed between the receptacle of undigested food (*āmāśaya-*) and "water" or substances closely connected with this. These links may explain Gayadāsa's remark in his commentary on SU,Ni 1,23 that the *kloman-*[734] is situated in or on the *āmāśaya-*,[735] for the *kloman-* is intimately connected, at least according to some authorities, with "water", and is the "seat of thirst". The *kloman-* is also equated with the *tila(ka)-*;[736] the *tila-* is mentioned by ŚĀR 1,5,8cd (= ĀY 4,241cd = ĀĀS 6871ab = BHĀ 1,3,215cd) as lying above the *agnyāśaya-*,[737] whereas below it lies the receptacle of wind.[738] According to Hariścandra as quoted by ĀY 4,239cd (= ŚĀR 1,5,45ab), the *tila-* is the *mūla-*[739] of the vessels (*śirā-* = ☞ *sirā-*). This is stated by Āḍhamalla in his commentary on ŚĀR 1,5,8 too, whereas Kāśirāma in his commentary on ŚĀR 1,5,45 has the problematic *tilavṛkkāv āhārajalavāhiśirāmūlau*; since *vṛkka-* means "kidney", one could explain *āhārajala-* as "ingested water", but does "water" in the case of the *tila-* have this meaning too? This is a problem which to my knowledge has not yet been examined. Āḍhamalla also equates *tila-* and *kloman-* (also in his commentary on 1,5,45), then quoting some verses describing the

---

732 ☞ *nāḍī-*, here plural, not dual, much to the discomfiture of GHĀṆEKAR 1977B, pp.21f.

733 On what is said here with regard to urine see also Excursus XI.

734 See on this organ MEULENBELD 1974, pp.457f.; cf. too HARIPRAPANNAJĪ and KAVAḌE.

735 *klomno hy āmāśayasthatvād vātakṛtaklomaśoṣeṇa pipāsā*.

736 On this latter see MEULENBELD 1974, p.458.

737 *tasyopari tilaṃ jñeyaṃ* (ĀĀS: *tasyoparitalaṃ*). On the *agnyāśaya-* see §§7.27f., especially n.595.

738 *tadadhaḥ pavanāśayaḥ*.

739 See on this term §6.18.

cooking of food clearly as a parallel to actual cooking in daily life, namely by means of water heated from below by fire, fanned by wind beneath.[740] It may also be noted that the arrangement of the substances mentioned here (food, water, fire and wind) is obviously taken by Āḍhamalla to correspond to that of the receptacles of the body mentioned in ŚĀR 1,5,7f. = ĀY 4,240f. = ĀĀS 6870f. (*śleṣmāśaya-*, *āmāśaya-*, *agnyāśaya-*, *pavanāśaya-*); note also BYT 2,51, where we have the arrangement *śleṣmaṇaḥ ... āśaya-*, *āmāśaya-*, *analasya* [*āśaya-*], *pakvāśaya-* and *pavanāśaya-*.

CA,Ci 15,7f. too tell us that fire cooks food (which according to 15,6 is in association with liquids[741]) from below to give ☞ *rasa-* and an impurity/impurities, just as rice and water situated in a receptacle above it are cooked by fire to give edible food; the fire is fanned by wind.[742] KIRFEL 1954, p.339 also cites some similar verses in his study of the *Garuḍapurāṇa*, but labels them secondary to this text (p.350). The first verse is more or less similar to that of the passage quoted by Āḍhamalla, but the rest differs.[743] According to KIRFEL 1954, p.350, these verses "beschreiben den Verdauungsprozeß in der Form des natürlichen Kochens. Von oben nach unten liegen übereinander: Speise (Erde), Wasser, Feuer und Wind. Es ist die gleiche Folge, die auch die Elementarschichten einnehmen, die im kosmographischen System des Hinduismus die Welt umschließen."

Such correspondences are interesting; however, the relation of ideas on digestion such as these to those usually found in our medical texts is

---

740 *agner ūrdhvaṃ jalaṃ sthāpyaṃ tad annaṃ ca jalopari*
agner adhaḥ svayaṃ vāyuḥ sthito 'gniṃ dhamate śanaiḥ.
vāyunādhamamāno 'gnir atyuṣṇaṃ kurute jalam
tad annam uṣṇatoyena samantāt pacyate śanaiḥ.

741 *tad dravair bhinnasaṅghātaṃ snehena mṛdutāṃ gatam.*

742 *samānenāvadhūto 'gnir udaryaḥ pavanodvahaḥ* (var: *udīrṇaḥ pavanena tu*)
kāle bhuktaṃ samaṃ samyak pacaty āyurvivṛddhaye
evaṃ rasamalāyānnam āśayastham adhaḥsthitam
pacaty agnir yathā sthālyām odanāyāmbutaṇḍulam.

743 The whole passage (verses 45-47 on pp.339f.) is given here in what is regarded loc.cit. as the critical reading; the numerous variants (none of any great import) are not reproduced:
ūrdhvam agner jalaṃ kṛtvā tad annaṃ ca jalopari
agneś cādhaḥ svayaṃ prāṇas tam agniṃ ca dhamec chanaiḥ
vāyunā dhamyamāno 'gniḥ pṛthak kiṭṭaṃ pṛthag rasam
dvārair dvādaśabhiḥ kiṭṭaṃ bhinnaṃ dehāt pṛthag bhavet
karṇākṣināsikā jihvā dantā nābhir vapur gudam
nakhā malāśrayā hy ete viṇmūtraṃ cety anantaram
śukraśoṇitasaṃyogād etat ṣāṭkauśikaṃ smṛtam.

clearly quite problematic. But the whole does remind one of a theory in Tibetan medicine, of which WALSH, p.1226 writes: "The theory of digestion is curious. The stomach is earth, the air blows, so that the fire burns and causes the water to boil, and so digests the food." Water, fire and wind are mentioned together in BH,Śā 4,21 too, which refers to the abdominal fire, arisen from water, acting together with/through the winds [in the body];[744] in 4,22f. this fire is then compared with a lamp floating on water. However, the connection between fire and water is here of quite a different sort from what we have seen above.

The verses quoted by Āḍhamalla could have confused Kāśirāma, who says that *tila-* refers to *pācakapitta-* "[food] cooking bile (☞ *pitta-*)", a common name for one of the five sorts of bile within the body (cf. n.642). On the other hand, Kāśirāma might also have had ŚĀR 1,5,30ab in mind, which tells us that bile is present in the form of fire and to the measure of a *tila-* (sesame seed)[745] within the *agnyāśaya-*;[746] Kāśirāma equates this with the bile called *pācaka-*.[747] I may also mention that right before the verses quoted above, KIRFEL 1954, p.339 has a verse (44) according to which digestion takes place in the *guda-*, which is usually taken to refer to the anus or rectum, but here could be used in the older meaning "intestine, entrails".[748] This shows how problematic even seemingly clear terms may be in medical contexts.

EXCURSUS IX

(in place of footnote 675)

Cakrapāṇidatta, commenting on SU,Sū 14,11,[749] says expressly that there are not two ☞ *rasa-*s (which would give a total of eight ☞ *dhātu-*s), but only one.[750] Here both *rasa-*s are termed ☞ *āhārarasa-*, but this is most

---

744 *jāṭharo jalasambhūtaḥ pāvakaḥ pavanaiḥ saha.*

745 On this *tila-* see also MEULENBELD 1974, p.558.

746 *agnyāśaye bhavet pittam agnirūpaṃ tilonmitam.*

747 *agnyāśaye yat pittaṃ bhavet tat pācakaṃ jñeyam.*

748 *āhāro bhuktamātras tu vāyunā kriyate dvidhā*
sampraviśya gude samyak pṛthag annaṃ pṛthag jalam.

Here too I have refrained from reproducing the variants. (On the separation of *āhāra-* into *anna-* and *jala-* here cf. §10.19, on *āhāra-* and *anna-* also §7.39.)

749 After which, by the way, he (in contrast to Ḍalhaṇa) comments on 14,13, which creates the impression that 14,12 might not have been a part of SU as he knew it.

750 *annapānād ekāhenotpanno raso 'nnapānarasaḥ. sarvadhātūnām* (cf. n.649) *ity atra sarvaśabdena sthāyirasam api gṛhṇāti. yad uktaṃ carake puṣyanti tv āhārarasād rasarudhira-*

probably due to the fact that Cakrapāṇidatta is commenting primarily on the *āhārarasa-*, saying that *this* is not different from the other *rasa-*.[751]

Indu in his commentary on AS,Sū 9,p.79b, which says that one of the reasons for the body being overwhelmed by sleep is that its ☞ *srotas-*s are enveloped by phlegm,[752] characterises this enveloping by phlegm as *āhāra-rasavaśāt* (p.80a):

āhārarasavaśāc chleṣmaṇā srotaḥsv āvṛteṣu

"when the *srotas-*s are enveloped by phlegm due to [its] origin from/on account of *āhārarasa-*."

Both "due to [its] origin from *āhārarasa-*" and "on account of *āhārarasa-*" are puzzling, as it is not clear what *āhārarasa-*, which is not mentioned in the text of AS, is doing here; it may be that Indu means to say obliquely that the body is often overwhelmed by sleep after a meal, when *āhārarasa-* is produced due to digestion of the food just consumed. But in any case what is clear is that phlegm is here linked to *āhārarasa-* and not, as is usual (see §7.34) to *rasa-*, which too could be taken to be proof that the two *rasa-*s were not clearly differentiated and that both could be referred to not only as *rasa-*, but also as *āhārarasa-*.

I may in this connection also draw attention to Cakrapāṇidatta's commentary on CA,Ci 22,16 (see §7.22), where he says:[753]

āhārarasāt sarvadhātupoṣako dhāturasa utpadyate

"The *rasa-* which is a *dhātu-*, the *poṣaka-* of all the *dhātu-*s, arises from *āhārarasa-*."

Does this not seem to be a contradiction of Cakrapāṇidatta's own statement, just mentioned, on SU,Sū 14,11? But "arises" could merely mean "replenished" or the like, in which sense *rasa-* could even be said to be "derived" from *āhārarasa-*. As regards *poṣaka-*, the fact that *rasa-* is here expressly

---

māṃsa (see n.658) *ityādi. nanu yady evaṃ tat kiṃ rasadvayam. tathā cāṣṭadhātutvaṃ śarīrasya.* (SU3, p.239 here adds, in contradistinction to SU4, p.104, the puzzling and maybe corrupt or out of place *sthānāgnyādivat tasya raktam adhikam.*) *maivam. ekam evāyam āhāraraso dviprasthānaḥ poṣyapoṣakabhedena ... tena sthāyipoṣakarasayoḥ karmabhedād bhedo 'bhidhīyate pākāntarasthānāntaragatyantarābhāvāc caikatāpi.*

751 Note also the reading given by Aruṇadatta, in his commentary on AH,Śā 1,17, of AS,Śā 1,p.275b (also AS1,Śā 1,p.195), where semen is described as the essence of fully digested food; Aruṇadatta has *tatrāhārarasasya ... sāraḥ* instead of *tatrāhārasya ... sāraḥ*. However, this does look like a scribal or editorial error, though we cannot be sure on this point.

752 *śleṣmāvṛteṣu srotaḥsu śramād uparateṣu ca*
indriyeṣu svakarmabhyo nidrāviśati dehinam.

753 Reminding us of what was said of Aruṇadatta in n.668.

called a *dhātu-* seems to signify that what Cakrapāṇidatta actually means is the *rasa-* which is *poṣya-* with respect to the *poṣaka- (āhāra)rasa-*, but which is taken not be unchangeable, but also the starting point for a metabolic transformation leading to the other *dhātu-*s, which thus derive from it and may perhaps therefore be said to be "nourished" or "satiated" from it.[754] Indeed, Śivadāsasena in his commentary on DR,Āh 21 might be mentioning this form of "nourishing" or "satiating", for he says that the nourishing or satiating function of *rasa-* with regard to all the *dhātu-*s can be alternatively explained by the fact that the *rasa-* situated in the heart[755] permeates the whole body,[756] from which it could follow that this latter too somehow nourishes or satiates the following *dhātu-*s (cf. also n.663).

In this connection I may also draw attention to Cakrapāṇidatta's commentary on CA,Ci 15,16,[757] where he says that *rasa-* is of two types, *sthāyin-* or *poṣya-*, and *poṣaka-*. Commenting on different theories on how *rasa-* nourishes/satiates the elements, he uses the term *rasa-* to describe the theory according to which *rasa-* is converted in full to blood, this then to flesh etc., but *āhārarasa-* when describing the theory according to which *rasa-* divides into different parts which then go to each of the elements simultaneously. But matters are not that easy, for he uses *rasa-* (not *āhārarasa-*) when relating the theory that *rasa-* as a whole nourishes or satiates each of the elements in succession, a part of it remaining with each element; it is to be noted that in this connection Cakrapāṇidatta mentions the nourishment or satiation of the elements beginning with *rakta-*, not with *rasa-*, which is a further complication. So is *āhārarasa-* used by Cakrapāṇidatta in a quite different sense? But whatever Cakrapāṇidatta may actually mean, this is basically a problem relating to Cakrapāṇidatta's own theories and thus of no primary importance for the 'classical' texts themselves.[758]

---

[754] Cf. also AS,Sū 19,p.138b, where each *dhātu-* brings about nourishment or satiation (*puṣṭi-/poṣaṇa-*) of the next.

[755] I.e. the *sthāyin-* or *poṣya- rasa-*. Śivadāsasena, loc.cit.: *raktādivat sthāyī yaḥ sa poṣyaḥ. tasyaiva hṛd avasthānam uktam.*

[756] *yat tu rasasya sarvadhātupoṣakatvam uktaṃ tad api raktādipoṣakatayā tathā hṛdayasthāyirasasya kṛtsnadehavyāpakatayā vopapannam.*

[757] Very similar in this respect to his commentary on SU,Sū 14,11 cited in n.750.

[758] On Cakrapāṇidatta's remark *sa ca raso dehapoṣako 'mbubhava iti āpya ity arthaḥ* in his commentary on CA,Ci 22,16 see §7.22.

EXCURSUS X

(in place of footnote 676)

In his commentary on CA,Ci 8,39 (*srotasā ca yathāsvena dhātuḥ puṣyati dhātutaḥ*), Cakrapāṇidatta explains the process of nourishing/satiating in more or less the same way as we have seen above (§7.38), i.e. that ☞ *rasa-* (actually called *poṣaka-* in the commentary on the subsequent verses) either nourishes/satiates all the elements as *rasa-*, or else changes to each of these in succession to nourish them. But the list of the nourished/satiated elements begins with blood; *rasa-* is missing. This seems a bit of a mystery. The explanation lies in the fact that 8,40 states that "kingly consumption" (*rājayakṣman-*) arises due to (along with other things) obstruction of the ☞ *srotas*-s and due to the waning/wastage of blood etc.; 8,29 too mentions the obstruction of the *srotas*-s of blood etc., due to which the ☞ *dhātu*-s cannot be nourished/satiated.[759] Obviously Cakrapāṇidatta has here only followed the text of CA. Why this does not mention the nourishment/satiation of *rasa-* is, however, puzzling, especially since 8,166f. expressly mentions the nourishment/satiation of all *seven dhātu*-s due to clearing of the *srotas*-ṣ through medication. See in this connection also AS,Sū 19,pp.141b-142a, which tells us that the *dhātu*-s of the body increase and decrease due to variouṣ factors, "heavy" (*guru-*) ones increasing (*upacīyante*) through "heavy" things, "light" (*laghu-*) ones through "light" things, from which it follows that blood, flesh, fat etc. are respectively increased or augmented better by blood, flesh, fat etc. than by "other substances". *rasa-* is missing here,[760] though it is mentioned in the next passage (19,p.142b). A similar passage is found in CA, namely Śā 6,10; in this too *rasa-* is missing, but the total number of *dhātu*-s is nevertheless seven (not six as in the AS passage), namely flesh, blood, fat, tissue-fat/muscle-fat (☞ *vasā-*), bones, marrow and semen! There is no mention of *rasa-* in the passages adjoining either. Cf. also the parallel to these passages in Ḍalhaṇa's commentary on SU,Sū 15,10 in §7.19 and note also the discussion on ☞ *guṇa-* in §7.19 (*guru-* and *laghu-*, mentioned in the AS passage, are two well-known *guṇa*-s). It is also not clear what "substance" (*dravya-*) refers to, whether to substances in general, or to *dravya-* in a sense approaching that of *dhātu-* (cf. ☞ *dravya-* in Appendix II).[761]

Nevertheless, one reason I can think of for *rasa-* not being mentioned

---

[759] According to 8,43 *rasa-* is hindered from flowing and hence causes the disease.

[760] *tasmād anyebhyo dravyebhyo 'pi sutarāṃ raktam āpyāyyate raktena māṃsaṃ māṃsena medo medasā ....*

[761] Though probably *dravya-* does not denote the gross elements (☞ *mahābhūta-*).

is that the passages just mentioned tell us that certain substances of the body are augmented (or the like) by certain substances ingested; now blood, flesh, fat etc. may be ingested according to the texts, but I have not come across any clear reference[762] to the ingestion of the *dhātu- rasa-* (formed by the breaking down of food) of an animal.[763] That *rasa-* is mentioned in AS,Sū 19,p.142b would not speak against this explanation, since this passage tells us of ways to overcome the *kṣaya-* (cf. §§5.2f.) of the substances of the body with the help of substances not the same as, but serving to build up the substance whose *kṣaya-* is noticed; *rasa-* is to be built up by infusing the broth of flesh (*māṃsarasa-*), alcohol and milk.

Anyway, Cakrapāṇidatta's commentary on CA,Śā 6,10 would appear to show that he was aware of the problem of the non-mention of *rasa-*, for it states:

> *rasas tu pradhānatvena prakaraṇāntarokta eveti nehoktaḥ*
> "*rasa-* is not mentioned here because (*iti*) it is really (*eva*) mainly mentioned in connection with a different topic."

However, this statement is made in connection with a discussion on whether or not the various perceptions are also implied in Śā 6,10 or not; since touch, sound, sight, smell are mentioned, it seems that the fifth, namely taste, is here implied by *rasa-*, so that Cakrapāṇidatta's statement seems actually of no use in the context we are concerned with. What further complicates matters is that Indu in his commentary on the AS passage says that *dhātu-* refers to everything in the body that sustains this: "all the *doṣa*-s etc.".[764]

Cakrapāṇidatta omits *rasa-* in quite another context too, namely commenting on CA,Ci 30,282;292, where he explains ☞ *dūṣya-* as *raktādīni* "blood etc."; one wonders why *rasa-* here is not considered to be a substance capable of being corrupted. This might not be directly relevant for the discussion above, but it is interesting nonetheless, especially since CA itself refers to a "cooking" of the "elements" blood etc. in Ci 15,219, where no *rasa-* is mentioned. Cf. also Cakrapāṇidatta's commentary on *raktādi-* in CA,Si 1,44: *raktādayaḥ dhātavaḥ*. AH,Ut 36,39 = AS,Ut 42,p.339b tells us that the bite of a poisonous snake effects the *dhātu*-s blood etc. (*dhātūn rudhirādīn*); here too *rasa-* is not mentioned.

All this clearly necessitates further investigation. As regards the pos-

---

[762] CA,Śā 3,4,1 and 3,4,5 in §§13.20f. are not clear.

[763] See also the list of ingestible *dhātu*-s beginning with blood in CA,Sū 27,337, SU,Sū 46,130, AH,Sū 6,71 (on which cf. n.935) and AS,Sū 7,p.50b.

[764] *dhātuśabdena dehadhāraṇasāmānyāt sarve doṣādaya ucyante.*

sibility of a non-'standard' list of *dhātu*-s being presupposed here,[765] this should not be precluded either (especially in the case of CA,Śā 6,10, which expressly speaks of seven *dhātu*-s, one of these being *vasā*-).[766] One may speculate whether such passages may be remnants of a theory in which *rasa*- maybe played no part as a *dhātu*-, but the explanation may lie somewhere else. In any case, statements such as these are interesting, and show how important more detailed research into the theories of digestion and metabolism in ancient India is.

EXCURSUS XI

(in place of footnote 686)

The remarks on *mūtra*- here apply of course to a possible older theory of its origin, not the one which has gained general acceptance in our texts. On this latter see also BHĀ 1,3,172f., which, while giving a fair example of this theory, at the same time does create the impression of trying ingeniously to solve the problem of "water" in the body by saying that the liquid impurity of food is at first "water", but later turns to urine in the bladder (*vasti*-).[767] This stands in contrast to e.g. SU,Ni 3,21,[768] which calls the fine tubes (☞ *nāḍī*-), leading from the *pakvāśaya*- to the bladder and augmenting (*tarpayanti*) the urine in the latter, *mūtravaha*- "urine-carrying", from which it follows that urine is not formed in the bladder according to this passage, but merely stored in it. Ḍalhaṇa in his commentary on SU,Ni 3,21 explains the text as referring to tubes arising from the *pakvāśaya*- filling the bladder with urine, though he has difficulty in reconciling the plural number of tubes with a dual number of urine-carrying ducts (☞ *dhamanī*-), by which he is obviously referring to SU,Śā 9,7, which mentions two such ducts.[769] Interestingly, in

[765] Cf. §10.29 and the remarks on the *Suvarṇa(pra)bhāsottamasūtra* on p.558.

[766] Though with regard to the commentaries the same considerations as detailed in n.663 apply.

[767] *evaṃ vipakvasyāhārasya sāro nigadito rasaḥ. śeṣo grahaṇīstho* (cf. ☞ *grahaṇī*- in Appendix II) *maladravaḥ. maladravasya jalabhāgaḥ śirābhir vastiṃ nīto mūtraṃ bhavati. uktaṃ ca.*

āhārasya rasaḥ sāraḥ sārahīno maladravaḥ
śirābhis taj jalaṃ nītaṃ vastiṃ mūtratvam āpnuyāt (= ĀĀS 6931).
śeṣaṃ kiṭṭaṃ ca yat tasya tat purīṣaṃ nigadyate
samānavāyunā nītaṃ tat tiṣṭhati malāśaye.

[768] Cf. on this p.212 in Excursus VII.

[769] That the same text may possibly (though not necessarily) have in different parts different theories on the same topic is of course not considered by him.

his commentary on SU,Śā 9,7 Ḍalhaṇa then tells us more or less the same as the BHĀ passage above, namely that the two water-carrying (*toyavaha-*) ducts (*dhamanī-*) leading downward in the body and not further elaborated upon in the text of SU actually carry a sort of pre-urine not actually water, but only looking like water; this pre-urine is called urine when it reaches the hollow of the bladder.[770] The difficulty of reconciling this with the statement of the text that two ducts which approach the bladder carry urine[771] seems to have been circumvented by Ḍalhaṇa through the vague explanation that two ducts carrying urine are attached to the bladder, without any specification on what this means.[772]

CA,Si 9,4 likens the bladder to the ocean and all the ☞ *srotas*-s in the body carrying water (*ambuvaha-*) to rivers, the former being the receptacle or base (*pratiṣṭhā-*) of the latter. The passage says too that the body is filled (like the sky with the sun's rays) with many *srotas*-s, called ☞ *marman*-s,[773] having the bladder as their *mūla-*,[774] though in this case the *srotas*-s are not expressly characterised as carrying water. This is quite problematic, for the *mūla-* of the *srotas*-s carrying water is *not* the bladder according to e.g. CA,Vi 5,8, SU,Śā 9,12, AS,Śā 6,p.331a, this being the *mūla-* of the *srotas*-s carrying urine. What makes the whole even more problematic is BHĀ 1,3,174, which tells us that wind in the *malāśaya-* forces the urine and faeces out of the body.[775] From this one would have to conclude that urine too is present in the *malāśaya-*, even though 1,3,172 (mentioned above) has just told us that it is in the bladder![776]

---

[770] *nāpi toyasya toyabhāvanā. tena toyam atra bhaviṣyato mūtrasya kāraṇabhūtaṃ pakvāhāravivekajaṃ dravamayam* (var: *dravam accham*) *udakam ivodakam. tad api dve eva vahataḥ. tad evodakaṃ bastivivaraprāptaṃ mūtram ity ucyate.*

[771] *mūtrabastim abhiprapanne mūtravahe dve.*

[772] *tadvahe dve mūtrabastisambaddhe.*

[773] This latter nomenclature, reminiscent of SU,Ci 7,37f. discussed in Excursus II at the end of chapter 6, causing the commentator Cakrapāṇidatta some difficulty. On *marman-* = *srotas-* see too the discussion on ☞ *marman-* in Appendix II.

[774] On *mūla-* cf. §6.18.

[775] *tatra malāśayasthenāpānavāyunā* (the variant *malāśayenāpānavāyunā* is, as the editor rightly points out, obviously a mistake) *preritaṃ mūtraṃ meḍhramārgeṇa purīṣaṃ gudamārgeṇa śarīrād bahir yāti.*

[776] According to ŚĀR 1,5,27 = ĀĀS 6890 too, as well as ĀĀS 6891, the *malāśaya-* is one of the particular abodes of wind within the body.

CHAPTER 8

The Suśrutasaṃhitā (3): Semen of Females and Other Problems

8.1. SU has, however, still more information to offer us regarding not only the origin of the procreatory-menstrual fluid, but also its relation to semen. Śā 2,36 tells us that during coitus, the *ārtava-* of a woman flows, just as clarified butter melts through the agency of fire;[777] both NYC and Ḍalhaṇa say that this is due to the heat of friction of the generative organs,[778] which could be correct. But in any case, the verse could state either that

---

[777] *ghṛtapiṇḍo yathaivāgnim āśritaḥ pravilīyate*
visarpaty ārtavaṃ nāryās tathā puṃsāṃ samāgame (var: *nāryāḥ puruṣasya samāgamāt*).

NYC (like ĀY 4,131) obviously reads *ghṛtakumbho* for *ghṛtapiṇḍo*; see the commentary in n.788. ĀY 4,131 also has *suvilīyate* for *pravilīyate*. Cakrapāṇidatta's and Śivadāsasena's commentaries on SU,Sū 14,14 and DR,Āh 22 respectively quote this verse with *ghṛtakumbho* instead of *ghṛtapiṇḍo* too, also reading *puṃsāṃ tathā strīṇāṃ* for *nāryās tathā puṃsāṃ* (i.e. "thus *ārtava-* flows during the coitus of men [and] women"). The verse given above has caused GHĀṆEKAR 1986, p.43 much headache, for it obviously causes difficulties if one insists, as is done loc.cit., that *ārtava-* here means "ovum". His solution is: the verse means to say that the excitement of sexual intercourse makes the Graafian follicles burst and *hastens* the liberation of ova (see §11.26 below)! It may here also be noted that on p.71 the state of maturity of the "ovum" (this includes the premature "ovum" too) is taken, in connection with SU,Śā 3,5 (see n.178), to be in actual fact the cause of the sex of an embryo. Pp.69f. say that *strībīja-* (= *ārtava-*, p.62) and *śukrāṇu-* (of which *śukra-* is made up, p.62) have two X and one Y chromosome(s) respectively. When the former divides, so do the X chromosomes; thus each new part has one. But as there is only one Y chromosome, we get *śukrāṇu-*s with or lacking this, the presence of the Y chromosome being the excess of semen of the Indian texts. The *strībīja-* has female and male determinants, in which context a theory attributed to Otto Schoner is cited (GHĀṆEKAR 1986 was first published at about the outbreak of the Second World War, as the date of the foreword shows); this theory relates to the discharge of different ovaries and is very reminiscent of what we will see below in §13.3. How all this relates to what is clearly a fluid seems immaterial in this apologetic context. The difficulties of those who insist on using the term "ovum" is also evident in the case of RAY/GUPTA/ROY, p.11, where we find that "It is stated that this act is the cause of activating the female secretion (*ārtava*)", without any further explanation, though one would very much have liked to have had an explanation in view of the fact that on pp.10f. the talk has all along been of the "ovum". Similarly, we read on p.36 of SINGHAL/GURU: "As a lump of ghṛta melts upon contact with heat, similarly ovum is discharged in woman on contact with man", "ovum" being explained as "*Ārtava*" and "contact" as "intercourse". Here too we find no further comments, though in this case we may at least point to the statement, showing some awareness of the problems involved, in §1.7 above.

[778] See n.788 and n.790 respectively.

ārtava- flows out into the uterus (to combine there with semen), in which case it cannot have been present there before, or else that *ārtava-* flows out of the vagina, in which case it would be some other fluid. As regards the latter possibility, it could maybe represent what is commonly regarded as the lubricating fluid of the woman,[779] or else be something different; whatever it may be is, however, only of secondary importance for our discussion at present, though we shall return to the problem later. We shall also shelve for later scrutiny (cf. §12.15ff.), since in the context here it would complicate matters even more, the possibility that the fluid mentioned does not leave the vagina at all, but merely the vulva, i.e., that it issues from some place within the vulva but different from the vagina; we shall thus confine ourselves for the present to the two cases of a flow into the uterus and a flow out of the vagina. In the latter case, if the fluid is simply said to flow out of the vagina, then it could either have been present throughout in the uterus before,[780] or else have flowed into this just prior to its emission, as in the first alternative.

8.2. If this first alternative is correct, then we have here a parallel to passages such as those mentioned in n.168, which say that semen has to collect from all over the body before being ejaculated into the woman, in the case of the woman herself the ejaculation then being within her own body, into the uterus. Indeed, Cakrapāṇidatta in his commentary on Ca,Ci 2,4,46-49, cited in n.168 as detailing the process of this ejaculation of semen, tells us that *śukra-* here subsumes *ārtava-*, which could be taken to point to his saying that a parallel exists between these regarding their flow. Actually, however, the *tertium comparationis* seems to be the role these two fluids play in giving form to the *ātman-*,[781] which means that Cakrapāṇidatta's state-

---

[779] Actually, there are at the very least two fluids, namely that secreted by Bartholin's glands, which have minute ducts invisible to the naked eye and open out in the vulva near the mouth of the vagina before its introitus (they became known to Western medical science in 1678), and that which is exuded from the inner epithelium of the vagina itself. Moreover, the cervical canal of the uterus too gives off a viscous and clear mucous secretion (on this see also n.1437). (Males too may discharge a fluid on arousal and before an orgasm; this fluid comes from the Cowperian glands. But the discharge is as a rule not copious and as such often not perceived by the males concerned themselves.)

[780] One could argue that the simile of clarified butter melting would imply that a solid in the uterus melts, but this does appear to be rather far-fetched. Indeed, if the state of the *ārtava-* prior to its flowing should be hinted at here at all, then it is rather more probable that a viscid state is meant.

[781] The text says:
śukraṃ ceha prakaraṇāgatatvenoktaṃ. tenārtavam apy ātmano rūpadravyaṃ

ment is of no use to us here. Nevertheless, the inference of a similar ejaculation of semen and the procreatory-menstrual fluid would be in perfect keeping with our deductions in §6.12. In this context I may be allowed to mention Śā 3,10 (cf. §6.3) too, which describes what is obviously menstruation by saying that the wind propels *ārtava-* to the ☞ *yonimukha-*, which could refer to the uterus. This description of propulsion is a parallel to that of semen (during intercourse) not only as given in Śā 3,4,[782] but generally as well (see n.168). From this it could be inferred that during menstruation too *ārtava-* is made to enter the uterus in the same manner as during intercourse, but in such a large quantity that it flows out via the vagina. However, there may be a difference between what is supposed to happen during menstruation and what is supposed to happen during intercourse; it is possible that, apart from the ejaculation into the uterus during the latter, the procreatory-menstrual fluid was held to trickle slowly into the uterus at all times, so that Śā 3,10 might be interpreted differently, as discussed already in §6.7. Note in this context also the accumulation in the uterus during pregnancy assumed in §6.17, though it could be that what was thought to happen during a pregnancy is different from what was thought to happen at other times.

8.3. If, however, the second of the alternatives in §8.1 is correct, then we will have to infer that *ārtava-* also comes out of the vagina during coitus, and this fluid can obviously not be blood-like, i.e. not like the menstrual fluid or like the procreatory fluid, provided this be blood-like too, as Sū 15,5,[3]

---

jñeyam.

This refers to CA,Ci 2,4,49:

aṣṭābhya ebhyo hetubhyaḥ śukraṃ dehāt prasicyate
carato viśvarūpasya rūpadravyaṃ yad ucyate.

The qualification of semen as *rūpadravya-* is interesting in the light of a similar terminology relating to procreatory fluids in some modern religious groups; cf. DAS,R.P. 1992A, p.406. This does not of course necessarily imply any direct relationship, but see nevertheless DAS,R.P. 1992A, p.390,n.17 (cf. too p.400,n.76).

[782] SU,Śā 3,4:

tatra strīpuṃsayoḥ saṃyoge tejaḥ śarīrād vāyur udīrayati. tatas tejo'nilasannipātāc chukraṃ cyutaṃ yonim abhipratipadyate saṃsṛjyate cārtavena.

This passage also tells us that the procreatory fluids of male and female combine in the ☞ *yoni-*. Since such combination is otherwise said to take place in the uterus, ☞ *yoni-* here must include or refer to this, as Ḍalhaṇa in his commentary too points out:

yonim abhipratipadyata iti yonim abhitaḥ sarvataḥ pratipadyate. tena yones tṛtīyāvartasthitagarbhaśayyāṃ pratipadyata ity arthaḥ. tatra prāguktavisarpitenārtavena ca saṃsṛjyate.

On the problematic "whorls" (or "folds"?) (*āvarta-*) of the ☞ *yoni-*, in the third of which the uterus is situated, see e.g. TIVĀRĪ, 1,pp.6-9 (with a lot of speculation).

leads us to assume (cf. §6.9). In this case too Śā 3,10 assumes importance, for it says, at least according to some textual traditions, that the *ārtava-* which flows is also *vivarṇa-*, which may mean "colourless" (cf. §§7.8f.), and that would obviously be in keeping with a discharge from the vagina during intercourse.[783] But then "accumulated/increased in a month" (*māsenopacita-*) becomes a problem, for this seems to refer to both states of the *ārtava-* mentioned in 3,10, namely *īṣatkṛṣṇa-* and *vivarṇa-*, implying that the latter too flows out of the body during menstruation, which is of course a very problematic supposition. One could counter that, if another fluid is really meant, then it too could, as an exact parallel to the procreatory-menstrual fluid, have been thought to be discharged after being collected for a month. Or else one could say that the collection for a month mentioned in the verse refers only to the other state (*īṣatkṛṣṇa-*) of *ārtava-*, though the wording does make it difficult to defend this proposition.

8.4. Nevertheless, according to the above *ārtava- could* refer to two different fluids. It could, however, also be that it only seems to us that different fluids are meant, whereas what is referred to is in actual fact only the same fluid in different states. Thus it is possible to suppose that the procreatory-menstrual fluid merely undergoes a change of colour, being either blood-like in the uterus (and capable of fertilisation) and later, when it is "old", coming out with this colour during menstruation, but colourless when it comes out of the vagina during intercourse, or else colourless when it comes freshly into the uterus and in part flows out via the vagina, but blood-like when it comes out later during menstruation, in which case it would *not* be fertile when like blood, which would, however, be in opposition to Sū 15,5,[3] (see §8.3)[784] as well as the general usage of "blood" to describe this fluid in its procreatory function too. However, it may be that there were different theories current on this score. Note in this respect the quotation from Gopura's work in ĀY 4,129f. already referred to in §6.11; after describing the quantity of the fluid *rajas-* in the body, the text says (4,130cd):

ahni (printed: *ahniś*) *caturthe prasṛtis tiṣṭhec chuddham athārtavam*
bālānāṃ vātiraktaṃ ca svacchaṃ snigdhaṃ ghanaṃ tathā

"On the fourth day [after the onset of menstruation] now, a *prasṛti-* [of] pure *ārtava-* of girls (scil. obviously: women) may

[783] In this context it is irrelevant what ☞ *yonimukha-* (cf. §8.2) actually refers to, for the fluid will in any case have to come out via the vaginal opening.

[784] Unless *ca* in *raktalakṣaṇam ārtavaṃ garbhakṛc ca* be taken to mean not "and", but "or", which is, however, problematic.

be taken to remain[785] [in the body],[786] extremely red indeed (*vā*), and clear [and] unctuous, further dense".

I suspect that there is a mistake in this verse as it is printed, and that we have to read either *nātiraktaṃ ca svacchaṃ* "is not too red, and clear" or *vātiraktaṃ vā svacchaṃ* "is either extremely red, or clear". But whatever the true reading be, we see here that *ārtava-* (a synonym of *rajas-* mentioned before it in the text?, or of *rajas-* in a certain state?) is, in one form at least, clear (*svaccha-*), which would fit *vivarṇa-* "colourless" in SU,Śā 3,10. The trouble is, however, that *vivarṇa-* can mean not only "colourless", but also "bad-coloured" and "pale". Moreover, "colourless" could also be taken to refer to the fact that the menstrual fluid is believed in SU not to stain clothes, but to be removable by washing (see §7.4). Complicating matters still further is that we have so far not even considered the possibility that Śā 2,36 could refer to a fluid flowing out of the vagina but bypassing the uterus, or to a fluid flowing first into the vagina and then into the uterus! Actually, I have come across no hints in Indian medical literature which could lead us to consider such possibilities, but I mention them here nevertheless because we know that discussions on such points are to be found in other medical systems, e.g. in Graeco-Arabic medicine. More plausible would be, however, a discharge into the vulva not coming through the vagina (of the sort secreted e.g. by Bartholin's glands; cf. n.779), but in the case of SU,Śā 2,36 the possibility of this being meant does not seem very great, as the context leads us to expect something which enters the uterus. In any case, considerations such as the above on the colour of the fluid we are here dealing with lead us to consider whether the question ĀY 4,68cd, cited in §2.2,[787] might not actually have to be understood as: "Is the *rajas-* of women red indeed, or is *rajas-* without blood/red[ness]?"

8.5. All this is confusing, but not only to us. Indeed, the commentators of SU,Śā 2,36 too have obviously had similar problems (though these may not tally in all with what we have discussed above). Thus NYC in the commentary (fols.20r;v) tells us that not only *ārtava-* (= *rajas-* and *śoṇita-*), but also a fluid called semen (*śukra-*) is discharged by women during intercourse. Will not, the author Gayadāsa asks, this latter be said to be capable of producing an embryo in a later passage (see §8.6)? But it is the former which, according to him, is meant here, this being capable of uniting with semen

785 My effort to translate the optative. On the use of the optative in general cf. e.g. RENOU, §292.

786 Cf. §6.11.

787 *raktam eva rajaḥ strīṇāṃ kiṃ vā raktaṃ vinā rajaḥ*.

from the male to form the embryo in the uterus, namely by being discharged due to the friction of the two generative organs.[788] From this it seems to follow that the semen of the female, even though capable of somehow producing an embryo (how, is as yet unclear), cannot do so in the same manner as the procreatory-menstrual fluid, which alone can produce what we might call a proper embryo by uniting with semen from the male. It may here be recollected that there is a chance, though probably rather slender, that both fluids of the female are discharged to the exterior through the same opening according to a certain reading of NYC on SU,Śā 4,24 given in Excursus III at the end of chapter 6. Whether NYC indeed makes such a statement on "semen" of women or not, in the case of the procreatory-menstrual fluid the discharge would in any case most probably refer only to menstruation, so that this would not necessarily contradict the statement that this same fluid is ejaculated into the uterus at another time, namely due to intercourse. But it is doubtful whether the passage in Excursus III is to be understood in this sense anyway. As regards Ḍalhaṇa, in his commentary on Śā 2,36 he first poses a question: how can *ārtava-* flow (when intercourse takes place just after menstruation), since the old *ārtava-* is gone, but new *ārtava-* has yet to take its place, this latter being present in only a small quantity and in a state like to that of viscousness and therefore incapable of flowing?[789] The answer, Ḍalhaṇa says, is that *ārtava-*, capable of uniting with semen to form a child, is made to flow into the womb through the heat of friction of the generative organs. Women, however, also secrete a fluid called semen (*śukra-*) during intercourse, but this cannot produce an embryo and is hence not relevant in the context.[790] Since NYC does not say that *śukra-* cannot produce

---

[788] *śukram apy āsāṃ samāgame visarpati* (ms: *visarpyeti* or *visarppeti*). *nanu tad garbhāyeti* (ms: *garbhoyeti*) *prāg eva vistāreṇābhihitam. rajas tu garbhāyeti. tadvisarpaṇenātra prayojanam. tad eva vyaktīkurvann āha ghṛtakumbho yathā hy agnim ityādi. ṛtubhavam ārtavam ṛtukālajaṃ śoṇitam. tat puṃsāṃ samāgame. dvayendriyasaṅgharṣajenoṣmaṇā vilīnam ārtavam itas tato dhāvitaṃ garbhāśaye* (ms: *garbhāgarbhāśaye*) *śukreṇa sahaikasthānastham ekajīvopagataṃ sthānadvayasthaṃ jīvadvayopagataṃ janmasambhavahetur bhavati* (the last line introduces Śā 2,37, which states that, if the *bīja-* in the womb is divided, each part will give rise to a separate child; on *bīja-* cf. §§2.5f.).

[789] I may add that the mention of the quantity of the fluid seems to presuppose that the old *ārtava-* issues due to its quantity; does this hint at something like the Tibetan theory as described by MEYER,F. 1983 in §6.15?

[790] *nanu purāṇam ārtavam upacayād dinatrayaṃ srutvā svayam eva vinivṛttaṃ nūtanaṃ tv alpaṃ styānībhūtam iva pravartitum akṣamam. tat katham ārtavasañcāro yena tatsaṃsṛṣṭaṃ śukraṃ garbhajananasamarthaṃ bhavatīty* (var: *śukraṃ garbhāśayam anupratipadyata ity agre vakṣyamāṇam upapatsyata ity*) *āśaṅkyāha ghṛtetyādi. puṃsāṃ samāgama indriyadvaya-*

an embryo at all, we seem to have different statements of the commentators of this passage here. But the commentators do agree in supposing that the procreatory-menstrual fluid enters the uterus during intercourse. This is also borne out by Ḍalhaṇa's commentary on SU,Śā 3,4 (which presupposes an ejaculation of semen from the male during intercourse in the manner mentioned in n.168), where it is said that "there (in the womb) it (semen) unites with *ārtava-* which has flowed in the manner described before (which can refer only to SU,Śā 2,36)";[791] from this it follows that according to Ḍalhaṇa the procreatory-menstrual fluid is not present in the womb before intercourse, but enters it subsequently. But there seem also to have been theories according to which *ārtava-* in SU,Śā 2,36 refers *only* to *śukra-* "semen";[792] though we seem to have no further statements on this seemingly not widely accepted explanation of the verse, it does appear unlikely that according to this explanation *ārtava-* in this verse is a procreatory fluid.

8.6. It must, however, be stressed here that what the commentators say on semen is not all based on the text of SU, for Śā 2,36 refers only to *ārtava-*.[793] In our discussion of the verse (not the commentaries) above we also considered the possibility (in part because of other statements in SU) that a fluid different from the one so far being considered might (also) be meant, but whether this is semen or not was not part of the discussion, as the SU passage does not use this term. However, the semen of women *is* mentioned by SU, but what is said of this is the exact opposite of what Ḍalhaṇa says in his commentary on Śā 2,36, since Śā 2,47 tells us that two women having sexual intercourse with each other[794] discharge semen (*śukra-*), out

---

saṅgharṣajenoṣmaṇā vilīnam ārtavaṃ visarpati. tac ca visarpitaṃ śukropagataṃ garbhāśayam anuprāptaṃ jīvopagataṃ garbhasambhavahetur bhavati. ārtavam ṛtukālajaṃ śoṇitam. śukram apy āsāṃ puṃsāṃ samāgame kṣarati. na tu tad garbhopayogīti tatkṣaraṇapratipādanaṃ na yuktam. tathā ca vṛddhavāgbhaṭaḥ. yoṣito 'pi sravanty ... (i.e. AS,Śā 1,p.286a, which we shall discuss in §9.7) *iti.*

[791] *tatra prāguktaprakāravisarpitenārtavena ca saṃsṛjyate.*

[792] See e.g. Śivadāsasena's remarks cited in §8.15 below.

[793] Cf. n.777.

[794] According to PLOSS/BARTELS/BARTELS 2,p.78f., such forms of tribadism are quite possible with an enlarged clitoris, such enlargement being characteristic of certain races, but also artificially brought about. In this regard see also the interesting statement of CHÖPEL, pp.196f.: "It is said that the women of Loṭiyaṇa in the Sindhu of Greater Persia are very passionate; that is well known throughout the world. They have a very large clitoris; at times it even emerges outside the labia. Some can even do the deed with another woman; it is almost the size of the male member. For some, it usually is clearly outside the female genitals." "Greater Persia" probably refers to the region of what is now Pakistan and Afghanistan or a

of which a boneless child is born.[795] Ḍalhaṇa, as in many other cases where the going gets tough, chooses to ignore this problem, stating merely that "boneless" means "with little [developed]/few and soft bones";[796] by thus choosing not to let us know his opinion on whether *śukra-* here is indeed a procreatory fluid he seems to evade the problem that this verse poses if set in relation to his own commentary on 2,36 (see §8.5). In his commentary on SU,Sū 14,14 (discussed in §8.15 below), however, he refers to this verse as a statement showing that women have *śukra-* with procreatory power, but in connection with a theory of "others" (*anye*), which thus does not seem to be his own. Gayadāsa's commentary in NYC on fol.23v is, however, in keeping with what we have seen in §8.5: he characterises "boneless"[797] in the same manner as Ḍalhaṇa, but also says expressly that the child is indeed born out of the semen of the women. According to him, begetting a child thus is, however, something as bad as begetting a child only out of the procreatory-menstrual fluid, as e.g. in Śā 2,48f.,[798] according to which a woman having in-

---

part of it; if so, it bears consideration whether this (modern) passage from CHÖPEL might be of help in localising the belief referred to in SU,Śā 2,47, even though the span of time between the two statements is very large.

795 *yadā nāryāv upeyātāṃ vṛṣasyantyau kathañcana*
muñcataḥ śukram anyo'nyam anasthis tatra jāyate.

A variant reads *muñcantyau* for *muñcataḥ* (thus also ĀY 4,158, BHĀ 1,3,57;187). Cakrapāṇidatta, in his commentary on CA,Sū 1,92ff., quotes this passage with a variant first line: *yadā nārī hi nāryā tu maithunāyopapadyate*, whereas ĀY 4,158 reads in its stead: *dve nāryau tu vṛṣayantyau kāmādhikyāt kathañcana*. Yaśodhara in his commentary (p.81) on KĀM 6(=2,1), p.80 quotes the verse as:

yadā nārī ca nārī ca maithunāyopapadyate
anyo'nyaṃ muñcataḥ śukram anasthis tatra jāyate.

Incidentally, *vṛṣasyat-* (*vṛṣasyantyau*) "in heat; randy, horny", used (if our dictionaries are to be believed) as a rule for *females* (originally: "longing for a bull"), is by O'FLAHERTY, p.38 translated as "acting like virile men", obviously to prove that we have here a "male analogy (or rather the male-female analogy with male predominance)"; this is thus another example showing how little the translations and explanations of O'FLAHERTY are to be trusted (see n.20).

796 *anasthir ity īṣadarthe nañ. tenālpakomalāsthiḥ.* BHĀ 1,3,57 is similar.

797 Which to him seems to hint at the lack of what a child usually obtains from its father.

798 This passage runs:

ṛtusnātā tu yā nārī svapne maithunam āvahet
ārtavaṃ vāyur ādāya kukṣau garbhaṃ karoti hi.
māsi māsi vivardheta garbhiṇyā garbhalakṣaṇam
kalalaṃ jāyate tasyā varjitaṃ paitṛkair guṇaiḥ.

According to Ḍalhaṇa, these lines might be spurious, since Jejjhaṭa (mostly known as Jejjaṭa or Jajjaṭa) does not quote them (*imau dvāv api ślokau jejjhaṭācāryeṇa na paṭhitau*).

tercourse in a dream during her fecund period produces out of her *ārtava-* an embryo called ☞ *kalala-* bereft of anything coming from a father.[799]

8.7. The problem of female semen also crops up in connection with SU,Śā 2,38-45, where five different abnormal persons are described as being produced due to abnormal intercourse of the parents. The abnormities of the first four refer to their sexual practices, the fifth, called *ṣaṇḍhaka-* in 2,41 and *ṣaṇḍha-* in 2,44, is according to 2,42 a male who acts and looks like a female (*strīceṣṭitākāra-*), and is, according to 2,44, without *śukra-*, whereas the other four have *śukra-*. Now the term ☞ *ṣaṇḍha(ka)-* in general is problematic, as it may refer to persons who are impotent, homosexual, otherwise sexually 'perverted', sterile etc., but fortunately the description given here is clear, so that this does not create any problems for us. What is problematic is Śā 2,43, which tells us that due to certain types of intercourse a girl may be produced who acts like a man.[800] Coming as it does after 2,42, this verse does give the impression of describing a female counterpart of the male *ṣaṇḍha-* in 2,42. According to both NYC (fol.22v) and Ḍalhaṇa (in their commentaries on 2,45),[801] there are both male and female *ṣaṇḍha-*s,[802] and NYC in the commentary on 2,43 states expressly that this passage refers to a female *ṣaṇḍha-*.[803] A breast-less,[804] man-hating female *ṣaṇḍha-* or *ṣaṇḍhī-* is indeed mentioned in AH,Ut 33,40 = AS,Ut 38,p.292a and CA,Ci 30,35 respectively, but not described further. Thus we could have a reference to a female *ṣaṇḍha-* in the SU passage.[805] If, however, a female *ṣaṇḍha-* should indeed be meant in Śā 2,43, then how would this *ṣaṇḍha-* be defined with respect

---

[799] NYC: *anasthimūrtir jāyate* (ms: *jājāyate*). *anasthiviśeṣeṇa pitur aṃśatvāt. tatrādheyabhūtāyā yoṣitaḥ śukreṇādhārabhūtāyā yoṣitaḥ śukraṃ garbham ādhatte. alpakomalāsthir anasthi[ḥ]. atiśayitapātāt strīśukreṇa garbhasambhavaṃ pratipādya tādṛśād* (ms: *nādṛśād*) *eva pāpād ārtavamātreṇāpi garbhasambhavaṃ nirdiśann āha. ṛtusnātetyādi* (which refers to the text given in n.798).

[800] *sā bhaven naraceṣṭitā.*

[801] Both also referring to CA,Śā 2,17, which is quite problematic; see p.582 in the discussion on ☞ *ṣaṇḍ(h)a(ka)-* in Appendix II.

[802] NYC and Ḍalhaṇa: *ete puruṣadoṣāḥ. yoṣitām apy eta eva draṣṭavyāś* (ms: *dṛṣ°*) *carake* (Ḍalhaṇa: *carakeṇa*) *naranārīṣaṇḍhāv ity ukteḥ.* (Only NYC:) *atrāpi naraṣaṇḍhā nārīṣaṇḍhāś coktāḥ.*

[803] *naraṣaṇḍhaṃ praviśya nārīṣaṇḍhaṃ darsayann āhartau puruṣavad vetyādi* (i.e. 2,43) (ms: *puruṣaṃdvetyādi*).

[804] This is explained by the commentaries variously as meaning without or with small breasts.

[805] Cf. also ZWILLING, p.208 on a female ☞ *paṇḍaka-*.

to 2,44? Would it not follow from the latter that the female *ṣaṇḍha-* too would be without *śukra-*, especially since such *śukra-* is expressly mentioned in 2,47 only a few verses later (see §8.6)? That this is problematic goes without saying, for 2,44 obviously refers to procreatory semen, whose absence in the female *ṣaṇḍha-* would then mean that it is present in other females. Or is *śukra-* in 2,44 seen not as "semen", but as a procreatory fluid? In that case the procreatory-menstrual fluid might be meant in the case of the female *ṣaṇḍha-*. But according to 2,47 female semen too can be procreatory! Thus we see that, if 2,43 does indeed refer to a female *ṣaṇḍha-*, we end up with a problem in any case.[806] This has obviously also been seen by Gayadāsa in NYC, for why else would he point out in his commentary on 2,45 that in the case of the female *ṣaṇḍha- ārtava-* is meant?[807] Indeed, SU,Ut 38,18, mentioned already in §6.1, tells us expressly that a female *ṣaṇḍī-* (= *ṣaṇḍhī-*) lacks *ārtava-*.[808] Aggravating the whole are CA,Sū 19,4,9, which mentions a disorder called *ṣaṇḍhayoni-*, i.e. a ☞ *yoni-* which is characterised as

---

[806] I may here also mention that 2,45 (see p.581) seems to be taken by GHĀṆEKAR 1986, p.47b (2,45 is 2,47 there) to refer to females too. As far as the vessels carrying semen are concerned, this could indeed be justified. But according to loc.cit. these vessels are taken to refer to the male only, for which one could indeed build a case, since 2,44 (loc.cit. 2,46) refers only to the 'abnormal' males, and 2,46 (2,48 in op.cit.) too expressly mentions only a son, i.e. a male (though it must also be pointed out that "son" might have been used in the sense of child, and that a variant has *garbho* "embryo/foetus" in the place of the generally accepted *putro* "son"):

āhārācāraceṣṭābhir yādṛśībhiḥ samanvitau
strīpuṃsau samupeyātāṃ tayoḥ putro 'pi tādṛśaḥ.

The problem seems to be seen by GHĀṆEKAR 1986 not in the mention of vessels carrying semen, but in that of an erection, and in this context we are told that females too have an erection during intercourse, namely that of the clitoris, intercourse, however, still being possible even without an erection of this, as women are passive (*niśceṣṭ*) during the act!

[807] After a quotation from CA,Śā 4,30f. detailing the reasons for certain "diseases" (*vyāpad-*) (among them sterility and non-defined gender) of female and male offspring, the commentary states (fols.22v-23r):

tatra doṣā[s] strīpuruṣavyāpattī ṣaṇḍhasaṃjñateti. ārtavavahaśukravahasrotovināśād (ms: °*śroto*°) *avi[dya]mānārtavaśukre ca tayo[ḥ] śaktyupapātād vāphalatvam.*

Cf. also the *strīṣāṇḍhyanidāna-*, BYT 147,27f.:

strīṇām ārtavam apy evaṃ pṛthag doṣair dviśas triśaḥ
raktena ca viduṣṭaṃ hi nāpatyajananakṣamam.
pūrvoktāny eva nāmāni vidyād eṣāṃ bhiṣagvaraḥ
bījadoṣā api klaibye gaṇitā niṣphalā yataḥ.

Here both *ṣāṇḍhya-* "*ṣaṇḍha-*-hood" and *klaibya-* "☞ *klība-*-hood" seem to mean "sterility".

[808] She also lacks breasts: *anārtavastanā ṣaṇḍī kharasparśā ca maithune.*

being *ṣaṇḍha-*,[809] and SU,Ut 38,8, where *ṣaṇḍākhyā* is clearly to be connected with *yoniḥ* in the same verse.

8.8. But whether the passages discussed in §8.7 refer to the semen of women or not does not alter our discussion in §8.6, from which it is quite clear that there was a theory that women do discharge something called semen, which is also fertile. Now the passage discussed in §8.6, namely SU,Śā 2,47, tells us only of a discharge of this semen during homosexual intercourse between women, but if it is discharged then, would we not be right in assuming a discharge during heterosexual intercourse too? Also, the procreating power of this female semen raises the question of its possible relation to the procreatory-menstrual fluid, more or less on the lines of the discussion in §§8.1-4, though in the case of this semen as described in the passage in question here we can be sure that it must on the one hand be present in the womb (since a child is produced out of it),[810] on the other flow out of the vagina (since the semen of the two women mingles), obviously coming out of the womb.

8.9. As if all this were not already problematic enough, we have yet another problem, namely the significance of the term *śukra-*. Is it used merely for reasons of parallelity, because the fluid which leaves the body of males is called so? Or does it, rather, even in the above contexts referring to females, primarily mean "white"? That would seem unlikely, for though the ☞ *dhātu- śukra-* "semen" is thought to be present in the body of females too (see §2.4)[811], in actual descriptions usually only male semen is described (cf. also §5.16),[812] and the discharge(s) we are here dealing with do(es) not

[809] That the *yoni-* itself is characterised thus is clear from the fact that this disorder is, along with several others, subsumed under the heading ☞ *yonivyāpad-* "disorder of the *yoni-*". Cf. in this respect also the other references to such terminological problems given in the discussion on *yonivyāpad-* in Appendix II. But even if *ṣaṇḍhayoni-* should refer to the woman, "she whose *yoni-* is *ṣaṇḍha-*" would still make *ṣaṇḍha-* a characterisation of *yoni-*.

[810] The semen of one woman must reach the womb of the other to mingle there with the semen of the latter.

[811] Cf. also n.20.

[812] This tendency of the texts is very obvious e.g. in CA,Ci 30,134, where only the *śukra-* of males is described, but referred to as *śukraṃ śarīriṇām* "*śukra-* of corporeal beings", in keeping with the great male-orientedness of this text remarked upon in §5.16. The use of *śukra-* in this way in our texts has played a role in a modern theory, often used to "prove" the mention of hormones in ancient Indian medical texts, that *śukra-* in these texts is of two kinds, namely "internal" (*antaḥśukra-*) and "external" (*bahiḥśukra-*) (cf. e.g. TIVĀRĪ, 1,pp.46;50, as well as the detailed chart on p.49, which is an interesting document on the development of modern Āyurvedic thought; incidentally, Gananath Sen seems to be the first to have postulated

seem to be the same as this latter. But we must nevertheless regard the explanation of *śukra-* of women as primarily referring to something white as a possibility until we know more, since our knowledge on the notions prevailing in ancient India in such connections is very meagre. Still, we have not so far come across any description which would actually allow us to regard this *śukra-* as being characterised as "white". SALOMON,C., p.274 does speak of a notion current in *modern* Bengal among a certain religious group about "*Strī bīrya*, the white female sexual fluid" different from menstrual blood, but does not give any references, so that it is unclear whether "white" is a characterisation by the members of this group, or her own characterisation. On the other hand, such a characterisation could be in some accord with the ejaculate of the female modern Western medicine has finally (re)discovered (see §12.15). But the relevance of all this for ancient Indian notions remains dubious unless we have concrete evidence that in ancient India too "semen" of women was (also) characterised as "white". If *śukra-*, however, is not primarily "white", then it remains to be examined whether we could have here an old term later fallen into disuse, for *śukra-* can also mean "pure, clear", and the latter of course would be a fitting term if *śukra-* should refer to or subsume the lubricating fluid(s) mentioned in §8.1. "Pure" would, however, theoretical-

---

two sorts of *śukra-* — cf. SEN 1,p.76). Priyavrat Śarmā carefully distances himself from such speculations (see TIVĀRĪ, 1,p.(10)), confirming what was said on him in §1.7. It must, however, be remarked that his statement (loc.cit.; cf. also 1,p.(8)) that the texts distinguish clearly between "manifest" (*vyakta-*, *prādurbhūta-*) and "unmanifest" (*avyakta-*) *śukra-* is not borne out by these texts themselves; of course they describe both *śukra-* present all over the body in general and the reproductive *śukra-* of the male which issues out of the body, and in a few sporadic cases even distinguish their actions (thus e.g. Hemādri in his commentary on AH,Sū 11,37-39, cited several times in the discussion on ☞ *ojas-* in Appendix II, where he distinguishes between *ojas-* as a *sāra-* of *śukra-* within the body, and *ojas-* as a ☞ *mala-* of ejaculated male *śukra-*, and also in §13.26; Hemādri lived around 1300 according to MEULENBELD 1974, pp.435f.), but that does not allow us to make a classificatory theory of this sort out of such statements and say that this classification is what the texts presuppose, for these do not show such classifying tendencies, seeming rather not even to see any problem here, so that we cannot be sure that, when the *śukra-* of males is mentioned, this is always to be seen expressly in its relation to the theory that both males and females have *śukra-*. (I must point out here that I am referring to true medical and particularly the 'classical' texts, not to such of a pseudo-medical nature as e.g. the text discussed in n.1086, in which one might on searching perhaps find material serving as a basis for subsequently developing a modern theory of the sort mentioned.) It also seems a pity that Śarmā too has not been fully able to resist the fashion prevalent in India today to twist statements of texts to fit modern medical theories; cf. his remark in TIVĀRĪ 1,p.(9) that SU,Śā 2,47 (on which see §8.6) does not tell us that an embryo is produced because it is known that actually only the semen of males with spermatozoa in it, coming from the testes, is a procreatory form of *śukra-* (see also n.15). Cf. too n.236.

ly apply equally well to male semen as the last, most refined and therefore "purest" part of the chain of substances of alimentary transformation. It is of course well-nigh impossible that such a meaning was responsible for the *original*, quite ancient nomenclature of male semen, but it is nevertheless worth considering whether such could not secondarily have been — because of the theory of the chain of alimentary substances — taken to be the meaning of *śukra-* in connection with male semen, and whether the term *śukra-* in this meaning might not then have been applied to some discharges of women too. But if this should indeed have been the case, then would it not mean too that the origin of these discharges was considered to be more or less the same as that of male semen? We thus see that there are several possibilities which we may explore in this regard. Unfortunately, we have no evidence so far that may prove or disprove any of these. Note in this context also the assumption of EIM, p.196, that the "semen" of women we are discussing above is *exactly* the same as that of males; EIM states that "it is said to be present in females also,[813] *being secreted during the orgasm*" (my italics). But that a *śukra-* is secreted in this manner would not automatically entail such a linkage with the *dhātu-*, thus making the *śukra-* of females and males ultimately the same in all respects, and I doubt very much indeed whether one could find clear evidence for this latter assumption, though I would be dishonest if I failed to draw attention to CA,Ci 15,34f. mentioned in §6.14.

8.10. Let us therefore turn our attention to another very interesting passage, namely SU,Sū 14,14, which says:

> *evaṃ māsena rasaḥ śukraṃ strīṇāṃ cārtavaṃ bhavati*[814]
> "Thus in a month *rasa-* becomes *śukra-*, and the *ārtava-* of women".[815]

---

[813] That the ☞ *dhātu-* semen was deemed to be present in women too is by now of course known to us.

[814] Variant (with no change of content of the statement):
evaṃ māsena rasaḥ śukrībhavati strīṇāṃ cārtavam.
This is also found in the commentary of Śivadāsasena on DR,Āh 22, and in ŚARĪRA 2,12 (with *strīṇām ārtavam*). On the version of this passage as quoted in BHĀ 1,3,187 see §11.22.

[815] ĀY 4,73 has the following verse stating the same:
annaṃ caturvidhaṃ samyak pācitaṃ pācakāgninā
māsena yāti śukratvaṃ tadvad strīṇāṃ rajo bhavet.
The text says that this is a verse of Suśruta, but actually that does not seem to be the case. Note, however, ŚĀR 1,6,11f.:
pācitāḥ pittatāpena rasādyā dhātavaḥ kramāt
śukratvaṃ yānti māsena tathā strīṇāṃ rajo bhavet.
We shall return to this latter verse in §11.9.

The mode of expression seems to point to a direct parallelism between the two fluids, which goes so far as to make the time necessary for the formation of both the same. It is, however, noteworthy that the time taken for ☞*rasa-* to become *śukra-* is a matter of great debate in medical literature. We have already mentioned CA,Ci 15,21, which gives us quite different spans of time for the formation of semen, in §5.16. Cakrapāṇidatta in his commentary on this mentions the opinions of different authorities on this matter, and it seems that the different spans of time presupposed were various, not only a month, as here (see also n.171).[816] More on this controversy is to be found in AS,Śā 6,p.336b = AH,Śā 3,65f.[817] (see also the respective commentaries of Indu, and Aruṇadatta and Śrīdāsa) and in the commentaries of Cakrapāṇidatta and Ḍalhaṇa on SU,Sū 14,14-16.

8.11. Now SU,Sū 14,14 creates a problem in that it seems to say that both the fluids mentioned represent the end of the chain of digestive breakup. This seems at variance with the statement of SU,Sū 14,6 that the procreatory-menstrual fluid is derived from "*rasa-* itself" (see §7.11), especially if this ☞ *rasa-* should be ☞ *āhārarasa-* (see §7.40), i.e. that form of *rasa-* which serves as a nourishing or satiating agent and of which it is unclear whether the changes it undergoes are the same as that of the *rasa-* nourished by it, or whether they are of a different sort (cf. §§7.24ff., especially §§7.38f.). But there need not be any contradiction, for it should be noted that Sū 14,14 does not tell us that both these fluids are produced in exactly the same way. It may be recalled in this context that, though our texts (SU too) seem to presuppose the existence of *rasa-* in two different functions, basically there is only one and the same *rasa-* (cf. §§7.38f.). Thus what might be meant could be that semen is the end product of the digestive process in which the seven ☞ *dhātu-*s are formed one after the other from what the commentaries call *poṣya-/sthāyin- rasa-*,[818] whereas the procreatory-menstrual fluid is the direct product of *rasa-* = *āhārarasa-*, both, however, taking the same time to evolve. If we allow this deduction to be valid, then it could follow that *ārtava-* is what is left of *āhārarasa-* when it has completed the process of nourishing or satiating all the seven *dhātu-*s. This would mean that semen and *ārtava-* have a different, but parallel development.[819]

---

[816] The process can be hastened by the use of aphrodisiacs; cf. e.g. SU,Sū 14,17, CA,Ci 15,20.

[817] = ĀY 4,36, BYT 2,78f., which have minor variants.

[818] Though what the commentaries mean in particular need not necessarily be presupposed by the texts commented upon too.

[819] This postulate does not *a priori* stand in contradiction to the first part of 14,14 (on

8.12. We must also ask ourselves whether it might not be possible that the text of SU,Sū 14,14 regards *śukra-* as something special to males, *ārtava-* as something special to females. But can we really suppose this? Now it is true that, though basically *śukra-*, the last element of the chain of alimentary breakup, was held to be present in both males and females, in most descriptions of it in our medical texts we usually find it spoken of with reference to males; it is also unmistakably this same *śukra-* which men are said to ejaculate (cf. also §8.9). We might thus ask with some justification whether this association with the male might not be what SU,Sū 14,14 primarily has in mind. I may point out that Sāyaṇa in his commentary on the passage from *Aitareyāraṇyaka* 2,3,7 discussed on p.522 in Appendix II says that *both* the "blood" (*lohita-*) of women and the semen (*retas-*) of men are the seventh element. On the other hand we must not overlook the fact that the passage SU,Sū 14,14 is part of a context dealing with metabolism in general; moreover, the very fact that something special to women is expressly mentioned might be taken to mean that everything else not expressly related to one of the sexes is common to both. If, however, *śukra-* here should refer to both the male and the female, then we would have an asymmetry, since only females would have, over and above this, *ārtava-*. But this assymetry would be there according to Sū 14,6 too and thus be consistent. Moreover, if the text should indeed suppose a derivation of *ārtava-* from *āhārarasa-*, then we would have great problems if we were to contend that it speaks only of the

---

which see also n.829), which gives us the time, measured in the unit ☞ *kalā-*, during which *rasa-* is present in (and this seems to mean: in the form of) each element:

> *sa khalu trīṇi trīṇi kalāsahasrāṇi pañcadaśa ca kalā ekaikasmin dhātāv avatiṣṭhate. evaṃ māsena rasaḥ* etc.

Even if this should apply only to the (in later terminology) *poṣya- rasa-* and not to the *poṣaka- rasa-* (*āhārarasa-*), that need not preclude the latter too being meant in the next line, as *rasa-* in each case is ultimately one and the same fluid. We have also seen (§7.38) that the commentators mention, among several conflicting theories, one in which the *poṣaka- rasa-* too is changed, so that the above line could also be taken to refer to this *rasa-*, though of course we have no way of knowing not only whether such a theory was presupposed by SU, but also if the theory of *poṣaka-* and *poṣya- rasa-*s as given by the commentators was subscribed to by the author(s) of SU at all, even though some similar theory at least does indeed seem to have been presupposed (cf. §7.38). It may also be pointed out that we could even analyse the locative *dhātau* as meaning not "in", but "with respect to", in accordance with a common usage of the locative; this would then mean that the *rasa-* remains static (*avatiṣṭhate*) for the time given with respect to each *dhātu-*, and that could refer to the time needed for each element to be nourished/satiated as well as to the time needed for each transformation of the transformational chain, i.e. need not necessarily presuppose a change of the nourishing/satiating *rasa-* itself, as its nourishing/satiating function could have been taken not to require such a change.

śukra- in males, for that would mean that we would have a loose end regarding the metabolism of the *rasa-* which does not function as *āhārarasa-* in the female, as she would have nothing corresponding to the seventh element of the male as part of the transformational chain, which would not be in keeping with the general theory of transformation of the alimentary elements. This would speak for *śukra-* as present both in males as well as in females being meant above. But since the text is not specific, we can nevertheless still not be fully sure that 14,14 does indeed refer to *śukra-* common to males and females. Since there is, however, nothing we can do about this, we shall simply have to accept this residual uncertainty. But be that as it may, we still have a problem on our hands, for, if our derivation of *ārtava-* from *āhārarasa-* is correct, then we may legitimately ask what happens to the residue of *āhārarasa-* in males, since these do not have *ārtava-*. We shall unfortunately never know whether the author(s) of SU gave no thought to what happened to the remnants of *āhārarasa-* in males, or whether he or they simply did not consider this worth commenting upon (provided, of course, that our deductions are correct in the first place).

8.13. These are, however, not the only problems connected with the above explanation, for the exact nature of what is then left over in the end as *ārtava-* too remains unclear. Somehow or other the *āhārarasa-* must be quantitatively reduced through its nourishment/satiation of the elements, as otherwise the quantity of each element in the body would soon diminish. But are we to look at it as somehow changing to form *ārtava-*, or is *ārtava-* to be viewed as the last part of *āhārarasa-* left over, more or less unchanged, after it has fully completed its function of nourishing/satiating? This could obviously be of importance also for the discussion of the colour of *ārtava-* (see §§8.3f.), for one can assume that *āhārarasa-*, if it be basically the same as *rasa-*, must have been conceived of as being colourless or pale like this (cf. SU,Sū 14,4f. in §7.11; see also §7.23). But we get no information which can help us in tackling these problems.

8.14. All this is, however, valid only if the procreatory-menstrual fluid indeed evolves from *āhārarasa-*. But it must be pointed out that we are by no means certain that this is indeed the case, since the other *rasa-* the commentators label *sthāyin-* or *poṣya-* might also be meant; I have remarked expressly on this possibility (see especially §§7.40f.). But if the latter is indeed what is meant here, and if the *śukra-* mentioned in SU,Sū 14,14 is that of both men and women, then *ārtava-*, as it is not a part of the regular chain of transformation leading from *rasa-* to *śukra-* (since the theory seems to leave no room for it), would appear to be a by-product produced from *rasa-* by some other means. Theoretically, one could also hold that *ārtava-* is derived ultimately from *śukra-* itself, or even that in the female both *ārtava-* and *śukra-* are produced as final products of the chain, whereas in the male the final product is only *śukra-*, but are there statements in the medical texts which could support this supposition? On the other hand, if *śukra-* should apply only to males (a perhaps slender, but nevertheless not negligible possibility), then we could conceive of a derivation of *ārtava-* from *rasa-* in exactly the same way as *śukra-*, which of course at once reminds us not only of the Tibetan theory in §6.15, but also of that of COMBA 1981 in §2.4.

8.15. It seems that ultimately SU,Sū 14,14 creates more problems than it helps solve, and it is therefore no wonder that later medical authorities had great difficulties in explaining it, as is evident from the commentaries of both Ḍalhaṇa and Cakrapāṇidatta. Thus Ḍalhaṇa makes the rather startling remark that *ārtava-* in this passage does *not* stand for *rajas-* (with which he seems to mean the procreatory-menstrual fluid), but for *śukra-*. Two alternative reasons for this statement are given. The first states that *rajas-* arises from *rasa-* like blood [in general] (*rakta-*),[820] the time taken for this being, according to Ḍalhaṇa, seven days[821] and hence not the same as that of the *ārtava-* of 14,14.[822] The second reason given is that, if *ārtava-* should not refer to *śukra-*, women would have one *dhātu-* less than men. The first argument seems to be in keeping with what we have remarked on in n.663, namely that Ḍalhaṇa may perhaps not subscribe to the theory of *rasa-* functionally being of two kinds, which would of course mean that *rasa-* would not be able to undergo more than one kind of transformation of the same sort (i.e., *śukra-* and *ārtava-* might come perhaps from the same substance, but then not in the same manner), whereas the second argument seems to be based on the notion (already referred to in §8.12 and §8;14) that the *śukra-* mentioned in 14,14 refers to that of males only. Ḍalhaṇa, however, also mentions the opinion of *anye* "others" that *śukra-* here subsumes the *śukra-* of females too (as a procreatory fluid)[823] and that *ārtava-* here does indeed refer to *rajas-* (i.e. most probably the procreatory-menstrual fluid), this being, however, produced in a week, but accumulated and thus manifest in a month.[824] Though

---

[820] This is obviously a reference to SU,Sū 14,4-6; see §7.11.

[821] I.e. not the month supposed by 14,14.

[822] Apart from pointing out the discrepancies in the times taken for formation, this argument also seems — even though Ḍalhaṇa does not say so expressly — to be based on the different positions in the chain of alimentary change, since if blood and *rajas-* should be taken to develop in a truly parallel fashion, then the latter cannot of course be an end-product of the same sort as *śukra-* is.

[823] SU,Śā 2,47 (see n.795) being cited in this context.

[824] The original wording of Ḍalhaṇa's commentary is:

he does not say so expressly,[825] the very fact that he labels the latter explanation that of "others" seems to show that he does not agree with it, which would be in keeping with his view on SU,Śā 2,36 that the *śukra-* of women is not procreatory (see §8.5). As a matter of fact, it is this very explanation which Ḍalhaṇa does not seem to agree with that Cakrapāṇidatta gives us in his commentary on 14,14 (SU3: 14,12) as his own, albeit *without* the part stating that the *śukra-* of women is a procreatory fluid, but also saying that *śukra-* in SU,Sū 14,14 refers to that of both males and females. He also mentions the explanation of others, who do not agree with his own explanation, that *ārtava-* here refers to the *śukra-* of women. But he points out that this *śukra-* cannot lead to the formation of a child, this being possible only by means of *rajas-*, on which he might be saying (see n.827) that it emerges during sexual intercourse to unite with the procreatory fluid of the male; there are, however, a few cases in which the *śukra-* of women, if it be conjoined/attained[826] (if this is what *āśrayībhūta-* means), may lead to the formation of an embryo, SU,Śā 2,47 being cited as an example.[827] Cakrapāṇidatta also says

---

> *atrārtavaśabdo 'yaṃ śukre vartate na tu rajasi. rajo hi rasād raktavat saptame 'hani jāyate. athavā yady ayam ārtavaśabdaḥ śukre na vartate tadātra śukrotpattyadhikāre strīṇāṃ śukrasyānuktatvāt ṣaḍdhātutvaṃ syāt. tasmād ārtavaśabdaḥ śukre vartate. anye tu punar ārtavaśabdaṃ rajasy eva vartayanti cakāreṇa strīśukraṃ samuccinvanti śukre 'pi garbhajananaśaktidyotanārtham. tathā coktaṃ* ... (here we have SU,Śā 2,47) *iti. asmin pakṣe māsenārtavasya bhavanam upacayo 'bhipretaḥ prakāśaś ca. yasmād ārtavasya raktavat saptāhenaivotpattir iti.*

A part of this commentary is also to be found in ĀY 4,74; the beginning up to *śukre na vartate* is the same, after which we have:

> *tadā strīṇāṃ śukrasyābhāvāt ṣaḍdhātutvaṃ bhavati. tasmād ārtavaśabdaś cātra śukre pravartate. strīṇāṃ śukram apy astīti vṛddhavāhaḍaḥ* (var: *vṛddhavāgbhaṭaḥ*).

Then AS,Śā 1,p.286a (see §9.7) is quoted in a variant form as ĀY 4,75 (see n.891).

[825] We have already seen in §8,6 that he prevaricates with regard to SU,Śā 2,47, which the "others" here give as proof for the correctness of their view.

[826] Or maybe, in this special case: received?

[827] The whole commentary runs (version of SU4, with variants of SU3):

> *yady api rasād ārtavaṃ māsād arvāg eva bhavati tathāpi rajasaḥ sūkṣmaviprakṛṣṭamārgatayā māsena bhavanam. yad uktaṃ viśvāmitreṇa.*
>
> *sūkṣmakeśapratīkāśā bījaraktavahāḥ sirāḥ*
> *garbhāśayaṃ tarpayanti* (SU3: *pūrayanti*) *māsād bījāya kalpate* (SU3: *kalpyate*)
>
> (on this verse see §10.23) *iti. atra bījabhūtaraktaṃ bījaraktaṃ ārtavam ity arthaḥ. atra yadi nāmārtavasya raktarūpatā stokenāpi kālena bhavati tathāpi tas-*

that even if *ārtava-* turns to blood (i.e. obviously: assumes the colour of blood) in only a little time, it nevertheless takes a month to reach the uterus. Most of this argument is repeated by Śivadāsasena in his commentary on DR,Āh 22 (= CA,Ci 15,17; see on this §§5.12f. and §7.41); Śivadāsasena also mentions the theory of "others" (*anye*) that *ārtava-* is used in the sense of *śukra-* in SU,Sū 14,14 and (this is particularly intriguing) Śā 2,36 (cf. on the latter §8.5).[828] Other authorities too have given us their opinions; we shall

---

ya garbhāśayaprāptir māsenaiva bhavati. garbhāśayagataṃ ca tan mukhyārtavaśabdavācyam. ato māsena tadbhavanābhidhānaṃ yuktam. anye māsenopacayād āvirbhāvam ārtavasya māseneti vadanti. ye tv ārtavaśabdaṃ strīśukre varṇayanti tanmata aśāstrīyeyaṃ vyākhyā (SU3: *aśābdīyā vyākhyā*). *yad api* ... (here follows a variant of SU,Śā 2,36; see on this n.777) *iti kaiścit paṭhyate tad yady ārṣaṃ tadā tatrārtavaśabdena raja evocyate. tad dhi maithunasamaye puruṣa[bīja?]grahaṇāya visarpati. strīśukraṃ tu śukraśabdena puruṣastrīsādhāraṇena māṃsādivat prāptam eva. yat tu strīśukrasya garbhānārambhakatvaṃ tadvacanād eva tathā nārīrajaso garbhārambhakatvād eva labdham. uktaṃ ca tantrāntare*

strīṇāṃ śukraṃ na garbhāya bhaved garbhāya cārtavam

(on this see §11.9) *iti. strīśukraṃ ca kvacid āśrayībhūtaṃ garbhārambhakaṃ vaktavyaṃ. yathā* ... (here SU,Śā 2,47cd is quoted) *iti*.

Special attention may here be drawn to the line *tad dhi maithunasamaye* etc. SU3 has *tad dhi maithunasamaye puruṣagrahaṇāya visarpati*, whilst SU4 has *bīja* within round brackets after *puruṣa* (*puruṣa(bīja)grahaṇāya visarpati*), this *bīja* seemingly being an insertion of the editor. With *bīja*, the line simply says that the procreatory-menstrual fluid issues to mingle with semen. Without *bīja*, however, the line says that *ārtava-/rajas-* issues for the purpose of uniting with the male, which does look very like a reference to the lubricating fluid(s) and leads to the same sort of problem as last discussed in §8.4.

828 *raktam iti strīṇāṃ rajaḥsaṃjñaṃ raktam apy upadhāturūpam. rasād eva raktaṃ prasādabhāgajanyam ity arthaḥ. uktaṃ ca suśrute* ... (here follows a variant of SU,Su 14,6ab; cf. n.413) *iti. etac ca rajorūpaṃ raktaṃ* (not *rakta*, as in the printed text) *rasajanyam api dhātuśoṇitavan na śīghraṃ jāyate kintu śukravan māsenaiva. tad uktaṃ suśrute* ... (then SU,Sū 14,14 in the variant of n.814 follows) *iti. anye tu* ... (here we have SU,Śā 2,36 with the variants given in n.777) *ity atra yathārtavaśabdena śukram ucyate tathā strīṇāṃ cārtavam ity atrāpy ārtavaśabdaḥ śukra eva vartate na tu rajasi. rajo hi rasād raktavat śīghram eva jāyate na tu māsenety āhuḥ. tan nety anye. visarpaty ārtavam ity ādāv api strīṇāṃ ārtavaṃ puṃsāṃ samāgame sati visarpatīti yojanayārtavaśabdasya rajovacanatvenāpy upapatteḥ. vastutas tv ārtavasya raktarūpatā* (printed: *raktarūpatayā*) *yady api māsād arvāg bhavati tathāpi tasyārtavarūpatā garbhāśayaprāptyaiva bhavati. garbhāśayaprāptis tu māsenaiva bhavatīti kṛtvā māsenārtavaṃ bhavatīti suśrutenoktam* (printed: *suśrutatenoktam*). *uktaṃ hi viśvāmitreṇa* ... (here we have the same verse as given in n.827, with *pūrayanti* and *kalpate*, but also with *sūkṣmāḥ* for *sūkṣma*) *iti. tatra bījabhūtaraktaṃ bījaraktam ārtavam ity arthaḥ*.

refer to some of these views below, as their discussion here would lead us too far off the track. But even the brief overview given here suffices to show that there was indeed quite some confusion, not only on the nature of *śukra*-, but also on that of *ārtava*-, which has already caused us difficulties in §§8.1ff.[829]

8.16. In this context I may also be allowed to draw renewed attention to the passage from Vṛddhātreya's work (ĀY 4,69-72) given in n.222, which differentiates between "blood" = *ārtava*-, and *rajas*-, the latter referring to the part of the former that remains in the body after the menstrual discharge, which discharge is incidentally the periodic voiding of a fluid that could (but need not) have been conceived of as always trickling into the uterus (periodically voided due to accumulating pressure on the uterine opening into the vagina?).[830] Now the "rest" is obviously referred to in 72ab:

tadā śukrasya saṃsargi jāyate malavarjitam.

Interestingly, though Śivadāsasena calls *rajas*- an ☞ *upadhātu*- here, which could, in accordance with the theory later current on the *upadhātu*-s, imply its derivation from blood, he nevertheless does not mention such a derivation.

829 Let me also add that Cakrapāṇidatta in his commentaries on CA,Sū 28,4 and CA,Ci 15,20f. quotes the passage from SU,Sū 14,14 given in n.819, in the former case as:

sa khalv āpyo rasa ekaikasmin dhātau trīṇi trīṇi kalāsahasrāṇi pañcadaśa ca kalā avatiṣṭhate. evaṃ māsena etc.,

in the latter case as:

sa khalv āpyo rasa ekaikasmin dhātau trīṇi trīṇi kalāsahasrāṇy avatiṣṭhati daśa (probably a mistake) *ca kalāḥ. evaṃ māsena* etc.

One does have the feeling that *āpyo rasa[ḥ]* here seems to be due to the incorporation of some commentary into the text; a feeling is, however, no proof, so that we cannot attach undue importance to it. But SU,Sū 14,4 and 14,5 tell us respectively that blood in general (and not the procreatory-menstrual fluid) is derived from "watery" (*āpya-*) *rasa-*" and "water" (*ap*-), so that if *āpya- rasa-* in the versions of SU,Sū 14,14 just quoted should indeed be the correct reading, then we obviously end up with a problem on our hands. And if that were not enough, we also have the problem of the meaning of these latter terms, a problem which busied us for most of the last chapter, chapter 7.

830 This follows from 4,70:

strīṇāṃ garbhāśaye vāyur yad raktaṃ labhate śanaiḥ
catvāro 'ñjalayo māsāc cyavate yonimārgataḥ,

which seems to say that the wind gradually (*śanaiḥ*) "seizes" (*labhate*) the "blood" in the womb, of which then four *añjali*-s flow out through the vagina after a month. The gradual seizing would seem to presuppose a presence of fluid within the womb even between discharges, i.e. probably a steady trickle or the like into the uterus. On the other hand, one could take *śanaiḥ* to refer to the flowing out of blood, since the menstrual discharge does take place over a period of some days; in that case, however, the entry of the fluid into the uterus need not be gradual too, but could occur as a one-time flow.

This can be translated as:

> "It (i.e. the rest), devoid of impurities, then becomes commingled with semen",

or else as:

> "It, commingled with semen, then becomes devoid of impurities."

The first alternative would tell us that menstruation is a process of cleansing of the procreatory-menstrual fluid, and that it is this pure fluid which must combine with semen to produce an embryo; the second would tell us that this fluid is not (necessarily?) pure even after menstruation, being purified through contact with semen. The editor of ĀY takes the line just quoted to have the latter meaning, and when one looks at the immediately following lines 72c-f, one must confess that this does make sense, for these run:

> *yathā malinakhaṇḍasya malaṃ ca harate balāt*
> *kṣīraṃ kṣāras tathā vidur garbha ārtavaśuddhikṛt*
> "As milk [and] a caustic forcibly takes (sic) away the impurity of soiled candied sugar, so do they know the embryo (*garbhaḥ*) to be a purifier of *ārtava-*/do they know [it] (scil. probably: semen) to be a purifier of *ārtava-* in the case of an embryo (i.e. probably: when the combination of the fluids has formed an embryo) (*garbhe*)."

The reasoning behind this could be that, since menstruation is a discharge of impure "blood", and since such "blood" does not issue during pregnancy, the "blood" during this period must be pure, the reason for this pureness being its contact with semen (and/or, if we take *garbha* to stand for *garbhaḥ*, with the embryo). The weak point of this deduction is that the passage in question does not say that menstruation is a discharge of impure fluid. Nevertheless, our deduction does make sense, even though we cannot be sure that it is indeed correct. But if it is correct, then we could, since I know of no other similar statements, have here a theory which either did not gain general acceptance, or else was superseded in time. On the other hand, the reasoning behind the statement above could be that the "blood" of women is not by nature pure, and requires purification before it can form the embryo. For this statement too I know of no parallels. The passage could, however, be understood in quite another way too, namely as:

> "As milk forcibly takes away the impurity of soiled candied sugar, so do they know the caustic embryo (*garbhaḥ*) to be a purifier of *ārtava-*/do they know the caustic (scil. semen?) to be a purifier of *ārtava-* in the case of an embryo (*garbhe*)."

But in this case I also know of no parallels. Unfortunately I can say no more on the subject. What, however, at any rate remains clear is that the differentiation of *ārtava-* and *rajas-* that the commentaries quoted in §8.15 mention has a striking parallel in the ĀY passage, even though the differentiation the commentaries mention is between *ārtava-* when it is used for *śukra-*, and between *rajas-*, it not being clear whether the two are two different fluids or two different forms of the same fluid, whereas Vṛddhātreya's differentiation obviously is of the latter sort. It must also be stressed that the differentiation of the commentaries applies only to (a) very special case(s), whereas that of Vṛddhātreya is of a general nature. Note that, even though *ārtava-* seems to be the general name of the procreatory-menstrual fluid according to Vṛddhātreya, one could in a narrow sense refer to it as the menstrual fluid to differentiate it from its form called *rajas-*, which is that part left over after menstruation and obviously procreatory. But even from this one could not deduce that the two fluids are by nature quite different, as some modern scholars have postulated (cf. §2.3).

8.17. But all this does not necessarily follow out of SU,Sū 14,14 itself. We shall thus have to see whether SU itself has more information to offer us. An intriguing statement is contained in Ni 5,27, which is part of a passage (5,20ff.) in which the effects of the skin-disease *kuṣṭha-*[831] on the seven ☞ *dhātu-*s[832] are described. Now 5,27 details the effects of *kuṣṭha-* when it "has gone to the place of *śukra-*" (*śukrasthānagate*). This must at once remind us of the discussion on the receptacle of semen in the male in §§6.13ff. Indeed, one must ask in this connection whether there was thought to be a correspondence to the receptacle of semen of the male in the body of the female, and, if there was, whether this was a receptacle of procreatory-menstrual fluid (cf. especially the Tibetan parallels in §6.15), or whether the female body too was thought to have a receptacle of semen, then probably over and above that of the procreatory-menstrual fluid.[833] The problem is, however, that SU,Ni 5,27 seems to describe the effects of *kuṣṭha-* in humans in general. Or could it be that we have here the by now familiar problem that the author may, nevertheless, have had only the male in mind as the

---

[831] On this see especially EMMERICK 1984 and EMMERICK 1986.

[832] Starting, interestingly, not with ☞ *rasa-*, but with ☞ *tvac-* "skin"; we shall return to this problem below in §10.10 and not discuss it here, as it is irrelevant in this context. I may, however, nevertheless point out here that the term *dhātu-* is expressly mentioned in 5,21.

[833] In this connection we would, however, quite pertinently have to ask how this correlates with e.g. SU,Śā 9,7, which lists ☞ *dhamanī-*s carrying semen in males and procreatory-menstrual fluid in females, saying nothing about any *dhamanī-*s carrying semen in females (see §6.8).

ideal person?[834] On the other hand, what may be meant is not a certain receptacle of semen, but the semen permeating the whole body (of males as well as females), the "place of *śukra-*" being then wherever the *śukra-* should happen to be, in which case *sthāna-* would be used more or less in the sense of "(up)to" (and not "place"). This would indeed be in keeping with the rest of the passage, which simply describes the effects of *kuṣṭha-* when it is in intimate contact with the other *dhātu-*s; in each case the name of the *dhātu-* is compounded with the word *samāśrita-*, except in the case of blood, where we have *śoṇitasaṃśrite*, which, however, means roughly the same. Thus one may opine that what is said with respect to *śukra-* must be more or less the same too, and Ḍalhaṇa and Gayadāsa actually do substitute *śukragata-* for *śukrasthānagata-* in their commentaries without any comment. On the other hand, the very fact that the mode of expression in this case is so different may be used as a counter-argument to this. What complicates matters even more is the fact that a Nepalese manuscript[835] inserts two verses in the text of SU between Śā 2,37 and 2,38, in which the fluid of the woman uniting with the procreatory fluid (here called *retas-*) of the male is quite clearly called *śukra-* (and nothing else).[836] These verses, which elaborate on 2,37, appear only in this ms, and do seem to be spurious, but they nevertheless clearly show that there must have been a tradition (of unknown age) which simply applied the label *śukra-* to the procreatory fluid of females. Since the verses do not say anything else on this *śukra-*, we do not know whether *śukra-* is here simply used as a synonym of *retas-* "flow",[837] or whether the word is used here in a meaning derived from "semen", i.e. as "procreatory fluid". Happily for us, the discussion so far has shown that this case mentioned here is a singular case as far as the evidence so far sifted goes, as it would have caused us difficulties even greater than the ones we are already tackling if we had had here a nomenclature in general use in medical works.

8.18. I shall not, however, go on with this discussion, as I cannot offer any definitive solutions. It suffices if we are aware of the problem here. What

---

[834] Cf. e.g. §5.16, also n.812.

[835] Abbreviated loc.cit. as *tā.*, which obviously stands for the *prācīnatāḍapatrapustaka-* mentioned on p.22 of the introduction of SU.

[836] *miśrībhāve yadā puṃsāṃ retas tu bahu nirmalam*
tadā prasūyate nārī garbhiṇī dārakadvayam
miśrībhāve striyāś caiva bahu śukraṃ yadā bhavet
kanyādvayaṃ tadā nārī sūyate nātra saṃśayaḥ.

[837] A sort of reverse synonymy due to the fact that *retas-* "flow" often is a synonym of *śukra-* "semen".

I would like even more to draw attention to is the following verse Ni 5,28, which tells us that the child of a woman and a man whose *śoṇita-* "blood" (obviously the procreatory-menstrual fluid here) and *śukra-* have been corrupted by *kuṣṭha-* is also stricken by the disease.[838] Whether or not 5,28 be intimately related to 5,27, this is a problematic statement, for the procreatory-menstrual fluid is not mentioned before this. But one may, if the two verses simply follow one another without any basic connection, say that the fact that this fluid of the female has not been mentioned before does not automatically mean that it is somehow to be subsumed under those substances that have been mentioned; in other words, what has gone before need not be relevant for 5,28, and *kuṣṭha-* may then affect the fluid irrespective of how it affects the seven *dhātus-* mentioned and not be in any way connected with this latter process of affecting. If, however, 5,27 and 5,28 are related, then we have a problem, for the end result of their combination would be that *śukra-* in 5,27 (then obviously in both males and females) subsumes the semen of males and the procreatory-menstrual fluid of females, which then at once reminds one of the discussion in §8.14. Unfortunately here too we have to make the frustrating experience that the text merely throws up a problem, but then leaves us in the lurch as to what it actually means to say.

8.19. For the problem of the relation of *ārtava-* to semen Śā 5,39 too is of interest. In it we find a list of muscles (☞ *peśī-*) which are present only in women. Of these, three are called *śukrārtavapraveśinyaḥ* "characterised by/connected with/belonging to the entrance of *śukra-* and *ārtava-*", i.e. guarding the entrance of *śukra-* and *ārtava-*,[839] and obviously within the uterus[840] (this is presupposed too by the variant in n.840, where *garbhāśaye*

---

[838] *strīpuṃsayoḥ kuṣṭhadoṣād duṣṭaśoṇitaśukrayoḥ*
yad apatyaṃ tayor jātaṃ jñeyaṃ tad api kuṣṭhitam.

[839] Thus also TIVĀRĪ, 1,p.17. According to PUROHIT, p.165 this refers to "three [Peśīs] at the place where Śukra and Ārtava enter inside", but that is actually begging the question, for this place is after all the uterus, quite apart from the fact that *praveśa-* (in *praveśin-*) refers either to the act of entering or the entrance, but not to that which has been entered, as PUROHIT assumes, unless his "at the place" is actually Indian English for "at the places", i.e. "openings". In SINGHAL/GURU, p.116 we find: "three more muscles are meant to bring together the sperm and ovum"; what is this supposed to mean?

[840] The relevant part of the passage:
apatyapathe (var: *bhagāpatyapathe*) *catasraḥ. tāsāṃ prasṛte abhyantarato dve. mukhāśrite bāhye ca vṛtte dve. garbhacchidrasaṃśritās tisraḥ. śukrārtavapraveśinyas* (var: *śukrārtavapraveśinyo garbhāśaye*) *tisra eva. pittapakvāśayayor madhye garbhaśayyā yatra garbhas tiṣṭhati.*

The three muscles for *śukra-* and *ārtava-* cannot be identical with the three attached to the

is expressly mentioned, and the parallel AS,Śā 5,p.325a,[841] whereas a passage from an unknown source quoted by Aruṇadatta in his commentary on AH,Śā 3,17f. does not expressly mention them as being within the uterus[842]). As regards *ārtava-*, this cannot but refer to the entrance of the procreatory-menstrual fluid into the uterus through two ducts or the like, as discussed earlier (see e.g. §§6.12ff.). This means that the third muscle is connected with an opening through which semen enters the uterus. The question now is: where does this semen come from? Is it the semen of the male? One could point in this connection to the periodic closure of the uterus mentioned in §5.6, especially since BHĀ 1,3,38 expressly says that this closure refers to the closure of the "entrance of the holder (scil.: womb)", which in all probability means the uterine cervix.[843] But we must not forget that

---

garbhacchidra- "opening for the embryo/foetus" mentioned just before them, as then we would obtain, together with the other muscles mentioned in this passage, seventeen additional muscles in women, and not twenty as the passage itself expressly states. ZYSK 1986, p.704 translates *śukrārtavapraveśinyas tisra eva* as "and three enter the passage where the semen and the menses unite", but this is a patently impossible translation (has he perhaps wrongly analysed *°praveśinyaḥ* as a genitive *°praveśinyāḥ*, deducing from this that the muscles belong to the place "characterised by the entrance of *śukra-* and *ārtava-*", and are not themselves called so?). Ḍalhaṇa, cited by ZYSK 1986, p.704,n.104 as saying "and of them, three muscles are in the passages where the semen and menses unite", says only *tāsāṃ* (i.e. of the twenty additional muscles of women, or, alternatively, of women) *śukrārtavapraveśinyaś ca tisraḥ peśyaḥ*, which is more or less a mere repetition of the SU text. NYC unfortunately has a lacuna here (cf. p.97).

841 This latter says:

daśa yonau. tāsām abhyantarāśrite dve. mukhāśrite vṛtte dve. tisro garbhamārgāśrayāḥ. yasyāṃ garbhas tiṣṭhati. yonis tu śaṅkhanābhyākṛtis tryāvartā. tasyās tṛtīya āvarte pittapakvāśayāntare garbhaśayyā. tasyāṃ śuklārtavapraveśinyas tisraḥ peśyaḥ.

Note *yasyām* in this passage. Could this be a mistake for *yasmin*, referring thus to *garbhamārga-*? Or is it maybe a mistake for the plural *yāsu* (which, however, creates difficulties, as it would mean that the embryo is situated within the muscles attached to the *garbhamārga-*)? Or is it a construction *ad sensum*, with a presumed *garbhaśayyā-* or the like as a synonym of *garbhamārga-*? Or do we rather have a transposition of the sentence *yasyāṃ garbhas tiṣṭhati* from an original position after *pittapakvāśayāntare garbhaśayyā* or after *daśa yonau*. Note too that Śrīdāsa in his commentary on AH,Śā 3,17f. quotes the AS passage *without* the sentence beginning with *yasyāṃ* (see the text in n.868).

842 *śaṅkhanābhyākṛtir yonis tryāvartā jāyate striyaḥ* (var: *jāyate tu yā*).
tasyās tṛtīya āvarte rohitasyākṛtir bhavet
garbhaśayyātha tisraś ca bhaveyuḥ sampraveśikāḥ.
śukrasya cārtavasyaivaṃ peśīs tantravido viduḥ.

This same passage is also quoted in the commentary of Śrīdāsa, who says that it comes from "another work" (*tantrāntare*).

843 *yonir atra dharādvāram*, which explains the line *ṛtau vyatīte nāryās tu yoniḥ saṃvriyate*

BHĀ is an extremely late text; the 'classical' texts do not elaborate on this closure, so that we do not know what exactly is meant; indeed, they might even be referring to a contraction of the uterus as a whole (which would not leave any enterable hollow space in it). As it is, Ḍalhaṇa in his commentary on SU,Śā 3,9 does tell us that the uterus as such contracts.[844] If, however, a contraction of the cervix or its opening should be meant, would we not expect such regulation, should it indeed be postulated, to be effected by the three muscles mentioned separately as attached to the "opening for the embryo/foetus"[845] or the "path for the embryo/foetus"[846]? True, Ḍalhaṇa in his commentary on SU,Śā 5,39 does say that these three muscles refer to the three *āvarta*-s within which the uterus is situated (cf. n.782), but the text quoted in n.841, which mentions these very *āvarta*-s separately, would rather speak against this. On the other hand, *garbhamārga*- of this same text could be taken to refer to the vaginal passage, though this then raises the question of whether *chidra*- in *garbhacchidra*- of the SU parallel (see n.840) could also be taken to mean an elongated passage. Taking the three muscles to be attached to the cervix or mouth of the uterus would not lead to such difficulties, though I admit that that in itself is no proof that this is indeed what is meant, for ultimately it remains unclear what these three muscles actually refer to, and even if they refer to the same in different texts.[847] Nevertheless, the chance that they are somehow attached to the cervix is not bad, and if they should indeed be so, then it does become difficult to conceive of the muscle guarding the entrance of *śukra*- to be connected with this too. And we also have passages listing, among the signs (*lakṣaṇa*-) of conception in a

---

tathā taken from SU,Śā 3,9.

[844] *yonir garbhāśayaḥ saṃvriyate saṅkocaṃ yāti.*

[845] *garbhacchidra*- (see n.840).

[846] *garbhamārga*- (see n.841).

[847] TIVĀRĪ, 1,pp.19-21;23f. has difficulties explaining these three muscles. In the course of her Hindi discussion she takes *garbhacchidra*- to refer to the cervix uteri (*garbhāśay-grīvā*), whereas the *garbhamārga*- is equated with the uterus (*garbhāśay*) itself (in which connexion she spares not a word on the problem caused by *yasyām*; cf. n.841), also presenting diagrams with two alternative explanations of the muscles mentioned, in one of these even clearly taking the *garbhacchidra*- to be the uterus. In the English version the cervix uteri is mentioned, but seemingly transformed into the uterus itself. As regards the muscle regulating the entry of *śukra*- (amongst the *śukrārtavapraveśin*- muscles), in 1,pp.22;24 she writes of several conditions, known to modern medical science, of the cervix of the uterus not being conducive to the entry of semen; she opines that this is what the passage in question means, but I cannot find any proof for such a statement in the texts themselves, and it indeed seems peculiar that the texts should here speak specifically of abnormal conditions.

woman, not only the [subsequent] non-visibility (*adarśana-*) of *ārtava-* (CA,Śā 4,16; ĀY 4,144: of the *puṣpa-*[848]), but also the obstruction/retention (*avabandha-*) of *śoṇita-* and *śukra-* (SU,Śā 3,13), the non-discharge (*apāta-* and *asrāva-* respectively) of *śukra-* and *rajas-* (ĀY 4,146) or of *śukra-* and *śoṇita-* from the ☞ *yoni-* (BHĀ 1,3,42) and the interrupted sequence (*ananubandhana-*) of [the issue of] *śukra-* and *asra-* (AH,Śā 1,36)[849] or *śukra-* and *ārtava-* (AS,Śā 2,p.287b)[850] (on the non-visibility of certain secretions see also Ḍalhaṇa[851] and Aruṇadatta[852] in their commentaries on the SU and AH passages respectively).[853]

8.20. Now the statements above (except that of CA) which pertain to the non-visibility of the fluids are part of the description of conception which has just occurred (i.e. this is a *sadyogṛhītagarbhalakṣaṇa-*). In this connection it may not be amiss to refer to the theory current in ancient Greek medicine that conception can be assumed to have occurred if no semen (of the male) reappears from the vagina after coitus has taken place, and that the non-appearance of semen is due to the contraction/closing of the uterus.[854]

---

[848] Cf. on this term §2.1.

[849] Quoted in BYT 2,6 with *śukrāsrasya tu bandhanam* "stoppage of *śukra-* and *asra-*" in the place of *śukrāsrānanubandhanam*; what is meant is, however, ultimately the same, namely the stoppage of the flow.

[850] Printed loc.cit. as *śuklārtayor anubandhaś ca*. In this reading *śuklārtayor* is quite obviously a mistake for *śuklārtavayor* (i.e. *śukrār°*); as regards *anubandhaś ca*, this too seems to be a mistake for *ananubandhaś ca*, not only because this is what ASI reads on p.202, but also because Indu in his commentary expressly presupposes it (see n.853), quite apart from the evidence of the AH parallel. One could, however, also translate *śuklārtavayor anubandhaś ca* as "the obstacle for semen and *ārtava-*", which would ultimately say the same.

[851] *avabandho 'pravṛttiḥ*.

[852] *śukrāsrayor apravartanaṃ śukrāsrānanubandhanaṃ śukrāsrayor yonimukhād avamanaṃ bahir aniḥsaraṇam*.

[853] Indu in his commentary on AS, however, sees here not a reference to the non-visibility of secretions, but to semen from the male and procreatory-menstrual fluid from the female not being able to unite, because of the obstruction due to the *vidāraṇa-* ("tearing asunder", i.e "obstruction"?) of (= by) the embryo/foetus etc.:

śuklasyārtavasya vā parasparaṃ anyenānanubandhanam. ananubandhanaṃ garbhavidāraṇādinā pratibandhaḥ.

This is repeated in more or less the same words in Indu's commentary on AH5,Śā 1,35 (i.e. AH,Śā 1,36), though with *garbhavidhāraṇādinā* "due to separating by/carrying of the embryo/foetus etc." instead of *garbhavidāraṇādinā*. I doubt, however, whether this meaning of *ananubandha(na)-* is justified here, as the parallel passages show.

[854] On this theory and the controversies connected with it, especially as discussed by

One oould therefore postulate that *śukra-* here refers to semen coming from the male; the theory may be that, since this has combined with procreatory-menstrual fluid to produce an embryo, none of it can flow out of the womb and thus out of the vagina again. As a parallel, one may point to ancient Greek notions of women having the power to retain or eject the seed from both partners after intercourse (cf. e.g. DEAN-JONES 1994, p.28). Note in this context the woman (or maybe the ☞ *yoni-*)[855] called *vāminī-*[856], who (or which) lets semen (*śukra-* or *bīja-*,[857] obviously from the male), together with procreatory-menstrual fluid, flow out again after a few days.[858] It is also obvious that some sort of mechanism (which in all probability would consist of or include the closure of the uterine opening to the vaginal canal) would have to be at work to prevent the semen and procreatory-menstrual fluid which have combined from flowing out of the uterus again. I have not been able to find any statement as to what this mechanism is, but one can surmise that it may somehow be triggered by the discharge of the procreatory-menstrual fluid into the uterus. But we should also not overlook the fact that, since the procreatory-menstrual fluid too is said not to flow any more, this can hardly refer only to the fluid of the time of conception, for any mechanism holding back the fluid released into the uterus should be expected to do so regardless of whether conception has taken place or not; were this not the fact, then we should have expected some sort of comment on the appearance of procreatory-menstrual fluid showing that conception has not taken place. Thus, whether the fluid of the woman be seen here primarily in its role as procreatory or as menstrual fluid, it seems to me that what is meant is that it is no more able to flow *as it usually does*, whatever be the reasons for this, and that would pertain to something which is easily seen or felt (by the woman), namely to menstruation. So we may deduce that this cessation applies to a longer period of time, since we are here dealing with a fluid which in the general course of events is voided periodically and actually *does* stop being visible after conception; note too what Hiraṇyākṣa (quoted by NYC, fol.29v) says:

"The non-visibility of the *puṣpa-* is in the first month a sign of

---

Galen, see NICKEL, pp.50ff.

855 But not the disorder, as TIVĀRĪ 2,pp.50f. presumes. See on the *yoni-* or woman here being what is characterised also the remarks on ☞ *yonivyāpad-* in Appendix II.

856 The feminine form of *vāmin-*.

857 On *bīja-* cf. §§2.5f.

858 On this see TIVĀRĪ, loc.cit.

the [presence of a] foetus."[859]

In that case the symptom we are here primarily concerned with is not necessarily one which pertains to conception which has just occurred, and this would fit in with the fact that SU,Śā 3,14, which quite obviously describes symptoms which occur later in a pregnancy,[860] is joined with 3,13 without any transition of subject-matter. We seem thus to have some sort of confusion here, which may have influenced the other, later texts quoted. Indeed, the latter is quite clearly the case as regards the ĀY passage, which calls the non-visibility of the *puṣpa-*, which can hardly but refer to the *menstrual discharge* (cf. §2.1), a sign of fresh conception! In this connection it is also significant that CA,Śā 2,23, which lists the signs of conception just occurred, does *not* list the symptom we are here discussing. Or is *sadyas* perhaps not to be taken (at least in the SU passage) to mean "recently" in the sense of "freshly"? Maybe it just means "a short while ago". But whatever *sadyas* does actually mean, HILGENBERG/KIRFEL, p.165 seem to be on the wrong track when they translate *sadyogarbhā-* of AH,Śā 1,35 as "Frau, die sogleich empfangen hat", seemingly influenced by Aruṇadatta's explanatory *sadyas tatkṣaṇāt*; in the light of the above it is doubtful whether the latter is to be understood as "at once", it seeming much more probable that on the level of AH "just" (German: "soeben") or something similar referring to a recent occurrence is meant.[861] (Interestingly, TIVĀRĪ has "sadyograhīta-garbhā [sic] or immediately conceived woman" in 1,p.199, but "woman having been recently pregnant" in 1,p.200. GHĀṆEKAR 1986, p.81b on the other hand solves the problem thrown up here by taking *sadyogṛhītagarbhā-* to refer to a pregnant woman until four to six weeks after conception.)

8.21. Whatever the answer to the problem posed by *sadyas* may be, since *śukra-* is here clearly treated parallelly to the procreatory-menstrual fluid, this means that the cessation of its flow is probably not a momentary matter and thus does not apply to the non-visibility of semen ejaculated by the male during a single act of intercourse only. As it also cannot apply to the non-ejaculation by the male, it must therefore be something the female secretes.[862] And that then leads us to deduce that the third muscle (☞ *peśī-*)

---

[859] *puṣpasyādarśanaṃ māsi prathame garbhalakṣaṇam* (cf. also the CA passage above).

[860] Ḍalhaṇa: *uttarakālīnaṃ gṛhītagarbhāyā lakṣaṇam āha*; cf. also DAS,R.P. 1991A, p.57,n. 101.

[861] Indu has the similarly ambiguous *sadyogarbhāyās tatkālaprāptagarbhāyā nāryā[ḥ]* in his commentary on AS,Śā 2,p.287b.

[862] TIVĀRĪ, 1,p.200: "absence of mens[t]rual and coital discharges". GHĀṆEKAR 1986, p.81f. opines that what is meant is the cessation of the flow of an adhesive fluid from the

mentioned by SU,Śā 5,39 is connected with an opening through which "semen" enters the uterus from within the woman's body. See in this context also what was said on *śukra*--carrying channels in §6.19 (though note that in Śā 5,39 only *one* opening for *śukra-* is mentioned). Now of course, if we simply equate this "semen" with the fluid lubricating the vagina,[863] we end up with problems, for such lubrication does not stop after conception, a fact which must have been generally known in ancient India too. Nevertheless, *śukra-* could be what we have been having to consider throughout this chapter, namely something else discharged by the female during coitus (and maybe capable of fertilisation too), whose relation to the procreatory-menstrual fluid is, at least for now, quite unclear.

8.22. Or are we to assume that, with respect to the signs of pregnancy, what we have been describing as the procreatory-menstrual fluid is actually something else?[864] That would of course complicate matters even more. *śukra-* could nevertheless refer to a female discharge different from this *ārtava-*, but, since a female discharge not the procreatory-menstrual fluid may be momentary, the possibility that it refers to male semen then becomes greater. In this context note what the problematic authority Bhoja[865] (quoted by NYC, fol.29v; see DAS,R.P. 1991A, p.57,n.101) says in his enumeration of the characteristics of a woman who is *sadyogṛhītagarbhā-*:

yonir nodvamate bījam

"The ☞ *yoni-* (uterus? vagina?) does not emit the *bīja-*."

Should *bīja-* here refer to the male semen (cf. §§2.5f.), then we might cite this as evidence that the above passages too refer to male semen, maybe in the sense that a woman not pregnant discharges semen received into the uterus after some time, a pregnant woman, however, not. Or is *bīja-* here used

---

uterus, this cessation even leading to a sort of dryness of the mouth of the uterus (*strī ke garbhāśay se jo ek prakār kā cip'cipā srāv hotā hai, vah srāv garbh'dhār'ṇā hone ke kāraṇ baṃd ho jātā hai. yahā̃ tak ki us'kī apravṛtti se strī ko garbhāśay'mukh meṃ ek prakār kī khuśkī mālūm hotī hai*).

[863] Cf. in this connection BAKKER, p.569, who holds that the "seed" flowing from the vagina of women after coitus might refer to sperm alone as well as to sperm together with uterine secretions (the original Dutch runs: "In tegenstelling met MÜLLER ben ik van meening, dat, wanneer er gesproken wordt van zaad, dat na den coitus vloeit uit de vrouwelijke geslachtsdeelen, daaronder zoowel sperma alleen als sperma gemengd met uterusslijm kan worden verstaan"; *sperma* must refer to the male procreatory fluid, as it is used in this sense several times in the essay, whereas in the case of women or in an undifferentiated sense the author uses *zaad* "seed; semen").

[864] Cf. the discussion on *ārtava-* in §§8.1ff. above.

[865] Cf. MEULENBELD 1974, pp.418f.

for female semen, in the same manner as *śukra-* may be? But *bīja-* can also refer to the embryo (or rather, the "germ", the stage in which it has just been formed), in which case Bhoja's statement cannot be used as evidence, as the passage would be saying that the newly formed embryo is retained. The parallel AS,Śā 2,p.287b lists *yonyā bījagrahaṇam* "acceptance/retention of the *bīja-* of (= by)/through the *yoni-*". Indu explains this as referring to the non-flowing of semen;[866] though Indu probably means male semen, he does not say so expressly. Moreover, he need not necessarily be right, for *bīja-* could refer to the embryo (the "germ") here too. And lastly, we have no guarantee that Bhoja and the AS mean the same. But in any case, even if ultimately *śukra-* in the passages dealing with the signs of fresh conception should refer to male semen, that does not necessarily tell us anything about the third muscle (☞ *peśī-*) within the uterus, for the other arguments in §8.19 against this referring to a muscle guarding the opening of the uterus into the vagina would remain. Nevertheless, there does remain a residual uncertainty as regards our explanation of this third muscle (see §8.21); though it seems most probable that this indeed guards an opening leading into the uterus from within the body and discharging something secreted by the woman and called *śukra-*, we cannot be fully sure on this point. Another difficulty is caused by Śrīdāsa in his commentary on AH,Śā 3,17f.,[867] as he quotes AS,Śā 5,p.325a to the effect that not three, but four muscles are mentioned.[868] Of course this is not at all in keeping with the other statements so far seen and the total number of twenty additional muscles in the woman the texts (AS too) presume (we would have twenty-one), and the editor too has realised this and suggested emending the text;[869] indeed, one can even explain the reading as a mislection of *tisraḥ* due to the *i* being dropped (leaving *tasraḥ*, later 'corrected' to *catasraḥ*). Thus the probability is very great that we have here only a mistake of no import; nevertheless, we will have to keep this passage in mind.

8.23. We are now in a position to sum up what we have learnt from SU in chapters 6-8. According to SU, the procreatory and menstrual fluids of women are one and the same. This procreatory-menstrual fluid is derived from ☞ *rasa-*, though whether this refers to ☞ *āhārarasa-* or to *rasa-* that is

---

[866] *yonyā bījasya śuklasya grahaṇaṃ prasravābhāvaḥ.*

[867] Note that 3,18 is wrongly numbered 3,17 too.

[868] *daśa yonau. tāsām abhyantaraprasṛte* (var: *abhyantarāśrite*) *dve. mukhāśrite vṛtte dve. tisro garbhamārgāśrayāḥ. yonis tu śaṅkhanābhyākṛtis tryāvartā. tasyās tṛtīya āvarte pittapakvāśayayor antare rohitamatsyamukhākārā garbhaśayyā. tasyāḥ śuklārtavapraveśinyaś catasraḥ peśyaḥ.*

[869] To *śuklārtavapraveśinyas tisraḥ*.

nourished or satiated by this[870] is unclear (if the two *rasa*-s are indeed always kept apart). By contrast, blood, which is a different fluid and is, contrary to the procreatory-menstrual fluid, always regarded as a ☞ *dhātu*- of the regular metabolic chain, is derived from the latter *rasa*- (and in any case in a different way); there is, however, some confusion with regard to the respective properties of the two fluids. The procreatory-menstrual fluid seems to be discharged into the uterus during intercourse due to the effects of heat (produced through friction, according to the commentaries) and/or wind (cf. §§8.1f.), though it could also be that the passages in question actually tell us that it is propelled out via the vagina, in which case it may or may not have been in the uterus before intercourse. If we are indeed told that it may (also) be propelled out via the vagina during intercourse, then we are faced with problems as regards its nature, as it then might actually be of two sorts (blood-like and non-blood-like); these problems are compounded by the fact that women also seem to discharge a fluid called "semen" in connection with intercourse, which may (at least in certain circumstances) or may not be capable of fertilisation. The exact nature of this fluid and its relation to the procreatory-menstrual fluid is not clear; at times the two could be identical, but then again they seem to be different, and it is unclear whether the terminology used for the procreatory-menstrual fluid is also used to refer to this "semen" (cf. e.g. §8.15) (there is even a case, though as a verse of SU most probably spurious, that simply calls the procreatory fluid of the female "semen"; see §8.17). As regards the discharge of procreatory-menstrual fluid into the uterus, the same discharging process as during intercourse might also account for the menstrual discharge;[871] in any case, there seem to be two tubular structures leading into the uterus, in all probability from some place within the body where the procreatory-menstrual fluid collects prior to discharge (see on this especially §§6.12ff., also §8.17), though the conclusion that there is such a place of prior collection is based more on circumstancial evidence than on express statements of SU. It is also unclear whether the "semen" of a woman collects in her body in the same way as semen collects in that of a man (see §8.17). With regard to this "semen", there seems to be evidence to assume that there is a third opening through which it enters the uterus from within the body of the woman.[872] As regards the possibility of

[870] This *rasa*- is called *sthāyin*- by the commentators, though their notions on it need not be those of SU (see §§7.38f.).

[871] The possibly different theory mentioned in §8.16 is not from SU.

[872] I may remark here that if CA,Ci 15,34f., mentioned in §6.14 and §8.9, should really presume a collection of semen of women into the urinary bladder prior to its discharge, then we

there being other openings for ducts, channels or the like leading to or from the uterus and carrying procreatory-menstrual fluid, SU offers very little material which could help us to tackle this problem, though cf. especially §6.25 for a discussion of the possibility of a variant theory postulating not two, but a plural number of passages carrying procreatory-menstrual fluid opening into the uterus. Finally, we also learn that milk is in all probability the same as or at least very intimately connected with the procreatory-menstrual fluid; its relationship to semen is unclear, though, but this problem lies outside the scope of our study (cf., however, §§6.23f. and §7.40).

8.24. But I must point out that much of the above cannot, as yet, be proved beyond the shadow of a doubt; in many cases all we have to go on is the degree of probability. Nevertheless, comparing the data thus gathered from SU with the data gathered from CA, we find that they do not contradict each other. The data from CA were, however, found to be quite meagre, whereas SU offers us a lot of material, which is, though, often very difficult to interpret properly. In this we find another confirmation for the epigram saying that (see ŚARMĀ, p.511):

> *śārīre suśrutaḥ śreṣṭho na ca naṣṭaḥ kathañcana*
> *vyākhyāne tu paraṃ kaṣṭam iti me niścitā matiḥ*
> "Suśruta is the best with regard to [the discussion of] the matter pertaining to the body, and is in no way corrupt. With regard to the explanation [of what the text says], however, [one has] extreme difficulty. Thus is my determined opinion."

It may also be pointed out that, even though the material we were able to glean from SU was not always consistent and is indeed not easy to interpret, on the whole there does not seem to be any basis for agreeing with the frustrated scholar who changed the verse given above, the altered form beginning with *śārīre suśruto naṣṭo* (see ŚARMĀ, loc.cit.) "Suśruta is corrupt with regard to [the discussion of] the matter pertaining to the body."

---

would of course have grave difficulties with regard to this assumption; this matter shall, however, be taken up later, in §§12.17ff.

CHAPTER 9

The Aṣṭāṅgahṛdaya and the Aṣṭāṅgasaṅgraha

9.1. Let us now turn our attention to AH and AS. It is convenient to discuss these together, for, whether they were composed by one Vāgbhaṭa or two different Vāgbhaṭas, and whether AS evolved out of AH or AH out of AS, they are obviously very closely related (see also n.91).

9.2. On the whole, AH and AS draw heavily on CA and SU, large portions, both prose (in AS) and verse, being literal or nearly literal borrowings from these texts. Indeed, from AH,Ut 40,84 it clearly follows that one of the aims of the text is to create a synthesis of CA and SU (or, to be more literal, to supply what is missing in one text from the other);[873] however, from AH,Ut 40,88 (= AS1,Ut 50,p.421) it also follows that the information from CA and SU has been supplemented by that from other texts which were, already at that time, not generally studied anymore — the texts mentioned are BH and others (*bheḍādyāḥ*). Even though it is possible that these verses might be later accretions, they very aptly state what one indeed spontaneously feels when one reads AH (and AS).[874] Thus what AH and AS have to say on the subject of the procreatory-menstrual fluid, blood and semen can be expected to be on the whole only a repetition of what we find in CA and SU, albeit more systematically arranged and harmonised and maybe at times more lucidly (though not necessarily always correctly) explained, and also supplemented by additional information now and then. And that is in truth what we do find, even with more or less the same terminology used as in CA and SU. Thus it does not seem necessary to give such detailed references from AH and AS or to discuss them in such detail as was done in the case of CA and SU; rather, it seems sufficient to remark only on passages of especial interest — and which have not already been referred to in the foregoing chap-

---

[873] Cf. also AH,Sū 1,3ff. and AS,Sū 1,pp.2f.

[874] HĀLADĀRA, p.275 mentions a verse current in South India:
aṣṭāṅgasaṅgrahe jñāte vṛthā prāktantrayoḥ śramaḥ
aṣṭāṅgasaṅgrahe 'jñāte vṛthā prāktantrayoḥ śramaḥ
"The labour of [studying] the two previous works (most probably CA and SU) is in vain if AS is known, the labour of [studying] the two previous works is in vain if AS is not known."
Eulogising statements adhering to this model are incidentally nothing uncommon; loc.cit. draws attention to several similar verses relating to various matters, such as certain grammatical works, worship of certain gods, etc.

ters, in which we have also discussed many passages not only from AH and AS, but also from other texts, in detail.

9.3. It goes without saying that in these two works too the procreatory-menstrual fluid and male semen are said to combine to give the embryo,[875] their relative quantities determining the sex of the child.[876] The procreatory fluid collects and then exits as the menstrual discharge, but the problems regarding the actual place of collection prior to discharge remain the same as in §8.23: AH,Śā 1,22cd-23ab = AS,Śā 1,p.280b has already been discussed in §6.16 and §7.9, whereas AS,Śā 1,p.276a tells us that the "blood" of women reaches the uterus monthly and is then discharged, being called *ārtava-*.[877] This latter statement comes right after a description of how semen is discharged, namely from all over the body;[878] it moreover begins with *tathā* "in that manner; thus". Thus Indu's commentary, which says that the "blood" is to be collected from all over the body,[879] does seem plausible. This does not, however, necessarily mean that it collects there directly from all over the body and not via a receptacle which serves as an intermediate place of collection, so that this problem remains. In this connection the translation of the passage regarding the semen of males cited in n.878 is highly intriguing; it can be translated as:

> "Agitated by excitement, however, [the semen of males] through friction/(sexual) union (*saṅghaṭṭana-*) [and?] due to extreme excitement (*āveśa-*) of the heart (*hṛdaya-*) (scil. emotive excitement), issues, united (i.e. in this way uniting), from limb to limb (= from all limbs/each limb)",

but also as:

> "Agitated by excitement, however, [the semen of males] through friction/sexual union, united due to entrance (*āveśa-*) into the heart, issues from limb to limb (= from all limbs/each limb)."

It is obvious that if the second translation should be the one actually apper-

---

[875] E.g. in AH,Śā 1,1, AS,Śā 2,p.286b; see also AH,Sū 1,9, AS,Sū 1,p.4b.

[876] E.g. in AH,Śā 1,5, AS,Śā 2,p.287a.

[877] *tathā raktam eva ca strīṇāṃ māse māse garbhakoṣṭham anuprāpya tryahaṃ pravartamānam ārtavam ity āhuḥ.*

[878] Literally "from limb after limb":

> *harṣodīritaṃ tu saṅghaṭṭanena hṛdayāveśāt piṇḍībhūtam aṅgād aṅgāt pravartate.*

Before this the passage states that semen permeates the whole body. Cf. also n.168 and §8.2.

[879] *garbhakoṣṭhaṃ yoṣito garbhasthānaṃ sarvataḥ śarīrāt prāpya.*

taining here, what would be meant is that due to sexual excitement the semen of males issues from all over the body and unites in the heart; the heart would in this case not only quite obviously be the receptacle of semen we have discussed in §§6.13ff., but the specification of the heart as being this receptacle would moreover correspond more or less exactly to the location of this receptacle in the Nepalese diagram mentioned in §6.14. And because of the seeming parallelity of male semen and female procreatory-menstrual fluid in the AS passage just referred to above, this could be taken to mean that the procreatory-menstrual fluid too was taken to collect in the heart prior to its discharge into the uterus. We must, however, be cautious in our deductions, for the relevant part of AS,Śā 1,p.276a may, as we have seen above, also be understood in quite another way as not referring to any receptacle at all,[880] and we can furthermore not be fully sure that a complete parallelity between male and female is indeed presupposed by the AS passage. But even if the AS passage should indeed be referring to the heart as the place of collection not only of semen of the male, but also of procreatory-menstrual fluid of the female, that still would not necessarily mean that we have here a theory acknowledged as valid by all.

9.4. Nevertheless, the matter is highly intriguing. But since no other information on the subject seems to be available for the present, we will turn to other matters. After what has been cited in n.877, AS,Śā 1,p.276a tells us, in much the same manner as SU,Śā 2,18-21 (see §6.9), that *ārtava-* may flow in too large a quantity (*atiprasaṅgena*; cf. n.208) or at the wrong time, that this disorder is called *asṛgdara-*, *pradara-* and *raktayoni-* (cf. n.161), and that it may be cured by means of the cure for "blood-bile" (☞ *raktapitta-*) (1,p.276b). This, however, contains nothing new for discussion here. AH,Śā 2,61 describes a false pregnancy much like CA,Śā 2,8f. in §5.5. As regards the origin and symptoms of *raktagulma-*, these are described in much the same way as in CA (see §5.9) and SU (see §6.22, also §7.3), though we find no information on the problem of how exactly the procreatory-menstrual fluid is blocked (cf. §6.22); see e.g. AH,Ni 11,33;49-51 and AS,Śā 1,p.276b and

---

[880] Indu's commentary is unclear on this point, for he says:

kimbhūtaṃ saṅghaṭṭanena vegāveśena hṛdayāveśāt piṇḍībhūtam ekatvaṃ yātaṃ ... .

Does *vegāveśena* "through the extreme agitation of vehemence/the emotions" explain *saṅghaṭṭanena* or *hṛdayāveśāt*? If it explains the former, then Indu would be begging the question of relevance to us, for he would not be telling us what *hṛdayāveśa-* means. If *vegāveśena*, however, explains the latter expression, then Indu would be explaining *hṛdayāveśa-* in accordance with our first translation, and we would according to him have no mention of the heart as a receptacle of semen here.

Ni 11,pp.58a;59b (some parts of the AH and AS passages have already been dicussed in §6.22). An interesting remark is, however, made by Indu in his commentary on AS,Śā 1,p.276b, for he says that the "blood", not able to flow out, accumulates in the ☞ *mahāsrotas*-.[881] Since the blood is also said to be there together with discharged semen, this does look very much as if *mahāsrotas*- were used here to refer to the uterus,[882] though one might also argue that the *mahāsrotas*- is referred to because it holds the uterus, something in the latter thus being within that which encloses this latter too. On the other hand, AS,Ni 11,p.58a tells us that *gulma*-s *in general* are caused by the morbific entities, alone, in various combinations or with blood, entering the *mahāsrotas*- and blocking its outlets on both sides.[883] Could Indu simply have taken up this terminology, even though the special *gulma*- of women has a different cause, thus leading to another cavity (the visceral cavity?) and the uterus being referred to by the same term? Or could he be telling us that "blood" accumulates first in the uterus, but then in the passages leading to this and via them in other parts of the body? But that would seem to suppose that this "blood" also enters the general visceral cavity by this means, and the passages remarking on *raktagulma*- cited above do not mention this. One could in this context point to the entry of the foetus/embryo into the visceral cavity referred to in §6.20 and opine that the same path is also used by blood to leave the uterus and enter this cavity. Could Indu (not AS!) have had something similar in mind? But since we have seen in §6.20 that the blocking of openings referred to with regard to the stopping of the flow of blood seemingly does not refer to openings leading from the uterus to the visceral cavity, what sort of openings to the latter could Indu have meant? Thus the problem remains.

9.5. The procreatory-menstrual fluid is differentiated from blood; blood causes disorders of the fluid (AH,Śā 1,11, AS,Śā 1,p.278b). Other differentiations (not always consistent) have been mentioned in §§7.5ff. In AH,Śā 3,80f. and AS,Śā 5,p.326b blood and the procreatory-menstrual fluid

---

[881] Indu says:

tad raktaṃ bahir apravartamānam antar mahāsrotasi niṣiktaśuklasahitaṃ kevalaṃ eva vā vardhate,

mahāsrotas- here obviously explaining *antar* of the text, which runs:

tad dhi vardhamānam antar vartamānaṃ saśuklam aśuklaṃ vā jīvarahitaṃ ... .

[882] Which is called *garbhakoṣṭha*- in AS,Ni 11,p.59b as well as Indu's commentary on this. Cf. in this regard also the remarks on similar terms used for both uterus and stomach in Western medicine made by LAQUEUR 1990, pp.27;251.

[883] *mahāsroto 'nupraviśyāvṛtyordhvam adhaś ca mārgam ... gulmam abhinirvartayanti.*

are both said to be present as separate entities in the body. Regarding their respective origins, I would like to draw attention first to AH,Śā 1,7 cited in n.421. Though this passage, actually only a variant of SU,Sū 14,6, tells us only that the procreatory-menstrual fluid (*rajas-*) comes from ☞ *rasa-* without specifying the nature of the latter, Aruṇadatta interestingly explains it as stating that the procreatory-menstrual fluid arises not simply from *rasa-*, but from ☞ *āhārarasa-*.[884] And in his commentary on AH,Śā 1,22cd-23ab (see

---

[884] Aruṇadatta says in his commentary:

> *kimbhūtaṃ rajaḥ. rasāj jātaṃ rasajam. nanu raktaṃ rasajam eva. vakṣyati hi rasād raktaṃ tato māṃsam ityādi. tat kimarthaṃ rasajam ity uktam. brūmahe. rasād āhārarasād pariṇamato* (var: *āhārarasapariṇamato*) *jātaṃ rasajam ity atra nirdiṣṭam. na tu rasadhātoḥ. ity acodyam etat*
>
> "Of what nature (literally: arisen from what) is *rajas-*? Arisen from *rasa-*, [therefore] *rasa-*-arisen (*rasaja-*). Is it not so that blood[, which is different from *rajas-*,] is already *rasa-*-arisen (*rasaja-*)? For [the author Vāgbhaṭa] will say [in AH,Śā 3,62 = CA,Ci 15,16; cf. n.408]: 'From *rasa-* blood, from that flesh' etc. Then why is [*rajas-*] called '*rasaja-*' [here]? We are saying [it]. [With] '*rasaja-*' is here meant: arisen from *rasa-*, [i.e.] from *āhārarasa-* undergoing change. Not, however, from the element (☞ *dhātu-*) *rasa-*[, the first of the seven *dhātu-*s of the chain of metabolic transformation]. Thus this [statement of Vāgbhaṭa] is not controversial/objectionable."

Śrīdāsa in his commentary derives *rajas-* from *āhārarasa-* too, using the same words as Aruṇadatta (*rasād āhārarasād* etc.). Before this he, however, gives us some information not found in Aruṇadatta's commentary:

> ... *tadutpattir iti kecid. jīvaraktaṃ raja iti kecid āhuḥ. śoṇitaṃ ca pakvaṃ caivam āpadyate raktadhātū rajaś ceti. tasmād raktam iti kecid.*

The dots above (..) are found in the edition and obviously signify a part of the text which is missing. This makes it difficult to say exactly what is being discussed here, but since just before this Śrīdāsa has remarked on the derivation of the procreatory-menstrual fluid from *rasa-* the text presupposes, it looks as if he is here referring to another derivation of the procreatory-menstrual fluid. We may translate:

> "'... is its origin', [say] some. Some say '*jīvarakta-* is *rajas-*/*rajas-* is *jīvarakta-*'. And blood, when cooked (*pakvaṃ ca*) (i.e. when metabolically converted), is changed thus: 'the *dhātu-* blood and *rajas-*'. Therefore 'blood', [say] some."

This passage gives the impression of being a reference to the later theory according to which the procreatory-menstrual fluid is derived not from *rasa-*, but from blood, and which will be discussed below (see §§11.5ff.); Śrīdāsa's mode of deriving *rajas-* and blood from blood is actually already found in Indu's commentary on the same passage (see on this §11.33). The equation of *jīvarakta-* and *rajas-* in this context is puzzling, as it does not seem to be a necessary part of the argument to say that *jīvarakta-* is another name of *rajas-*; but we do not know what the missing part states. One could also opine that *jīvarakta-* here actually refers to blood in general (cf. §7.12), in which case we might have a statement equating blood in general with the procreatory-menstrual fluid. However, that would not be in keeping with the line immediately following, which does not equate the two, but derives one from the other; thus, if *jīva-*

§9.3) he explains *upacita-* "accumulated/increased", which in the verse refers to the procreatory-menstrual fluid,[885] as meaning *āhārarasato vṛddhiṃ prāptam* "having obtained augmentation/growth from *āhārarasa-*". Now what a commentator says is of course not necessarily that which the text commented upon means to say, especially if the two are separated by a span of several centuries,[886] but it must be conceded that this interpretation of the AH passages does tally with what we have independently deduced above as being possible (see §§7.40f.).

9.6. However, this alone is of course no proof that Aruṇadatta indeed gives us the opinion of AH on the origin of the procreatory-menstrual fluid, as the same discussion in §§7.40f. just mentioned has also shown us that this fluid could, on the basis of the same material, also be taken to evolve from *rasa-* different from *āhārarasa-*; since it is also unclear whether the 'classical' texts really have any sort of systematic differentiation between *rasa-* and *āhārarasa-*, it is therefore methodically unsound to let Aruṇadatta's statements on the origins of the procreatory-menstrual fluid overly influence our discussion of AH and AS.[887] This must also hold true for a discussion of AS,Śā 5,p.326b, where female milk and the procreatory-menstrual fluid are both called the pure part (☞ *prasāda-*) of *rasa-*;[888] we can say no more on this here than what has been said on similar passages in §§7.40f. But frustrating as this unclearness must be, we at least have a confirmation here for the close connection between milk and the procreatory-menstrual fluid. Indeed, AS,Śā 2,p.288a, which we have already mentioned in connection with the

---

rakta- should indeed refer to blood, then subject and predicate of the nominal sentence would most probably not be in a relation which makes them fully equal, but would refer to one being the same as the other, but in an altered state, which would be a parallel to the mode of expression we have already discussed with reference to Gayadāsa in §5.13 and Cakrapāṇidatta and Ḍalhaṇa in §7.12. We must also not neglect the possibility of an asyndetic expression for: "Some say '*jīvarakta-* and *rajas-*'" (then probably referring to the derivation of both). On the other hand, that *jīvaraktaṃ raja[h]* is expressly said to be the opinion of "some" (*iti kecid āhuḥ*) might mean that Śrīdāsa is here indeed quoting an opinion held by some authorities equating the two fluids, but then not accepted by him. We will have to wait for the missing portion of the commentary to turn up before coming to a final decision on this point.

[885] *māsenopacitaṃ raktaṃ* ... .

[886] Aruṇadatta probably lived in the twelfth century (cf. MEULENBELD 1974, pp.389ff.).

[887] On Indu's remarks in this regard see §11.33.

[888] The AS passage states:
strīṇāṃ [śarīre] rajaso 'ñjalayaś catvāraḥ. stanyasya dvau. ubhayam api caitad rasaprasādātmakam iti.
On the quantity of fluid mentioned here cf. §6.11.

parallel passage SU,Śā 4,24 (see §§6.17ff.), leaves no doubt that the procreatory-menstrual fluid and milk are directly connected, for it states not only that the procreatory-menstrual fluid whose flow is blocked by the foetus accumulates in ever higher regions of the body, not only forming, among other things, the ☞ *aparā-*, and making the breasts swell, but also being coloured in the breasts due to the action of phlegm and thus giving rise to milk (see n.272). The text further tells us that this milk is subsequently increased/augmented by *āhārarasa-* (cf. also §7.37),[889] but this is of course something quite different from saying that milk is derived from this latter, for *(āhāra)-rasa-* has such a "nourishing", "satiating" or similar function with regard not only to many parts, but ultimately also to the whole or at any rate[890] most of the body (see especially §§7.37ff.), in the course of which function it might undergo such changes as make no longer it directly, but some substance derived from it the direct agent of "nourishment" or the like of another substance, even though the root source remains the *(āhāra)rasa-* of the ultimate source, namely ingested food and drink.

9.7. As to the relation of the procreatory-menstrual fluid to semen, both male and female, we do not get much new information. The problem of three passages leading to the uterus, mentioned in AS,Śā 5,p.325a, has been discussed in §§8.19ff. From AS,Śā 1,p.286a we learn that women ejaculate "semen" during intercourse, and what is said on this seems the exact opposite of what SU,Śā 2,47 seems to say (cf. §8.6 and §8.8). The AS verse states:

yoṣito 'pi sravaty eva[891] *śuklaṃ* (= *śukraṃ*) *puṃsaḥ*[892] *samāgame*
garbhasya tan na kiñcit tu[893] *karotīti na cintyate*[894]

---

[889] The whole relevant portion runs:

stanāśrayam eva ca kaphoparañjitaṃ stanyatām upagataṃ prasūtāyāḥ punar āhārarasenāpyāyate.

Indu in his commentary on the colouring:

tataś cordhvapratipannād raktād yat stanaprāptaṃ tat sthānasthitena śleṣmaṇā śvetena kṛtoparāgaṃ stanyatāṃ yāti.

[890] Cf. e.g. AS,Śā 6,p.334a.

[891] AS1,Śā 1,p.201, Ḍalhaṇa (see n.790), Aruṇadatta (commentary on AH,Śā 1,5) and Hemādri (commentary on AH,Sū 1,9f.) read *sravanty* for *sravaty*. In ĀY 4,75 (cf. n.824) pāda a runs: *striyo api* (printed against the metre as *śtriyo 'pi) sravanty eva.*

[892] Ḍalhaṇa, ĀY and Hemādri (var): *puṃsāṃ*.

[893] Ḍalhaṇa: *tan na garbhasya kiñcit tu*, Aruṇadatta and Hemādri, also Śrīdāsa in his commentary on AH1,Śā 1,5: *garbhasya tu na tat kiñcit*, BHĀ 1,3,157: *tatra garbhasya kiñcit tu na* (against the metre) (on the verse in BHĀ see also §11.22).

[894] ĀY: *tan na garbhopakāri* (printed: °*ri*) *syād atɔ nātraiva kathyate.*

> "The semen of a woman too does flow during intercourse with a man.[895] But it does not do anything for[896] the [creation of the] embryo; thus [it] is not thought about[897] (= cared for, of no consequence)."

The problems caused by varying theories on the fertilising power of female semen have already been discussed in §§8.5ff.; the above does not seem to add any really new material for discussion. But it could be that *garbhasya tan na kiñcit tu karoti* is not to be translated as we have done. "Of the embryo" (*garbhasya*), taken to mean "for the embryo", could also be taken in the sense of "to the embryo", i.e. the semen does not harm the embryo. This could imply that the female *śukra-* may unite with female *ārtava-* and/or male/female *śukra-* to form an embryo outside the normal process of fertilisation, but not interfere with this normal process. In that case the above verses would not actually be in contrast to SU,Śā 2,47. BHĀ 1,3,187, which explains *garbhasya* in AS,Śā 1,p.286a as *śuddhasya* (*śuddha-* "pure", probably = "normal") and then quotes SU,Śā 2,47 with an introductory "The same (i.e. female *śukra-*) is the cause of an abnormal embryo",[898] seems to imply such an explanation. But the AS verse could, if *garbhasya* should mean "to the embryo", also imply that the "semen" of women does not harm the embryo in some other way, i.e. not necessarily through any power of fertilisation, which would imply that the fluid is basically of no consequence for conception, the end effect of which would be more or less the same as taking *garbhasya* to mean "for the [creation of the] embryo". Thus here again we end up with a *non liquet*.

9.8. Whatever the correct interpretation may in actual fact be, however, we see that here again women are thought to lose "semen" during intercourse. But this does not help us further in determining what this fluid actually is. But I may point out that Aruṇadatta in his commentary on AH,Śā 1,5 calls *rakta-*, which here denotes the procreatory fluid,[899] *strīretorajas-*.[900] Later in the commentary he glosses *rakta-* with *strīretas-*, but this latter gloss is not found in all mss, which means that we cannot be certain that it is authentic. In any case, it is unclear whether *strīretorajas-* means "female *retas-*,

---

[895] Vars: "Women too emit semen during intercourse with a man (or: men)."

[896] ĀY: "It is not helpful towards".

[897] ĀY: "talked about here".

[898] *vikṛtasya tu garbhasya kāraṇaṃ tad api bhavati.*

[899] The passage relates the theory on the relative quantities of the male and female procreatory fluids determining the sex of the child (see n.103), the female's fluid being *rakta-*.

[900] *strīretorajo'lpatvāt pumān jāyate na strī.*

[i.e.] *rajas*-", or "female *retas*- and *rajas*-". If the former, then we would have one and the same fluid, and in this case *retas*- would probably not be used here in its common later meaning "semen", but in the older meaning "flow" (i.e. flowing secretion),[901] though there is also the possibility that Aruṇadatta may here be calling the procreatory fluid of the female "semen" in the same way as the additional SU verse in n.836 does (cf. also §8.23). If, however, the second explanation of *strīretorajas*- should apply, then we would have two different fluids (cf. §§8.3ff.), both capable of fertilisation, in which case *retas*- could mean either "semen", or else "flow", the latter, however, then referring to something *not rajas*-. Of course, the explanation of Aruṇadatta, whatever it may actually mean, is not necessarily what AH wants to say, but it is nevertheless interesting to note that it leads to problems of a sort already discussed in the last chapter. These remarks also hold good for what Aruṇadatta says later in his commentary. After quoting AH,Śā 3,62f. (see on this n.408), which tells us that an embryo is formed from *śukra*-, he asks whether it is not true that *śukra*- here does *not* refer to the *śukra*- of males alone, since women too have seven *dhātu*-s. Then why does not a woman who discharges *retas*- through sexual excitement without cohabiting with a man beget a child, since to this end *śukra*- and *ārtava*- must combine, and both these are already at hand here, i.e. discharged without intercourse with a male?[902] To this Aruṇadatta counters that this is because *śukra*- of a male is absent, and it is this which combines with *strīretorakta*- (obviously the same as the *strīretorajas*- mentioned above) to form a child. Finally he quotes AS,Śā 1,p.286a (see §9.7) to show that women too ejaculate *śukra*-, which is, however, of no use in procreation.[903] This argument creates great

---

[901] Cf. the discussion on ☞ *dviretas*- in Appendix II.

[902] It may be pointed out that this argument does not seem to take into account theories like that found in SU,Śā 2,48f. discussed in §8.6, according to which two women can beget a child together.

[903] Aruṇadatta's commentary is also more or less literally reproduced by Śrīdāsa: *nanu rasād raktaṃ tato māṃsaṃ ... śukrād garbhaḥ prajāyate* (see n.408) *ity aviśiṣya nirdiṣṭaṃ na tu puṃśukrād garbhaḥ prajāyata ity evaṃ nirdiṣṭam. ubhayor api ca* (Śrīdāsa: only *api*) *strīpuṃsayor avaśyaṃ saptadhātukatvam abhyupagantavyaṃ na puṃsa eva. tad evaṃ taruṇīnāṃ kusumaśarākrāntamānasānāṃ* (Śrīdāsa: *kusumaśaraśarākrāntamānasānāṃ*) *tathāvidhena puruṣasaṃyogena vinā* (variant in Aruṇadatta's commentary: *saṃyogenāpi*) *kevalāt smṛtisaṃsparśadarśanāc calitaprasrutaretasāṃ* (Śrīdāsa: *smṛtidarśanāc calitasrutaretasāṃ*) *kim iti garbho na jāyate. śuklārtavaṃ hi garbhakāraṇam. tac ca sannihitam eveti kecid. tān brūmahe. puṃśukrābhāvāt. puṃśukraṃ hi strīretoraktayuktaṃ garbhakāraṇam. na ca* (Śrīdāsa: *strīretoraktasaṃyuktaṃ na*

problems. Aruṇadatta seems to begin by telling us that the *śukra-* of *both* males and females can form an embryo, but says at the end that the *śukra-* of women plays no part in procreation. This might mean that he uses *śukra-* in two different meanings here. But we seem to have no evidence to substantiate this deduction, so does he actually mean that the *śukra-* of women, though in principle capable of creating a child, does not do so either because it cannot combine with the semen of a man, or because this particular combination is infertile? This is indeed a possible interpretation of his statement. But Aruṇadatta also does not tell us what the *strīretorakta-* with which the *śukra-* of a male combines is; are *retas-* and *rakta-* (obviously = *rajas-* in *strīretorajas-* above) here synonyms, or do they refer to two different substances? If the latter should be the case, then this would mean that the semen of males can combine with both. This deduction could be valid if the alternative deduction made previously that the combination of the *śukra-* of male and female is infertile be valid too. And it would also fit the statement that a woman discharging *retas-* without intercourse discharges both *śukra-* and *ārtava-*, which, however, cannot combine to give a child. One could then object that the mention of only *retas-* being discharged would speak against two fluids being present, but against this one can hold that *retas-* might simply mean "flow", without differentiating between what flows (so that the emendation mentioned in n.903 is not necessary). All this would of course not hold true if *retas-* and *rakta-* in *strīretorakta-* should refer to one and the same fluid. In this case there would be no problem with regard to the *retas-* discharged by the woman without intercourse with a man (there would also be no problem if the emendation were effected, as *raktaretas-* would simply be a synonym of *retorakta-*). There would, however, be a problem as regards the argument that both *śukra-* and *ārtava-* are already present in the discharge without cohabitation with a man, as this refers to two fluids. We thus seem to have a contradiction here. But actually there is a way out, hinging on a different translation of the relevant part of the commentary, namely of:

śukrārtavaṃ hi garbhakāraṇam.

Our paraphrase above was based on the translation:

"for *śukra-* (of males) and *ārtava-* are the cause of the embryo".

---

ca) tatrāsti. tadabhāvād garbhasyānutpattiḥ. tathā (Śrīdāsa adds *ca*) *saṅgrahe 'py adhyagīṣṭa* (Śrīdāsa: *'dhyagīṣṭa*). *yoṣito 'pi sravanty* (Śrīdāsa: *sravaty*) ... *na cintyate* (see §9.7). *iti*.

The commentary of Aruṇadatta as printed in AH has *calitaprasruta[rakta]retasāṃ*, but *rakta* seems to be an addition by the editor, who might have been irritated by the terminology used. We shall, however, see below in this paragraph that even if we accept the emendation, the commentary still says the same.

But one could also translate this as:

> "for *śukra-* (of females) which is *ārtava-* is the cause of the embryo".

In that case we would have to understand the objector's argument as saying that this *śukra-* of women, which is the same as *ārtava-*, is the counterpart of the *śukra-* of males and capable of creating a child, so that it should create one even when it issues without actual intercourse. To this Aruṇadatta would then counter that a child cannot be created unless this *śukra-* = *ārtava-* combines with semen (*śukra-*) from a man. This would be a smooth explanation and do away with the difficulties drawn attention to, but leave us with another, namely that in this case metabolically there would be no distinction between *śukra-* (of males) and *ārtava-*, even though they would quite obviously be different substances. Thus we see that whichever way we look at the matter, we are faced with problems. I am unable to offer any solutions at this point, but shall return to the subject in another context in §11.16 and §11.26.

9.9. These are, however, special problems, relating, moreover, to commentaries and not to the texts right now concerning us themselves. All in all, we see that, as we suspected at the outset, AH and AS say more or less what we have already deduced from CA and SU and do not seem to add any really new material, except for the statement of AS,Śā 1,p.276a (see §9.3) on the heart maybe being the receptacle of semen in the male and of procreatory-menstrual fluid in the female; this potentially very important statement is, however, unfortunately ambiguous.

CHAPTER 10

Other Ancient Medical Texts

10.1. Apart from the 'classical' texts CA, SU, AH and AS there are a few other ancient texts which, though they have been preserved in full or part, are not so well known. Two of these ancient works have material bearing on our problem;[904] they are texts which have not been fully transmitted and which, though maybe still studied even by some scholars who lived after the age of the 'classical' texts (cf., however, §9.2), gradually became neglected and nearly forgotten until modern times, when interest in them revived. They could quite possibly be even older than at least the later of the 'classical' texts, but the redactions which we today possess are probably of a much later date. Nevertheless, it is very probable that these redactions contain ideas which belong to quite ancient traditions and are not influenced by the four 'classical' texts.

10.2. The first of these two texts is BH, which is quite ancient, according to some scholars even more ancient than CA, but seems in the redaction known to us to stem from about the seventh century A.D. Large parts are missing, and the remaining portions are often in a corrupt state. Nevertheless, it is possible to glean something of interest with regard to our subject-matter from it. This is, however, very little. We find the usual terminology for the procreatory-menstrual fluid being used;[905] in Śā 3,6 and 5,6 this fluid is called *ārtava-*, in Sū 16,3 and Śā 5,6 *rajas-*[906], otherwise "blood"[907].

---

[904] These are not the only ancient texts of this sort. Thus the YOGAŚATA(KA), ascribed traditionally to Nāgārjuna or Vararuci, probably dates to the seventh century A.D. at the latest; it may even be older than AH and AS. But it contains nothing relevant for our subject. Ravigupta's SIDDH was composed around (probably slightly after) the time of (the older of?) AH and AS; it too does not (though quoted a few times in this study) contain any really relevant material, and neither does RAS, which may even have been composed in the fifth century A.D. — the commentary of Narasiṃha on the latter (itself relatively old, though its exact date is difficult to ascertain), however, contains some material of interest to us and has been utilised a few times in the course of this investigation. As regards HĀ, though the work as such seems to be ancient too, the form in which it is available today seems not to be very old, which is why it is discussed in the next chapter.

[905] The text offers no grounds for doubting that the procreatory and menstrual fluids are one and the same, but, as in the case of AH and AS in the last chapter, it seems needless to go into details here.

[906] Passages with *rajasvalā-* (e.g. Ci 16,27) have not been noted (cf. §4.1).

[907] E.g. *śoṇita-* in Śā 3,8, *lohita-* in Śā 5,6, *rakta-* in Śā 3,10, *rudhira-* in Śā 5,5. In Śā

In Śā 5,5[908] menstruation is described as a flowing out of "old blood" (*purāṇarudhira-*); we learn basically nothing more from this than from CA,Śā 4,7.[909] In BH,Ni 3,21-28 we find a description of *lohitagulma-* = *raktagulma-* which tallies with that found in other texts (see §9.4); in v.23 the fluid which is pent up (how exactly, is unclear here too) is called *rakta-*, and in v.24 *śoṇita-*. Then v.26 tells us that the breasts are full of milk, which too tallies with the descriptions of this disease already discussed.

10.3. Śā 5,6 says:

"The blood of women, it should be known, dries up the body, therefore they experience/see no *rajas-*.[910] When, however, their body is full of the element (☞ *dhātu-*)[911], then each month they make known/offer blood which is [their] water/fluid of distinction (?)[912]. And, as something whose entering

---

3,8;10 we find the same theory on procreation as the one mentioned in n.103.

[908] See n.123.

[909] See n.94, also §5.16.

[910] *rajas-* can also mean "passion", and this would fit here, were it not for the following.

As regards the translation "experience" for *paśyanti*, a similar usage of √*paś* could be found in KĀM 7(=2,2),p.99, where it might be said of a woman being embraced that she "experiences something *rāmaṇīyaka-* (scil.: connected with pleasure)" (*kiñcid rāmaṇīyakaṃ paśyet*; KAN 2,2,14 has *ramaṇīyatvaṃ paśyet* — on the nature of KAN see n.1267). Yaśodhara, who on p.99 says that the embrace described serves to generate and heighten passion (*rāgasya jananaṃ vardhanaṃ ca*), on p.100 seems to explain the statement of KĀM as meaning "experiences *rāmaṇīyaka-* of the nipples [which are] to be scratched, etc., by the left [of the lover] or [which are] marked with the marks of [his] teeth" (*vāmalekhyādeḥ stanamukhasya daśanapadāṅkitasya vā rāmaṇīyakam ... paśyet*), from which it would follow that *rāmaṇīyaka-* refers to sexual excitement or its signs, in this case erection of the nipples. This interpretation could be substantiated by reference to what Yaśodhara seems to say a few lines later, namely that the woman moans or coos (√*kūj*) for the sake of kissing, not to show *rāmaṇīyaka-*: *cumbanārtham eva na rāmaṇīyakadarśanārtham*. However, it is not clear whether the latter sentence is really an explication of the remark on √*kūj*, or of the remark of the KĀM text saying that, while in his embrace, the woman strives to climb upwards on the body of the man, in which latter case it would seem to make more sense to take *darśana-* in the sense of "looking", which again would have repercussions on the translation of √*paś* in the first statement of Yaśodhara (though this need not necessarily be the correct interpretation of the KĀM passage itself). As it is, both the KĀM passage and Yaśodhara's explanation of it are difficult to understand, as is also evident from the various translations of them we have, translations which mostly do not make proper sense.

[911] *rajas-*? Blood? Or is a plural number of *dhātu-*s meant, which could stand for the seven elements *rasa-*, blood etc.?

[912] This probably refers to that which distinguishes women from men. I must admit, how-

> (into the uterus?) has returned (i.e. taken place again), it returns back again in a month as *ārtava-*."[913]

As the rather awkward translation makes clear, this passage is quite difficult to understand, and maybe corrupt. Nevertheless, one is of course rather inclined to see in the first "blood" mentioned (which also seems to be the "it" returning as *ārtava-*) the procreatory-menstrual fluid, especially since menstruation is obviously mentioned in the passage too, and combined with "blood". But the next passage, Śā 5,7, says:

> "But if that blood, corrupted, attains the passage/channel, then women have *pradara-*. One should treat that [*pradara-*][914], drying up the body, by means of the remedy for blood-bile."[915]

On *pradara-* and blood-bile (☞ *raktapitta-*)[916] cf. especially §5.14 and also §6.9; we have seen in §5.14 that *pradara-* may arise due to the combination of excessive blood with the procreatory-menstrual fluid. Does the corrupted "blood" of the BH passage (Śā 5,7) refer to this blood (in general), which is the root cause of the disorder, or does it refer rather to the fluid which is made to flow in an abnormal manner by this blood? We could thus have a problem, since the "blood" mentioned in Śā 5,7 seems to be the same as that in Śā 5,6. One may of course argue that the two passages may not have belonged together originally, so that "blood" in them need not be the same. But we have also seen in §6.9 and §9.4 that SU,Śā 2,17-21 and AS,Śā 1,p.276 respectively refer to an excessive and untimely flow of *ārtava- only*, calling this *asṛgdara-* or *asṛgdara-/pradara-/raktayoni-* respectively (cf. n.164), and saying that it is to be treated *like raktapitta-*. The BH passage could thus be referring to the same as the SU passage, irrespective of whether *pradara-* and *asṛgdara-* be exactly, or not quite, the same (cf. n.161).

---

ever, that this translation seems rather strained, but I cannot translate *vivekajalalohita-* in any other way. As it is, it should be considered whether we might not have here a mistake of the ms (in Telugu script), originating probably from a confusion of letters of some northern or eastern script, in which *r* and *b/v* are very similar, for *virekajalalohita* = "blood which is the water/fluid of evacuation", or maybe even *virekajalohita-* "blood arising from evacuation".

913 *strīṇāṃ khalu śoṇitaṃ śarīraṃ śoṣayati tasmān na tā rajaḥ paśyanti. paripūrṇadhātuśarīrās tu yadā bhavanti tadā vivekajalalohitaṃ māse māse prativedayanti. pratigatapraveśaṃ ca tat punar māsena samāgacchaty ārtavam.*

914 That *pradara-* is to be supplied follows from the masculine *taṃ*.

915 *yadā tu tac choṇitaṃ duṣṭaṃ mārgaṃ pratipadyate tadā strīṇāṃ pradaro bhavati. taṃ śarīraṃ śoṣayantaṃ lohitapittabheṣajenopakramet.*

916 Here *lohitapitta-*.

10.4. BH, at least that portion of the text we today possess, offers us no information on the origin of the procreatory-menstrual fluid. We also find nothing (apart from Śā 5,6f., just discussed) on its relation to blood in general and their differences. This, as well as the general lack of information on the procreatory-menstrual fluid, may be because the text has not been transmitted to us in full, but what seems more probable is that the author(s) simply did not give the matter much thought or deem it worth mentioning. This is substantiated by Śā 7,6, where, as in CA,Śā 7,15 (see §5.11), we have a list of the quantities of different substances present in the body, which includes semen (*śukla-*) and blood (*śoṇita-*), but not the procreatory-menstrual fluid. Thus BH is in this respect very similar to CA (cf. §5.16f.), which is not surprising, as the two are known to be so close to each other that they are often taken to represent different branches of the same school.

10.5. But even if the text does not tell us anything directly on the origin of the procreatory-menstrual fluid, it may still contain some information in this regard. BH,Śā 5,8 says that in a pregnant woman ☞ *rasa-* divided into three parts is responsible for (i.e. obviously: for the growth or nourishment of) the embryo/foetus, milk and *rasa-* (probably her own) respectively.[917] The same theory seems to be hinted at in Śā 4,31, which says that the embryo/foetus is nourished and made to grow by the mother's ☞ *anna-rasa-*.[918] We have seen several times above that milk is in all probability the same or at least very intimately connected with the procreatory-menstrual fluid;[919] we have also seen that there was a theory deriving both fluids from ☞ *āhārarasa-*,[920] though in other cases just *rasa-* is mentioned, which need not be *āhārarasa-*. Now when Śā 5,8 tells us that *rasa-* serves to nourish *rasa-*, then the nourishing *rasa-* here performs exactly the same function the *āhārarasa-* performs according to the theory seemingly considered 'standard' in later times (cf. §§7.37ff.), and Śā 4,31 actually mentions *annarasa-*, which seems but a synonym of *āhārarasa-*. This does make it look as if BH is of the opinion that milk, and, by implication, the procreatory-menstrual fluid, is derived from *āhārarasa-*. Of course we must be cautious in making such deductions, for not only do we have no surety that *rasa-* and *āhārarasa-* were really differentiated thus by BH too (cf. §9.6), but we must also not lose sight

---

[917] *garbhiṇyās tu tridhā kāye raso 'bhinirvartate garbhatvāya stanyatvāya rasatvāya ceti.*

[918] *nābhyāṃ nāḍī pratiṣṭhitā tasyām aparā mātur hṛdayam āśritā tayā mātur annaraso 'bhivahan garbhaṃ prīṇayaty abhivardhayati. tad yathā kulyāḥ kedāram abhisaṃśrayantyo bhāvayanti tadvat.*

[919] Cf. §7.40, §8.23, §9.6, as well as the description of *raktagulma-* in §10.2.

[920] Cf. §§9.5f., with further references.

of the possible differentiation between *āhārarasa-* and *annarasa-* discussed in §7.39. But despite these caveats we could build a good case for BH,Śā 5,8 deriving milk and therefore maybe also the procreatory-menstrual fluid from *āhārarasa-*. Since this statement is part of a description of the nourishment of the embryo or foetus, it might not be amiss here to see if we can find parallels in similar descriptions in other texts. I must, however, stress that, since we are not primarily interested in the nourishment of the embryo or foetus here, we shall touch upon this subject only insofar as it is directly concerned with our problem.[921]

10.6. AH,Śā 1,2, AS,Śā 2,p.291a and Cakrapāṇidatta's commentary on CA,Śā 4,21 tell us that the embryo or foetus is nourished by the mother's ☞ *āhārarasa-* (according to SU,Śā 3,31 it is the *āhārarasavīrya-*, on which see §§13.20f.), without mentioning the tripartition BH does. But in his commentary on SU,Śā 3,31f. Ḍalhaṇa quotes Bhoja,[922] who states that the "essence" (*vīrya-* "strength") of food, having become *rasa-*, is in the pregnant woman divided into three parts, one part nourishing/satiating the body, one increasing the milk, and one leading to growth of the embryo or foetus.[923] Since *(āhā-*

---

[921] On the nourishment of the embryo or foetus see e.g. COMBA 1981, pp.204-208, TIVĀRĪ 1,pp.141ff. (add: ĀY 4,147-153), HARA 1987, p.145,n.23 and DAS,R.P. 1984A, p.238. In this context see also the material on ☞ *ojas-* in connection with the formation of the embryo in COMBA 1981, p.206,n.38 and TIVĀRĪ 1,pp.155-157 (add ĀY 4,177).

[922] On whom cf. MEULENBELD 1974, pp.418f.

[923] This passage is also found in ĀY 4,147-150. Ḍalhaṇa quotes:

garbho ruṇaddhi srotāṃsi rasaraktavahāni vai (ĀY: var, see below)
raktāj jarāyur bhavati nāḍī caiva rasātmikā
sā nāḍī garbham āpnoti tayā (ĀY: *tathā*) *garbhasya vartanam* (ĀY: *vardha°*)
yad yad aśnāti mātāsya bhojanaṃ hi (ĀY: *ca*) *caturvidham*
tasmād annād rasībhūtaṃ vīryaṃ tredhā pravartate (ĀY: *yat sampravartate*)
bhāgaḥ śarīraṃ puṣṇāti stanyaṃ bhāgena vardhate
garbhaḥ (ĀY: *°aṃ*) *puṣyati bhāgena vardhate ca yathākramam*
garbhaṃ kulyeva kedāraṃ nāḍī prīṇāti tarpitā (ĀY: var, see below).

NYC in the commentary on SU,Śā 4,24 gives us only the second and third lines of the above passage (*raktāj jarāyur ... garbhasya vartanam*), with the same reading as Ḍalhaṇa (see Excursus III at the end of chapter 6). The first line is different in ĀY, namely:

garbho ruṇaddhi (printed *ruṇāddhi*) *srotāṃsi rajaḥśukravahāni vai*,

which poses a problem which we shall take up later (in §13.14), as it is not relevant here. The last line in ĀY reads:

garbhaṃ kulye ca kedāre (var: *kedāraṃ*) *nāḍī prīṇāti tarpati*,

the first part of which looks like a corruption. Note that here as well as in the BH passage in n.918 we have the simile of the irrigation ditch(es) (*kulyā-*) and the field (*kedāra-*) (we find this simile in AH,Śā 1,56 too, which simply states that the foetus is nourished by the mother through the umbilical cord as a *kedāra-* through a *kulyā-*). The comparison in this case seems

ra)rasa- does indeed ultimately nourish or satiate the whole body, the nourishment or satiation of *rasa-* being the first step in this process, this statement of Bhoja seems to tally with what BH tells us.[924] KĀ,Sū,Lehādhyāya, p.2 says much the same as BH,Śā 5,8, except that instead of the nourishment of *rasa-* and milk that of the woman and the breasts respectively is mentioned, whereas CA,Śā 6,23 says more or less the same as Bhoja.[925] True, the passages mentioning the tripartition of *rasa-* do not expressly mention *āhāra-rasa-*, as the other passages (in which the tripartition is not mentioned) cited before them in this paragraph do, so that we cannot be fully sure that these are to be linked. If, however, they are, then we do have grounds for assuming that the notion that milk (and, by implication, the procreatory-menstrual fluid) was derived from *āhārarasa-* was wide-spread, even if we should have different theories on the nourishment of the embryo. A word of caution would, however, not be amiss here: the passages discussed in §§10.5f. deal with pregnant women, and it may be (though the chances seem slim) that the metabolic processes in pregnant women were with regard to the procreatory-menstrual fluid thought to be different from those in other women. But on the other hand it must be brought to mind that we do have express statements of Indian medical authorities linking the procreatory-menstrual fluid

---

to be merely between the irrigation of a field through a ditch or ditches and the nourishment of the embryo through a ☞ *nāḍī-*, i.e. the umbilical cord (cf. also p.102), which both Bhoja and the BH passage mention. However, a similar simile belongs to the standard repertoire of the commentaries commenting upon the role of ☞ *āhārarasa-* as nourishing or satiating agent of the elements (see §§7.37f.); what is meant in that context is that *āhārarasa-* performs its function in the same way as the water of an irrigation ditch, of which a part first enters the nearest field, after that a part the one further removed etc. (in like manner, first *rasa-*, then blood, then flesh etc. are nourished or satiated). It seems probable that the mention of this simile in connection with the foetus refers to the same mode of nourishment in the body of this. Incidentally, this mode is only one of the alternative modes discussed by the commentaries mentioned above as being possible (though they usually seem to prefer this explanation).

[924] As regards the other parts of Bhoja's statement, particularly the first verse of the passage, these, though of relevance to our study as a whole, are not of immediate relevance to the particular problem we are here dealing with, and shall thus be discussed in §§13.14ff. below.

[925] The KĀ passage says:
yad annapānaṃ prāyeṇa garbhiṇī strī niṣevate
raso nirvartate tādṛk tridhā cāsyāḥ pravartate
mātṛpuṣṭyartham ekāṃśo dvitīyo garbhapuṣṭaye
tṛtīyaḥ stanapuṣṭyarthaṃ nāryā garbhas tu puṣyati.

The CA passage states: *striyā hy āpannagarbhāyās* (Cakrapāṇidatta in his commentary on CA, Ni 3,13f.: *āpannasattvāyās*) *tridhā rasaḥ pratipadyate svaśarīrapuṣṭaye stanyāya garbhavṛddhaye ca*. See on this passage, however, also §§13.18ff.

and *āhārarasa-* in women in general (see §§9.5f.). Nevertheless, we cannot be definite that *rasa-* as distinct from *āhārarasa-* may not after all be meant (see also loc.cit.), provided that the two were differentiated at all in the first place. In this context it is interesting that BHĀ 1,3,175 says that *rasa-* (which, as this and the following passages of the text, especially 178ff., make clear, functions like that *rasa-* also termed *āhārarasa-*) goes to the heart,[926] the seat of *rasa-* which is *śarīrārambhaka-*, and mingles with this there; the latter *rasa-* would thus seem to be that called *sthāyin-* or *poṣya-* by the commentators. Now *śarīrārambhaka-* means "causing the body to begin" or "the beginning/origin of the body". Could this refer to the nourishment of the embryo we are here discussing? Then of course we would have clear evidence for *rasa-* not *āhārarasa-* being meant in such a context, though that would not necessarily have to apply to the passages discussed above too. But I may also point to passages such as CA,Ci 22,16 (quoted in §7.22, with parallels) saying that the body arises from *rasa-*, which is not part of a context like the one we are discussing here, but stands in connection with the theory permeating Indian medical literature that *rasa-* is of utmost importance for the upkeep of the body; the BHĀ passage could also be seen in this light. See, however, Aruṇadatta's commentary on AH,Śā 3,6, in which verse the birth, growth etc. of the body are said to arise from *rasa-* (*rasaja-*);[927] Aruṇadatta says that *rasa-* here refers to *āhārarasa-*. This is followed by Śrīdāsa, commenting on the same passage.[928] Then again, BHĀ 1,3,179(p.57) and 1,3,181 speak of semen which is *śarīrārambhaka-*, and which according to the former passage is expressly said to be situated all over the body (cf. §6.13). One could of course see semen here in its function as causing new life (see also §8.2), but it is also to be considered whether *śarīrārambhaka-* might not refer to the being situated all over the body, on the basis of the characterisation of semen as being present throughout the body and because it is a commonplace in our texts that *rasa-* too circulates throughout this. Of course this would entail ascribing a meaning "foothold; support" or the like to *ārambha(ka)-* (from *ā√rabh* "take hold of; keep fast, cling to; attain") which would to my knowledge not be otherwise attested (though one may also consider whether we might not actually have a mistake e.g. for *śarīrarambhaka-* "support of the

---

926 Cf. on this the discussions on ☞ *rasa-* and ☞ *hṛdaya-* in Appendix II.

927 *rasajaṃ vapuṣo janma vṛttir vṛddhir alolatā.*

928 Indu in his commentary says, however, that the body arises from a plural number of *rasa-*s (*rasajāni yāni rasebhyaḥ pariṇatebhyaḥ sampadyante*), the plural seemingly referring to characteristics of the body which are unknown to us, as the commentary has lacunae, so that it is unclear what these *rasa-*s refer to.

body" here), but the matter seems worth considering nonetheless. However, it does not seem as if we can say much more on this subject at present, so that we may shelve it for now, hoping to gather new information in the course of this study for a fresh look at the matter.[929]

10.7. We do not seem to be able to garner any other information relevant to the nature of the procreatory-menstrual fluid from BH, so we shall turn our attention to the other old text, KĀ, which in its present redaction dates probably to about the seventh century. Known basically through one manuscript, it is unfortunately extremely fragmentary as regards the chapters most interesting to us. As it is, the text says little directly on the procreatory-menstrual fluid, which it seems more often than not to call simply "blood", e.g. *rakta-* in Śā,Asa, p.46 (on this passage cf. n.401), *śoṇita-* in Śā,Gar 3 and Śā,Jā 4. All we learn from these passages is that this "blood" unites with semen (in the first two instances called *bīja-*; cf. on this §§2.5f.), which is by now nothing new to us. Nevertheless, KĀ does contain other extremely interesting material. Thus Sū 28,p.38 contains the startling statement that the first of the seven ☞ *dhātu*-s, so well known to us by now, is *not* the substance *rasa-* (a major part of our discussions on the origins of the procreatory-menstrual fluid so far), but ☞ *tvac-* "skin"![930] This is followed in KĀ,Śā,Jā 10 by the statement that the "elements" skin (*tvac-*) to semen are strengthened/nourished by food,[931] reminding us of CA,Vi 8,102,[932] AH,Śā 3,117 and AS,Śā 8,p.350a, where we have a list of eight "essences" (*sāra-*), which is the same as the usual list of "elements" (*dhātu-*), except that *tvac-* "skin" takes the place of *rasa-* at the beginning and *sattva-* "(true) essence" follows semen.[933] We obviously have a problem here, one which the commentators of these passages do not remark upon, except for Cakrapāṇidatta, who says in his commentary that *sāra-* here means a very pure *dhātu-*,[934] after which he says no more on the problem either.[935] This does not, however, mean that the com-

---

[929] This will indeed be the case in §13.21.

[930] *tvagasṛṁmāṃsamedosthimajjaśukrāṇi dhātavaḥ.*

[931] *tena* (= *āhāreṇa*) *tvagādayaḥ śukrāntā dhātava āpyāyante.*

[932] To which DAS,R.P. 1984A, p.243,n.24 has already drawn attention.

[933] *sattva-* here takes the place of ☞ *ojas-*.

[934] *sāraśabdena viśuddhataro dhātur ucyate.* See also SHARMA,P.V. 1985, p.344.

[935] Hemādri in his commentary on AH,Sū 6,69-71 cites AH,Śā 3,117, but does not tell us anything on the actual problem of this passage. In Sū 6,71 we find the line *śoṇitaprabhṛtīnāṃ ca dhātūnām uttarottaram* in the context of which parts of the body of animals are heavier (*guru-*) than others; Aruṇadatta explains this as meaning that flesh is heavier than blood, fat

mentators were unaware of the problem; see on this §§10.9f. The problem is present in SIDDH 21,10+19, CA,Ci 28,92-95 and AS,Ci 23,p.291b too, where we find the mention of diseases due to wind, present in what are obviously the seven *dhātu*-s (though the word is not mentioned here) ending with semen, but beginning with "skin" instead of *rasa*-. The same is the case in AS,Sū 19,p.143b, the diseases mentioned, however, being those due to bile and phlegm. In CA,Sū 18,7 we twice find a list consisting of "skin, flesh, blood etc." (*tvaṅmāṃsaśoṇitādīni*),[936] it refers to substances overwhelmed by the corrupted morbific entities. Narasiṃha in his commentary on RAS 1,6 (p.15) too quotes a list (upto ☞ *majjan*- "marrow") beginning with "skin",[937] whereas in SU,Ni 9,4 (= BYT 110,1 = MĀDH 40,1) we find, in connection with *vidradhi*- (cf. §7.2), a similar list which goes up to *asthi*- "bone".[938] In CA,Sū 11,48, AH,Sū 12,44 and AS,Sū 22,p.166a it is stated that, of the three paths by which diseases enter the body, the first[939] consists of the chain of *dhātu*-s beginning with *blood*, and *tvac*- "skin", which latter thus, though not itself called a *dhātu*-, is intimately linked with blood etc.[940] And in CA,Ni 5,3 and CA,Ci 21,15 (= BYT 123,4 = MĀDH 52,4cd-5ab) we find, in a list of

---

than flesh, etc. Hemādri, however, analyses the line as referring to animals whose essence (*sāra*-) is composed of blood etc., saying that the flesh of animals whose essence is flesh is heavier than that of those whose essence is blood, the flesh of animals whose essence is fat is heavier than that of those whose essence is flesh, etc. Then quoting AH,Śā 3,117, he asks why *tvac*- and *sattva*- are not mentioned, with which he clearly gives to understand that he sees these substances as being on a par with the other substances (blood etc.) mentioned in Sū 6,71. His answer incidentally is that the flesh of animals whose essence is skin is unfit for consumption, whereas there are among animals (*tiryakṣu*) none whose essence is *sattva*-. (On the chain of *dhātu*-s beginning with blood see also Excursus X at the end of chapter 7. Parameśvara says that *rasa*- is not mentioned here because it is not *mūrtimat*-, i.e. "solid" or "material", like flesh etc., whereas Indu explains the passage — AHS,Sū 6,72 — as if it spoke of all the seven *dhātu*-s including *rasa*-.)

936 Once we also have only *tvaṅmāṃsaśoṇitāni*, which is probably an editorial or printing mistake.

937 This verse is from an unidentified source:
tvacchoṇitāntare pittaṃ māṃsamedo'ntare kuphaḥ
asthimajjāntare vāyur evaṃ tristhūṇam ucyate.

938 The edition of SU has:
tvagraktamāṃsamedāṃsi pradūṣyāsthisamāśritāḥ.
Gayadāsa in his commentary on 9,26-28 quotes this with *asthi*- twice:
tvagraktamāṃsamedo'sthi pradūṣyāsthisamāśritāḥ .

939 Technically called *śākhā*- "branch[es]".

940 Cakrapāṇidatta and Ḍalhaṇa in their commentaries on SU,Sū 45,56 say that *śākhā*- refers, according to some, to the *dhātu*-s blood etc., *tvac*- not being mentioned by them.

seven *dhātu*-s of which three are the morbific entities,[941] a series consisting of skin, flesh, blood and serous fluid (☞ *lasīkā*-);[942] this series is found also in AH,Ni 14,2 and AS,Ni 14,p.70a, where the word *dhātu*- is absent. Gayadāsa in his commentary on SU,Ni 9,10f. quotes a passage of Bhoja in which "*tvac*- etc." (*tvagādīni*) are said to be corrupted by the morbific entities; it is unclear what *tvagādīni* actually refers to, but in the light of the above there is the possibility that a chain of *dhātu*-s might here be meant.

10.8. Even though some of them call ☞ *tvac*- a ☞ *dhātu*-, none of the passages quoted in §10.7 go as far as KĀ, which expressly calls *tvac*- one of the *dhātu*-s (clearly taking the place of *rasa*-) of the known metabolic chain.[943] We nevertheless see that there seem to be traces of a similar theory in them too. It is obvious that this creates a problem as regards the origin of the procreatory-menstrual fluid too, for we have so far as a rule discussed this in connection with its assumed origination from *(āhāra)rasa*-. As it is, the list beginning with "skin" is most probably the remnant of a very ancient list going back to Vedic, and maybe even to Indo-European times. JAM-

---

[941] *trayo doṣā[ḥ]* and *doṣās trayaḥ* respectively.

[942] On these passages see §10.10 below, and also the discussion on ☞ *dhātu*- in Appendix II.

[943] According to AH,Ni 14,4f. = AS,Ni 14,p.70a *kuṣṭha*- eventually attains all the *dhātu*-s, permeates the interior (of them? or of the body?), wets them through and through (obviously as the first step of putrefaction) and brings about "worms" (☞ *kṛmi*-):

prapadya dhātūn vyāpyāntaḥ sarvān saṅkledya cāvahet
sasvedakledasaṅkothān kṛmīn sūkṣmān sudāruṇān.

It is unclear what "all the *dhātu*-s" (*dhātūn ... sarvān*) refers to here. Aruṇadatta and Indu give no explanation, whereas Toḍaramalla in his commentary on AH (see the editor's note on loc.cit.; on this commentary see n.1006) says that this refers to "*rasa*- etc." (*sarvān samagrān dhātūn rasādīn*). But in the passage immediately preceding the above, *tvac*- is mentioned, but not *rasa*- (see §10.7). Is Toḍaramalla therefore referring to another chain of substances different from the one beginning with *tvac*-, or is he presupposing that *tvac*- and *rasa*- are the same? In this context SU,Ni 5,3, which tells us that *kuṣṭha*- first attacks the skin, then moves into the body and attacks the *dhātu*-s there, is of interest:

evaṃ samutpannas tvaci doṣas tatra tatra ca parivṛddhiṃ prāpyāprati-
kriyamāṇo 'bhyantaraṃ pratipadyate dhātūn abhidūṣayan.

Though this passage too does not specify what the *dhātu*-s are, they seem at least not to include *tvac*-. But we do not know whether the chain of the *dhātu*-s presupposed here begins with *rasa*-, or whether it begins with the element following on *tvac*-/*rasa*-, i.e. blood. And then of course there is always the problem of whether SU,Ni 5,3 and the passages mentioning *tvac*- (and not *rasa*-) in connection with *kuṣṭha*- (see §10.10) actually presuppose the same theory; this problem is compounded by the fact that the arrangement of the vv. AH,Ni 14,2-5 = AS,Ni 14,p.70a, which suggest's a connection between them, need not necessarily reflect a homogeneous and homogenous theory, given the known syncretistic tendencies of these texts.

ISON 1986, pp.172ff. has not only drawn attention to ancient (and also later non-medical) lists, of some — though not all — of which "skin" is a part, but has also briefly traced the development which led to the formulation of the doctrine of the seven *dhātu*-s (beginning with *rasa*-) which was later considered 'standard' in Indian medical theory; see also MÜLLER 1934, pp.27ff.[944] That all this takes us well beyond the sphere of only medical texts is made abundantly clear by HEILIJGERS-SEELEN (pp.51;62;65f.), where attention is drawn to the confusion, in Tantric texts, caused by two different lists, one being the known list of the seven *dhātu*-s, but with skin in the place of *rasa*-, the other a list of six substances, which are the same as in the list of the seven *dhātu*-s (with skin as the first element), but without the last element, namely semen; according to the texts, the constituents of the list of six substances are found in six "sheaths" or "repositories" (*kośa*-).[945] All this points to a development which is obviously quite complex; much of this development is still unclear, but fortunately we are not concerned so much with the historical development as with the problems occasioned in the context of our study.

10.9. That the commentators too, even though unaware of this diachronic development, were aware of the problem occasioned by ☞ *tvac*- and ☞ *rasa*- is obvious from Cakrapāṇidatta's commentary on CA,Sū 11,48 (see §10.7), which says that *tvac*- here subsumes *rasa*- residing in the former, to differentiate it from *rasa*- in the heart;[946] see also CART on the same passage.[947] ĀĀŚ introduces 4591 (p.659) by stating that it will describe the disease *masūrikā*- situated in the *rasa*-, but the verse 4591 actually describes the disease as situated in the skin (*tvac*-); after 4591 the text then tells us in its commentary that *tvac*- here actually refers to *rasa*-, as it is *rasāśraya*- "having the seat (i.e. situated) in *rasa*-".[948] In his commentary on SU,Śā 9,12, Ḍal-

---

944 An interesting list is contained in MBH 12,293,35. See also SHARMA,P.V. 1972, p.28,n.5 and CA,Śā 6,10 (see ☞ *rasa*- in Appendix II), AS,Ut 40,p.314a. Sāyaṇa in his commentary on the passage from *Aitareyāraṇyaka* 2,3,7 quoted on p.522 expressly mentions the "repositories; sheaths" (*kośa*-) skin (*tvac*-), blood, flesh, fat, bones and marrow.

945 Cf. on the latter Sāyaṇa's statement given in n.944.

946 Cf. on this latter the discussions on ☞ *rasa*- and ☞ *hṛdaya*- in Appendix II.

947 For a precise translation of Cakrapāṇidatta's commentary see MEULENBELD 1974, p.72,n.42 (cf. also SHARMA,P.V. 1985, pp.123f.). Interestingly, Śrīdāsa in his commentary on the AH parallel Sū 12,44 expressly remarks that *rasa*- is *excluded* here: *na rasadhātuḥ śākhā tac ca bāhyarogāyanaṃ bāhyānāṃ rogāṇām ayanaṃ sthānam*; Śrīdāsa also calls *tvac*- an ☞ *upadhātu*-.

948 *tvakśabdenātra rasa ucyate rasāśrayatvāt.*

haṇa says of the statement that the *mūla-*[949] of the ☞ *srotas*-s carrying flesh includes the skin (*tvac-*):

> *tvakśabdena raso 'tra tātsthyāt*
>
> "By the word *tvac-*, *rasa-* is here [meant], because of [its (i.e. the *rasa*-'s)] being situated in it (i.e. *tvac-*)."

This is remarkable, as the passage commented upon does not mention *rasa-* in the connection in which *tvac-* is mentioned at all:

> *māṃsavahe dve. tayor mūlaṃ snāyutvacaṃ raktavahāś*[950] *ca dhamanyaḥ.*

Similarly, the commentary *Madhukośa* on MĀDH 22,15 (= CA,Ci 28,30), where the effects of corrupted wind in the skin (*tvac-*) are described, says that this refers to *rasa-*, since *tvac-* here is an additional or secondary element (☞ *upadhātu-*), obviously derived from *rasa-*.[951] Here too *rasa-* is not mentioned in the text commented upon, but one could opine that it may have been felt to be implied because the following verses (16ff.) deal with wind in the substances blood to semen, i.e. the remaining chain of ☞ *dhātu*-s.[952] Similar considerations may have made the *Madhukośa* equate *tvac-* in MĀDH 54,14 with *rasa-*,[953] since, after describing the effects of the disease *masūrikā-* when it attacks the skin, the text goes on (54,15ff.) to describe the effects as regards the substances blood to semen, followed by the effects as regards vessels (☞ *sirā-*), cords (☞ *snāyu-*) and junctures (☞ *sandhi-*).

---

[949] On this term see §6.18.

[950] Var: *raktavāhinyaś.*

[951] The *Madhukośa* says:

> *tvaggata iti upadhāturūpāṃ tvacaṃ prāpte. candrikākāras tu tvakśabdena rasam āha. tena rasagata ity arthaḥ.*

(On the Candrikākāra cf. MEULENBELD 1974, p.403.)

[952] The commentary also mentions an ingenious theory by Kārtika (on him see MEULENBELD 1974, pp.394f.), who opines that, as *rasa-* is situated in the heart, which is in the vicinity of the *āmāśaya-*, which has already been mentioned in connection with the disorders due to wind in it, a renewed mention of *rasa-* is superfluous:

> *hṛdayasthasya ca rasasyāmāśayasāmipyād āmāśayagatavātalakṣaṇenaiva tad-adhigate rasagatasyānabhidhānam iti kārtikaḥ.*

(On *rasa-* in connection with both heart and *āmāśaya-* see, apart from n.946, also n.420.) Kārtika is known as a commentator on SU, so that what he says probably refers to SU,Ni 1,25 (referred to below too), the *āmāśaya-* being mentioned in Ni 1,22. But since the context in MĀDH is similar to that in SU, and since MĀDH 22,15 and 22,13 correspond to SU,Ni 1,25 and 1,22 respectively, it is understandable that to the *Madhukośa* Kārtika's arguments are relevant for MĀDH too.

[953] *tvakśabdenātra raso 'bhidhīyate dhātugataprastāvāt.*

SU,Ni 1,25 too, like MĀDH 22,15, describes the effects of corrupted wind situated in the skin, the following verses (26ff.) detailing the effects with regard to the substances blood to semen, the chain, however, being interspersed with a few other substances,[954] namely vessels, cords and junctures.[955] It comes as no surprise to see that Ḍalhaṇa in his commentary on the SU passage simply equates *tvac-* here with *rasa-*.[956] Gayadāsa too equates *tvac-* and *rasa-*, saying that the latter is in the place of the former.[957] But he then goes on to ask what the other substances interspersed among the *dhātu-*s in the passage are doing here, as they are not *dhātu-*s. In reply, he points to CA,Ci 15,17 with the reading mentioned in §5.13. Here milk and *ārtava-* are *abhinna-* from blood, he says, and the skins (plural) *abhinna-* from *rasa-*, blood and flesh, which means that the wind situated in them (i.e. obviously milk, *ārtava-* and the skins) behaves similarly (i.e. like that situated in *rasa-*, blood and flesh), from which it follows that the diseases of wind in skin are also those of wind in *rasa-*. The appropriateness of the mention of the diseases of wind situated in the vessels, cords and junctures also follows.[958] Gayadāsa's statement (including the significance of *abhinna-*) has already been discussed in §5.13. Here it creates difficulties, for the CA verse as quoted derives the six skins (☞ *tvac-*) from flesh, which itself is derived ultimately from *rasa-* (this derivation obviously being presupposed here too); as such, simply equating these skins thus derived with the skin mentioned in

---

[954] Which in the MĀDH passage follow *after* semen. Note that MĀDH is much later than SU, and is known to have rearranged much of the material it has taken over from the 'classical' texts.

[955] In AS,Sū 19,p.143b too the effects of corrupted wind and phlegm are detailed, in two separate lists, with regard to skin, the substances blood to semen, and, following these, vessels, cords and *koṣṭha-*, i.e. the visceral region.

[956] *tvakstho rasasthaḥ.*

[957] *tvakstho rasasthaḥ ... rasasya tvaksthānatvāt.*

[958] The full wording of Gayadāsa's commentary is:

kuryāt sirāgataḥ śūlam ity aprastutaṃ dhātusthavātādhikārād asthigatavātādhikārāt (var: °*sthavātādhikārāt sirāgatavātānadhikārāt*). *naitad asti. vyākhyātam evaitad doṣadhātumalakṣayavṛddhivijñānīye.*

rasād raktaṃ tathā stanyaṃ asṛjaḥ kaṇḍarāḥ sirāḥ
māṃsād vasā tvacaḥ ṣaṭ ca medasaḥ snāyusandhayaḥ.

ity atra. tatra stanyārtavayo raktābhinnatvāt tvacāṃ ca rasaraktamāṃsair abhinnatvāt tatrasthavāyoḥ saiva kriyā rogāś ca. ata eva ya eva tvaksthasya vātasya rogās ta eva rasasthasya vātasyoktāḥ. atas tebhyaḥ śeṣāṇāṃ sirāsnāyusandhīnāṃ āśrayabhūtānāṃ āśritavātarogāvasaraḥ.

A part of this has already been quoted above in n.154.

SU,Ni 1,25 is highly problematic, since this latter skin stands in exactly the place at the head of the chain of substances where *rasa-* usually stands. And of course, Gayadāsa simply assumes, in traditional manner, that CA and SU must refer to the same, which is not at all certain. What is therefore much more probable is that in SU,Ni 1,25ff. we have a non-'standard' series of the sort mentioned in §10.8, since some of the elements not fitting into the 'standard' list are found in other known lists, Indian as well as Iranian (cf. JAMISON 1986, loc.cit.); see also GARBH 2, where we find the series *rasa-* ("nutrient fluid"?, or "taste"?; cf. §7.15), blood, flesh, fat, cords, bones, marrow and semen.[959] Fortunately, we are concerned not with the correctness or incor-

---

[959] The relevant part of GARBH is:

saptadhātukam (scil. *śarīram*) *iti kasmāt[.] yadā* (GARBH1: *yathā*) *devadattasya dravyādiviṣayā jāyante. parasparaṃ saumyaguṇatvāt ṣaḍvidho raso rasāc choṇitaṃ śoṇitān māṃsaṃ māṃsān medo medasaḥ snāyavaḥ snāyubhyo 'sthīny asthibhyo majjā majjātaḥ śukram* ... .

Lakshmi Kapani (GARBH, p.14) translates *yadā devadattasya* etc. as: "Lorsque naît chez Devadatta le domaine des substances matérielles, etc., [alors se produit] le sextuple *rasa*, fait de la propriété saumique, d'òu successivement [se produit] le sang, du sang la chair ..." (the square brackets in the translation are Kapani's own). After the word "saumique", we find a footnote which tells us that what is meant here is phlegm, which is "rapporté au *soma* qui lui-même est rapporté à la lune, astre froid". Now of course phlegm is a ☞ *saumya-* or "somic" substance, but Kapani does not tell us in what manner it is supposed to fit into the picture, and I fail to see how it possibly could. As it is, Kapani has not seen the problem occasioned by the fact that *ṣaḍvidho rasaḥ* "*rasa-* of six types" obviously refers to the tastes, whereas the following *rasa-* (*rasāt*) looks as if it were the first ☞ *dhātu-* of the chain of metabolic change (this in a variant form, with *snāyu-* inserted between fat and bones). One would also have liked an explanation on what "fait de la propriété saumique" is actually supposed to mean with respect to the *rasa-*s according to Kapani. Moreover, it is unclear whether *parasparam* and *saumyaguṇatvāt* belong together or not (Kapani obviously holds the latter to apply), and whether the period behind *jāyante* is correctly placed or not (one could consider placing it behind *parasparam* or *saumyaguṇatvāt* too), or even whether *ṣaḍvidho raso* might not in actual fact be a later insertion, or whether there should not be a period before *rasāc*. In all of these considered cases, what is actually meant remains problematic. Matters are not helped by what is supposed to be a translation in GARBH1 (p.68): "Why this body is said to possess as many as seven constituent tissues, just as *Devadatta* [note the peculiar footnote loc.cit. on this name!] has manifold possessions? Due to the mild blending of the six fluids (*rasa*), is formed the blood ...". As regards *saumyaguṇatva-*, I may also draw attention to the apparent use of ☞ *soma-* to denote the gross element water in n.515 (see §7.19), which leads to the problem of water and *rasa-* in its different meanings, in which connection we have already cited the above GARBH passage (see §7.15). And in this context we may also ask what "le domaine des substances matérielles" means; is ☞ *dravya-* here used in a manner approaching its Vaiśeṣika usage for the nine elementary substances, which include the five gross elements, the ☞ *mahābhūta-*s (see too CA,Sū 1,48)? Or is it maybe used in the sense found in §§10.10f., i.e. approaching that of *dhātu-*? And what exactly does the formation of the "abodes/spheres of the *dravya-*s etc." of Devadatta

rectness of Gayadāsa's arguments, but with the fact that he obviously saw difficulties. Moreover, the mention of milk and *ārtava-* in this context shows again that we are justified in discussing the problem of *tvac-* here in the course of our study.

10.10. I may also draw attention to BH,Śā 5,1, which first tells us that ☞ *ojas-* and ☞ *tejas-* are present in twelve places (*sthānāni*) of the body, namely *tvac-* "skin", the six substances blood to semen in the known chain, sweat, bile, phlegm, urine and faeces, but then tells us that the three morbific entities (wind, bile and phlegm) can bring about disorders of "*rasa-* etc.". The whole is complicated by the fact that the morbific entities are said to act thus when they themselves are corrupted through the disordered/injured (*vyāpanna-*) forms of the twelve substances just mentioned.[960] "*rasa-* etc." cannot but refer to the ☞ *dhātu-*s *rasa-*, blood, flesh etc.; is there now a difference here between blood etc. as "places" (*sthāna-*), and as *dhātu-*s? If so, how does this relate to *tvac-* (which precedes blood in the first list) on the one hand, and *rasa-* on the other? I am unable to offer a solution to this problem. Cf. in this regard also the discussion of the disease *kuṣṭha-*,[961] in which *tvac-* obviously takes the place of *rasa-*, in SU,Ni 5,22-29, AH,Ni 14,33-36 and AS,Ni 14,pp.72b-73a, and also CA,Ni 5,3, AH,Ni 14,2 and AS,Ni 14,p.70a, which, as well as the remarks in the respective commentaries on the problem of *tvac-* and *rasa-* here, have been discussed by EMMERICK 1984, p.98 and EMMERICK 1986, p.188.[962] Here too the commentaries debate on the relation of *tvac-* and *rasa-*, according to some the latter being subsumed by the former.[963] In his commentary on SU,Ni 5,21, which says that *kuṣṭha-* first arises in the skin (*tvaci*) and then spreads to the other *dhātu-*s in succession,

---

(= N.N.) refer to? Does this refer to the formation of the corresponding sites in the body? Or is *viṣaya-* to be understood here in the sense of "object" or "property" relative to the five sense objects and/or the five gross elements? But what of *ādi-* "etc." then?

[960] *iha khalv ojas tejaḥ śarīre nitye ca bhavataḥ. tayoḥ sthānāni dvādaśa bhavanti. tad yathā tvakśoṇitamāṃsamedo'sthimajjāśuklasvedapittaśleṣmamūtrapurīṣāṇīti. tāny avyāpannāni sukham ity ucyate. vyāpannais tu vātapittaśleṣmāṇaḥ praduṣṭā rasādiṣu vikārān upajanayanti.*

[961] See also §§8.17f.

[962] See also n.943 and MÜLLER 1964, p.24.

[963] I may in this context also draw attention to the statement of Gayadāsa in his commentary on SU,Ni 5,17 that *kilāsa-* is an affliction of *rasa-* (*rasagatam eva kilāsam*), an added complication in the discussion on the relationship between *kuṣṭha-*, *kilāsa-* and *śvitra-*, which hinges to a large extent on the affliction of the skin through *kilāsa-* (on this discussion see EMMERICK 1984, pp.99f. and EMMERICK 1986, pp.189f.; see also the *Madhukośa* commentary on MĀDH 49,37-39).

Ḍalhaṇa expressly says that "skin" (*tvac-*) stands for *rasa-*:

> *tvacīty atra tvakśabdena rasa ukto rasārabdhatvāt. tena rasastham ity arthaḥ*
>
> "'*tvaci*': here through the word *tvac- rasa-* is mentioned because of [the chain of *dhātu*-s'] beginning with *rasa-*. Through this 'situated in *rasa-*' is the meaning."[964]

Similarly in his commentary on 5,22 he says that some opine that *tvac-* mentioned here is *rasa-* according to "others":

> *apare tvakśabdena rasam āhuḥ.*[965]

And in his commentary on SU,Sū 21,33, where both *visarpa-* and *kuṣṭha-* are said to be caused by the three corrupted morbific entities situated in the skin (*tvac-*), flesh and blood,[966] Ḍalhaṇa simply equates *tvac-* and *rasa-*:

> *tvakśabdenātra raso vivakṣitaḥ.*

(Cf. his commentary on Śā 9,12 mentioned in §10.9.) Interesting too is the fact that YOG 2,Kuṣṭhanidāna,30 (p.215), which deals with *kuṣṭha-* in the skin (*tvakstha-*), is prefaced by a line saying that this verse deals with *kuṣṭha-* in *rasa-*:

> *tatrādau rasagatam āha.*

In their commentaries on SU,Ni 5,17, both Ḍalhaṇa and Gayadāsa quote a passage by Viśvāmitra explaining the difference between the diseases *kuṣṭha-* (in the version of Gayadāsa) or *śvitra-* (in the version of Ḍalhaṇa) and *kilāsa-*: the latter is in the skin, the former in the *dhātu*-s;[967] it is unclear what *dhātu-* here means (according to both Ḍalhaṇa and Gayadāsa it would be the list of elements of the chain of metabolic transformation beginning with *blood*), though it is clear that the skin (*tvac-*) is not a *dhātu-* here, in

---

[964] The verse commented upon runs:

> *evaṃ kuṣṭhaṃ samutpannaṃ tvaci kālaprakarṣataḥ*
> *krameṇa dhātūn vyāpnoti narasyāpratikāriṇaḥ.*

[965] See also the *Madhukośa* on MĀDH 49,25cd (= SU,Ni 5,22cd in a variant reading); this commentary is also referred to in §10.12 below.

[966] No *kuṣṭha-* in fat and bones is, however, mentioned. KAL 11,82-84 refers to *kuṣṭha-* in skin, blood, flesh, and fat, and to *anyad*, i.e. "other" (seemingly referring to any other sort of *kuṣṭha-* in other substances, which the Hindi translation takes to be bones, marrow and semen).

[967] The whole passage says:

> *tvaggataṃ tu yad asrāvi kilāsaṃ tat prakīrtitam*
> *yadā* (Gayadāsa: *yadi*) *tvacam atikramya tad dhātūn avagāhate*
> *hitvā kilāsasaṃjñāṃ ca śvitrasaṃjñāṃ labheta tat*

(the last line according to Gayadāsa:

> *hitvā kilāsasaṃjñāṃ tu kuṣṭhasaṃjñāṃ labhet tadā*).

contrast with the passages described in §§10.7f. In AH,Ni 14,2 and AS,Ni 14,p.70a, just cited, *tvac-*, ☞ *lasīkā-* "serous fluid", skin and flesh are enumerated, and KĀ,Ci,Ku, p.82 similarly says that all *kuṣṭha-*s are situated in skin, flesh, blood and *lasīkā-*, whereas we find a list of seven substances, called ☞ *dhātu-* as well as ☞ *dravya-* in the CA passage Ni 5,3: the three morbific entities (☞ *doṣa-*), and skin, flesh, blood and *lasīkā-* (cf. §10.7) (the latter four are mentioned again in Ni 5,10); this nomenclature is highly reminiscent of the use of *dravya-*, detailed in n.522, to describe another list of *dhātu-*s. The same list as in CA,Ni 5,3 (the substances being here called *dhātu-*s too) is also mentioned in CA,Ci 21,15 (cf. §10.7), in connection not with *kuṣṭha-*, but with the disease *visarpa-*;[968] see also SU,Ni 5,11, where skin, blood and flesh are mentioned.

10.11. EMMERICK 1984, p.98 and EMMERICK 1986, p.188 hold that the mention of ☞ *lasīkā-* in connection with *kuṣṭha-* is the reason for the commentators including *rasa-* in the system too, leading to its being connected with *tvac-*, which seems to usurp its place in the system, since *lasīkā-* is a ☞ *mala-* of *rasa-*. This is, however, problematic, not only in the light of what we have seen above (§10.8), but also because the connection between *lasīkā-* and *rasa-* referred to by EMMERICK 1984 and EMMERICK 1986 seems to have been made relatively late, found, of the four 'classical' texts, only in AS (Śā 6,p.334b).[969] The whole is actually much more complicated, for in CA,Ci 7,9 we have the same list of seven substances as in §10.10 cited in connection with *kuṣṭha-* (and called ☞ *dravya-*s), but with an important difference: instead of *lasīkā-* we find *ambu-* "water"!

10.12. This latter brings yet another component into the already thorny problem of the nature of the first ☞ *dhātu-* of the metabolic chain, for in §7.26 and §7.30 we have already discussed KĀ,Śā,Śa, pp.51f., where we found water taking up the nourishing or satiating function of ☞ *(āhāra)rasa-* as regards the other *dhātu-*s. Now this passage[970] also tells us that the ten *añjali-*s (the quantity present in the body) of this "water" are *tvagāśraya-*.[971] *tvagāśraya-* could mean "having the seat (i.e. situated) in the skin", but it could also mean "having the seat in a skin". If the former, then what is meant is clear: "water" is situated in the skin as such. In the latter case the refer-

---

[968] On this cf. also the *Madhukośa* commentary on MĀDH 49,1-6.

[969] *rasasya sāro raktaṃ malaḥ kapho lasīkā ca* (Indu says that "*lasīkā-* is like water": *lasīkā jalasadṛśī*). (On the passage from which this is extracted see n.1112.) In the other three texts only ☞ *kapha-* occurs (cf. MEULENBELD 1974, pp.488f.).

[970] The text of which is given in n.616.

[971] *daśaivāñjalayaḥ proktā udakasya tvagāśrayāḥ*.

ence could be to one of the six (or seven) "skins" discussed in Appendix II (☞ *tvac*-). Now there seems to have been, apart from the controversy on the exact number of "skins", a difference of opinion regarding the first of these, for, though the SU, ŚĀR, BYT and BHĀ passages mentioned in the discussion on ☞ *tvac*- are silent on this point, the CA, BH and AS passages tell us that the first "skin" is *udakadharā*, i.e. holds water (the AS passage giving this as an alternative theory).[972] (On the corresponding ĀY passage see §10.13.) But the "skins" here seem to be the layers of the outer covering of the body, i.e. of the skin as such, which means that actually it is immaterial how we translate *tvagāśraya*- in the KĀ passage, for in either case the passage would be in keeping with the theory of the latter set of texts, namely that water is situated in the skin.[973] If we now bear in mind our long discussion on the possible connection of ☞ *rasa*- and water in most of chapter 7,[974] we see that this could have been the bridge to make the commentators equate *tvac*- and *rasa*- (§§10.9f.). Of course the commentators may have felt no more than that some sort of connection existed between water and *rasa*-, for we can hardly assume that they held the two to be fully identical, as they cannot but have been aware of those passages in which both these fluids are mentioned side by side, a case in point being the passages mentioned in n.144 (along with CA,Śā 7,15), where KĀ,Śā,Śa, pp.51f. was also mentioned. Indeed, in its comments on MĀDH 49,25-30,[975] where the effects of *kuṣṭha*- on the *dhātu*-s of the chain of metabolism are described, this chain beginning (as in the passages cited in §10.10) with *tvac*- and not *rasa*-, the *Madhukośa* commentary, while expressly saying that *tvac*- subsumes the missing *rasa*-,[976] also differentiates between *rasa*- and water (*udaka*-).[977] This of course leads to problems, which the *Madhukośa* solves by saying that the *tvac*- here meant is not the first or outermost skin containing water, but the second one, which holds blood,[978] the mention of blood implying that of *rasa*- too.[979] Gaya-

[972] The AH passage is very brief and non-committal.

[973] One could debate whether there might or might not be a slight difference if KĀ should not differentiate between the individual layers of the skin, but we seem to have no evidence to decide on this one way or the other.

[974] See §§7.13ff., also n.829.

[975] = A variant to SU,Ni 5,22 = ĀĀS 4135.

[976] Substantiating this by a quotation from Bhoja — also found in Gayadāsa's commentary on SU,Ni 5,3 — which has *rasa*- in the place of *tvac*- in connection with *kuṣṭha*-. (On Bhoja cf. MEULENBELD 1974, pp.418f.)

[977] Cf. too n.615 and n.965.

[978] The theory taking the first skin to hold water regards the second one as holding blood;

dāsa in his commentary on SU,Ni 5,9-16 obviously differentiates between *tvac-* and *udaka-* in this context too, for, commenting on *tvagraktādīni* in Ni 5,11 (obviously a variant to the reading *tvagraktamāṃsāni* given in SU), he says that this excludes flesh and water.[980]

10.13. But the fact that "water" as such enters the picture does not mean that ☞ *lasīkā-* does not, for the same KĀ passage just cited (Śā,Śa, pp.51f.) not only tells us that various diseases and disorders of the skin arise when the water is "corrupted" or spoilt" (*duṣṭa-*) and that this same water issues forth as sweat,[981] but also that, "when a wound is present [in the skin], serous fluid (*lasīkā-*) and pus (*pūya-*) and slimy discharge (*picchā-*) arise

---

cf. the texts mentioned above.

979 The *Madhukośa* says:

idānīm uttarottarasaptadhātugatakuṣṭhasya lakṣaṇaṃ krameṇāha tvaksthe vaivarṇyam (MĀDH 49,25) *ityādi. tvakśabdenātra raso 'bhidhīyate dhātuprastāvāt. tvakśabdena rasasyābhidhānaṃ tātsthyāt. jejjaṭas* (on Jejjaṭa see MEULENBELD 1974, pp.406-408) *tu tvacam evāha. bāhyāyās tvacaś caraka udakadharatvenokte* (this refers to CA,Śā 7,4) *rasagrahaṇasyānarhatvād dvitīyāsṛgdharātroktā. udakāc ca raso bhinnaḥ. tathā hy uktam udakaṃ daśāñjalimitaṃ raso navāñjalimita iti* (this is not found in CA,Śā 7, as the editor opines; it could, however, be a paraphrase of CA,Śā 7,15). *rasaṃ vihāya raktādyuktiś ca prāg eva rasaṃ kroḍīkṛtya tiryaksirāgatair doṣaiḥ kuṣṭhajananāt. sirāsu raktavyavasthiteś ceti tasyāśayaḥ. sākṣād rasa ity anabhidhānād bṛhatkuṣṭhaṃ rasagataṃ na bhavatīti nāśaṅkanīyam. yataḥ suśrutenaivoktaṃ teṣāṃ tu mahattvaṃ sarvadhātvanusāritvād* (SU,Ni 5,7, however, says: *teṣāṃ mahattvaṃ kriyāgurutvam uttarottaraṃ dhātvanupraveśād asādhyatvaṃ ceti*; also note that SU,Ni 5 mentions *tvac-*, but not *rasa-* — see §10.10) *iti. sarvaśabdena ca rasasyāpi grahaṇam. bhoje 'py uktaṃ*

praduṣṭāḥ pracyutā doṣā rasāsṛṅmāṃsasaṃśritāḥ
kuṣṭhāni janayanty āśu śarīreṣu śarīriṇām

iti (quoted by Gayadāsa in his commentary on SU,Ni 5,3 too, with the first line: *praduṣṭopacitā doṣā rasāsṛṅmāṃsasaṃsṛtāḥ*). *tvaksrāvo romaharṣaś ca svedasyātipravartanam* (MĀDH 49,26ab, which, however, has *tvaksvāpo*) *ity etat kecic choṇitagatasyaiva liṅgaṃ manyante. anye rasagatasyeti* (MĀDH 49,26cd itself assumes the former of these two alternatives: *kaṇḍūr vipūyakaś caiva kuṣṭhe śoṇitasaṃśrite*).

The ĀĀS, commenting on the same line (ĀĀS 4135), says very laconically:
tvakśabdenātra rasa ucyate. dhātuprastāvāt tvaksthāc ca.

980 *tvagraktādīnityādiśabdān māṃsāmbunor avarodhaḥ.*

981 *bhavanti tasmin* (scil. *udake*) *duṣṭe ca dadrukaṇḍūvicarcikāḥ*
tvagāmayāḥ kilāsāni pāmā keśavadhas tathā
tad agnimārutodviddhaṃ [kū]pakaiḥ sveda ucyate.

from/through it".[982] Now "it" could refer to the water, in which case we would have a connection between water and *lasīkā-*; it may be thus no wonder that Cakrapāṇidatta in his commentary on CA,Sū 20,8 says:

> *lasīkodakasya picchābhāgaḥ*
> "*lasīkā-* is the *picchā-* part of water."[983]

"It" could, however, also refer to the wound; in that case the text would not expressly say that water and *lasīkā-* are connected, but the context would make this a probability to be seriously considered nevertheless. That there was indeed a feeling that there was a connection between the two in this context is borne out by ĀY 4,259, which tells us that the first "skin" (☞ *tvac-*) is a holder of *both* water and *lasīkā-* (*lasīkāvāridhārin-*). On the other hand, this same passage could also be held to show that water and *lasīkā-* were, since *both* are mentioned here, differentiated. Cf. in this connection Indu's commentary on AS,Sū 20,p.146b:

> *lasīkā jalasadṛśas tvaci snehaḥ*
> "*lasīkā-* is *sneha-*[984] in/on the skin like water".

Interesting is also CA,Ci 25,28, where both *lasīkā-* and water (*jala-*) are listed amongst the possible discharges of a wound (along with pus, blood and other discharges with various colours and characteristics). Even more interesting is SU,Sū 22,8. This passage first tells us that a discharge having the appearance of water (*salilaprakāśa-*) and also slightly musty (*kiñcidvisra-*) and of yellowish appearance (*pītāvabhāsa-*) appears if the skins (on the plural cf. §10.9) of the body are damaged (see on this also §10.19).[985] Later this same passage says that, if a wound reaches up to the viscera, it discharges blood, urine, faeces, pus and water;[986] what this water is is unclear, but its mention in connection with the viscera does make us wonder whether it may not be different from the water mentioned in connection with the skins, and the commentators too seem to have had difficulties explaining it, for Cakrapāṇidatta in his commentary says:

> "The flow of water, however, is due to the influence of the beverage[s], having the nature of flowing (var: the influence of the

---

982 *vraṇe lasīkā pūyaṃ ca picchā cātaḥ pravartate.*

983 Cf. also the commentaries of Ḍalhaṇa and Cakrapāṇidatta on SU,Sū 21,7.

984 On this term in this context cf. §10.18 below.

985 Shortly after this statement we have a line in which water (*toya-*) is mentioned again, but here it is just part of a comparison in which the issue of pus is compared to that of water:
pakvāsu ca [cchinnāsu sirāsu] toyanāḍībhir iva toyāgamanaṃ pūyasya.

986 *koṣṭhagato 'sṛṅmūtrapuriṣapūyodakāni sravati.*

> characteristic, of the imbibed beverage[s], of having the nature of flowing)."[987]

According to Ḍalhaṇa's commentary as given in SU, water here refers to a discharge of *rasa*--wetness or to what has been drunk.[988] According to SU3, p.394, however, Ḍalhaṇa says that water here is *rasa-* in the meaning of "wetness" or as something drunk.[989] Whatever it be that Ḍalhaṇa actually says, note too that there is no mention of ☞ *lasīkā-* in SU,Sū 22,8, which further complicates the problem. On the other hand, CA,Ni 4,7 has a list in which we find both *rasa-*[990] and *lasīkā-*, as well as *śarīrajakleda-* "wetness originating in the body" (which turns to urine according to 4,8f.); this is part of the aetiology of *prameha-* due to phlegm, in connection with which CA,Ci 6,5 (= MĀDH 33,2) mentions only *śarīraja- kleda-*, but no *rasa-* or *lasīkā-*, whereas these (the former in *rasaujas-*) are mentioned in 6,8 (= MĀDH 33,4), along with water (*ambu-*). The parallels in the other 'classical' texts too are interesting: AH,Ni 10,4 has *vapuḥkleda-* "body-wetness", sweat (*sveda-*) and *rasa-* (var: ☞ *vasā-*), but no *lasīkā-*,[991] AS,Ni 10,p.51b has *kleda-* and *rasa-* only, whereas SU,Ni 6,9 mentions none of all these.

10.14. ☞ *lasīkā-* is also mentioned in CA,Śā 7,15, which deals with the water present in the body and is of great interest. We have already referred to this passage in a different context (§7.26). Right after the portion given in n.564, the text goes on to say that this water, present in the body, fills the outermost skin (cf. §10.12), and is known as *lasīkā-* when it is in a wound in the skin, as sweat when it issues through the pores of the skin.[992] This is a

---

[987] *udakasrāvas tu pānīyasya saraṇātmakasya prabhāvāt* (SU3, p.394), and *udakasrāvas tu upayuktapānīyayasya saraṇātmakatvaprabhāvāt* (SU4, p.171).

[988] *udakam atra rasakleda eva gṛhyate pītaṃ vā.*

[989] *udakam atra rasaḥ kleda eva gṛhyate pito vā.*

[990] This is also called ☞ *ojas-* here (*rasaś caujaḥsaṅkhyāta[ḥ]*).

[991] *vapuḥkleda-* is translated as "Aussehen, Feuchtigkeit" (i.e. as two separate words) by HILGENBERG/KIRFEL, p.249. In this they follow Aruṇadatta, who seems to take *vapus-* as a word separate from *kleda-* (*vapurādīn dūṣayitvā*). This explanation, clearly quite problematic, is, however, not followed by the other commentators; Śrīdāsa mentions it as an opinion held by "some" (thus obviously disagreeing with it) — interestingly, in his own explanation he subsumes *lasīkā-* under *vapuḥkleda-* (*vapuḥkledaḥ śarīrakledo lasīkādidravaviśeṣaḥ. vapuḥ śarīrakāntir ity eke*). Śrīdāsa also cites an AH variant taught him by his teachers (*asmadguravaḥ*), in which semen (*śukla-*) takes the place of *sveda-*; this tallies with the reading of AS,Ni 10,p.51b.

[992] *yat tu sarvaśarīracaraṃ bāhyā tvag bibharti. yat tu tvagantare vraṇagataṃ lasīkāśabdaṃ labhate. yac coṣmaṇānubaddhaṃ lomakūpebhyo niṣpatat svedaśabdam avāpnoti. tad udakaṃ*

very clear statement, and it is thus no wonder that Gayadāsa in his commentary on SU,Ni 5,3 (see §10.12) expressly refers, after quoting CA,Ci 7,9 and 21,15, to CA,Śā 7,15 as proof that *tatrāmbu lasīkodakam eva*, i.e. that *ambu-* "water" in CA,Ci 7,9 (cf. §10.11 above) is the same as *lasīkā-* in CA,Ci 21,15 (cf. §10.7) and as *udaka-* (which is the term used for "water" in CA,Śā 7,15). In the same passage Gayadāsa also clearly refers to CA,Ni 5,3,[993] which mentions skin, flesh, blood and *lasīkā-* (see §10.7 and §10.10), saying that this refers to *rasa-* (here thus obviously equated with *tvac-* "skin"), blood, flesh and water (*ambu-*). We may also recall that Aruṇadatta and Śrīdāsa in their commentaries on AH,Śā 3,60-63 (see n.506) also say that *lasīkā-* is but transposed water (*udaka-*).[994]

10.15. On the other hand, AS,Śā 6,p.334b (see n.1112) tells us that serous fluid (☞ *lasīkā-*) is derived from *rasa-* as its impurity (☞ *mala-*) (cf. §10.11). We must therefore ask whether the author of AS saw some connection between water and *rasa-*, or whether he has merely given us a variant theory not connecting water and *lasīkā-*. At any rate, he definitely saw some problems, for in AS,Śā 5,p.326a we find more or less the same stated as in CA,Śā 7,15 with respect to *udaka-* (see §10.14), but with a puzzling difference: at the beginning of its description the AS passage tells us that what follows pertains to *dravāḥ* "fluids" (water is not mentioned). This plural is then, however, connected with the singular as CA uses it (cf. n.564 and n.992) by means of a relative clause (*dravāś tu yat tad* ... "But that which is fluids ..."), the text telling us that this "that" (*tat*) is also *lasīkā-* and sweat (as in the CA passage). The passage ends by stating laconically:

tad udakam
"That is water."[995]

This can only mean that what has been described as *dravāḥ* is in fact water, which is extremely puzzling and could maybe be taken to show that the au-

---

daśāñjalipramāṇam.

993 *ataḥ prabhavāṇy abhinirvartamānāni* etc.

994 Note also that what CA,Śā 7,15 says on sweat is remarkable too, since in all the four 'classical' texts sweat is regarded as the ☞ *mala-* of fat. It may also be pointed out that in some of the enumerations of the ☞ *srotas*-s different *srotas*-s are mentioned for water and for sweat (e.g. in CA,Vi 5,7f., CA,Ci 13,20, CA,Ci 20,16, AS,Śā 6,p.331), so that at least in such cases water and sweat seem to be differentiated.

995 *dravās tu yat tad atiyogena cyavamānaṃ purīṣam anubadhnāti. tathā mūtraṃ rudhiram anyāṃś ca śarīre dhātūn. yac ca sarvaśarīragataṃ bāhyā tvag bibharti. yac ca tvagantare vraṇagataṃ lasīkāśabdavācyam. yac coṣmānubaddhaṃ romakūpebhyo 'bhiniṣpatat svedavācyam. tad udakam.*

thor was not quite comfortable with what CA (very obviously excerpted here, for the wording of the AS and CA passages is more or less the same) says on the characteristics of "water", and may have tried to circumvent this difficulty by subsuming the fluids of the body in general under this heading, even though his text (see n.995) also makes it clear that the fluids urine and blood are *not* subsumed under "fluids".[996]

10.16. Indu in his commentary on this passage says that *dravāḥ* implies "water, urine (!) etc.",[997] but then takes the whole passage to refer to water (*udaka-*) only (thus following the text of AS), without in the least commenting upon the difficulty thus caused; indeed, his subsuming "water" (together with urine etc.) under "water" only adds to the difficulties. But he does make a very interesting observation, for he interprets the statement of the passage (see n.995) that excessive *udaka-* adheres (*anubadhnāti-*) to faeces (*purīṣa-*), urine (*mūtra-*) and blood (*śoṇita-*) in the sense that the first applies to *atīsāra-* (cf. §§7.25ff.), the second to *prameha-* "urinary disease" and the third to *raktapitta-* "blood-bile". Regarding the adherence to urine, note that CA,Ni 4,37 tells us that urinary disease is caused not by water, but by ☞ *lasīkā-* adhering to urine![998] As regards *raktapitta-* mentioned by Indu, the origin of the disease ☞ *raktapitta-* is not clear in all aspects; excited or corrupted bile somehow combines with blood to cause the disease, but the exact circumstances under which this happens are a bit problematic; this is perhaps best exemplified by AÑJ 63, which simply says laconically that corrupted bile corrupts blood, giving no other particulars.[999] As it is, the causes for the corruption or excitation of bile are variously given as different foods, emotions, heat etc. (the texts are not consistent), the exact process of this corruption or

---

[996] Cf. also Excursus XI at the end of Chapter 7.

[997] *te codakamūtrādayaḥ.*

[998] It is thus no surprise that Cakrapāṇidatta in his commentary on the CA passage tells us that *lasīkā-* is water situated between skin and flesh, citing in this context a passage to this effect which seems to be a variant or a paraphrase of CA,Śā 7,15 (see n.992):

lasīkā māṃsatvagantara (for °*tare*) *udakabhāgaḥ. vakṣyati hi śārīre yan māṃsatvagantara udakaṃ tal lasīkāśabdaṃ labhate.*

(In his commentary on SU,Sū 21,7 Cakrapāṇidatta quotes the CA passage more or less as given in n.992.)

[999] We might in this context also point to other diseases, which are not *raktapitta-*, but also originate due to a combination of corrupted bile with blood, as e.g. in CA,Sū 18,23-25. However, in this passage the combination is a special one affecting only the skin and causing diseases of this, and does not cause haemorrhaging; as such such diseases at least are not a further complicating factor.

excitation and the mode of combination with blood not being clear.[1000] According to CA,Ni 2,4, in *raktapitta-* the excited or corrupted bile combines with blood that exceeds its own measure (*svapramāṇam ativartate*). This is a puzzling statement. According to CA,Ci 4,7 the blood is increased by the bile, but Ni 2,4 seems to say that the blood is increased before the bile reaches it.[1001] What then increases the blood? In this context Indu's explanation of AS,Śā 5,p.326a does make sense, i.e. it would be "water" which causes the blood to increase. On the other hand, amongst the factors bringing about the excitation of bile and/or blood according to the passages giving the aetiology of the disease, as well as, in CA,Ni 2,4, the increase of blood, are different food-stuffs, in which context it is striking that SU,Ut 45,4 tells us expressly that bile, which subsequently corrupts blood, is corrupted by corrupt *rasa-*.[1002] Though their functions seem, according to these different theories, not the same, we nevertheless find once again that water and *rasa-* are mentioned in similar contexts.

10.17. This similarity is even more pronounced in the descriptions of thirst and its causes in the 'classical' texts,[1003] which state that thirst (*tṛṣ-*

1000 On the possibility of "blood-bile" maybe being due to an independent corruption or excitation of blood, which then either corrupts the bile, or else combines with independently corrupted or excited bile, see the discussion on ☛ *raktapitta-* in Appendix II.

1001 CA,Ni 2,4 states:

> *...pittaṃ prakopam āpadyate. lohitaṃ ca svapramāṇam* (var: *cāśu pramāṇam*) *ativartate. tasmin pramāṇātivṛtte pittaṃ prakupitaṃ śarīram anusarpad yad eva* (var: *yadaiva*) *yakṛtplīhaprabhavāṇāṃ lohitavahānāṃ ca srotasāṃ lohitābhiṣyandagurūṇi mukhāny āsādya pratirundhyāt* (var: *pratipadyate*) *tad eva* (var: *tadaiva*) *lohitaṃ dūṣayati.*

This is translated by MEULENBELD 1974, p.322 as:

> "... one's bile enters into a state of excitement and one's blood exceeds its own measure. When this (blood) has exceeded its measure, the excited bile spreads through the body, approaches the openings, which are heavy with the effluxion of blood, of the blood-carrying channels originating from the liver and the spleen, and (the bile) may become checked; this same (bile) then corrupts the blood."

The variants do not perceptibly alter the meaning as regards our problem.

1002 The verse is a continuation of 45,3; both together run:
krodhaśokabhayāyāsaviruddhānnātapānalān
kaṭvamlalavaṇakṣāratīkṣṇoṣṇātividāhinaḥ (3)
nityam abhyasato duṣṭo rasaḥ pittaṃ prakopayet
vidagdhaṃ svaguṇaiḥ pittaṃ vidahaty āśu śoṇitam (4).

Cf. also AS,Sū 10,p.89b and Indu's commentary on AS,Sū 19,pp.143f. mentioned in §7.30, according to which *rasa-* can corrupt all the morbific entities.

1003 CA,Ci 22, SU,Ut 48, AH,Ni 5,45-57½, AS,Ni 5,pp.36b-37b. On thirst as an affliction

ṇā-) is caused by the drying up (√*śuṣ*) in the body of water (*jala-*, *ap-*) or the element (of) water (*apāṃ dhātu-*, *abdhātu-*)[1004] (CA,Ci 22,7;11;13;19;21;25) as well as of the "somic" (☞ *saumya-*) substance(s)[1005] (obviously not a clear term, as it is variously explained by the commentators)[1006], by the loss/diminution (*kṣaya-*; see §§5.2f.) of *rasa-* (CA,Ci 22,16,[1007] SU,Ut 48,13, AH,Ni 5,57, AS,Ni 5,p.37a)[1008] and by the outflow (*nirgama-*) of blood (SU,Ut 48,12). CA,Ci 22,6 states that bile and wind dry up (*sam*√*śuṣ*) the *rasa*--carrying tubes[1009] (var: ducts[1010]), as well as the root of the tongue (*jihvāmūla-*), the throat (*gala-*),the palate (*tāluka-*) and the *kloman-*[1011] and cause thirst. According to SU,Ut 48,5, however, thirst is caused by wind and bile drying up the channels (☞ *srotas-*) carrying water.[1012] This is made even more intriguing by Ḍalhaṇa's commentary, which tells us that the channels carrying water refer to those arising from tongue, throat, palate and *kloman-*, and to the ducts (☞ *dhamanī-*) carrying *rasa-*,[1013] even though there is no mention of the latter in the passage commented upon. AH,Ni 5,47f. and AS,Ni 5,p.36b tell us that the vessels (☞ *sirā-*) of the root of the tongue (*jihvāmūla-*), of the throat, of the *kloman-*, and of the palate, (and?) carrying water,[1014] dry up and cause thirst.

---

categorised more or less like a disease see also MÜLLER 1961D.

1004 On this cf. §§7.25ff.

1005 CA,Ci 22,5: *saumyān dhātūṃś ca śoṣayataḥ*; AH,Ni 5,46, AS,Ni 5,p.36b: *saumyadhātu-praśoṣaṇāt*.

1006 Cakrapāṇidatta: phlegm, *rasa-* and water; Aruṇadatta: phlegm, *rasa-* etc.; Toḍaramalla (see the editor's note on the AH passage; on the ms of Toḍaramalla's commentary see no.32 on p.(15) of the Sūcipatram batch of numbered pages in the AH edition): phlegm or *rasa-*; Śrīdāsa (in AH1): *rasa-*; Indu (on AS and AH): *rasa-*.

1007 On which cf. §7.22.

1008 SU,Ci 26,11 tells us that *klaibya-* "☞ *klība*--hood" is a *kṣaya-* of the *saumyadhātu-*:
saumyadhātukṣayo dṛṣṭaḥ klaibyaṃ tad aparaṃ smṛtam.
Since semen is both "somic" (cf. n.423) and a ☞ *dhātu-*, *saumyadhātu-* in this context probably refers to semen.

1009 *nālī-* = ☞ *nāḍī-*.

1010 ☞ *dhamanī-*.

1011 On this see Excursus VIII at the end of chapter 7.

1012 The drying up of palate, lips, throat and mouth (*āsya-*) is mentioned in 48,7 as a prodrome of thirst.

1013 *ambuvāhīni srotāṃsi jihvāgalatāluklomajāni rasavāhinyaś ca dhamanyaḥ.*

1014 *jihvāmūlagalaklomatālutoyavahāḥ sirāḥ.*

10.18. Now the *mūla-*[1015] of the channels (☞ *srotas-*) carrying water consists according to SU,Śā 9,12 of the palate and the *kloman-*, according to CA,Vi 5,8 and AS,Śā 6,p.331a of the root of the palate (*tālumūla-*) and the *kloman-*, according to Hemādri in his commentary on AH,Sū 12,3 of the *kloman-*.[1016] Damage to the channels causes thirst (var: and incurable brownness of the body) and immediate death according to SU, drying up of the tongue, palate, lips, throat and *kloman-* and excessive thirst according to CA, and excessive thirst, drying up of the mouth, buzzing in the ears and seeing darkness (*tamodarśana-*)[1017] according to AS. No ducts, channels or the like carrying *rasa-* are connected with water in these three passages. Actually, the *mūla-* of the channels carrying *rasa-* consists of the heart and the ducts carrying *rasa-* according to SU, of the heart and ten ducts according to CA and AS,Śā 6,p.331b. Damage to these channels causes the disorders listed, as the text of the CA passage tells us, in CA,Sū 28,[1018]9f.; similarly, AS,Śā 6,p.331b refers us to AS,Sū 19[1019],pp.139a;140a. But in the passages referred to there is no mention of thirst as a disorder caused by *rasa-*. The SU passage says that the disorders connected with *rasa-* are drying up and the same disorders as those connected with the channels carrying *prāṇa-* (which incidentally have in this passage exactly the same *mūla-* as the channels carrying *rasa-*)[1020], here too there is no mention of thirst. However, SU,Sū 15,9 tells us that the *kṣaya-* (see above) of *rasa-* causes, among other disorders, thirst. ĀĀS, p.255 expressly connects the mention of two *srotas*-s carrying water in SU (i.e. obviously SU,Śā 9,12) with CA,Ci 22,6 linking thirst with *rasa-* (see §10.17),[1021] thus clearly linking water and *rasa-* or even equating them. CA,Ci 8,4;25 complicate matters still further, for these passages relate of disease arising due to the *kṣaya-* of *sneha-* in the body. Now

---

1015 On this term cf. §6.18.

1016 See also *tila-* in Excursus VIII at the end of chapter 7.

1017 "Seeing darkness" and "entering darkness" (*tamaḥpraveśa-* and the like) are often referred to in medical works. On the latter cf. the *Madhukośa* commentary on MĀDH 2,54:

tamaḥpraveśanam andhakārapraviṣṭasyevāsaṃvittiḥ

"Entering darkness is non-perception like [that] of [one who] has entered murkiness."

1018 Referred to in the Vi passage as the *vividhāśitapītīya-* [chapter].

1019 Referred to in the Śā passage as [the chapter called] *doṣādivijñānīya-*.

1020 Cf. also p.526 in Appendix II, where the *prāṇa*-s are connected with the theory of "fiery" and "somic".

1021 In a form showing extreme variation.

sneha- means "oleaginous substance", but it might also refer simply to moisture (cf. especially *upa√snih* and its derivatives); does *sneha-* therefore refer to water here? Cakrapāṇidatta in his commentaries loc.cit., however, glosses *sneha-* with "semen and ☞ *ojas-*" (*śukraujas-*) and calls it the essence (*sāra-*) of the body, in his commentary on 8,25 further saying that *ojas-* actually refers to ☞ *rasa-*. CA,Ci 8,24 does actually mention a diminution of semen and *ojas-* in the body, after which 8,25 enigmatically tells us that disorders arise

tataḥ snehakṣayāt.[1022]

This can be translated as:

"due to the *kṣaya-* of *sneha-* due to this",

but also as:

"thence, due to the *kṣaya-* of *sneha-*";

in the latter case *sneha-* on the one hand and semen and *ojas-* on the other would indeed be the same. On *sneha-* see also Indu's commentary on AS,Sū 20,p.146b in §10.13 above. Cf. in this connection also CA,Sū 6,27, which tells us that in summer the sun drinks up (*pepīyate*) the *sneha-* of the world; Cakrapāṇidatta, however, holds *sneha-* to refer to "essence" (*sāra-*).[1023]

10.19. These connections between water and *rasa-* are intriguing, but unfortunately no clear-cut picture emerges from all this. As it is, we have already seen in §7.22 that the theories on the origin of thirst may in part be connected with an otherwise discarded theory of Vāryovida which states that all beings derive from *rasa-*, and that water is in some way characterised by (contains?) this *rasa-*; the nature of this *rasa-* is, however, not clear. Attention may here also be drawn to CA,Ni 1,20;23;26, where as part of the aetiology of fever we find the mention of channels (☞ *srotas-*) which are *rasasvedavahāni*. Does this refer to channels which carry *rasa-* and sweat, or to different channels carrying *rasa-* and sweat respectively? The matter is complicated because, of the 'classical' texts, CA,Vi 5,8 and AS,Śā 6,p.331b mention separate *srotas-*s for *rasa-* and sweat in their lists of the different *srotas-*s of the body, whereas SU,Śā 9,12 mentions only those of *rasa-*; moreover, we have seen in §§10.14f. that sweat seems also to have been regarded as a form of water. Another knotty problem is caused by SU,Ut 47,70-72, which tell us that suppression of thirst (*tṛṣṇānirodha-*) weakens the element (of) water

---

1022 The whole passage in question runs:
īrṣyotkaṇṭhābhayatrāsakrodhaśokātikarśanāt
ativyavāyānaśanāc chukram ojaś ca hīyate (24)
tataḥ snehakṣayād vāyur vṛddho doṣāv udīrayan
pratiśyāyaṃ jvaraṃ kāsam aṅgamardaṃ śirorujam (25) ... .

1023 The matter of *sneha-* and *ojas-* will be taken up again in another context in §13.21.

(*abdhātu-*) in the body and consequently strengthens the fiery energy (☞ *tejas-*), which has to be counteracted by the intake of water (*ambhas-*) to strengthen the former and weaken the latter. This leads us to the problem of the classification of substances as "somic" and "fiery", on which see especially ☞ *āgneya-* in Appendix II and §§7.11ff.[1024] And it also brings us back to the very beginning of our problems regarding the relation between water and *rasa-*, namely §7.11, where water and *rasa-* as well as the terms "somic" and "fiery" first appeared together in connection with the origins of and differentiation between blood in general and the procreatory-menstrual fluid, leading to the problem of whether the latter was thought to be derived from ☞ *rasa-* as such or from ☞ *āhārarasa-* (see the summaries in §7.44 and §8.23, also §§8.11ff., §§9.5f. and §§10.5f.).[1025] It also brings up again the problem of the gross elements (☞ *mahābhūta-*) and the relation of *rasa-* in its various meanings to the gross element water, more so since CA,Ci 22,16 not only states that *rasa-* is derived from water, but also that the diminution/loss[1026] of this *rasa-* causes one to feel thirsty, i.e., this passage, seemingly to be connected with Vāryovida's theory just mentioned and already discussed in §7.22, not only differentiates between the two liquids, but also does so in the context of thirst. Cakrapāṇidatta says that due to the *kṣaya-* of *rasa-* a *kṣaya-* of water takes place, which then leads to thirst,[1027] which shows that here he too differentiates between the two fluids (cf. also §7.22). In his commentary on SU,Sū 22,8 the same Cakrapāṇidatta says that the skins men-

---

1024 Particularly §§7.15f. and §§7.19f.

1025 In this context I may also remark upon the puzzle caused not only by CA,Sū 20,8 and AH,Sū 12,2 telling us that *rasa-* is a receptacle of bile, the "fiery" substance par excellence (whereas according to AS,Sū 20,p.146b and AH,Sū 12,3 it is, in keeping with its "somic" character, that of phlegm, the "somic" substance par excellence) (cf. §7.35), but also by the statement of all three texts (loc.cit.) that not only blood, but also ☞ *lasīkā-* and sweat (which, as we have seen, are very closely related to "water") are amongst the receptacles of bile. And of course it is well-known (cf. e.g. EIM, p.150) that bile was thought to be located in the skin too (thus also stated in the passages just quoted). (Cf. also n.642.)

1026 *kṣaya-*; see §§5.2f.

1027 Cakrapāṇidatta then quotes SU,Sū 15,29, which says:

> *doṣadhātumalakṣīṇo balakṣīṇo 'pi vā naraḥ* (Cakrapāṇidatta: *'pi mānavaḥ*)
> *svayonivardhanaṃ yat tad annapānaṃ prakāṅkṣati*
> "The man whose morbific entities (☞ *doṣa-*), elements (☞ *dhātu-*) and impurities (☞ *mala-*) are diminished, or else whose vigour is diminished, desires that food and drink which increases its own matrix."

Cf. on this §7.16.

tioned there (see §10.13) are the receptacle of water *or rasa*-,[1028] which means that he seems to differentiate between the two here too. Ḍalhaṇa states in his commentary that the skins are the receptacle of the two clear fluids water and *rasa*-,[1029] which too seems to differentiate between the two. But in his commentary on another portion of the same passage he also seems to equate water and *rasa*- (see §10.13), though it is unclear whether the two "water"s he comments upon are one and the same. Aruṇadatta and Śrīdāsa in their commentaries on AH,Śā 3,90, which says that a person of bilious nature is characterised by intense thirst and hunger, say that this is because in him the *dhātu*- (of) water and the *dhātu*- *rasa*-[1030] dry up, which not only differentiates between the two, but also seems to connect the former with thirst, the latter with hunger respectively. AH,Ci 7,29 = AS,Ci 9,166b (see also SU,Ut 47,70f.) say that the element (of) water (*abdhātu*-), made feeble by the heat of too much alcohol, dries up the throat, palate and lips (cf. §10.17); since a few lines later the effects of the heat on the skin (☞ *tvac*-) are detailed, the two are obviously differentiated here, which might (though it need not) also imply a differentiation between water and *rasa*-. I may also draw attention to the explanations of both Cakrapāṇidatta and Ḍalhaṇa, which might be understood to the effect that "water" refers to ingested beverages (§10.13). We have already briefly touched on water being a part of what is ingested in §7.15 (see also §§7.29ff.) and also discussed a problem connected with "water" in the alimentary system in Excursus XI at the end of chapter 7; cf. also the statement of KIRFEL 1954, p.339 in Excursus VIII in the same chapter on the separation of *āhāra*- "food" into *anna*- "(solid) food" and *jala*- "water" right after entering the body, as well as the Vedic theory distinguishing between ingested "food" and ingested "water" referred to in §7.43.[1031] That the problem is, however, historically even more complica-

---

[1028] *atra salilaprakāśatā srāvasya tvacām udakasya rasasya vādhāratayā.*

[1029] *salilaprakāśa ity udakarasayor acchayos tvacām adhiṣṭhānatvāt.*

[1030] *abdhātū rasadhātuś ca.*

[1031] It is most interesting to see how the commentators on *Chāndogyopaniṣad* 6,5,1-3 quoted in n.690 try to come to grips with the problem caused by the theory stated there being so obviously at variance with that of the transformation of alimentary substances in medical texts. Thus Śaṅkara, commenting on the statement that the middle *dhātu*- (on this see n.689) derived from "food" is flesh, says that this is flesh derived through the chain of *rasa*- etc.:

> *yo madhyamo 'ṃśo dhātur* (printed: *dhāturar*) *annasya tad rasādikrameṇa pariṇamya māṃsaṃ bhavati.*

Since blood is derived from "water" and bones and marrow from fiery energy (☞ *tejas*-) according to the same *Chāndogyopaniṣad* passage, this is clearly problematic, and the difficulty is only magnified by Ānandagiri's explanation that *rasa*- etc. subsumes blood etc. (*rasādīty ādi*-

ted than this is shown e.g. by the passages from the *Śatapathabrāhmaṇa* discussed by HARIPRAPANNAJĪ, pp.1f.

10.20. The mention of ingested substances in this context brings us back to the starting point of our discussion on ☞ *tvac-*, ☞ *rasa-* and water, namely to §10.7, for KĀ,Śā,Jā 10, discussed there because it mentioned *tvac-* where we would have expected *rasa-*, also ascribes the nourishing or satiating function we have come to associate with ☞ *(āhāra)rasa-* (§§7.37ff.) to *āhāra-* "food" (see §10.7). Cf. in this context also §7.29 on the impression that *āhārarasa-* may have been subsumed under "food" (but see also §7.42). If now, as we have just seen, there is a possibility that there were theories current according to which "water" referred to the liquid part of ingested food (or to ingested "water" as distinct from ingested "food"[1032]), then, if we also consider the discussion in Excursus XI of chapter 7 on "water" referring to a state of ingested food following digestion, this brings us straight *in medias res* as regards the problem of water and *āhārarasa-* discussed above in §§7.29ff., shelved in §7.43, and very briefly referred to in §10.12 (in connection with KĀ,Śā,Śa, pp.51f.). As already mentioned in §10.19 (where further references are given), this must then willy-nilly also lead to the problem of the metabolic origins of the procreatory-menstrual fluid. But the passages quoted in §7.43 tell us that the internal channels (☞ *srotas-*) §6.18) of the body carry food *and* water (and *rasa-*); the same is said of the ducts (☞ *dhamanī-*) of the body in SU,Śā 9,5;7 and AS,Śā 6,p.330. One is tempted to ask not only just what is meant by the statement that "food" and "water" (if ingested liquids should be meant by the latter) are carried through the body,[1033] but also how "food" and "water" (if this should refer to substances liquid before ingestion, as opposed to solids) could be differentiated after ingestion, not only because they are mixed, but also because solid food itself is liquefied within the body. What is, however, worth considering too is the solid and liquid parts of the ingested substances separated through digestion (irrespective of what state they were in when ingested); it is actually these that were mentioned in §7.15.[1034] In this connection it is noteworthy that in his commen-

śabdena rudhirādi gṛhyate).

1032 Cf. especially the separation of *āhāra-* into *anna-* and *jala-* mentioned in n.748.

1033 Since we do after all know that the ingested substances are actually broken down in the alimentary tract according to the medical texts themselves.

1034 Cf. also the verse from the *Garuḍapurāṇa* (KIRFEL 1954, p.339), quoted in Excursus VIII at the end of chapter 7 (see n.748).

tary on CA,Sū 17,63-75, which describe the effects of the *kṣaya-*[1035] of different substances of the body, Cakrapāṇidatta remarks expressly on the absence of *udaka-* "water", *svara-* "voice" etc., with respect to the former drawing attention to CA,Vi 5,8, where in the list of the "internal" *srotas*-s and their *mūla*-s[1036] *udaka-* too is mentioned.[1037] Since Sū 17,63ff. mention the ☞ *doṣa*-s, the seven ☞ *dhātu*-s, their impurities (☞ *mala-*) and ☞ *ojas-*, it seems as if to Cakrapāṇidatta "water" is something on a par with these, i.e. not something ingested.[1038] One may also ask in this connection how these statements correlate with Cakrapāṇidatta's own remarks given in §10.13.

10.21. On the other hand, KĀ,Śā,Śa, pp.51f., which we last discussed in §§10.12f.,[1039] does seem to ascribe the nourishing or satiating function we have come to associate with *(āhāra)rasa-* to water. If we now compare this with the statement in the same text just mentioned in §10.20 ascribing this same function to food, and when we also consider the statements pertaining to skin in §10.7, the starting point of this long discussion, we find that the whole conglomerate of problems related to what *(āhāra)rasa-*, skin, water, ☞ *lasīkā-* and food are seems to be referred to in KĀ, and I for one am unable to unravel the tangle of this skein either in general or specifically as regards KĀ. Clearly, the whole matter calls for an independent, in-depth study encompassing Vedic, medical and other relevant texts, in which context one will also have to consider statements such as that of ŚB 6,7,3,3, where ☞ *agni-* "fire" and ☞ *rasa-* are equated, which might refer to or have given rise to relevant speculations too, since bile (☞ *pitta-*), which is closely related to fire, is, as is well known, taken to be situated in the skin in its *bhrājaka-*function[1040] (and is e.g. in SU,Sū 21,10 expressly called *agni-*). But I am afraid that I will have to leave it at that, even though it should by now be

---

1035 Cf. on this §§5.2f.

1036 See §6.18.

1037 *ete cāṣṭādaśa kṣayā āviṣkṛtatamatvenoktāḥ. tenodakakṣayasvarakṣayādyanabhidhānaṃ nodbhāvanīyam. uktaṃ hi. svarakṣayam urorogam iti. tathodakakṣayalakṣaṇaṃ yathā. jihvātālvoṣṭhakaṇṭhaklomasaṃśoṣaṃ pipāsāṃ ... jānīyād* (as in Vi 5,8, with variants) *iti.*

1038 Note that Cakrapāṇidatta does *not* mention food, which is found listed in Vi 5,8. As remarked already above, what exactly the statement that "food" "flows" through *srotas*-s within the body means is a problem too — and not only because of the difficulties just mentioned of differentiating between "food" and "water". I may in this context also refer to the remarks on food and *(āhāra)rasa-* made above, though these are ultimately no more than speculation, as a detailed discussion of this problem lies outside the scope of our study.

1039 See there for references to further passages.

1040 Cf. n.1025.

clear that somehow or other all this has a bearing on the theory of the origin of the procreatory-menstrual fluid, though KĀ itself does not comment upon this origin — indeed, in the list of substances in Śā,Śa, pp.51f. blood and semen are mentioned, but no procreatory-menstrual fluid, which leads to the same deductions regarding KĀ as in the case of CA and BH as regards the text's general lack of interest in this topic (see §10.4).

10.22. This seems a rather disappointing conclusion after such a long and intricate discussion, but that cannot be helped. However, KĀ does have some more interesting information on matters concerning the procreatory-menstrual fluid in Kh 9,16f. The list of the receptacles of the body in 9,16 (cf. on this also §7.28) lists the uterus (☞ *garbhāśaya-*) as the special receptacle of women, of which receptacle 9,17 then says:

> "... into which vessels (☞ *sirā-*) carrying *rajas-* pour forth *rajas-*. Due to this, this [*rajas-*] issues forth indeed, *daivāt*, as the *puṣpa-* (see §2.1) month after month; due to the opposite (i.e. if it does not flow) this very [*rajas-*] is here, however, collected in that very place[1041]."[1042]

This passage is particularly interesting for two reasons, the first of which is *daivāt*. Much like its English equivalent "providentially", this very often has the connotation of "by chance/happenstance"; should this be meant here too, then we would have an extremely interesting statement, since this would be opposed to the regularity of the process otherwise stated to be (as we have seen several times) its characteristic, and also referred to in this very same line. However, *daivāt* can, again like its English equivalent, also mean "by fate/destiny"; if this is what it means here, then the above would state simply that it is the fate or destiny of women to menstruate,[1043] and that would not be in conflict with the rest of the sentence, which is why it seems more

---

[1041] This would seem to refer to the uterus, though one could also see here a reference to the visceral cavity (*koṣṭha-*) (which holds the uterus and is mentioned in 9,16).

[1042] Together with the here untranslated portion of 9,16, the whole passage runs:
viṇmūtrakrimipakvāmakaphavātāśayāḥ pṛthak
saptaite dehināṃ koṣṭhe strīṇāṃ garbhāśayo 'ṣṭamaḥ (16)
rajovahāḥ sirā yasmin rajaḥ pravisṛjanty ataḥ
puṣpabhūtaṃ hi tad daivān māsi māsi pravartate
viparyayāt tad eveha tatraiva tu nicīyate (17).

[1043] Cf. in this connection especially the mythological origins of menstruation related by SCHMIDT 1922, pp.273ff., MEYER,J.J. 1952, pp.225;228 and also *passim* in chapter 13 (pp.250-292) of GHURYE (cf. in this context also LESLIE,I.J. 1989, pp.251;254f. on menstruation and its symbolism, though whether the symbolistic relationships supposed loc.cit. are always correct I do not know).

probable that *daivāt* here means "by fate/destiny" and not "by chance/happenstance".

10.23. The second interesting point is the mention of a plural number of vessels (☞ *sirā-*) carrying the procreatory-menstrual fluid into the uterus. This reminds us of the discussion in §6.25 (see also §8.23) on a possible variant theory holding that not two, but a plural number of passages carrying procreatory-menstrual fluid opens into the uterus, especially on the basis of CA,Ci 30,207, which too speaks of (a plural number of) vessels (*sirā-*). In this respect I would also like to draw attention to Cakrapāṇidatta's commentary on SU,Sū 14,14 and Śivadāsasena's commentary on DR,Āh 22 (= CA,Ci 15,17), which quote the following verse of Viśvāmitra:[1044]

> *sūkṣmakeśapratīkāśā* (var: *sūkṣmāḥ keśapra°*) *bījaraktavahāḥ sirāḥ*
> *garbhāśayaṃ tarpayanti* (var: *pūrayanti*) *māsād bījāya kalpate* (var: *kalpyate*)
>
> "Vessels resembling fine hairs (var: Fine vessels resembling hairs), carrying seed-blood/germ-blood (blood with the seed/germ?, or, blood which is the seed/germ?, or, blood for the seed/germ?)[1045] satiate (var: fill) the womb in a month. It (or: the womb. In a month it) is fit for (or: effects/becomes?) (var: is made fit for) the seed/germ."

Whether "it" here refers to the womb or the procreatory-menstrual fluid, it is on the whole clear that the verse speaks of the latter entering the former to cause conception. Since the vessels through which the fluid enters the womb are, however, in the plural here too, it does look as if this verse too speaks of a theory at variance with the one according to which the fluid enters the womb through two ducts (or the like).

10.24. Interesting are also the verses following on KĀ,Kh 9,17, i.e. 9,18ff.[1046] They tell us of the origin and symptoms of *raktagulma-*, a disor-

---

[1044] Already cited above in n.827 and n.828; the verse in the version of Śivadāsasena is also found, without any ascription, in the commentary of Śrīdāsa (AHI, p.121) on AH,Śā 3,65f.

[1045] On *bīja-* "seed/germ" cf. §§2.5f.

[1046] *anena hetunā strīṇāṃ raktagulmo hi jāyate*
tad āśayasya cābhāvāt puruṣāṇāṃ na jāyate (18)
hīnayonyās tu bālāyāḥ kāyaṃ gacchati śoṇitam
atha pūrṇasvabhāvāyāḥ kāyaṃ yoniṃ ca gacchati (19)
garbham aṅge bhāvayati kiñcit stanyāya kalpate
paktaye śoṇitādes (ms: *°ādyas*) *tu śeṣaḥ kāyaṃ samindhati* (20)
tathaiva garbhasūtāyāḥ (printed *garbhaḥ sū°*) *sadyaḥ stanyāya kalpate*
śeṣaṃ tu rudhirībhūtaṃ kāyaṃ yoniṃ ca sarpati (21)
dhātuṣu pratipūrṇeṣu śarīre samavasthite

der which we have already discussed several times above.[1047] From this we learn that this disease occurs only in women, since it is only they who have a womb (18). Then vv.19-22 seem to refer to healthy women, possibly to show how they differ from women having *raktagulma-*. We are told that in an immature girl whose ☞ *yoni-*[1048] is immature (*hīna-* "wanting, deficient, incomplete") the blood (*śoṇita-*) goes into the body, in a woman who is in full possession of her (womanly) characteristics (*pūrṇasvabhāvā*), however, into her body and her womb.[1049] This blood of the latter in part fosters/furthers (*bhāvayati*) the embryo or foetus in the body/in the [respective] body-member (*aṅga-*; in the case of the latter translation obviously the womb) and furthers/is turned in part to milk,[1050] whilst the rest "ignites" (*samindhati*)

---

sañcitaṃ rudhiraṃ yoniḥ punaḥ kālena muñcati (22)
yadā raktavahā raktaṃ pradoṣān nānupadyate
vimārgād yonim anveti [vikṛti]s tena jāyate (23)
tathaiva raktagulmo 'pi hetunānena jāyate
yadā ṛtumatī nārī prāptān vegān vidhārayet (24)
hriyā trāsād vyavāyād vā vartamānān adhogatān
evamādibhir apy anyair udāvṛttaiḥ prakopitaḥ (25)
vāyuḥ śoṇitam ādāya pratisrotaḥ prapadyate
garbhāśayam udāvṛttas tasyā vahati śoṇitam (26)
mārutaś cyutagarbhāyā yadā mithyopacaryate
tasyāḥ sa vāyur udvṛttaḥ pratighātāt saśoṇitaḥ (27)
gatvā garbhāśayaṃ ruddhaḥ sthiratvam upapadyate
saṃvṛttaṃ śoṇitaṃ tatra māruto viṣamaṃ gataḥ (28)
rajovahāḥ samāvṛttaḥ saṃstambhayati garbhavat
sa gulmaḥ spandate 'bhīkṣṇaṃ mārutena samīritaḥ (29).

(On the emendation of the printed *garbhaḥ sūtāyāḥ* to *garbhasūtāyāḥ* in v.21 see n.1052.) The following verses tell us that the woman afflicted by *gulma-* shows the same external signs as a pregnant woman, including flowing breasts (*payodharau prasicyete* in 34).

1047 See §5.9, §6.22, §7.3, §9.4 and §10.2.

1048 Here probably the uterus or the reproductive tract as a whole.

1049 KĀ,Kh 9,4, in the form of a student's question to his teacher, lets us know that "young men and young girls" do not have *raktagulma-*:

raktagulmaḥ kathaṃ strīṇāṃ jāyate durupadravaḥ
atha kasmāt kumārāṇāṃ kanyānāṃ ca na jāyate.

In the light of what we have just seen, "young men" could refer to males in general, "young girls", however, to immature girls, which does seem a bit strained. TIVĀRĪ, 1,p.402,n.5 has *kumārīṇāṃ* in the place of *kumārāṇāṃ*. This does appear more probable, though it must also be pointed out that *kumārī-* can then hardly mean simply "virgin" (as a virgin too may menstruate), but must mean "immature virgin"; whether it is then simply a synonym of *kanyā-*, or whether there is some difference (of age) between the two, I do not know.

1050 *kiñcit stanyāya kalpate.*

the body (of the woman, so it seems) to bring about the "cooking", i.e. transformation, of blood (*śoṇita-*) etc. (or, if the reading of the ms[1051] should be correct, to bring about metabolic transformation as, i.e. in the successive new forms of, blood etc.), which obviously refers to the chain of elements by now well-known, but excluding the first member. In the same way, it furthers/is turned to milk in a woman who has delivered.[1052] The rest turns to blood (*rudhira-*)[1053] and goes into the body and the *yoni-*; when the elements are full and the body is ready (?; or: is steady, i.e. in equilibrium?),[1054] the *yoni-* liberates the accumulated blood (*rudhira-*). Then vv.23-24ab say:

> "If the blood does not follow/enter (*anupadyate*) the ones carrying blood (*raktavahā[ḥ]*) due to [morbific] corruption, then it goes along/seeks (*anveti*) the *yoni-* by a wrong course. Through this a disorder arises; *raktagulma-* too arises through this cause just in this manner."

The following verses then tell us that the wind in a menstruating woman may be agitated due to the suppression of natural urges[1055] (due to shame, fear or intercourse) and other swollen/pent up (*udāvṛtta-*)[1056] substances, and may, having seized the blood (*śoṇita-*) against the stream (*pratisrotas*) (i.e. through being forced into instead of out of the body), attain the womb; swollen/pent up (*udāvṛtta-*), it bears/carries (*vahati*)[1057] her blood (*śoṇita-*). In

---

[1051] See v.20 in n.1046.

[1052] The printed text makes *garbhaḥ* "foetus, embryo" the subject of the verb *kalpate* in the line *tathaiva garbhaḥ sūtāyāḥ sadyaḥ stanyāya kalpate*, which is not only logically problematic, but also metrically out of place, as we thus have a heavy syllable where we should have a light one. Thus we must in all probability read not *garbhaḥ sūtāyāḥ* but *garbhasūtāyāḥ*, i.e. "of one who has given birth to a foetus (or, here better: to the contents of the womb, i.e. to a child)". I would not be surprised to learn that the latter is indeed what the ms reads, and that the former is a 'correction' of the editor (since the *visarga* is often omitted before sibilants in manuscripts), but unfortunately I have been unable to consult the ms (on which cf. §10.7). As it is, TIVĀRĪ 1,p.571 too has emended the printed text in the way proposed, though she has not given any reasons for this.

[1053] *śeṣaṃ tu rudhirībhūtam*.

[1054] *dhātuṣu pratipūrṇeṣu śarīre samavasthite*.

[1055] *vega-*, which in such a context usually refers to urine, faeces and wind; especially (and obviously) the latter would fit here.

[1056] MONIER-WILLIAMS, p.185c lists a meaning "to secrete; to retain" for *ud-ā/√vṛt*, which is not quite the same. But the prefixation of *ud* and *ā* to *√vṛt* gives us a basic meaning "revolve up" (or something similar), which would, as "swell up", fit well here. In fact *udāvṛttā-* (or the like) is the name of a disorder of the female genital tract whose causes show a marked similarity to the causes of the disorder discussed here (see TIVĀRĪ 2,pp.34f.).

[1057] From v.28 it follows that the wind somehow stops the blood, i.e. seemingly prevents

a woman who has miscarried, the wind, if attended to improperly, is swollen (*udvṛtta-*), and, obstructed (and thus unable to leave the body), attains, together with blood (*śoṇita-*), the womb, where it is rendered immovable (cf. too n.1057). The blood (*śoṇita-*) is conglomerated/made to grow (*saṃvṛtta-*) (obviously because it cannot flow out). The wind, in a disproportionate state (*viṣamaṃ gataḥ*) and turned back (*samāvṛttaḥ*) (since it cannot leave the body), blocks (*saṃstambhayati*) the ones carrying *rajas-* (*rajovahāḥ*), like a foetus, causing *gulma-*.

10.25. It would seem that "the ones carrying blood" in 9,23 and "the ones carrying *rajas-*" in 9,29 mentioned here are the same as the vessels carrying *rajas-* to the womb mentioned in 9,17 (see §§10.22f.). What the above therefore seems to be telling us is that in *raktagulma-* the "blood" in the womb mixes with wind and swells up, which blocks the ingress of further "blood", leading to this bringing forth the signs of pregnancy in other parts of the body (as Kh 9,30ff. expressly tell us); this tallies with what we have tentatively deduced to be a possible explanation for SU,Ut 42,13-15 in §6.22.[1058] Thus SU,Ut 42,15 might indeed be telling us that the abdomen does not distend, for neither can the *gulma-* formed by the effects of "blood" and wind in the womb be nourished by a fresh influx of fluid, as it itself blocks the ingress of this, nor can it grow in the same way as an embryo or foetus does, since this is nourished by other means (which we shall discuss in more detail in §§13.14ff. below). But this does not do away with all the difficulties of this passage, for, if we were to assume that one *na* of *na spandate nodaram eti vṛddhiṃ* should indeed refer to a non-distension of the abdomen,[1059] then the other *na* would have to be construed with *spandate*, i.e.

---

it from flowing. But can *vahati* "bear, carry", even though at times used in the sense of "support, (up)hold" (i.e. not only in the sense of "convey, carry away"), really be used to mean "hinder, stop, block"? I must admit to feeling uneasy about translating it in this manner. If, however, *vahati* in v.26 should be understood in the sense of "convey", what would be meant is that the swollen wind (or: the wind pent up in the body) forces the blood into the womb. In v.27 we would then have another reason for wind behaving in this way (after a miscarriage), and the statement in v.28 that blood is rendered immovable would apply to both cases. To me this makes sense, especially since the meaning of *vahati* would then cause no difficulty. TIVĀRĪ, 1,p.405 too understands 26cd in this way (the English explanation on p.407 is not as clear as the Hindi explanation on p.405).

[1058] In this respect there could (but need not) eventually be a difference to the theory presupposed by CA and similar texts, depending on how the passages discussed in §§6.21f. are to be understood.

[1059] The non-distension itself poses a problem too, though not so much for the SU passage as for KĀ,Kh 9,33, which tells us that in *raktagulma-* a *romarājī-* too appears (*romarājī ca*

the SU passage would say that the *gulma-* does not palpitate, which would be the exact opposite of what KĀ,Kh 9,29 tells us,[1060] though there could be a difference of opinion between the texts in this regard (as also with regard to growth of the *gulma-* as opposed to that of the abdomen; cf.n.1059). However, since SU,Ut 42,13-15 actually do not even mention obstruction (even though they must obviously presuppose some sort of obstruction), we really have no means of proving or disproving the correctness of our deductions on non-distension of the abdomen, as well as on how the procreatory-menstrual fluid actually is blocked, so that ultimately the KĀ passage does not help us to gain surety in this regard.

10.26. It might, however, help us with regard to another problematic passage, namely SU,Śā 4,24, which we have discussed in detail in §§6.17ff., only to break off the discussion, unresolved, in §6.25. According to what we have just seen above, and given the fact that ☞ *srotas-* may subsume both ☞ *sirā-* and ☞ *dhamanī-*,[1061] there can, I think, now be little doubt that what this SU passage refers to is firstly a plural number of channels ("vessels") which bear procreatory-menstrual fluid *to* the womb, and secondly a blocking of these channels by the foetus (note that the blockage through the *gulma-* of the channels carrying *rajas-* is expressly said to be like that occasioned through a foetus in KĀ,Kh 9,29). But in *gulma-* the vessels are blocked, according to KĀ,Kh 9,29, by the wind, the inference, however, being that, since the wind makes the blood swell up, it is the blood, mixed with wind, which fills up the whole womb and thus blocks all openings. The same

---

mūrchitā; on this cf. DAS,R.P. 1991A, p.65). According to DAS,R.P. 1991A, pp.38f., there is the possibility that the *romarājī-* was explained as being formed due to the distension of the abdomen. That according to CA,Ni 3,14 (on which see also DAS,R.P. 1991A, pp.36-38) the *romarājī-* appears due to *raktagulma-* could be in keeping with this, for we have seen in §6.22 that according to CA the abdomen is distended due to *raktagulma-*. But does KĀ too presuppose such a distension of the abdomen? If it does not (see §6.22), then we obviously have a problem on our hands. Kh 9,58f., according to which a foetus grows fast, a *gulma-*, however, slowly (cf. §6.22), could be taken to mean that the abdomen is indeed distended. But the text does not say so expressly, so that we seem not to be able to gain any surety as to whether a distension is presupposed by it or not. However, a distension would be in keeping with the appearance of the *romarājī-*, provided the reasons deduced in DAS,R.P. 1991A for the appearance of this are indeed correct. As regards the growth of the *gulma-* referred to in KĀ, it must be asked what can cause this if the ingress of fluid into the uterus is blocked. Could the growth be due to wind, which causes the *gulma-* to palpitate (see n.1060)? Or does the text presume that, regardless of the blockage due to the *gulma-*, some fluid still manages to seep into the uterus in minute quantities?

[1060] *sa gulmaḥ spandate 'bhīkṣṇaṃ mārutena samīritaḥ.*

[1061] See the relevant discussions in Appendix II.

can, however, logically be effected only by a foetus in a stage in which it is large enough to fill the womb completely; what happens when it is still too small? I have already referred to this in my discussion of Su,Śā 4,24 above. As regards the formation of the ☞ *aparā-*, we have discussed in §6.25 whether this might have been taken to be formed due to the pressure of fluid outside the womb and pushed into this through the wall of the womb, opining too that Su,Śā 4,24 and its parallel As,Śā 2,p.288a (see §6.17) might mean to say that the procreatory-menstrual fluid first of all gradually fills the womb, whilst the foetus gradually grows and blocks the ingress of further fluid. The data from KĀ do rather seem to speak for the correctness of this assumption.[1062] With regard to the uterine orifice leading to the vagina, in the light of our discussion in n.277 and §6.25 (see also §§8.19f.) it seems most probable that in a pregnancy this is presumed to be blocked anyway, at least in the early stages of pregnancy; afterwards the foetus (or its enveloping membranes) probably account for the blockage of this orifice, if necessary. In the case of the *gulma-* the question may be taken not to arise, since no fluid reaches the uterus; if one, however, should hold that small amounts of fluid might yet reach the uterus,[1063] then one could always say that at those times when the uterus periodically opens (cf. the references on pregnancy just cited) the conglomerate of blood swollen by wind within the womb stops any fluid from leaving via the vagina.

10.27. KĀ,Kh 9,18ff. are interesting in another respect too. We have seen above (§10.24) that the verses also mention what is clearly a tripartition of blood in the healthy pregnant woman.[1064] This theory is very similar to the one referred to further above in §§10.5f. (and found also in KĀ,Sū,Lehādhyāya, p.2), in which the *rasa-* of a pregnant woman divides into three parts nourishing the embryo or foetus, her milk (or breasts) and her body (or *rasa-*) respectively; what is, however, quite different is that instead of *rasa-* we here have "blood". It is difficult to see in this "blood" anything else but the procreatory-menstrual fluid. Indeed a quotation from Kṛṣṇātreya in ĀY 4,151-153[1065] has more or less the same theory, but uses the word *rajas-*;

[1062] On the possibility of the blockage being due to a membrane being formed within the uterus cf. n.301, where this possibility was viewed sceptically.

[1063] E.g. to make the *gulma-* grow; cf. n.1059.

[1064] That the woman referred to is pregnant follows from the mention of the embryo or foetus.

[1065] The quotation from Kṛṣṇātreya (*tathā ca kṛṣṇātreyāt*) runs:
daivād gṛhītagarbhāṇāṃ rajo bhāgās tridhā bhavet
stanayoḥ pīnatāṃ stanyam eko bhāgaḥ prayacchati (151)

this *rajas-* of providentially (*daivāt*)[1066] pregnant women in part causes swelling of the breasts and milk, in part nourishes the embryo,[1067] and in part nourishes the mother. This parallel does seem to substantiate our deduction regarding the nature of "blood" mentioned in the KĀ passage. However, the ĀY passage also causes some problems, for after the above the text says: *tato na hi*. One could take this to mean "after that no more at all", which could refer either to the tripartition after pregnancy (i.e. this does not take place any more), or else to no part of the *rajas-* remaining after it has nourished milk, embryo and mother as mentioned. The next verse would then be saying that either (*athavā*) *rajas-* of women issues, or (*vā*) the passages of the ☞ *srotas*-s carrying *ārtava-* are surely blocked (*rudhyante*) by the foetus. The first part of this statement would be extremely problematic. It is therefore

---

dvitīyaḥ svabalenaiva garbhagaṃ poṣayaty api
tataḥ paraṃ poṣayati garbhiṇīṃ ca tato na hi (152)
rajaḥ pravartate (var: *pradṛśyate*) *strīṇām athavārtavavāhinām*
srotāṃsy asyā hi rudhyante mārgā garbheṇa vā dhruvam (153).

The last verse seems to be corrupt. *ārtavavāhinām* cannot qualify *strīṇāṃ* if it presupposes *vāhin-*, as this is not feminine here. One could argue, though, that it stands *metri causa* for *ārtavavāhīnām*, in which case it might come from *vāhi-* or *vāhī-*. But what would that mean? Would it refer to women who possess *ārtava-*? Or would it rather refer to women whose *ārtava-* flows (scil. can flow), but is not allowed to do so because it is obstructed? Or do we actually have *ārtavavāhin-* qualifying *srotas-*? But then we would expect *srotasām*, which would obviously give more sense than *srotāṃsy*, and with which *mārgā[ḥ]* could be construed without any problem. I would therefore suggest reading this line as:

srotasām asyā rudhyante mārgā garbheṇa vā dhruvam.

This is of course not in the 'normal' śloka metre, but then neither is 152c, so that that is no serious argument against the emendation, provided, of course, that *ārtavavāhin-* does refer to the *srotas*-s. One could also try to avoid an emendation by construing *ārtavavāhinām* with *mārgā[ḥ]*, which would thus stand in apposition to *srotāṃsy*; we would then have "the passages of those carrying *ārtava-*, [namely] the *srotas*-s". But *mārgā[ḥ]* would in this case be so far removed from *ārtavavāhinām* that this is problematic, especially since this is no kāvya (where such things might yet be expected), but nearly prosaic verse for practical use and obviously with no high literary aspirations. But whether we emend the verse or not, what is ultimately meant would be the same. Thus, though the explanation of this line given below follows its emended form, the arguments remain the same in any case.

[1066] On *daivāt* cf. §10.22, though note that the word is used there in connection with menstruation and not pregnancy.

[1067] Literally: "nourishes/satiates that which attains/is situated in the womb" (*garbhagaṃ poṣayaty*). This translation is of course possible only if *garbha-* here means "womb" (cf. also COMBA 1981, p.202,n.28); otherwise the translation is: "nourishes/satiates that which attains the embryo/foetus", which would probably refer to the nourishment (by whatever means) of this latter, this nourishment thus being said to be augmented by the *rajas-*.

much more probable that *tato na hi* is to be joined to the next verse; the whole would then be saying that there is either (*athavā*) no flow of *rajas*- of women therefore/after that (*tatas*) (because all the fluid is used up), or (*vā*) the passages of the *srotas*-s carrying *ārtava*- are surely blocked by the foetus. It would then seem that the passage ĀY quotes is telling us either that no fluid reaches the exterior because it is all used up internally for satiating or nourishing, or else, alternatively, that it does not reach the exterior because the foetus blocks certain openings. The text would leave it unclear whether the two explanations given refer to different stages or types of pregnancy, or whether we have here in general two different, variant explanatory traditions. I must, however, admit that I do not feel fully at ease with this explanation. The text is not clear, and, apart from the difficulties mentioned in n.1065, we also have problems like the construction with *athavā ... vā*.[1068] The above is but an attempt to come to grips with the verse as it is. In any case, if the explanation given should indeed be correct and the deduction that no procreatory-menstrual fluid is seen because it is used up internally be valid, then that does not mean that we can simply generalise this as having been accepted by all authorities. But be all that as it may, it does not change the fact that the verses quoted by ĀY say expressly that the procreatory-menstrual fluid performs the same nourishing functions in a pregnant woman that according to other texts are performed by *rasa*-.

10.28. But let us return to KĀ,Kh 9,18ff. In the light of the above, this passage appears to state in a manner whose clearness is unmatched in anything from the 'classical' texts we have seen so far that not only is the procreatory-menstrual fluid present throughout a woman's body (on which see also §5.16 and §8.23), but that in an immature girl it flows only throughout the body (cf. also §6.23), whereas in a woman able to bear a child it flows through the body and into the womb.[1069] Though this may in actual fact not be in keeping with the theory or theories of other texts in all respects, it does seem, from what we have seen so far, to be reconcilable with what they seem on the whole to say. But what is really difficult to reconcile with what we have seen so far is the statement on the tripartition of the procreatory-menstrual fluid in the same manner as that of ☞ *rasa*- (see §10.27), the for-

[1068] Do we actually have an original *atha ārtavavāhinām*, with a glide (*v*) inserted to remove the hiatus? Or should we actually read *athacārtava*°?

[1069] Given the connection between milk and the procreatory-menstrual fluid, one could deduce from this that, like the latter, the former too is always present in the female body. However, this matter requires a rather detailed discussion, for which a footnote is not the proper place; the discussion has therefore been shifted to Excursus XII at the end of this chapter (pp.309-311).

mer even seeming to take up the function of *rasa-* in the chain of alimentary transformation. This is highly intriguing, especially since *rasa-* is nowhere mentioned in the whole passage quoted in n.1046. We have seen above that there were theories deriving the procreatory-menstrual fluid from *rasa-* (or ☞ *āhārarasa-*) (cf. §10.19). So could the procreatory-menstrual fluid in the body of a woman according to some theories have somehow taken up functions of *rasa-*, or even been ensconced in its place in certain metabolic functions? In the light of what the passage would then say about this fluid (see §10.24), this seems to cause great difficulties.

10.29. As regards the non-mention of *rasa-*, particularly in v.20, which speaks of "blood etc." being "cooked" (see §10.24), attention may be drawn especially to CA,Sū 11,48, AH,Sū 12,44 and AS,Sū 22,p.166a in §10.7, which list the chain of ☞ *dhātu-*s beginning not with *rasa-*, but with blood, and do not place ☞ *tvac-* "skin" at the beginning of the whole chain, but link it with this chain by means of *ca* "and", showing that, though linked with the chain, it is here not an actual part of this chain as in other cases mentioned in §§10.7ff. It may also not be amiss to refer to §7.30 (also n.663, and Excursus IX and X at the end of chapter 7). Of course we cannot simply postulate a chain of elements beginning with blood on the basis of this, but we can at least consider the possibility of such a chain having existed. Furthermore, we may also consider the seeming confusion between blood in general and the procreatory-menstrual fluid as detailed e.g. in §§7.2f. and §§7.5ff. (cf. also §5.14). This does allow us to ask whether the KĀ passage being discussed may not in actual fact bear testimony to an otherwise lost theory in which blood was not only the first element in a chain of *dhātu-*s, but also the same as the procreatory-menstrual fluid, as it is in many cultures (cf. §2.2). But even if the first part of this question regarding the list of *dhātu-*s should in actual fact be justified (only further research can hope to bring surety in this regard), what the KĀ passage tells us apparently does not allow us to uphold the second part of the question, for since one of the fluids seems to turn into the other in KĀ,Kh 9,21 (*rudhirībhūta-*), the two cannot be quite the same. If now it is the procreatory-menstrual fluid that is here said to turn into blood, how does that fit into the general picture when we consider the male too? Clearly, postulating such a transformation causes immense difficulties, especially in the light of the statement in v.21 that this newly created blood goes into the body and the ☞ *yoni-*, and is then, according to v.22, liberated from this after accumulation. Is thus v.21 actually telling us something quite different? We have seen above that the procreatory-menstrual fluid might possibly have been taken (at least in some theories) to exist in two forms within the body, one more or less without colour or pale, the other blood-like (see §§8.3ff.; cf. also §9.8). Could *rudhirībhūta-* "turned to blood" in v.21 actually refer not to a metamorphosis to blood in general, but simply to a change of colour? Then of course the whole question of a oneness with or transformation to blood becomes irrelevant. But v.20 also tells us:

> *paktaye śoṇitādes tu śeṣaḥ kāyaṃ samindhati*
> "The rest, however, ignites the body for the cooking[1070] of blood etc."

This need not contradict what was said above, though, as the fluid itself is not said to be changed into blood, but only to give what seems to be the energy to "cook" this and other substances;[1071] but even if a more direct connection with blood etc. and the fluid in question should be meant, then this latter could here be taken to nourish or satiate blood etc. without itself being blood. Things of course get more difficult if the reading of v.20 should in actual fact be:

> *paktaye śoṇitādyas tu śeṣaḥ kāyaṃ samindhati*
> "The rest, however, ignites the body as (i.e. in the form of) blood etc. for cooking (see §10.24)."

Though the exact meaning of "cooking" then would admittedly be unclear, we would nevertheless have to postulate an actual transformation to blood etc. here, with all the problems that this entails.

10.30. We should, however, also not overlook the possibility of the fluid meant in the KĀ passage being not the procreatory-menstrual fluid, but blood in general. Several things, however, speak against this. Firstly, Kh 9,17 speaks of *rajas-* being collected if it is hindered from flowing out (see §10.22), and v.18 then says that it is through this very reason (*anena hetunā*) that *raktagulma-* arises. This does make one suppose that the fluid mentioned in 18ff. is the same as in v.17. On the other hand, v.29 tells us that the *gulma-* arises due to "those carrying *rajas-*" being blocked.[1072] Now the agent which causes this blocking seems to be wind mixed with "blood" (see §10.26), and it is possible that this latter is not the same as the *rajas-*, i.e. that it is blood in general, which too reaches the uterus via "the ones carrying blood" (v.23). This would then mean that we would here have a cause for *gulma-* due to blood which is very similar to that of *vidradhi-/makkalla-* (cf. §7.2) and *pradara-* (cf. §10.3). But not only would that clash with what we have already seen on the causes of *gulma-* (see the references in §10.24), but

1070 "Cooking" obviously refers to the metabolic transformation of blood etc.

1071 We may in this case point to the essence of semen, namely ☞ *ojas-* "vital fluid/force" and ask whether we might not have a sort of energy similar to this being postulated here as the final essence of the procreatory-menstrual fluid.

1072 *rajovahāḥ samāvṛttaḥ saṃstambhayati garbhavat.*

also with KĀ,Kh 9,24, for this verse refers expressly to a menstruating woman, which does make it most probable that the "blood" the agitated wind mingles with and forces into the uterus (see §10.24, especially n.1057) is the procreatory-menstrual fluid. One could nevertheless hold that this need not apply to the "blood" mentioned in connection with healthy women in vv.19-22, and that *śoṇita-* here refers to blood in general, which is according to 21 turned to the procreatory-menstrual fluid (*rudhirībhūta-*), especially since according to 22 it is *rudhira-* too, and not *śoṇita-*, which is periodically voided. This would indeed be in keeping with the later theory of the derivation of the procreatory-menstrual fluid from blood we shall discuss in §§11.5ff. Still, it does strike one as queer that we should then have blood in general mentioned in the middle of a passage which otherwise seems to speak of the procreatory-menstrual fluid. Nevertheless, the possibility cannot be fully discounted. But it must be pointed out that this explanation also occasions difficulties inasmuch as according to 19 this blood itself would go to the ☞ *yoni-* in a healthy pregnant woman, but have to be changed to the procreatory-menstrual fluid first before going to the same place in a woman who has delivered (22). If, however, the procreatory-menstrual fluid and not blood in general were meant, no problem would arise, provided *rudhirībhūta-* referred to a change of colour (cf. §10.29): the procreatory-menstrual fluid according to this would turn red before leaving the body, but not in the other case, in which it remains within this. Such a change of colour could not be postulated in the case of blood in general, since this is always red. Thus the evidence does rather seem to speak for the procreatory-menstrual fluid and not blood in general being meant in the KĀ passage we have been discussing, though, as in most of the problematic cases so far, we have no full surety that this is indeed so (cf. also the variant translated in §10.29, though one may very well ask of what evidential worth it really is).

10.31. To sum up the discussion on KĀ,Kh 9,18ff.: the procreatory-menstrual fluid here appears in functions in which we have so far found only *rasa-*, seemingly even in metabolic functions where we would not have expected it. The quotation from Kṛṣṇātreya in ĀY 4,151-153 cited in §10.27, in which such functions are expressly ascribed to *rajas-*, seems to confirm this. Since, however, we cannot obtain full surety with regard to the KĀ passage, as this speaks only of "blood" in the relevant contexts, we cannot fully discount the possibility (though this does seem meagre) that what functions here in a manner similar to *rasa-* is blood in general. But even if that should indeed be the case, we would still have the verses quoted by ĀY to show that the procreatory-menstrual fluid too could have a similar function; moreover, the remarks on *gulma-* in the KĀ passage would most probably nevertheless refer to the procreatory-menstrual fluid.

10.32. We thus see that, even though BH does not offer us much by way of information, KĀ compensates for this amply, since it contains much that is either not to be found in the other texts studied so far, or else hidden in these in such a way as to be quite difficult to collect. As it is, we will probably not go amiss in saying that especially KĀ bears testimony to the fact that ancient Indian medicine is neither a homogeneous nor a homogenous system. It is, however, a fact that the variant theories to which KĀ seems to adhere do not seem to have made much headway against what we may on the basis of the 'classical' medical texts call the 'mainstream' of Indian medical theory as it has come to be generally accepted and is still accepted today. Though traces of such other theories are, as we have by now amply demonstrated, found in all the 'classical' texts too, they are not as easily discernible. The reason for this is probably that texts such as KĀ and BH (and others; cf. n.904) have been spared the fate of repeated 'modernisation' at the hands of scholars through the ages, because they were not as extensively studied as the other texts (and this applies even more to KĀ than BH). This, however, also means that they could not claim for themselves the authoritativeness of the four 'classical' works CA, SU, AS and AH, which have influenced the development of later Indian medical literature massively.

EXCURSUS XII

(in place of footnote 1069)

KĀ,Kh 9,18ff. appears to state that not only is the procreatory-menstrual fluid present throughout a woman's body, but that in an immature girl it flows only throughout the body, whereas in a woman able to bear a child it flows through the body and into the womb (cf. §10.28). I may draw attention here to a possible deduction based on the fact that one part of the fluid nourishes the embryo or foetus in a pregnant woman (see §10.24). In a woman who has delivered, this part can obviously not nourish any embryo or foetus. So what happens to it? Does it simply flow out of the body? Or does it, in full or at least in part, go to the breasts? Should such be the case, then it would imply that more of the fluid reaches the breasts in a woman who has delivered than in one who is pregnant, which would imply that a woman who has delivered has more milk in her breasts than one who is pregnant — and that is of course exactly what we see in reality. A confirmation for the validity of this assumption could be contained in BHĀ 1,4,10, which tells us that the quantity of milk of women diminishes due to a second pregnancy;[1073]

[1073] *strīṇāṃ stanyaṃ bhavet svalpaṃ garbhāntaravidhāraṇāt.*

this obviously relates to a woman breast-feeding her first child.[1074]

We may also look at the KĀ passage from another angle, namely the fact that the text refers (apart from an immature girl) only to a pregnant woman and to one who has delivered. Neither does the passage touch upon what happens in a woman past her menopause, nor do we learn what happens in a woman who is of child-bearing age and also able to bear a child, but has neither already given birth nor is pregnant; could this be because in the society of that time, in which it was common for women to be married off even before puberty[1075] and to have one pregnancy after the other through most of their lives before the menopause, it was considered unusual for a healthy woman (unless she be widowed or an ascetic) to be in any other state? When we also consider the fact that the passage could be taken to imply that a woman always has milk in her breasts,[1076] then we do gain the impression that for the author(s) a mature woman was either always pregnant or else breast-feeding. Of course this may be over-stressing the wording of the passage, for after all it is quite possible that in the given context women in other states were simply not considered relevant enough to merit discussion. Nevertheless, and quite apart from the question of what the normal state of healthy adult women was considered to be, we seem to have other evidence that points to a notion that women always have milk in their body irrespective of the state they may be in; see §5.16 and n.1622, as well as, e.g., the lists on the constituents of the body listed in n.144, at least some of which list milk too. Such an assumption could explain the wide-spread notion in ancient India that the mere sight of a child (or even thinking about it) may make milk shoot into or even from *any* mature woman's breast if it stirs her motherly feelings enough.[1077] I myself can testify from personal observation that a milk-filled breast may begin to flow not only when the other breast is sucked upon by an infant, but even when the mother hears her child cry. May not such and similar observations have, under the circumstances

[1074] However, the same text may also have contradictory data; see the passage quoted in n.1115.

[1075] Healthy and normal, yet unmarried, women were probably extremely rare, at least in the social strata to which the medical authorities probably belonged.

[1076] Or maybe in some other receptacle of milk; cf. e.g. CA,Ci 30,238;243;246;249.

[1077] Cf. MEYER,J.J. 1952, p.211, also SU,Ni 10,21-23, BHĀ 1,4,9, KĀ,Kh 9,45cd-46ab (which speaks of the mere thought of being pregnant — even if no true pregnancy exists — making milk flow), and n.300. According to BHĀ 1,4,10 adverse emotions, among them absence of love for the child, cause the flow of milk to cease; one can only speculate on what distress this notion must have caused women having difficulty in breast-feeding.

prevailing, been generalised?[1078] Trying to find an answer to this question is unfortunately something that falls outside the scope of this study, but it is nevertheless well worth looking into.

It must, however, also be pointed out that at least one passage (ŚĀR 1,5,16) expressly remarks on the periodicity of milk, and that of the procreatory-menstrual fluid, though it is not clear whether this refers to their appearance and disappearance at puberty and the menopause respectively[1079] (from which we would then deduce that they are always present in the body at least as long as a woman has reproductive power), or at shorter (irregular) intervals during her adult life between the two extremes of puberty and menopause;[1080] should the former be meant (and verses such as SU,Sū 14,6 — also quoted by Āḍhamalla — in §7.11 could let one suppose so), then that would obviously not be in conflict with our tentative deductions above. Cakrapāṇidatta in his commentary on CA,Ni 3,14 might be remarking on the theory that milk is present only during pregnancy,[1081] saying that this is not true,[1082] as it is also present in the *gulma-* due to blockage of the procreatory-menstrual fluid (cf. on this §10.24). However, it is not clear whether this theory (and Cakrapāṇidatta's refutation) refer to milk in the body in general, or only milk in the breasts; obviously an absence of milk in the latter except at certain times would not preclude its being present in the body as such at other times. Moreover, *garbha* in *garbha eva* (see the text in n.1082) could stand not for *garbhe*, as supposed above, but for *garbhaḥ*; should this indeed be the case, then Cakrapāṇidatta's remarks would be directed against a theory holding that the embryo itself (or probably rather: the combination of the procreatory fluids from which the embryo is formed) gives rise to milk. This does seem unlikely, but in the course of this study we have met and will meet with so many theories not in keeping with our own notions on similar matters that instead of issuing a categorical denial we should rather regard the issue as not yet resolved.

1078 Cf. also n.298.

1079 As e.g. the commentary of Āḍhamalla would make us suppose; see p.107.

1080 As the quite late commentary of Kāśirāma would make us suppose; see n.1098. See also n.1115.

1081 The proponents of this theory basing this notion on CA,Śā 6,23, on which see §10.6.

1082 *garbha eva stanyaṃ bhavatīti tan nirastaṃ mantavyam.*

CHAPTER 11

Later Medical Works

11.1. Given the authoritativeness of the four 'classical' medical works (cf. §10.32), it is no wonder that these in course of time came to be supposed to contain all the necessary theoretical discussion on anatomy and physiology, and it is thus not surprising that later works have occupied themselves more and more primarily with diseases, their symptoms and cures. This is of course not to say that other theoretical considerations became irrelevant, or that no more works like the 'classical' texts treating on all branches of medicine were composed, but on the whole it will not be wrong if we *a priori* do not expect to glean much really new material on our specific problems from later works. The material that they do offer has to a large extent already been utilised in various contexts above; this holds good also for the commentaries on the 'classical' and other medical works, which often have a lot to say that concerns us directly, in contrast to independent medical works composed at about the same time as these commentaries. Nevertheless, it would be methodically unsound if, based solely on the just mentioned considerations, we were not to examine later works and commentaries any more than has already been done.

11.2. It has, however, been impossible for me to procure *all* later works or to go through the works I could procure more than cursorily, and this is even more true of the often extremely voluminous commentaries on medical works in general. Thus what I have to say below on the opinions of later works and commentators over and above that which has already been said is not much, and may not encompass all the material relevant to our study. Nevertheless, I did strive for comprehensiveness. I must, however, point out that I did not consider Nighaṇṭus. Nor did I usually consider works on Pākaśāstra or Sūdaśāstra (i.e. works such as the *Bhojanakutūhala*), on Dravyaguṇa (e.g. *Annapānavidhi* of the *Āyurvedamahodadhi*) (DR too seems to contain nothing relevant, though its commentary, already utilised several times above, does), on Pathyāpathya (e.g. the works listed in IOL, p.1907), on *jvara-* (e.g. *Triśatī*, *Jvaranirṇaya*), on Nāḍīparīkṣā or Nāḍīcikitsā (e.g. the works listed in IOL, pp.1680-1683),[1083] on *unmāda-* or *graha-* (e.g. *Kumāra-*

[1083] Works on Nāḍīcikitsā are mostly relatively late anyway, as is this whole branch of Indian medicine (see e.g. ROŞU 1982, pp.62-64, MEULENBELD 1995, p.6, KURIYAMA; cf. also MEYER,F. 1990, pp.251-254). I have, however, drawn attention *passim* to a few passages from the *Nāḍījñānadarpaṇa* which are to be found in a small chapter (vv.248-252) on the examina-

tantra), and on medicinal properties and/or extraction and/or preparation of certain substances such as iron or plant products (e.g. *Lohasarvasva*, *Bheṣajakalpa*, *Śataślokī*), as works of such a nature, of which I consulted quite a number (this includes those cited as listed in IOL), showed that they could not as a rule be expected to have any bearing on our topic of interest;[1084] I also refrain from giving the names and particulars of *all* such works I consulted here, partly because I did not always make full bibliographic notes when I had the opportunity to consult them, and partly because I do not wish to needlessly bloat my bibliography. I have, however, mentioned a few representative works above, so that the reader may by their perusal gain an idea of why I did not consider such works further.

11.3. As a rule, I have also neglected medical chapters or chapters containing material bearing on medical matters in works not purportedly of a medical nature (e.g. encyclopaedias,[1085] epics, Purāṇas and the like), as their relationship to 'real' medical texts (irrespective of whether they actually agree with such texts — 'classical' or otherwise — in most respects or not) is not always clear and thus creates additional problems; such texts should better be studied along with other works containing medical information, but not of a strictly 'medical' nature.[1086] It also goes without saying that mod-

---

tion of *ārtava-* (*ārtavaparīkṣā-*); to my knowledge, other works of this genre do not have similar chapters.

[1084] Texts of an alchemical-medical or iatrochemical nature were, after four (*Āyurvedaprakāśa*, *Rasaratnasamuccaya*, *Rasārṇavakalpa* and *Rasahṛdayatantra*) had been consulted, totally neglected, as it seemed that nothing of use for us could be expected from them. Neither did I, after having consulted the *Madanamahārṇava* — "which discusses the causes of and cures for diseases mainly in terms of what may be called *karman* and *adharma*, and *prāyaścitta* respectively" (WEZLER 1990, p.146,n.104), i.e. discusses *karmavipāka-* — as well as the *Vīrasiṃhāvaloka*, try to find and utilise other similar works (on such works cf. also RAO, p.14, PINGREE).

[1085] This applies of course only to general encyclopaedic works which also have chapters on medicine, not to medical works which have the nature of compilations or encyclopaedias (such as BHĀ and ĀY). The Prakrit *Haramekhalā*, however, was consulted.

[1086] This is valid also for that class of texts termed 'medical' which are usually — though not always — fairly recent and seem to be primarily concerned with extolling the benefits of Yogic and Tantric practices for a 'healthy' life and are cloaked in a medical garb, so that I would call them 'pseudo-medical' works. There are undoubtedly some references to Yogic practices in medical works (also, though sporadically, in old ones, though their exact relationship to the other matter of these medical works remains to be examined and is in any case problematic), but the connection between Yoga (as a system, not necessarily as individual practices also termed *yoga-*) and Āyurveda is quite peripheral, it only being in fairly recent times that the connection became more intimate, so that today we often find Yoga and Āyurveda (i.e. the Indian medical system proper, though on the difficulties of defining this exactly see DAS,R.P. 1993B) treated on a par or even as being more or less the same, which is of course not to be justified (cf. on this subject e.g. ZYSK 1993, pp.211-213, also DAS,R.P. 1992B, pp.184ff., 1993A, pp.21ff., and 1993B, pp.69f., and MEULENBELD 1995, p.2), especially when such unhistoric subsumption goes on to encompass Tantrism too, as is often the case nowadays (on 'medicine' in Tantric texts see also ROŞU 1986B, pp.123-125 and DAS,R.P. 1995). As regards later 'pseudo-medical' texts of the sort just mentioned, their theories and explanations are on the whole highly esoteric and sectarian, often giving the impression of being either secret knowledge only for the initiated, or else of being ad-hoc formulations to explain away difficulties. Despite the modern tendency just mentioned to reckon such texts to be on a par with the texts discussed in this study (indeed, often all that a Westerner otherwise ignorant of Indian medicine gets to know of 'Āyurveda' is such texts), I have not deemed it necessary to discuss them here, though I do admit that by not doing so I may have inadvertently overlooked valuable information. However, I had to draw the line somewhere. Such texts should be studied along with other basically non-medical texts dealing with medical matters too, and which too may contain information valuable for our topic of interest, but not of the same nature as that of the texts actually discussed in this study. As an example of the class of 'pseudo-medical' texts just mentioned we may take the ĀYS (for a characterisation of which see ZYSK 1993, p.212) with its in many respects even more interesting commentary by Yogānandanātha, both according to the editor (see p.xv) "not older than the 16th century A.D.". Here we find an ingenious theory combining ☛ *rasa-* "taste" with the seven elements (☛ *dhātu-*) of the alimentary chain, each of the six tastes being (as parts of the ingested food) taken to "nourish" or "satiate" one or more of the elements or even form it/them. The theory is, however, not consistent throughout, cf. e.g. 1,73f.;88, 2,97, 4,40-61 (passim), 5,44-49 and the commentary on 5,33. According to the commentary on 2,56f. both semen and the procreatory-menstrual fluid come from *svādurasa-* "sweet taste". The elements *rasa-* and blood are considered to be one, c.f. e.g. 2,98f. and the commentary on 1,74;85; for other passages and an apologetic explanation of this see DASH 1971, p.101, and cf. too Śāyaṇa's commentary on *Aitareyāraṇyaka* 2,3,2 (*prāṇabhṛc-charīreṣu raktarūpo raso dṛśyate* "in the bodies of those bearing life, *rasa-* is seen in the form of blood"), from which it follows that this notion is not confined to the ĀYS. But because of the oneness of blood and *rasa-* in this text it is only logical that the skins (☛ *tvac-*), which according to some medical texts derive from blood (see on this also §11.8), are said in this text to evolve out of *rasa- and* blood (1,79). According to the commentary on 2,2f. the skins have the same function of "holding" (*dadhati*) as the "receptacles" (*āśaya-*; on these see n.581). (All this is further proof that, as a large part of chapter 10 has shown, the theory of the *dhātu-*s is actually replete with difficulties, which have, however, mostly been smoothed over in course of time.) In the commentary on 1,87 the procreatory-menstrual fluid (*raktadhātu-*; cf. p.555 in Appendix II) is said to be the combination/sum (*samāhāra-*) of *rasa-*, blood, flesh and fat, semen (*śukla-*), however, of bones, marrow and *retas-*. This is very interesting, the more so since the semen discharged during intercourse and that present all over the body as the seventh element (here: *retas-*) are obviously differentiated (cf. on this §8.9). The commentary on 2,63 states that from the first day on of the period beginning with menstruation to the fourteenth day all the seven elements combine together and flow as "blood" (*sapta dhātavo raktarūpeṇa sravanti*); this "blood" is obviously the same as *rajas-* in 2,60. I may also mention that the development of the different body parts of the embryo is, according to the commentary on 2,5 (see also that on 2,6;19), dependent on certain fluids originating from the mother (on a similar

ern medical works of a traditionalistic nature, be they written in Sanskrit or in another language, have not been consulted as primary, but as secondary sources, being as a rule products of the "renouveau contemporain de l'Āyurveda" (on which see e.g. ROŞU 1982) and as such full of extraneous influences, both Indian and non-Indian; moreover, they seem mostly to be based not on the tradition(s) we are here chiefly concerned with, but on (a) parallel tradition(s) of the sort we find not only in non-medical works nevertheless dealing with medical matters, but also in the practical tradition as it still exists today.[1087] Since Indian medicine seems to have suffered a general loss of vitality and gone through a period of stagnation and finally decay since the Middle Ages before having been rejuvenated in modern times, premodern independent works of a relatively late date too are to be utilised with caution; even for instance BHĀ, which is basically not an independent work, but a compilation from various, ancient, sources, shows the need for this, for the presentation of the compiled passages, especially the connecting or explanatory portions written by the compiler himself, show that he in

---

Hippocratic theory, but relating to both parents, see n.1651); thus the head arises when the essence of the mother's bile flows (*mātuḥ pittasāraḥ sravati*), the hands when [the essence of] the mother's hands (i.e. a fluid containing this) flows (*mātuḥ pāṇiḥ sravati*), the feet when the essence of the mother's feet flows (*mātuḥ pādasāraḥ sravati*) etc. Could this theory have ultimately been derived from statements such as those relating to the *bīja-* (on which see §§2.5f. and Excursus I at the end of chapter 1)? Cf. also §5.16. In chapter four, various *padma*-s/*kamala*-s "lotuses" (i.e. the plexuses known from Tantric systems) are mentioned and connected with the tastes as well as the elements of the chain of digestive change; in this connection the *padma-* with a hundred petals situated below the navel mentioned in 4,66 is of great interest, as it is according to the commentary the "lotus" of the digestive fire (*jaṭharānalapadma-*) (on the hundred-petalled "lotus" see also DAS,R.P. 1992A, pp.410f.).

1087 Though traditional practitioners today purport to be part of the tradition as it is found in the 'classical' works and those drawing on them, their knowledge seems in actual fact to be based on (a) parallel tradition(s), for all we know maybe quite ancient (though probably greatly altered in course of time), that has/have been handed down in certain areas, schools and/or families and not necessarily written down. The general tendency to see Āyurveda as a homogenous and homogeneous mass (cf. §1.7) has not allowed this to be generally perceived, and we today have a peculiar situation in which ancient texts are taught and learnt by heart (which does not say anything about the full understanding of their contents), whereas the actual practice of traditional medicine is not necessarily based on them; but what it is actually based upon has to my knowledge not been studied in detail so far, though some relevant material is found in KAISER (we can, however, get a preliminary idea, as regards a part of South India, from ZIMMERMANN 1989A; there is though in this case too the problem of what is authentically traditional and what may have been re-introduced into the living system in recent times on the basis of the 'classical' texts). We seem to have a similar state of affairs in traditional Tibetan medicine as it is practised today.

many cases cannot have understood the passages he has culled from other sources properly.

11.4. Of the later works not falling in the above categories and which I was able to peruse, the *Cakradatta*, *Vaṅgasenasaṃhitā*, *Ārogyacintāmaṇi*, *Vaidyamanorama*, *Vaidyamanoramā*, *Kūṭamudgara*, *Rājamārtaṇḍa*, *Parahitasaṃhitā*, *Vaidyāmṛta*, *Siddhayoga*, *Cikitsākalikā* and *Cikitsātilaka* seemed not to offer anything (or anything new) bearing on our discussion. Other works, containing relevant information (often only mere scraps) have already been cited in the relevant contexts above; their names need not be repeated here. I shall now discuss other matters found in later texts and not yet touched upon.

11.5. In ŚĀR 1,5,16-17[1088] we find the well-known list (lacking in the 'classical' texts, SIDDH, KĀ and BH) of "additional elements" (☞ *upadhātu-*) in the form today regarded as being 'authoritative', apparently for the first time. According to this list, milk[1089] is the *upadhātu-* of ☞ *rasa-*, the procreatory-menstrual fluid that of blood;[1090] the other ☞ *dhātu-*s give tissue-fat/muscle-fat (☞ *vasā-*), sweat, teeth, hair and vital fluid (☞ *ojas-*) respectively.[1091] This *upadhātu-*-list seems, however, to be preceded in time by the list in KAL,Hi p.734,[1092] which states that *rasa-* gives rise to milk and

---

[1088] *stanyaṃ rajaś ca nārīṇāṃ kāle bhavati gacchati*
śuddhamāṃsabhavaḥ snehaḥ sā vasā parikīrtitā (16)
svedodantās tathā keśās tathaivaujaś (printed *tathaivojaś*) *ca saptamam*
iti dhātubhavā jñeyā ete saptopadhātavaḥ (17).

16cd is more or less the same as SU,Śā 4,13ef (see n.1114), where it, however, appears in quite a different context (see n.1969).

[1089] Maybe, since it is mentioned along with similar substances, at least originally held to be present in a woman not only at certain times; on this cf. Excursus XII at the end of chapter 10.

[1090] See also n.884; cf. too §10.30.

[1091] We have already seen that milk is connected with *rasa-* in an older theory too, according to which the procreatory-menstrual fluid, which is the same as (or gives rise to) milk, is itself a product of *rasa-* (see §§10.5f., with further references, and §10.19, also §10.9, §10.24 and §§10.27f.), whatever the exact nature of *rasa-* may be (i.e. whether it be ☞ *āhārarasa-* or not). It is interesting to note that BHĀ 1,3,206 contains a paraphrase of SU,Śā 4,24 (on which see §6.17), but that the author does not remark on the obvious implications of this passage (regarding milk and the procreatory-menstrual fluid in their relationship to each other) and the problems it creates when compared with the later *upadhātu-*-lists. Milk is derived from *rasa-* according to other BHĀ passages too; see §11.8.

[1092] KAL was probably composed in the first half of the ninth century; this is discussed in the unpublished work mentioned in n.91.

blood (in general), blood gives *rajas-* and flesh, flesh gives ☞ *sirā-*s and fat, fat gives cords (☞ *snāyu-*) and bones, bones give tissue-fat/muscle-fat (☞ *vasā-*) and marrow, marrow gives marrow and semen, the latter giving rise to the embryo. This statement,[1093] attributed to Kṣārapāṇi, an ancient medical authority,[1094] is followed by a verse referring only to the *dhātu-*s (cf. n.1095), and then by the laconic sentence that thus the *dhātu-*s and *upadhātu-*s have been described, which does lead one to conclude that the first member of each pair of products (i.e. milk, *rajas-*, *sirā-*s etc.) is meant by the latter term.[1095] The list of KAL is not in keeping with the list of ŚĀR, today accepted as the 'standard' list, except as regards milk and *rajas-*. Note also that the term *upadhātu-* is not a part of Kṣārapāṇi's statement, but of the comment of KAL on this. There is also a special problem attached to the KAL-list, for it does seem puzzling that no *upadhātu-*-list of the same highly systematised nature is found until several centuries later, i.e. in ŚĀR; even Cakrapāṇidatta (probably older than the author of ŚĀR)[1096], who uses the word *upadhātu-*,[1097] does not mention anything of a similar nature (see §11.7 below). The statement of KAL is contained in a chapter (Hi) which is an appendix to the main work, for, since the chapter immediately preceding it is expressly called a *pariśiṣṭa-* "addendum", it too must be seen as such. In contrast to all the preceding chapters (which are in verse), Hi is composed mainly in prose; it purports to be the text of a discourse of the author Ugrāditya in defense of vegetarianism (with arguments drawn mostly from medical literature) in the court of the King Nṛpatuṅgavallabha (and is in fact one of the main supports for the supposed date of KAL). Actually, a perusal of this chapter does create the impression that it is a later addition to the main

---

[1093] Note that no ☞ *mala-* "impurity" is mentioned in this connection.

[1094] On Kṣārapāṇi see MEULENBELD 1974, pp.395f.

[1095] The whole relevant passage runs:
kṣārapāṇināpy uktam. raso bhūtvā dvaidhībhavati stanyaṃ śoṇitaṃ ca. śoṇitaṃ bhūtvā dvaidhībhavati rajo māṃsaṃ ca. māṃsaṃ ca bhūtvā dvaidhībhavati sirā medaś ca. medo bhūtvā dvaidhībhavati snāyv asthi ca. asthi bhūtvā dvaidhībhavati vasā majjā ca. majjā bhūtvā dvaidhībhavati majjā caiva śukraṃ ca. śukrād garbhaḥ sambhavatīti. tathā coktam.
rasād raktaṃ tato māṃsaṃ māṃsān medaḥ pravartate
medaso 'sthi tato majjā tasyāḥ śukraṃ tataḥ prajā.
iti. evaṃ dhātūpadhātuniṣpattir āptair upadiṣṭā ... .
On the verse quoted here see n.408.

[1096] Cf. MEULENBELD 1974, pp.400ff.;428f.

[1097] Thus e.g. in his commentaries on CA,Ci 15,17 and 15,38-41 and SU,Sū 14,6.

body of the text not by the author of KAL, or at least a later revised version of something the author once may have said or written. Of course I have no proof for this statement except for a general feeling (the chapter has not yet been critically studied in this regard), and that is as a rule not much to go on, but it would tally with the fact that both lists (of ŚĀR and KAL), though they do not fully agree, are highly systematised none the less, i.e. create the impression that they are not greatly separated in time. But there is also the possibility that the passage found in ŚĀR is not an original composition there, but merely a loan from or a paraphrase of some older authority. Thus the problem remains unresolved, though it is very intriguing. But even if the list cited by KAL should be old, it could still be that the text it is imbedded in is not, and it is this text, and not the list itself, which uses the term *upadhātu-*.

11.6. But be all that as it may, it is a fact that the above lists do not tally with the probable oneness of milk and the procreatory-menstrual fluid in theories which we have already discussed (see n.1091). As it is, one very much gains the impression that these lists are attempts to explain not only a relation between milk and the procreatory-menstrual fluid no longer clearly understood, but also the problem of the relation between blood and the procreatory-menstrual fluid and the origin of the latter, at the same time maybe also serving as an answer to the problems with regard to the nature of *rasa-* and milk (also in non-medical contexts) drawn attention to by DAS,R.P. 1984A, pp.237ff. We could have, in these lists, old theories on the derivation of the procreatory-menstrual fluid at variance with that found, remarkably consistently, in the 'classical' texts, and not utilised by these, but they could also be a reaction from a later time in which the theory found in the 'classical' texts was no longer generally known or accepted.[1098] The former

[1098] It is not without interest to note in this connection that a commentator on ŚĀR 1,5,16-18, Kāśirāma (of the early seventeenth century; see MEULENBELD 1974, p.429), remarks on the dissymmetry in the developmental chain in males and females (as we too have done with regard to different substances in §8.12). His problem stems from the list of *upadhātu-*s as it is in ŚĀR: if milk and the procreatory-menstrual fluid are *upadhātu-*s, he asks, what do males have in their stead? Do they not have corresponding *upadhātu-*s? His answer:

> *apakvo vikṛtim āpanno raso yadā nirgacchati sa eva puṃsāṃ rasopadhātuḥ. vikṛtim āpanno raktaḥ yadā pittaprakopata ūrdhvaṃ gacchati sa eva raktopadhātur nānyaḥ*
>
> "When/If undigested *rasa-*, having undergone [irregular] change, exits (= appears), that indeed is the *upadhātu-* of *rasa-*. When/If blood, having undergone change, goes upwards [in the body] due to excitement of bile, that indeed is the *upadhātu-* of blood, nothing else."

The substances Kāśirāma lists are quite obviously the result of disorders and as such hardly what one would expect in the body in the regular course of events, but since ŚĀR 1,5,16 (see

would of course be more probable if we could be sure that the ascription of the KAL passage as it is to Kṣārapāṇi is correct. Unfortunately, such ascriptions need not necessarily be in keeping with the historical facts. Moreover, KAL,Hi pp.733f. deals with a matter in which the question of the *upadhātu*-s is actually irrelevant. The argument brought forward there is that eating flesh is not necessary for building up flesh in the body, as food is first broken down so as to form *rasa*-, after which other substances (including flesh) are formed; ingested flesh too having to undergo this process and thus not being treated in any way differently from other foodstuffs. In this context the verse *rasād raktaṃ tato māṃsaṃ* ... quoted in n.1095 after the statement in prose attributed to Kṣārapāṇi is of course important, for this verse seems to be very old, found as it is in various forms in a variety of texts (including SU and CA; see n.408), and can in traditional scholarly discourse thus safely be used as an *ārṣavacana*- to underpin a point. Also, the fact that it is found in a number of — also old — texts, and the fact that in KAL Kṣārapāṇi is mentioned in its vicinity, should make us consider whether it may not actually have been part of the today lost treatise of the very ancient authority Kṣārapāṇi. The intervening prose passage in KAL, from *raso bhūtvā* to *tathā coktam*, on the other hand has none of these arguments in its favour, and it is thus well worth considering whether it really could be taken to stem from Kṣārapāṇi, or even whether the author of Hi indeed actually meant it to be understood as such, for one must in this context also reckon with the possibility of the verse originally having followed directly on *kṣārapāṇināpy uktam* and the intervening prose passage, ending with a final *iti*, having been inserted later, with a corresponding addition after the verse, since the argument on flesh not being needed to build up flesh would follow directly from the verse too. I do not mean to say that things must have been so, but the whole matter does have so many peculiarities that we cannot simply neglect the possibility of their having indeed been thus. The outcome of which is that we

---

the text in n.1088) expressly states that milk and the procreatory-menstrual fluid are substances which periodically appear and disappear in the female body (though it must be emphasised that this statement is valid only on the level of ŚĀR and does not necessarily say anything on older or variant notions regarding these substances), and since Kāśirāma remarks on this, he seems to see the substances he presupposes to be peculiar to the male under the aspect of their periodicity (here then obviously pertaining to recurring appearances and disappearances all through life, which would thus hold good for the corresponding substances of the female too; cf. in this context Excursus XII at the end of chapter 10 and n.1089), though that still does not do away with the fact that they refer to abnormal conditions. Moreover it is not clear why the substances mentioned should be present only in males (see in this respect also §10.16 on *atīsāra*- and ☞ *raktapitta*-).

are not in a position to state with certainty whether the list of additional elements found in the KAL text is really old or not. This does not, however, preclude the possible existence of an ancient tradition indeed deriving the procreatory-menstrual fluid from blood or else seeing a close connection between the two (though not necessarily as part of systematised lists of derivations as in the cases we have mentioned here).

11.7. The term *upadhātu-* itself seems, however, to be old, not necessarily because of the KAL passage mentioned, but because Cakrapāṇidatta[1099] in his commentary on CA,Ci 15,17 (on this passage see §5.12), Gayadāsa in his commentary on SU,Ni 10,18[1100] and Śivadāsasena in his commentary on DR,Āh 22 (= CA,Ci 15,17) quote a list of *upadhātu*-s mentioned by Bhoja:[1101]

> *sirāsnāyurajaḥstanyatvaco gativivarjitāḥ*
> *dhātubhyaś copajāyante tasmāt ta upadhātavaḥ*
>
> "Vessels (☞ *sirā-*), cords (☞ *snāyu-*), the procreatory-menstrual fluid, milk and the skins (☞ *tvac-*), devoid of movement,[1102] arise from the elements (☞ *dhātu-*). Therefore they are additional elements (*upadhātu-*)."

This list is not systematic and does not ascribe any particular *upadhātu*-(s) to any particular *dhātu-* — it does not even mention enough *upadhātu*-s with which to build a system corresponding to the seven *dhātu*-s; as such, it is far removed from the lists of ŚĀR and KAL. Indeed, even the list of CA,Ci 15,17 (see §5.12), which seems much more archaic than these last two, shows greater systematisation, even though the CA list does not mention the word *upadhātu-*. Not only does the CA list have some substances (☞ *kaṇḍarā-* and

---

[1099] Who, as we have seen in n.440, uses the word *upadhātu-* to refer to the procreatory-menstrual fluid.

[1100] Where milk, arisen from that whose cause is "cooked" food, is called the ☞ *prasāda-* "pure part" of *rasa-*: *rasaprasādo madhuraḥ pakvāhāranimittajaḥ*. (On Gayadāsda's — and Ḍalhaṇa's — comments on *prasāda-* here see the discussion of this term in Appendix II.)

[1101] On Bhoja see MEULENBELD 1974, pp.418f.

[1102] According to Śivadāsasena, loc.cit., "devoid of movement" means that they cannot move (scil. act) to nourish/satiate other ☞ *dhātu*-s and so on:

> *atra gativivarjitā ity anena dhātvantarapoṣaṇādyā gatir niṣedhyate.*

Whether this is indeed what the verse wants to say I do not know, but even if it is not, then this statement at least makes us ponder whether *gati-* might not here refer to changing into other substances, which then according to the verse quoted the substances mentioned would not do. Unfortunately, we do not seem to have material allowing us to come to a definite conclusion.

☞ *vasā-*) in addition to those in Bhoja's,[1103] but it shows the beginning of some systematisation in that it ascribes each substance individually to one of four *dhātu*-s listed. All this seems to point to a need gradually felt to account for certain important constituents of the body which could be classified neither as *dhātu*-s nor as ☞ *mala*-s, but were felt to be in need of being made part of the system incorporating these; the attempt which looks most archaic is that of Bhoja (or, more correctly, of the verse attributed to him), followed by that of CA, and it is interesting to note that Bhoja uses the term *upadhātu-* to refer to these substances somehow connected with the *dhātu*-s, whereas CA does not, which shows that the agreement on the use and meaning of this term one finds today cannot simply be extrapolated backwards in time. This is also borne out by the use of the term by commentators; thus I may point out that Ḍalhaṇa (who uses the term *upadhātu-* several times) while paraphrasing the CA passage in his commentary on SU,Sū 14,10 (cf. n.151) uses *upadhātu-* to refer to the substances the CA passage enumerates (even saying expressly that these are formed only from the elements *rasa-* to fat and not from bones and those following),[1104] i.e. not in the sense of ŚĀR.[1105] In his commentary on SU,Sū 15,10, Ḍalhaṇa says that the passage he quotes there[1106] refers to *dhātu*-s, ☞ *mala*-s and *upadhātu*-s[1107] (even though in actual fact the passage itself says expressly that it refers only to *dhātu*-s and *mala*-s), which must refer here to *ārtava-*, sweat and milk, mentioned in the passage with no sort of derivation even hinted at! Śivadāsasena in his commentary cited in n.828 too refers to the procreatory-menstrual fluid as an *upadhātu-*, but derives it from *rasa-*, i.e. not in the same manner as ŚĀR. And in their remarks on AH,Sū 12,44 and MĀDH 22,15 respectively, Śrīdāsa and the *Madhukośa* commentary call *tvac-* "skin" an *upadhātu-* (see §10.9, includ-

---

1103 As regards °*tvacas* (°*tvaco*) in the verse attributed to Bhoja, this may refer to a plural or a singular number of "skins", since as the last member of a copulative compound *tvac-* here stands in the plural in any case. I have translated it as a plural because of the parallel CA,Ci 15,17, but must also draw attention to the passage of the *Madhukośa* commentary mentioned in §10.9 and drawn attention to below, which says that the skin (singular) as such is an *upadhātu-* of *rasa-*.

1104 *prajāyata iti kriyāpadaṃ rasādīnāṃ medaḥparyantānāṃ viśiṣṭakāryāntarotpādadarśanārtham. ebhya evopadhātavaḥ svabhāvād utpadyante na punar asthyādibhya iti. tathā hi rasāt stanyam ārtavaṃ ca raktāt kaṇḍarāḥ sirāś ca māṃsād vasātvacau medasaḥ snāyusandhī iti.*

1105 It is unclear whether Ḍalhaṇa is younger or older than the author of ŚĀR (see MEULENBELD 1974, pp.408f.;428f.).

1106 Which is given in n.515.

1107 *tatra svayonidravyāṇām avabodhārthaṃ dhātumalopadhātuṣu ślokāḥ kathyante.*

ing n.947), which is not at all in agreement with the list of ŚĀR.

11.8. It may, however, not be amiss to bear in mind that the identity and date of Bhoja is a most vexing problem. There may have been several Bhojas, older as well as younger than CA. Thus, though the Bhoja verse does create the impression of being more archaic than that of CA, we cannot simply assume that it is in fact older, for it could just as well be merely part of a variant tradition (with more archaic elements), which, though using the term *upadhātu-*, would then not necessarily be older than the CA verse. It must in this context be remarked that it is intriguing that none of the other 'classical' texts (SU, AS and AH) has anything corresponding to CA,Ci 15,17 (cf. §7.41), quite apart from the fact that none of the 'classical' texts (CA included) mentions the word *upadhātu-*, this latter clearly having caused some commentators discomfiture.[1108] I must in this connection also draw attention to PUROHIT, p.4, which states that CA has, just after Ci 15,17, another verse stating:

asthno dantās tathā majjñaḥ keśā ojaś ca saptamāt
dhātubhyaś copajāyante tasmāt ta upadhātavaḥ

"From bone teeth, further from marrow hairs, and ☞ *ojas-* from

---

[1108] Cf. for instance Ḍalhaṇa's commentary on SU,Sū 15,1, which, like 15,3 (*doṣa-dhātumalamūlaṃ hi śarīram*), mentions only the morbific entities (☞ *doṣa-*), the ☞ *dhātu-*s and the ☞ *mala-*s:

athāto doṣadhātumalakṣayavṛddhivijñānīyam adhyāyaṃ vyākhyāsyāmaḥ.

Ḍalhaṇa says that *dhātu-* here subsumes the *upadhātu-*s:

dhātavo rasādaya upadhātavaś ca.

This echoes Cakrapāṇidatta's commentary:

atra ca dhātuśabdenopadhātūnām ojasaś ca grahaṇāt tatkṣayādyabhidhānam.

Ḍalhaṇa adds further:

anye tu malaśabdāgra ādiśabdo lupto draṣṭavyaḥ. tena stanyārtavagarbhādayo 'pi dhātavo gṛhyanta iti vadanti

"Others, however, say: 'After the word *mala-* the word *ādi-* (i.e.: etc.) is to be regarded as elided/elliptically implied. Through this the *dhātu-*s milk, *ārtava-*, the embryo etc. too are taken (i.e. included or implied)'."

(Note the use of the term *dhātu-* here! Cf. on *ārtava-* p.555 in Appendix II.) Now it is true that 15,5;12 mention milk, *ārtava-* and the embryo, and 15,21ff. *ojas-*; thus there is some justification in saying that an "etc." is to be presumed as standing after *mala-*, though of course this presupposes that the text of Sū 15 is homogeneous. This is, however, something quite different from holding, as Ḍalhaṇa and Cakrapāṇidatta do, that *dhātu-* subsumes a class of substances classed as *upadhātu-*s. (It is also interesting to see that both Cakrapāṇidatta and Ḍalhaṇa seem to see no problem in the fact that their claim would also make the embryo an *upa-dhātu-*.)

> the seventh. And they originate[1109] from the ☞ *dhātu*-s, therefore they are *upadhātu*-s."

However, I have not been able to find this verse in any edition of the text, and also not in any commentary on the text; the only place where I found it[1110] was in the commentary of Āḍhamalla on ŚĀR 1,5,16-18, where it is quoted just after CA,Ci 15,17, both verses together, however, said merely to be taken from "elsewhere" (*anyatra*). Since Ḍalhaṇa too seems to know nothing about the verse quoted by PUROHIT (cf. §11.7), this verse appears to be spurious as far as CA is concerned. As it is, it would only add one more problem to the problems we already have, since the substances enumerated here are in full agreement with the list today regarded as 'authoritative' (see §11.5), and it would thus be rather peculiar if the verse immediately preceding it and listing products of the other *dhātu*-s were *not* to presuppose this list. Indeed, AS,Śā 6,p.334b[1111] makes it rather improbable that the above verse could be part of CA. Here we find the "essences" (*sāra*-) and impurities (☞ *mala*-) of the *dhātu*-s listed: the essence of *rasa*- is blood, of blood flesh (☞ *māṃsa*-), large cords (☞ *kaṇḍarā*-) and vessels (☞ *sirā*-), of flesh fat (☞ *medas*-), the skins (☞ *tvac*-) and tissue-fat/muscle-fat (☞ *vasā*-), of fat bone, cords (☞ *snāyu*-) and junctures (☞ *sandhi*-), of bone marrow (☞ *majjan*-), of marrow semen, and of semen either ☞ *ojas*-, or nothing, or the embryo.[1112] Except for *rasa*-, we find here derived from each of the *dhātu*-s up to fat, beside the following *dhātu*- of the chain, exactly the same substances derived from the *dhātu*-s up to fat in CA,Ci 15,17, but no "essences" (apart from the following *dhātu*-) of the remaining *dhātu*-s bone to semen. This passage from one of the 'classical' texts, clearly a parallel of the CA verse, shows firstly that the list of CA is most probably complete as it is, the verse PUROHIT quotes being then but an attempt to "complete" it in accordance with much later views, and, secondly, that to the author(s) of the AS there was no such thing as an *upadhātu*-, he (they) even going so far as to expressly put the other substances mentioned as "essences" on a par with the *dhātu*-s, even though

---

[1109] *upajāyante*, *upa* expressing contiguity, togetherness, and maybe also subordination.

[1110] After Gerrit Jan Meulenbeld had given me the necessary information.

[1111] On this passage see also §10.11, §10.15 and n.609.

[1112] *athānnakiṭṭam accham mūtram ghanaṃ śakṛt. rasasya sāro raktaṃ malaḥ kapho lasīkā ca. raktasya sāro māṃsaṃ kaṇḍarāḥ sirāś ca malaḥ pittam. māṃsasya sāro medas tvaco vasā ca kiṭṭaṃ karṇākṣināsikāsyaromakūpaprajananamalāḥ. medasaḥ sāro 'sthisnāyusandhayaḥ kiṭṭaṃ svedaḥ. asthnaḥ sāro majjā kiṭṭaṃ keśaromanakhāḥ. majjñas tu sāraḥ śuklaṃ malo 'kṣiviṭ tvacāṃ snehaḥ. śuklasya sāram ojaḥ. atyantaśuddhatayāsya malābhāvaḥ. anye punar ata eva tasya necchanti pākam. apare punaḥ śuklasāraṃ garbham evāmananti.*

they are not called so! Why, however, milk and the procreatory-menstrual fluid are not mentioned here is a puzzle that I am unable to find an answer to, though one could surmise that the reason is that these two substances are peculiar to women, whereas all the other substances mentioned (semen of course included) are found in both men and women. But be that as it may, this passage from AS is but further proof that the history of the usage and significance of the term *upadhātu-* is quite a confusing problem; that this confusion has actually lasted until fairly recent times is proved by BHĀ 1,3,95, which enumerates *ārtava- and* the *upadhātu*-s, thus making it very probable that the procreatory-menstrual fluid is here regarded as something not an *upadhātu-*, quite in contradistinction to not only the KAL and ŚĀR passages discussed, but also to the list of Bhoja.[1113] Moreover, in 1,3,94-98 we find substances enumerated which are obviously (like the procreatory-menstrual fluid) not *upadhātu*-s there, though they are *upadhātu*-s according to one or more of the other passages just mentioned, have an unknown status in CA,Ci 15,17, whilst AS,Śā 6,p.334b accords them a status similar to that of the *dhātu*-s, without calling them so. The problem is compounded by BHĀ 1,3,212, which lists the *upadhātu*-s in exactly the same order (and thus with the same derivation in accordance with the order of the *dhātu*--list) as the today 'authoritative' list (also found in ŚĀR) given in §11.5, and BHĀ 1,3,210f., which (like BHĀ 1,4,7 too) derive milk from *rasa-*, but are not sure whether ☞ *vasā-* is derived from flesh[1114] or from fat, thus contradicting the statement of 1,3,212.[1115] Moreover, according to BHĀ 1,3,279 the skins (☞ *tvac-*) are de-

[1113] BHĀ 1,3,94-98 state:
atha doṣāḥ pravakṣyante dhātavas tadanantaram (94cd)
āhārāder (printed *ahārāder*) *gatis tasya pariṇāmaś ca vakṣyate*
ārtavaṃ cātha dhātūnāṃ malās tadupadhātavaḥ (95)
āśayāś ca kalāś cāpi marmāṇy atha ca sandhayaḥ
śirāś ca snāyavaś cāpi dhamanyaḥ kaṇḍarās tathā (96)
randhrāṇi bhūri srotāṃsi jālaiḥ kūrcāś ca rajjavaḥ
sevanyaś cātha saṅghātāḥ sīmantāś ca tathā tvacaḥ (97)
lomāni lomakūpāś ca deha etanmayo mataḥ (98ab).

This text is also found, with numerous corruptions and variants, on the left of the head of the figure on the cover of SIMH and COMBA 1991 mentioned in §6.14. The relevant portion (2ab) of this passage corresponding to 95cd of the BHĀ passage too is corrupt: *ārttavaś cātha dhātūnāṃ malā jadrayadhātava*; however, the corruptions in 2b are easily explainable in the Nāgarī script as those of an original *malās tadupadhātavaḥ*.

[1114] The line containing this derivation = SU,Śā 4,13ef (*śuddhamāṃsasya yaḥ snehaḥ sā vasā parikīrtitā*), which latter occurs in a very different context (see n.1969, cf. also n.1088).

[1115] 1,3,210f. run:
vanitānāṃ prasūtānāṃ dhamanībhyāṃ stanau gatāt

rived not from flesh[1116] or from blood[1117],[1118] but from semen and the procreatory-menstrual fluid (just as the creamy skin derives from boiled milk, a common simile for their formation);[1119] this seems, however, just a restatement of SU,Śā 4,4,[1120] in which text, however, the list of *upadhātu*-s BHĀ presupposes is not found. According to BYT 2,85, the skins are formed from the *piṇḍa*-; it is, however, unclear whether this refers to a particular stage of the foetus in the womb called thus,[1121] or whether *piṇḍa*- is here simply a clump of matter, in which case it might refer to the embryo/foetus too, but in a more general manner (probably simply as the procreatory fluids united). But be that as it may, the descriptions mentioned pertaining to the formation of the skins can all be linked (except for those of AH and AS in n.1117), for if the notions on the formation of the skins were originally part of embryology, then this formation from the conglomerate of semen and procreatory-menstrual fluid could with a shift of focus also have been explained as a formation from what this conglomerate at some stage represents, and that is of course a clump of flesh or similar matter. This connection with embryology might not have been felt anymore later, the skins then being linked with metabolic processes in the adult too. In that case the formulation of a theory deriving them from flesh (as in the case of CA and AS,Śā 6,p.334b) would make sense if viewed as a development of the theory pertaining to the clump of matter just mentioned. But the descriptions of AH,Śā 3,8 and of

---

rasād eva hi jāyeta stanyaṃ stanayugāśayam.
śuddhamāṃsasya yaḥ snehaḥ sā vasā parikīrtitā
medasas tāpyamānasya (var: *medasaḥ sravamāṇasya*) *sneho vā kathitā vasā.*

It is most interesting to see here that milk is mentioned only in connection with women who have [recently?] delivered (*vanitānāṃ prasūtānām*) (cf. on this Excursus XII at the end of chapter 7).

1116 As in CA,Ci 15,17, AS,Śā 6,p.334b; the Bhoja verse gives no exact derivation (see §11.7).

1117 As e.g. in AH,Śā 3,8 and AS,Śā 5,p.318a.

1118 On the derivation of the skins cf. also n.1086 and n.1103.

1119 The relevant portion of 1,3,279 states:
kṣīrasya pacyamānasya yathā santānikā bhavet
pacyamānasya śukrasya rajasaś ca tathā tvacaḥ.

This verse, with the last pāda corrupt, is found in the text at the right shin of the figure on the cover of SIMH and COMBA 1991 (see §6.14).

1120 *tasya khalv evampravṛttasya śukraśoṇitasyābhipacyamānasya* (*śukraśoṇitasya* is missing in some mss) *kṣīrasyeva santānikāḥ sapta tvaco bhavanti.*

1121 On this *piṇḍa*- see e.g. SUNESON, pp.114ff., COMBA 1981, p.197. On *piṇḍa*- in Tantric texts see HEILIJGERS-SEELEN, pp.62f.

AS,Śā 5,p.318a cannot be so easily integrated in this explanation. One could opine that "blood" in this context originally meant the procreatory-menstrual fluid, but was reinterpreted subsequently to mean blood in general when the connection of the formation of the skins with the embryo was lost. Or one could see here an attempt to link the skins with something within the body of the adult when a linkage with a procreatory fluid (in the womb) was no more feasible, and blood might have suggested itself due to its similarities with the procreatory-menstrual fluid.[1122] But of course all these are musings which the evidence does not allow us to substantiate. Nevertheless, we have here a good example of the sort of problems the theories on the *upadhātu*-s can cause, and this is only one problem related to one of the substances mentioned by the texts. The whole matter of the *upadhātu*-s, which is so obviously of relevance for the procreatory-menstrual fluid too, is thus much more complicated than most modern handbooks would make us believe; note also that there seems to be *no* reference to the theory of *upadhātu*-s even in a very late text, namely ĀY. I may also draw attention to the commentary of Cakrapāṇidatta on CA,Ci 15,17, which states expressly that both milk and the procreatory-menstrual fluid (both seemingly subsumed under the term *upadhātu*- by Cakrapāṇidatta and seemingly regarded as separate substances) derive from *rasa*-,[1123] which is at direct variance to the theory found also in the ŚĀR and KAL passages cited in §11.5 deriving the procreatory-menstrual fluid from blood as an *upadhātu*-.[1124]

11.9. ŚĀR 1,6,11cd-12ab (quoted already in n.815) says the following:

[1122] One might also be tempted to see in this theory of AS and AH the remnant of a notion holding blood and the procreatory-menstrual fluid to be the same substance; blood would thus have been taken to be present in the womb and, originally probably together with semen, form the skins, this notion later being transferred to the body of the adult. But this would be, as in the case of n.169 and n.402, ultimately nothing but speculation. The idea cannot be categorically rejected, but our evidence is too meagre to build up any sort of convincing case.

[1123] Cf. the references in n.1091.

[1124] Cakrapāṇidatta says:

> *dhātūnāṃ poṣaṇam abhidhāyopadhātupoṣaṇam āha rasāt stanyam ityādi. rasāt stanyaṃ prasādajam. tathā raktam api rajaḥsaṃjñaṃ rasād eva prasādabhāgajanyam. ... etac ca rajo rasajanyam api sūkṣmatayā cireṇaiva jāyate.*

It may here be noted that in his commentary on Ci 15,19 Cakrapāṇidatta draws attention to the fact that there is a discrepancy between Sū 28,4, in which nails (*nakha*-) are classed under the impurities (☞ *mala*-) of the chain of metabolic breakdown, and Śā 7,6, where nails are said to be bones (*asthi*-), i.e. a *dhātu*-. (Cakrapāṇidatta incidentally quotes CA,Sū 28,4 as *kiṭṭāt keśanakhādayaḥ puṣyanti*; the reading of CA is much longer, so Cakrapāṇidatta has probably reduced it to its for him relevant portion.)

pācitāḥ pittatāpena rasādyā dhātavaḥ kramāt
śukratvaṃ yānti māsena tathā strīṇāṃ rajo bhavet
"Cooked (i.e. transformed) by the heat of bile (☞ *pitta-*), the elements (☞ *dhātu-*) ☞ *rasa-* etc. in succession turn to semen (lit.: go to the state of semen-hood) in a month. Likewise the *rajas-* of women comes[1125] into existence."

Here we have essentially the same as in SU,Sū 14,14, which we discussed in §§8.10ff. We have no way of knowing how Śārṅgadhara understood the verse, which is exactly as ambiguous as the corresponding prose of SU. However, we have an additional problem,[1126] for the derivation of what seems to be the procreatory-menstrual fluid from *rasa-* (even though here it seems to take place via the intermediary stages of the other *dhātu-*s) is clearly difficult to reconcile with the theory of its being an ☞ *upadhātu-* of blood, which ŚĀR itself presupposes (see §§11.5ff.), quite in contrast to SU, which derives the procreatory-menstrual fluid from *rasa-* (see §7.44), even though in this case too there are difficulties, as §§8.10ff. amply demonstrate. Kāśirāma in his commentary understands the verse to mean that *rajas-* is made in the manner of semen (*retaḥprakāreṇa*) in one month (without signifying what this "manner" is); he either sees no problem here, or is begging the question. What we also do not find is any statement on the problem that is created by the seeming parallelism of *rajas-* and *śukra-*.[1127] Āḍhamalla in his commentary too does not remark on the *upadhātu-* problem. Of relevancy for us is that he relates the opinion of "some" (*eke*) who, basing themselves on SU,Sū 14,6 (see §7.11), according to which the procreatory-menstrual fluid is derived from "*rasa-* itself" (*rasād eva*), postulate the same derivation for the procreatory-menstrual fluid and for semen.[1128] Āḍhamalla then opines that

---

1125 I see no other way of translating the optative here except by using the present indicative. On the use of the optative in general cf. e.g. RENOU, §292. Note that here metrical considerations have probably played a role too.

1126 Similar to that of TIVĀRĪ in §2.4.

1127 For a detailed discussion on this see §§8.10ff.; cf. also §9.8.

1128 This commentary states:
pittatāpena pittatejasā pācitāḥ paripācitā rasādyā dhātavaḥ krameṇa śukratvaṃ māsena yānti prāpnuvanti. rasādyā ity atra rasagrahaṇād rasa eva śukratvaṃ prāpnoti. ata eva yātīty ekavacanaṃ kṛtam. ādigrahaṇād raktamāṃsamedomajjāsthīnīti. eke rasādyā dhātavo māsena śukratvaṃ yāntīti paṭhanti vyākhyānayanti ca. teṣāṃ mate tu kāryakāraṇayor abhedopacārān na kevalaṃ śukratvaṃ strīṇām api rajo bhavet. tad uktaṃ rasād eva striyā raktaṃ ... (SU,Sū 14,6) *iti*
"The elements *rasa-* etc. cooked, [i.e.] fully cooked/brought to maturity,

the use of *tathā* "likewise" signifies that women have *śukra-* "semen", as is shown by methods causing it to flow etc., as otherwise how could women obtain [sexual] delight (*saṃharṣa-*)?[1129] And then he quotes a verse from an unidentified source:

> *saumyaṃ tvagāśrayaṃ svacchaṃ snigdhaṃ yonimukhodgamam*
> *strīṇāṃ śukraṃ na garbhāya bhaved garbhāya cārtavam*[1130]
>
> "Somic (☞ *saumya-*), having its seat (abode?; origin?) in/on the skin (*tvagāśraya-*), clear, oleaginous, arising from the ☞ *yonimukha-*, the semen (*śukra-*) of women is[1131] not for (i.e. does not bring about) an embryo; and [it is] *ārtava-* [which] is for an embryo."

11.10. Āḍhamalla seems to have allowed himself to be sidetracked here, but what he says is very interesting nonetheless. There are several difficulties in the verse quoted, notably those regarding the exact meaning of *tvagāśraya-* and of *yonimukha-*. Nevertheless, we do have here quite a vivid description of female "semen". It brings to mind the discussion on female "semen" above in chapter 8 and §§9.7f., and has many points of similarity with some of the views mentioned there. The fluid described in this verse is clearly colourless and also different from the procreatory-menstrual fluid. This description makes it look as if the *śukra-* of women here referred to is the lubricating fluid (actually: fluids; cf. n.779), though the discussion just

---

> through the heat of bile, [i.e.] the fiery energy of bile, in succession go to, [i.e.] attain, the state of semen-hood in a month. '*rasa-* etc.': here through the taking (i.e. mention) of *rasa- rasa-* itself attains semen-hood. Therefore the singular '*yāti*' (goes) is effected (i.e. used)/proper. (This seems to show that Āḍhamalla actually read *śukratvaṃ yāti* instead of *śukratvaṃ yānti*.) Through the taking (i.e. mention) of '*ādi-*' 'blood, flesh, fat, marrow and bone' [are implied]. Some recite and explain '*rasādyā dhātavo māsena śukratvaṃ yānti* (the *dhātu*-s *rasa-* etc. go to semen-hood in a month)'. (This seems to refer to an alternative reading *yānti* instead of *yāti* mentioned above.) In their opinion, however, not only semen-hood, but also the *rajas-* of women should arise, because of the carrying over/application of [the principle of] the absence of difference with regard to action/effect and cause. Thus it is said: 'The blood of a woman [is] from *rasa-* itself ...'."

What the last part obviously means is that since in SU,Sū 14,6 the procreatory-menstrual fluid is derived from *rasa-* too, therefore its derivation must be the same as that of semen.

[1129] *tatheti grahaṇāc chukratvam api strīṇām eva bhavati. yato dṛśyate hi tatsrāvaṇādibhiḥ prayogaiḥ. anyathā tāsāṃ saṃharṣaḥ kathaṃ syād iti.*

[1130] The second line (pādas c and d) of this verse has also been quoted by Cakrapāṇidatta in his commentary on SU,Sū 14,14; see n.827.

[1131] On the optative cf. n.1125.

alluded to (and in whose course we also came across a clear fluid) has shown that such an assumption is not without its problems. Still, it is very tempting to see in *yonimukhodgama-* a reference to the fluid arising from Bartholin's glands at the mouth (*mukha-*) of the vagina (*yoni-*), and in *tvagāśraya-* to the fluid exuded by the inner epithelium of the vagina;[1132] *tvagāśraya-* could even be taken to refer to the site of Bartholin's glands, and thus to that of what they discharge, too. In any case, these are fluids which are discharged steadily on arousal, irrespective of actual stimulation due to intercourse or other means. Āḍhamalla might be alluding to them, even though he may be seeing in them one and the same fluid, arising from the inner epithelium near the introitus of the vagina. This deduction is facilitated all the more by the fact that ŚṚṄG 3,29 does indeed seem to refer expressly to the lubricating fluid:

> "[I shall relate][1133] as [follows]: if the member (= penis) of men enters slowly, with distinct/bright/pleased tip,[1134] into the abode of Manmatha (= Kāma) of the lotus-faced one without pain, then gradually *retas-* flowing from the pair of openings situated at its (i.e. the abode's) margin/end (*prānta-*)[1135] makes the abode of Madana (= Kāma) wet there/in that case."[1136]

Where the two openings mentioned are situated is not quite clear. If *prānta-* means "margin", then of course we have to think of somewhere near the introitus of the vagina, either just within or just outside it (which then could eventually be taken to be a reference to the openings of Bartholin's glands). On the other hand, *prānta-* may mean "end", and that then could refer either to the end of the vagina, or to the end of the uterus, since "abode of Kāma" (or synonyms) is a common kenning for ☞ *yoni-*, which may also subsume the uterus;[1137] thus in this case the fluid would come from somewhere deep inside the vagina, in which connection one could even consider the interior of

---

[1132] See on this n.779.

[1133] Cf. 3,28: *vakṣye 'haṃ*.

[1134] This seems to refer to the glans penis being clearly visible as the result of its being pushed out from under the prepuce due to a firm erection.

[1135] I am unable to translate the *ca*.

[1136] *yathā puṃsāṃ liṅgaṃ kamalavadanāmanmathagṛhe*
prasannāgraṃ mandaṃ viśati yadi pīḍāvirahitam
tatas tasya prānte sthitavivarayugmāc ca śanakaiḥ
sravad retaḥ sāndraṃ madanasadanaṃ tatra kurute.

[1137] Cf. the discussion in Appendix II.

the uterus. The latter is, however, problematic. We by now know of the theory of discharge of procreatory-menstrual fluid into the uterus through two tubular structures (cf. e.g. §8.23), and of the theory maybe also current of an additional tubular structure for the discharge of "semen" into the uterus (cf. §9.7). The *retas*- presupposed by ŚṚṄG 3,29 may be something different from the procreatory-menstrual fluid, or but another form of this;[1138] in any case it can hardly be the procreatory-menstrual fluid serving for conception. That, however, creates difficulties in that it would either have to be discharged into the uterus via the two passages of reproductive procreatory-menstrual fluid (should no other passages be taken to open into the uterus), or else through two passages otherwise not mentioned (and also not in accord with the possible theory of three passages opening into the uterus, since in this "semen" has only one passage). Both possibilities cannot be dismissed out of hand, but it has to be admitted that they do seem to be singularly problematic, so that it is perhaps not feasible to consider an emission into the uterus. This then leaves an emission near the introitus or at the cervical end of the vagina as feasible alternatives. But wherever it may come from according to the ŚṚṄG verse, to all appearances *retas*- in it does seem to describe the lubricating fluid (actually fluids).

11.11. If we now assume that it is this fluid or these fluids which are meant by Āḍhamalla, then we run into difficulties due to the context in which he places the verse, since he says that women would have no "delight" (*saṃharṣa*-) if *śukra*- did not flow, this *śukra*- to him obviously being that described in the verse. Now *saṃharṣa*- might be taken to refer to sexual arousal here, which would fit the interpretation of *śukra*- as the lubricating fluid. Alternatively, the mention of methods causing this *śukra*- to flow so that the woman has *saṃharṣa*-, the obvious parallelism to male semen implied and the fact that we have seen above (in chapter 8 and §§9.7f.) that the "semen" of women seems to have been connected with an ejaculation too, leads one to wonder whether what Āḍhamalla (who may not necessarily be interpreting the verse correctly) means when he uses the word *saṃharṣa*- is not in fact an orgasm of the woman. Should this be the case, then Āḍhamalla's statement would be proof that there was a theory current that women ejaculate a clear fluid in connection with their orgasm; since the lubricating fluid is not necessarily linked with this or even with physical stimulation, but with arousal as such, we obviously have a problem if we regard the fluid Āḍhamalla refers to as the lubricating fluid. But the deduction that the lubricating fluid is indeed meant is actually by no means certain. Thus "arising from the ☞ *yonimukha*-",

---

[1138] Cf. on this latter also §§10.29f.

which could be taken to be a reference to the fluid of Bartholin's glands near the vaginal introitus, could simply refer to the fluid's coming out of the opening ("mouth") of the vagina or even the vulva (Āḍhamalla clearly refers to a fluid which can be seen) and not necessarily produced there, and one could explain this as referring to a fluid discharged further within the body and then flowing to the exterior. Moreover, *yonimukha-* could, as is discussed in Appendix II, refer to the uterus or its opening into the vagina, and in that case a fluid ejaculated into the uterus or arising from some part of it (such as e.g. its cervix), and then entering the vagina could be meant. As regards *tvagāśraya-* "having its seat (abode? origin?) in the skin (☞ *tvac-*)", I may draw attention to the fact that the term ☞ *kalā-* is also used for a layer of skin,[1139] i.e. ☞ *tvac-*. Could we deduce from this that some authorities used *kalā-* and *tvac-* as synonyms? If we now recall[1140] that one of the *kalā-*s, permeating the whole body, carries semen (*śukra-*), and that this must refer to both males and females, then we could arrive at quite another explanation of *tvagāśraya-*, namely one based on the parallelism of semen in the male and female body and in keeping with some of the arguments of native authorities we have already seen. The deduction would then be that Āḍhamalla is referring to semen in women, permeating the whole body as in men, which comes out (probably due to an orgasm) and is colourless. But it must be borne in mind that all this is conditional upon *tvac-* here being used in the sense of *kalā-*. We have no clear proof, though, that this is indeed the case, and that *tvac-* is not to be taken in its primary meaning "skin", then most probably as the internal epithelium of the female genital tract. We have, however, also mentioned the possibility of an ogasmic ejaculate arising from the uterine cervix. But this is actually within the vagina, and if we were to suppose that what is meant is something coming from its surface (i.e. not from within the uterus), then this would be *tvagāśraya-* as well as *yonimukhodgama-*. This explanation could even be taken to apply to the lubricating fluid(s), given the possibility that this/these is/are actually meant in the verse irrespective of what Āḍhamalla sees in the "semen" he mentions.[1141] On the other hand, even in the case of such (a) fluid(s) the explanation of *tvac-* = *kalā-* considered above might still be valid, so that what might be meant could be a discharge of the lubricating fluid(s) believed to come from further inside the genital tract, but not by means of an orgasm. In this case even (the cervical part of?) the uterus might be considered as the place of

---

1139 See the discussion on this term in Appendix II.

1140 Cf. §6.13.

1141 Cf. on this also n.779.

origin, since the verse Āḍhamalla quotes does not mention any openings (cf. against this the ŚṚṄG verse in §11.10). Indeed, one could argue that the fact that Āḍhamalla differentiates *śukra-* from *ārtava-* according to their power of fertilisation shows that the former too must somehow be associated with the uterus, otherwise there would be no point in making this statement.

11.12. However, the commentary of Śrīdāsa on AH1,Śā 3,67 (pp.124f.) leads to the deduction that we may be on the right track as regards the connection of clear, visible *śukra-* with an orgasm. Śrīdāsa repeats the argument of his and Aruṇadatta's commentaries on Śā 1,5 (see §9.8) that women too have seven ☞ *dhātu-*s and therefore *śukra-*, and then says:

> "If [it be asked:] 'what is the place/receptacle of *śukra-* (*śukra-sthāna-*)[1142] of a woman, of what sort is, moreover, the *śukra-*, or, what is its result?', [then the answer is as follows]. For it is stated in the [lost] work of Bharadvāja: 'Just as men have seven *dhātu-*s, so indeed do women too. For the *śukra-* of women [too] appears through [their sexually] delighted (*hṛṣṭa-*) mind. And when delight (*harṣa-*) is generated, the embryo arises. And the delight of women does not always arise. The abode of the[ir] *śukra-* is the ☞ *yonimukha-*.[1143] And that *śukra-*, resembling water, does not lead to the construction of an embryo because of its being somic (☞ *saumya-*). The *bīja-* (see §§2.5f.), the *ārtava-* of women, however, is fiery (☞ *āgneya-*). The *śukra-* of men is somic.'"[1144]

In the theory propounded here the "semen" of women is obviously a clear fluid, since it resembles water. The problem here, similar to that in §11.11,

---

1142 On this see §§6.13ff. and (especially with regard to women) §8.17, also §8.23.

1143 See on this statement also §6.14.

1144 This is a translation of:
kiṃ striyāḥ śuklasthānaṃ (śukla- = śukra-) kīdṛśaṃ punas tac chuklaṃ kiṃ vā tasya phalam iti cet. uktaṃ hi bhāradvājīye. yathā puruṣāṇāṃ sapta dhātavaḥ strīṇām apy evam eva. strīṇāṃ hi śuklaṃ hṛṣṭena manasā sambhavati. utpanne ca harṣe garbhaḥ sambhavati. na ca harṣaḥ strīṇāṃ sarvatropajāyate. śuklasthānaṃ yonimukham. tac ca śuklam udakaprakhyaṃ saumyatvān na garbhabandhanāyopakalpate. nārīṇāṃ tu bījam ārtavam āgneyam. puruṣāṇāṃ śuklaṃ saumyam itīti.
As regards the two *iti*s closing this passage, it is unclear whether one of them marks the end of the quotation and the other is used in the sense of "thus" ("For it is stated thus in the [lost] work of Bharadvāja:"), or whether one *iti* (the last one) marks the end of the quotation from the work mentioned, while the other one in its turn marks a quotation from yet another source. As regards Bharadvāja, see on him e.g. HĀLADĀRA, pp.120-123, also p.(44).

is whether *hṛṣṭa-* and *harṣa-* refer to arousal (i.e. "excited" and "excitement" respectively) or to an orgasm (i.e. "satisfied" and "satisfaction" respectively). In the first case the lubricating fluid could be meant, but in the second something else would probably be meant. Now here there are several points to consider. The first is the parallelism here not only with respect to male semen, but also, so it seems, the procreatory-menstrual fluid. The second is the fact that the "delight" mentioned seems to be linked with the formation of an embryo. And the third is the statement that this "delight" does not always arise. One could hold that the latter refers to a woman who is forced to have sexual intercourse even though she is not prepared for this, i.e. even though she discharges no lubricating fluid. However, the other points mentioned do rather seem to speak for an orgasm being meant, during which the female "semen" is ejaculated like male semen; this orgasm, the text points out, does not always occur, which would tally with the fact that sexual intercourse does not necessarily lead to an orgasm of the woman, far more commonly than in the case of the man. Should an orgasm indeed be meant, then because of its link with conception another deduction would be valid too, namely that an orgasm is also a pre-condition for the discharge of the procreatory-menstrual fluid (which would fit with the notion of the ejaculation of this fluid we have already discussed). We would then further have to deduce that what is implied is that a woman discharges both the procreatory-menstrual fluid and "semen" during an orgasm; however, the former would seem to be discharged internally, whereas the place of discharge of the latter is unclear. Since the text characterises the latter fluid as clear, and since there seem to be no other cases except sexual acts[1145] in which it can be seen,[1146] it would seem a valid deduction that one can actually see it when it is discharged, i.e. that it reaches the exterior of the body. That must raise the question of what the ☞ *yonimukha-* is here. It may be the vaginal orifice or the uterine orifice (cf. §11.11), but in both cases it is difficult to reconcile this with *śukrasthāna-*, unless *sthāna-* here be taken to mean not "receptacle", i.e. the place where *śukra-* is stored, but merely "place", i.e. the place where *śukra-* occurs or arises, scil. issues, in which case we might also consider *yonimukha-* to refer to the vulva, into which *śukra-* issues in the manner considered most probable with regard to *retas-* in ŚṚṄG 3,29 in §11.10. The actual "receptacle" of *śukra-* might then be, just as in the male, in some other place

[1145] Would this include masturbation? Cf. too §9.8.

[1146] The procreatory-menstrual fluid, by contrast, can of course be seen during menstruation.

in the body;[1147] the same might even be the case if *yonimukha-* should refer to the uterus as the place from which the *śukra-* issues. As it is, since in the passage of Śrīdāsa quoted the question of the procreatory power of "semen" plays a role, it would seem to follow that it must issue from the uterus, as otherwise the matter of procreation would probably not have arisen. If, however, the uterus be meant, then the question arises whether the interior or some other part of this is meant, leading ultimately to the same sort of argument as in the case of Āḍhamalla's remarks (see §11.11). If, however, the interior is meant (cf. also §8.5), then naturally we must ask why, if "semen" is discharged from this and can then flow out of the body, this does not hold true for the procreatory-menstrual fluid too.[1148] A possible answer could be that the former fluid is discharged in greater quantities than the latter, so that it does not stay within the uterus, though in that case one may also ask what prevents it from diluting and mixing with the latter. It may be that these are problems which were simply not seen, but can we be sure in this regard? And of course we also encounter difficulties when we consider §§8.19ff., for we might ask, if the clear fluid mentioned here should be the same as the "semen" considered there, whether it was indeed believed that a pregnant woman did not discharge this fluid when she had an orgasm. But here one must also raise the question of whether a pregnant woman was at all credited with being able to have an orgasm; we must also not forget the strong injunctions against having sexual intercourse during a pregnancy (though this prohibition may not have been as general as is widely supposed; cf. n.1368).

11.13. There is yet another possibility which we will have to consider. STIFTER (e.g. pp.59ff.;134f.;155f.;160;165) has drawn attention to the fact that Skene's ducts, which are today known to be responsible for the discharge of a fluid connected with the very peak of sexual excitation (see §12.16), were, and maybe still are, believed by some to open not into, but just outside the urethral opening of women. The location of the urethral opening at the upper end of the vulva could be brought into connection with *prānta-* in ŚṚṄG 3,29, which we discussed in §11.10, and thus the pair of openings referred to in this passage could be taken to refer to Skene's ducts. However, this leads to difficulties in connection with Āḍhamalla's statement in §11.9, where we find the characteristic "arising from the ☞ *yonimukha-*" (*yonimukhodgama-*). If we consider taking *yonimukha-* to refer to the opening of the vagina (see

---

[1147] Cf. the references in n.1142.

[1148] This question is relevant to all cases of postulated discharge of another fluid beside the procreatory-menstrual fluid into the uterus (also in the case of Āḍhamalla's commentary discussed in §11.11).

§11.11), then one could hold that this makes sense because this would serve to differentiate the place from which *śukra-* of the female exits from other organs within the vulva. But if we take the term to refer to the opening of the vulva, as we would have to do to correlate Āḍhamalla's statement with a possible reference to Skene's ducts, then we would have to ask ourselves why a special reference to the "mouth of the vulva" as the place where *śukra-* arises (in this case: issues from) is at all necessary in this regard, since there seems to be no other way for it to reach the exterior of the body. We would also have difficulties with the *yonimukha-* being the "abode of *śukra-*" (*śukla-sthāna-*) according to Śrīdāsa's statement quoted in §11.12. So this line of reasoning too ultimately does not lead to any clear conclusions.

11.14. In any case, the whole discussion above (beginning in §11.9), with all its problems, is valid only if what is meant here is indeed an orgasm. This does seem to be probable, but it is not certain, so that we must still go on considering the lubricating fluid. We must also always bear in mind that there seem to have been various theories with regard to female "semen" current, and that these seem not only to have been at times mutually contradictory, but that individual theories may not have given thought to all eventualities and objections, so that we in any case cannot take clarity and comprehensiveness for granted. Thus Āḍhamalla, even though the verse he quotes speaks of "somic" *śukra-* (see §11.9), need not necessarily have the same fluid in mind as Śrīdāsa has, and both Āḍhamalla and Śrīdāsa may not have the same in mind as the authors whose verses they quote. The outcome of all of which is that once again we end up with tantalising bits of information which may fit into a certain pattern, but about which we cannot be absolutely sure. We can do nothing but wait for other information to crop up, in the light of which we might be able to look at the passages of Āḍhamalla and Śrīdāsa a-new.[1149]

11.15. In connection with female "semen", attention may be drawn to the following verse from HĀ (Śā 1,p.414):

> *āmrapuṣpaṃ phalaṃ tadvad bījaṃ śukramayaṃ viduḥ*
> *strīṇāṃ rajomayaṃ reto bījāḍhyam indriyaṃ nare*
> "The flower of the mango[-tree], [and its] fruit, are like this.[1150] The seed they know as made up (= the same as)/ full of semen. The *retas-* of women is made up (= the same as)/full of *rajas-*, in the man the *indriya-* is filled with seed."

This is part of a passage (= SC 4,27-32) which says that unlike for instance

---

[1149] This will indeed be the case in §12.17 and n.1554.

[1150] For the explanation of this statement see n.1151.

plants, in which the male and female principles are already combined in the seed, in humans the male and female principles first have to be combined to form the seed.[1151] It is not quite clear whether the "seed" which is "made

---

[1151] The whole passage runs (pp.413f.; the variants of SC are given in brackets):

dṛśyate ca rajaḥ strīṇāṃ vinā yogena putraka
dṛśyate na vinā yogāt phalaṃ strīṇāṃ tu putraka
saṃśayād vismitaś citte hārītaḥ paripṛcchati

hārīta uvāca

saṃyogena vinā prājña kathaṃ garbho na jāyate
saṃyogena vinā puṣpaṃ phalaṃ vā na kathaṃ bhavet
vṛkṣavan na kathaṃ strīṇāṃ phalotpattiḥ pradṛśyate
etat pṛṣṭo mahācāryaḥ provāca ṛṣipuṅgavaḥ

ātreya (*atrir*) *uvāca*

viruddhānāṃ ca vallīnāṃ sthāvarāṇāṃ ca putraka
tatra dhātusamaṃ bījaṃ saha yogena vartate
na bhinnadṛṣṭis tasyeva (*teṣāṃ vai*) *dṛśyate śṛṇu putraka*
sthāvarāṇāṃ ca sarveṣāṃ śivaśaktimayaṃ viduḥ
niścalo 'pi śivo jñeyo vyāptiśaktir mahāmate (*vyāptaśaktimahāmate*)
tatra strīpuruṣaguṇā vartante samayogataḥ
āmrapuṣpaṃ phalaṃ tadvad bījaṃ śukramayaṃ viduḥ (not in SC)
strīṇāṃ rajomayaṃ reto bījāḍhyam indriyaṃ nare
tasmāt saṃyogataḥ putra jāyate (printed: *jātate*) *garbhasambhavaḥ*.

This passage is not quite clear in all respects (though *viruddhānāṃ*, which might appear problematic, is most probably to be explained as "[having the] opposite [characteristics from humans]"), but the gist seems to be clear. It tells us why, even though the menstrual fluid of women (*rajas-*) can be seen without any union (*yoga-*, *saṃyoga-*; for *yoga-* one could also consider "undertaking", i.e. "specific effort", but see *saha yogena vartate* in Ātreya's reply, where this meaning does not fit), the embryo (*garbha-*) or fruit (*phala-*) of women is not formed without any union, even though there is a *puṣpa-* "flower" (a common term for menstruation; cf. §2.1), whilst in trees the fruit is [formed without a union (at times even without a flower; see on this e.g. VṚK, pp.130-132, DAS,R.P. 1997, pp.212f.)]. This is because in immovable beings (clearly: as opposed to beings capable of motion) the seed (*bīja-*; on this term cf. also §§2.5f.), in which the elements (☛ *dhātu-*, here probably used for reproductive substances) are together/in equilibrium (*dhātusama-*), exists with [their] union [in it] (*saha yogena vartate*), viz. contains both Śiva and Śakti, the latter obviously referring to the male and female principles and as such highly reminiscent of 'Tantric' terminology. Thus too are the flower and fruit of the mango, i.e. the mango-fruit develops out of the flower without any sort of union. This obviously is because the male and female principles are already combined in its *bīja-* (in accordance with what has been said before), in which connection we may however ask whether *bīja-* refers to the seed as the kernel (or the like) out of which the plant develops, this (and therefore all its parts, including flowers etc.) thus having both male and female principles within it, or whether *bīja-* is actually the "germ" that develops out of the flower to give the fruit; though ultimately the outcome is the same, in the latter case the rest of the plant would not necessarily harbour the combination of male and female principles. (As regards the rest of the passage, on this see the translation above.)

up/full of semen" refers to humans, but if it does,[1152] then it would at first glance seem to refer to the male, since the *indriya-* of the male is said to be filled with "seed". What, however, does *indriya-* here refer to then? The meanings "power, might; power of the senses; virile power" do not fit here, and are, moreover, Vedic, so that it is highly debatable whether they should be considered at all. The derived meaning "faculty/organ (of sense/action)", so common in later times, too does not fit here in this general sense. But what could fit here are the derived (and also common) meanings "generative organ"[1153] and "semen"; note that according to the Hindi translation of the parallel passage SC 4,32 the *indriya-* of the man is quite clearly the penis, which is filled with *śukra-* = *bīja-*.[1154] The parallelism to *retas-* may, however, also point to *indriya-* being used in the HĀ passage as "flow", or (perhaps better) "ejaculated fluid" (which would not be too far removed, and derived from, "semen"), in the same manner as *retas-*, which can mean both "semen" and "flow". We must then, however, debate on whether the generative organ, semen or flow of the male is here referred to by *indriya-*, as well as on whether *maya-* means "made up of (= the same as)" or "full of". Now one could hold that, if *bījaṃ śukramayaṃ viduḥ* should indeed refer to the male, then *indriya-* could not refer to semen, as *bījāḍhyaṃ indriyam nare* would tell us that it is the semen which is filled with seed, i.e. the opposite of the first statement. This leads to four possible explanations of the HĀ verse with regard to the male:

(1) the generative organ (*indriya-*) is filled with seed (*bīja-*) which is

---

Śiva is here also characterised as he whose Śakti is all-pervasiveness (*vyāpti-*), even though he himself is unmoving. However, TIVĀRĪ 1,pp.104;106 (clearly not taking it as a Bahuvrīhi qualifying *śiva-*) emends *vyāptiśaktir* to *vyāptaśaktir* (which one could translate as "whose Śakti is all-pervading") and takes this to mean "permeated energy" (p.106; p.104 has *vyāpt śakti*), but in that case a better emendation would probably be *vyāptā śaktir*, i.e. Śiva is immovable, but the Śakti is all-pervading. The reading *vyāptaśaktimahāmate* of SC makes the term in question part of the vocative with which Hārīta is addressed, which spoils the whole explanation and is therefore definitely incorrect. (Note too that here we have further proof that the sexual mode of reproduction of plants through pollination etc. was unknown to the ancient Indians; cf. on this e.g. VṚK, pp.246;249ff., DAS,R.P. 1997, pp.214, also n.123 above.)

I may also draw attention to HEILIJGERS-SEELEN, p.63, where the combination of Śiva and Śakti to form the embryo (*piṇḍa-*, on which cf. n.1121) is mentioned.

[1152] For a different interpretation see §11.17.

[1153] Which is not only one of the faculties/organs of action, but also (and then of course the male is meant) a symbol of manly power.

[1154] *aur strīyoṃke rajasvalārūp vīrya hai aur puruṣ'kī iṃdriy'meṃ jo śukr hai vah bīj hai ....*

the same as semen (*śukra-*),

(2) the ejaculated fluid (*indriya-*) is filled with seed (*bīja-*) which is the same as semen (*śukra-*),

(3) the generative organ (*indriya-*) is filled with seed (*bīja-*) which is full of semen (*śukra-*),

(4) the ejaculated fluid (*indriya-*) is filled with seed (*bīja-*) which is full of semen (*śukra-*).

In the light of the discussion in §§2.5f., *bīja-* could refer to the generative fluid, and this meaning would of course fit cases (1) and (3) well. However, we actually have no guarantee that the two *bīja-*s in the verse do refer to one and the same substance. But if they do not (and according to §2.6 this possibility is always to be reckoned with), then the first *bīja-* might, for all we know, be a synonym of *indriya-*, both of which would then of course refer to the ejaculated fluid. This would lead to two more possible explanations:

(5) the ejaculated fluid (*indriya-/bīja-*) is filled with seed (*bīja-*) as well as semen (*śukra-*), and

(6) the ejaculated fluid (*indriya-/bīja-*) is filled with seed (*bīja-*) and is the same as semen (*śukra-*).

As regards option (5), the parallelism of "seed" and "semen" would seem to point to their being the same, i.e.:

(5) the ejaculated fluid (*indriya-/bīja-*) is filled with seed/semen (*bīja-/śukra-*).

But if the two *bīja-*s should not denote the same here, then there is also no necessity to assume that *bījaṃ śukramayaṃ viduḥ* refers only to the male. Should this indeed be the case, then we obtain the following possible explanations for *bījāḍhyaṃ indriyam nare:*[1155]

(A) the semen (*indriya-*) is filled with seed (*bīja-*),

(B) the ejaculated fluid (*indriya-*) is filled with seed/semen (*bīja-*).

Options (A) and (B) would, however, say the same as options (6) and (5) respectively, so that ultimately we are left with the options (1) to (6). However, the parallel construction of *strīṇāṃ rajomayaṃ reto* and *bījāḍhyam indriyaṃ nare* does let us wonder if *indriya-* can really refer to something not a fluid, i.e. if the options (1) and (3) are at all viable (cf. too n.1155). Moreover, the correspondence of *bījāḍhya-* to *rajomaya-* also raises the question of whether *āḍhya-* and *maya-* may not be semantic equivalents not primarily in the meaning *āḍhya-* has, but in that which *maya-* has, i.e. whether *āḍhya-* could be seen here as a variant of *maya-* (and not *maya-* as one of

---

[1155] In this case there would be no need to consider *indriya-* "generative organ", which the parallel use of *retas-* makes problematic anyway.

āḍhya-), in which case one could, even though our dictionaries do not seem to list such a meaning, also consider the meaning "made up of (= the same as)" for *āḍhya-*. Applying this to the options above, we would then deduce that the ejaculated fluid is here the same as semen (*bīja-*, *śukra-* and *indriya-* all referring to this, or else only *bīja-* and *śukra-*, if *indriya-* should refer to the generative organ, which, however, seems problematic).

11.16. But if *bījaṃ śukramayaṃ viduḥ* does not refer to the male, then, since this option was considered only due to the mention of *bīja-* twice, and since here neither male nor female is mentioned, whereas in the following line both expressly are, it seems to follow that we would here have a reference to both male and female. From this it would then, however, also follow that *bīja-* is used for the ejaculated fluid of both, and that *śukra-* is used for females too in a sense similar to that discussed in §9.8 (where *śukra-* could be = *ārtava-*), either as a synonym of *bīja-*, or else as something within the *bīja-*. Similar considerations apply to the statement expressly referring to women: the *retas-* may be the same as *rajas-*, or else it may contain *rajas-*, *retas-* either meaning "semen" (on which then cf. *śukra-* just mentioned) or simply "flow" (in this case internal, or at any rate with an internal component; this would also apply to *bīja-*, should this really refer to women too). All this brings to mind the problems with regard to the different meanings of *bīja-* discussed in §§2.5f., especially the problem of whether the *bīja-* might not be something within the procreatory fluid, or better: the procreatory part of the fluid emitted, though if we apply this to our case here we seem then to have to deduce that in *bījaṃ śukramayaṃ viduḥ* the *bīja-* itself has a procreatory part (i.e. is not itself this procreatory part), and thus have a problem with regard to the terminology employed. As regards the fluid of the female, which is after all our main subject of study here, we would then be faced with the question of whether *rajas-* might not be the procreatory part of the fluid emitted (i.e. not the whole fluid). That this is problematic goes without saying, as such a notion seems not to have played a role in the material we have sifted so far, except maybe in KAL 23,18 discussed in §7.7, in which case, however, this notion followed from only one of several equally probable, and different, possibilities considered. It might, however, be of relevance in Tantrism; I myself have drawn attention to problems which might be connected with similar notions in the context of a study of a modern syncretistic religious group.[1156] Unfortunately, I know of no other relevant studies, and so can offer no new material here. But that a Tantric connection should at least be considered follows also from the remarks in n.1151. However, even

---

1156 See DAS,R.P. 1992A, pp.402ff., also p.413.

apart from the question of any possible Tantric connection, the possibility that the verse quoted could be referring to fluids of both male and female of which not the whole, but only a certain part is procreatory needs to be considered. It must be emphasized, though, that this is only one of several possibilities discussed above, that all these possibilities may be leading us onto the wrong track, and that they are based on a text which is not only (like most Indian medical texts) still in need of critical editing, but also obviously in bad textual shape. These problems become really acute when we consider the fluid not only of the female, but also of the male, the discussion above having shown how absolutely confusing the matter is; particularly difficult in this regard is the question of which of the different terms used with reference to the male (or the male and the female) would refer to the ejaculatory fluid, and which to its procreatory part (note especially the difficulty occasioned by the two *bīja*-s.[1157] And, moreover, there may actually be no need at all to postulate such a procreatory part of the ejaculatory fluid, as we have seen above that the various terms used may also be explained as being synonyms; thus all the remarks on a procreatory part of the fluid emitted must be seen as *merely going through various possibilities*. This is, however, all I can offer by way of explanation for the present. But it may in this connection (as regards the woman) be interesting to draw attention to a problem already touched upon, namely that of *strīretorajas-* and *strīretorakta-* in Aruṇadatta's and Śrīdāsa's commentaries on AH,Śā 1,5 in §9.8; here too we have the problem of whether we have two different fluids or one fluid, and how, if different fluids are meant, they might be connected. But in this case too we do not yet have any further information; we shall return to it in §11.26, though.

11.17. We may, however, also consider an alternative explanation for the problematic verse HĀ,Śā 1,p.414 reproduced in §11.15, namely that *bījaṃ śukramayaṃ viduḥ* does not refer to humans, but to the mango[-tree] (*āmra-*) mentioned immediately before this, whose flower and fruit are given as examples for the reproduction due to the inherent combination of the male and the female principles in the immovable. Then

---

[1157] Attention may also be drawn to the discussion in §2.5 on the temptation to translate *bija-* with "sperm", which would however be highly problematic not only because there seems to be no evidence to assume that there was a theory postulating the existence of spermatozoa or something similar (cf. especially n.54), but also because that would obviously lead to problems as regards the female procreatory fluid. (In the latter case translating *bīja-* with "ovum", in the manner of modern apologists arguing as detailed in §1.7, would be no help, as then we would have to assume that this "ovum" is part of a fluid similar to semen, and would moreover, as detailed in §11.18, lead to contradictions with modern views on reproduction, and that is the last thing a modern apologist would want!)

āmrapuṣpaṃ phalaṃ tadvad bījaṃ śukramayaṃ viduḥ
could be translated as:
"The flower of the mango[-tree], the fruit, [and] the seed they know as like this (i.e. in this manner) made up/full of semen."
One could, however, also retain the translation of §11.15:
"The flower of the mango[-tree], [and its] fruit, are like this. The seed they know as made up/full of semen."
The two differing translations would reflect the different considerations on *bīja-* in n.1151. If, however, *bīja-* should refer to the seed (whatever "seed" may actually imply) of the mango, then we would seem to have a contradiction to what this example is supposed to demonstrate, since according to the context we would expect the seed to have not only semen (*śukra-*), but also a female procreatory part. One could, however, counter that, since an asexual mode of reproduction is presupposed here (cf. n.1151), therefore the *bīja-* would be full of a female procreatory substance anyway, so that what would merit special mention would be the presence of a male procreatory substance in it too. But it must be admitted that all this requires quite some amount of explaining, and thus is not fully convincing; *bījaṃ śukramayaṃ viduḥ* thus *might* refer to the preceding *āmra-*, but then again it might not, in which case we cannot but explain it as pertaining to humans, with all the accompanying problems discussed in §§11.15f.[1158] In this context it is interesting to see that TIVĀRĪ 1,p.105,n.1 reproduces the HĀ passage given in n.1151 *without* the problematic *āmrapuṣpaṃ phalaṃ tadvad bījaṃ śukramayaṃ viduḥ*,[1159] which is also how SC gives the passage (see n.1151). This quite clearly saves the authoress a lot of bother, but is not a course we can simply follow.

11.18. However, HĀ contains more interesting information. Śā 1,p.415

[1158] One might also consider seeing in the verse an agricultural comparison found in many cultures all over the world, namely that of the male semen being the actual seed out of which the child grows, the woman only serving as the nourisher, like the earth. The verse would then be saying: the ejaculated fluid (*indriya-*) of the male is full of seed (*bīja-*), which is the same as/full of semen (*śukra-*), whereas the (internal) flow of the female is the same as/full of *rajas-*, the latter then representing the earth (cf. *rajas-* "dust"). This explanation does not, however, seem feasible, not only because everything we have seen until now seems to presuppose a formation of the child out of a *combination* of substances from both father and mother, but also because the HĀ passage of which this is a part expressly remarks on the need to combine the male and female principles (see n.1151).

[1159] Also without the preceding *tatra strīpuruṣaguṇā vartante samayogataḥ*. It is quite clear that the authoress has cited the text selectively. However, since I had no access to the edition she used I do not know whether the particular line in question too has been left out by her, or was not found in the edition she consulted.

has a peculiar pair of verses:

> *narareto'dhikatvena tathā śukrādhikena tu*
> *hīnarasendriyair vāpi jāyate puruṣādhikaḥ*
> *strīretaso 'dhikatvena*[1160] *hīnaśukrendriyād api*
> *rajaso 'py*[1161] *adhikatvena strīsambhūtiḥ prajāyate*
>
> "Due to an excess of male *retas-*, but likewise due to a surplus of semen, or else due to insufficient/weaker *rasendriya-*, a surplus of males[1162] arises. Due to an excess of female *retas-*, also due to insufficient/weaker *śukrendriya-*, also due to an excess of *rajas-*, an increase/a production of females is produced."

Part of this passage gives us the known theory (see §1.5) of males being born due to an excess of semen (here *śukra-*), females due to an excess of the procreatory-menstrual fluid (here *rajas-*). But we also learn that an excess of male *retas-* (*nararetas-*) and female *retas-* (*strīretas-*) respectively have the same result. What are these? The use of *retas-* would imply that they are some sort of fluid, but the way they are mentioned here does not make it seem as if they were identical with *śukra-* and *rajas-* respectively. From the discussion in §§11.15f. we could maybe deduce that *nararetas-* is a fluid containing *śukra-*, *strīretas-* a fluid containing *rajas-*. It goes without saying that this creates problems. A scholar arguing in the manner we have drawn attention to in §1.7 would now probably say that there is actually no problem, as *śukra-* denotes spermatozoa and *rajas-* the ovum here, but that of course would not solve any problems at all, particularly with regard to the female, since it would not explain what the *strīretas-* is, and moreover would have to assume that in actual fact the female sex of a child is determined by a quantitative excess of the ovum, which is patently absurd.[1163] Moreover, since

---

[1160] Printed as *strīretasādhikatvena*.

[1161] Printed as *rajasāpy*.

[1162] *puruṣādhikaḥ* is problematic, as the noun *adhika-* is to my knowledge not masculine, but neuter, so that we would have expected *puruṣādhikam*. Or do we here have a compound "[that which is] excessive as regards males" or "[that] with respect to which the surplus is males (i.e. having an excess of males)" qualifying some masculine noun meaning "progeny" or the like which we have to supply? Since this is very problematic, I suspect that we have here a mistake for *puruṣādhikam* or maybe *puruṣādhikyam* "surplus/excess of males", and have translated accordingly. What could, however, also be possible is *jāyate puruṣo 'dhikaḥ* "a male is born more numerously (i.e. more often)". But be that as it may, what is intended seems clear, even though the actual translation poses problems.

[1163] See also n.1157.

according to the verse *nararetas- and śukra-* and *strīretas- and rajas-* seem to be capable of bringing about conception, a fluid composed of one member of each pair and having the other member of that pair as a constituent, procreatory part, does not make sense, unless one assumes that there are two ideas at work here, one seeing only the absolute quantities of the procreatory parts of male and female, the other assuming that if the quantity of one aggregate of procreatory and non-procreatory parts is larger, the procreatory part of that aggregate would automatically be larger than that of the other aggregate too. It is clear that all this is a bit difficult to swallow. Nor can we simply equate *strīretas-* and *rajas-* with female "semen" and the procreatory-menstrual fluid respectively (or even the procreatory-menstrual fluid and female "semen" respectively), since what about *nararetas-* as distinct from *śukra-* then? The same difficulty would crop up if we were to differentiate between the procreatory-menstrual fluid as such and that part that is procreatory under certain conditions (remains in the body after menstruation?).[1164] As it is, it does not look as if we could for the present come to grips with these problems except by being wildly speculative, which is why it seems most prudent to shelve them until more relevant material crops up (if, that is, such material can be found at all).

11.19. But there is yet another problem in the passage just quoted, namely the meaning of the terms *rasendriya-* and *śukrendriya-*. As for *indriya-*, the same chronological considerations (regarding Vedic usage) of course apply as in §11.15. Nevertheless, we could still consider taking *indriya-* to mean "power, might" here, which would mean that a weakness (obviously relative) of the power or might of (or: power/might whose manifestation is) *rasa-* or *śukra-* results in a male or female child respectively. This would, however, be very problematic, as it would place *rasa-* of the female on a par with *śukra-* of the male; whether *rasa-* be the ☞ *dhātu- rasa-* or some other fluid within the body, it is obvious that we end up with a terminological problem here. We would also end up with a similar problem if we took *indriya-* to mean "virile power" and *rasendriya-* and *śukrendriya-* to mean either "virile power of *rasa-*" and "virile power of *śukra-*" or "*rasa-* and virile power" and "*śukra-* and virile power" respectively; similar would be the case should we consider *indriya-* to have the meaning "flow" (i.e. "ejaculated fluid", cf. §11.15), with *rasendriya-* meaning "flow of/with *rasa-*" and *śukrendriya-* meaning "flow of/with *śukra-*". The problem of *rasa-* cannot simply be explained away by substituting

---

1164 In this case we would have the same fluid in different forms, and not a fluid made up of different constituents. Cf. in this regard also the differentiation of *ārtava-* and *rajas-* described in §§8.15f.

raja for *rasa* in *hīnarasendriyair* (i.e. assuming *hīnarajendriyair*), as TIVĀRĪ 1,pp.171-173 does, for *raja-* as an alternative form of *rajas-* is, according to our dictionaries, attested only lexically. True, we do have the compounds *vi-raja-*, *araja-* and *nīraja-* (in which *rajas-* more often than not stands for "dust" or "passion" or the like),[1165] but these are Bahuvrīhi compounds in which special factors regulate the loss of the final *s* (cf. e.g. AIGR II,1,§41bα) and as such cannot simply serve to explain a similar postulated loss in the simplex. It is methodically extremely unsound to base an emendation, especially one of such consequence, on such flimsy evidence, just because one has difficulties in understanding a text as it is. *rajas-* on the other hand would not fit metrically (*hīnarajaïndriyair*), unless one were to assume an incorrect sandhi (*hīnarajendriyair*), which would then give us the same form as TIVĀRĪ postulates, but in this case as a *wrong* form, which does not make arriving at her emendation via this line of reasoning methodically any sounder either; even though the dictionaries do list sporadic attestations of such sandhi forms of *rajas-* (cf. also n.1322), we cannot simply replace a to us difficult word form with another one which is actually wrong grammatically.[1166] So is *rasa-* of the female indeed an exact correspondence to *śukra-* of the male, in which case it would refer to the procreatory-menstrual fluid? One could in this case point to parallels in New Indo-Aryan, in which *ras* may, for instance, refer to the menstrual fluid and to semen in certain esoteric terminology[1167] (cf. the Tantric influences referred to in n.1151). If, however, *rasa-* should indeed refer to the procreatory-menstrual fluid and *rasendriya-* thus mean either "flow of (scil. ejaculated fluid which is) *rasa-*" or "*rasa-* and virile power" or even "virile power of *rasa-*"[1168] (*śukrendriya-* would be similar, with *śukra-*

1165 Apart from what our dictionaries list, I also found the compounded form *araja-* "one whose menstrual fluid is not [yet visible]" (here as a feminine accusative: *arajām*) in a passage from the *Gṛhyasaṅgraha* (not available to me) cited by THIEME 1971, pp.433f. Cf. also the name *Arajā-* (see e.g. RAY, p.25), as well as the denominative √*arajāy* "to become dustless/ menstruation-less" (seemingly only lexically attested).

1166 Cases of removal of correct hiatus do sporadically occur in medical texts, and I myself have remarked on this (DAS,R.P. 1990, pp.56f.), but these are anomalies, which even the commentators have as a rule felt they had to remark upon. Moreover, these are anomalies of the texts themselves, not emendations, and that is a very great difference.

1167 See e.g. DAS,R.P. 1992A, p.391,n.22. There may quite possibly be parallels, of which I know nothing, in Old and Middle Indo-Aryan literature too for *rasa-* referring to a fluid of the woman. In this connection attention may also be drawn to the use of *rasa-* listed in our Sanskrit dictionaries to sometimes describe male seminal fluid, the best known example being the name *rasa-* for mercury, explained as Śiva's seed, in alchemical texts.

1168 See above; "virile" would probably have to be understood as "procreatory" here.

for *rasa-*), then the problem of the meaning of *hīna-* crops up. In §11.18 we have considered both "insufficient" and "weaker". The former seems to make better sense, since it refers to the known theory of the relative quantities of the procreatory fluids determining the sex of the child, whereas the latter would refer to the relative strength or power of the fluids, which is something quite different, and apparently creates difficulties (this would also apply to "insufficient virile power of *rasa-/śukra-*", for this too would be a statement on a difference not relating to the quantity of the fluids concerned).

11.20. But as a matter of fact the notion of the relative strength or power of the procreatory fluids being a factor determining the sex of the child is not only attested, but even expressly commented upon — albeit in an as yet unpublished text. NYC, commenting on SU,Śā 3,12, says (fol.29v):

> *śukrādhikyaṃ vīryato mānato 'pi. ubhayathāpi bāhulyam abhipretam. tad uktam.*
>
> > *bahutvād vīryavattvāc ca pumān śukrā[t] sañjāyate*
> > *alpadurbalaśukratvān nārī syād iti gotamaḥ*
>
> "Excess of semen is by strength/potency, by measure too. In *both* cases abundance (viz. excess over procreatory-menstrual fluid) is meant. That/Thus is said:
>
> > "Due to [its] abundance and due to [its] being potent/strong, a man is born of semen. Due to semen being scant and weak a woman is born, thus [says] Gotama[1169]."

Cf. also KOHLBRUGGE, p.42(no.20), BṚSAṂ 67,15[1170] (with a parallel in Utpala's commentary); according to these a man with thin (*tanu-*) semen begets daughters. However, it is unclear whether *tanu-* "thin" here refers to the consistency of the semen (which would be a qualitative statement) or the thickness of the stream (which would be a quantitative statement). But be that as it may, we still have the statement of NYC, and this could point to our being on the right track if we take *hīna-* in HĀ,Śā 1,p.415 to stand for "weaker". We would, however, have a parallel to what NYC says only if *rasa-* in this verse does indeed stand for the procreatory-menstrual fluid, something which, though not impossible, is unfortunately not yet substantiated by any sort of firm evidence, though I must also draw attention to a possible similar usage described in n.1182. But actually *rasa-* and *śukra-* need not necessarily be exact correspondences here if we were to take *indriya-* to mean "facul-

---

[1169] Probably the same as Gautama, on whom see MEULENBELD 1974, p.400.

[1170] In most other editions 68,15.

ty/organ". The *śukrendriya-* would then be the "faculty/organ of *śukra-* (i.e. semen)", or, in other words, the male generative organ. Now semen is a fluid which the male visibly ejaculates and whose connection with procreation is obvious; the female does not have any easily discernible fluid of exactly the same nature (which has, as we have seen several times above, caused ancient Indian medical authorities a great many problems). But if *indriya-* should indeed mean "faculty/organ", then there is no need to take *rasa-* here to be on a par with *śukra-*, as we would have to assume in the case of the other translations of *indriya-*, for *rasa-* here could mean "delight",[1171] i.e. the *rasendriya-* would be the "faculty/organ of delight", that is, the female generative organ.[1172] *indriya-* "generative organ" too could be considered, as it would ultimately mean the same ("generative organ of *śukra-*" and "generative organ of delight"). What *hīna-* "insufficient; weaker" in this connection should, however, mean, is not quite clear. One could consider the relative sizes of the organs, or their relative position with respect to each other (e.g. above or below each other), or else their action, i.e. the relative vehemence of their use, or the vehemence of the result achieved (e.g. an orgasm or the like; this would also have to be considered in the case of the translation "weaker/insufficient *rasa-*/semen- and virile power" considered in §11.19). However, we see that this explanation seems to be an invitation to unsubstantiated speculation, which does make it seem better to break off the discussion here. But in any case one thing remains clear: should *rasendriya-* and *śukrendriya-* refer to procreatory fluids, then we would have an insufficiency (qualitative or quantitative) of *rasendriya-* having the same effect as an excess of *nararetas-* and of *śukra-*, and an insufficiency (qualitative or quantitative) of *śukrendriya-* having the same effect as an excess of *strīretas-* and *rajas-*, from which it could follow that *śukrendriya-* is the same as either *nararetas-* or *śukra-* (which seem not to be the same), *rasendriya-* as *strīretas-* or *rajas-* (which too seem not to be the same). If only *rasa-* and *śukra-* in *rasendriya-* and *śukrendriya-* should refer to procreatory fluids, then we would have to reason similarly; but since *śukra-* in *śukrendriya-* would then in all probability be the same as *śukra-* in excess, and not *nararetas-* (should this be different from *śukra-*), *rasa-* would probably be the same as *rajas-* and not as *strīretas-* (should this be different from *rajas-*). All this is of course conditional upon

---

1171 One could also consider the meaning "fluid", but then one would have difficulties explaining why especially the fluid(s) discharged by females during intercourse should be termed so and not that discharged by the male.

1172 The description of the female generative organ as a seat of delight (especially from the point of view of the man) is a commonplace, not only in India.

rasa- or *rasendriya-* indeed referring to a procreatory fluid. Ultimately — and frustratingly — we have in this case too no evidence which would allow us to come to a definite conclusion regarding the possible correspondences, as also not as regards the exact meaning of the terms *rasendriya-* and *śukrendriya-* (we will return to this passage in §13.10 and §13.22).

11.21. Let us now turn to one of the youngest medical texts, BHĀ (composed about the middle of the sixteenth century), which is at the same time one of the most interesting, as it gives the impression, at least as regards the parts of especial interest to us, of being a sort of compendium of different views of old authorities somehow sought to be united. Of course other texts too show such syncretistic tendencies, but BHĀ seems to have been composed primarily to serve as a sort of encyclopaedia of medical knowledge and not, as most of the other later medical texts, to explain medical subject matter, general or special, in a new and more up-to-date manner (usually as regards the method of presentation or arrangement) and incorporating a few novelties which the authors obviously felt were improvements over what their predecessors had done, at the same time leaving out much that seemed not to fit into the system as envisaged (cf. also §11.3). Thus BHĀ here and there presents us with snippets of information which mediaeval texts of the sort just mentioned cannot by their very nature be expected to have. I have already quoted BHĀ several times, but shall now do so in greater detail.[1173]

11.22. BHĀ 1,3,187 states that the gross (*sthūla-*) part of *rasa-* turns to semen in men in a month, in women, however, to semen *and ārtava-*.[1174]

[1173] ĀY too is an encyclopaedic work (a few decades younger than BHĀ) containing much interesting information, but this is of a sort which I have found it more convenient to discuss in connection with other matters throughout this study, whereas BHĀ throws up some special problems which to me seemed to merit separate discussion.

[1174] This quite obviously is to be connected with 1,3,177-181, which tell us that ☞ *(āhāra)-rasa-*, while nourishing/satiating the other ☞ *dhātu-*s of the body (cf. on this §§7.37f.), divides into three parts, a gross (*sthūla-*) part nourishing a *dhātu-*, a fine (*sūkṣma-*) part nourishing the next *dhātu-* of the chain, and a part left over which is the corresponding ☞ *mala-*. With respect to the next *dhātu-*, the same happens, i.e. the *(āhāra)rasa-* again divides into three parts, and so on. This theory of tripartition is quite common in medical commentaries and in later medical texts, either in the above form, or else saying that *rasa-* changes (and not merely nourishes/satiates) to the other *dhātu-*s in succession in the manner described. Incidentally, BHĀ 1,3,177 also contains a verse which it ascribes to Caraka:

rasas tu tatra tatra tridhā vibhajyate. uktaṃ ca carake.
sthūlāḥ sūkṣmās tanmalāś ca tatra tatra tridhā rasaḥ
svaṃ sthūlo 'ṃśaḥ paraṃ sūkṣmas tanmalo yāti tanmalam.

But as far as I could ascertain this verse is not found in CA. Whether BHĀ made use of a dif-

Then SU,Sū 14,14, relating to the formation of semen and *ārtava-* from *rasa-* in a month, is quoted (on this see §§8.10ff.), followed by the explanation that this means that women too have semen.[1175] The difficulties this Su passage has caused the native commentators have been described above, especially in §8.15. BHĀ apparently adds another difficulty, for the edition lists a variant which has *strīṇāṃ ceti* instead of *strīṇām ārtavaṃ ceti* (cf. the text in n.1175), according to which the passage would be saying that in both males and females *rasa-* finally turns to semen, without mentioning *ārtava-*. However, this is probably just a (scribal?) mistake, since the statement with which the author of BHĀ has prefaced the SU passage would be meaningless if the passage did not mention *ārtava-*, so that it will probably not be wrong not to attach too much importance to this variant. The BHĀ text then quotes AS,Śā 1,p.286a[1176] (given in §9.7) with the variant mentioned in n.893, but as-

---

ferent redaction of the text in which these lines were included, or whether this is a plain mistake, I cannot tell, though the absence (as far as I could ascertain) of this theory of tripartition in older texts in general could speak for the latter. Maybe the quotation is from a commentary on CA, but ascribed to the text commented upon, as is often the case (it bears remembering that usually quotations seem to have been made from memory, not from written texts, which were not easily obtainable before the introduction of the printing press). Ḍalhaṇa says in his commentary on SU,Sū 14,10:

pūrvokta evārthaḥ ślokābhyāṃ kathyate.
sthūlāṇvaṃśamalaiḥ sarve bhidyante dhātavas tridhā
svaḥ sthūlo 'ṃśaḥ paraṃ sūkṣmas tanmalaṃ yāti tanmalaḥ
svāgnibhiḥ pacyamāneṣu malaḥ ṣaṭsu rasādiṣu
na śukre pacyamāne 'pi hemanīvākṣaye malaḥ
iti.

Ḍalhaṇa quotes from an anonymous source. The editor of SU in a footnote asks us to compare CA,Ci 15,18-19ab, which list the ☞ *mala-*s; is this his own idea, or is he here following some explanatory tradition linking the verses quoted by Ḍalhaṇa with the CA verses? If the latter, then could the author of BHĀ have been following some similar tradition in which the ascriptions were mixed up? It is interesting to see too that ĀY 4,44 quotes the first verse (*sthūlāṇvaṃśamalaiḥ ... tanmalaḥ*) of Ḍalhaṇa's quotation (with *tanmalaḥ* for *tanmalaṃ*) and also ascribes it to Caraka (*tatra carakaḥ*); the editor of ĀY expressly remarks on the nonexistence of this verse in the known editions of CA (see p.425,n.23) and also says that it is found in Ḍalhaṇa's commentary (from which he quotes it with *sūkṣmo viṇmalaṃ* for *sūkṣmas tanmalaṃ*).

1175 The BHĀ text runs:
tataḥ sthūlo bhāgo raso māsena puṃsāṃ śukraṃ strīṇāṃ tv ārtavaṃ śukraṃ ca bhavati. uktaṃ ca suśrute. evaṃ māsena rasaḥ śukro bhavati strīṇām ārtavaṃ ceti. cakārāt strīṇām api śukraṃ bhavati.

1176 Which tells us that a woman emits "semen" during intercourse, but that this is of no consequence for the formation of the embryo.

cribing the whole to Suśruta, and adding the explanation that the embryo to which the verse refers (see n.1176) is a "pure/genuine" (*śuddha-*) embryo, whereas an unnatural (*vikṛta-*) embryo can be formed from the "semen" of women, after which follows SU,Śā 2,47[1177] (quoted without any ascription) with the variant listed in n.795. 1,3,188 then says:

> "And thus [too is said]:
>
> > 'The *ārtava-* of women should be [considered] agreed upon by all as appropriate/serving for the embryo. *śukra-* indeed occasions their strength, colour and nourishment/growth.[1178]'
>
> Through this [the deduction] 'the seventh *dhātu-* of women is *ārtava-*, the eighth *śukra-*' is made known. [This excess of a *dhātu-* is just] like the excess of a receptacle[1179] (i.e. the womb)."[1180]

After this we have, in 1,3,189, SU,Sū 14,10, describing the metabolic chain of ☞ *dhātu-*s. with the reading given in n.408 (and with no source being stated), followed by a summarising statement saying that thus *rasa-* turns to semen and *ārtava-* in a month.[1181] Unfortunately, all this does not tell us how exactly *ārtava-*, i.e. the procreatory-menstrual fluid, is formed; do we have a parallel development of semen and the procreatory-menstrual fluid, or is the latter formed from the former?[1182] In this context attention may also be

---

1177 Referring to two women begetting a child together.

1178 In the same way, it seems, as it does in men; cf. e.g. Narasiṃha's commentary on RAS 1,6,p.11.

1179 On the receptacles in the body cf. n.581.

1180 This is the translation of:
evaṃ ca.
strīṇāṃ garbhopayogi syād ārtavaṃ sarvasammatam
tāsām api balaṃ varṇaṃ śukraṃ puṣṭiṃ karoti hi.
etena strīṇāṃ saptamo dhātur ārtavaṃ śukram aṣṭamam iti bodhitam. āśāyādhikyavat.

1181 This part runs:
evaṃ rasa eva kedārakulyānyāyena sarvān dhātūn pūrayan māsena navadaṇḍottareṇa śukram ārtavaṃ bhavatīti siddhāntaḥ.

1182 I would here also like to draw attention to the extremely problematic
kāmāt sañjāyate śukraṃ śukrāt sañjāyate rajaḥ
"from *kāma-* semen arises, from semen *rajas-* arises"
from the *Mahābhārata*, which MEYER,J.J. 1952, p.362,n.1 quotes and discusses; see on this also O'FLAHERTY, p.34. The matter is complicated by the fact that MBH 14,24,7ab cites the line in question as:

drawn to the line introducing 1,3,204 (≈ SU,Sū 14,6; cf. n.415) and 1,3,205 (≈ SU,Śā 3,10; cf. n.395), namely:

> *strīṇāṃ rasa eva māsenārtavaṃ bhavatīty uktvā punar āha suśruta eva,*[1183]

of which we have a variant *śukrata* for *suśruta*, which the editor asks us specially to consider;[1184] according to this variant *ārtava-* would arise from semen. The question of course is whether this 'variant' or whether the reading printed as 'correct' is in actual fact the original reading; however, if the original reading of BHĀ should indeed be *śukrata*, then this would only tell us that the author of this line regarded the verses of SU quoted as saying that *ārtava-* derives from semen. But as even that is not sure, we ultimately end up basically with the same problem of apparent parallelity of male semen and the procreatory-menstrual fluid as in §11.9, which here is compounded by the mention of "semen" of women which has the power to create an embryo, and maybe even by a derivation of the procreatory-menstrual fluid from semen.[1185] Added to this is the problem of the exact translation of *śukram*

> *kāmāt sañjāyate śukraṃ kāmāt sañjāyate rasaḥ*
> "from *kāma-* semen arises, from *kāma- rasa-* arises",

listing as a variant not only the line as quoted above, but also as:

> *kāmāt sañjāyate śukraṃ kāmāt sañjāyate rajaḥ*
> "from *kāma-* semen arises, from *kāma- rajas-* arises",

and as:

> *kāmāt sañjāyate śukraṃ śukrāt sañjāyate rasaḥ*
> "from *kāma-* semen arises, from semen *rasa-* arises".

Not only do we have the problem of different meanings of *rajas-* here (on the one hand "procreatory-menstrual fluid", on the other "emotion, passion"), but also of *rasa-*, which may mean "flavour; pleasure", but also refer to a fluid, in which connection we are reminded of the problem of *rasa-* maybe referring to the procreatory fluid of the woman discussed in §§11.19f., especially since the context in which the line quoted above is embedded pertains to reproduction, and since in 14,24,7cd semen and the procreatory-menstrual fluid (*śukraśoṇite*) are expressly mentioned.

[1183] This we may translate as:

> "Having said: 'verily the *rasa-* of women becomes *ārtava-* in a month', he (the author) again says in [the text of] Suśruta indeed" (or: "[The author] having said: 'verily the *rasa-* of women becomes *ārtava-* in a month', Suśruta indeed says again").

Whether "the author" refers to Suśruta or to someone else (the author of BHĀ?) is unclear.

[1184] *śukrata iti pāṭhāntaraṃ cintyam*. The translation (n.1183) then is: "... he again says: from semen indeed", with no alternative translation.

[1185] I may in this connection also draw attention to TA 16, where we find a variant of SU,Ci 37,74cd:

ārtavaṃ bhavatīti in n.1181 (BHĀ 1,3,189), which we have taken to refer to *rasa-* giving both *śukra-* and *ārtava-* in a month. But since the text does not have a word for "and", one could also consider whether it says that *rasa-* gives *śukra-* which is *ārtava-* in a month, even though this would ultimately be at variance with the differentiation between *śukra-* and *ārtava-* in 1,3,187f. Clearly, this can serve only to increase our difficulties!

11.23. BHĀ 1,3,28-30 too are of interest. First we learn that during sexual intercourse semen of the male is collected from all over the body and then ejaculated.[1186] The text than tells us that this semen combines with the procreatory-menstrual fluid (*ārtava-*) in the uterus, this latter fluid having entered the uterus in the same manner as (male) semen, which means that it has been collected from all over the body and ejaculated internally.[1187] This statement is of course to be seen in the light of the discussion in §§8.1ff. (see also §8.23, §9.3 and §11.12); cf. too the commentaries on SU discussed in §§8.5f., on internal ejaculation into the uterus. Indeed, that the compiler of BHĀ adhered to such a theory is made no less certain by BHĀ 1,3,40(5), which says first that due to various factors *śukra-* (here obviously that of the male) sometimes flows into the uterus in large, sometimes in small quantities, and then that the same holds good for *ārtava-* too.[1188]

11.24. This is, however, all the relevant information that BHĀ seems to offer us (over and above what has already been garnered in other parts of this study). Turning now to other later sources, we find that Aruṇadatta and Śrīdāsa in their commentaries on AH,Śā 1,5 quote a passage ascribed to Dāruvāhi(n)[1189]:

---

tathā śuklāsragān doṣān navāpi nava vastayaḥ

(instead of *evaṃ śukragatān doṣān dviguṇaḥ sādhuḥ sādhayet*). This refers to nine different clysters (*vasti-*/*basti-*; cf. n.393) subduing different substances, the fourth to ninth of which are the ☛ *dhātu*-s ☛ *rasa-* to marrow (☛ *majjan-*), when the morbific entities (or, depending on what ☛ *doṣa-* here actually means, disorders) have corrupted semen (according to SU: *śukra-*) or semen and the procreatory-menstrual fluid (according to TA: *śukla-* and *asra-*). Here too we see a parallelism between the two procreatory fluids.

1186 The process is the same as in n.168.

1187 30ef states:

tatra śukravad āyātenārtavena yutaṃ bhavet

"there (i.e. in the uterus) it (i.e. *śukra-* from the male) becomes[, when it has been discharged into the uterus (on the optative cf. VṚK, pp.508f.),] joined with *ārtava-* which has come like *śukra-*".

1188 *śukraṃ kadācid ... garbhāśaye bahu sravati ... kadācid ... alpam iti. evam ārtavam apīti.*

1189 This is evidently the same as the old medical authority Dāruvāha mentioned by MEULENBELD 1974, p.410.

strīpuṃsayos tu saṃyoge yady ādau visṛjet pumān
śukraṃ[1190] *tataḥ pumān vīro*[1191] *jāyate balavān dṛḍhaḥ.*
atha ced vanitā pūrvaṃ visṛjed raktasaṃyutam
tato rūpānvitā[1192] *kanyā jāyate dṛḍhasaṃhatā*[1193].

"If, however, during intercourse of male and female the male emits semen (*śukra-*) first, then a male is born, heroic (var: resolute), strong, firm. If now the woman discharges that together with/containing/consisting of blood first, then a girl is born, possessed of (var: with exceeding) beauty, firmly compact (scil. strong-limbed?)."

GHĀṆEKAR 1986, pp.43b;71b quotes the above as beginning with *strīpuṃsayoḥ susaṃyoge* "If during excellent/right intercourse ...". The same lines, with a very minor variant,[1194] are also found in ĀY 4,133f., where they are ascribed to Aruṇadatta. ĀY 4,137f. seem to be a continuation, not quoted by Aruṇadatta or Śrīdāsa, of the verses quoted above; their gist is that a simultaneous ejaculation of both male and female gives rise to ☞ *ṣaṇḍha*-s; and that the child born is otherwise according to (or: the same as) the one who ejaculates first.[1195] Whether these latter lines too actually do derive from the same source as that quoted by Aruṇadatta or not, they do not alter the deduction possible from the lines quoted by Aruṇadatta and Śrīdāsa, namely that they contain a clear reference to an orgasm of the woman, for the parallel treatment of male and female can hardly lead to any other deduction. Thus these lines show that the discussions in §§11.12ff. and §11.23 (with further references) were quite justified, and that, in a certain tradition at least, the discharge of the procreatory-menstrual[1196] fluid was obviously

---

[1190] Śrīdāsa: *śuklaṃ*.

[1191] Aruṇadatta, var: *dhīro*.

[1192] Aruṇadatta, var: *rūpādhikā*.

[1193] Aruṇadatta, var: *dṛḍhasaṃhitā*.

[1194] This is: *strīpuṃsayoḥ samāyoge* instead of *strīpuṃsayos tu saṃyoge*. The last line (*tato* ...) is found only in one ms (see p.431,n.111 of the edition).

[1195] *strīpuṃsayor visṛṣṭiś ced ekadaiva bhaved yadā*
ṣaṇḍhās tadā prajāyanta iti me niścitā matiḥ.
tayor atyutsukākrāntaḥ (printed: °*āntā*) *pūrvaṃ hi visṛjec ca yaḥ*
tadrūpaṃ janayed garbhaṃ nānyathā samprapadyate.

[1196] This is clearly also what "together with/containing/consisting of blood" (*raktasaṃyuta-*) in the passage quoted above implies, irrespective of the exact meaning of the Sanskrit term, which is a subject for debate and will be dealt with in §11.26 below.

thought to be an ejaculation due to orgasm.[1197] Moreover, we also see that the sex of the child conceived was taken to depend on which of the two sexual partners has an orgasm first.[1198]

11.25. This deduction also helps us to understand BH,Śā 5,13, which states:

> "In this regard, when, of the two come together, the man enjoys [his] object/aim first, the woman last, in that case here a male, similar as regards body and limbs [to the man having intercourse], is born. But where the woman attains [her] object/aim first, the man last, there conversely (*vā*) a woman, similar as regards body and limbs, is born."[1199]

In the light of the above, it is clear that this too must refer to orgasms, on which the sex of the embryo depends.

11.26. There, however, remains a problem regarding the verses of Dāruvāhi(n) quoted in §11.24. It is not clear whether *raktasaṃyutam* is elliptical for *śukraṃ raktasaṃyutam* or not, i.e. whether "that" refers to *śukra-* or not. In the first case we would have the statement that the woman ejaculates "semen" either consisting of "blood", containing "blood" or together with "blood". Following the first option, we would have to deduce that the "semen" (*śukra-*) mentioned is actually the procreatory-menstrual fluid (here *rakta-*), in which case we would have a terminology of the sort discussed particularly in §9.8 and last referred to in §11.16. The last option would lead to the conclusion that the woman discharges both "semen" (whatever that may actually be) and procreatory-menstrual fluid together, leading again to the problem of simultaneous discharge already discussed in §§11.12. The second option, however, would mean that "semen" is a composite fluid containing a procreatory part ("blood"), which would lead to a problem similar to that drawn attention to in §§11.15ff. If, conversely, *raktasaṃyuta-* should not qualify *śukra-*, then it would follow that what the woman ejaculates is either a fluid consisting of "blood", a fluid containing "blood", or a fluid together with "blood". In the

---

1197 The use of *vi√sṛj* "emit" and (in the text in n.1195) *visṛṣṭi-* "emission" to refer to an orgasmic ejaculation is to be noted. We shall see in chapter 12 that this narrow meaning does not always seem to be implied (e.g. in §12.3), though very often, probably even mostly, it is.

1198 Attention may in this respect also be drawn to the discussion on *hīnarasendriya-* and *hīnaśukrendriya-* in HĀ,Śā 1,p.415, in the course of which we too considered the orgasms (see §11.20); however, the difficulties connected with that explanation l.c. must also not be overlooked.

1199 *tad yadā sannipatitayor yatra puruṣaḥ pūrvam arthaṃ nandayati jaghanyaṃ strī tatra pumān ihāṅgapratyaṅgaiḥ sadṛśo jāyate. yatra tu strī prathamam arthaṃ sādhayati jaghanyaṃ puruṣas tatra strī vāṅgapratyaṅgaiḥ sadṛśī jāyate.*

first two cases we would have only one fluid, but the second of these would tell us that the fluid does not consist wholly of this "blood", but is made up of this only in part. In the last case something else beside "blood" would obviously be discharged too. All this brings us right back into the middle of a conglomerate of problems out of which so far no way seems to have led out. That we are not the only ones who have seen problems here is shown by the explanation of Dāruvāhi(n)'s verse by GHĀṆEKAR 1986, pp.43f., which may also be mentioned in this context, as it is quite interesting. First he opines that intercourse makes the ovum[1200] mature and ovulation take place faster than it would in the normal course of events.[1201] This *ārtava-* (clearly not a fluid to him!), he says, is also called *śukra-*, as in the verse of Dāruvāhi(n) (l.c.: Dāruvāha) quoted; this would seem to mean[1202] that according to his analysis of the verse the woman discharges *śukra-* which is *raktasaṃyuta-*, which would thus, we must deduce, have to be analysed as "consisting of (or eventually: containing) *rakta-*". After this we are told that *śukra-* may, however, also refer to a fluid which is not procreatory. He then refers to the discharge of Bartholin's glands (cf. on these §11.10), which he likens to that in the male due to the prostate and the Cowperian glands, and which, he says, is further followed by an orgasmic ejaculation of fluid from the uterus, which, however, plays no role in procreation.[1203] (To buttress the latter statement, he quotes a Western work, which, though, expressly mentions "real ejaculation of fluid" which "comes largely from glands, situated near the mouth of the vagina"![1204]) Finally, he says that all this shows that *ārtava-* is used for *strībīj* ("ovum" to him), and that *śukra-* may refer to this or to a

---

[1200] As he identifies what we have seen to be the procreatory-menstrual fluid, here *ārtava-*.

[1201] Cf. n.777. He quotes a Western source which says: "In general, I believe I admit that in the majority of cases the ovum is set free from the Graafian follicle on the twelfth day after the commencement of the last menstruation. While it may perhaps happen more frequently that, if an orgasm in copulation happens to take place, the follicle bursts prematurely. I would regard it as probable that the said boy births came from premature ovulations provoked by coitus which produce ova not yet quite mature. *Ideal birth*." (The English bibliography, no.30, says on the work quoted only: "Ideal Birth by Von de Velda".) From this it seems to follow that most women do, indeed should, *not* have an orgasm during intercourse!

[1202] No explication is to be found, but we may deduce this because *rakta-* here must also refer to the "ovum" according to the author.

[1203] P.44a: *thoṛī der pīche jab maithun'janya ānand paramocc koṭi tak pahũc'tā hai, tab puruṣoṃ ke samān strī ke garbhāśay se dūs'rā kṣārīya cip'cipā padārth utsargit hotā hai. is taral kā garbhotpādan se koī saṃbaṃdh nahīṃ hai*. Cf. also n.862.

[1204] Havelock Ellis, *Psychology of Sex* (no other particulars are given). The "real ejaculation" from near the mouth of the vagina will be taken up again in §12.16.

vaginal discharge (*yonigat srāv*). These various statements are, specially in the light of his statement already referred to in §2.3,[1205] full of contradiction, showing quite clearly the problems modern scholars cause themselves when they feel that they have by all means to bring statements of ancient texts into line with modern notions. But we also see here an awareness of the problems that have been bothering us too, and parts of the explanation of GHĀṆEKAR 1986 are in striking agreement with our own deductions. Be that, however, as it may, the context in which Dāruvāhi(n)'s verses are mentioned, namely the commentaries of Aruṇadatta and Śrīdāsa on AH,Śā 1,5, must bring to mind the problem of *strīretorajas-* and *strīretorakta-* in Aruṇadatta's and Śrīdāsa's comentaries already referred to in §9.8 and §11.16. Are we to assume that Aruṇadatta and Śrīdāsa saw in these verses something relevant to what they meant in using the expressions just cited? Of course even if they did, this does not necessarily mean that they analysed the verses in the same way as Dāruvāhi(n) meant them to be understood, but maybe one could get at what *strīretorajas-* and *strīretorakta-* mean if one knew how they understood Dāruvāhi(n)'s statement. Unfortunately neither the original intent of Dāruvāhi(n) nor how Aruṇadatta and Śrīdāsa understood him (and even whether Aruṇadatta and Śrīdāsa understood the same) is clear. We are thus, as regards the statements of Aruṇadatta and Śrīdāsa, basically no further than in §11.16: they may have meant two fluids, or they may have meant one. But if they should have meant two, then we should now also consider the discussion in §§11.17ff., according to which HĀ,Śā 1,p.415 seems to refer to male *retas-* as distinct from *śukra-* and female *retas-* as distinct from *rajas-*, all these fluids playing a role in determining the sex of a child, just like *strīretorajas-* and *strīretorakta-*. Are there parallels here? But if there are, then they do not help us further in determining the nature of the fluids, so that all we can do for now is shelve the matter again and wait for any new information that may crop up.

11.27. In connection with female orgasm a certain female disorder described by our texts is of interest. This disorder is *acaraṇā-*, commonly taken to be the name of the disease, though in actual fact it seems originally to be an adjectival qualification (the feminine of *acaraṇa-*) either of the woman afflicted or her ☞ *yoni-*.[1206] This use is still evident in SU,Ut 38,16ab, where it seems as if the woman were meant (the context would lead to this deduction)[1207] (though note that the *yoni-* seems to be meant in the case of

---

[1205] *ārtava-* consists of two parts, one being the menstrual fluid (maybe also functioning as the lubricating fluid), the other consisting of the "ovum".

[1206] On this problem of nomenclature see the discussion on ☞ *yonivyāpad-* in Appendix II.

[1207] The *Madhukośa* on the similar passage MĀDH 62,9 says:

several other disorders in this chapter[1208]). In SU,Ut 38,16ab we find:

> *maithune 'caraṇā* (var: *'caraṇī*) *pūrvaṃ puruṣād atiricyate*
> "The *acaraṇā-* (var: *acaraṇī-*) overcomes/surpasses/goes over (or, probably better: is quite freed)[1209] before the man during intercourse" (also possible: "the *acaraṇā-* surpasses/overcomes the man first during intercourse").[1210]

MĀDH, BHĀ and YOG follow the description of SU (for details see TIVĀRĪ 2,p.26,n.1; cf. also pp.5-10), whereas BYT 137,9 and SC 2,8 have *maithunācaraṇāt pūrvaṃ* etc. and ĀĀS 5836 *maithune 'caraṇāt pūrvaṃ* etc. (these readings will be discussed in n.1220 below).[1211] This disorder obviously refers to an orgasm of the woman before that of the man, which the *Madhukośa* on MĀDH 62,9 explains by saying that the woman "ceases/comes to an end" (*viramati*), which is an apt description of the post-orgasmic state.[1212] The explanations in the texts on BHĀ 2,70,12(p.765) and ĀĀS 5836 are even more explicit by explaining *atiricyate* through *rajo muñcati* "discharges *rajas-*",

---

atrācaraṇaśabdenopacārāt tadvatī strī bhaṇyate.

[1208] Cf. especially *imā yonayaḥ* in 38,20.

[1209] The basic meaning of √*ric* is "leave; release", whereas *ati* conveys the sense of surpassing. Cf. in this connection also VS 23,24:

> *mātā́ ca te pitā́ ca té 'graṃ* (the metre requires *te ágraṃ*) *vṛkṣásya rohataḥ*
> *prátilāmī́ti te pitā́ gabhé muṣṭím ataṁsayat*
> "Your mother as well as your father climb to the tip of the tree. Saying/thinking, 'I (am) pass(ing) over,' your father moved his fist (= penis, cf. OERTEL, pp.20f.) to and fro in the cunt."

(Note *l* in the place of the usual Vedic *r* in *prátilāmi*; this is in keeping with the postulated location of the Vājasaneyins in the east.)

[1210] The two translations depend on whether *puruṣāt* is construed with *pūrvam* (which seems more probable) or with *atiricyate*.

[1211] CA,Ci 30,18 too lists a disorder called *acaraṇā-*, but this is not the same, for it refers to nymphomania (on nymphomania see TIVĀRĪ 2,pp.25-27;60f.). This is caused due to "creatures" (*jantu-*), which arise when the genital tract is not washed, causing itching. On this explanation of sexual arousal, often explained as due to "worms" (☛ *kṛmi-*), cf. e.g. WEZLER 1971, p.282,n.33 (note that here the place in KĀM where the quotation in the commentary of Yaśodhara appears is wrongly given as "79.10-12"; it should be "78.2-3"), SCHMIDT 1922, p.257, JAINI, p.190,n.36 (though note on the latter also n.54 above); add to their references: RATIK [1],85f., RATIR 4,19f., NĀG 14,14f., KEL 6,28ff. (with an interesting discussion) and the editor's note on KAN 2,1,24, and see also §12.10.

[1212] A scholion in one ms of the text (see the editor's note l.c.) has obviously not understood this, for it states that this means that the woman pauses until the man has attained his pleasure:

> *ayam arthaḥ. yāvat puruṣas tu sukhaṃ nāpnoti tāvad asau viramati.*

which, on the basis of our knowledge by now, is a reference to an orgasmic ejaculation (cf. §11.24).[1213] Why, however, an orgasm of the woman prior to that of the man should be considered to be a disorder is unclear. The remaining part of SU,Ut 38,16 (16cd) does not seem to give us an answer, for it merely tells us that

> "And another [woman] does not find/acquire the *bīja-* (on this see §§2.5f.) because of frequent excessive practice (i.e. intercourse) (*aticaraṇa-*)[1214]"[1215]

In this connection attention may be drawn to AS,Śā 2,pp.292b-293a saying that if the woman is successful/contented/satisfied (*kṛtārtha-*) (this obviously refers to an orgasm) before the man, then the wind of the now disinterested woman makes the semen of the man ejaculated later become ineffective and damages his (this seems to refer to the child) seminal ☞ *srotas*-s, and thus a child called *vātendriya-*, ejaculating only wind during intercourse,[1216] is produced.[1217] Attention may also be drawn to the statement of Kāśyapa given in n.126, according to which a woman without passion/desire gives birth to a *kumbhila-*; though here an orgasm is not mentioned, disinterestedness is, and it is worth considering whether a similar theory is presupposed. But even should this passage, in which no orgasm is mentioned, not presuppose the

---

[1213] Ḍalhaṇa in his commentary on SU,Ut 38,16, however, says (obviously commenting upon the reading *maithune 'caraṇī* (though the other reading too is mentioned afterwards):

> *pūrvācaraṇy* (wrongly printed as: *pūrvā caraṇī*) *atiricyate maithunācaraṇe 'dhikā bhavati. ... puruṣāt pūrvam atiricyate. atyartham kaṇḍūyata ity arthaḥ*
> "The *acaraṇī-* overcomes/surpasses/goes over/is quite freed as the first one [of the two indulging in intercourse], [i.e.] is excessive/superior in the performance of intercourse. ... *atiricyate* before the man. 'Itches excessively', is the meaning".

This can hardly mean simply that the woman is terribly aroused (cf. n.1211), unless one wants to maintain that Ḍalhaṇa did not understand the verse, which is, however, so clear in saying that the woman does something before the man during intercourse that one would probably be doing him great injustice then. What he thus seems to mean is that she is so aroused sexually that she has an orgasm before the man, even though this does seem a peculiar mode of expression. My suspicion is that Ḍalhaṇa has explained the SU verse in this manner to bring it into line with the statement of CA on *acaraṇā-* mentioned in n.1211.

[1214] Mostly we find a disorder called *aticaraṇā-* mentioned, which also seems to be an original adjectival qualification. On this disorder see TIVĀRĪ 2,pp.28f.

[1215] *bahuśaś cāticaraṇād anyā bījaṃ na vindati.*

[1216] Cf. on this also p.583 in Appendix II below.

[1217] *yadā strī prathamaṃ kṛtārthā bhavati tataḥ puruṣeṇa paścāc chuklam utsṛṣṭaṃ harṣānavasthitacetasaḥ striyā vāto viguṇīkaroti puṃstvavāhīni cāsya srotāṃsi copahanti. tadā vātendriyaṃ bhavati. tan maithune vātam evotsṛjati.*

same theory, the question nevertheless crops up whether something similar to what has been said in AS,Śā 2,pp.292b-293a may not be the reason for *a-caraṇā-* being a disorder, namely because it leads to progeny which is not normal.

11.28. Or may the line quoted in n.1215 be taken to refer to *acaraṇā-* in the line immediately preceding (*maithune 'caraṇā* etc.) too? In this connection it may be pointed out that Ḍalhaṇa, commenting on the SU passage, gives us another reading of the line quoted in n.1215, and this is also the reading found in the other texts mentioned in §11.27, except for BYT and ĀĀS (on which see n.1220). This reading states:

> *bahuśaś cāticaraṇā* (Ḍalhaṇa: *cāticaraṇī*) *tayor bījaṃ na tiṣṭhati* (var:[1218] *vindati*)
>
> "And the *aticaraṇā-* (Ḍalhaṇa: *aticaraṇī-*) frequently; their (dual) *bīja-* (or: the *bīja-* in them) does not remain (var:[1219] one does not find their *bīja-*)".[1220]

"Their" here seems to refer to both the *acaraṇā-/acaraṇī-*[1221] and the *aticaraṇā-/aticaraṇī-*[1222]. The combination of 16ab and 16cd makes it seem

---

1218 MĀDH 62,9, YOG 2,Yonirogādhikāra,13 (p.405).

1219 See n.1218.

1220 BYT 137,9 and SC 2,8, which correspond to SU,Ut 38,16, read:
maithunācaraṇāt (var: *maithune 'caraṇā*) *pūrvaṃ puruṣād atiricyate*
bahuśaś cāticaraṇāt tayor bījaṃ na vindati (SC: *tayor vīryaṃ na tiṣṭhati*).

According to this, the disorders would be *maithunācaraṇa-* ("Due to *maithunācaraṇa-*, [the woman] overcomes ...") and *aticaraṇa-* respectively. One could also consider translating as: "Due to the performance/behaviour (*ācaraṇa-*) of/during intercourse ...", but in the given context that does not seem feasible, as the disorder would not be named then. The Hindi translation of SC says that the *yoni-* is called *caraṇā-* in the first case, and *praticaraṇā-* in the second, even though the text reads differently! It also says that in both these *yoni-*s the semen of the man cannot stay: *jo strī maithun kar'te samay puruṣ'ke pah'le skhalit ho jāve us'kī yoni-kā nām caraṇā hai. aur jo anek bār bahut jyādā skhalit ho vah praticaraṇā yoni kahī hai. in donoṃ yoniyoṃmeṃ puruṣ'kā vīry nahīṃ ṭhahar'tā hai. skhalit honā* (literally: "fall [from a pinnacle?]", but maybe understood rather as "be discharged") of the translation clearly refers to an orgasm.

ĀĀS 5836 is similar to BYT, but with *maithune 'caraṇāt* for *maithunācaraṇāt*, and *tiṣ-ṭhati* for *vindati*; according to this reading, the disorders would be *acaraṇa-* ("Due to *acaraṇa-* during intercourse, [the woman] overcomes ...") and *aticaraṇa-* respectively.

1221 On the latter see n.1213.

1222 CA,Ci 30,105 also has the form *aticāriṇī-*. This is combined with *prākcāriṇī-* (usually called *prākcaraṇā-*) to form a compound *prāgaticāriṇī* (Cakrapāṇidatta: *prāgaticāriṇīti prāk-caraṇāticaraṇe*, obviously referring to *prākcaraṇā-* and *aticaraṇā-* in 30,20 and 30,19 respectively); as the adjectives qualifying it show, this is not a dual (with an archaic ending -*ī*), but

as if the *aticaraṇā-/aticaraṇī-* has frequent orgasms, and this is also how BHĀ 2,70,12(p.765) explains the text.[1223] Ḍalhaṇa and the *Madhukośa* commentary, however, take it to refer to frequent intercourse,[1224] the outcome of which may ultimately be the same, but need not necessarily, since intercourse need not always lead to an orgasm of the woman. But how the commentaries understand the text is ultimately of secondary import; what interests us primarily is that, if SU,Ut 38,16cd holds good for 16ab too, then it would tell us that the *acaraṇā-/acaraṇī-* cannot conceive, and that of course would not be the same as what AS,Śā 2,pp.292b-293a says. As regards frequent orgasms, we might hold that this could have been thought to lead to too much procreatory-menstrual fluid (and/or some other fluid; cf. §11.12, also §11.26) being discharged and washing out semen or the embryo, which would thus be a parallel to what happens due to the menstrual discharge in the case of intercourse during menstruation (cf. §5.7) or during abnormal bleeding (cf. §5.14). But a similar explanation would be of little use in the case of a prior orgasm of the woman hindering conception, for if this should result in the washing out of the semen or embryo, then the same would logically also happen if the woman had her orgasm after the man had his[1225] — not the point of time

---

a feminine singular, which is peculiar:

> *vātaghnaiḥ śatapākaiś ca tailaiḥ prāgaticāriṇī*
> *āsthāpyā cānuvāsyā ca svedyā cānilasūdanaiḥ.*

Or does *prāgaticāriṇī* actually refer to only one disorder, in spite of 30,19f.?

As regards the disorder referred to by *aticaraṇā-*, this consists according to CA,Ci 30,105 of swelling, numbness and pain in the ☛ *yoni-* due to wind agitated by excessive intercourse. (Actually, *prākcaraṇā* and *aticaraṇā* seem to refer to the *yoni-*; cf. e.g. 30,22: *yonir upaplutā*, and on the nomenclatural problem ☛ *yonivyāpad-* in Appendix II. In other words, we have here the feminine forms of adjectival *prākcaraṇa-* and *aticaraṇa-* because of the feminine *yoni-*.)

1223 *bahuśo vāraṃvāram atiricyate.*

1224 Ḍalhaṇa:

> *bahuśo 'ticaraṇān maithunād anyāticaraṇā bījaṃ garbhāṅkurajananaṃ na vindati na prāpnoti*
> "Due to frequent *aticaraṇa-*, [i.e.] intercourse, another one, the *aticaraṇā-*, does not find, [i.e.] does not obtain, the seed (*bīja-*) producing the sprout, the embryo."

Madhukośa:

> *bahuśo maithunācaraṇād aticaraṇā ... bahumaithunācaraṇād bījaṃ na dhatte*
> "The [woman who is] *aticaraṇā-* due to frequent practice of intercourse ... does not hold the *bīja-* due to the practice of much intercourse."

Cf. the explanation of *aticaraṇā-* found in CA (see n.1222).

1225 A variant to the *Madhukośa* commentary on MĀDH 62,8-10 says, seemingly with

of orgasm, but the quantity of fluid discharged would to be of relevance. So the explanation for too frequent intercourse (or orgasms) being a hindrance to conception may be entirely different. I may here draw attention to the fact that medical texts say in connection with the *aticaraṇā-/aticaraṇī-* that too much intercourse aggravates the wind in the genital tract.[1226] Cf. in this context AS,Śā 2,pp.292b-293a in §11.27 (where, however, not conception as such, but conception of a normal child is the issue), and BHĀ 2,24,15(p.228), where the *kṣaya-*[1227] of semen and the loss/destruction (*nāśa-*) of the procreatory-menstrual fluid and of the embryo or foetus are subsumed under disorders due to wind. It may be debated whether too frequent intercourse (or orgasms) may not have been taken to hinder conception for similar reasons.[1228] But even if this should ultimately turn out to be the correct reason for a woman having many orgasms not being able to conceive (I must stress that I am here only mentioning a possibility), would this also apply in the case of a woman who has her orgasm before the man? We may try to circumvent this difficulty by holding that *tayoḥ* "their" in SU,Ut 38,16cd refers

---

reference to the *aticaraṇā-* in 9:

> *iyaṃ puruṣād atiricyate paścād dravati tena bījaṃ na gṛhṇāti*
>
> "She overcomes/surpasses/goes over/is quite freed (see §11.27) after the man [during intercourse], [i.e.] becomes fluid/flows (scil. ejaculates?; cf. on this §12.10 below); through this [she] does not hold/accept the *bīja-*",

or:

> "She surpasses/overcomes the man (first) [and] becomes fluid/flows afterwards; through this [she] does not hold/accept the *bīja-*."

Regardless of which translation is correct, the explanation given would, from the look of it, better fit the *acaraṇā-*, who is also described in 62,9. But be that as it may, the first translation could be telling us that the semen ejaculated by the man is washed out if the woman has an ejaculation after him. The second translation is more difficult to explain. If "afterwards" refers to an ejaculation after the man, then what may be hinted at would be the same, but the mention of the orgasm before the man would leave us with a very difficult puzzle. On the other hand, "afterwards" might also refer to an ejaculation right after the orgasm, but before the male has his. In that case there would be a similarity with SU,Ut 38,16ab, but the explanation of why the *bīja-* is not retained would have to be different from that in the case of the first translation.

[1226] Cf. TIVĀRĪ 2,pp.28f., also n.1222.

[1227] Cf. on this §§5.2f.

[1228] Cf. also KĀ,Kh 9,24-26, already mentioned in §10.24, which say that agitated wind in a menstruating woman may block the flow of "blood", and SU,Śā 10,30, which says that aggravated wind makes *śukra-* and *śoṇita-* in the womb (Ḍalhaṇa: pent up in the womb; *vāyunoparuddham*) swell up and distend the womb. Of course here the contexts are different, but the blocking function of wind is nevertheless evident.

not to the *acaraṇā-/acaraṇī-* and the *aticaraṇā-/aticaraṇī-*, but to the woman and the (otherwise not mentioned) man having intercourse, in which case the explanation could be taken to pertain only to the *aticaraṇā-/aticaraṇī-* and not the *acaraṇā-/acaraṇī-*. This would, however, leave us in the dark as to the reason why the latter should be a state requiring treatment.

11.29. But be all that as it may, we find that the above is clearly at variance with what we discussed in §§11.24f., namely that the sex of the child may depend upon which of the partners having intercourse has an orgasm first, for §§11.27f. tell us that an orgasm of the woman before that of the man is a disorder (to be treated), whether it in actual fact was taken to hinder conception or not (in the latter case the production of an abnormal child might be considered; cf. §11.27). We obviously have different theories here. The theory of the sex of the child depending on which of the partners during intercourse has an orgasm first would also seem to be at variance with that according to which the relative quantities of the procreatory fluids is the determining factor, but it may be speculated that with heightened passion (of which the time of orgasm could have been taken to be an indication) more fluid is ejaculated than through less passion. Though we have so far found no material buttressing such a premise, it is obvious that the matter nevertheless merits consideration, if only as a possibility to be kept in mind.

11.30. I would here also like to draw attention to AS,Śā 2,p.292b, which says:

> "If the object/aim of the woman and the man is effected equally (*samam*), or their *bīja-* (i.e. obviously: respective procreatory fluid)[1229] [is] of equal parts and favourable (i.e. not corrupted), and if the *bīja-*[1230] is corrupted in part,[1231] then a ☞ *napuṃsaka-*, [who is] *dviprakṛti-*[1232] [and] ☞ *dviretas-*, orig-

---

1229 On *bīja-* cf. §§2.5f.

1230 Procreatory fluid? Pre-embryonal state of the combined fluids? (Cf. §§2.5f.)

1231 Indu says that this refers to a corruption of the *bīja-* which is *strībhūta-* (from the woman?; in the form of a female?) in that part from which the *bīja-* (here seemingly the procreatory fluid) of (i.e. within?) the embryo originates (or: a corruption of the *bīja-* of the embryo which is *strībhūta-*, in that part from which the *bīja-* originates), due to which the sex of the *garbhabīja-* (most probably: "the *bīja-* which is the embryo/foetus") is undifferentiated:

> *yadā ca tasmin strībhūtaṃ bījaṃ garbhasya bījārambhake bhāge duṣyati yenaikarūpatā puṃviśeṣeṇa yoṣidviśeṣeṇa vā garbhabījasya na sampadyate tasmin strīpuruṣabījabhāge kenāpi doṣakopakāraṇena duṣṭe dviprakṛtir* (printed as *dviḥ prakṛtir*) *eko 'py ubhayasvarūpo dviretā napuṃsako bhavati.*

1232 Cf. on this term the discussion on ☞ *dviretas-* in Appendix II.

inates."[1233]

We have here seemingly two theories, one in keeping with the passages cited in §§11.24f. and stating that the sex of the child depends upon who ejaculates first; however, "equally" (*samam*) might have been used in the sense of "with equal vehemence" (and not of "simultaneously" here), so that we cannot be quite sure as regards this AS passage.[1234] The other theory is in keeping with the parallel notion of the sex of the child depending on the absolute quantities of the fluids mingling[1235] (the text further says that damage to the *bīja-* may also play a role). How these two theories are related, if at all, is not mentioned, so that we are left to ponder how AS intends this passage to be understood. Moreover. it comes just before the passage from the same text quoted in §11.27 referring to an abnormal child called *vātendriya-* being born if the woman has an orgasm first, which is really puzzling. Has the author of AS simply juxtaposed two passages referring to orgasm here without heed to whether they agree with each other or not, or is there some link between them which is not at first glance perceivable?

11.31. The answer would seem to be the latter according to NYC, fol.22r. This part of the commentary on SU,Śā 2 comes after some comments on 2,42, and clearly contains an alternative theory on the development of the abnormal children which are described in SU,Śā 2,38-42;[1236] the beginning of this passage is also followed by *iti*, which might mark a quotation — maybe thus referring to something originally between 2,42 and 2,43, but no longer found in our editions.[1237] Be that as it may, we find here a link between

---

[1233] *yadā strīpuṃsayoḥ samam evārtho niṣpadyate bījaṃ vā tayoḥ samāṃśaṃ sammukhaṃ ca. yadā ca bījaṃ bhāge duṣyati tadā dviprakṛtir dviretā napuṃsakaṃ bhavati.*

[1234] Cf. in this connection also §§11.19f.

[1235] See on this latter n.103.

[1236] 2,38 tells us that an ☞ *āsekya-* is born due to both "fathers" (i.e. parents) having only a minute quantity of *bīja-* (cf. on this also n.20). A *saugandhika-* (2,44: *sugandhin-*) is born according to 2,39 from a putrid ☞ *yoni-* (*pūtiyoni-*). 2,40f. describe the *kumbhīka-* and the *īrṣyaka-*, but do not say how they are produced. 2,41f. say that if a man behaves like a woman or a woman like a man during coitus, a ☞ *ṣaṇḍha(ka)-* looking and behaving like a woman or a girl behaving like a man is respectively born (the behaviour during coitus probably refers to the man lying below the woman; see on this also below). On the various abnormal children mentioned here see n.1246.

[1237] The commentary runs:
taccyutiviśeṣa (or: *tac cyutiviśeṣa*?) *iti. anye tv āsekyādicaturṣu* (ms: *āsekādy-ādicaturṣu*) *samāṃśam alpaṃ* (or: *samasamam/samaṃ samam alpaṃ*?; ms: *sama2malpaṃ*, the *2* like a *śa* without the right vertical stroke) *ca bījaṃ puraścaraṃ paścāc ca cyutiviśeṣa iti. tatra striyā[ḥ] puruṣānte cyuter āsekyaḥ.*

different times of ejaculation and the formation of different sorts of abnormal embryos; the translation of the relevant portion of the text in n.1237 is:

> "But others [say]: 'With regard to the four, the ☞ *āsekya-* etc., *bīja-* whose parts are equal (or: *bīja-*, equal [to the other *bīja-*]),[1238] [of] minute [quantity], is first-going,[1239] and the difference/distinction according to ejaculation[1240] is afterwards.' With regard to that, from the ejaculation of the woman after the man an *āsekya-* [is born]. From the ejaculation before an *īrṣyaka-*. From the ejaculation of man and woman together a *kumbhīka-*. In the case of inverted/reversed coitus,[1241] a *saugandhika-* when the ☞ *yoni-* is corrupted in the case of equality during coitus.[1242] And when the *bīja-*, all parts of the *bīja-*,[1243] is corrupted, ☞ *ṣaṇḍhaka-*s in case of ejaculation at the same time of both indeed. In this regard Bharadvāja[1244] [says]:
>
> > 'Due to equality of pure *śukra-* and *rajas-*, [of] minute [quantity], an *āsekya-* arises[1245] here

---

pūrvacyuter īrṣyakaḥ. strīpuṃsayor yugapac cyute[ḥ] kumbhīkaḥ. rataviparyaye sāmye rate duṣṭe yonau saugandhikaḥ. praduṣṭe ca bīje sakalabījabhāge tayor eva samakālacyutau ṣaṇḍhakāni. atra bharadvājaḥ.

> *alpayoḥ sāmyataḥ śukrarajasoḥ śuddhayo[ḥ] striyāḥ*
> *paścāc cyuter ihāsekyo bhaved īrṣyaḥ purā cyuteḥ.*
> *yugapac cyavanāt kumbhī ṣaṇḍho vai duṣṭayos tayoḥ*
> *sāmye rate viparyāsād duṣṭayonau sugandhika*

iti. atra duṣṭayonir ity atiśayena (ms: *ātiśayena*) *tac* (ms: *tā*) *chakyāṃ śuddhayonir ity asyaitadvyapadeśāt.*

1238 In the case of the former translation, *bīja-* looks rather as if it referred to the germinal embryo (though another interpretation may be possible), whereas in the case of the latter translation the procreatory fluid of male and female respectively would definitely be meant.

1239 This obviously means that the main cause consists of the procreatory fluids equal in quantity to each other, but of minute quantity. What then follows is an additional cause.

1240 Literally: "falling; coming/streaming forth" (*cyuti-*); it is quite clear that ejaculation is meant.

1241 Most probably: when the woman is above the man. Cf. on this also n.1368.

1242 This looks as if it too referred to ejaculation at the same time.

1243 Does the latter statement explicate the former, or are they to be joined by "and" or "or"? On *bīja-* and *bījabhāga-* see Excursus I at the end of chapter 2.

1244 On him see n.1144.

1245 On the optative cf. VṚK, pp.508f.

> from the ejaculation of the woman later, an *īrṣya-* from ejaculation before. From ejaculation together a *kumbhī-* (or *kumbhin-*?), a *ṣaṇḍha-* indeed when they are corrupted. In the case of equality during coitus a *sugandhika-* in the case of a *yoni-* corrupted due to interchange'."[1246]

In other words, one will, if the procreatory fluids be equal in quantity, in any case have a child neither male nor female, but only if the quantity of the fluids is minute will this be abnormal in other respects too, the regulating factor in such cases being the orgasm. One could see in the passages in §§11.27f. something similar, and this would then also hold true for the AS passage quoted in §11.27. However, the passages in §§11.27f. do not qualify their statements, i.e. these seem to relate to a woman having an orgasm before the man in general. In other words, the AS passage quoted in §11.27 remains at variance with that quoted in §11.30 in spite of the statement of NYC; note also that the abnormal child mentioned in §11.30 is different, at least as regards its name, from those mentioned in the passage of NYC.[1247] Of course we can nevertheless not neglect the possibility that the statement in §11.27 has been taken out of a context such as the one we have in the passage of NYC, but that would not do away with the other passages in §§11.27f. Thus we still seem to be faced with different theories here. But whatever the answer to this problem may be, from the discussion so far we see once again that there were theories in vogue stating that the sex of a child depended on the orgasm(s) of the parents.

11.32. I must here draw attention to another very interesting statement in NYC, namely the commentary (3rd last folio,r) on SU,Śā 10,54, which latter says that if a man younger than twenty-five begets a child in a woman younger than sixteen the child dies either in the womb or soon after birth, or else remains weak. NYC says that this is because *ārtava-*, the second ☞ *dhātu-*,

---

[1246] On the ☞ *āsekya-* and ☞ *ṣaṇḍha(ka)-* see Appendix II. The other abnormal children here and in n.1236 are described in SU,Śā 2,39-42 as follows; the *saugandhika-* is excited by the smell of genitals, the *kumbhīka-* is a bisexual male, homosexually passive, and the *īrṣyaka-* a voyeur. The commentaries contain additional material. I have not gathered other relevant information. It is nevertheless interesting to see the importance given to sexual inclination here, in which context I may also point out that the question of sexual inclination divorced from natural gender plays a very important role in Jaina discussions on who is capable of being liberated; for some details cf. the data in JAINI, pp.76ff.;119;133f.;144;163ff.;173ff.

[1247] AS,Śā 2,p.293 also mentions the *āsekya-*, *īrṣyaka-* and *saugandhika-*, but their formation depends on factors other than an orgasm (the quantities of the fluids combining may also play a role, but not in each case). On the *kumbhīka-* see also n.126 on the *kumbhila-*.

is complete in a woman only at the age of sixteen, as also *bīja-* (here clearly semen), the seventh *dhātu-*, is complete in a man only at the age of twenty-five.[1248] We have already seen[1249] that the procreatory-menstrual fluid is sometimes called a *dhātu-*,[1250] so that in itself does not constitute a problem. What is, however, a great problem is that this *dhātu-* is expressly called the second *dhātu-*, which (since semen is here called the seventh *dhātu-*) must automatically entail a reference to the chain of alimentary transformation, in which the second *dhātu-* is, however, blood in general. The obvious conclusion from this would be that to Gayadāsa blood in general and the procreatory-menstrual fluid are one and the same, and in this context one could point to his opinion that both fluids are "fiery" (see §7.12). But Gayadāsa also says several times[1251] that what we have come to know as the procreatory-menstrual fluid is procreatory; is he here really of the opinion that blood in general too can be procreatory? Taking his statement at face value would mean that Gayadāsa adhered to a theory, otherwise lost or at least not provable beyond doubt,[1252] that blood and the procreatory-menstrual fluid are one and the same. This deduction may be buttressed by his remarks on SU,Śā 4,25. This passage, already discussed in §7.10, speaks of the development of certain organs from blood, which blood, however, seems to be blood in general. Śā 4,24, however, refers to the procreatory-menstrual fluid. We have seen in Excursus III (at the end of chapter 6) that at the end of his commentary on 4,24 and before his commentary (which says nothing relevant for our problem) on 4,25, Gayadāsa has a problematic

ārtavasya raktatvāt

"Because of *ārtava-* being blood."

Several possible interpretations of this have been considered in Excursus III;

[1248] The relevant passage says:

tatra striyā dvitīyo dhātur ārtavaṃ tathā ṣoḍaśabhir eva varṣaiḥ sampūrṇaṃ bhavat[i. pu]ruṣasya tu saptamo dhātur bījam. tac ca pañcaviṃśatibhir varṣair eva sampūrṇaṃ bhavati.

This need not be in contradiction to SU,Sū 14,6, which tells us that the procreatory-menstrual fluid of a woman begins to flow after the age of twelve (see §7.11), as even a flow of the fluid would not necessarily mean that the body already has the full quantity of the fluid within it. But of course we do not know if Gayadāsa himself was aware of this problem or tackled it with this argument.

[1249] E.g. in n.1086, n.1108 and §11.22.

[1250] Cf. p.555 in Appendix II.

[1251] An explicit passage is quoted in n.788.

[1252] Cf. n.169, n.402 and n.1122.

except for one, these were unproblematic. The problematic explanation considered whether the line quoted here may not have served to lead from the discussion of 4,24 (with *ārtava-*) to that of 4,25 (with blood), from which one could then deduce that to Gayadāsa both passages describe the same substance. In this connection one may also point to his commentary on SU,Ni 1,25-29, where he calls *ārtava-* and blood *abhinna-*. But we have seen in §5.13 (cf. also §10.9) that this most probably does *not* mean that the two are not different; cf. also the use of *abheda-* by Cakrapāṇidatta and Ḍalhaṇa in their commentaries on SU,Sū 14,7 and 14,6 respectively quoted in §7.12, as well as the mode of expression of Śrīdāsa in his commentary on AH,Śā 1,7 in n.884. Can we therefore really credit Gayadāsa with the opposite view here? Though it does appear difficult to swallow, especially since *ārtavasya raktatvāt* might also be explained differently (see loc.cit.), one cannot categorically exclude the possibility that Gayadāsa is either inconsequent or waffles with regard to the relationship between blood and the procreatory-menstrual fluid. Or is "second" in his commentary on SU,Śā 10,54 to be taken to mean that women have a *dhātu-* over and above semen, i.e. a second procreatory *dhātu-* (cf. §11.22)? But then we would have problems with the reference to semen as the seventh *dhātu-*. Thus whichever way we look at it, NYC causes us difficulties.

11.33. A similar problem seems to be caused by Indu's commentary on AS,Śā 1,p.276a. This AS passage has already been referred to in §9.3; it says that the blood of women enters the uterus month after month and flows out, and is then called *ārtava-*. After this the disorder called *asr̥gdara-*[1253] is described (see §9.4). The parallels in this discussion to SU,Śā 2,18-21 (discussed in §6.9) are quite clear and very probably no coincidence. Now the AS passage begins with:

tathā raktam eva ca strīṇāṃ ... (see n.877)
"In that manner the blood of women indeed ... ."

This can hardly but refer to the procreatory-menstrual fluid. What Indu says on this (as well as in his commentary on AH,Śā 1,1), though, is, problematic:

raktam eva[1254] *sarvasya dhātutvena jāyate*[1255]. *tad eva viśiṣṭaṃ sad ārtavam ity ācāryā āhuḥ*[1256]

[1253] *asr̥gdara-* = *pradara-*, or *asr̥gdara-* as well as *pradara-*, = *raktayoni-*, according to the text.

[1254] Commentary on AH: *eva yat.*

[1255] Commentary on AH: *jñāyate.*

[1256] Commentary on AH: *sad ārtavasaṃjñam.*

> "Blood indeed is produced as the *dhātu-* of every one. That indeed, being particular/distinct, is *ārtava-*: thus say the preceptors (var: Blood indeed, which is known as the *dhātu-* of every one, is indeed, being particular/distinct, called *ārtava-*)."

It is to be noted how "the blood of women indeed" here appears only as "blood indeed". When one reads these lines, one is very much inclined to think that according to Indu it is blood in general which is meant, "every one" referring to all humans, male as well as female; this blood present in all then is called *ārtava-* in its special function of reaching and flowing out of the womb of women. From this it would follow that it is blood which reaches the womb and flows out, only being called *ārtava-* then. Clearly, this would be a very startling statement, and so we may also be permitted to ask whether the particular or distinct nature of blood may not refer to a *derivation* from blood, which would not be the same as an equation with blood. Could other parts of Indu's commentary help us to reach a decision? AS,Śā 5,p.326b says clearly that both milk and *rajas-* are derived from the pure part (☞ *prasāda-*) of ☞ *rasa-* (cf. §9.6). Indu merely rephrases this statement and says no more; since blood too is derived from *rasa-*, this would not speak against his equating blood and the procreatory-menstrual fluid. His commentary on AH,Śā 3,80f. (cf. §9.5) is similar (see AH5). But in his commentary on AH,Śā 1,7 (cf. §9.5) he is quite explicit (see AH5): in women, blood arisen from *rasa-* "cooked" by its fire divides into two parts due to a reason peculiar to women; one part is the *dhātu-* blood, while the other is *rajas-*. This *rajas-* then gradually collects in the channels (☞ *srotas-*) of *rajas-*, enters the uterus each month, flows for up to three days and is called *ārtava-*.[1257] According to this, the *dhātu-* blood and the procreatory-menstrual fluid have the same origin, but develop differently. Both arise from a fluid called blood too, and this latter blood arises from *rasa-*. From this it is clear that the *dhātu-* blood (blood in general) and the procreatory-menstrual fluid are *not* the same fluid; thus it cannot be used to buttress the statement of NYC in §11.32. On the other hand, the latter fluid is according to Indu's statement not derived directly from the former fluid, but from something which is obviously very similar. As such, this statement is very interesting, as it seems to represent an intermediate stage between the derivation of the procreatory-menstrual fluid directly from ☞ *(āhāra)rasa-* and from blood (cf. §§11.5ff.). This would be in

---

[1257] *raja ārtavam. tatra rasāgninā pakvād rasād udbhūtaṃ raktaṃ strīviśiṣṭena hetunā dvidhā bhavati. tayor ekaṃ dhāturūpeṇa pariṇamati. anyad rajorūpeṇa. rajorūpeṇa pariṇataṃ rajovāhiṣu srotaḥsu sthitiṃ kurvan kālakrameṇopacīyamānaṃ dvādaśavarṣāyāḥ striyāḥ pratimāsaṃ sadā tridināvadhi pravartate. tad ārtavasaṃjñāṃ labhate.*

keeping with the fact that Indu lived before 1050.[1258] But if what Indu says indeed represents an intermediate stage with regard to the two different concepts on the origin of the procreatory-menstrual fluid, then it becomes difficult to see an old theory in its derivation from blood (see §11.6). It should be noted that Śrīdāsa in his commentary on AH,Śā 1,7 has the same argument; Śrīdāsa is, however, later than Indu[1259] and clearly only reproduces the older tradition (though not necessarily directly from Indu). Attention may also be drawn to Indu's commentary on AH,Śā 3,81, according to which both milk and the procreatory-menstrual fluid arise from the pure part (☞ *prasāda-*) of *rasa-*; the relationship of this statement to the other statements on the derivation of the procreatory-menstrual fluid quoted above is not clear.

11.34. Let us now sum up what we have discussed in this chapter apart from what might be a — very problematic — equation of blood and the procreatory-menstrual fluid in NYC (§11.32). The first subject to be discussed was the different theory regarding the metabolic origin of the procreatory-menstrual fluid, which we here saw to be derived from blood (in keeping with the today authoritative theory of ☞ *upadhātu-*s) and not from ☞ *(āhāra)rasa-*. Though the theory of derivation from blood seems to be younger, we cannot be fully sure that it is not in actual fact an old theory which was neglected by older texts, but came to the fore again later; however, the remarks of Indu discussed in §11.33 would seem to speak against this latter. We also took up the vexing problem of female "semen" again, coming, however, to no definite results as to the nature or derivation of this substance or substances. In this connection special mention may be made of passages from HĀ which complicated the matter even more by mentioning fluids whose relationship to the fluids so far discussed is quite unclear (see §§11.15ff.). We also found further confirmation for our so far tentative deduction that during intercourse an internal ejaculation of procreatory-menstrual fluid takes place in the female (as a parallel to the external ejaculation of males), and we also deduced that it is most probably the orgasm of the female that brings about the internal ejaculation of procreatory-menstrual fluid into the uterus, and maybe (though this is by no means certain) of another fluid (though the place of ejaculation is problematic), namely "semen" (which is visible and may be a clear fluid), on which see especially §§11.10ff. (also §11.26).[1260] In connection with this "semen" we also discussed the problem of whether it

---

1258 See MEULENBELD 1974, pp.392f.

1259 But before 1300, according to MEULENBELD 1974, p.430.

1260 Cf. also the statement of EIM in §8.9, though this seems to see no difference between the semen of males and females ejaculated.

was thought to be collected in a special place within a woman in the same way as in a man (cf. §11.12), and in the same way as the procreatory-menstrual fluid might be too (cf. on this latter §11.23). The "semen" alluded to, especially in §§11.10f. might, however, rather refer to the lubricating fluid(s) of women emitted without an orgasm; in this case it is unclear whether an emission from some place near the vaginal introitus is described, or from deeper within the genital tract. As regards the sex of the child conceived, there seem to have been theories current linking this with which of the parents has an orgasm first (see §§11.24f., also §11.29); we also found that there was seemingly a theory stating that an orgasm of the woman prior to that of the man (in general, and not in the special case referred to in §11.31) resulted not only in an abnormal child, but might even have also been taken to be a hindrance to conception, though we do not have any clear evidence for this latter assumption. In any case, this theory clashes with the theory according to which the sex of the child depends on which of the sexual partners has an orgasm first, this latter theory on its part leading to difficulties when confronted with the theory that the sex of the child depends on the amount of fluid liberated (also in the case of the female most probably due to an orgasm) by the partners in the sexual act. Apart from these, we also found evidence for another theory according to which the relative strength or potency (and not the quantity) of the procreatory fluids of man and woman may be responsible for the sex of the child conceived (§11.20), and also discussed the possibility of the heftiness of the passion or orgasm maybe being a factor too (§§11.19f., §11.30).

11.35. One of the things that the above shows is the importance of ejaculation, brought about by an orgasm, for the creation (and also sex) of a child, this applying not only to males, but to females as well; indeed, if such an ejaculation is necessary for bringing the two procreatory fluids of the male and the female into contact within the uterus, then it follows that conception is not possible if not only the man, but also the woman has no orgasm. This means that the orgasm of the woman too must have been considered of paramount importance for begetting progeny. If this is so, why is the subject of orgasm not discussed in more detail in the medical texts, which after all contain detailed discussions on other matters relating to conception and embryology? Ancient Indian scholars certainly did not have any taboos in this respect, and it is also hard to believe that they would simply have shelved the matter or circumvented it either as irrelevant or problematic, for it is obviously not irrelevant, and as regards its being problematic for some reason unknown, so many other problematic matters have been taken up and discussed in detail in medical texts that it would be extremely curious if such shelving or circumventing should have taken place only in this case. Could we therefore deduce that in actual fact the discussion of orgasm in medical texts was regarded to be superfluous because it was to be found elsewhere? And if this should indeed be the case, where could such a discussion have been found?

11.36. The answer any Indologist must give to this is: sexological literature, i.e works on Kāmaśāstra. And in truth, this branch of learning has a great antiquity in India,.going back probably to at least some centuries before the Christian era. It may here also be pointed out that AH,Ut 40,41 contains an express reference to this branch of learning; it says that those who have taken aphrodisiacs[1261]

[1261] These are a means for begetting progeny according to Ut 40,2 (= AS,Ut 50,p.450a) (see also CA,Ci 2,1,4, SU,Ci 26,39; I have not looked for similar passages in other medical works). Attention may also be drawn to BH,Ni 2,14:

> *ratimūlaṃ śarīraṃ hi śarīrasya ratiḥ phalam*
> *tasmāt phalārthī mūlārthaṃ striyaḥ seveta yuktitaḥ*
> "Verily, the body has its root in sexual enjoyment; sexual enjoyment is the fruit of the body. Therefore may he wishing to attain the fruit duly [sexually] attend upon women for the sake of (or: upon women, the cause of) the root."

SMAR, p.271 states:

> *tasmād ratisukhaṃ labdhvā dadyāt piṇḍaṃ sa yoṣitaḥ*
> *ātmajanmotsukaḥ kuryāt pitṛṇām ṛṇanāśanam*
> "Therefore, having obtained the delight of sexual enjoyment, may [he] give the [offering of the] lump [of food for the manes, which only the son can offer] from [out of] the woman. May the one anxious for the birth of his [own] self [through his son] bring about the destruction of the debt of/to the fathers (i.e. manes)."

KĀD 120-132 too are of interest, especially 128f., which state that sexology is as necessary as the study of the Vedas, since through it one obtains pure (*pūta-*) progeny. And the end of the commentary on KĀDS has verses telling us that through the study of sexology one gives women pleasure and also obtains this oneself, thus obtaining the dual result of personal pleasure and progeny. See also RATIŚ, p.11 (the Sanskrit text has unfortunately not been given loc.cit.). (Cf. too ŚRĪVĀSTAVYA, pp.38ff. for an apologetic justification of the chapters on aphrodisiacs in Indian medical works.) In the light of all this one might question whether ROCHER, pp.528f. is fully right in remarking on KĀM: "*Sukha* may well be the basic concept which sets the Kāmasūtra apart from the texts of dharma. ... The Kāmasūtra operates, as far as possible, within the concepts and guidelines of the dharmaśāstras, but it also has a purpose of its own: it aims at providing *sukha*." Though ROCHER does not say so explicitly, from the tone of op.cit. it does seem as if here *sukha-* were deemed inferior to 'real' *dharma-*. But in the light of what we have seen above, one could actually hold that, since pleasure (*sukha-*) (also of the woman) is necessary to ensure progeny, sexual *sukha-* not only does not stand in opposition to *dharma-*, but is on the contrary an integral part of or prerequisite for the latter (see too n.1669). What one could however see as an opposition to *dharma-* is the fact that the *sukha-* of KĀM appears basically in a context not concerned with procreation, i.e. a context in which the seemingly original purport of studying sexology is lost, inasmuch as it shows a development ultimately emancipated from the needs of *dharma-*, this development thus resembling that of oth-

"should engage in all not censurable practices of sexual enjoyment ordained by the Kāmasūtra (or, according to the variant: Kāmaśāstra)[1262]".[1263]

This shows that the author of AH had a knowledge of sexological literature, and since we know that such knowledge was common among cultured persons in ancient India, there seems no reason to assume that the authors of other medical treatises did not have it or that they should assume that those studying their treatises did not have it,[1264] all the more so since the prepa-

er sciences such as grammar, mathematics, astronomy etc. (see also DE, pp.89f., as well as the arguments for the superiority of *kāma-* over *dharma-* and *artha-* referred to in n.1671). In any case, note the connection between *sukha-* and *dharma-* made by NĀṬ 22,148 (as usual, from the point of view of the male):

> *dharmārthaṃ hi tapaścaryā sukhārthaṃ dharma iṣyate*
> *sukhasya mūlaṃ pramadās tāsu sambhoga iṣyate*
> "The practice of austerities is assuredly for the cause of *dharma-*; *dharma-* is prescribed/acknowledged as for the cause of *sukha-*. Women are the root/basis of *sukha-*; with regard to them, sexual enjoyment is prescribed/acknowledged."

In a similar vein, Yaśodhara in his commentary on KĀM 2(=1,2),p.25 says:

> *dharmamūlaḥ smṛtaḥ svargas tatrāpi paramāḥ striyaḥ*
> *gṛhasthadharmo durvāro narāṇāṃ dharmayatnajaḥ*
> *hitāś cāpatyasantānaiḥ striyas tv iha paratra ca*
> *paraṃ sampratyayo bhogaprakarṣārthāya* (var: *bhogaḥ prakar°*) *vai striyaḥ*
> "The celestial world is authoritatively taught as having *dharma-* as its root/basis; there too/with respect to it too, women are the chief [thing]. The *dharma-* (here maybe in the sense of: duties/station) of the householder (is this a euphemism referring to sexual intercourse, used in the same manner as *grāmyadharma-* '*dharma-* of the villager', which is common in medical texts?), irresistible to men, is born of the endeavour after *dharma-*. And [both] here and yonder (i.e. in the afterworld) women are, however, beneficial through [the production of] offspring and descendants (or: through [the production of] uninterrupted series of offspring). It is absolutely [my] firm conviction: women are certainly for the sake of intense enjoyment."

And quite apart from that, the life of men with an aversion to *sukha-* is futile like that of grass, as Yaśodhara says loc.cit. (*tṛṇānām iva hi vyarthaṃ nṛṇāṃ janma sukhadviṣām*). On *kāma-* as great good see also STERNBACH 1953, pp.9f.

1262 This might be a reference to KĀM, but *kāmasūtra-* might equally be a synonym of *kāmaśāstra-* (thus also Śivadāsasena in AH3: *kāmasūtravihitām iti kāmaśāstravihitām ity arthaḥ*; similarly the *Kairalī* commentary in AH4: *kāmasūtravihitāṃ kāmaśāstrakṛtām*) and as such refer to this branch of literature in general. Conversely, the variant of this reading, looking like a reference to the branch of literature, might also be taken to be a reference to KĀM, as this may also have been known under the name "Kāmaśāstra" (see SCHMIDT 1922, pp.14;21).

1263 *ācarec ca sakalāṃ raticaryāṃ kāmasūtravihitām* (var: *kāmaśāstravihitām*) *anavadyām*.

1264 Most sexological works too, especially in the chapters concerned with recipes and the

ration of aphrodisiacs is one of the eight parts of traditional medical knowledge in India. And even if the deliberations above should be off the mark and medical authorities not have expected those interested in more details to turn to sexological works, the fact that these works by their very nature may be expected to at least partly deal with topics of interest to us justifies our turning to them for further information.

---

like, show at least some acquaintance with medical works. Some such works even quote from medical texts. Cf. also KĀM 35(=7,1),p.367 = BĀ 7,1:

āyurvedāc ca vedāc ca vidyātantrebhya eva ca
āptebhyaś cāvaboddhavyā yogā ye (BĀ: *ye yogāḥ*) *prītikārakāḥ*,

KAN 7,1,58:

āyurvedād vedād vidyātantrāt tathāptavākyebhyaḥ
yogā ete jñeyāḥ prīte prītipradā nityam,

and RAT 14,1:

dṛṣṭvānekavidhāni manmathakalāśāstrāṇi śabdārṇavaṃ
soḍḍīśaṃ haramekhalāṃ ca nikhilāṃ yogāvalīnāṃ śatam
tisro vaidyakasaṃhitā bahuvidhān ālocya śaivāgamān
yogādhyāyam ahaṃ mahājanamataṃ vakṣyāmi saṅkṣepataḥ.

Whether the *Haramekhalā* referred to in the RAT passage is in Prakrit or in Sanskrit (cf. SHARMA,P.V. 1985-1986, p.143) is unclear. With regard to the work *Uḍḍīśa*, SHARMA,P.V. 1985-1986, loc.cit. writes: "Uḍḍīśatantra is also mentioned as Haramekhalā in some of the Mss."; according to the RAT passage, the two would be different works. The expression *manmathakalāśāstrāṇi* has also been discussed below in the context of the ☛ *candrakalā-*.

According to Madhusūdanasarasvatī's *Prasthānabheda*, the Kāmaśāstra is a part of Āyurveda (see e.g. DEUSSEN, pp.59f.).

CHAPTER 12

Sexological Works: Orgasm and Ejaculation

12.1. The best known sexological treatise, which is also regarded as the oldest work on this subject that has come down to us, is KĀM, by Mallanāga Vātsyāyana.[1265] This work contains in chapter 6(=2,1),pp.77ff. a description of the theories of older authorities[1266] on the matter of female orgasm and their refutation by Vātsyāyana,[1267] thus allowing us not only to learn about and analyse these theories, but also to trace in part the historical development of the notions on this subject. I shall, however, not give a detailed account of the not in all particulars clear arguments and counter-arguments here, but merely a résumé with the salient points relevant for our discussion.[1268]

12.2. The first (and seemingly oldest) theory is that of Auddālaki (Śvetaketu),[1269] who held that women have no ☞ *bhāva-* like men,[1270]

[1265] Following the usual practice, I refer to the author by his Gotra, and not by his personal name. According to SASTRI, pp.73ff., the Vātsyāyanas originally hailed from Magadha, which need not, however, necessarily apply to Mallanāga himself, whom most hold to have been from Western India (on Mallanāga see e.g. CHAKLADAR, pp.11ff., especially pp.68ff., SCHMIDT 1922, pp.9ff., MYLIUS 1987, pp.14f.;195, DANIÉLOU, p.10).

[1266] The work BĀ purports to be made up of the concise statements in verse (*kārikā-*) of Bābhravya, whom Vātsyāyana quotes. However, it seems in actual fact most probably to have been made up later of verses culled from KĀM (and maybe other sexological treatises); see especially BĀ 7, which is merely a repetition of the last verses of KĀM (including Vātsyāyana's account — with the mention of his name — of the circumstances of and reasons for the composition of his treatise!). Note also that BĀ 2,2 = KĀM 6(=2,1),p.80 is attributed by Yaśodhara in his commentary on KĀM to Auddālaki, and that it indeed makes sense only in this case, *not* if it comes from Bābhravya.

[1267] This description, in prose interspersed with verse, is recast in verse form in KAN 2,1,24ff. (KAN is indeed regarded by many as being but a commentary, in verse form, on KĀM).

[1268] Several problematic points of the discussion have been discussed in detail by WEZLER 1971, pp.277ff. On the whole discussion see also SCHMIDT 1904, pp.230ff., RAI, pp.233ff., HAMPIHOLI, pp.78f. That the matters discussed continue to be of relevance even for modern South Asia is evinced by CHÖPEL, pp.57-60;198-201.

[1269] If this is the same as the famous Vedic sage (and there do seem to be grounds to presume so), then the theory unfolded here must belong to the late Vedic period at the very latest. See also CHAKLADAR, pp.14-16, DE, pp.89f., AGRAWALA, pp.14ff.;19ff., RAI, p.vi; cf. (on Uddālaka Āruṇi, the ancestor of Auddālaki Śvetaketu — according to BRONKHORST 1996,

and that all that happens during coitus is that their "itching"[1271] is taken away through the continuous action of the man. This delight[1272] is unfathomable to men,[1273] but inferable from the fact that the man stops after completing the sexual act without reference to the woman, this not being so in the case of a woman, she on the contrary longing for lengthy coitus and being enraged when things are otherwise.[1274] But from this it does not follow that women have *bhāva-* like men, since all this can be explained satisfactorily by the desire of the woman to have her itching removed. The school of Bābhravya (Pāñcāla)[1275], however, held that women do have *bhāva-*, and that this takes place during the whole period of intercourse, in the male, by contrast, however only at the end. The reason for saying that women do have a *bhāva-* is in striking agreement with what has been said in §§11.34f., namely (p.80):[1276]

---

p.595 his father) also FIŠER, p.115, CHAKLADAR, p.15 (on Śvetaketu's time cf. also BRONKHORST 1996, p.599).

1270 This does seem to refer to an orgasm. According to Yaśodhara this means that women do not have the delight connected with ejaculation that men have, and that this is due to their lack of semen:

> *yādṛśaṃ sukhaṃ visṛṣṭiprabhavaṃ puruṣo 'nubhavati tādṛśam eva na strī śukrābhāvāt.*

visṛṣṭi- "emission" here clearly refers to an ejaculation, like *vi√sṛj* and *visṛṣṭi-* in §11.24, though the word is not always used in this narrow meaning ("ejaculation") in all cases; cf. e.g. §12.3 (especially n.1281). Indeed, it is often difficult to say what exactly *visṛṣṭi-* refers to, as e.g. in Yaśodhara's commentary on KĀM 2(=1,2),p.14, where he says that the genitals generate joy (☞ *ānanda-*) in the state of *visṛṣṭi-*:

> *tatra strīpuṃsayor yad adhovyañjanaṃ sambādhakādi tanmātrasvabhāvaṃ tat tvagindriyam eva. tasya kaścid eva pradeśa upasthendriyam ucyate. yo visṛṣṭyavasthāyam ānandakarma janayati.*

In the case of the male, this would refer to an ejaculation, but that need not be the case as regards the female.

1271 On this see n.1211.

1272 The characterisation of this delight by KĀM leads to a problem which is discussed in Excursus XIII on pp.438-442.

1273 On Yaśodhara's explanation that this is due to the fact that women have no semen (*śukrābhāvāt*) and hence no ejaculation (*visṛṣṭi-* "emission") like men (*puruṣasukhaṃ tu visṛṣṭibhāvitvāt*), see also n.1270.

1274 She therefore obviously has pleasure, which is, however, not curtailed through a single act like that of a man, so that one cannot gauge its pinnacle.

1275 On him and his school see e.g. CHAKLADAR, pp.12-14, DE, p.93, AGRAWALA, pp.17ff.;22ff. Bābhravya has (as *pāñcāla-*) even given his name to one of the *nāyaka-*s in sexological literature; cf. ŚR, pp.34-42 of the Introduction, RAI, pp.144;150, UPADHYAYA, p.13.

1276 On statements such as the following giving the intent, but not necessarily the original

na hy asatyāṃ bhāvaprāptau garbhasambhavaḥ[1277]
"for there is[1278] no production of an embryo if there is no attainment of *bhāva-*."

Against the objection that that cannot be because the passion of the woman slowly builds up during intercourse and reaches a peak, after which she wishes to stop, it is said that this does not show a special delight at the end, but only a different degree of the same delight that is there from the beginning, the wish to stop being occasioned by the woman's "loss/diminution of the element" (*dhātukṣaya-*) (p.82).[1279]

12.3. The term *dhātukṣaya-* reminds one of the problematic *rajaḥ-kṣaya-* in §5.2 and *dhātukṣaya-* in §5.3, also of the use of the term *dhātu-* to describe the procreatory-menstrual fluid.[1280] But these do not really help us much in determining what the expression as used here in KĀM refers to. Yaśodhara's commentary on p.81 explains the beginning of the argument of the school of Bābhravya by saying that both male and female experience the delight of "emission" (*visṛṣṭi-*), the woman, however, from the very beginning of intercourse, for she gradually becomes, as can be perceived, wet like a broken water vessel. Her delight is conjoined with an emission (*visṛṣṭi-*) like that of the man, accompanied by ☞ *bhāva-*, from the beginning, while the man's *bhāva-* is obtained at the end, due to the voiding (*visarga-*) of semen (*śukra-*). Their delight of emission (*visṛṣṭi-*)[1281] is the same, though not the time. And this is a precondition for conception in the woman to take place, such conception not being possible without the satisfaction (*tṛpti-*)[1282] (which here seems the same as or the cause of *bhāva-*) due to the emission (*visṛṣṭi-*) of semen (*śukra-*).[1283] Explaining *dhātukṣaya-*, Yaśodhara says

---

words of an authority or his school, cf. WILHELM 1979, pp.403f. (also SCHMIDT 1922, p.56).

1277 Var: *garbhasambhava upapadyate.*

1278 Var: "takes place".

1279 *dhātukṣayāc ca virāmābhīpseti.*

1280 Cf. p.555 in Appendix II.

1281 On *visṛṣṭi-* cf. n.1270. It is clear that in the passage being discussed here *visṛṣṭi-* is used with regard to the woman in the more general sense of "emission" and not only of the narrower "(orgasmic) ejaculation", though with regard to the man the latter meaning pertains. Note too that Yaśodhara uses the word *visarga-* "voiding" for the discharge of semen by the man; is he thus attempting to differentiate between "ejaculation" and mere "emission" by avoiding *visṛṣṭi-*, which here is used in both senses?

1282 On this term see also RAI, pp.112f.

1283 The whole passage commented upon (p.80) runs:

(p.83) that when the sentiment/emotion (*bhāva-*) called *kāmitā-*[1284] arises, a ☞ *dhātu-* called *śukra-*, leaving its own place, reaches its own tube (var: tubes)[1285] and gradually flows out from the very beginning of coitus, its gradual loss or diminution bringing about a cooling down until finally the woman wishes to stop.[1286] From this it follows that he understands the fluid called *dhātu-* to refer to the lubricating fluid(s)[1287] of the female, its (their) flow causing *bhāva-*, which ends when a loss or diminution (*kṣaya-*) has taken place. Whether other fluids possibly released to the exterior too are subsumed under this is unclear. I must confess that I find that this makes very good sense, and that I cannot think of any other way in which we could understand the theory of the school of Bābhravya. It must, though, be pointed out that the term *śukra-*, which is according to Yaśodhara what flows out, is extremely problematic and, moreover, not used in the argument of Bābh-

---

> *sātatyād yuvatir ārambhāt prabhṛti bhāvam adhigacchati. puruṣaḥ punar anta eva. etad upapannataram* (vars: *tad upapannataram* and *tad upapannatarakam*). *na hy asatyāṃ bhāvaprāptau garbhasambhava* (var: *garbhasambhava upapadyata*) *iti bābhravīyāḥ.*

The relevant portion of the commentary of Yaśodhara says:

> *dvāv api visṛṣṭisukham adhigacchataḥ. strī tv ārambhād yantrayogāt prabhṛti sātatyān nairantaryeṇa. sā hi puruṣeṇopasṛpyamāṇā prabhinnajalabhāṇḍavac chanaiḥ klinnasambādhā bhavatīti pratyakṣasiddham etat. sukhaṃ ca puruṣasyeva visṛṣṭyanuviddham ity ārambhāt prabhṛti bhāvam adhigacchati. puruṣaḥ punar ante bhāvam adhigacchati. tadānīṃ śukravisargāt. etad iti yathoktam upapannataram. pramāṇasiddhatvāt. tataś ca tayor bhinnakālatvān na sādṛśyam iti na kālato nava ratāni* (on *nava ratāni* cf. n.1332). *bhāvatas tu santi. visṛṣṭisukhasādṛśyāt. nanu sambādho vraṇasvabhāvatvād upanudyamānaḥ klidyatīty āha na hīti. rasaprāptau visṛṣṭisukhādhigame tṛptā hi strī garbhaṃ dhatte. yathāha carakakāraḥ.* (After this we have CA,Śā 2,23, with some minor variants; on this — to us in this context uninteresting — description of conception which has just occurred cf. also §8.20.) *tṛptiś ca* (var: *tṛptis tu*) *bhāvaḥ. sa ca na śukravisṛṣṭiṃ vinety abhiprāyaḥ. ārtavaṃ visṛjati na śukram iti kecit. yad āha. kāmāgnitaptastrīpuṃsayor anyonyadehasaṃsargād araṇīdaṇḍābhyām iva vahniḥ śukrārtavamathanād iti.* (The portion from *ārtavaṃ visṛjati* to *śukrārtavamathanād iti* will be discussed below in §12.6.) *asti tāvat tṛptinibandhanaṃ. kiṃ tad iti cintyate. yadi tan na śukraṃ kathaṃ yoṣito garbhasambhava utpadyate.*

[1284] On this cf. the discussion on ☞ *bhāva-* in Appendix II.

[1285] ☞ *nāḍī-*.

[1286] *sātatyena bhāvasya pravṛttatvāt kathaṃ virāmābhīpsety āha dhātukṣayāc ceti. samutpanne kāmitākhye bhāve yaḥ śukradhātuḥ svasthānāc cyutaḥ svanāḍīṃ* (var: *svanāḍīḥ*) *pratipadyate tasyārambhāt prabhṛti śanaiḥ śanaiḥ syandanāt kṣaye nivṛttarāgatvād virāmābhīpsā.*

[1287] Cf. n.779.

ravya's school. What must also be drawn attention to is that the fluid itself is nowhere by the school of Bābhravya said to form the embryo; this is said to be formed due to the *bhāva-*.[1288] This latter is, however, according to Yaśodhara dependent on the flow of *śukra-*.[1289] But the exact wording of his statement is puzzling:

> "There is indeed a basis/cause for satisfaction. 'What is it?' is pondered. If it is not *śukra-*, how does the production of the embryo of a woman take place?"[1290]

This looks as if to Yaśodhara *śukra-* (of the woman) were responsible for conception too, the more so since he then quotes SU,Śā 2,47 (in the form mentioned in n.795) to show that two women too can beget a child through *śukra-*. This is rather confusing, though it does remind one of the case last referred to in §11.26 (with further references) in which *śukra-* could maybe refer to the procreatory fluid. One might of course hold that Yaśodhara is here giving not his own arguments, but the arguments he supposes the school

---

[1288] KAN 2,1,37 too says simply:

> *aratau tāsāṃ na garbhaḥ syāt*
> "If their (feminine) *rati-* (on this cf. ☞ *bhāva-* in Appendix II) were not, there would be no embryo",

the whole verse running:

> *sātatyād ārambhāt prabhṛti mṛgākṣīṣu jāyate bhāvaḥ*
> *prānte tu nāyakānām aratau tāsāṃ na garbhaḥ syāt*,

with a variant *amatau* ("if their intention/acquiescence were not") for *aratau*, which in the given context is a feasible alternative only if it means that a woman unwilling to have intercourse cannot be expected to attain *bhāva-*.

[1289] The mention of a place of *śukra-* from which this *śukra-* flows of course reminds one of the receptacle of semen of women last mentioned in §11.12 (cf. also §11.33). This is here clearly connected with a tube or tubes through which the *śukra-* flows, ultimately reaching the exterior. Whether this tube or these tubes lead(s) directly to the exterior is not clear. But the discussion in the rest of this chapter, if I be allowed its anticipation here, does rather speak for the discharge being first internal (not necessarily uterine) and only then external; only if a discharge via the urethra should be meant (see §12.16; on the connection of semen with the urinary system cf. also §§6.13f.) could one consider seeing in the tube mentioned (if the singular is indeed the correct reading) the urethra leading to the exterior, but it is obvious that this is a very doubtful proposition. Moreover, we have the added problem that *śukra-* here seems to be a procreatory fluid of the woman (see below); could it thus be the same as the procreatory-menstrual fluid? In this case the ejaculation of course would have to be into the uterus, at least at first. But whether *śukra-* be identical with the procreatory-menstrual fluid or not, this does not alter the fact that *svasthāna-* in the text in n.1286 seems to presuppose a receptacle of this *śukra-* somwhere within the body, which would definitely not be the uterus if we had here an internal discharge into the latter.

[1290] From *asti tāvat tṛptinibandhanaṃ* to *yoṣito garbhasambhava 'utpadyate*; see n.1283.

of Bābhravya would give, showing the difficulties they would occasion, without explicitly saying so; this assumption may be substantiated by pointing to the commentary on the theory of Auddālaki (see §12.2), in which Yaśodhara closely follows the arguments given, explaining these from the point of view of their proponents.

12.4. But be that as it may, the difficulties just described arise from the commentary of KĀM, not from the text of KĀM itself. As regards this, after having stated the theories of his predecessors Vātsyāyana states his own theory (which seems to be the one generally accepted in later times). According to this the manifestation of delight of men and women is exactly the same,[1291] which must mean that both experience an orgasm due to ejaculation at the end of their coitus, as Yaśodhara too holds.[1292] Later (p.90) Vātsyāyana says that intercourse[1293] is carried on until the *dhātukṣaya-* of both, which in this context must refer to the woman too. Moreover, since ☞ *dhātu-* is clearly used for the semen of the male, it seems very probable that it also signifies the "semen" of the female,[1294] even though Vātsyāyana gives us no explicit statement on the substance the latter ejaculates.[1295]

[1291] Vātsyāyana states (p.83):

tasmāt puruṣavad eva yoṣito 'pi rasavyaktir (vars: *'pi rasasya vyaktir* and *'pi ratir*; on *rasa-* and *rati-* cf. ☞ *bhāva-* in Appendix II) *draṣṭavyā*.

Later he says (p.86):

tenobhayor api sadṛśī sukhapratipattir iti.

[1292] In his commentary on the passage from p.83 in n.1291 he says: *yathā puruṣasya visṛṣṭir ante ca tadvad eva yoṣito 'pi draṣṭavyā*. Note that *visṛṣṭi-* is here obviously used in its narrower sense (cf. n.1270).

[1293] The characteristic of which is that the male is hefty during the first coitus and ejaculates fast, subsequently doing the opposite, whereas the behaviour of the female is the opposite of this. Similar statements are found in many later sexological works.

[1294] Cf. in the context of this terminology also the discussion on ☞ *dhātu-* in Appendix II.

[1295] The original words of the KĀM are:

prathamarate caṇḍavegatā śīghrakālatā ca puruṣasya. tadviparītam uttareṣu. yoṣitaḥ punar etad eva viparītam. ā dhātukṣayāt. prāk ca strīdhātukṣayāt puruṣadhātukṣaya iti prāyovādaḥ.

Yaśodhara (p.91) glosses *dhātukṣaya-* here with *śukradhātukṣaya-*. As regards *dhātu-*, attention may also be drawn to RAT 3,26, where the *vaḍavā-* type of woman is characterised as *prathamacaramadhātvadhikā*. SCHMIDT 1922, p.156 translates this as "besitzt zu Anfang und zuletzt überreichen Samenstoff", but Kāñcīnātha in his commentary explains this as meaning that the woman has an excess of the first *dhātu-* wind and the last *dhātu-* phlegm (*prathamadhātur vāyuḥ. caramadhātuḥ śleṣmā. tābhyām adhikā vātaśleṣmaprakṛtir* ...) (on ☞ *dhātu-* denoting the morbific entities or ☞ *doṣa-*s cf. Appendix II); Kāñcīnātha's explanation does seem to give more sense.

In any case, *kṣaya-* here must refer to a loss/diminution brought about by the ejaculation (as Yaśodhara too seems to opine), i.e. it is used in the same meaning (loss/diminution of a fluid) as that in which we have supposed it to be used in the case of the theory of the school of Bābhravya, with the difference that the latter theory speaks of a gradual loss or diminution right from the beginning of coitus (thus referring to the lubricating fluid?), the theory of Vātsyāyana, however, of a loss or diminution brought about by an ejaculation (which may, however, occur several times; cf. n.1293) (thus referring to something different from the lubricating fluid?); cf. §12.6.

12.5. It is in this connection perhaps not without interest that Vātsyāyana is of the opinion that the man should ensure that the woman obtains her orgasm[1296] first,[1297] because, since both are of the same nature (as regards the character of their orgasm), their delight during intercourse too is desired to be (or, probably better: is regarded as) (*iṣyate*) of the same nature.[1298] It is obvious that with this remark Vātsyāyana is in direct contrast to those medical authorities who see a prior orgasm of the woman as a grave disorder to be avoided (see §11.34), though it should not be forgotten that these authorities on their part seem in conflict with others according to whom a prior orgasm of the woman is no disorder. But we must remember that Vātsyāyana is not primarily concerned with procreation as the medical authorities are, but with sexual pleasure (cf. too n.1261), his reasoning seeming to be (as Yaśodhara too remarks on p.87)[1299] that a man cannot as a rule carry out intercourse after he has had his orgasm, which leaves the woman who has not already obtained her orgasm unsatisfied.[1300] Of course

---

1296 The term used here is *rati-*, on which see also ☞ *bhāva-* in Appendix II.

1297 Yaśodhara adds: if the orgasm cannot be simultaneous. On the simultaneity of orgasm see e.g. KĀMS, p.23, where this is said to be what one should strive for.

1298 This is stated in a verse on p.87:
jāter abhedād dampatyoḥ sadṛśaṃ sukham iṣyate
tasmāt tathopacaryā(var: *°cāryā*) *strī yathāgre prāpnuyād ratim.*

1299 In his commentary on 13(=2,8),p.161 Yaśodhara says too that an orgasm of the man before that of the woman results in intercourse whose savour is interrupted (*rataṃ vicchinnarasam*).

1300 Attention may in this context also be drawn to the last line of the passage in n.1295:
"'And the *kṣaya-* of the *dhātu-* of the man is before the *kṣaya-* of the *dhātu-* of the woman', is a general assertion".
In other words, a man is unable to carry on because of his having no more *dhātu-* before the woman (see §12.4 on *kṣaya-*). The outcome is the same as when he has an orgasm before her, namely that the woman remains unsatisfied, in which case a remedy of the sort Vātsyāyana prescribes makes sense.

Vātsyāyana's advice (which many sexological treatises repeat) is clearly in conflict with the theory that a prior orgasm of the male ensures that an (as a rule probably desired) male child is born, as also the notion of a child being born at all if the male has his orgasm first could conflict with the theory of the medical texts that an orgasm of the female too is needed for conception (see §11.34), but it does look as if these different theories were formulated quite independently of each other, and so we would probably be wrong in trying by all means to look for agreement among them. Nevertheless, it must be pointed out that the medical texts seem to have ignored the problem drawn attention to by the sexologists, namely the difficulty of carrying on intercourse to ensure an orgasm of the woman too if the man has already ejaculated.[1301] This is a most interesting example of what may be called the fragmentation of knowledge in traditional India into different Śāstras which need not necessarily be connected, i.e. of a specialisation of the sort deemed often to be a very recent Western phenomenon.

12.6. In any case, it seems as if at least Vātsyāyana refers to the same phenomenon we have already discussed a few times (last in §11.34; cf. also n.1289), namely to the ejaculation by the woman of "semen" (☞ *dhātu*- "element" in Vātsyāyana's terminology). Since this is emitted at the time of orgasm, we cannot simply equate it with the lubricating fluid(s). On the other hand, the fluid mentioned by Auddālaki and the school of Bābhravya would fit this latter fluid. The whole is complicated by the use of the same word (*dhātu*-) for the fluid mentioned both by the school of Bābhravya and by Vātsyāyana, it appearing as if different substances were meant (for we face great difficulties should the same fluid be meant, especially due to the connection with orgasm made by Vātsyāyana), though it must also be pointed out that this applies only if Yaśodhara's explanation of the former's theory is correct. Further complicating the matter is Yaśodhara's usage of *śukra*- "semen" to refer to the emission the theory of the school of Bābhravya mentions. This is then made even more complicated by a passage in the same part of Yaśodhara's commentary on p.81 as that given in n.1283. Comment-

---

[1301] CA,Śā 8,6 tells us that one of the causes for non-conception is too great lust of the woman:

maithune cātikāmā vā na garbhaṃ dhatte.

This Cakrapāṇidatta explains in his commentary as referring to a woman who craves for intercourse even when the man's desire has ceased:

maithune cātikāmeti nivṛttecche 'pi puruṣe maithunam icchantī.

Regardless of what CA actually means (cf. in this context also §11.28), Cakrapāṇidatta's remarks could be taken to refer to a woman who has not had her orgasm and cannot have it any more because the man breaks off intercourse (scil.: after his orgasm).

ing on the *dhātukṣaya-* the school of Bābhravya presupposes, Yaśodhara says that according to some authorities a woman discharges (*visṛjati*)[1302] not semen (*śukra-*), but *ārtava-*! In this context he then quotes a passage from an unknown source giving basically the same cause for this flow as in §8.1.[1303] Since the emission we are dealing with seems external, the theory of *ārtava-* being emitted clearly causes problems of the sort we have already discussed, which, as we have seen, some commentators have "solved" by e-quating *ārtava-* with *śukra-* and differentiating it from *rajas-* (see especially §§8.3ff. and §8.15, also §9.8 and §11.34). It must, however, also be pointed out that, irrespective of how Yaśodhara understands it, we can by no means be sure that the passage he quotes does indeed presuppose an external emission of *ārtava-*, for it could very well be that all it refers to is the internal ejaculation of procreatory-menstrual fluid into the uterus; cf. the two alternatives mentioned in §8.1. This then makes us wonder whether Vātsyāyana too might not have had an internal ejaculation in mind, and not an external ejaculation as we have hitherto supposed; we shall return to this problem in §12.10.

12.7. Yaśodhara himself[1304], though, does not seem to adhere to the theory that it is *ārtava-* and not *śukra-* which is emitted (see §12.3). In this connection he also tells us that *ārtava-* is the ☞ *dhātu-* blood, derived from ☞ *rasa-*, in some situations/conditions/states, while the *dhātu-* semen (*śukra-*) derives from the *dhātu-* marrow (☞ *majjan-*).[1305] The statement on *ārtava-* is maddeningly imprecise. Does Yaśodhara mean that blood in general and *ārtava-* are basically one and the same, the former appearing as the latter in certain situations (such as menstruation)? That would, needless to say, be extremely problematic (cf. also §11.32). Or is he rather saying that *ārtava-* is blood in a different condition, i.e. either in the same manner as supposed by Indu in §11.33, or else as in the theory we have discussed in

---

1302 According to the context, which is the same as in §12.3, *visṛjati* should mean "emits" and not "ejaculates (orgasmically)" (cf. n.1281). Whether the passage (mentioned below) Yaśodhara quotes (and which does not use *vi√sṛj* or any nominal derivative of this) presupposes a mere emission or an ejaculation is unclear, as is indeed also how Yaśodhara actually understands it.

1303 *ārtavaṃ visṛjati na śukram iti kecit. yathāha kāmāgnitaptacittastrīpuṃsayor anyo'nya-dehasaṃsargād araṇidaṇḍābhyām iva vahniḥ śukrārtavamathanād iti.*

1304 Or, if the possibility broached in §12.3 should turn out to be correct, the school of Bābhravya as Yaśodhara represents it.

1305 *tasmād rasadhātor utpanno 'sṛgdhātur eva kasyāñcid avasthāyām ārtavam. śukradhātus tu majjadhātor utpadyata iti.*

§§11.5ff., and as such in both cases unproblematic as regards the derivation of *ārtava-*? I may here point out that Yaśodhara was obviously acquainted at least somewhat with Indian medical theories and terminology, so that it would not be surprising if he had known of a theory deriving the procreatory-menstrual fluid from blood or closely linking the two with regard to their formation.[1306] Moreover, someone acquainted with Indian medical theory would also probably have known that *ārtava-* was held to combine with semen from the male to form a child. But can we be certain that Yaśodhara really did know of *these* theories? There is, for instance, the problem that he seems to hold only "semen" to be procreatory, not *ārtava-*. Or is he merely telling us that this "semen" too is procreatory (as many of the medical authorities we have quoted too hold), without necessarily supposing that *ārtava-* is not? For it could be that the main reason for his mentioning the difference between the two fluids is that he might hold *ārtava-*, connected with blood as it is, to be basically red, which seems not to hold good for the *śukra-* emitted during intercourse, whatever it actually be.[1307] Apart from this we also have the problem of whether *ārtava-* is emitted only internally or not (cf. §11.34), as a substance not coming out of the body during intercourse would obviously not fit here. And then there is also the statement deriving *śukra-* from marrow, which here seems to refer to *śukra-* of the female, from which it would follow that this would then be the same as the *śukra-* of the male. All in all, we find ourselves in a maze of problems, which is, however, — unfortunately — nothing new by now. The whole is compounded by the possibility that Yaśodhara may not be giving us his own opinion here (see n.1304): note in this regard also the reporting or quotative *iti* which ends the statement in n.1305. But can we be sure that his manner of argumentation in that case is indeed that which the adherents of the position he would be representing followed?

12.8. But be that as it may (and irrespective of whether the way in which Yaśodhara may represent others is the way in which these themselves would have represented themselves or not), the commentary of Yaśodhara

---

[1306] This does not necessarily mean that he also was acquainted with the ☞ *upadhātu-* theory as today current, though he may have known the term; Yaśodhara (and the oldest post-KĀM works we today know), could belong to the period between the eleventh and thirteenth centuries (cf. SCHMIDT 1922, pp.21ff.; MYLIUS 1983, p.314), i.e. the time when the *upadhātu-* theory as we today know it seems not yet to have gained general acceptance (cf. §§11.5ff.).

[1307] I have written "seems" because ŚṚṄG 2,7 says that the *anaṅgajala-* (on this see n.1346) of the *hastinī-* type of woman is red (*rakta-*). I do not know what to make of this statement, which is not only peculiar, but to my knowledge also singular — at least I could find no parallel.

on the remarks of Vātsyāyana himself seems to make the possibility that Yaśodhara all along plays the *advocatus diaboli* of the authorities whose views he explains very much a probability, for while commenting on the first line of the verse given in n.1298, which says, among other things, that the orgasmic delight of the "couple" (*dampati-*) is of the same sort, he says that women have an emission at the end of intercourse just like a man (which is in keeping with the theory given in §12.4), this emission consisting of *śukra-*. This seems to be something different from the *śukra-* Yaśodhara mentions in connection with the theory of the school of Bābhravya (see §§12.6f.), showing that here Yaśodhara is now putting himself in the place of Vātsyāyana and arguing for him. But a closer look at what Yaśodhara says makes our joy turn out to be short-lived, for we find it said (on p.87) that:[1308]

> "Of the couple [means] of the woman and the man. The meaning is: of the two by whom the same goal is obtained. [When this is so, then] it is,[1309] however, of such a measure: [that] which is the different delight due to the removing (var: departure) of her itching[1310] and [that] which is the flowing (*syandana-*) of semen when the vagina is being subjected to friction is according to the respective difference of type of the wom-

---

[1308] The passage translated below states:
dampatyoḥ strīpuṃsayoḥ (var: *strīpuṃsoḥ*). *ekārthābhiprapannayor ity arthaḥ. etāvat tu syāt. avāntarastrījātibhedād yad aparam asyāḥ kaṇḍūtyapanoda-sukhaṃ* (var: *kaṇḍūtyapagamāt sukhaṃ*) *yac copamṛdyamāne* (var, giving no sense: *copamṛdyamānena*) *sambādhe syandanaṃ śukrasya. visṛṣṭisukhaṃ tu puruṣavad anta eveti* (var: *antarā ceti*). *yathoktam* (var: *yathā coktam*)

kaṇḍūtyapagamāt strīṇāṃ kṣaraṇāc ca sukhaṃ dvidhā
syandanaṃ ca visṛṣṭiś ca śukrasya kṣaraṇaṃ dvidhā
klinnatā kevalasyandād visṛṣṭer mathanāt sukham
ante tv ākṣiptavegāyā visṛṣṭir naravat smṛtā.

The last three lines also occur as ĀY 4,139, in the following garbled form:

syandanaṃ ca visṛṣṭiś ca śukrasya kṣaraṇaṃ tathā
klinnatā kevalaṃ syandāt visṛṣṭaṃ madhunā śubhā
ante tv (var: *sujāyate tv*) *ākṣiptavegāyā visṛṣṭir naravat smṛtā.*

This is translated by the editors as: "*Syandana* (melting), *visṛṣṭi* (elimination) and *kṣaraṇa* (external ejaculation) — these are the three characteristic features of semen. If there is only *syandana* (melting) then this produces stickiness. If there is *visṛṣṭi* (elimination) then this gives pleasant feeling. *Kṣaraṇa* (ejaculation) is the terminating point of the sexual urge." This is a valiant effort, but not quite what the verses say; the translators themselves have appended after this: "The meaning of these verses is not clear."

[1309] On the optative cf. VṚK, pp.508f.

[1310] On this see n.1211.

an.[1311] But the delight of emission (*visṛṣṭi-*) is, as [in the case of] the man, only at the end.[1312] As (var: And as) is stated:

> The delight of women is twofold due to the removal of itching and due to streaming (*kṣaraṇa-*). Streaming is twofold: flowing (*syandana-*) as well as emission (*visṛṣṭi-*) of semen. Moistness is through a mere flow(ing) (*syanda-*), the delight of emission (*visṛṣṭi-*) due to churning/friction. But the emission (*visṛṣṭi-*) of the one (feminine) in whom an orgasm is manifested/effected[1313] is authoritatively taught (*smṛta-*) as being at the end, as [in the case of] the man."

(It is obvious that *visṛṣṭi-* here refers to an ejaculation; cf. on this n.1270.)

12.9. From this it is clear that Yaśodhara calls both the fluid which seems to be the lubricating fluid(s) and the fluid ejaculated at the end of intercourse "semen", which is also an elegant way of getting around the termi-

---

1311 This could refer to different sizes of the generative organ, or to different temperaments, or to times required to obtain orgasm, and the like (cf. on different types of women SCHMIDT 1922, pp.153ff.;261ff., RAI, pp.143ff.;154f., KAPP, and KEL 5,1ff.;7,1ff., also KĀMS, pp.19ff., WAYMAN, pp.184f. and PLOSS/BARTELS/BARTELS 1,p.358; see also n.362). However, these differences can hardly be of relevance as regards the *basic* procedure of attaining an orgasm (i.e. the final ejaculation), and which is discussed here as being common to all women, as clearly also irrespective of the actual circumstances accompanying or leading to it. Thus it seems more likely that the "respective difference of type of the woman" refers not to different types of women, but to the difference of a woman with respect to a man, as this is what Vātsyāyana has discussed in the foregoing passages and remarks concludingly on in the passage here being explained by Yaśodhara. "Itching", it follows, is the characteristic of a woman, not of a man; the "friction" of the vagina too is of a different sort (internal) from that experienced by the penis. Nevertheless, men and women attain orgasm in the same manner according to the text. So it could also be debated whether *avāntara-* ("between" > "included in" > "according to") could not be translated here as ("included in" > "subordinate to" >) "irrespective" or "irrelevant", a meaning which is attested in Sanskrit and in some New Indo-Aryan languages.

1312 *iti* here ends the actual explanation of Vātsyāyana's words, what follows being an amplification. As regards the variant

> "But the delight of emission is as [in the case of] the man and in between/and now and then (*antarā*)",

this is clearly quite out of place here, unless multiple orgasms of the woman are meant during the time the man takes to reach his orgasm, which could, however, hardly have been taken to be what usually happens.

1313 On *vega-*, mentioned also in the discussion on ☞ *bhāva-* in Appendix II, see WEZLER 1971, p.277,n.26; the term seems to have different technical meanings.

nological difficulties drawn attention to in §12.6. We must of course ask ourselves what Yaśodhara bases this on; is it only the verse he quotes, or is it something else besides? If the former, then the line

syandanaṃ ca visṛṣṭiś ca śukrasya kṣaraṇaṃ dvidhā,

translated above as:

"Streaming is twofold: flowing as well as emission (scil.: ejaculation) of semen",

would be the main prop of Yaśodhara's statement. This verse is, however, as the translation too tries to make clear, ambiguous: "semen" may be connected with both "flowing" and "emission", in which case we would have a flow of semen and an emission of semen, or it may be connected only with "emission", in which case we would have only an emission of semen, but a flow of something else. One may even translate the whole line differently, namely as:

"The streaming of semen is twofold: flowing as well as emission."

We thus cannot be sure what the line actually means to say, and whether Yaśodhara has misunderstood it or not. If, however, Yaśodhara also bases his remarks on other notions known to him, then it is immaterial what the verse actually says, for we would in any case have a theory, calling both sorts of fluids *śukra-*, in actual currency and not merely based on a possible misrepresentation of a single line by a single individual.

12.10. These are, however, not the only difficulties Yaśodhara occasions us. In his commentary (on p.91) on the passage cited in n.1295 he tells us that males and females behave differently during intercourse[1314] because the female has eight times the quantity of the ☞ *dhātu-* (i.e. the fluid ejaculated) of the male:

yat puruṣasya dhātor ekaguṇatvād yoṣitaś ca paścād aṣṭaguṇatvāt tad āha prāk ceti[1315]

"Because/Since the male's [way to *dhātukṣaya-*] is due to the ☞ *dhātu-*'s[1316] being one-fold and the female's [different way to *dhātukṣaya-*] afterwards is due to the [*dhātu-*'s] being eight-fold, therefore he says: 'and before'[1317]."

1314 Which ideally lasts, as we have seen in §12.4, until *dhātukṣaya-* occurs.

1315 Var: *aṣṭaguṇatvād yad āha prāg avocad iti.*

1316 On this term cf. also §§12.3f.

1317 This refers to:

prāk ca strīdhātukṣayāt puruṣadhātukṣaya iti prāyovādaḥ

(on which see also n.1300). As regards the variant in n.1315, the translation would be:

"... being eight-fold, as/because he says: 'before, he said'."

This is extremely intriguing in view of the fact that the AH, AS and BYT passages as well as KIRFEL 1954 mentioned in n.144 (but not the ĀY passage of §6.11) say that the quantity of procreatory-menstrual fluid in the body is eight times that of semen. Is Yaśodhara seeking a link with this? Or does the ratio of the two fluids rather stem from the notion, found in the same passage of Yaśodhara's commentary on pp.90f. and also on p.82, of lust (*kāma-*) and passion (*rāga-*) of women being eight times that of men, which is also a cliché in Indian literature?[1318] But even if Yaśodhara might have

---

Is this just a mistake, or could it refer to a variant of the KĀM passage?

[1318] Cf. e.g. the (in)famous verse (STERNBACH 1977 5714 on p.1368), whose commonest form is:

āhāro dviguṇaḥ strīṇāṃ buddhis tāsāṃ caturguṇā
ṣaḍguṇo vyavasāyaś ca kāmaś cāṣṭaguṇaḥ smṛtaḥ.

See also AN 4,29, PAU 16 with commentary, PA 2,39, MEYER,J.J. 1952, p.380,n.2, JAINI, p.169, LESLIE,I.J. 1994, p.65. It is thus no wonder that women were considered to be sexually insatiable, cf. the verse Yaśodhara cites on p.79 and refers to several times (e.g. on pp.82;91):

agnis tṛpyati no kāṣṭhair nāpagābhiḥ payodadhiḥ
nāntakaḥ sarvabhūtaiś ca na puṃbhir vāmalocanā.

An only slightly different version is cited by BÖHTLINGK 1966 3547; a similar verse is found in PA 2,40. This verse is only the beginning of a passage, not very complimentary to women, in MBH 1,1189* (with variants given loc.cit.):

nāgnis tṛpyati kāṣṭhānāṃ nāpagānāṃ mahodadhiḥ
nāntakaḥ sarvabhūtānāṃ na puṃsāṃ vāmalocanāḥ
evaṃ tṛṣṇā tu nārīṇāṃ puruṣaṃ puruṣaṃ prati
agamyāgamanaṃ strīṇāṃ nāsti nityaṃ śucismite
putraṃ vā kila pautraṃ vā kāsāṃścid bhrātaraṃ tathā
rahasīva naraṃ dṛṣṭvā yonir utklidyate tataḥ
etat svābhāvikaṃ strīṇāṃ na nimittakṛtaṃ śubhe.

(This passage is also quoted, with a few variants, in SMAR, pp.267f., one of these variants being *utsvidyate* for *utklidyate*; *utsvidyate* is also found in several of the variants given in MBH, loc.cit. See on this n.1441. A not quite so drastic version of the last part is found in MBH 4,Appendix I,15,ll.28-30 on p.311: this refers only to unchaste women.) That women were considered to be by nature so randy that even the sight of their nearest male kin made their vagina wet is also stated by the verse BÖHTLINGK 1966 7107:

sundaraṃ puruṣaṃ dṛṣṭvā bhrātaraṃ pitaraṃ sutam
yonir dravati nārīṇāṃ satyaṃ satyaṃ janārdana.

SMAR, p.268 adds another verse to this and says that the whole is extracted from the *Mahābhārata*; I could not find it in MBH:

saundaryapuruṣaṃ dṛṣṭvā bhrātaraṃ pitaraṃ sutam
yonir dravati nārīṇāṃ tathyaṃ me śṛṇu keśava
pañca me (var: *pañcamāḥ*) *patayaḥ santu ṣaṣṭho mama virocate*
lokeṣu puruṣābhāve sarvā nāryaḥ pativratāḥ.

Mānavadharmaśāstra 2,213-215 says:

svabhāva eṣa nārīṇāṃ narāṇām iha dūṣaṇam

this and not a medical theory in mind, he could still be referring to the procreatory-menstrual fluid, for PAU 28f. explicitly attributes sexual power (*utsāha-*) and sexual appetite (*ruci-*) to the accumulation of procreatory-menstrual fluid (*lohita-* "blood") and semen (*dhātu-*) in the body of female and male respectively.[1319] PAU and its commentary give no reason,[1320] but

---

ato 'rthān na pramādyanti pramadāsu vipaścitaḥ
avidvāṃsam alaṃ loke vidvāṃsam api vā punaḥ
pramadā hy utpathaṃ netuṃ kāmakrodhavaśānugam
mātrā svasrā duhitrā vā na viviktāsano bhavet
balavān indriyagrāmo vidvāṃsam api karṣati.

Cf. in this regard too KAKAR, pp.18f., who also gives modern parallels (also on pp.50f.); for more modern material see KAISER, pp.103f. In this regard cf. also CHÖPEL, pp.144;264: "If a pregnant woman has strong desire for copulation, it is a sign the child dwelling in the womb is a girl."

That other ancient cultures too had a similar view of women is evinced e.g. by DEAN-JONES 1992, whence we learn that in ancient Greece too women were considered to have an enormous sexual appetite, over which they have no control. See also ADRADOS, and KING, pp.30;40,n.5.

1319 The PAU passage states:
28: *prakṛtivikārasya lohitopacayaḥ ratyutsāhavardhane hetuḥ.*
29: *ubhayor upacayaḥ manmathasaṅgarasya rucyutpādakatve hetuḥ.*

See also 25f. (Note that intercourse is called *saṅgara-* "combat; battle" here, and similarly *āyodhana-* in 25. Similarly, intercourse is called *ratisaṅgrāma-* in AS,Ut 50,p.466b. Such a characterisation is common in sexological works; on it cf. SCHARFE, also O'FLAHERTY, p.378 under "sexual: aggression". Cf. also a parallel from Lucretius cited by LAQUEUR 1990, p.47, the battle even being internalised according to Hippocrates, as stated by op.cit., p.39. HARA 1986, p.78 quite aptly writes : "In other words, the act of loving is in a way single combat between male and female, where all the items in actual fighting can be appropriately applied." On the parallelism between sex and a battle in classical European sources see too KING, pp.29f.)

1320 PAU 27, however, states that introducing the tip of the penis (*udañjimukha-*) into the ☛ *yonimukha-* causes the breasts to grow sooner (than usual):
yonimukha udañjimukhapraveśasyācireṇorojodbhave paramakāraṇatvam.

This the commentary interprets as referring to an insertion up to the hymen:
yonicūcukasyādhastātpradeśe (note "*yoni*--nipple" for the clitoris here) *yāvatsthānaparyantaṃ carmamayasya* (scil. probably: *yonimukhasya*) *mukhapidhānaṃ tāvanmātram udañjimukhaṃ tasya praveśaḥ praveśanam ... .*

A reason for this is not given, though the commentary also quotes a line to the same effect which it attributes to Vātsyāyana (or his school?):
tathā ca vātsyāyanīyaṃ sūtram. pañcaghasrebhya upariṣṭāt karkandhuvad urojau paridṛśyete sati yonimukhe vighāṭite.

I have, however, been unable to find this line in KĀM. But there seems to be independent evidence for the belief that sexual intercourse causes the signs of womanhood to appear in a young virgin girl; thus Mukunda(rām) Cakrabartī's Middle Bengali *Caṇḍīmaṅgal* describes how Khullanā has intercourse with her husband, the merchant Dhanapati, and then says that

one could speculate that it might be because the accumulation must seek an outlet.[1321] In this context we might point to Āryasaumillaka's monologue drama (*bhāṇa-*) *Pādatāḍitaka*, which could be referring to a woman bereft of sexual passion because her *rajas-* has been obstructed; however, the passage in question is not clear.[1322] In the Jaina texts quoted by JAINI, though,

---

through this, carried out for one month, her first menstruation occurs (*Caṇḍīmaṅgal*, p.170: *āyāsa alasa ghume, premālāpe bāsadhāme, kutūhale gela eka māsa; sādhu saṅge sahabāse, puruṣaparaśarase, sbayambhu kusuma parakāś* — I have neglected pronunciation in transliterating the text).

As regards the word *udañji-* above, this term, which has caused MEYER,J.J. 1937 3,p.242,n.2 such difficulties, is a common word for "penis" in sexological works (thus often in the texts in *Kāmakuñjalatā*), whereas *añji-* is used for "vagina" (see e.g. PAU 10, KĀD 56). I refrain, however, from entering upon the discussion of whether these terms mean the same in Vedic, especially in the problematic passage VS 23,21 and its parallels discussed in §12.36 below (cf. too MEYER,J.J. 1937 3,p.242,n.2, JAMISON 1996, p.68). *gudá-* in this VS passage may incidentally be a mistake for *gṛdá-*, which several parallel passages (see on these BLOOMFIELD, p.257a) have (see MEYER,J.J. 1937, loc.cit., HOFFMANN 1975-1976, p.570,n.2), even though words for "anus" are often used to denote the vagina/vulva too (and vice versa), as e.g. in the passage in n.1480, but also otherwise in Indo-Aryan: see KATRE 1968, pp.93ff. (cf. also p.45), as well as nos.8321;8379;9047;9291 in TURNER 1966 and TURNER 1985 (I may add that in Bengali *gud*, though in many areas meaning "anus", is in some areas, e.g. Calcutta and its environs, a common vulgar expression for "vulva; vagina"); cf. too *Pöli* "vulva; vagina" in German children's language, a derivation of *Po(po)* "buttocks, bum", also the vulgar German *Fut* "vagina", in some dialects used in the meaning "anus", as well as the American usage of *ass* "arse" for "vagina" (LAQUEUR 1990, p.96), and Old Indo-Aryan *bhaga-*, as a rule "vagina; vulva", in *bhagandara-* "anal fistula(; fistula of the perineum?)". (The use of *vagina* for "anus" in Latin referred to by LAQUEUR 1990, p.258,n.58 is, however, to be viewed differently, as this term, originally only "sheath", seems only subsequently to have been used in the modern sense; cf. also LAQUEUR 1989, p.126,n.51.)

1321 A variant of NĀṬ 22,103 tells us that the woman of godly disposition (*devaśīla-*) has little semen and (depending on how the compound is analysed) either little or else great inclination for sexual intercourse (*svalpaśukraratapriyā*). Most mss, however, read *svalpabhuk suratapriyā*, i.e. she has little appetite and loves sexual intercourse. Since in vv.100-145 the food habits of women of various dispositions are referred to again and again, but the subject of semen of the women not raised once, the variant appears to be spurious. Nevertheless, it shows that someone, be it a copyist or a redactor, must have associated semen with the nature of a particular type of woman, though what form exactly this association had is unclear.

1322 *Pādatāḍitaka*, p.106:
athavā na tasyāḥ pariṣvaṅgeṇa prayojanam. sā hi tapasvinī nivṛttakāmatantrā rajo'parodhāt kevalaṃ kuṭumbatantrārthaṃ śabdakāmam anuvartate. gamyaś cāyam asyāḥ. apumāñ chabdakāma iti dāttakīyāḥ.

GHOSH (Sanskrit part, p.98) does not read *rajo'parodhāt*, but *rajoparodhāt*, and anyone who has read Sanskrit manuscripts knows that such forms (i.e. without any *avagraha*) are quite common in them. But if *rajo'parodhāt* should indeed be a secondary editorial emendation, then

---

it becomes unclear what *rajoparodha-* stands for; it could be a mistake for *rajaüparodha-* (a scribal error then, for the learned author can hardly be credited with such a blunder, unless we assume an archaic sandhi form), in which case it would mean "obstruction of *rajas-*". On the other hand, *rajoparodha-* is the correct form if the second part of the compound is not *uparodha-*, but *aparodha-* (i.e. *rajo'parodha-*). What could then be meant is "exclusion/hindering of *rajas-*", which ultimately works out to the same. The difficulty is that *uparodha-* is very common, whereas *aparodha-* is only rarely attested. On the other hand, the attestations of *aparodha-* are quite old, the word seeming to have fallen into disuse later; since the work in which the above appears is also quite ancient, this fact might rather speak not against, but for *aparodha-* having to be read here, the more so since it seems the grammatically correct form anyway. In any case, the meaning does not change, whatever the correct analysis may be.

Unfortunately, the whole passage is so problematic that its interpretation is not certain. Thus the translation of the passage quoted above is given by GHOSH (English part, p.141) as:

> "But it is not necessary for her to embrace him. That poor woman who due to menopause has stopped amorous intercourses, pursues a man who loves (only) the word. For the followers of Dattaka say, 'A man without virility loves the word "amorous".'"

Now the context of the above remarks is that a woman has left one lover for another, who happens to be pot-bellied. The speaker wonders why she has done so, why the fat fellow is embraced by her soft arms (*tathā cāsyā[ḥ] komalābhyāṃ bhujābhyāṃ pariṣvajyate*; the lines quoted above follow right after this). This does not look like a description of a woman past her menopause, so that we should leave the question of what has caused the flow of *rajas-* to discontinue open. Apart from this we also have the problem of the last part of the passage (*apumāñ chabdakāma iti dāttakīyāḥ*), in which *śabda-* is explained by GHOSH not as "word" or "sound" alone, but by "the word 'amorous'", which is problematic, to say the least. One could for instance just as well consider translating *śabdakāma-* as "one who loves [only] (the) word(s)" (referring to the "non-man", *apuṃs-*), which would be less problematic; other translations too might be possible.

SCHOKKER/WORSLEY on the other hand translate the passage as follows (p.26):

> "Or rather, there is no point in embracing her. For this unfortunate girl has given up love-making because she is menstruating and only makes love verbally for the sake of the support of her family. Nevertheless he is accessible to her. Dattaka's followers teach that love which consists only in word is unmanly."

Not only is the translation of the last line doubtful here, but the context also does not seem to speak for a woman who is only momentarily not able to make love (i.e. for taking *rajo'parodhāt* — if indeed we are to read so — in the sense of "because of the hindrance due to *rajas-*"), though of course this is no hard and fast proof against the correctness of the interpretation given here.

Quite apart from the quoted translations, we may also consider taking *rajas-* here to mean "emotion, passion", in which case the whole discussion would of course take quite another turn. We must thus conclude that the passage quoted could be taken to speak for the assumption that libido and the accumulation of the procreatory-menstrual fluid were linked, this being, however, only one of several possibilities.

(pp.142f.;166;184;192,n.49), a connection is indeed made between a woman's libido and her menstrual flow,[1323] there even being a theory that the latter stops when she overcomes her sexual urge (pp.184;192,n.49).[1324] JAINI, p.166 mentions an ingenious line of thought linking the theory that various minute beings are generated in the genitals of a woman due to the menstrual flow[1325] with the theory cited above (n.1211) that her sexual arousal is due to tiny creatures in her vagina. Could Yaśodhara thus indeed be talking about the procreatory-menstrual fluid when he says that women have eight times more *dhātu-* than men? But if so, then he creates a new problem, for we have seen in §12.7 that he does not seem to adhere to the view that what women discharge is *ārtava-*. So could his *dhātu-* mentioned above really refer to *ārtava-* or some synonym of this? On the other hand, his remark in §12.7 was in connection with the commentary on the theory of the school of Bābhravya and referred to a fluid not ejaculated at the end, but discharged from the very beginning of coitus. Moreover, there is the possibility (which we, however, cannot call more than this; cf. §§12.8f.) that Yaśodhara may just have been playing the *advocatus diaboli*. But in connection with the ejaculation postulated by Vātsyāyana Yaśodhara speaks of *śukra-* "semen" too (loc.cit.). Is he then equating this and the procreatory-menstrual fluid?[1326]

[1323] This is according to the translations; whether it is indeed (only) the menstrual flow that is presupposed, I do not know (on p.184 e.g. only a flow of *rudhira-* "blood" is mentioned), as I did not check the original texts quoted.

[1324] That sexual intercourse and the signs of womanhood are connected seems to be a common belief in at least certain parts or sections of modern South Asia; cf. e.g. KAKAR, p.112: "For Gandhi believed that in a really celibate woman menstruation stopped completely, the monthly period being but a stigmata of *vikara*, of the sexual distortions of a woman's soul."

[1325] They are also said to be generated in her navel and armpits, and between her breasts; cf. JAINI, pp.35;143;166 (the matter is discussed in n.54).

[1326] In this context attention may be drawn to NĀG 12,13, which says that a woman makes a man her slave by smearing her eyes with an ointment of *bīja-* and *rajas-* liberated during intercourse:

sambhogajaṃ bījarajo varāṅge talliptasaṃśuṣkasitārkavartiḥ
tadañjanenāñjitalocanāyāḥ sadaiva dāsatvam upaiti kāntaḥ.

Vv.8 and 15 show that *bīja-* (on which cf. §§2.5f.) here refers to male semen. *rajas-* must refer to something the female emits to the exterior, though what exactly it is is unclear. But since *rajas-* and *bīja-* seem to be used parallelly here, one may surmise that what the former means is a fluid ejaculated at the end of intercourse. The Hindi commentary of NĀG 12,9, which too speaks of *rajas-* that is emitted by the woman and thus probably anticipates the use of *rajas-* in 12,13, seems to interpret *rajas-* similarly. The verse commented upon, in which *rajas-* is not mentioned, says that a man who smears certain substances on his penis makes a woman flow (√*syand*) for a very long time (*suciram*) like a pitcher full of holes:

Or does he postulate the internal (but not necessarily uterine) ejaculation of (one form of?) the latter accompanying an orgasm which medical texts seem to presuppose?[1327] But I have not been able to find any express reference to such a notion in his commentary, as also in the other sexological works I consulted, in which one of the commonest means of describing a woman's orgasm is to use a form of the verb √*dru*, for instance to say that she "runs" (i.e. "flows profusely") (*dravati*).[1328] This of course would speak for an ex-

---

pāradaṭaṅkaṇavyoṣakākamācīmadhuliptaṃ yat
nārī syandati suciraṃ bahucchidram iva kalaśam.

This seems to refer to a discharge of what we would call the lubricating fluid(s); cf. in this context too the similar simile of a broken vessel in §12.3, as well as (since √*syand* is used in the verse quoted here) the usage of *syanda(na)-* in the texts in n.1286 and n.1308, and in §§12.8f. and §12.12, which too speaks for a fluid of this sort. One could of course postulate that what is meant in the NĀG passage is an ejaculation of the woman for a long time, which would imply multiple orgasms in very short succession, but is this really feasible, especially with regard to the parallel usage of *syanda(na)-* just referred to? The Hindi commentary (actually only an annotated translation) says: *in'kā lep yadi indriy'meṃ lagākar sambhog kare to pah'le strīkā hi śīghr rajaḥpāt ho* "if one puts a paste of these on the penis and has intercourse, then the *rajas-* of the woman indeed swiftly flows first". It is clear that the commentator here assumes (most probably wrongly) an ejaculation of the woman before that of the man and brought about by a *drāvaṇa(yoga)-* (see on this n.1328) (the verse in question is indeed prefaced by the heading *yonidrāvaṇam*, but a comparison with other verse headings shows that this probably stems from the editor), and that he calls this ejaculation *rajas-*. But whatever 12,9 may actually refer to, it does seem as if the commentator analysed *rajas-* as referring to an ejaculatory fluid.

[1327] See on this last §11.34, also n.1289.

[1328] Mere cursory reading turned up occurrences of √*dru* in AN 2,4;13(with *sam*), 3,23, 4,35, RATIK [1],95, PA 2,4;26;34, RATIR 1,36;44;68, 4,18f., 7,52;56;59;63;65, RAT 3,6f., 5,1f., 14,5;8;13, ŚṚÑG 3,31-33, 4,14. AN 2,8 has *dravatām* √*nī*. RATIR 3,3f. uses *drava-* for an ejaculation of the man, saying that, since this is undesirable (cf. §12.5), the woman should be stimulated by other means first; we similarly find *druti-* used in KAN 2,1,29 for an ejaculation of the male in connection with the theory of Auddālaki (see §12.2), *ciradruti-* and *śīghradruti-* being the respective equivalents of *ciravega-* and *śīghravega-* in KĀM 6(=2,1),p.79 (on these latter terms see n.1313). Cf. also *dravati* in n.1225, *dravanti* in n.1337, and *drāvayati* and *drava-* in *Bālatantra* 3,54f. Note too the word *apadravya-* "dildo" (see p.439), which we might analyse as "connected with the flow(ing) away" or "fit for the flow(ing) away" (cf.AIGR II,2,§652, particularly §652b), though other etymologies theoretically remain possible. Recipes for hastening a woman's orgasm are called *drāvaṇa(yoga)-*; see on these RAI, pp.130f. (where, however, RAT 15,1-8 have been forgotten), SCHMIDT 1904, pp.244f., SCHMIDT 1922, pp.641ff. — the passages referred to there also have several occurrences of √*dru*. See also ŚIVAT 6,13,123ff., BYT 147,461f., ŚĀRPA 3209-3213 (3209 says of a woman that a certain recipe *retaḥsrotaḥ pravartayati* in her, i.e. makes a stream of her *retas-* issue). A chapter with similar recipes is also found in SHARPE; the authoress gives it the heading "Ut prius emittat mulier" (p.43), and in fact

ternal ejaculation. But even if an internal ejaculation should be taken to occur, the fluid might also be held to flow out after that and not stay within the body (cf. §§11.11ff.), so that it would ultimately be external too. I must confess that I am unable to untangle this skein of problems. Conversely, even if Yaśodhara and the later sexological works[1329] should refer to an external ejaculation (whatever this actually may be in the case of Yaśodhara), we in actual fact have no definite proof that this is also presupposed by KĀM itself, as the text does not say anything on this point. Thus, though it does look as if Vātsyāyana meant an external ejaculation (and the school of Bābhravya too an external flow), we cannot be fully sanguine that an internal ejaculation (or flow) is not meant after all. But even an internal ejaculation would not automatically entail that the fluid discharged is procreatory; cf. on this §12.3. All this is admittedly frustrating, but it cannot be helped; as it is, we have seen above (cf. §§11.34f., also §12.6) that some (not necessarily all) medical texts seem to presuppose a discharge of two fluids due to orgasm, one of these being the procreatory-menstrual fluid. Since the sexological works studied are not primarily interested in the medical and procreatory aspects of sexual intercourse and as such touch upon this subject only peripherally, we should not expect too much of them in this regard.

12.11. However, KĀM and its commentary, as well as KAN (see n.1267), are not the only texts that have something to say on the subject of what happens during intercourse. Let us therefore examine what other, and

---

several of the passages on such recipes in Sanskrit works too expressly say that the methods described make a woman have her orgasm or ejaculate before the man (cf. on this §12.5). It must, however, also be pointed out that *dravati* in the verse BÖHTLINGK 1966 7107 given in n.1318 seems rather to refer to wetness of the vagina due to sexual excitement, and not to an ejaculation; it may, however, be that we have to differentiate between a general and a technical usage of √*dru* (the latter then as a rule in sexological works). The modern work KEL (cf. n.1330) at any rate clearly distinguishes between the two in 5,17, the fluid due to excitation being termed *jala-* "water", whilst √*dru* is used to denote what is obviously an ejaculation of the woman:

> *mandaṃ cūcukam oṣṭhataḥ paricaran sītkārapūrvaṃ kucaṃ*
> *vaktreṇaiva kṛṣan manojabhavane strīṇāṃ jalaṃ vāhayan*
> *dvāviṃśākṣaramantrarājam atulaṃ kāmaṃ japan kāmuko*
> *nārīṃ drāvayate ciraṃ vitanute jegīyate hastinīm.*

The relevant portion of the Hindi auto-commentary of the author states; *striyoṃ ke ānandodbhūt guhyasthān se ek prakār kā srāv ho jātā hai aur usī avasthā meṃ sambhog kartā huā ... strī ko dravit kar detā hai. arthāt · us'ke vīrya ko vahā detā hai, aur cir'kāl tak sambhog kar'tā rah'tā hai.* Note too that the action described by √*dru* is here explained as being the flowing of *vīrya* (of the woman), a known expression for semen.

[1329] On their dates see n.1306.

that means later,[1330] sexological works have to say.[1331] Seemingly elaborating on KĀM 6(=2,1),p.87, which states that there are nine different types of coitus depending on the time taken,[1332] KAN 2,1,20 says that this is due to the time taken by the man and the woman to achieve their emission (this clearly refers to an ejaculation[1333] — of both).[1334] NĀG 14,15 refers to the removal of the "itching"[1335] of a woman and her quieting down (*śama-*) by the word *cyuti-* "falling; coming/streaming forth",[1336] this must refer to an ejaculation, as is also evinced by NĀG 12,1, where *cyuti-* is used for the man's premature ejaculation and the consequent difficult orgasm of the woman,[1337] as well as by the text in n.1237 (cf. n.1240). In AN 3,11 we find a

---

1330 See also n.1306. I have also considered the modern work (composed most probably in the late forties of this century) KEL, as it clearly draws heavily on older sexological works, even though it at the same time also contains much that is obviously quite modern or belongs to other traditions. In this connection I may draw special attention to chapter 1 of KEL, which is a diatribe against masturbation, in the course of which we find the interesting verse 1,37, which tells us that a masturbator's progeny is weak and timid, and asks how such progeny can make India prevail over others:

apakvavīryasya tavāpi santatir vidhāsyate hā kim u śauryasāhasam
tataḥ katham nāma videśinām jaye svakīyalipsā na samāpyate tvayā.

These verses may refer to victory over the British colonialists, but if they do not, rather referring to victory over other nations, then the other nations of the world should now probably pray fervently that Indian males never let up on their masturbation!

1331 I shall, however, disregard texts such as the *Kāmakalāvilāsa*, which, despite its name, is merely a Tantric text of a non-technical nature carrying no information of use to us. Moreover, sexological works seen, but not containing any relevant matter, are not mentioned in this study.

1332 *kālataḥ pramāṇavad eva nava ratāni.*

1333 The word used for "emission" is *visṛṣṭi-*; cf. on this term especially §§12.2f. The verse in question runs:

strīpuṃsayor visṛṣṭeḥ śaighryaṃ dairghyaṃ ca mādhyasthyam
ity evaṃ navadhā syād ratam iha bhedena kālasya.

1334 AN 3,12 is similar; see also the terminology in 3,15. The same is found in RAT 3,11 = ŚIVAT 6,12,42. All these verses speak of *visṛṣṭi-*.

1335 On this see §12.2 and §12.8.

1336 *kaṇḍūtyapanayanapaṭoḥ kāntācyutiśamahetubhūtasya*
prāṇeśvarasya dayitā sahate na muhūrtavicchedam.

1337 *cyuti-* in this verse has a double syntactical function, as the translation shows:
mṛduhrasvadhvajo yatra priyo 'śakto drutacyutiḥ
tatra striyāś ca kāṭhinyāt sa tu nīcaratodbhavaḥ
"Where the beloved is one whose banner (= penis) is soft and short, [where he] is incapable and one whose *cyuti-* is fast, there the woman's [*cyuti-*] too

description of the orgasm of a woman, of which it is said that she then emits "the water of the one whose arrow is of flowers (i.e. of Kāma)".[1338] This too clearly refers to an ejaculation.

12.12. An interesting description of women's discharges during sexual intercourse is also found in RAT 3,9f.[1339] = RATIR 4,20cd-23ab:[1340]

---

is [achieved] through difficulty. This, however, arises from 'low' coitus." On 'low' (and other sorts of unrecommendable) coitus see e.g. SCHMIDT 1922, pp.261ff. Note that several of the verses cited loc.cit. speak of *tuṣṭi-* "satisfaction" of the woman arising through the proper form of coitus; this does seem to refer to an orgasm, or the post-orgasmic state. This being so, the statement in RAT 3,6f. saying that women do not "run" (*dravanti*) due to improper coitus would seem to refer to an orgasmic ejaculation (cf. n.1328).

[1338] The whole verse (for the variants of which see SCHMIDT 1922, p.268) says:

nārī visṛṣṭakusumeṣujalā ratānte
nṛtyaṃ karoti bahubalganarodane ca (var: *bahuvandanarodanena*)
vaikalyam eti mukulīkṛtacārunetrā
śaknoti no kim api soḍhum atiprayāsam (var: *atipriyā sā*)

"The woman who has emitted the water of the one whose arrow is of flowers at the end of coitus dances (scil. jerks) with much jumping and crying (literally: makes dancing and much jumping and crying) (var: dances with much extolling and crying). She attains, her lovely eyes closed, [a state of] weakness, and is not able to bear any further endeavour (var: to bear anything [else], she, the very dear one)."

Before this, AN 3,9 speaks of the man alleviating the woman's "itching" (cf. n.1335) and making her "flow" again and again (*śamayan mūlakaṇḍūtiṃ syandayaṃś ca muhur muhur*), thus, according to 3,10, making her, as one whose "Kāma-water has been emitted" (*visṛṣṭakāmasalila-*) through the application of various means, subject to his will. The latter emission must be of the same sort as that of 3,11. What, however, "flow" √*syand* refers to is not clear, as it may according to the context refer either to an orgasmic ejaculation or to a flow not connected with an orgasm (i.e. probably the lubricating fluid); cf. in this context n.1326 and n.1400. As regards the use of *kāma-* (or synonyms) in the nomenclature of the fluid emitted (see especially §12.13), though I have translated this as a proper noun ("Kāma"), it must also be borne in mind that *kāma-* "lust; passion" too is implied, it being unclear whether we can actually postulate a clear-cut demarcation between the abstract and the "personified".

[1339] There are three verses in this passage, the first two *both* numbered "9" in the edition quoted.

[1340] The version given here is that of RAT (SCHMIDT 1922, p.267 and RATIR have a few minor variants, which are, however, of no import, except for the one variant mentioned in n.1344):

caṇḍadhvajābhighātena kaṇḍūvyapanayād ataḥ
kṣaraṇāc ca sukhaṃ tāsāṃ visṛṣṭeḥ syandanātmanaḥ
prārambhāt prabhṛti syandaḥ kledātmā sukhaleśadaḥ
ante tu naravat tāsāṃ visṛṣṭer mūrchanāsukham
kṣaṇaṃ raṭanti nṛtyanti rudati cātivihvalā

"Now because of the removal of itching[1341] through the impetuous striking[1342] of the penis and because of the streaming [thus caused] they (feminine) have delight: from emission [and from that] whose nature is flowing (i.e. which is of a flowing nature).[1343] From the beginning [of coitus] onwards the flow is moist (i.e. no gushing flow), giving [only] a small portion of [the ultimate] delight,[1344] but at the end they have,

---

niḥsahatvaṃ tadā yāti mukulīkṛtalocanā.

1341 On this see §12.2 and §12.8.

1342 It is unclear whether this refers to the movement of the penis in general or its striking special parts within the vagina such as those which will be discussed in §§12.18ff. below.

1343 One could see in *visṛṣṭi-* and *syandanātman-* not two parallel nouns (in the latter case a substantivised adjective), but rather in the latter a qualification of the former, which would then lead to the translation:

"... have delight from emission whose nature is flowing (i.e. of a flowing nature)."

But since the following verse itself differentiates between *visṛṣṭi-* and *syanda-*, this does not seem feasible; see also §12.13.

1344 The RATIR parallel has *kleśātmā* in the place of *kledātmā*. This would say that the flow at first is painful or causes discomfort, which is patently absurd, especially as the line just preceding has remarked on the delight caused by the flow, and also in view of the discussion in §12.3 and §12.8. Nevertheless, there seems to have been such an explanatory tradition in vogue as regards RAT too, though it is unclear whether the reading *kleśātmā* of RATIR is a result of this explanatory tradition or whether the tradition arose out of a similar variant of RAT (which, however, the mss and printed text then do not seem to follow). This explanatory tradition is not evinced by the commentary of Kāñcīnātha; nevertheless, the modern translators I consulted have followed it, whether influenced by some oral explanation in vogue, New Indo-Aryan or Dravidian versions of the text, or simply a misunderstanding of the commentary of Kāñcīnātha, I do not know. Thus UPADHYAYA, p.29 translates: "In the beginning of the union between man and woman, there is a feeling of pain, resulting in meagre pleasure. Both the man and the woman experience pleasure only towards the end." Obviously the translator has seen the discrepancy between his "feeling of pain" and the delight mentioned in the previous verse, and has therefore supressed the word *syandaḥ* and made it look as if the line referred to the beginning of coitus, clearly in a situation in which the woman's vagina is not lubricated enough. This is all the more puzzling since he prefaces the translation with the following explanation: "While the process of oozing does yield some pleasure, the actual climax is experienced by the woman only when the fluid flows freely"; this is more or less a summary of Kāñcīnātha's commentary on 9f.:

ataś caṇḍadhvajābhighātena caṇḍadhvajadaṇḍamardanena kaṇḍūvyapanayāt kaṇḍūter abhāvāt syandanātmano dravasvarūpatvāt kṣaraṇāt tāsām aṅganānāṃ visṛṣṭer visargāt sukham. satatakṣaraṇād visṛṣṭer eva sukhaṃ vidyata ity arthaḥ (i.e., in accordance with what follows, both *satatakṣaraṇa-* and *visṛṣṭi-* give pleasure). *kṣaraṇād alpasukhaṃ jāyate tato visṛṣṭer adhikasu-*

like men, the delight, causing swooning, of emission. [The woman,] shouting, dancing (scil. jerking) and crying very agitatedly for a short while, then attains a state of weakness/of not tolerating [any other stimulation], with eyes shut."

12.13. It is clear that this is a parallel to the verses Yaśodhara quotes in his commentary on KĀM and which are reproduced in n.1308. It leaves no doubt that secretions reaching the exterior are indeed meant, one of these obviously the result of an orgasm and of the nature of an ejaculation, for which the expression *visṛṣṭi-* is used, as in the passage in n.1308.[1345] But in contrast to this latter passage, no *śukra-* is mentioned here; so we have in this case no cause for a discussion of the sort found in §12.8. What the discharges reaching the exterior are remains unclear. A similar uncertainty also pertains with regard to that flow of the female which sexological texts often refer to variously as "Kāma-water" (synonyms for "Kāma" are also employed; see too n.1338), "coitus-water" "passion-water" or something similar,[1346] or

---

khaṃ jāyata iti bhāvaḥ. satatakṣaraṇena tu sukham alpam iti pratipādayitum āha satatakṣaraṇe 'dhikaṃ sukhaṃ nāsti (is this a — to us unknown — part of the passage commented upon, or, rather, just a paraphrase?). *ante visṛṣṭau sukham adhikam aṅganānāṃ vidyata iti bhāvaḥ. prārambhāt surataprārambhāt prabhṛti prārambham ārabhya syandaḥ kṣaraṇaṃ kledātmā klinnatāsvarūpaḥ kledakārīty arthaḥ. tasmāc ca sukhaleśadaḥ surate 'lpasukhadaḥ. prārambhakāle yādṛśaṃ sukhaṃ madhye 'pi tādṛśam alpam eva sukham ity arthaḥ. ante tu naravat puruṣasyeva paścād visṛṣṭau satyāṃ mūrchanāsukhaṃ pāravaśyam ātyantikaṃ sukham astīti bhāvaḥ. yathā narasya prārambhāt prabhṛti sravaṇaṃ nāsti tathaivānte visṛṣṭau satyām aṅganānāṃ mahat sukhaṃ vidyate. prārambhāt prabhṛti syandaś ced alpasukhaprada ity arthaḥ.*

It looks as if UPADHYAYA has understood the verse correctly, but then felt that he also had to bow to the (postulated) explanatory tradition described above (this tradition incidentally looking as if it reflected the frequent and sordid fact of the initiation of intercourse by the man even though the woman is not ready for it). Similarly, RAI, p.291 translates: "At the outset this sensation is unpleasant and brings them little satisfaction, but at the climax they experience a discharge like that of the man, which renders them practically senseless with pleasure"; what the "sensation" here is is unclear, as the line before this refers to "the vigorous thrusting of the penis and the flowing of their love-juice" as well as "the need for Visṛṣṭi" which "women experience" (all translations are those of RAI). LIENHARD 1960, p.47 writes: "Zu allererst ist die Ausscheidung schmerzhaft und gewährt nur geringe Wonne. Gleich den Männern erfahren sie aber am Ende die Ohnmachtswonne der Wollustauflösung"; SCHMIDT 1922, pp.267 too says: "Von Anfang an ist dies Fließen mit Unbehagen verbunden und gewährt nur geringen Genuß; am Ende aber haben sie, wie die Männer, infolge der Ergießung Wonneempfindung bis zur Ohnmacht". Here the flow itself is made painful!

[1345] See also *retaḥsrotas-* "flow of *retas-*" in RAT 15,3 and ŚĀRPA 3209 as a result of applying *drāvaṇa-* (see n.1328) means.

[1346] Examples are e.g. *kāmasalila-*, *kāmāmbhas-*, *anaṅgajala-*, *ratyambu-*, *ratyambhas-*, *rati-*

simply *mada-* "passion; passion-fluid". We have seen in §12.11 that this may refer to the ejaculatory fluid. That this is indeed so follows quite clearly from the *Kairalī* commentary on AH,Ut 33,50f. (= AS,Ut 38,p.294a), which quotes the following verses:

> *kāścid ratānte kāmāmbucchalād vañcayituṃ priyān*
> *mūtraṃ sṛjanty akālaṃ*[1347] *vā madanāmbv eva tad dvayam*
> *akālavāhanasyārtham iti vyācaṣṭa sārakṛt*
> *dvedhāpy anicchā surate dvaitībhāvaḥ*[1348] *pratīyate*
>
> "To deceive the beloveds by means of the deceit of [emitting] Kāma-water some (feminine) emit urine at the end of intercourse or at the wrong time.[1349] These two, Madana-water (Madana = Kāma) indeed,[1350] are the meaning of *akālavā-*

---

jala-, *ratisalila-*, *ratipānīya-*, *ratāmbu-*, *ratodaka-*, *ratanīra-*, *ratadrāva-*, *ratasalila-*, *ratajala-*, *madoda-*, *madanodaka-*, *madanajala-*, *madanasyanda-*, *madanāmbu-*, *smarajala-*, *smarasalila-*, *smarodaka-*, *smarāmbu-*, *kandarpanīra-*, *suratāmbu-*, *suratapayas-* (cf. on *payas-* also DAS,R.P. 1984A, pp.237ff.) and so on (see also MÜLLER 1955, p.20,n.78). Examples of passages mentioning such a fluid are RAT 1,11;14-17, 3,21;26;28, RATIR 1,3;13;16, 6,46;49f.;52 (these latter describing cunnilingus), AN 1,13-15, 2,9-10;12;18, 3,10;21;23;25, ŚṚṄG 2,7;11;15, ŚĀRPA 3096;3099;3102;3104. See too SCHMIDT 1922, p.169, where we find, from a recension of the *Smaradīpikā*, an example (*kāmadhārā-*) not found in SMARAD. Other examples have been mentioned in §12.11 or will be found below (passim, especially §§12.18ff.). AN 1,17 speaks simply of "water" (*toya-*) evolved due to *rati-* (*ratija-*; on *rati-* cf. ☞ *bhāva-* in Appendix II), while in 4,8 we find *gharmasalila-* "heat-water". In ŚIVAT 6,13,115 *madanāṃbhas-* is ambiguous; the verse says:

> *patanān madanāmbhasaḥ purā paritoṣo na bhaved vilāsinīnām*
> *suratotsavasiddhihetave tad ayaṃ stambhavidhir nirūpyate.*

One could understand *patanān madanāmbhasaḥ purā* as "due to falling (i.e. discharge) of *madanāmbhas-* before [the partner]" (i.e. women are not satisfied if the man discharges this fluid before them), and this clearly has been done so by the person (the editor?) who gave the passage 6,13,115-118, which deals with methods to retard the ejaculation of the male, the heading *madanāmbhaḥstambhanam* "retardation of *madanāmbhas-*", obviously taking the latter to refer to semen. The three words could, however, also be taken to mean "before the falling of *madanāmbhas-*", i.e. women are not satisfied before discharge of this fluid, which would therefore refer to a fluid of the female (in this case too it would make sense for the man to retard his orgasm until the woman has achieved hers).

1347 Variant in the extract, obviously from the *Kairalī*, appended to Indu's commentary on AS,Ut 38,p.294a by the editor: *akāle*.

1348 Variant in loc.cit.: *drutibhāvaḥ*.

1349 Does the latter refer to the time even before an orgasm is or could be expected to be attained, or does it just mean that urine is discharged when it should not be?

1350 Is this ironical? Or are we to understand that the fluids are like or wrongly taken to be Madana-water, or something similar?

> *hana-*; thus the Sārakṛt[1351] has explained [in the passage here commented upon].[1352] The unwillingness with regard to intercourse [shown] in *both* ways[1353] is recognised as the state of duplicity (var: of flowing)."[1354]

This clearly shows that what the terms discussed refer to is a fluid discharged at the end of coitus (i.e. obviously: due to an orgasm) — and at least in such a quantity[1355] that it resembles a flow of urine. In many of the cases mentioned above, though, what is meant is not clear, the context not elaborating further, or else fitting both the ejaculatory fluid as well as the lubricating fluid(s) (on the latter see too n.1338, with further references). But that does not alter the deduction made on other evidence, namely that the sexological works in general do presuppose an external ejaculation of some fluid accompanying a woman's orgasm, on a par with the orgasmic ejaculation of men, as is also suggested by AN 3,16;19, where we find *madanasyanda-* and *manmathajala-* (i.e. expressions such as those mentioned in n.1346) respectively for male semen, and as is conclusively proved by the passage from the medical commentary *Kairalī*.

1351 The author of the *Sāra*; cf. on him ☞ *ulba-* in Appendix II.

1352 AH,Ut 33,50f. states:
akālavāhanād vāyuḥ śleṣmaraktavimūrchitaḥ
karṇikāṃ janayed yonau rajomārganirodhinīm (50)
sā karṇinī ... .
akālavāhana-, literally "making flow at the wrong time", also means "labouring at the wrong time", and it and similar expressions in the parallels to the passage commented upon seem generally to be seen in connection with labour at childbirth, an explanation obviously not followed here. (As the text also tells us, due to *akālavāhana- karṇikā-*(s) arise(s) in the genital tract; on this matter see TIVĀRĪ 2,pp.36ff.)

1353 In urinating at the end of or during intercourse.

1354 I must admit that the verses cause difficulties, and that my translation may not be fully correct. But that does not detract from their worth as proof that an ejaculation due to orgasm was indeed postulated. TIVĀRĪ 2 says on p.37: *kuch striyā̃ maithun ke ant meṃ kāmāmbu ke vyāj (bahāne) se priyoṃ ko ṭhag'ne ke lie akāl meṃ mūtr yā madanāmbu kā srāv yā donoṃ (kā hī srāv) kar'tī haiṃ. akāl-vāhan kā sārāṃś yahī hai. anicchā-pūrvak maithun meṃ drutibhāv (sravaṇ kar'nā) do prakār (kāmāmbu yā mūtr) se pratīt hotā hai*, and on p.38: "Premature ejaculation of fluid brought about by untimely straining refers to either normal excretion of reproductive system of woman coming at the end of orgasm or else deliberate discharge of urine by woman to deceive her husband. In a woman undesirous for coitus above both the discharges i.e. secretions from reproductive system and urine are seen." I have great difficulties in understanding this.

1355 This need not be very large, but it must be more than just a few drops, since it must be noticed by the man and also differentiated from the lubricating fluid and similar secretions.

12.14. This must at once remind one of the several decades old and acrimonious debate among Western sexologists on the subject of female ejaculation, which has only recently been conclusively settled in favour of ejaculation on irrefutable evidence gathered under clinical conditions (the expulsion of fluid has even been filmed). It is needless to go into the details of this discussion here, as one can find all the needed information in STIFTER and LADAS/WHIPPLE/PERRY. WEISBERG, with this discussion in mind, has written: "Years from now I am sure that a medical school lecturer will joke about how it wasn't until the 1980s that the medical community finally accepted the fact that women really do ejaculate." This statement will now have to be modified to "the medical community in the West"; clearly, medical professionals in ancient India were better observers.[1356]

12.15. The female ejaculatory fluid is often described by STIFTER as being discharged with great vehemence, at times in a stream more forceful than that of an ejaculation by the male.[1357] It has by now been analysed in the laboratory, and has been found to be made up of something akin to prostatic fluid. It is very often confused with urine (which is in striking agreement with the verses found in the *Kairalī* commentary and quoted in §12.13); though mostly described as colourless,[1358] it seems that it may also be — at least somewhat — milky.[1359] Even though it appears that such milkiness is not something always apparent, one is of course at once reminded of the term, among various appellations, we have seen to be applied to this fluid in

---

[1356] The existence of an ejaculate of women linked with orgasm seems to have been known to or was at least postulated in Western medical theory too, at least from the time of the ancient Greeks onwards (see e.g. STIFTER, *passim*, LAQUEUR 1990, *passim*, also LADAS/WHIPPLE/PERRY, p.27). As it is, its existence has also been vaguely known to many later Westerners, yet even those who have witnessed or even experienced it seem often to have had very little idea of what actually happens (see e.g. LADAS/WHIPPLE/PERRY, pp.24ff.).

[1357] STIFTER, p.16 cites an Indian doctor who observed a woman, excited by a gynaecological examination, ejaculating with such force that the jet hit a burning light-bulb in the chamber and made it explode.

[1358] Thus also in STIFTER, *passim*.

[1359] Thus also in the case WEISBERG observed. WHIPPLE, p.16 says: "The phenomenon of female ejaculation refers to expulsions of fluid from the urethra. Many women reported having surgery to correct this 'problem' and others reported that they stopped having orgasm to prevent 'wetting the bed'. The fluid was described as looking like watered down skim milk, tasting sweet and usually about a teaspoon in volume." STIFTER, p.179 opines that in some cases women may, like some men, have a so-called retrograde ejaculation in which the ejaculated fluid does not leave the body but enters the bladder. The author writes that this may be why some women feel an urge to urinate after an orgasm; what leaves the body, he continues, is, however, not urine, but a small amount of a different fluid which is colourless or whitish.

Indian texts, namely *śukra-* "semen". Against this we, however, have several descriptions of it as "clear" (on which see last §11.34); nevertheless, the matter is intriguing (see also §8.9).

12.16. But I must also draw attention to the fact that the orgasmic ejaculate of women is actually not vaginal, but urethral; it originates in the paraurethral or periurethral glands and enters the urethral passage through Skene's ducts. The texts we have so far studied do not, however, seem to refer to a urethral ejaculation — at any rate, we find no statement that links the urethra with the ejaculate,[1360] if the place of discharge can be deduced at all. Even the verses of the *Kairalī* quoted in §12.13 only tell us that males may be deceived to think that what is actually urine is the fluid, which says nothing about where this fluid actually comes from, except that it obviously flows out of the vulva. But all this does not necessarily mean that the ejaculation the texts describe is not the same as the one observed by Western sexologists. As it is, unless one really does observe the ejaculation minutely, as seems to have been done under the clinical conditions mentioned in §12.14, the ejaculated fluid, coming out as it does via the vulva, most probably would appear to be vaginal.[1361] This is borne out by several statements in STIFTER and LADAS/WHIPPLE/PERRY showing ignorance about the actual place of ejaculation among not only doctors, but also the women concerned.[1362] On the other hand, women having such an ejaculation often wrongly seem to see in it an unwanted discharge of urine, so that in these cases at least an awareness of the actual place of discharge seems to be present.[1363] This, however, does not necessarily mean that those describing the ejaculation in

---

[1360] This also holds true for SU,Śā 2,36 mentioned in §8.1.

[1361] Cf. also the text speaking of "glands, situated near the mouth of the vagina" in §11.26.

[1362] Thus e.g. on p.70 of LADAS/WHIPPLE/PERRY we learn of a woman who ejaculated regularly, but was not sure where actually the discharge came from, i.e. whether it was urethral or vaginal. On p.74 we find a woman saying: "I asked my doctor about it and his answer was that doctors don't understand or even know where this fluid comes from ...". And p.85 states: "Perry, Whipple, and their colleagues have examined some women who have reported the expulsion of fluid from the vagina. In these women, the opening of the urethra was found to be located in or close to the opening of the vagina. This may account for their reporting that the fluid was felt to be expelled from the vagina." Cf. also STIFTER, p.180f.: "Beim Geschlechtsverkehr kommt nämlich die äußere Harnröhrenöffnung von der Richtung her nahezu 'innerhalb' der Vagina zu liegen. Das könnte zur Folge haben, daß viele Frauen gar nicht merken, daß sie eigentlich auch ejakulieren, vor allem wenn der eingeführte Penis oder auch eine geringe Sekretmenge das 'Ausspritzen' verhindert."

[1363] WHIPPLE, p.17 states: "Some women expel a fluid that is different from urine during sexual activities and orgasm and some women may expel a little urine."

our texts knew about this too; we must remember that they were males, unable to gain first-hand knowledge of this phenomenon. One could of course say that all they would have had to do would have been to ask the women concerned, but this apparently simple mode of proceeding might, as is often the case, not necessarily have been the one actually followed;[1364] cf. in this regard also the discussion on whether women actually do have an orgasm or not (§12.2). Also, not all women may have realised where the ejaculate comes from (see above). Furthermore, there is the problem that the women concerned might not have wanted to or been able to divulge such matters to men; indeed, commenting on KĀM 13(=2,8),p.160 Yaśodhara expressly remarks on the unwillingness or inability of women to let others know what causes them pleasure (see n.1558). Thus if, as remarked above, the male observers most probably saw only a discharge that came out via the vulva,[1365] it would seem most probable that they would have classified it as vaginal, especially since it was clearly connected with (vaginal) sexual intercourse. But here again we cannot be sure. It is for instance possible that the discharge was felt to come from some place within the vestibule before the introitus of the vagina, but not the urethra, though it must be pointed out that, since the vaginal and urethral openings are well visible, this would presuppose a theory postulating a discharge through one or more minute opening(s) not visible to the eye, or else a discharge through the mucous epithelium.[1366] Nevertheless, since a similar theory of just such a minute opening to the exterior seems to exist in Greek medicine,[1367] we cannot neglect this possibility fully. It may also not be amiss here to point to STIFTER, pp.182-186, which ask whether there may not in actual fact be more places than the urethra from which women ejaculate (cf. also n.1366). What complicates the matter even more are passages like AH,Sū 7,69, in which, with reference to a woman,

---

[1364] Note too what ROUSSELLE, p.26 writes — hyperbolically — on ancient medical authorities in the West: "Women's bodies were shrouded in such secrecy that Galen learned more from dissecting female monkeys, insofar as their organs were the same as human ones, than all the men of ancient times learned through their sexual contacts with their own wives."

[1365] Very interesting in this regard are the photos reproduced by STIFTER, p.152.

[1366] One could also think of Bartholin's ducts in this connection, as in our times too they are, even though usually taken to produce only the lubricating fluid, sporadically believed to discharge an ejaculate on orgasm (cf., e.g., STIFTER, pp.155f.;183).

[1367] See DEAN-JONES 1994, p.158, according to which one ancient Greek author refers to a minute duct, a *mikròs póros*, opening out just above the urethra (on this theory see also §12.24). The statement is found in a text whose authorship is not clear; it is part of the corpus of Aristotle's biological canon, but DEAN-JONES 1994, p.15 holds it to be actually "the work of a later Hippocratic doctor writing under the influence of Aristotle".

mehana- "urinary organ" is obviously used for "vagina".[1368] A similar terminology is found e.g. in HAṬH 3,85,[1369] and probably even in a much earlier

---

[1368] The verse is part of a passage which lists women unfit for intercourse, among these also one whose *mehana-* is impure or (too) narrow. Vv.69f. say:

grāmyadharme tyajen nārīm anuttānāṃ rajasvalām
apriyām apriyācārāṃ duṣṭasaṅkīrṇamehanām
atisthūlakṛśāṃ sūtāṃ garbhiṇīm anyayoṣitam (var: *anyayoṣitām*)
varṇinīm anyayoniṃ ca gurudevanṛpālayam.

Aruṇadatta, Śrīdāsa, Hemādri, Parameśvara and Candranandana (7,51 in AH2) too all explain *mehana-* as *yoni-*, Parameśvara also remarking that here the openings for urine and menstrual fluid have been combined.

As regards the prohibition, mentioned in the passage cited, of having intercourse with a pregnant woman (*garbhiṇī-*), this is a commonplace; nevertheless, RATIŚ, p.69 says that intercourse is allowed if the pregnancy has not progressed too far and the woman is lustful:

śāstroditāṃ doṣahīnāṃ sakāmām anṛtāv api
upeyāt tu janaḥ kāmī sadā te kathitaṃ śṛṇu
apraudhagarbhā yā nārī bhaved atyantakāmukī
tām arhati jano gantuṃ tatkāmasādhanāya tu.

That intercourse during pregnancy is allowed is also stated by several other texts, the oldest among these being TS 2,5,1,5, which says that it is allowed until the time of delivery:

tā́ abruvan vā́raṃ vṛṇāmahā ṛ́tviyāt prajā́ṃ vindāmahai kā́mam ā́ víjanitoḥ sáṃ bhavāméti. tásmād ṛ́tviyāt stríyaḥ prajā́ṃ vindante kā́mam ā́ víjanitoḥ sáṃ bhavanti.

See also n.133. The matter has been discussed by LESLIE,I.J. 1989, pp.251;289, where other, similar passages too are mentioned. Add to these VĀS 12,24, which says that according to the Kāṭhaka school lying with women is allowed even if they are to deliver on the next day (var: on the same or the next day):

api ca kāṭhake vijñāyate. api naḥ śvo vijaniṣyamāṇāḥ (var, not in keeping with the verb: *adya śvo vā vijaniṣyamāṇābhiḥ*) *patibhiḥ saha śayīrann iti strīṇām indradatto vara iti* (the other variants are meaningless).

Since the lines immediately before this (12,21-23) speak of intercourse, it seems clear that 12,24 too refer to this. Finally, KĀM 13(=2,8),p.165 says that intercourse with the woman above the man (*puruṣāyita-*) should not be carried out in certain cases, which includes pregnancy (and also puerperium):

na tv evartau na prasūtāṃ (var: *prajātāṃ*) *na mṛgīṃ na ca garbhiṇīm*
na cātivyāyatāṃ nārīṃ yojayet puruṣāyite.

RAT 10,51, which speaks of "inverse infatuation (scil. intercourse)" (*viparītamohana-*) is similar:

yojayen na viparītamohane nūtanaprasavapuṣpayoginīm
garbhiṇī[ṃ] ca hariṇīṃ ca pīvarāṃ kanyakām api kṛśāṃ ca varjayet.

Cf. also CHÖPEL, pp.84;250. From all this it follows that 'normal' intercourse during pregnancy is not prohibited. (On inverted coitus cf. also §11.31.)

[1369] This verse describes the sucking in of fluids through the *mehana-* of male and female:

mehanena śanaiḥ samyag ūrdhvākuñcanam abhyaset
puruṣo 'py athavā nārī vajrolīsiddhim āpnuyāt.

In the case of the male, it is clearly the penis which is meant (cf. on this too DAS,R.P. 1992A,

text, namely AV 1,11,5,[1370] so that the notions of the AH passage may have been presupposed by other medical authorities too; note also that Gayadāsa, commenting on SU,Ni 1,86, explains *upastha-* (usually "lap; (female) genitals") of the text as *mūtramārga-* "path of urine".[1371] As it is, the common theory (very much alive even today) of nine openings of the body,[1372] viz. eyes, nostrils, ears, mouth, anus and urino-genital opening (probably with the male as the standard, but applied to females too), as well as passages such as KAL 3,11, which refers to urine *and* menstrual discharge as well as semen and faeces as being discharged from only two openings,[1373] seem to show that it was the vulva as such, and not the vaginal and urethral openings separately, which was often taken to be the exterior opening, even though the urethral opening was of course known to be a separate opening (see on this latter e.g. MEULENBELD 1974, p.517, to the passages cited there add BHĀ 1,3,269f., ĀY 4,263)[1374 1375] But even if we must therefore reckon with

---

p.391b); in the case of the female, what is meant is, as the context makes clear, the vagina. Obviously, the terminology applied to the male organ has simply been transferred to the female, even though in her case the urinary and procreatory organs, though they have a common opening to the exterior, can be seen to have different openings within. The commentary seems to go one step further by explaining *mehana-* "urinary organ" through *meḍhra-* "penis", at the same time remarking expressly that this applies to females too. One could now opine that *meḍhra-* is used here to mean "urinary organ" (irrespective of sex) too, which, though to my knowledge a usage not noted by our dictionaries, could follow from the derivation of this word from √*mih* "urinate"; however, in the case of the female it would nevertheless describe not the urinary organ, but the vagina. See also Yaśodhara's comment on KĀM 6(=2,1),p.84: *ekasya nimnaṃ mehanam aparasyonnatam* (var. *anyasyonnatam*).

1370 This is part of an incantation supposed to lead to an easy delivery:
ví te bhinadmi méhanaṃ ví yóniṃ ví gavī́nike
ví mātáraṃ ca putrám ca ví kumārám jarā́yuṇā́va jarā́yu padyatām.

Here the *méhana-* and the ☞ *yóni-* are clearly differentiated. But since it would be absurd to postulate that giving birth was taken to require a "splitting apart", i.e. opening wide, of the urethra or its opening, *méhana-* here must refer to something part of the reproductive tract. Therefore, whatever ☞ *yóni-* may here refer to (most probably either the uterus or the vagina), it does very much look as if *méhana-* here denotes the vulva or the opening of the vaginal vestibule to the exterior. (On ☞ *gavī́nikā-* see Appendix II.)

1371 This seems to be in keeping with general medical theory, as e.g. CA,Śā 1,26 too sees the *upastha-*, one of the *karmendriya-*s according to 1,25, only in the role of an organ for voiding excrement (*pāyūpasthaṃ visargārthaṃ*), sexual action not being mentioned. On Gayadāsa's remark cf. also the discussion of his statement on SU,Śā 4,24 discussed in Excursus III at the end of chapter 6.

1372 Or ten, if the fontanelle too is counted.

1373 *mūtraṃ saretaḥ sapurīṣaraktaṃ sravaty adhastād vivaradvaye ca.*

1374 In the passages mentioned the urinary opening is described as common to males and

some medical authorities at least not bothering to differentiate clearly between the organs for urinating and for intercourse of a woman,[1376] can we simply generalise this?[1377] The problem ultimately boils down to a simple question: did the Indian sexologists and medical authorities know about the urethral ejaculation, or did they not?

12.17. The answer to this question is vital for the problem of the relationship between the procreatory-menstrual and the "seminal" fluids. For if it was known that the discharge of the latter is urethral, then the parallels to male semen would probably have been even more obvious; on the other hand, a connection with the procreatory-menstrual fluid, which was of course perceived as coming out of the vagina, would have been more difficult to construe. Conversely, if the ejaculation was taken to be vaginal, then a connection with the procreatory-menstrual fluid would have been easier to construe, but some of the parallelism with male semen would have been lost. In this connection attention may be drawn to Cakrapāṇidatta's commentary on CA,Sū 1,92ff. where the notion of the "heaviness" of the urine of males (that of females is "light") being due to its connection with semen (*śukrasambandhād guru*) is refuted by means of SU,Śā 2,47, which states that two women having intercourse with each other release "semen",[1378] and from which it is deduced that the urine of females too is comparable to that of men with

---

females, the vagina, often called the passage through which blood (i.e. procreatory-menstrual fluid) flows, as an additional opening.

[1375] Cf. in this regard also WECK, p.98 on Balinese medical notions: "Der Austritt des Urins erfolgt durch den Geschlechtsteil; daß es bei den Frauen zwei verschiedene Wege für die Abscheidung des 'Samens' und des Urins gibt, ist nur aus den Schriften bekannt [i.e. plays no role in oral traditions]."

[1376] See in this context also Excursus III at the end of chapter 6.

[1377] Note that we are here talking of medical authorities, not common people, among whom there quite well may have been many who really did not know of the difference between the urethral and vaginal openings — or even worse. In this context attention may be drawn to such modern notions as cited by KAKAR, p.20: "Ignorance, of course, thrives in the socially generated pall of silence. One college-educated patient believed well into her late teens that menstrual blood, urine and babies all came out through the urethra. Another woman, brought up in a village and presumably more familiar with the 'barnyard' facts of life, realized with deep consternation only when giving birth to her first child that babies were not born through the anus as she had believed. Writing on Indian sexuality some two thousand years after the *Kamasutra*, I still cannot say to its author — 'Elementary, dear Vatsyayana.'" Some ancient Greek medical authorities too probably did not differentiate between urethra and vagina, the most prominent amongst them being Aristotle (cf. DEAN-JONES 1994, pp.80ff.).

[1378] The verse is quoted without any ascription and with the variant given in n.795.

regard to being connected with semen:

śukrasambandhasya strīṇām api śukravattvena tulyatvāt.[1379]

However, this does not necessarily mean that Cakrapāṇidatta has a urethral ejaculation of "semen" of women in mind, for it is quite possible that all he means is that urine and semen could come in contact within the vestibule before the vagina or in the vulva (between the labia). Thus this passage cannot supply us with a solution to our problem. One could, however, point to the passages from the commentaries of Āḍhamalla and Śrīdāsa discussed in §§11.10ff. In the light of what we have just seen on female ejaculation through the urethra in connection with an orgasm, we may take a fresh look at them and ask whether their statements regarding the ☞ *yonimukha-* may not be taken to refer to an orgasmic ejaculate of the woman which issues out of the vulva (here then the "mouth of the *yoni-*"), but does not come through the vagina, which could point to an urethral ejaculation (though the authorities whose views are here quoted may simply have perceived an ejaculation through the vulva but not coming from the vagina, without actually linking this with the urethral opening; cf. §12.16). This would neatly circumvent the objections raised loc.cit. regarding the possible explanation of *yonimukha-* in the manner considered, as these pertained to a discharge of lubricating fluid not connected with an orgasm. Unfortunately, there are, as we have already seen, other plausible explanations for the passages from the commentaries of Āḍhamalla and Śrīdāsa, so that we still can say nothing definite on whether an urethral ejaculation was presupposed by the ancient Indian sexologists or not. As regards CA,Ci 15,34f. discussed in §6.14 (see also n.872), here we do have an express connection between semen and the urinary tract, but we cannot be sure that this does not pertain only to males, as is or seems to be the case as regards the passages linking semen and the urinary tract mentioned in §6.13. Let us therefore turn to some passages in sexological works in which we find anatomical descriptions of those parts of the female body responsible for discharge during coitus, in the hopes of finding some relevant information.[1380]

12.18. RAT 10,6-8 say:[1381]

1379 Śivadāsasena in his commentary on DR,Madyādivarga 30 is even more laconic in his refutation, saying simply that the urine of women too is connected with semen, without giving any reason:

śukrasambandhasya strīṣv api praśaktatvāt (obviously = *prasaktatvāt*).

1380 The first two descriptions cited in §§12.18f. below are also to be found in SCHMIDT 1922, pp.258-261; see too SCHMIDT 1904, p.197.

1381 This passage is also found in SMAR, pp.269f.

nivasati bhagamadhye nāḍikā liṅgatulyā
madanagamanadolā[1382] *dvyaṅgulakṣobhitā*[1383] *sā*
sṛjati madajalaugham sā ca kāmātapatraṃ
dvayam iha yuvatīnām indriyaṃ nirdiśanti (6)[1384]
madanasadanarandhrād ūrdhvato nāsikābhaṃ
sakalamadaśirāḍhyaṃ[1385] *manmathacchattram āhuḥ*
vasati madanarandhrasyāntare nātidūrāt
smarajalaparipūrṇā pūrṇacandreti nāḍī[1386] (7)
nivasati bahunāḍīcakram anyat pradhānaṃ
tritayam idam ihoktaṃ hastaśākhāvimarde
karikaraphaṇibhogārdhendukāmāṅkuśādyair
alam iha karaśākhāyogabhedābhidhānaiḥ (8)[1387]

6: "Within the vagina there dwells a (small) tube[1388] comparable to the penis, the *madanagamanadolā-*[1389]. Excited by

---

[1382] SMAR: °*ḍolā*.

[1383] SCHMIDT 1922, p.259,n.1 mentions a variant *dvyaṅgulā kṣobhitā*. SMAR has *dvyaṅguli-kṣobhitā*.

[1384] This and the next verse are interchanged in SMAR.

[1385] SMAR: °*sirāḍhyaṃ*.

[1386] SMAR: *pūrṇacandrākhyanāḍī*.

[1387] RATIR 4,35-40 has nearly the same in a different metre:
yonimadhye nivasati nāḍikā liṅgasannibhā (var: *candrasannibhā*)
madanāgāraḍolā sā dvyaṅgulakṣobhitā satī (35)
madodakaughaṃ sṛjati sā ca kāmātapatrakam
tad etad dvayam evāhur indriye yuvatījane (36)
madanāgārarandhrād apy ūrdhvato nāsikānibham
sirābhir madayuktābhir yuktaṃ kāmātapatrakam (37)
madanāgārarandhrasya cāntare nātidūrataḥ
pūrṇacandrābhidhā nāḍī pūrṇā smarajalair bhavet (38)
bahunāḍīcakram anyad asti mukhyaṃ tadantare
etat tritayam evāhuḥ karaśākhābhimardane (39)
phaṇīndrabhogārdhacandrakarihastasmarāṅkuśaiḥ
tarjanīmadhyamābhyāṃ ca kṣobhayet smaramandiram (40).
The only difference worthy of note is that according to the variant *candrasannibhā* above the *madanāgāraḍolā-* "swing of the abode of *madana-*" (= the *madanagamanadolā-* in RAT) is not like the penis (*liṅgasannibhā*), but like the moon (*candrasannibhā*) (full?, or a segment?); see on this also n.1434.

[1388] My attempt at translating *nāḍikā-* (cf. also ☞ *nāḍī-* in Appendix II). Whether *-ka-* (in this case feminine) here is actually a diminutive or pleonastic suffix is unclear.

[1389] "Swing for passionate/lustful (*madana-*) intercourse/moving (*gamana-*)", "swing for the

two fingers,[1390] it gives forth a flood of passion-water (see §12.13), it and the umbrella of Kāma (*kāmātapatra-*). They designate both here as the sensual organ of young women.
7: "They call [that which is] above the aperture of the abode of Madana (= Kāma), resembling a nose, filled with all the vessels (*śirā-* = ☞ *sirā-*) of passion(-fluid) (*mada-*; see §12.13), the umbrella of Manmatha (= Kāma) (*manmathacchattra-*). Within the aperture of the abode of Madana abides, not too far removed (from where?), a tube filled with Smara-water (Smara = Kāma), the '*pūrṇacandrā-*' (i.e. full-moon [tube]).
8: "Another wheel of many tubes, the essential/principal one, abides (where?). This triad is mentioned here with regard to rubbing with the limbs of the hand (= fingers). With [the enumeration of] the names of the varieties of finger-application,[1391] [namely] elephant's trunk, serpent's coil (or: serpent-eater, i.e. peacock), half-moon [and] goad of Kāma, enough [is said] here."

12.19. In AN 4,32-35 (= ŚIVAT 6,12,116-119, also SCHMIDT 1922, p.259) we find:

yonimadhye 'sti nāḍy ekā kāmāṅkuśasamā hi yā[1392]
liṅgena kṣobhitā[1393] *saiva madavāri nirantaram* (32)
kāmātapatrāt sṛjati sa syanda iti kīrtyate
varāṅgarandhrād ūrdhvaṃ tu nāsikābhaṃ yad asti tat (33)
manmathacchattram ity āhur āḍhyaṃ madaśirācayaiḥ[1394]
yonimadhye[1395] *nātidūrāt pūrṇacandrāsti nāḍikā* (34)

---

intercourse of (i.e. deriving from, caused by) Madana (= Kāma)", or "swing for Madana's moving (upon)". On *madana-* cf. what was said on *kāma-* in n.1338. According to the commentary of Kāñcīnātha, some explain *madana-* here as referring to the organ for urinating (i.e. obviously: the penis):

> *athavā madanagamanadoleṭy atra eke* — printed *eka* — *madanaśabdena mehanaṃ vivakṣitam iti kathayanti.*

1390 Var: "Two *aṅgula-*s/fingers [in length], excited". Note that *aṅgula-* in *dvyaṅgulakṣobhitā* can hardly be translated adequately if one takes *aṅgula-* to be a measure of length (on which see e.g. VṚK, p.169).

1391 On such application of the fingers in general see e.g. RAI, pp.344f.

1392 ŚIVAT and SCHMIDT 1922: *sā*.

1393 ŚIVAT: *śobhitā*.

1394 ŚIVAT and SCHMIDT 1922: *madasirā°*.

1395 ŚIVAT and SCHMIDT 1922: *yonirandhre*.

manojavārisampūrṇā strīṇāṃ tiṣṭhati sarvadā
tadvisarge[1396] *drutā nārī procyate pūrvasūribhiḥ* (35)
32: "Within the vagina[1397] there is a tube (☞ *nāḍī*-) which (var: tube. It) is indeed like the goad of Kāma[1398] (= penis). Excited[1399] by the penis, passion-water it verily constantly(?)/at once(?)/compactly (=profusely)(?) (*nirantaram*)
33: "emits from the umbrella of Kāma (*kāmātapatra*-). It (i.e. the emission) is authoritatively called (*kīrtyate*) 'flow' (*syanda*-).[1400] But that which, resembling a nose, is above the aperture of the best (scil. genital) organ,
34: "filled with a multitude of vessels (*śirā*- = ☞ *sirā*-) of passion(-fluid) (*mada*-; see §12.13), [that] they call 'the umbrella

---

[1396] ŚIVAT and SCHMIDT 1922: *tadvisṛṣṭyā*.

[1397] This is obviously what ☞ *yoni*- refers to here.

[1398] On *kāma*- see also n.1338.

[1399] "Adorned, embellished" according to the variant in n.1393, which is obviously out of place here.

[1400] That *sa[ḥ]* must refer to the flow follows from §§12.8f. and n.1326; *syanda*- here is thus obviously a technical term describing a certain type of discharge. RAI, p.227 translates: "This is called *syanda* or the secretive organ", but that does not seem feasible, not only because of loc.cit., but also because *syanda*- is to my knowlege a nomen actionis and not a nomen agentis. For the same reason, "Fließer" for *syanda*- in MEYER,J.J. 1937 3,p.242,n.2 does not seem feasible either. SCHMIDT 1922: p.260 reads *sasyanda iti kīrtyate* instead of *sa syanda iti kīrtyate* and translates: "und wird bezeichnet als sasyanda", i.e. he takes *sasyanda*- "[that which is] together with flowing" (an intensival derivation of √*syand* is highly improbable, both on morphological grounds and because no intensive forms of the verb are known) to be the name of the tube (☞ *nāḍī*-). But in general (as the medical texts bear ample testimony to, and as *pūrṇacandrā*- below and in §12.18 too shows) the names of *nāḍī*-s are feminine, whereas *syanda*- is masculine and would as such fit the pronoun *sa[ḥ]* without any difficulty; there is thus no reason to postulate a problematic *sasyanda*- (or *saspanda*-, as in SCHMIDT 1904, p.197) here. As regards the explanation of RAI, the argument regarding the feminine names of *nāḍī*-s would also hold against this if it should refer to the *nāḍī*-; the same applies to MEYER,J.J. 1937. However, RAI may have an alternative name of the *kāmātapatra*- in mind — MEYER,J.J. 1937 definitely does. In this case the convention applying to the nomenclature of *nāḍī*-s would not apply. But *kāmātapatra*- is neuter and not masculine like *syanda*-; moreover, the other arguments against the explanations of RAI and MEYER,J.J. 1937 also remain. As it is, the explanations of RAI, MEYER,J.J. 1937 and SCHMIDT 1922 needlessly complicate a statement which is quite straight-forward in the light of the fact that *syanda(na)*- is often used for an emission; see e.g. §§12.8f., §12.12, and *madanasyanda*- in AN 3,16 mentioned in §12.13 (cf. also n.1326 and n.1338). The problem of the exact nature of the emission described here will be taken up in §12.32.

of Manmatha (= Kāma) (*manmathacchattra-*)'. Within (var: In the opening of) the vagina[1401], not too far removed (from where?), the (small) tube[1402] *pūrṇacandrā-*
35: "of women is present, always full of Manoja-water[1403]. With regard to its voiding/When its voiding [takes place][1404] (var: Through its emission)[1405] the woman is [over]flowed (*druta-*; cf. §12.10), is taught by the bygone masters."[1406]
12.20. RATIK [1],93-95 says:
madanasṛnisamānā nāḍikā kāpi madhye
sravati madanavāri kṣelitā yā dhvajena
kṣarayati madapūrṇā sā sarojena tulyaṃ
yad api bhavati madhye khaṇḍam ity āha kāmaḥ (93)
ūrdhvaṃ yad bhagarandhrato madaśirācandrāṃśuśṛṅgāṭaka-[1407]
prakhyaṃ tan madanātapatram iti ca vyākhyānam ācakṣate
kiṃ caikā nikaṭe manobhavajalaiḥ pūrṇāsti yā nāḍikā
cāndrīṃ tāṃ kathayanti bhūrisukhadāṃ yoṣinmadadrāviṇīm (94)
yad garbhaśayyāmadanāntikasthaṃ
ciñcāprasūnapratimaṃ bhagāntaḥ
tad brahmasaukhyāspadam āhur āryāḥ
sparśāt paraṃ drāvakam aṅganānām (95)

1401 Cf. n.1397.

1402 Cf. n.1388.

1403 See §12.13 and n.1338 (Manoja = Kāma).

1404 *tadvisarge [sati].*

1405 On a possible attempt to differentiate between *visarga-* ("voiding") and *visṛṣṭi-* ("emission") cf. n.1281. The commentary of Kāñcīnātha quoted in n.1344, however, equates the two (*aṅganānāṃ visṛṣṭer visargāt sukham*).

1406 Later sexological works such as AN and RAT have been popularised in various languages in a variety of translations the quality of which often leaves much to be desired. A case in point is the following translation of AN 4,34f. from an unknown source quoted by STIFTER, p.27: "Der Madama-chatra (die Klitoris) am oberen Eingang zum Yoni ist dasjenige Organ, das hervorragt wie das *Rohr eines Springbrunnens* aus der Erde; es ist verbunden mit der Arterie Mada-vashi (der Samenquelle) und bewirkt die Ausstoßung des Spermas. Endlich gibt es noch eine Arterie, Pūrṇa-chandrā genannt, die voller Kama-salila ist und der die Gelehrten u.a. den Monatsfluß zuschreiben." Such translations are bound to transmit erroneous conceptions about the contents of the Indian sexological works.

1407 The printed text reads *madaśirācandrāśu śṛṅgāṭaka°*, which gives no sense.

93: "Kāma[1408] says: A certain (small) tube[1409], like the goad of Madana (= Kāma) (i.e. the penis) within [the vagina], which, filled with passion(-fluid) (*mada-*; see §12.13), like a lotus,[1410] lets, shaken by the penis, Madana-water[1411] flow, makes whatever [piece of] candied sugar is within [the genital opening] melt (i.e. would make it melt if it were placed within).

94: "And that which is above the opening of the vagina, resembling the peak/point of convergence (*śṛṅgāṭaka-*) of the passion(-fluid)-vessel[1412]-moon-rays,[1413] they give the explanation 'umbrella of Madana (= Kāma) (*madanātapatra-*)'. Moreover, one (small) tube[1414], which, filled with Manobhava-water[1415], is in the vicinity (of what?), that they call *cāndrī-* (i.e. lunar [tube]), giving much pleasure, making the passion(-fluid) (*mada-*; see §12.13) of the woman flow.

95: "The honourable ones/Aryans call that within the vagina which is situated near and is (or maybe: that which is situated near that which is) impassioning/exciting/intoxicating in connection with the womb (or maybe: that which is impassioning/exciting/intoxicating in connection with the womb, and is situated at the end[1416] [of the vagina?]),[1417] like the flower

---

[1408] Does this refer to a real authority, or to the mythical Kāma?

[1409] Cf. n.1388.

[1410] This could refer to the stalk of a lotus, which is firm and fleshy. The picture would fit the description of the penis-like tube. Or is the peculiar shape of the lotus-bud meant? One can, however, also think of the standard topos in Indian literature of the inability of water to cling to a lotus leaf and to drip off this rapidly.

[1411] Cf. n.1403; Madana = Kāma.

[1412] *śirā-* = ☞ *sirā-*.

[1413] This means that the organ described is in the centre of rays of vessels emanating from it like the rays of the moon from the latter. The likeness could also be taken to show that the organ described is round in form.

[1414] Cf. n.1388.

[1415] Cf. n.1403; Manobhava = Kāma.

[1416] With *antika-* derived (as "connected with the end") from *anta-* "end", not from *anti* "near".

[1417] The compound *garbhaśayyāmadanāntikastha-* is problematic, but I could not think of any other translations. The problematic nature of the compound might be due to the relatively

(or maybe: fruit)[1418] of the tamarind, the abode of supreme enjoyment, the unsurpassed agent of flowing, through touch, of women."

12.21. These descriptions have caused modern scholars examining them no little headache, for they have, all in all, not been able to understand them.[1419] One of the organs mentioned variously as the "umbrella of Kāma" or the like is, however, easily identifiable (and is the only organ modern scholars have been able to identify so far): it is the clitoris, as AN 4,33f. (see §12.19) too make quite clear.[1420] I may in this connection also draw attention to RATIK [1],94, where we might have an untranslatable play on words,

---

late date of composition of this text. The date is given in the last verse of RATIK ([2],193), which runs:

bāṇendrapṛthvīdharaśītapāṇivarṣe tathā mādhavaśuklabhūte
vyākhyād imāṃ dīkṣitasāmarājo ratyādikallolavatīṃ rasena.

The first expression of this verse gives the year in which the work was composed. The editor seems — judging by the superscribed numerals in the edition — to have identified the numeral values as 5 (*bāṇa-*), 1 (*indra-*), 5 (*pṛthvīdhara-*) and 1 (*śītapāṇi-*), i.e. gives the date as 1515. To my knowledge, however, *indra-* has the value 14, and *pṛthvīdhara-* = *bhūdhara-* = *parvata-* the value 7. But this causes difficulties. So could the first part of the compound actually be *bāṇendra-* (and not *bāṇa-* and *indra-* as separate words)? *bāṇendra-* = *bāṇeśvara-* = *śiva-* = *rudra-* would have the value 11, giving us the date 1711. Or is *indra-* to be taken as standing for *indriya-*, which would have the value of 5? *bāṇa-* and *indra-* would then give 55; i.e. we would obtain the date 1755.. However, such difficulties concern mostly the last two numerals (denoted by *bāṇendra-*); we can be relatively certain that the work was composed in the eighteenth century of some era. In all probability this latter is either the Vikrama or the Śaka era; in the former case we would obtain a date of the middle of the seventeenth century A.D. at the earliest, in the latter the end of the eighteenth century A.D. at the earliest.

[1418] The usual meaning of *prasūna-* is "blossom, flower", but the native lexicographers also give the meaning "fruit".

[1419] Cf. e.g. SCHMIDT 1922, pp.258f., MEYER,J.J. 1937 3,p.242,n.2.

[1420] On the *smarātapatra-*, as he calls it, cf. also Ḍalhaṇa's commentary on SU,Śā 5,10, which fits the descriptions given above. Commenting on *adhastād raktavahaṃ* (scil.☞ *srotas-*), he says, quoting a verse which calls the clitoris the *madanātapavāraṇa-* "umbrella of Madana":

adhastād raktavahaṃ smarātapatrasyādha ārtavavahaṃ. smarātapatraṃ bhagasyoparitane bhāge. uktaṃ ca.

vipulapippalapattrasamākṛter avayavasya śirastalam āśritam
sakalakāmasirāmukhacumbitaṃ nigaditaṃ madanātapavāraṇam

iti.

According to this description too, many vessels (☞ *sirā-*) converge at the organ mentioned. Note also the statement that the organ in question is situated at the head of an organ resembling a huge leaf of the holy fig tree (*Ficus religiosa* Linn.), which latter fits the labia of the female generative organ.

as *śṛṅgāṭaka-* is also the name of plants of the *Trapa* species and their fruit, namely the water chestnut, this latter resembling the clitoris;[1421] at the same time, this organ is said to be the juncture (or peak) (*śṛṅgāṭaka-*) of a number of vessels,[1422] a description which fits what is said in the other texts mentioned in §§12.18f.[1423]

12.22. The other organs described above in §§12.18ff. have, however, not been identified so far. However, one of these could, on the basis of the descriptions given, be identified as the Gräfenberg spot, long disputed in Western medicine. The Gräfenberg spot is an approximately 1×2 cm. firm,

---

[1421] Ḍalhaṇa in his commentary on SU,Ci 3,61 explains *śṛṅgāṭaka-* as a three-thorned aquatic fruit (*śṛṅgāṭakaṃ trikaṇṭakaṃ jalajaṃ phalam*). Usually *śṛṅgāṭaka-* is identified as *Trapa bispinosa* Roxb. which, though basically triangular with thorns at the tips, seems to have as a rule only two thorns (as its name implies); fruits of other *Trapa* species are, however, known to have different numbers of thorns. In any case, water chestnuts usually do seem to have a prominent mound, looking indeed much like an unsheathed clitoris.

[1422] Note in this connection also Indu on AS,Ut 5,p.46a, who explains *śṛṅgāṭaka-* as *tripatha-*, i.e. meeting point of three roads, which fits the context loc.cit. Both the mention of the number three as well as the characterisation as a mound also fit in with n.1421.

[1423] Note also NĀG 18,1, which puts the number of (small) tubes (*nāḍikā-* cf. n.1388) at twenty-four:

> *mukhaṃ caturviṃśatināḍikānāṃ madāvahānāṃ madanātapatram*
> *nārīvarāṅgodaramadhyavarti pracodayet tryakṣarabījayuktam.*
> "One should excite the umbrella of Madana (*madanātapatra-*) within the cavity (*udara-*) of the best organ of the woman, the [umbrella which is the] mouth (or: face/front) of twenty-four (small) tubes conveying passion(-fluid) (*mada-*; see §12.13), accompanied by the incantation consisting of three syllables (*oṁ āḥ aṃ* according to 18,2)."

According to this, the clitoris would be within the genital opening, which here cannot but refer to its position between the labia. Or is the above actually a clumsy way of stating that the (small) tubes mentioned are within the vagina, but have their peak in the clitoris? This is how the Hindi commentary understands the verse; this commentary, however, also sees in the clitoris the place of discharge of the mentioned (small) tubes, which is of course problematic if an *opening* to the exterior be meant, but might perhaps be taken to refer to a discharge internally within the clitoris, which discharge would then ultimately emerge via the vulva, as discussed in §12.24 (*strīke varāṃgʼke madhyameṃ saṃbhogecchā prakaṭ karʼnevālī caubīs nāṛiyāṃ haiṃ, unʼkā mukh yā nirgam sthān madanʼcchatr kahātā hai*). The commentator then says that the stimulation mentioned in the verse refers to stimulating the clitoris with the fingers (*aṃguliyoṃkā gharṣaṇ karʼnā cāhiye*). This is in keeping with 18,3, which refers to stimulation of the clitoris both with the fingers and with the penis, provided *liṅgāṅgulibhyāṃ* is to be construed with *pracodanīyaṃ*:

> *samprāpya bālāṃ karaśākhayaiva liṅgena cātmīyasamāṃ sametya*
> *liṅgāṅgulibhyāṃ tu tathābhirūḍhāṃ pracodanīyaṃ madanātapatram.*

(On *pracodanīya-* cf. also n.1557.)

oval area different from the rest of the vagina; it swells on stimulation, and is mostly detectable only then, though in some women it may also be detectable as a small bean-like lump in its non-stimulated form. It is located about 1 to 2 cm. deeper within the vagina than the inner border of the pubic bone (other sources describe it as lying about half-way between the back of the pubic bone and the uterine cervix), not directly on the upper vaginal wall (it is between vagina and urethra), but detectable through the epithelium.[1424] This could have led to its having being described as resembling a small penis.[1425] What would make the identification even more credible is the fact that stimulation of the Gräfenberg spot leads to an ejaculation, just as the texts seem to describe as taking place in the case of the penis-like organ.[1426] It must, however, also be pointed out that the ejaculate due to stimulation of the Gräfenberg spot is actually a fluid discharged through the urethra; we have already discussed both this fluid and the problems its urethral ejaculation poses in §§12.16f.

12.23. In this context RATIK [1],93 (see §12.20) deserves special consideration, for its description makes it appear as if a fluid were discharged into the vagina from the very spot where what we have taken to be the Gräfenberg spot is situated. Against this we, however, have AN 4,32f. (= ŚIVAT 6,12,116f.) (see §12.19), which tell us that the discharge of or due to what seems to be the Gräfenberg spot takes place via the clitoris. On the other

---

1424 WHIPPLE, p.16 gives the following description: "This sensitive area is usually located about halfway between the back of the pubic bone and the cervix, along the course of the urethra and near the neck of the bladder. It swells when it is stimulated, although it is difficult to palpate in an unstimulated state. Other researchers have also reported anterior vaginal wall sensitivity and orgasm from vaginal stimulation ... . The Grafenberg spot has not been found universally by all researchers who have conducted sexological examinations of the vagina. It may be that not all women have this distinct area, or the lack of universality may be due to different methods of stimulating or different criteria for identifying this area ... ."

1425 In this connection cf. also LADAS/WHIPPLE/PERRY, p.66, where a woman states: "Whenever sexually excited this spot begins to swell to the point of protruding, where it actually comes out to reach my husband's penis". This is, however, a rather special case and cannot be generalised.

1426 This has of course led to comparisons with organs of the male; thus in LADAS/WHIPPLE/PERRY, p.42 we find: "One controversial question is whether the G spot can be considered a homologue of the male prostate. This controversy goes back to ancient Rome, where Galen, a physician writing in the second century A.D., raised the question and voted yes. But William Masters, commenting on the Perry and Whipple research in April 1981, insisted that the word *prostate* is not appropriate." It is in this context interesting that according to LADAS/WHIPPLE/PERRY, p.134 the ejaculation of males may be triggered by stimulation not only of the penis, but also of the prostate (by women of the clitoris and the Gräfenberg spot).

hand, RAT 10,6 (also in SMAR) and RATIR 4,35f. (see §12.18) seem to be telling us, if I understand them correctly, that both the seeming Gräfenberg spot and the clitoris emit a fluid when stimulated. All this does seem puzzling, and remarks of this sort have actually led MEYER,J.J. 1937 3,p.242,n.2 to conclude that Indian sexologists regarded the clitoris as an exact equivalent of the penis,[1427] which included the notion of an ejaculation "durch die Klitoris". This then seems to have led him to make several wrong deductions, in particular to identify the penis-like organ, mentioned as lying within the vagina (i.e. what seems to be the Gräfenberg spot), as the clitoris.[1428]

12.24. We must not, however, do MEYER,J.J. 1937 injustice, for "durch" (i.e. "through") is ambiguous, since it does not tell us where the discharge goes to, and may even be understood in the sense of "by means of; because of". Though we can of course not be sure, nevertheless, what "durch" in the statment of MEYER,J.J. 1937 could actually be referring to might be a discharge into the body (where, is unclear) via the clitoris (then probably: its root),[1429] which subsequently emerges to the exterior via the vulva. If this is indeed what MEYER,J.J. 1937 presupposes, then that could hit the nail on the head. The texts quoted in §§12.18ff. give no details on the discharges due to stimulation of the clitoris. One could nevertheless assume that this discharge issues directly from the clitoris to the exterior, pointing as a parallel to notions common in Europe only a couple of centuries ago (and probably based on older beliefs) that the clitoris has a very minute opening which may open up and discharge to the exterior (see on this STIFTER, pp.52ff.). However, this is quite circumstantial evidence as regards India, especially ancient

[1427] Cf. in this regard also BṚSAṂ 69,3 (chapter 69 is chapter 70 in most other editions), in which *maṇi-*, a term common for the glans penis, is used for the clitoris. The commentator Utpala also quotes a verse from a mantic work which uses *maṇi-* in the meaning "clitoris". See also CHÖPEL, pp.56;196, who holds that the clitoris is the internalised female penis.

[1428] It goes without saying that this has caused MEYER,J.J. 1937, loc.cit. some headache, since the clitoris is after all an organ outside and not inside the vagina. But modern Indian scholars have been no less puzzled as regards the description of the clitoris itself; thus the editor of AN says on p.3,n.2: *madanacchattraṃ yoniḥ*! (Such an explanation of this and similar names of the clitoris can also be found in various commonly used dictionaries of Sanskrit; Kāñcīnātha in his commentary on RAT 1,12 too seems to equate *yoni-* and *kāmātapatra-*.) It is not only modern Indian scholars who have been thus nonplussed; a good example of this is SCHMIDT 1991, where we find e.g. *kāmātapatra-* explained as "Vulva" and "Organ in der Vulva" (p.143), the synonym *madanātapatra-* (misprinted loc.cit.) as "Vulva" (p.290) and another synonym *manmathacchattra-* simply as "ein Organ in der Vulva" (p.293)!

[1429] Cf. in this connection also the characterisation of the clitoris as an internalised penis by CHÖPEL (see n.1427) — does this mean that it is taken to discharge internally?

India (as the Indian texts only refer to a discharge and give no details at all), and the older Greek evidence, which could be taken to be a stronger parallel, is also not very solid,[1430] since the Greek theory referred to in §12.16, mentioning a minute duct opening to the exterior above the urethra,[1431] seems not only not to have been common, but also says merely that this structure serves as a respiratory duct (no discharge via it is seemingly mentioned) for a tube (*kaulós*) which discharges a woman's "semen" *within* her body, into the vagina in front of her womb.[1432] Hence, until we have better evidence, we should be careful of simply assuming that the Indian authorities were unable to observe the clitoris properly (though we cannot discount this possibility) and thus thought that it actually discharged something to the exterior. So maybe we should try to understand our texts by assuming that what they mean is an internal discharge that then flows out of the body. The internal discharge could take place somewhere in the vicinity of the clitoris (one could think, e.g., of an internal opening at the end of its shaft — but where to?); the texts, however, seem to have no clear statements which would prove the validity of such a deduction either, so that it even remains possible that the discharge due to stimulation of the clitoris occurs somewhere else, namely via the place where what we have taken to be the Gräfenberg spot is situated, which would do justice to what RATIK seems to say (see §12.23). But wherever the discharge due to the clitoris actually takes place, according to AN 4,32f. (= ŚIVAT 6,12,116f.) (see §12.23) one could deduce that stimulation of the Gräfenberg spot too produces an ejaculation at this same place. But even if the AN passage is to be understood in this way, it does not automatically follow from this that this must then apply to all the other texts too, for these may adhere to a different theory. In this context we must, however, point out that the clitoris and the other organ being discussed here can hardly be taken to be stimulated in the same manner during 'regular' intercourse,

---

[1430] Even though STIFTER, pp.36f. tends to assume that what is meant is an opening in and ejaculation via the clitoris.

[1431] Just above the urethra, if DEAN-JONES 1994, p.158 is correct, but I doubt if we should place too much emphasis on the translator's "just", quite apart from the fact that this term is rather ambiguous.

[1432] On this see also ROUSSELLE, p.29, where it is stated that according to a theory in the Hippocratic Collection female semen passes from the body into the uterus through a small orifice situated next to the place from which urine flows. Rousselle says on this: "It was *pneuma* which was drawn in through this canal, and clearly it is the clitoris which was held to be its orifice. The *pneuma* reached the woman's spermatic secretions, which came down through the neck of the womb and were then drawn back up into the uterus together with the male sperm to bring about conception."

since the clitoris is situated outside the vagina and the penis thus cannot as a rule stimulate it directly, though other kinds of friction may.[1433] Now the AN passage expressly speaks of stimulation with the penis in connection with the discharge "from the umbrella of Kāma", and it should also be noted that the RATIK passage mentions no discharge due to stimulation of the clitoris, whereas RAT (and SMAR) and RATIR, which do mention a discharge, also expressly allude to stimulation with the fingers. One could thus hold that the discharges due to stimulation of the clitoris and the penis-like organ (tentatively identified as the Gräfenberg spot) were actually taken to be of the same sort and at the same place, both organs acting in the same manner, but due to different causes. But this would also seem to presuppose that the discharge due to stimulation of the clitoris is also an ejaculation (connected with an orgasm), and not a non-orgasmic discharge of fluid. In any case, this is just one postulated possibility; we have as yet no means of convincingly proving or disproving it or the other possibilities. One of these latter is that, if some texts indeed state that an ejaculation comes out from or near the Gräfenberg spot itself (see §12.23), then that could be taken to refer to an ejaculation from where it is situated within the vagina into the vaginal passage and thus to the exterior. On the other hand, since the Gräfenberg spot lies internally between vagina and urethra, a discharge via the latter might also have been presupposed (which would tally with what actually happens) (cf. §12.22), though there seems to be no statement which could be taken to clearly refer to the discharge as being urethral. Finally, we must also take into account the possibility of the ejaculate coming out from some place other than the urethra within the vestibule before the introitus of the vagina, in accordance with the discussion in §§12.16f. and above in this paragraph, where we have, however, also drawn attention to the difficulties connected with such an assumption.

12.25. The identification of the penis-like organ as the Gräfenberg spot is, however, not unproblematic. Thus one may very well ask whether the fact that a (rather small) part of the vaginal lining swells on stimulation would be enough to qualify the swollen portion as similar to a penis, specially since it is controversial whether such swelling can indeed be observed in *all* women. Of course swelling and becoming firm is what the penis does too, but one must still ask in what way the Gräfenberg spot is, in its swollen state,

[1433] Modern Western sexological works sometimes describe women in whom the clitoris is situated at such an angle that it actually touches the shaft of the penis during intercourse, but since such descriptions are expressly called exceptions, we cannot use them to build a generalising theory here.

different in form from various other glands and knots of the body that can be easily felt under the skin, and are sometimes even clearly visible to the eye. One of the characteristics of the penis, which it shares for instance with the clitoris, is missing, namely that it clearly juts out and is not only a rounded bulge. This does not, of course, do away with the swelling and becoming firm which both penis and Gräfenberg spot have in common, which might indeed have led to the latter being called a small penis, but it shows that we cannot be fully sure that we have a water-tight case for assuming so. This is all the more so when we consider that RAT, SMAR and RATIR call the penis-like organ a "swing" (*dolā-/ḍolā-*); it is rather difficult to explain this with reference to the Gräfenberg spot. In other words, we are not absolved from the necessity of considering further parts of the female genitals in this connection. Then, however, one finds that there is indeed another organ which might be compared with a penis, namely the portio vaginalis of the uterus, that is, the vaginal part of the uterine cervix; not only does this perceptibly jut like the neck of a bottle with a conical tip[1434] into the vagina at its rear end, but it even has an opening which can be felt (quite clearly especially in women who have given birth), just like the penis.[1435] Now there is a debate among Western sexologists about whether the cervix (portio vaginalis) is or is not particularly sensitive to stimulation and can or can not trigger an orgasm. Might it not therefore be that the ancient Indian sexologists too adhered to the view that this area was indeed very sensitive? Whether they also thought that it could give rise to an orgasm and an ejaculation is still to be debated, but it is not unlikely that they knew that the uterine cervix emits a clear fluid[1436] associated with arousal, and/or orgasm[1437] (cf. also n.779); I may

---

[1434] Could, if the identification as the portio vaginalis of the uterine cervix is correct, the round shape of this tip from a frontal perspective be the reason for the variant of RATIR 4,35 likening the *madanāgāraḍolā-* to the moon (see n.1387)?

[1435] Actually, the uterus lies at an angle to the vaginal canal, and the cervix juts into the rear of the latter at a slight angle. With heightened sexual arousal the uterus assumes a more or less upright position, until it and the vaginal canal ultimately nearly form a straight line. This fact might help solve some of the problems LAQUEUR 1990, p.45 mentions.

[1436] On this cf. also §12.15.

[1437] On this latter point PLOSS/BARTELS/BARTELS 2,p.265 is quite definite: "Hat das geschlechtliche Empfinden, das von Wollusterscheinungen begleitet ist, seinen Höhepunkt erreicht, dann tritt der O r g a s m u s ein. Mit ihm ist der z w e i t e B e g a t t u n g s - r e f l e x verbunden, der sich beim Manne in der E j a k u l a t i o n d e s S a m e n s , beim Weibe in der E j a k u l a t i o n d e r S e k r e t e d e s C e r v i x teiles der Gebärmutter äußert. Diese erscheinen hauptsächlich als ein feiner Faden aus glasigem Schleim, den man den *Kristeller*schen Pfropfen nennt."

mention too that the Japanese sexologist Atsushi Oshikane, basing his deductions on personal observation, is quite definite that the ejaculate of women originates from the uterus, in particular its cervix (cf. STIFTER, pp.147;149). Here one might now point out that the texts refer to the penis-like organ as a ☞ *nāḍī-* or *nāḍikā-*, i.e. a (small) tube. Whatever these terms may actually mean here, it does seem as if *nāḍī-* and *nāḍikā-* in general refer to structures which are hollow and tubular (like e.g. a pipe, the inside of a reed, an artery or vein, or the like). Would the cervix qualify as such a structure? This is a question I feel unable to answer, for, though one could call the cervix of the uterus a tubular structure, I am not certain what other criteria a structure had to fulfil to be called so. One could in this connection, however, also point to the comparison made with the lotus in RATIK [1],93, since the cervical portio vaginalis could be taken to look like a lotus bud (whether it could also be taken to feel or look like a lotus stalk — cf. n.1410 — I cannot say). On the other hand, this comparison is so vague (see n.1410) that it would be methodically most unsound to base such a far-reaching deduction on it. But it must be pointed out too that we have seen in §12.23 that the same RATIK passage might be understood to say that an emission takes place at the very spot where the penis-like organ is situated, and this would of course fit the cervix much better than the Gräfenberg spot. This must then bring to mind the problem last discussed in §12.10, namely whether we may presuppose a theory according to which there is an internal ejaculation accompanying orgasm, the product of which then comes to the exterior. Such an ejaculation could be uterine, and thus a parallel to the ejaculation of the procreatory-menstrual fluid; on the other hand, it might also appear from some place in or on the cervix or near it within the vagina (cf. on this especially §§11.10ff.). From all of which it follows that an identification of the penis-like organ as the portio vaginalis would indeed be in keeping not only with data already discussed by us, but also with modern knowledge (such as e.g. the discharge of fluid from the uterine cervix). What would seem to clinch the issue is the fact that in the West too both the vaginal passage and the uterine cervix were regarded as an internal penis until quite recently. See on this LAQUEUR 1990, pp.33f.;79ff. also p.45 and (on the vagina) *passim*. LAQUEUR 1989 too draws attention to Western descriptions of the uterine cervix as a penis (see especially pp.104f.), and ROUSSELLE, p.27 quotes Oribasius as saying that in women "the internal genital organs surround the neck of the womb just as, in men, the foreskin grows around the glans". In this connection an illustration of the reproductive organs of the female in SEN 2,p.251 is extremely interesting, as the uterus looks much like a penis with its tip jutting out into the vagina from above; this illustration is reproduced here:

गर्भाशयः (बीजाधार-बीजवाहिन्यादियुतः)।

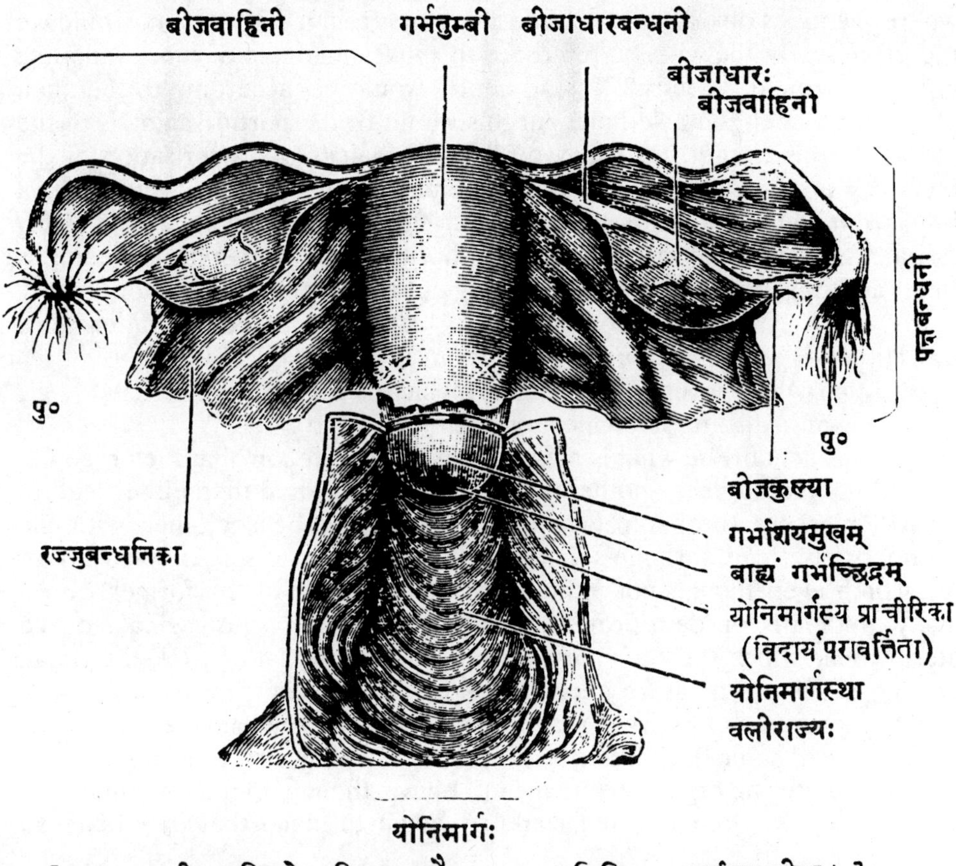

[पु०-पु०—बीजवाहिन्योः पुष्पितप्रान्तौ। ×-×—चिह्निता—गर्भाशय-ग्रीवा।]

12.26. But the portio vaginalis of the uterus might also fit the description of quite another of the organs described above, namely the organ which according to RATIK [1],95 might (according to alternative translations proposed in §12.20) be situated near the womb or at the end of the vagina, which in any case could apply to the uterine cervix.[1438] The RATIK passage

[1438] In this connection I may also mention the comparison with the *prasūna-* (on this see

also remarks on the importance of this organ, in which connection we must, because of the obvious parallelism in the descriptions, also consider the organ called the principal one in RAT 10,8[1439] (*pradhāna-*) and RATIR 4,39 (*mukhya-*) in §12.18. This latter organ is called a wheel (*cakra-*) of many tubes, which seems similar to the description of the clitoris in RATIK [1],94, where the tubes obviously spread out like rays from a single point. Could we therefore see in the wheel a reference to spokes made up of tubes radiating out from around a centre?[1440] This centre would be, according to the analogy of the wheel, the axle-hole. All this would fit the portio vaginalis of the uterus. In this connection I may point to Ḍalhaṇa's characterisation of the uterus as *alpamukha-* "whose mouth is minute" in his commentary on SU,Śā 5,44, from which it follows that according to Ḍalhaṇa (and, we may safely assume, at least some other medical authorities too) the cervical opening is small and thus does not spoil the analogy of the wheel by being too large.

12.27. We have, however, in §12.20 considered an alternative translation of *garbhaśayyāmadanāntikastha-* in RATIK [1],95, namely:

"that which is situated near that which is impassioning/exciting/intoxicating in connection with the womb".

Should this actually be what is meant, then it would follow that the organ described is situated near another organ. This latter would then either lead to sensual pleasure (probably: when stimulated) and be conjoined with the womb, or else lead, through some sort of connection, to sensual pleasure of the womb even though not joined with or near this. In the former case a likely candidate for identification would again be the portio vaginalis of the uterus, near which the other organ would lie. In the latter case, the organ leading to sensual pleasure need not be joined directly with the womb, but only bring about some sort of reaction in or on this; one could, e.g., think of a discharge of some fluid into the uterus, but one may also consider the possibility of a uterine orgasm being meant. I have, though, found no remarks of ancient Indian sexologists that could be taken to mean that they believed

---

n.1418) of the tamarind in the RATIK passage. The fruit of the tamarind is longish, tubular and firm, and as such resembles an erected penis. On the other hand, the blossom of the tamarind is elongated and more or less oval, the outer petals enclosing the inner part like a boat. Or are not these shapes, but something else responsible for the mention of the *prasūna-* of the tamarind? These considerations naturally apply only if the organ mentioned is indeed the cervical portio vaginalis.

[1439] See also SMAR, loc.cit. in n.1381.

[1440] Cf. in this context also SU,Śā 7,5 on the navel:
nābhisthāḥ prāṇināṃ prāṇāḥ prāṇān nābhir vyupāśritā
sirābhir āvṛtā nābhiś cakranābhir ivārakaiḥ.

that or debated whether a woman could have a uterine orgasm (a subject which is much debated in modern Western sexology, and in which connection the contractive spasms of the uterus during an orgasm play no small part); on the other hand, we have found a great deal of evidence to let us deduce that the notion of a discharge of some sort of fluid in connection with the uterus (into or within it, and/or in or near the cervix) and connected with sexual pleasure (scil. an orgasm) was not rare (see also §12.25). But be that as it may, the organ we are here concerned with would then lie somewhere near the organ bringing about such a result. Which means that we end up with the possibility that, if the alternative translation considered above is what is actually meant, the organ (wheel-like, according to the RAT, SMAR and RATIR parallels) said to give extreme pleasure by RATIK lies near the portio vaginalis of the uterus, or else near some other organ which has some sort of effect on the uterus, but is not necessarily directly joined with this. Also, since the organ giving extreme pleasure is said to be situated within the vagina, it most probably would, according to this explanation, be a spot or some similar small area somewhere on the vaginal wall.

12.28. I am, though, at a loss as to what such a spot or the like could be, specially if the organ near which this is situated be the portio vaginalis of the uterine cervix. In that case it would have to be something as yet unknown to Western medical science, as the Gräfenberg spot, though within the vagina, seems generally to be too far removed from its end to be referred to as being near this (cf. §12.22); at any rate I could not find any modern descriptions which describe it as being near the cervix. If the other organ connected with the pleasure of the uterus is, however, not the portio vaginalis of the uterus, then things become even more problematic. One could consider it to be the Gräfenberg spot (in this case then considered to lead to a discharge into the uterus; cf. §12.24), near which the organ we are looking for would lie. Or the organ connected with the uterus might be some unknown organ near which the Gräfenberg spot lies, which latter would then be what we are looking for. Or both organs may be (as yet?) unknown to Western medical science. One may in this connection also consider translating *cakra-* not as "wheel", but as "circle", and then think of identifying the organ this describes as a ring running around the inner lining of the vagina and perpendicular to its axis of length. We today know that the inner lining of the vagina emits droplets of a lubricating fluid, sort of "sweating" these out (cf. n.779); is it not possible that the ancient Indian sexologists too explained this phenomenon in a similar manner?[1441] They might have explained it as being due to a

---

[1441] It may even be that *sveda-* "sweat" was used by some authorities to denote the lubricating fluid, though not necessarily with any differentiation as to the actual place of discharge. RATIR 1,21 mentions a discharge called *ratisvedajala-* (as the characteristic discharge of the *hastinī-* woman). Now *sveda-* can mean "heat; toil", so that *ratisvedajala-* could be merely "water of the heat/toil of intercourse"; it could, however, also mean "sweat-water of intercourse". Unfortunately, we cannot in the latter case be fully sure that we then indeed have a reference only to the lubricating fluid and not to an orgasmic ejaculation, especially since *rati-* can refer to the orgasm (see ☛ *bhāva-* in Appendix II; on the RATIR passage see also KAPP, pp.194f.), which latter would incidentally seem to suit the former of the meanings of *ratisvedajala-* considered above. But *yonir utsvidyate* in the passage SMAR, pp.267f. (as well as in several variants of MBH 1,1189*) quoted in n.1318 clearly refers to wetness due to sexual arousal, not an orgasm. On the other hand, SMAR, p.271 has a passage (allegedly from the *Dattātreyopaniṣad*, but not to be found in the edition consulted) that says that the man should insert his penis into the vagina and work on the woman until *prasveda-* arises (*yāvat prasvedasambhavaḥ*), in this way both he and the woman attaining the fulfilment of pleasure (*sukhasiddhi-*); this looks like a reference to an orgasmic ejaculation (a reference to simple sweating seems unlikely). I may also mention in this context that Narasiṃha in his commentary on RAS 1,21 and Ḍalhaṇa in his commentary on SU,Ka 1,6 quote a verse on the "poison-maiden" (*viṣakanyā-*) from an unidentified source stating that she kills through her "sweat", intercourse too being mentioned in this verse (Ḍalhaṇa's readings, when they differ from Narasiṃha's, are given within brackets; these readings are also found in the quotation, without a source being given, in BANDYOPĀDHYĀY/DEBĪ, p.255, and BANERJI/BANERJI, p.148):

> *spṛśantī hanti (hanti spṛśantī) svedena gamyamānā ca maithunam (°ne)*
> *pakvaṃ vṛntād iva phalaṃ praśātayati mehanam.*

However, the relevant first line (the second line means: "she makes the penis fall off like a ripe fruit from the stalk") is ambiguous, as it could mean:

> "touching [one], she kills (Ḍalhaṇa: she kills, touching [one]), through sweat and [when] had recourse to for (Ḍalhaṇa: with regard to) intercourse",

or:

> "and, touching [one], she kills (Ḍalhaṇa: and she kills, touching [one]), through sweat [when] had recourse to for (Ḍalhaṇa: with regard to) intercourse".

Since in the first alternative "sweat" and intercourse are not connected, one could conclude that "sweat" need not necessarily refer to a fluid in the vagina here, especially since it is known that a "poison-maiden" was believed to be able to kill even through touch too. In this context I may also draw attention to HAHN 1974, p.213,n.40 and HAHN 1987, p.143, where it is said that the "poison-maiden" kills her husband through the emission of poisoned sweat whilst circumambulating the sacred fire with him during the wedding rites. Though no references supporting this statement are given, I suspect that the author had in mind the story found in Hemacandra's *Pariśiṣṭaparvan* to which PENZER, pp.16f. draws attention. One could now take this as an argument against "sweat" in the passage cited above referring to a vaginal fluid. Unfortunately, there seems to be no other information of relevance in the literature on the "poison-maiden" such as PENZER, pp.3-71, KOHL and SCHMIDT 1904, pp.565f. (with further bibliographic references; some meagre information is also found in BANDYOPĀDHYĀY/DEBĪ, pp.255f., BANERJI/BANERJI, pp.148ff., MÜLLER 1942, p.23,n.2). I may also mention that according to MEYER n.d., p.59,n.2 *sveda-* also means "lust, rut". Actually, this is a conjecture to fit one single occur-

sensitive ring running round the inner vaginal wall. But one may also think of the inner side of the very sensitive ring at the mouth of the vagina, especially since the ducts of Bartholin's glands (cf. n.779) open out to the exterior at the mouth of the vagina (cf. also §11.10). But two things speak against considering these as the *cakra*--like organ we are looking for. The first is that *cakra-* seems only very sporadically to be used for a circle with nothing enclosed within it (i.e. basically only the circumference, as would be the case here — in other words: for a ring); we would rather have expected an expression like *valaya-* or the like, or at least some further explication to make us hit upon such a not so common meaning of *cakra-*. The second is that the rings considered here would be connected with the lubricating fluid(s), whose discharge is not even necessarily connected with sexual acts, let alone an orgasm, while the contexts in which we find the organs we are trying to explain mentioned do point to stimulation with the fingers or during intercourse and maybe even an orgasmic ejaculation.

12.29. Ail this makes it not very probable that rings of the sort just considered could be meant. But even if we took the organ mentioned to be some sort of spot or similar small area within the vagina, we would still have problems inasmuch as we would have in this context to assume the existence of one or even two organs as yet unknown (one of which was taken to be connected with sensations relating to the uterus). When we consider all these difficulties, we must ask ourselves whether the alternative translation of *garbhaśayyāmadanāntikastha-* considered in §12.27 is at all feasible. If it is not, that would mean that the organ we are looking for is "situated near and impassioning/exciting/intoxicating in connection with the womb", which then would mean that we would be left with the portio vaginalis of the uterine cervix considered in §12.26. Of course I am aware of the fact that it is not impossible that the ancient Indian sexologists are here actually referring to an organ or organs (as yet?) unknown to modern Western science, but on the other hand it must also be pointed out that, if a known organ does seem to fit the description given, bringing in some as yet unknown organ or organs seems like a needless complication. Nevertheless, we cannot categorically exclude the possibility of some such organ or organs indeed being presupposed here, which is why the identification as the portio vaginalis of the cervix had best be regarded as a mere working hypothesis at this stage.

---

rence of the word, but it could nevertheless fit in the passage quoted above too:

> "and, touching [one], she kills (Ḍalhaṇa: and she kills, touching [one]),
> [when] had recourse to, with lust, for (Ḍalhaṇa: with regard to) intercourse".

In that case of course *sveda-* would have nothing at all to do with any fluid; moreover, *ratisvedajala-*, cited above, would simply be "water of the lust of intercourse".

12.30. The comparison with the wheel, however, also brings to mind the comparison with the lotus referred to in §12.25. If not a bud, but a flower were meant, then the petals, structures radiating out from a centre, could be regarded as analogous to the spokes of the wheel.[1442] But the comparison refers to a different organ, namely the penis-like structure, in identifying which we have considered both the Gräfenberg spot (see §§12.24f.), as well as the portio vaginalis of the uterus (§12.25), the latter with a greater degree of probability. Should the reference to the lotus indeed have this implication, then it would fit perfectly with what we have deduced with regard to the latter organ in connection with the picture of the wheel (cf. §12.26). But that would then mean that we would end up identifying two different organs mentioned as one and the same organ. If, however, the penis-like organ should, in spite of the problems connected with this identification (see §12.25), be the Gräfenberg spot, then the conclusion would be that this too lies at the centre of several tubes. Then, however, the wheel-like organ could refer to the Gräfenberg spot too (cf. §12.28), in which case we again end up with identifying two different organs mentioned as the same organ. When we consider all this together with the possibility that the penis-like and the wheel-like organs may be quite different from each other, it becomes obvious that we are on the best way to losing ourselves in a maze of problems, and thus should maybe consider trying another approach.

12.31. Now when we look at the descriptions of all the organs mentioned in the texts above, we find that RAT (and SMAR), RATIR and RATIK mention four organs,[1443] whereas AN (and ŚIVAT) speak of three. But RAT (and SMAR) and RATIR (§12.18) also mention a triad of organs, which consists of the clitoris, the wheel-like organ we have discussed above, and the tube called *pūrṇacandrā-*, said to be not too far removed (from where?). The description of this triad follows right after the description of a pair of organs, namely the penis-like organ also discussed above, and the clitoris. The mention of the clitoris twice does very much create the impression that we actually have two different theories side by side here, one speaking of only two, and one of three organs; especially the statement of RATIR, expressly calling the first group (of two organs: "this pair, verily") the "organs" in young wom-

---

[1442] In this connection I may also draw attention to Yogic and Tantric terminology, in which the mystical plexuses of the body may be synonymously referred to as *cakra*-s or *padma*-s, though at times it seems as if these two terms referred to different structures of which one is conceived of as being present within the other.

[1443] In the first three texts mentioned the clitoris appears twice.

en,[1444] while at the same time (like the others cited) simply going on then to describe the mentioned triad, does lend credibility to this assumption. AN (and ŚIVAT) (§12.19) list the clitoris, the *pūrṇacandrā-* and the penis-like organ. If we be allowed to combine these descriptions, then we would, based on the discussion just above on the nature of these organs, unfailingly have to deduce that the penis-like organ and the wheel-like organ in the two different groups are one and the same, which would leave us with three organs in all, of which one organ (the *pūrṇacandrā-*) would not figure in the alternative theory considering only two organs. Further, if these deductions be valid, then it would also follow from the preceding detailed discussion that the penis-like and wheel-like organs seem rather to refer to the portio vaginalis of the uterus, than to the Gräfenberg spot. But the latter would fit the description of the third organ, the *pūrṇacandrā-* (which we have not considered so far), very well. This organ[1445] is connected with an ejaculation (see especially AN 4,35 in §12.19[1446]), and is also said to be within the vagina and not too far removed. Should the latter refer to its distance from the exterior opening, then that would of course exactly describe the site of the Gräfenberg spot, which, though within the vagina, is yet near enough to the opening to be easily reached, at least according to the descriptions I could find (cf. §12.22); note too that §12.18 tells us that the *pūrṇacandrā-* is one of the organs which can be stimulated by the fingers, which would also apply to the Gräfenberg spot. As regards the place of discharge of the ejaculation due to the Gräfenberg spot, since the connection with the clitoris considered in §12.24 pertains to the penis-like organ which we have now identified with the portio vaginalis of the uterine cervix (and no longer with the Gräfenberg spot), we have hardly any material in this regard. One could, however, opine that an ejaculation through the urethra (cf. §12.22) would have been expressly mentioned, as it is not something one (at least most men, and probably also women) would a priori associate with sexual stimulation of a woman. Indeed, AN 4,34f. (see §12.19) seem to state that the ejaculation occurs from the *pūrṇacandrā-* itself, and since this organ is obviously situated within the vagina according to the text,[1447] one could take this to mean that the ejac-

---

[1444] 4,36: *tad etad dvayam evāhur indriye yuvatījane* (see n.1387).

[1445] Of which STIFTER, p.24, basing his deductions on translations of the Sanskrit texts, says that it almost certainly ("mit einiger Sicherheit") refers to the urethra; I do not know on what evidence.

[1446] In connection with n.1270 and n.1281.

[1447] Even though the Gräfenberg spot is not located within the vagina, but behind its wall, it can be felt through this wall from within the vagina (cf. §12.22); as such, it may be taken to

ulate too is vaginal. It must be admitted, though, that this is very scanty evidence, and as such cannot lead to any surety in this regard. Nevertheless, and even though in the light of the discussion in §§12.16f. we cannot be fully certain with regard to our deductions on this topic, we could therefore on the basis of §§12.18ff. at least tentatively deduce that a vaginal ejaculation could have been assumed, perhaps at the very spot where the Gräfenberg spot is located, or maybe somewhere nearer the mouth of the vagina (cf. in this regard also ŚṚṄG 3,29 in §11.10, though this probably refers to lubricating fluid). If, however, an ejaculation in the vicinity of the vaginal opening should indeed be presupposed by the texts, then we may also consider whether the statements of Āḍhamalla and Śrīdāsa last discussed in §12.17 might not pertain to just such an ejaculation. But that is about as far as we seem to be able to get with regard to the place of ejaculation due to stimulation of the Gräfenberg spot. But whatever the answer to this problem be, it does not affect the deduction regarding the existence of two different, partly overlapping, theories on particularly sensitive parts of the female genitals. All this means (provided, of course, all our deductions are valid) that the two systems of classification mirror exactly the difficulties Western sexologists have faced, and which are due to the same fact: two organs (the portio vaginalis and the clitoris) are easily perceivable, whereas the third organ (the Gräfenberg spot) is not. That we are indeed justified in at least considering the possibility of there being different theories on parts of the female genitals causing orgasm follows from the fact that, while commenting on KĀM 13(=2,8),p.160,[1448] Yaśodhara explicitly mentions a debate on whether one or more places (in the female genitals) cause an orgasm[1449] [1450].

12.32. Against all this one may hold that RATIK (§12.20) simply mentions four organs, namely the clitoris, the penis-like organ, the *cāndrī-* (obviously the same as the *pūrṇacandrā-*) and a tamarind-like organ (corresponding to the wheel-like organ), without any dyadic or triadic classification. There are, however, two arguments against giving this too much weight. The first is the very late date of composition of this text (see n.1417); the (nearly modern) author might simply have misunderstood older descriptions. The second is the fact that the text of [1],95, where the tamarind-like organ is mentioned, actually only says that this organ (on which see also n.1438) is

be located within the vagina, even though strictly speaking it is not.

1448 On the passage of which this is a part cf. also n.1558.

1449 *rati-*; cf. on this ☞ *bhāva-* in Appendix II.

1450 *atra ca ratibandhanam* (var: *ratisambandhanam*) *eko bahava iti keṣāñcit* (var: *kecit*) *pradeśavivādaḥ*. In connection with this passage cf. also the discussion in n.1558.

the organ of supreme enjoyment, which is ambiguous inasmuch as it may be taken to refer to a new organ not already mentioned,[1451] or else to a description of one of the organs already mentioned, but here specially characterised. If one of these counter-arguments holds good, then the description of RATIK cannot speak against the deduction of §12.31. It may, however, help to solve the problem of what sort of discharge is connected with the uterine cervix, for [1],95 refers to the tamarind-like organ as "the abode of supreme enjoyment, the unsurpassed agent of flowing, through touch, of women" (see §12.20). This does rather look like a reference to orgasmic ejaculation. If this should indeed be meant here, then the RATIK passage might also serve as a hint to our understanding of AN 4,32f. in §12.19, whence we learn that the discharge of the penis-like organ, which we have identified as the portio vaginalis of the uterine cervix, is (in 4,33) called a *syanda-*. Now √*syand-* and its nominal derivatives seem, as we have seen above, to be in general used for a fluid or fluids discharged due to sexual excitement or stimulation, but not connected with an orgasm, scil. the lubricating fluid(s) (see n.1326). In AN, however, √*syand* and its derivatives seem to be used differently, either as referring to an orgasmic ejaculate, or else ambiguous; see n.1338 as well as *madanasyanda-* (AN 3,16) for the male ejaculate in §12.13. This could speak for AN 4,33 referring to an orgasmic ejaculation of the woman, provided (and that is the crux) that *nirantaram* in 4,32 does indeed mean "at once" or "compactly (= profusely)" (see §12.19). Both the RATIK and the AN passage could thus be answering the question as to the nature of the discharge of the uterine cervix thrown up in §12.25 in favour of an orgasmic ejaculation from this organ. Now of course this does not automatically imply that the same must also be presupposed by all the other texts which mention what we have taken to be the portio vaginalis of the uterine cervix in §§12.18ff., but it must be admitted that the possibility is not small that this is indeed the case. That does not, however, solve the problem of where exactly the ejaculate in that case comes from, as it may come from inside the uterus and exit via the cervical opening, or else come from or near the cervix (portio vaginalis) itself (cf. §12.25). Wherever it may come from, however, it could with regard to the commentaries of Āḍhamalla and Śrīdāsa lead to a deduction different from the one just made in §12.31, namely that they too postulate an ejaculation from the cervix, a possibility already discussed in

[1451] Referring to RAT, RAI, p.207 writes: "There is also another nerve-centre which controls all the other three mentioned earlier (namely, Madanadola, Madanachatra and Pūrṇa Candra) and which is also susceptible to manipulations by fingers", but since the organ he means is in RAT 10.8 obviously part of the triad mentioned there (see §12.18), this explanation is hardly feasible.

§§11.11f. But the commentaries of Āḍhamalla and Śrīdāsa themselves are ultimately too vague to help us much. As regards other information on ejaculation connected with the portio vaginalis, the connection between this and the clitoris which AN 4,32f. seems to assume could speak for the discharge due to stimulation of the clitoris being orgasmic too, and originating in the same place as the discharge due to the portio vaginalis (cf. §12.24). But here too we can have no surety, as, even if the AN passage should indeed presuppose this, we cannot simply assume that other sexological works do so too. The characterisation of the discharge as "passion-water" or the like is no help either; though this can indeed refer to an orgasmic ejaculate, it may perhaps refer to what we would regard as the lubricating fluid(s) too (cf. §12.13).

12.33. Be all that as it may, we in any case seem to have two — overlapping — systems of classification in the texts extracted, one mentioning the Gräfenberg spot, and one not mentioning it. Now of course I am aware of the fact that this deduction is under the given circumstances anything but water-tight. However, it must be admitted that it does explain matters quite easily and is plausible when we take into consideration both what we today know on the organs being discussed and on the means of observation the ancient scholars we are concerned with had. The three organs discussed are all actual organs that can be perceived, their stimulating function is known, and their discharging or even ejaculatory function is either known or at least a matter of discussion to modern medicine, or part of folklore. And the descriptions of the texts, even though they are connected with difficulties, fit these organs without our having to resort to flights of fancy. Moreover, the fact that we would have a mention of the Gräfenberg spot only in one system would fit the difficulty in perceiving this very organ, in contradistinction to the other two organs presumably mentioned in both systems. Thus this explanation seems to be, if we consider all factors, the only one so far that even comes near making sense of what the ancient Indian sexologists have said; even if it should in future be shown that it is not correct, it can at least for now serve as a working hypothesis on which to base further research.

12.34. This working hypothesis can be put to use to solve a problem that has vexed several generations of Indologists, particularly Vedologists. The way to this leads through another sexological work, namely KĀD. The following is found in KĀD 85-88:

paścāt praveśitas tena śiśno dṛḍhataraḥ pumān
abhyantaraṃ jagāmoccaiḥ yāvat sardigṛde 'vadhi[1452] (85)

[1452] Printed: *sardigṛdebadhi*. One could consider emending this to *sardigṛdāv adhi/sardigṛdāvadhi* or *sardigṛder adhi*. See on this n.1455.

śukrapātasya yat sthānaṃ tena sthānena saṅgataḥ
tasmin kāle priyāyās tu ānando vavṛdhe param (86)
bhāvukasyāpy aviṣayam anirvācyaṃ rasaplutam
tadā cakāra sā yoṣā ānandasyātirekataḥ (87)
netre nimīlya bhaṇitaṃ mandaṃ mandaṃ cakāra ha (88)

85-86ab: "Thereafter the very firm penis was inserted by him. The man went in, upwards, until hitting took place on the *sardigṛda-* (or: until [she] was struck on the *sardigṛda-*),[1453] which is the place of emission of semen. [He was] united with (i.e. met) that place[1454] (alternatively: ... on the *sardigṛda-*. [He was] united with that place which is the place of emission of semen).[1455]

[1453] According to the emendations considered in n.1452, we could translate as follows: "until [he was] upon (*adhi*) the *sardigṛdi-*", or: "until upto (*avadhi*) the *sardigṛda-*".

[1454] One could also consider translating 86ab as: "[which] is united with that place which is the place of emission of semen", but this is valid only if *sardigṛdi-* or *sardigṛda-* is a masculine, which is, however, a very problematic assumption (cf. n.1462). Alternatively, the translation could be: "[where he] was united with (i.e. met) that place which is the place of emission of semen".

[1455] The emendation of *sardigṛdebadhi* to *sardigṛdāv adhi* or *sardigṛdāvadhi* has been considered as a possibility in n.1452 to remove the clumsy passive of the text as it has been edited and printed. It is based upon the assumption of the not uncommon misreading of superscribed *ā* as *e* (on such misreadings cf. e.g. VṚK, pp.328), and of *b* standing for *v* (the two are regularly mixed up, and in many scripts not even differentiated). The emendation to *sardigṛder adhi*, considered for the same reason, is based upon the confusion of *b* and *r* in many Indian scripts. One could also point out that *√vadh*/*√badh* seems more often to be used as "destroy", particularly "slay", than as "strike", even considering that the verb is used suppletively in the aorist for *√han*, the forms of which latter are not rare in the use "strike". Especially the passive of *√vadh*/*√badh* is, to my knowledge, as good as always used in the sense of "slay". This does not, at first sight, seem to fit the context either. But the use of *párā√vadh* in the text of n.1477 shows that in principle the use of the verb does not conflict with the given context, so that the argument of the meaning of the verb does not apply. (*√han* is sporadically used as "obstruct; hinder", so that we might also consider translating: "... until he/it was obstructed/ hindered at the *sardigṛda-*", "it" referring to the penis. But can we simply assume this meaning of the verb, which is after all not very common, here? Matters would be easier if we could postulate a form of *√bandh* "bind; restrain". To my knowledge, however, no aorists with this form have been found yet; a merely theoretically possible aorist passive *abadhi* from *√bandh* is no convincing argument.) But in any case, even if all my conjectures be wrong, there can I think hardly be any doubt that the verse somehow or other describes an insertion of the penis up to the organ named, whether this be called *sardigṛdi-* or *sardigṛda-*, and that this is somehow or other linked with the place of emission of semen in the woman. (On the form of the term used, i.e. *sardigṛdi-* or *sardigṛda-*, see also §12.35 below.)

86cd: "At that time the pleasure (☞ *ānanda-*) of the beloved (feminine), however, increased to the utmost.
87: "The woman then, due to an excess of pleasure (☞ *ānanda-*), made a deluge of juice (i.e. liquid) (or: made a floating/submersion in feeling), outside the sphere [of understanding/imagination] of, [and] indescribable for even one capable of appreciating/imagining (i.e.: a thinker or sensitive person)[1456].
88: "Having closed [her] two eyes, [she] indeed made, faintly/softly/sluggishly, [inarticulate] speech (scil. sounds)[1457]."

12.35. Here we find that the penis of a man may be inserted into the vagina until it reaches something called the *sardigr̥da-* (or *sardigr̥di-*)[1458], which is stimulated by the penis and thus causes what is obviously an orgasm accompanied by an ejaculation. The text further clearly states that the *sardigr̥da-(/sardigr̥di-)* is (or is attached to?)[1459] the place of seminal emission. The question of course at once arises: what is this *sardigr̥da-(/sardigr̥di-)*? It seems to be an organ the sexologists considered quite important, for in the prose part of NAR 1 it is stated that sexual intercourse culminates in ejaculation (*reta[ḥ]secanāntaḥ*) brought about by reaching up to the *sardigr̥di-* (*sardigr̥dim abhivyāpya*),[1460] to which the editor appends an explanation from the *Vaijayantī*[1461] saying that the *sardigr̥di-* is the place of ejaculation of semen.[1462] And in PAU 12[1463] and 21[1464] striking the *sardagr̥di-* (obvi-

[1456] This meaning of *bhāvuka-* is quite common in New Indo-Aryan and very late Sanskrit.

[1457] Editor's note: *bhaṇitaṃ ratikūjitam iti koṣaḥ*.

[1458] More on this latter form, which would result from some of the emendations considered in n.1452, below in this paragraph.

[1459] Cf. the first translation in n.1454.

[1460] *sardigr̥dim abhivyāpya reta[ḥ]secanāntaḥ puruṣakartr̥kavyavasāyayukto narmavyāpāra iti yāvat*. On this see also n.1468.

[1461] Which I have not been able to verify, for I could not find the word *sardigr̥di-* in the only edition of the lexicon of this name available to me.

[1462] This line says:

retaḥpātasya yat sthānaṃ tad vai sardigr̥di smr̥tam.

The line as printed makes *sardigr̥di-* neuter. Do we simply have a mistake (as is common in mss; cf. also n.1601) for *sardigr̥diḥ smr̥tam* here? That would make the name of the organ either masculine or feminine, which would be in keeping with the other mentions of this name where the gender of the word is visible; in such places the name is not neuter, though whether it is masculine or feminine is unclear, except in the passage quoted in n.1479, where it is feminine (which could be taken to speak against the first assumption made in n.1454). Against the

ously a variant name of the *sardigṛdi-*) is described as giving great pleasure (☞ *ānanda-*).[1465] According to Jayakṛṣṇadīkṣita's commentary on PAU 12,

---

assumption that the neuter above is a mistake one could, however, hold that one would rather (though not necessarily) have expected *sa vai sardigṛdiḥ smṛtaḥ* or *sā vai sardigṛdiḥ smṛtā* if *sardigṛdi-* were respectively masculine or feminine here (the correlative pronoun agreeing with the noun it refers to, not the relative pronoun). One could also hold that *sardigṛdi smṛtam* is simply a sandhi form for *sardigṛdiḥ smṛtam* (cf. AIGR I,§287c), but such a sandhi seems to be archaic and moreover mostly internal to a word, so that it does not really seem a viable alternative.

[1463] *sardagṛdivyāghātaḥ cidābhāsasyānudarśane paramakāraṇam.*

[1464] *sardagṛdivyāghātajanyasyānandasyānubhūter ubhayaniṣṭhatvaṃ rasānandānubhūter ubhayatra saṃśliṣṭatvāt.*

[1465] In his commentary on 12, Jayakṛṣṇadīkṣita says that the *male* experiences great pleasure on striking the *sardagṛdi-*:

> *evaṃ sthite diṣṭaprābalyānurodhena yadi tatra vyāghātuṃ samartho bhavati tadā sa puruṣaḥ cidābhāsadarśanajanyasyānandasyānubhavaṃ niṣprayāsenaiva karotīty arthaḥ.*

In his commentary on 22 he, however, says that both experience this pleasure:

> *pūrvapakṣamate vyāghātakartur evānudarśanam. uttarapakṣe tu vyāsajyavṛttitvād ubhayor apy anudarśanam. tatra hetur uttararūpavyāpāre pūrvarūpasya prayojakatvaṃ pūrvarūpābhāva uttararūpavyāpārasyāśrayāsiddhidoṣaprasaṅgāt. tasmād cidābhāsasyānudarśanaṃ rūpadvayasyaiveti tāttvikarāddhāntanidarśanam.*

(On *pūrvarūpa-* and *uttararūpa-* pertaining to the woman and man respectively see below.) Does the statement of the commentary on PAU 12 therefore pertain only to the *pūrvapakṣa-* mentioned in the commentary on 22? (All these statements of the commentary are not necessarily what the text commented upon presupposes, but they are interesting nonetheless.) As regards *cidābhāsa-*, Jayakṛṣṇadīkṣita speaks of five types of *ānanda-* in his commentary (on 13 and 22, for instance), obviously referring to *paramānanda-* in 7, *brahmānanda-* in 8, *lekharṣabhānanda-* in 9, *avāṅmanasagocara- ānanda-* in 11 and [the *ānanda-* due to] *cidābhāsa-* in 12; the *ānanda-* of the actual ejaculation, *sāyujyānanda-* in 22, seemingly being of another sort to him.

The definiton of *pūrvarūpa-* and *uttararūpa-* is given in the commentary on 14:

> *tasyāḥ* (scil. *yuvatyāḥ*) *pūrvarūpatvena prādhānyāvagamāt puruṣasyottararūpatvāt. ādheyasvarūpeṇānuṣaṅgikatvād utsāhavardhane yuvatyabhilāṣāyāḥ prādhānyam. tathā ca śrutiḥ. mātā pūrvarūpaṃ pitottararūpaṃ prajā sandhiḥ prajananaṃ sandhānam iti śruteḥ. tasmād yuvatyabhilāṣaiva ratyutsāhavardhane hetur iti.*

Cf. SMAR, p.274:

> *yajurvedopaniṣadi. athādhiprajaṃ mātā pūrvarūpaṃ pitottararūpaṃ prajā sandhiḥ* (printed *prajāsandhiḥ*) *prajananaṃ sandhānam. aitareyakopaniṣadi. athātaḥ prajāpatisaṃhitā jāyā pūrvarūpaṃ patir uttararūpaṃ putra[ḥ] sandhiḥ* (printed *putrasandhiḥ*) *prajananaṃ sandhānam.*

I have, like the editor of SMAR, been unable to trace the second quotation to its source; neith-

this *sardagr̥di-*, situated four *aṅgula*-s beneath the navel, is "the flower/blossom of the female generative organ" (*upasthendriyakusuma-*),[1466] reminding one of *ciñcāprasūnapratima-* in RATIK [1],95 (see §12.20 and n.1438).[1467] But if this (i.e. ultimately the penis-like organ) and the *sardigr̥di-/sardagr̥di-* are the same organ, then, since the latter can obviously be reached only by the penis inserted very deeply, it would seem to follow that what is meant must be the portio vaginalis of the uterus.[1468] That this was taken to be liable to be struck during intercourse follows from BHĀ 1,3,41, where it is said that after intercourse one should wait until after the next menstruation before having intercourse again, as otherwise one might occasion a(n) miscarriage/abortion by striking the portal of the womb.[1469] From the attestations of the word given above it further appears that the noun in question ends on *-i*, whether this be *sardigr̥di-* or *sardagr̥di-*, except in KĀD 85 in §12.34 above. This should make us consider whether the emendation *sardigr̥dāv adhi* or *sardigr̥der adhi* should not be preferred over the reading *sardigr̥de 'vadhi* or the alternative emendation *sardigr̥dāvadhi* (cf. n.1452).

---

er in *Aitareyopaniṣad* nor in its commentaries could I find the passage mentioned (even with the help of JACOB), though of course I might simply have overlooked it. As regards the first quotation (the same as the quotation in the commentary of PAC 14), *yajurvedopaniṣad-* refers to *Taittirīyopaniṣad*, in which I was able to find the quotation (*Taittirīyopaniṣad* 1,3,3).

[1466] The commentary states:

sardagr̥diśabdenopasthendriyakusumaṃ lakṣyate. nābheś caturaṅgulād adhastāt sardagr̥disthānam iti śāstrakārair upavarṇyate. śiśnasya sati paṅktisaṅkhyāṅgulitve tasmin kusuma āghātaṃ kartuṃ puruṣaḥ śakto bhavati netarathā. evaṃ sthite etc. (as in n.1465).

According to this, the *sardagr̥di-* can be reached only by a penis *paṅkti- aṅguli*-s in length. The editor takes *paṅkti-* to be "ten". The commoner numerical value of the term is, however, "five". It is unclear which Jayakr̥ṣṇadīkṣita means.

[1467] Is there any connection with the *kamala-* (and eventually *puṣpa-*) mentioned in §6.2?

[1468] Neither the passage of NAR nor the passages of PAC quoted speak expressly of the pleasure or ejaculation of the woman (in contrast to the KĀD passage in §12.34), which could be taken to indicate that some later sexological authorities were of the opinion that any pleasure and/or ejaculation connected with this organ pertained to the male (Jayakr̥ṣṇadīkṣita's commentary at times gives this impression). (Note here that *puruṣakartr̥kavyavasāyayukta-* in the NAR passage quoted in n.1460 refers to intercourse in which the male plays an active role, qualifying the word for "intercourse", i.e. *narmavyāpāra-*, in the same way as *reta[ḥ]secānta-* does, and thus cannot be used to show that what the text presupposes an emission of semen by the male. As it is, the line quoted leaves unclear whose emission of semen is referred to.)

[1469] *garbhadvāravighaṭṭanād garbhacyutiprasaṅgaṃ syāt. garbha-* in *garbhadvāra-* most probably means "womb" (a common meaning), but even if it should mean "embryo", the "portal of the embryo" would obviously be the portal of the womb.

12.36. But be that as it may, if the identification of this organ as the portio vaginalis of the uterus be correct, then we have also managed to solve a problem which has vexed Vedic scholars for over a century. The references of the sexological works just quoted do not seem to have been widely known, which explains why so far *sardigṛdi-* (*sárdigṛdi-*) has been regarded as a hapax legomenon occurring only in the Vedic TS 7,4,19,2 = KS,Aśvamedha 4(= 44),8.[1470] The context in which the word appears here seems to show that it must mean the same as in the sexological works, for it occurs in connection with the Horse Sacrifice,[1471] when the chief queen of the sacrificer lies down beside the dead horse and, under a cover, inserts its penis[1472] into her vagina,[1473] to the accompaniment of to us — and possibly also to people then — obscene verses, one of which, the verse in question, says of the "hastener to/cleaner of the hole" (*biladhā́vanaḥ*), the "secret/hidden dear one of women" (*priyā́ḥ*[1474] *strīṇā́m apīcyàḥ*[1475]):

> "who has struck down [upon]/crushed the *sárdigṛdi-*[1476] within their black [characteristic] sign (scil. obviously genitals)"[1477] [1478]

[1470] The editor of NAR in a note on NAR 1 quoted in §12.35 refers to this Vedic passage too, though without giving an exact reference and wrongly writing *sardigṛdhi-*.

[1471] See on this also RĀM 1,14,30ff.

[1472] That the penis is referred to is clear from the context and from various epithets. Whether *añji-* (*añjí-*), referred to here and in the parallel passages, too denotes the penis is a matter I do not feel competent to comment upon (cf. n.1320).

[1473] Cf. on this JAMISON 1996, p.68, which makes it quite clear that this action is anything but "symbolic". Loc.cit. also states: "We should also remember that death by suffocation induces 'reflex-conditioned tumescence and emission,' and since she is dealing with an animal not long dead, the Mahiṣī's experience may have been even less symbolic than we tend to hope."

[1474] Actually, KS and TS1 have *priyás*, a variant form of *priyā́ḥ* before the dental sibilant, whereas TS has *priyá*, which, however, is not a vocative (this would be *príya*), but a sandhi form due to the dental sibilant being followed by a mute surd; on this latter sandhi see AIGR I,§287b (with additions in the "Nachträge").

[1475] On this word see MAYRHOFER 1992- 1,p.87.

[1476] WATKINS, p.274 explains this as "clitoris?", but without giving a reason for this tentative identification: "which has hit the sardigṛdi (clitoris?) in their black(haired) mark".

[1477] *yá āsāṃ kṛṣṇé lákṣmaṇi sárdigṛdiṃ parā́vadhīt*. The exact meaning here of *párā √vadh* (literally something like "destroy away"; cf. n.1455) is unclear. It obviously signifies great violence. Our dictionaries usually translate it as "split/tear asunder", which could make one think of the hymen. However, the contexts in which *sardigṛdi-* etc. are used both in the Vedic and in the later passages quoted here clearly preclude defloration being meant.

[1478] The prose portion of NAR 1 too makes the connection with the horse sacrifice: *añjyudañjyoḥ parasparaṃ gharṣaṇarūpo vyāpāraḥ sambhogaḥ. añjim udañjim*

Now other Vedic passages leave no doubt that *gr̥dá-* means "penis" (see HOFFMANN 1975-1976, p.570,n.2, also WATKINS, p.274,n.11). Does it seem far-fetched to connect this, as HOFFMANN 1975-1976, loc.cit. does, with *sárdigr̥di-*? Though the meaning of the first member of the compound *sárdigr̥di-* is then still unclear, we are of course at once reminded of the penis-like organ inside the vagina discussed above; this too would confirm the deduction that what is meant is the portio vaginalis of the uterus. The commentary of Sāyaṇa-Mādhava on the TS passage too confirms this, as it says that the penis strikes against the place of emission of "semen" to make the latter come out, also equating the "holder of the embryo/foetus" (*garbhadhāriṇī-*) with the *sardigr̥di-*.[1479] The commentary of Bhaṭṭabhāskaramiśra given in TS1 (p.96) calls the *sardigr̥di-* the upper portion of the vagina, saying too that this portion is characterised by "flowing".[1480] This would seem to mean that the

---

anv ajetīti (TS 7,4,19,1 = KS,Aśvamedha 4(= 44),8; cf. also BLOOMFIELD, p.39b) *aśvamedha aśvamaithune tathā dr̥ṣṭatvāt.*

On *añji-* and *udañji-* see n.1320.

1479 The commentary of Sāyaṇa-Mādhava says:

śukrapātasthānaviśeṣaṃ vegena tāḍayāmāsa. yena tāḍanena śukrapāto bhavati tat karoti. antarbahiḥpradeśabhedena sthānatrayaṃ strīvyañjanaṃ vidyate. ata evāmnātam trivr̥t prajananam upastho yonir madhyameti. tatra ye 'yam madhyamā garbhadhāriṇī sā sardigr̥diḥ.

Even should we split hairs and take *śukrapātasthāna-* as the place of ejaculation of the man within the woman, it would, as the penis is to be thrust in as deeply as possible according to the text, still be somewhere at the end of the vagina, i.e. where the uterine cervix is situated.

1480 Bhaṭṭabhāskaramiśra says:

tatra sardigr̥dim saraṇaśīlam gudasyoparibhāgam parāvadhīt parāhanti.

Note the use of *guda-* for "vagina" here; cf. on this n.1320. The above explanation as *saraṇaśīla-* seems to include an etymology of *sárdigr̥di-*, connecting it with the root √*sr̥* "run flow". Indeed, further on we read:

sr̥ gatau. gr̥ secane. ābhyām auṇādiko 'pratyayaḥ. sāraṇam utsāraṇaṃ puṃsa utthānasthānīyaṃ (var: *puṃsañjana/na/sthānīyaṃ*?, printed as *puṃsaṃ janasthānīyaṃ*) *yad yonibr̥hmaṇam iti prasiddhaṃ tādr̥śam gr̥dam iti karmadhārayasamāse dāsībhārāditvāt pūrvapadaprakr̥tisvaratvam. yad vā saraṇaśīlo gr̥dir yasmin sthānaviśeṣe yac chuklena sicyate tādr̥śaṃ sthānaṃ sardigr̥diśabdenocyate.*

Parts of this are not clear (what, especially, is *yonibr̥hmaṇa-*; could it have anything to do with "swell"? — cf. on this MAYRHOFER 1956-1980 2,pp.415f.;446f. and 1992- 2,p.212-214), but it does seem probable that *puṃsa utthānasthānīyaṃ* ("occupying the place of coming forth of man") and its possible variant *puṃsañjana/na/sthānīyaṃ* ("occupying the place of production of man") refer to the womb as the place where man originates. But be that as it may, the etymology of *sárdigr̥di-* is still unclear, though one obviously should consider a root meaning "run; flow" or the like (cf. too *sŕ̥dākuḥ* in MAYRHOFER 1956-1980 3,p.499); however, a first member

orgasm of a woman is according to the commentaries linked with the flow of a fluid from out of the uterus. However, we have seen above (see §12.32) that a discharge from or very near the portio vaginalis itself seems more probable, though we can have no certainty on this count. It should also be noted that, since the portio vaginalis is after all a part of the uterus (jutting into the vagina), an emission from there is ultimately an emission from (but not from out of) the uterus too; we will, however, probably never know whether this is what the commentaries too have in mind. In any case, the descriptions of the texts quoted in §§12.34ff. do speak very much for the assumption that stimulation of the portio vaginalis indeed leads to an orgasmic ejaculation. Whether we can assume the same as applying to the clitoris remains, however, as unclear as in §12.32.

12.37. Let us now summarise our findings in this chapter. Our discussion of KĀM and Yaśodhara's commentary showed that the sexual pleasure of a woman appears as a rule to have been linked with the emission of some fluid, which seems according to an old theory (see §§12.2ff.) to be the lubricating fluid(s), whereas the common theory in later times appears to have seen in it something ejaculated during an orgasm.[1481]. This latter fluid is often called "semen", especially by Yaśodhara, but this term seems also to be used for the lubricating fluid(s), the confusion being increased not only by the unclear relationship between the two and the fact that Yaśodhara speaks of the "own place" (*svasthāna-*; see n.1286) which this *śukra-* of unclear nature leaves (see §12.3, especially n.1289), but also by the mention of a theory that what is ejaculated during orgasm is actually *ārtava-* (cf. too n.1326 on *rajas-*), which of course reminds us of medical theories which seem to speak of two sorts of *ārtava-* (see especially §§12.6ff.), and also brings up the question of procreation and the role of "semen" in this regard. Turning to later texts, we found that they too refer to some sort of ejaculation connected with orgasm, the fluid ejaculated — and reaching the exterior — being termed "Kāma-water", "passion-water" or the like.[1482] However, this term might also have been used to describe non-orgasmic discharges of the woman connected with sexual excitation (cf. §12.13). According to the texts several organs appear particularly susceptible to stimulation. In this regard there seem

---

of the compound meaning "small" might also be considered (MEYER,J.J. 1937 3,p.242,n.2 offers us only a lot of speculation on the etymology of *sárdigṛdi-*).

[1481] An orgasm of the female is, according to §12.2, also a prerequisite for conception, which tallies with what we have deduced from the medical texts.

[1482] An exception is KĀD 86, which uses the term *śukra-* "semen" (see §12.34), in connection with what seems to be the portio vaginalis of the uterine cervix.

to have been two theories current, one presuming only two such organs, namely the clitoris and the portio vaginalis[1483] of the uterus, the other in addition the Gräfenberg spot (see §12.31). This Gräfenberg spot leads to an ejaculation connected with orgasm. We cannot, however, say with surety whether this was taken to be from the Gräfenberg spot itself (i.e. vaginal; cf. too n.1447), or whether one surmised a urethral ejaculation; the former was taken to be more probable, but not on any conclusive evidence (see §12.31). As regards the other two organs mentioned, stimulation of the portio vaginalis of the uterine cervix seems linked with an ejaculation connected with this organ. Whether the ejaculate first enters the interior of the uterus (this ejaculation thus being a parallel to the ejaculation of procreatory-menstrual fluid into the uterus) and then comes out via its opening to the vagina, or whether it is discharged in, on or near the cervix itself, is unclear; on the evidence collected we have tentatively deduced that the latter is more probable, but here again no hard and fast proof seems available (cf. §12.32 and §12.36). Whether the clitoris too is connected with an orgasmic ejaculate is unclear, as is also whether such an ejaculate was taken to appear in the same place as that due to the portio vaginalis (i.e. the interior of the uterus or, maybe with a greater degree of probability, the region of the portio vaginalis). The evidence could allow one to make an affirmative deduction, but in this case too we have no solid evidence (cf. §12.32). There is also the possibility that the ejaculate due to the clitoris issues directly to the exterior (cf. §12.24), but such an assumption is connected with grave difficulties. We will probably have to reconcile ourselves to the fact that our material is simply not explicit enough to allow definitive conclusions.[1484]

12.38. The texts discussed in this chapter have as good as no material on the problem of internal ejaculation and the procreatory-menstrual fluid, as also the role of orgasm in procreation; this might be a result of their not being primarily interested with the procreatory aspect of sexual intercourse (cf. also §12.10). But with regard to this internal ejaculation we might also take up the subject of ejaculation of another fluid or other fluids not connected with procreation, but also the result of an orgasm. §11.12 and §11.26 (see also §11.34, and cf. §11.28) have already drawn attention to the problem caused by postulating that in this case an ejaculation into the interior of the

[1483] One of whose names seems to be *sardigṛdi-/sardagṛdi-/sardigṛda-*(cf. §12.35). On the "penis" of women cf. also the Western parallels mentioned in §12.25.

[1484] Whether clitoral stimulation can, according to *modern Western* medicine, indeed produce an ejaculation or not, is of course quite another matter (opinions vary), and the same holds true as regards the problem of sensitivity to stimulus of the cervix of the uterus.

uterus takes place, as what is thus ejaculated ultimately reaches the exterior of the body; since, however, the procreatory-menstrual fluid too is ejaculated into the interior of the uterus at the same time, one must ask why this too does not reach the exterior. If, however, the ejaculation not connected with procreation takes place not into the interior of the uterus, but in the cervical region, then no problem arises, since the contents of the uterus would stay within this. A theory postulating an extra-uterine discharge would also not lead to the problem referred to in §8.21, namely that a closure of the womb after conception would not allow anything, be it procreatory-menstrual fluid or something else, to exit. It would thus seem only logical that the fluid ejaculated by a woman due to an orgasm and reaching the exterior is indeed not ejaculated into the interior of the uterus, but somewhere in the region of the portio vaginalis of the uterine cervix (cf. §12.37).[1485] But logical as all this may seem, we cannot simply overlook §§8.19ff. (also §9.7), which have discussed the possible theory of SU,Śā 5,39 and AS,Śā 5,p.325a on three passages (guarded by muscles) opening into the uterus, two bearing procreatory-menstrual fluid and one "semen". From this it follows that we once again have to face the fact that our texts present us with various and often contradictory views, without any regard for the discomfiture they may thus be causing us.

[1485] On actual cervical discharge cf. also n.779. The modus of the discharge postulated here is, however, anything but clear. One could in this connection think of the two openings (*vivara-*) at the *prānta-* of the female genital passage described by ŚṚṄG 3,29 in §11.10 (and last referred to in §12.31, along with the possibly related commentaries of Āḍhamalla and Śrīdāsa discussed in §§11.10ff.), which according to one possible interpretation might refer to openings at or in the vicinity of the portio vaginalis. Though in the passage in question the fluid discharged seems to be a lubricating fluid, the theory of two openings in this region nevertheless bears serious consideration. But other modes of discharge are possible too, e.g. through a single opening or several (more than two) openings, or an oozing out through the epithelium itself (on such "sweating" cf. also §12.28). It also bears investigation whether the receptacle of "semen" of women discussed in §8.17, §11.12, §11.34, §12.3 and n.872, and in §11.12 expressly linked with the ☞ *yonimukha-*, might not refer to the portio vaginalis (cf. n.1559), i.e. some place within its fleshy walls (and thus not within the uterine cavity itself), in which case this receptacle would obviously be something different from both the receptacle of semen of males and of that of procreatory-menstrual fluid of females. I must, however, stress that this is only a thought that occurred to me in the given context as a subject worth looking into, which basically means keeping one's eyes open for new relevant information, as what we have discussed so far is of little help for proving or disproving the validity of this idea.

EXCURSUS XIII

(in place of footnote 1272)

According to the theory of Auddālaki (Śvetaketu), discussed in KĀM 6(=2,1),pp.77ff., on the delight felt by women during sexual intercourse (see §12.2), this is complete only together with a delight over and above that of the mere act of coitus:

> *sā punar ābhimānikena sukhena saṃsṛṣṭā* (var: *sukhenānuviddhā*) *rasāntaraṃ janayati. tasmin sukhabuddhir asyāḥ* (p.78)
>
> "She, furthermore, connected (var: intermixed) with *ābhimānika-* delight, creates a different savour[1486]. In this is her perception/impression of delight."

The term used to describe the other delight (*sukha-*) not consisting of the mere removal of "itching",[1487] *ābhimānika-*, is derived from *abhimāna-* (see p.80), whose meaning shall be discussed on p.441 below. The pleasure (*prīti-*) which is *ābhimānika-* is defined in KĀM 6(=2,1),pp.92f. as follows:

> *anabhyasteṣv api purā karmasv aviṣayātmikā*
> *saṅkalpāj jāyate prītir yā sā syād ābhimānikī* [1]
> *prakṛter yā*[1488] *tṛtīyasyāḥ striyāś caivopariṣṭake*
> *teṣu teṣu ca vijñeyā cumbanādiṣu karmasu* [2]
>
> "The pleasure, not of the nature of the sphere/object [of the specific affair/of the senses] (*viṣaya-*), which arises from the notion/volition (*saṅkalpa-*) even with regard to actions not (habitually?) executed (*abhi√as* "apply oneself to") before (or: that arises beforehand from the notion/volition, even when actions have not [yet] been — habitually? — executed), should be [known as] *ābhimānika-*, which is cognisable in the case of oral intercourse of [persons of] the third nature and indeed of women, and with regard to these and those actions, kissing etc."[1489]

We have here, in this pair of verses, first a general statement delimiting *ābhimānika-* pleasure from the foregoing *ābhyāsika-* pleasure, which latter consists of the mere carrying out of an action (*karmābhyāsa-*), such as hunting etc.[1490] Now *abhyasta-* may mean "practised, habitually performed", but

---

[1486] *rasa-*; is this used in the sense given in the discussion on ☞ *bhāva-* in Appendix II?

[1487] On this "itching" see n.1211 and §12.2.

[1488] Vars: *prakṛtir yā* and *prakṛteḥ sā*.

[1489] Cf. also KAN 2,1,76f., RATIR 7,44, SCHMIDT 1922, pp.81-83, SCHMIDT 1904, p.84, RAI, p.13, MYLIUS 1987, p.42; this passage often seems to have occasioned difficulties to translators.

[1490] Yaśodhara adds various fine arts; for other actions cf. also RATIK [1],96-99, where (as

since the actions mentioned may actually give pleasure even if performed only once or sporadically, the emphasis on the habitual carrying out would then seem to highlight either that they are habits (or hobbies), and as such conjoined with a special addictive pleasure, or that they give real pleasure only when one is proficient in them. In any case, one could think that the characteristic of *ābhimānika-* pleasure is that it does not require practice, such as sexual intercourse. However, one of the chief characteristics of intercourse as described in Indian sexological works is that it is *not* an affair that gives full pleasure just like that; one has to learn the correct ways and means, and one may very well ask if mere theory suffices in such a case. But all this is of course irrelevant if habitual actions are not meant, and indeed, it does seem as if what is important in the context of *ābhyāsika-* pleasure is that the pleasure derives out of the action itself, not of something outside this, so that habituality need not necessarily be implied. From this it would follow that the pleasure which is *ābhimānika-* is derived not only from the action itself (and in this context it becomes irrelevant if we take this to be habitual or not); indeed, in the context of Auddālaki's statement (see p.438 above), the mention of *ābhimānika-* pleasure only makes sense if we assume that what is here presupposed is that, in the context of sexual intercourse, coitus is a mere mechanical act which cannot give full pleasure without the appropriate framework not consisting of the act of coitus itself. It is thus understandable that Yaśodhara tells us in his commentary on p.78 that Vātsyāyana mentions the additional requirement for pleasure because otherwise women would need only an *apadravya-* or dildo[1491] to remove their "itching"[1492].[1493]

It is unclear what "actions" (not — habitually? — performed) in the first verse translated above actually refers to. If this refers to actions of the specific affair (in this case, the act of coition), then the verse would refer to the pleasure one derives from the notion/volition regarding this affair even when one does not — habitually? — perform it (i.e. even if one practises it only sporadically or not at all); this pleasure would be neither the pleasure of actual coitus nor perceptible through the sense organs,[1494] and as such the

---

in AN 4,25-28) the term *ābhyāsika-*, but not *ābhimānika-*, is found. The original wording of KĀM (p.92) is:

> *śabdādibhyo bahirbhūtā yā karmābhyāsalakṣaṇā*
> *prītiḥ sābhyāsikī jñeyā mṛgayādiṣu karmasu.*

[1491] Cf. on this also n.1328.

[1492] On "itching" cf. p.438 above.

[1493] *apadravyeṇāpi sā svayam apanayatīti ced āha* (var: *caivāha*).

[1494] I.e. not perceivable as sound, touch, taste etc.

actual meaning of *viṣaya-* (see the translation) is in this case irrelevant. If, however, actions not of the specific affair (in this case then, not coitus itself) are meant, then, should *viṣaya-* mean "sphere/object [of the specific affair]", the verse would refer to the pleasure one derives from actions (even if they are not usually carried out) which serve to build up the notion/volition regarding the specific affair (i.e. coitus). Should *viṣaya-* mean "sphere/object [of the senses]", i.e. sound, touch, sight etc.,[1495] then the pleasure felt must, since it results from some specific action, in its origin obviously derive from the sphere of the senses, but be seen here with regard to its object; i.e. the generated pleasure (here due to actions not consisting of coitus) induces a state of mind leading to pleasure directed towards the specific affair (here: coitus), and this state of mind is something not part of the sphere of the senses, since it exists only in the mind. So whatever the actual meaning of the verse may be, it would seem that what it refers to is pleasure that is not derived directly from the primary affair, in this case coitus. This pleasure[1496] could be regarded not merely as an accompanying or even substitutory pleasure, but also as an anticipatory pleasure, as the alternative translation[1497] of the first verse above tries to make clear.

As regards the examples given, the feeling of coital pleasure can of course be purely imaginary, and as such substitutory, in the case of persons unable to have real coitus or choosing other forms of sexual intercourse to the exclusion of this, but with regard to oral intercourse it could also be anticipatory; I may point out too that the persons mentioned here in connection with oral intercourse[1498] are probably not seen in their function of obtaining pleasure from it, but merely as those carrying it out on the man experiencing pleasure,[1499] for whom the pleasure thus derived may, but need not be, a substitute for true sexual intercourse (the pleasure may of course also be anticipatory with regard to a woman carrying out fellatio). As regards kissing etc., these can hardly be meant to be substitutory. It, however, looks

---

[1495] This is how Yaśodhara (p.93) understands the term; see the excerpt in n.1500.

[1496] One is tempted to call it pleasure derived from the imagination. I may add that an Arabic text refers to imagination as being a prerequisite to intercourse according to the Indians (SIGGEL 1950, p.21 = 1115): "Sie [i.e. the Indians] haben gelehrt: Wenn im Menschen drei Dinge erregt werden, die sich in ihm befinden, so wird die Begierde zum Geschlechtsverkehr in Bewegung gesetzt; diese sind das Liebessehnen [this is wrong; see n.168], die Einbildungskraft und die Freude."

[1497] Given in brackets.

[1498] On the "third nature" see the discussion on ☞ *napuṃsaka-* in Appendix II.

[1499] Cf. e.g. RAI, pp.47ff. and the references there.

as if Vātsyāyana does not make such clear distinctions, and we should also examine whether the ambiguity of the "actions" of the second verse (especially as regards the example of oral intercourse) may not in fact be intentional. But in any case, we see that the *ābhimānika-* pleasure is in actual fact that derived from actions somehow connected with sexual intercourse, but not consisting of coitus. As for the temptation to see in *ābhimānika-* pleasure the pleasure generated due to sexual acts which one does not usually perform or has not yet previously performed (oral intercourse might perhaps be seen as an example here), this line of thought seems to be blocked by the mention of kissing, which hardly falls into such a category, unless what we have here is a reference to kissing (and the like) of a sort not usually performed.

Regarding the meaning of *ābhimānika-*, Yaśodhara (p.93), after saying that the coming into existence (*ātmalābha-*) of the pleasure is derived not from the *viṣaya-*s as explained by him (see n.1495), but from the notion/volition (*saṅkalpa-*), takes *abhimāna-*, from which *ābhimānika-* derives, to mean *ahaṅkāra-* "perception of the individuality (also: self-conceit)", this latter being the cause or motive (*prayojana-*) for its coming into existence.[1500] This seems to mean that, since the pleasure does not arise unless the individuality in the form of the individual notion/volition is active, so its nomenclature is derived from this individuality. This makes sense. However, one may also consider whether *ābhimānika-* may not in actual fact derive from *abhimāna-* in an older meaning, namely "intention; desire", which would fit very well here.

Yaśodhara seems here also to be making a connection with another occurrence of *ābhimānika-*, namely in KĀM 2(=1,2),pp.13f.:

> *śrotratvakcakṣurjihvāghrāṇānām ātmasaṃyuktena manasādhiṣṭhitānāṃ sveṣu sveṣu viṣayeṣv ānukūlyataḥ pravṛttiḥ kāmaḥ. sparśaviśeṣaviṣayāt tv asyābhimānikasukhānuviddhā* (var: °*viṣaye tv ābhimānika*°) *phalavaty arthapratītiḥ prādhānyāt kāmaḥ*
> "*kāma-* [in a wide sense] is the engaging/activity of ears, skin, eyes, tongue and nose — [which are] governed by the mind, [which is] connected with the self/individuality/individual perception (*ātman-*) — with regard to [their] own respective spheres/objects (i.e. sound, touch etc.) in a conforming/an agreeable[1501] manner. But *kāma-* [in a narrow sense] is chief-

---

[1500] *aviṣayātmiketi. nāpi viṣayebhyaḥ śabdādibhya ātmalābho 'syā ity arthaḥ. kutas tarhīty āha saṅkalpāj jāyata iti. manasaḥ saṅkalpātmakatvān mānasīty arthaḥ. sā caivaṃvidhābhimānikīty ucyate. abhimāno 'haṅkāraḥ sa prayojanam asyā iti.*

[1501] An ambiguous term.

> ly its/his[1502] (var: the) cognisance/perception of an object that is characterised by a result (i.e. is attained), intermixed with *ābhimānika-* delight, resulting from (var: with regard to) a sphere/object [of sense] which is a particular touch."

Yaśodhara says that what is meant here is above all touch with regard to the genitals, which generate joy during the state of *visṛṭi-*.[1503] He goes on to say that the above excludes pleasure felt through touching the genitals during sleep, as the object (*artha-*), even if reached then, is not real, but false.[1504] Note that, though *artha-* itself may perhaps not denote this,[1505] what is meant here does seem to be an orgasm(ic ejaculation), since later Yaśodhara says explicitly that *sukha-* "delight" of the text refers to the streaming (*kṣaraṇa-*) of semen (*śukra-*), and to the "fruit" (seemingly a reference to *phala-* in *phalavat-* "fruitful") which is the joy (☞ *ānanda-*) lasting as long as this lasts.[1506] As to *ābhimānika-*, he says that the use of this precludes the application of the term *kāma-* to orgasms obtained through coitus with animals (*viyonau*) or openings not vaginas (*ayonau*, which could refer to other body openings, but also to artificial vaginas), since in such cases kissing, scratching, biting etc. on the correct places of the body are missing and cannot induce the appropriate passion and desire of real human coitus.

1502 Does *asya* refer to the *ātman-* or to the person who has *kāma-*?

1503 On Yaśodhara's text as well as on *visṛṣṭi-* see n.1270. Cf. also the discussion on ☞ *bhāva-* in Appendix II.

1504 *arthagrahaṇāt svapnavyañjanasparśārthasyālīkatvāt* (var: °*vyañjanapuruṣārthasyālīkatvāt*) *phalavaty api na kāmaḥ*.

1505 In contrast to its use in the text discussed in §11.25.

1506 *śukrakṣaraṇaṃ tattulyakālam eva cānandākhyaṃ phalaṃ sukham ity uktam*.

CHAPTER 13

Tubular Structures and Vital Fluids

13.1. PA 7,1-4 (also to be found in BHĀ 1,3,17-20 and SCHMIDT 1922, pp.260f.; 7,2-4 also in ŚARĪRA 4,14-16) say:[1507]

manobhavāgāramukhe 'balānāṃ
tisro 'tha nāḍyaḥ pramadājanānām[1508]
samīraṇī[1509] *cāndramasī*[1510] *ca gaurī*
viśeṣam āsām upavarṇayāmi (1)
pradhānabhūtā madanātapatre
samīraṇī[1511] *nāma viśeṣanāḍī*
tasyā mukhe yat patitaṃ nṛvīryaṃ[1512]
tan niṣphalaṃ syād iti candramauliḥ (2)
yā cāparā cāndramasī ca nāḍī
kandarpagehe bhavati pradhānā
sā sundarī yoṣitam eva sūte[1513]
sādhyā bhaved alparatodgamena[1514] (3)
gaurī ca[1515] *nāḍī yadupasthamadhye*[1516]
pradhānabhūtā kathitā tu tajjñaiḥ[1517]
putraṃ prasūte ca varāṅganā[1518] *sā*

[1507] Before this passage there is an introductory verse, without any own numbering.

[1508] Variant (thus also SCHMIDT 1922):
manobhavāgāramukhe tu nāḍyas
tisro bhavanti pramadājanānām.
BHĀ 1,3,17 has *'balānāṃ* for *tu nāḍyas.*

[1509] Var: *samīraṇā*; thus too SCHMIDT 1922 (also SCHMIDT 1904, p.197) and BHĀ.

[1510] BHĀ: *candramukhī.*

[1511] SCHMIDT 1922, BHĀ, ŚARĪRA: *samīraṇā.*

[1512] BHĀ, ŚARĪRA: *tu vīryaṃ.*

[1513] Wrongly printed *sūtre* in the PA edition.

[1514] SCHMIDT 1922: °*ratotsave 'pi*, BHĀ, ŚARIRA: °*ratotsaveṣu.*

[1515] SCHMIDT 1922, BHĀ, ŚARĪRA: *gaurīti.*

[1516] SCHMIDT 1922: °*sthagarte*, BHĀ: °*sthagarbhe.*

[1517] Var: *pradhānabhūtā bhavati svabhāvāt*; thus also SCHMIDT 1922 (also presupposed by SCHMIDT 1904, p.198), BHĀ, ŚARĪRA.

[1518] Var: *bahudhāṅganā*; thus also SCHMIDT 1922 (also presupposed by SCHMIDT 1904,

kaṣṭopabhogyā[1519] *surate pradiṣṭā*[1520] (4)
1: "In the mouth[1521] of the abode of Manobhava (= Kāma) of women, of womenfolk, are three tubes (☞ *nāḍī-*) (vars: But in the mouth of the abode of Manobhava of womenfolk are three tubes/In the mouth of the abode of Manobhava of women, of womenfolk, are three): *samīraṇī-* (var: *samīraṇā-*) ('windy/connected with wind' or 'stimulating'), *cāndramasī-* ('lunar') (var: *candramukhī-*, 'moon-faced') and *gaurī-* ('brilliant'). I am relating their distinction[s].
2: "The distinctive tube called *samīraṇī-* (var: *samīraṇā-*) is the chief one in the umbrella of Madana (= Kāma) (*madanātapatra-*)[1522]; that semen of a man (var: that semen, however) that falls in its opening would seem to be fruitless: thus [says] Candramauli[1523].
3: "And[1524] which other one, [namely] the tube *cāndramasī-*, is the chief one in the abode of Kandarpa (= Kāma): the beautiful one [in whom this is so] verily brings forth a maiden; [she] would seem to be mastered/attained with little production (scil. pleasure?, exertion?) (var: during scant pleasures/even during scant pleasure) of intercourse.
4: "And in the midst of whose generative organ the tube *gaurī-* is, however, said to be the chief one by those who are versed in this (var:[1525] In the hole/midst of whose generative organ the tube '*gaurī-*' is the chief one by nature), that excellent woman (var: that woman often) brings forth a son; [she] is declared to be enjoyable with difficulty during intercourse (var: [she] is enjoyable with difficulty [when] engaged in intercourse)."
13.2. What this passage states is quite clear: three tubes open into the

---

p.198), BHĀ, ŚARĪRA.

[1519] SCHMIDT 1922: *kaṣṭopabhojyā*.

[1520] BHĀ, ŚARĪRA: *suratopaviṣṭā*.

[1521] On *mukha-* see §13.2.

[1522] This of course refers to the clitoris. The Hindi translation of ŚARĪRA takes this to mean "within the vagina" (*bhag'madhye*)!

[1523] An authority otherwise seemingly unknown. Candramauli is also a name of Śiva.

[1524] I am unable to translate the second *ca* in this line.

[1525] SCHMIDT 1922 and BHĀ.

"mouth" of the "abode of Kāma" of women, one of these being unproductive, the other two leading to the birth of a daughter and of a son respectively, the sex of the child depending on which of the two tubes is the "chief one". 7,2 tells us that semen reaches one of the tubes, and from this it most probably follows that in the process of conception presupposed here semen must reach the other corresponding tubes too. But from all we have seen so far we may assume, unless we be of the opinion that a different mode of conception is presupposed here (the chance of which cannot but be extremely slender), that this semen has to combine with some sort of fluid from the female to produce an embryo, and from this it automatically follows, since this combination can take place only in the womb, that the three tubes must open into the womb; in other words, the "mouth" of the "abode of Kāma" corresponds to one of the meanings of ☞ *yonimukha-* discussed in Appendix II. It also follows that the tube connected with the clitoris ("umbrella of Madana") must open internally into the womb, which would be in accord with one of the alternatives mentioned in §12.37 on where the discharge due to this organ is emitted.

13.3. In the light of all we have seen and discussed so far, the PA passage will most probably have to be understood in the sense that the two tubes leading to conception discharge something which combines with semen from the male (and this discharge would then be the procreatory-menstrual fluid), but that there is some sort of difference in the discharge due to each tube, leading to a difference in the conceived child's sex. Of what sort this difference is, is, however, unclear: the tubes could each discharge different amounts of the same fluid, in which case we would end up with the same effect as in the case of the general theory we have encountered all along, according to which the relative quantity of procreatory fluid determines the sex of the child; or else the quality of the fluid discharged by the two tubes could be different, one being stronger than the other, or else one giving rise only to males, the other to females (by containing a different substance or a mix of different substances?) (cf. in this context also the discussion of the HĀ passages in §§11.15ff.). Further, the predominance of one or the other tube has an effect on the reaction of the woman during intercourse, since it leads to a difference in the way she is "mastered" (*sādhya-*) or "enjoyed" (*upabhogya-*). Now §12.5 makes it probable that the sexual act was not considered to be perfect unless the woman too had an orgasm;[1526] could we deduce from

---

[1526] Cf. too the verse of Padmaśrī quoted by ŚĀRPA 3141 (I could not find this verse in NĀG):

no ramate ratam tāvad yāvan notkaṇṭhitā priyā

this that, as regards the male, a lack of that which thus makes the sexual act perfect was considered (by the true connoisseur) to hamper his own sexual enjoyment too? According to this line of reasoning too, *sādhya-* and *upabhogya-* might hint at the orgasm of the woman. But if her orgasm is indeed referred to in §13.1, then that must remind us of the theories linking orgasm with procreation and the sex of the child (see §§11.34f.; cf. also §12.10); we would then have to deduce that due to the woman's orgasm, brought about by different degrees of stimulation, depending on which of the procreatory tubes predominates over the other, a procreatory fluid is discharged from the predominant tube, leading to a child of the sex specifically linked with this tube. In which particular manner the sex of the child is linked with that particular discharge is, however, still unclear. We have mentioned some possibilities above. Could the relative heftiness of the respective orgasms due to the different tubes also play a role in this regard (which would remind us of a possibility considered speculatively in §§11.19f., namely of the sex of the child perhaps depending on the heftiness of the orgasm; cf. also §11.30 and §11.34); could it maybe influence the quality of the discharge (cf. §11.20, also §11.34), or else the quantity (cf. §11.29)? Or may we assume that the degree of excitability is instrumental in making the woman have an orgasm with varying degrees of ease, i.e. before or after the man, thus influencing the sex of the child (see §11.24f., also §11.29 and §11.34), the possible causes for this which we may consider being the same as in the case of the relative heftiness just discussed? There are thus numerous possibilities which may be considered. But unclear as the actual factors determining the sex of the child may be, the fact that this is somehow linked with a particular tube (and consequently its discharge) might help us understand a puzzling passage in KAL which states that a woman desiring a son should lie down[1527] pressing her right side for a moment, a woman desiring a daughter her other side.[1528] Could it be that the pressure is somehow supposed to stimulate the respective tube or the like? For from YOG 2,Yonivyāpaccikitsā,10 (p.407) we learn

anyathā na sukhotpattir iti kāmānuśāsanam.

1527 After intercouse, according to the Hindi translation, but the text says nothing of the sort, simply referring to lying down.

1528 KAL 2,45:

niśi prayāyāt kuśalas tadaṅganāṃ sute 'bhilāṣo yadi vidyate tayo[ḥ]
prapīḍya pārśvaṃ vanitā svadakṣiṇaṃ śayīta putryām itaraṃ muhūrtakam.

This verse is also quoted (as coming from KAL) in SMAR, p.275; the readings of KAL and SMAR are identical (the change of *tadaṅganāṃ* of the ms of SMAR to *tadāṅganāṃ* by the editor does not seem justified in the light of the reading of KAL).

that there actually was a theory linking the right tube with male, the left with female offspring.[1529] Unfortunately, we have so far no means of determining whether the KAL passage too presupposes such a theory. Nevertheless, it should not be overlooked that in Indian belief, particularly as found in astrological and mantic literature, the right and left sides are indeed usually connected with males and females respectively;[1530] this notion is so common (even today) that references are not needed (cf. also n.1549). Actually, the link between sex and side of the body is nothing specifically Indian (cf. p.506 in Appendix I).[1531]

13.4. Such deliberations also might be relevant for a problem we have so far not touched upon. Apart from the common theory of the child's sex being determined by the relative quantities of the two procreatory fluids combining, as also from the seemingly much less known theories in which the sex depends on the time or (maybe) heftiness of the orgasm or quality of the fluid discharged,[1532] there is a theory, not so wide-spread, but nevertheless well-known, of the sex of the child depending on the days of the period of

---

[1529] This verse runs:

vāmanāḍyā bhavet kanyā putro dakṣiṇayā bhavet
rakte 'dhike bhavet kanyā putraḥ śukre 'dhike bhavet.

10cd has the known theory of the relative quantities of the fluids determining the sex of the child. One asks oneself whether the notions of 10ab and 10cd are two different theories, or whether they are somehow linked. Incidentally, 11ab is clearly a continuation of 10cd:

śukraśoṇitamiśreṇa bhaved yo 'sau napuṃsakaḥ.

Since it would be absurd to say that a *napuṃsaka-* (on this term cf. n.103) is born merely through the mixture of *śukra-* and *śoṇita-* (since this would make all children *napuṃsaka-*s), *miśra-* must be understood here as "mixture of equal parts", which is an interesting meaning of this word. This would not be the case only if *śukra-* and *śoṇita-* should refer to secretions of the *woman* (which would then unite with semen from the man), in which case we would probably have to assume that the right *nāḍī-* discharges the former and the left *nāḍī-* the latter; this assumption regarding the meaning especially of *śukra-*, however, quite obviously leads us into a field quite different from that of the medical texts mostly being discussed here (cf. e.g. n.1549), and is in any case not unproblematic.

[1530] On this cf. also MÜLLER 1955, pp.28ff.;52 (especially the latter page with unsubstantiated speculation), ANGOT, p.21.

[1531] See also WINDISCH, p.19,n.1, PRINTZ, p.125. ALTHOFF, pp.94;230 draws attention to the theory of Aristotle that the right side of the body is warmer (and drier) than the left, and it is intriguing that according to him males are characterised by greater heat/warmth than females. On this matter, and its connection with male and female nature, see also DEAN-JONES 1994, p.44. In more recent times, Otto Schoner's theory alluded to in n.777 actually links the sex of a child to the discharge of the ovum from the left (female) and right (male) ovaries.

[1532] See the references in §13.3.

fecundity[1533] on which intercourse takes place. The general theory is that boys are born of intercourse on even days, girls if it occurs on odd days. There are also lesser known variants of this general theory setting particular days or the junctions of day and night apart for the birth of those of the "third sex"[1534].[1535] Why different days have such effects is only rarely stated; of the 'classical' medical texts, only AS gives a reason: according to Śā 1,p.281b, only a little quantity of *ārtava-* is present on even days, whereas the quantity increases on uneven days.[1536] Indu and Aruṇadatta in their commentaries on AH,Śā 1,26f.[1537] say that on the even days *ārtava-* is less, on the odd days *śukra-*, according to Aruṇadatta seemingly (if I understand him aright) due to unknown reasons.[1538] Śrīdāsa in his commentary on the same passage (AH1,Śā 1,27f.) says that on the even days *śukra-* increases, on others *rajas-*.[1539] ĀY 4,123 tells us that the increase of *śukra-* and *rajas-* on even and odd days respectively is due to providential causes.[1540] Ḍalhaṇa and NYC (fol.29r-v) in their commentaries on SU,Śā 3,12, and ĀY 4,121f.,

---

1533 *ṛtu-*; cf. §2.1 on this term.

1534 Usually termed ☞ *napuṃsaka-*.

1535 References are to be found in TIVĀRĪ 1,pp.82f. (add BYT 1,8, ĀY 4,119f. and Cakrapāṇidatta's commentary on CA,Śā 8,5, on which latter see also n.1545) and COMBA 1981, p.195,n.10. See also JOLLY, p.51, SCHMIDT 1922, p.286,n.2, MEYER,J.J. 1952, p.368, KIRFEL 1954, pp.337;344;351.

1536 The passage actually speaks of even and odd nights:
tatra yugmāsu rātriṣv alpībhavaty ārtavam ayugmāsv āpyāyyate. tasmāt tāsu kramāt putrasya duhituś ca janma.

1537 *ṛtus tu dvādaśa niśāḥ pūrvās tisro 'tra* (var: *tisraś ca*) *ninditāḥ* (26cd)
ekādaśī tu yugmāsu syāt putro 'nyāsu (var: *'nyatra*) *kanyakā* (27ab).

1538 The text commented upon and the commentaries all speak of nights. Indu says:
yugmāsu samāsu ... rātriṣv ārtavālpatvāt putraḥ syāt. anyāsv ayugmāsu ... ārtavāpyāyanād duhitā syād ity arthaḥ.

Aruṇadatta remarks:

yugmāsu ... niśāsu saṅgame putraḥ syād acintyatvād ṛtubhāvasya (var: *hetubhāvasya*). *tāsu hi rātriṣv ārtavam alpībhavati. anyāsu ayugmāsu niśāsu ... kanyā jāyate. tadā hi śukram alpībhavati pūrvoktād dhetoḥ.*

The "inconceivability of the condition of *ṛtu-* (var: of cause)" (*acintyatvād ṛtubhāvasya*, var: *hetubhāvasya*) mentioned in the first line (and also obviously referred to in the last line) is difficult to explain. It seems to mean that *ārtava-* and *śukra-* issue in small quantities due to inconceivable reasons.

1539 Śrīdāsa too, in accordance with the text of AH, actually refers to nights:
yugmāsu śukravṛddhir anyāsu rajaso vṛddhir iti.

1540 *atra yugmāyugmatithiṣu śukrarajovṛddhau daivahetuḥ.*

quote a pair of verses, which all attribute to Videha (on whom see MEULENBELD 1974, p.426), to the effect that *rajas-* is present in a small quantity on even, a large quantity on uneven days.[1541] NYC also quotes a verse from "somewhere else" (*tathānyatrāpi*) that says that during coitus (*rahas-*) on uneven days *rajas-* increases, whereas on even days something unspecified (clearly semen) is more in quantity.[1542] See also COMBA 1981, p.195,n.10. Another reason is given by ĀĀS, p.994, in an explanation of vv.6941f.: a girl is conceived if *rajas-* is more in quantity on uneven nights, a son if semen is more in quantity on even nights. If, though, semen is more on uneven nights, then a male is conceived looking like a female, whilst if *rajas-* is more on even nights, then a female is conceived looking like a male.[1543] Now all these explanations may merely be later attempts to explain an old theory on the sex of the child depending on the days of intercourse and clashing with the better-known theory of the determining factor being the relative quantities of fluids mingling;[1544] if this is indeed so, then we can only note yet

---

[1541] The version of NYC runs (with the variants of Ḍalhaṇa):

yugmeṣu divaseṣv (Ḍalhaṇa: *tu dineṣv*) *āsāṃ bhavaty alpataraṃ rajaḥ*
saṃyogaṃ tatra gacched yā (ms: *gacchethā*, Ḍalhaṇa: *yā gacchet*) *sā pumāṃsaṃ prasūyate*
ayugme divase tv (ms: *divaseṣv*, Ḍalhaṇa: *ayugmeṣu dineṣv*) *āsāṃ bhaved bahutaraṃ rajaḥ*
saṃyogaṃ tatra yā gacchet sā strī (Ḍalhaṇa: *tu*) *kanyāṃ prasūyate.*

ĀY gives us the following:

ayugme divase tv āsāṃ rajaś cājaptarī bhavet
saṃyogaṃ tatra yo (var: *yā*) *gacched dhruvaṃ kanyā[ṃ] prasūyate*
yugmeṣu tu tataḥ strīṇāṃ rajaḥ svalpataraṃ bhavet
saṃyogaṃ tatra yo (var: *yā*) *gacched dhruvaṃ putraṃ prasūyate.*

What *rajaś cājaptarī bhavet* is supposed to mean, I have no idea. The English translation on p.357 says: "the ovum in the woman becomes *ajaptari* (more effective)"; I know not on what grounds. Even the translators obviously felt uneasy, for p.431,n.97 says that we should read *bahutara-*, i.e. *rajo bahutaraṃ bhavet.*

[1542] This is found on fol.29v:

tryahāt tryahād ayugmeṣu raha[ḥ]sv āpyāyyate rajaḥ
teṣu yugmeṣu bhūyiṣṭhaṃ jāyate sa pumān bhavet.

What *tryahāt tryahād* signifies, is unclear, but I suspect that it means "regularly after three days [from the onset of menstruation]", which would fit the fact that the first three days of menstruation are generally considered unfit for intercourse. As regards the second line, this seems corrupt, and I suspect that we actually have a lacuna here, the last part belonging to another line.

[1543] *yady api rajobāhulyādikaṃ strītvādau hetutayoktaṃ tathāpy ayugmāyāṃ rajobāhulye yugmāyāṃ śukrabāhulye ca tāvadvidha eva. yadā tv ayugmāyāṃ śukrabāhulyaṃ tadā tu pumān eva bhavati kintu stryākṛtiḥ. evaṃ yugmāyāṃ rajobāhulye 'pi stry eva puruṣākṛtir bhavatīti bodhyam.*

[1544] Note in this context also the statement of AS,Śā 1,p.281b that through food semen

another variant tradition about whose background we can say nothing. There were also notions current on the characteristics of children conceived on different days being different;[1545] it bears examination whether such ideas might not have their origins in astrological spheres (I myself do not feel up to the task).[1546] The notions of sex and character respectively being linked

---

can be made to appear in greater and smaller quantity on uneven and even days respectively, the children thus born, however, having the characteristics of the opposite sex and being moreover weak or defective:

yadi tv āhārānurodhād ayugmāsu śuklasyādhikatā yugmāsu ca nyūnatā syāt tataḥ pumān stryākṛtir durbalo hīnāṅgo vā jāyate strī ca puruṣākṛtir durbalā hīnāṅgā va.

Aruṇadatta and Śrīdāsa in their commentaries on AH,Śā 1,26f. say more or less the same (the effects being due to *āhārādi-* "food etc."), differing only in that they mention an excess of semen on uneven days and an excess of procreatory-menstrual fluid on even days, Śrīdāsa over and above this also a lesser quantity of procreatory-menstrual fluid and semen on these days respectively. Statements like these could (though they need not) also be attempts to synthesise different traditions, obviously making use of notions that the quantity of the procreatory fluids *in general* may depend on factors such as food, degree of excitation etc. (cf. e.g. Ḍalhaṇa's commentary on SU,Śā 4,5, BHĀ 1,3,40(5); I have not searched systematically for references).

1545 See for instance TIVĀRĪ 1,pp.84f., COMBA 1981, p.196,n.12, as well as the quotation from Hārīta in Cakrapāṇidatta's commentary on CA,Śā 8,5 (I could not find this quotation in HĀ) and BṚSAṂ 77,23 (78,23 in other editions). Cf. too CHÖPEL, pp.261f.

1546 Cf. on astrological notions connected with procreation for instance SCHMIDT 1922, pp.297f., MEYER,J.J. 1952, p.368 (the latter also telling us that the sex of the child may depend on the day on which the manes are worshipped). Note also BṚSAṂ 77,25 (78,25 in other editions), and the remarks of BHAT, p.728. According to a most interesting notion found in some texts the quantity of ☞ *rasa-* in the body increases on the days of full-moon and new-moon, which is why a child conceived on such days is incomplete. This seems to mean that the increase in *rasa-* is somehow to the detriment of the procreatory fluids. I have not searched systematically for passages with such notions, and thus can reproduce here only what I found by chance. RATS, p.69 states:

pūrṇimāyām amāvasyāṃ ramaṇīṃ yāti cen naraḥ
rasādhikyaṃ bhaved gātre apūrṇo jāyate sutaḥ.

In RATIŚ, p.75 we find:

pūrṇimā ca mahādevi rasādhikyavidhāyinī
amāvasyā tathā caiva ratiśāstre prakīrtitā
ramaṇīgamane tatra rasādhikyaṃ bhaved dhruvam
durbalo vikalāṅgaś ca tadgarbhe jāyate śiśuḥ
akāle mriyate caiva jānīhi varavarṇini.

ĀY 4,124 cites a verse from Vaikhānasa which, since it is part of a description of conception and comes right after the line 4,123 (cited above) on the different quantities of procreatory fluids on even and uneven days, is obviously placed in a similar context of increase of human fluid in connection with procreation, even though it speaks of the increase of the fruits of

with certain days have amalgamated in course of time; we even find them applied not only to individual days, but also to individual parts of the day (i.e. as a rule: night).[1547] Be all that as it may, whether the notion of dif-

---

mātuluṅga- (on the identification of *mātuluṅga-* cf. MEULENBELD 1974, p.589, VṚK, pp.235;331;447):

> *yathā bahulapakṣeṣu* (var: *°pakṣe tu*) *mātuluṅgo 'dhikāyate*
> *na tathā jāyate śukle svabhāvaś cātra kāraṇam.*

A variant (see p.431,n.103) has *mastuluṅgo* in the place of *mātuluṅgo*. *mastuluṅga-* denotes the brain, and could lead one to speculate on Tantric theories according to which semen has a cranial location (cf. e.g. DAS,R.P. 1992A, pp.390f.), but a simple scribal error is more probable.

Cf. also n.1875. Note too the links between the notion of the sex of the child depending on the day of intercourse and notions similar to that of the ☞ *(candra)kalā*-s in mediaeval esoteric Bengali texts (see e.g. CASHIN, pp.162f.).

[1547] I have not searched systematically for such occurrences, and thus can here refer only to RATIŚ, pp.53-63, RATS, pp.70-76. These texts are representative of a large number of compilations with similar data; most compilations of this sort seem to be rather modern (though the contents need not be), many dating to the turn of or early part of the twentieth century, and usually surviving only in libraries. Some, though, continue to be reprinted regularly, but it is difficult to procure them through ordinary book-shops. From medical texts I can offer only the data collected by TIVĀRĪ 1,pp.84f. (BHĀ 1,3,16 cited there is also found as ĀĀS 6942), as well as a passage from the commentary of Śrīdāsa on AH,Śā 1,26f.:

> *tantrāntare 'py uktam.*
> *śuddhasnātā caturthyāṃ tu janayet putram uttamam*
> *pañcamyāṃ putriṇīṃ nārīṃ daśamyāṃ puruṣeśvaram*
> *ekādaśyāṃ rājapatnīṃ dvādaśyāṃ tu janeśvaram.*

I am, however, certain that more material is present in medical texts (especially commentaries) and only waiting to be turned up by a systematic search; texts of a sexological and astrological nature too should have much information. This also holds good for theories on the effects on the man due to intercourse at certain times, and for theories regarding the astrological prognoses connected with the month, day or even day time of the very first menstruation of a girl; stray references I myself found are RATS, pp.66-69 and RATIŚ, p.50, and RATS, pp.48-64, RATIŚ, pp.35-49 and RASAR,Rasendrakhaṇḍa,Yonivyādhi 38-45 (p.786) respectively (the last passage attributed to Nārada; I could not find it in *Nāradasmṛti*).

Apart from the notions mentioned above, there were also such that attributed the sex of the child to diet or certain medicines; I have not noted references to these when I have come across them. Other notions made certain actions (ritual or other) before or during intercourse responsible for the child's sex; on such notions see e.g. WEZLER 1993B, also *Bālatantra* 4,4, as well as, on modern notions, POKARNA, pp.126-128. Incidentally, WEZLER 1993B, p.289 refers to both AH,Śā 1,17 and Aruṇadatta's commentary as proof for the notion of a correspondence "between the particular colour of the sperm and the (skin?) colour of an embryo". This is not quite correct, as the AH verse merely gives the characteristics of semen guaranteeing good progeny (*sadgarbhāya*). However, the commentator Aruṇadatta does indeed link the colour of the semen with that of the child.

ferent amounts of fluid discharged on different days be an original part of the theory relating the sex of the child to even and uneven days or only a later explanation, the fact remains that such a notion existed; it is thus legitimate to ask what may have given rise to it. One could of course think of a periodic increase and decrease of the fluid within the body, maybe even due to astronomical/astrological influences (cf. n.1546),[1548] but one may also consider the possibility that the difference in quantity is due to different tubes discharging on different days, though then one would have to assume that only the procreatory-menstrual fluid is affected (and some of the explanations above do refer *only* to this), or else that the tubular structures discharging semen in the male behave in similar fashion, in which connection we could be entering the realms of Tantric theories too (cf. n.1549 below). Unfortunately, apart from speculation we can offer no answer to the question raised.

13.5. But what about the third tube in the PA passage, the one linked with the clitoris? Does this too then discharge something, or does it have no discharge? One of the deductions from the last chapter was that the relevant texts discussed might presuppose an ejaculation due to clitoral stimulation (see §12.37). If now the two procreatory tubes in the PA passage do indeed presuppose a discharge (i.e. an internal ejaculation) through them, could not the third tube too have been, in the light of the material discussed in the last chapter, credited with a discharge too? But if this should indeed be so, then the question of what this discharge was taken to consist of automatically crops up: it is obviously nothing that leads to conception, but what actually is it?

13.6. In this context we may recollect what we have seen in §§8.19ff. (see also §9.7 and §12.38) on the possible theory in SU,Śā 5,39 and AS,Śā 5,p.325a of three passages (guarded by muscles) opening into the uterus, two of these bearing procreatory-menstrual fluid, and one "semen". The PA text does lend credibility to our tentative deduction that there was a theory stating that three such passages exist,[1549] and that the reading of Śrīdasa men-

---

[1548] Cf. in this regard also the link between menstruation, conception, the phases of the moon and the heat of the atmosphere made by Aristotle (ALTHOFF, p.202, also pp.232f.).

[1549] It is obvious that this opens up highly interesting perspectives with regard to the problem of the three main tubes presupposed by Tantric theories (Iḍā, Piṅgalā and Suṣumnā, in the commonest terminology) and their openings within the body. Usually they are taken to open somewhere at the perineum, but in the case of women the uterus might also be meant, especially in connection with theories linking menstrual blood with them (see e.g. DAS,R.P. 1992A, p.413). In this regard attention may also be drawn to SATA 2,23, which tells us that there are two tubes within the *yoni-*, the left carrying "semen" (*śukra-*) and the right

tioning four such passages, discussed in §8.22, is indeed nothing more than a mistake of no import. It is also most probable that SU, AS and PA all agree in presupposing two of these three passages to carry procreatory-menstrual fluid (cf. §§13.2f.). We have, however, just seen (§13.5) that it is unclear whether PA presupposes a discharge of the third passage too, and also what the nature of such a discharge would be. Even if the comparison with the medical texts makes it more probable (though not sure) that such a discharge was indeed presumed, a discharge which would then have to be of non-reproductory character, it does not automatically follow that such a discharge would have to be "semen" too, especially when we consider the fact that the other sexological works discussed so far use the term "Kāma-water", "passion-water" or the like when referring to discharge due to stimulation of the clitoris (cf. §12.37) — though note that one text does speak of *śukra-*, but in connection with the portio vaginalis of the uterus (see n.1482). Though it does look as if both "semen" and "passion-water" ultimately referred to the same, we cannot be sure, added to which there is not only the problem of what "semen" actually signifies in the medical as well as in the sexological texts (cf. on the latter the summary in §12.37, also the remarks on the "tube" in §12.3), but also of what "Kāma-water", "passion-water" or the like means (see §12.13 and §12.37).

13.7. With regard to the discharge of this "passion-water" or the like

---

"blood" (*rakta-*):

dve nāḍyau yonimadhye tu vāmadakṣiṇayos tathā
vāme śukraṃ vijānīyād dakṣiṇe raktam eva ca.

The translator seems not to have understood this, since he appears to speak of *four* tubes ("veins" to him): "There are two veins in the middle of the *yoni* (the female organ) and likewise to the left and the right. One should know that semen is on the left and the menstrual fluid on the right" (p.242). From what we have seen so far, it seems reasonable to assume that ☞ *yoni-* here refers either to the uterus or to the whole female genital tract including the uterus, and not only the vagina. But be that as it may, the text does seem to refer to *two* tubes, one carrying *śukra-*, the other *rakta-*, both opening into the *yoni-*. The notion that the left of the two outer tubes (of the three main tubes) carries semen and the right a fluid connected with the female is found in so-called "Buddhist" Tantrism (cf. e.g. DAS,R.P. 1992A, p.403; in "Hindu" Tantrism right and left are usually interchanged); see also n.1529. Mediaeval Bengali esoteric texts too seem to consider the sex of the child to be linked with different tubular structures; see e.g. CASHIN, p.88,n.98 and p.135,n.63. We must, however, content ourselves here with drawing attention to such interesting examples regarding the Tantric aspect of the matter, as pursuing this further would lead us into realms too far removed from the subject immediately concerning us (on left and right see also §13.3). It is, however, also worth noting that there seem to be theories, for instance among certain groups in Bengal, postulating a discharge of menstrual blood from a plexus in the head of the woman (see e.g. SALOMON,C., p.297,n.18, DAS,R.P. 1992A, p.409).

discussed throughout chapter 12, I may draw attention especially to the possibility of there being theories postulating a simultaneous discharge of both procreatory-menstrual fluid and another fluid (mostly "semen") or else of two forms of what is basically procreatory-menstrual fluid, usually termed *ārtava-* (one colourless, one coloured); see e.g. §§11.12ff., §11.26, §11.34, §§12.6ff. and §12.37 (with further references). If there was indeed a theory of three tubes (or the like) discharging into the uterus, two with procreatory-menstrual fluid and one with another fluid, then that would seem to fit well with this theory of simultaneous discharge. But does the theory of three tubes, even if it presupposes a discharge of all three tubes, also presuppose their simultaneous discharge? The PA passage at least seems to hold that the two tubes discharging a procreatory fluid do not do so at the same time. One could, of course, opine that actually the text is telling us that, though they discharge simultaneously, only one of them actually leads to conception, depending on near which one the semen of the man is ejaculated. But that would not only raise the question of what then happens to the ejaculate of the other tube within the womb, but also lead to problems with regard to the obvious link of the different tubes with different orgasms (cf. §13.3). But if the two tubes were not held to discharge together, might not the same have been presupposed of the third tube too — if an ejaculation from this was presupposed at all? Another problem is occasioned by SU,Śā 3,10, which, though it does seem to speak of two forms of *ārtava-*, at the same time speaks not of three, but of *two* ducts (☞ *dhamanī-*) (cf. §7.8 and §§8.3f.). Thus we see that here too there are no easy equations.

13.8. This, however, makes us turn once again to the medical texts speaking of only two ducts discharging the procreatory-menstrual fluid, and seemingly opening into the uterus (see §§6.3ff., §§6.12ff. and §8.23). Now in the descriptions of the various ducts within the body in general, no third duct, corresponding to the *samīraṇī-* of the PA passage, seems to be mentioned among those opening into the uterus.[1550] Does this mean that we have two different theories here, one postulating only two openings for fluids entering the womb, whereas the other postulated three? This would be in keeping with the general confusion reigning on the presence and nature of female "semen", if the third opening be indeed connected with this controversial fluid (see §§13.5f.). On the other hand, there is also the possibility that the medical authorities mentioning only two ducts did know of the third passage, but considered it to be not a ☞ *dhamanī-*, but something else, and thus not worthy of mention in the given context, whereas the sexologists used the

[1550] Cf. for relevant passages on the ducts in the body MEULENBELD 1974, p.470.

word ☞ *nāḍī-* in a loose sense, subsuming that which the medical authorities called a *dhamanī-* too, which is why they could mention three "tubes".

13.9. Whether the latter possibility is also plausible is, however, a matter of debate. If it is not, it could still be that the medical authorities mentioning only two ducts nevertheless also assumed that there was a third opening into the uterus, but for some reason or the other did not comment upon this. But it could also be that there were indeed two different theories current, one assuming three openings to exist, and one two, which would be in accord with what was said in §12.38. In this context we may once again turn our attention to HĀ,Śā 1,p.414f. discussed in §§11.15ff. (and already briefly referred to in §13.3), which speaks of the seed (*bīja-*) made up/full of semen (without it being clear whether this indeed refers to humans), the *retas-* of women made up/full of *rajas-*, and the *indriya-* of men filled with (or: made up of?) seed (see also §11.26). It bears examining whether here two theories, one assuming an internal discharge of "semen" of the woman into the uterus and one not assuming this, might not have coalesced and given rise to a new theory doing justice to both these theories, this new theory then also, for reasons of symmetry, being applied to the male. It must, however, be pointed out that this is just a possibility that presents itself in this context, no proof or solid evidence being available, at least as yet.

13.10. Nevertheless, a mixing up of theories may also be considered for explaining the puzzling passage HĀ,Śā 1,p.415 discussed in §§11.18ff. (also §11.26), which links the production of males with an excess of male *retas-* and also a surplus of semen, as well as insufficient/weaker *rasendriya-*, the production of females, however, with an excess of female *retas-* and also an excess of *rajas-*, as well as insufficient/weaker *śukrendriya-*. The difficulties connected with this statement have already been discussed loc.cit. Now according to the PA passage, we find that the sex of the child seems somehow to be linked with the particular discharge of one of the two tubes *gaurī-* and *cāndramasī-*, though in what way exactly is unclear (cf. §13.3). It is worth considering whether, in the case of the HĀ verse, this or a similar theory (perhaps even in some variant or in an interpretation not originally intended) might not have been mixed up with the more general theory of the sex of the child depending on the relative quantities of procreatory fluid from the father and the mother, with the result that four different fluids were conceived of. If we indeed have a mix-up of theories here, then this would seem to pertain to male *retas-* and female *retas-* on the one hand, and male semen and female *rajas-* on the other; this is indeed worth considering, though of course it is very doubtful whether any sort of proof for the validity of such an assumption will ever be forthcoming; unfortunately — and frustratingly — we can once again do no more than raise the possibility detailed above, without being able to confirm or discount it. But in any case it is an explanation which one cannot dismiss out of hand. As to *rasendriya-* and *śukrendriya-*, however, we are as much in the dark as before.[1551]

13.11. In any case, it should be pointed out here that if we can really assume that both procreatory-menstrual fluid and "semen" were thought to be discharged internally into the uterus, irrespective of the details of such a theory or theories, then the assumption that they were also thought to be, before their discharge, stored in some receptacle,[1552] assumes more plausibility, though of course that is far from making it a certainty. On the other hand, the assumption that two fluids are discharged, of which only one (namely "semen") as a rule reaches the exterior during intercourse, must also bring to mind the problem, discussed last in §12.38, of how this can be, as also that of possible dilution (cf. also §11.28). All in all, a most frustrating situation!

13.12. Let us therefore for the present turn to other matters. A very interesting description of (small) tubes of passion(-fluid)[1553] is found in NĀG 19,1ff.; according to this there seem to be *six* such tubes, which would seem to be evidence for another theory over and above those postulated above with two or three tubes. These six tubes are called *satī-* ("chaste"), *asatī-* ("unchaste"), *subhagā-* ("blessed/charming"), *durbhagā-* ("unfortunate/repugnant"), *putrī-* (here probably "connected with/characterised by a son") and *duhitr(iṇ)ī-* ("connected with/characterised by a daughter"). They are situated *yonimātre*, which is unclear, though we could surmise that it means "only in the *yoni-*" (and not anywhere else, as e.g. postulated by 19,11 quoted in n.1557). The *satī-* and *asatī-* are on the skin (*tvaci*), on the left and right side respectively;[1554] an unchaste woman is excited (*prakupyati*) through stimu-

---

[1551] Thus, though one might also consider six different fluids according to the wording of the HĀ passage, one cannot say more, even by way of weighing probabilities, unless one knows what these two terms actually denote. But cf. nevertheless the theory of six tubes in §13.12.

[1552] See on these receptacles of procreatory-menstrual fluid and semen §8.23, §9.3, §11.23 and §12.3 (also §11.34, §12.37), and n.1485 (with further references) respectively.

[1553] *madanāḍikā-*; on *mada-* cf. §12.13, on *nāḍikā-* n.1388.

[1554] Are these tubes within the vagina, or outside it? In the latter case, they could be within the vulva near the mouth of the vagina, but might also be somewhere within the labia. Even if they are within the vagina, they cannot be too far inside it, since the next pair is said to be situated just a little within, and the order of mention of the pairs of tubes seems to be in the succession of deeper situation. It is in this connection tempting to consider once again the various passages (of ŚṚṄG and of the commentaries of Āḍhamalla and Śrīdāsa) discussed in §§11.10-14. But there are various difficulties connected with these, as well as the possibil-

lation of the *satī-* and has an orgasm through stimulation of the *asatī-*, whereas a chaste woman is exhausted (?)[1555] by stimulation of the *asatī-* and has an orgasm through stimulation of the *satī-*. The *subhagā-* and *durbhagā-* are situated a little within the vagina, on the left and right respectively; stimulation of the former makes the woman beautiful and attractive, that of the latter makes her ugly and sickly. The *putrī-* and *duhitr(iṇ)ī-* seem to be situated in the uterus,[1556] again on the left and right respectively; stimulating the former gives rise to a son, stimulating the latter gives rise to a daughter, and stimulating both gives rise to a ☞ *klība-* (on whom nothing was said in the PA passage).[1557] Basically, in spite of various tubes being mentioned,

---

ity that different matters are discussed in these passages and that of NĀG. In any case, what is said in all the passages mentioned is ultimately so unclear and problematic that comparing them or using them to explain each other seems, at least at this stage, speculative and rather rash. But be that as it may, it is very tempting to see in the first pair of tubes of the NĀG passage a reference if not to, then at least to something akin to Bartholin's glands (even if these may not have been actually observed, but only postulated theoretically due to observation of female discharges in this area); here too we can, however, not have full certainty.

1555 The text has *khidyate* "is depressed/distressed/tired/exhausted"; if the reading is correct, this probably has a sexual connotation, e.g. "is exhausted through sexual activity". But it seems improbable that this refers to an orgasm, as that would spoil the symmetry of the description of the effects of the *satī-* and the *asatī-* on chaste and unchaste women. Thus it should be considered whether *khidyate* is not actually a mistake for *svidyate* "sweats", which looks very similar in the Nāgarī script; on "sweating" due to sexual excitement see §12.28 (especially n.1441).

1556 The text says that the tubes are situated at the sides of the *khagamukha-*. This is highly reminiscent of the ☞ *yonimukha-*, and, as the description of the effects of these two tubes are to determine the sex of the progeny (see below), it appears most probable that *khagamukha-* refers to the uterus or womb. Linguistic arguments for this are to be found in the discussion of ☞ *khagamukha-* in Appendix II, but whether any of these explanations of *khagamukha-* be correct or not, quite apart from them the non-linguistic arguments just adduced do rather speak for taking *khagamukha-* to be a synonym of *yonimukha-*, whatever the exact meaning of *khagamukha-* ultimately may be.

1557 NĀG 19,1-10:
satī caivāsatīsaṃjñā subhagā durbhagā tathā
putrī duhitriṇī ceti ṣaḍ etā madanāḍīkāḥ (1)
yonimātre 'pi santy etāḥ kathitā madanāḍikāḥ
satī vāme 'satī savye dve varāṅgatvaci smṛte (2)
kiñcid abhyantare caiva randhre savyāpasavyayoḥ
durbhagā subhagā caiva dve nāḍyau samudāhṛte (3)
sarvato'bhyantare caiva pārśve khagamukhasya vai
vāme putrī samākhyātā dakṣiṇena duhitriṇī (4)
satīsañcodanāc caivāsatī rāmā prakupyati

only these last two appear to be relevant for the purpose of reproduction and seem to open into the uterus; as such, the NĀG passage, though it may at first sight seem to speak of a number of relevant tubes over and above those mentioned in the two theories (postulating two and three tubes respectively) we have discussed above, appears ultimately basically not different from one of these.[1558] But here caution is called for. In the PA passage, semen was

---

asatīcodanān nārī khidyate ca satī sadā (5)
satīsañcodanād eva satī hṛṣyati nirbharam
asatīcodanāc caivāsatī tuṣyati nityaśaḥ (6)
subhagācodanān nārī subhagā priyadarśinī
bhavet snigdhānanāṅgī ca śyāmā pīnapayodharā (7)
durbhagācodanenāpi vanitā durbhagā bhavet
rūkṣā kṛśā vivarṇāṅgī jarāroganipīḍitā (8)
putrīsañcodanād eva priyā putravatī bhavet
duhitrīcodanenāpi duhitā jāyate dhruvam (9)
ubhayoś codanāt sūte klībam eva na saṃśayaḥ
nipuṇaṃ codayan nāḍīṃ yadīcched dhitam ātmanaḥ (10)

priyadarśinī in 7 seems to be a correction for *priyadarśanī*, which is also printed as an alternative. Note also the use of √*cud* here (see also *pracodanīya-* in n.1423), which is very reminiscent of the New Indo-Aryan usage of this verb (which in Old Indo-Aryan is also used in very sacred contexts) to mean "fuck; screw" (see in this regard MÜLLER 1963B, p.421; the author is, however, unaware of the New Indo-Aryan usage); cf. also *mātaraṃ codayāmi* (note the extremely common Bengali *māke cudi*!) always in the mouth of the Bengali student in the eleventh century Kashmiri poet Kṣemendra's *Deśopadeśa* 6,44 (see SALOMON,R., p.30).

NĀG 19,11 mentions what appears to be a variant theory (though the Sanskrit text does not say this, these verses make no sense otherwise), according to which the tubes mentioned above are situated not in the female genitals, but on or in the breast (*kuca-*), armpit (*kakṣa-*), lip (*oṣṭha-*), loin/hip (or: part between the shoulder blades) (*trika-*), belly/navel (cf. *tuṇḍi(kā)-* and *tunda-*) and hip/buttock (*nitamba-*; cf. on *nitamba-* also DAS,R.P. 1991A, p.54, n.49) respectively (cf. also 18,4 and 25,1):

satī kuce 'satī kakṣe subhagauṣṭhe ca durbhagā
trike tuṇḍe sthitā putrī nitambe tu duhitriṇī (printed *dahitriṇī*).

In this connection attention may also be drawn to the theory of the ☞ *(candra)kalā*-s.

[1558] NĀG 19,11 may incidentally help in explaining a problematic passage in KĀM, namely 13(=2,8),p.160. After stating that one has to find out how to gratify the woman (*enāṃ katham anurajyate*; the verb *anurajyate* seems, however, to fit the variant *eṣā katham anurajyate* "how she is gratified" better) by means of *pravṛtti-* (which is a difficult word, explained by Yaśodhara as *ceṣṭā-* "action, effort"; on this line cf. also BODEWITZ 1989, p.258,n.6), Vātsyāyana quotes the opinion of Suvarṇanābha that one should press (*pīḍayet*; not "küssen", as MYLIUS 1987, p.64 translates) her during intercourse in that place where/from which (this can have various meanings) she turns her gaze (*yato dṛṣṭim āvartayet*):

yuktayantreṇopasṛpyamāṇā yato dṛṣṭim āvartayet tata evaināṃ (vars: *tatra evaināṃ, tatra tatraināṃ* and *tatraināṃ*) *pīḍayet. etad rahasyaṃ yuvatīnām iti suvarṇanābhaḥ.*

expressly said to be able to reach one, and thus probably all, the tubes, from which it followed that they would have to be within the womb (see §13.2). Furthermore, the tubes themselves were not said to be stimulated, but to determine, according to their predominance within the body, through factors unknown to us the excitability of the woman. The NĀG passage, however, speaks of stimulating the tubes, which would seem to mean that they would have to be outside the uterus so as to be able to be stimulated by the penis or some other means. With regard to the first two pairs of tubes this poses no problems, but in the case of the last pair, viz. *putrī-* and *duhitr(iṇ)ī-*, this is problematic; they would either have to be responsible for some other passages or the like ejaculating procreatory fluid into the womb, or else they themselves would have to be so placed that they are susceptible to stimula-

---

Yaśodhara, who is of the opinion that all this is knowledge unavailable [to males], as women do or can not divulge such matters (*etad rahasyam. strībhir aprakāśyatvāt* "This is a secret, due to [its] being inexpressible through women") obviously takes *yato dṛṣṭim āvartayet* to refer basically to an aversion of the woman's gaze, explaining it as referring to a rolling of the eyes whenever the penis within the vagina presses a portion that gives extreme pleasure and rapidly leads to an orgasm (*rati-*; see on this ☞ *bhāva-* in Appendix II):

> *yata iti yatra sambādhasyāntaraṃ* (printed *saṃvāda*°) *bhāgaṃ lakṣīkṛtya* (var: *kakṣīkṛtya*) *sādhanenopasṛpyamāṇā tatsparśasukhād dṛṣṭim āvartayed dṛṣṭimaṇḍalaṃ bhramayet tata eveti tam āśritya pīḍayet sādhanenātyartham upasarpet. tatra hi pīḍanād drutaṃ ratim adhigacchati.*

(Note that this seems to have been misunderstood by several translators who refer to the woman fixing her gaze on parts of her vagina!) But NĀG 19,11 could point to not internal, but external parts of the body (to be pressed probably by the hands or lips) which the woman glances at. Whatever may in actual fact be meant here, however, the theory of there being certain parts of the body giving more pleasure than others need not presuppose any sort of special tubes connected with such parts. Yaśodhara, however, is of the opinion that such tubes are indeed meant here, namely tubes referred to in other works (var: elsewhere):

> *yasminn* (*upasṛpyamāṇā* from the previous line is obviously to be added) *atyarthaṃ dṛṣṭim āvartayet tasminn atyartham eva pīḍayed iti boddhavyam. etena nāḍīpradeśā apy anyatantroktā* (var: °*deśā anyatroktā*) *vyākhyātāḥ. teṣām anenaiva prakāreṇa jñānamānatvāt.*

Yaśodhara of course must mean internal tubes, but according to the NĀG passage one could just as well postulate external tubes, if such postulation should turn out to be justified. I may add that, though the above is part of the chapter describing coitus with the woman above the man (*puruṣāyita-*, pp.156;163ff.), in this part the actions of the man in general (*puruṣopasṛpta-*, p.158) are described, these being what the woman in the case of such a union emulates. Thus one cannot argue that the lines quoted above refer to a passive man lying on his back, who does not use his penis for thrusting (which would then speak for the particular parts of the body of the woman mentioned being external). On the other hand, one may ask how the woman could function in exactly the same way as a man with regard to him if the particular parts of the body mentioned are internal. (Cf. on Yaśodhara's commentary also §12.31.)

tion, but open into the womb. In the latter case a situation on or near the portio vaginalis of the uterine cervix would be an obvious explanation, also doing justice to the system seeming to underlie the order of mention (see n.1554), as well as to their characterisation as at the side (*pārśve*) of the ☞ *khagamukha-* (if this should indeed be a synonym of ☞ *yonimukha-*; cf. n.1556), as this could refer not only to a position on the walls in the interior of the uterus, but also to one on the two sides of the portio vaginalis.[1559] As such not their function (whichever of the two possible ones considered this actually be), but at least their position would be in accord with what we have already seen in other texts.[1560] Now we have already seen above that we might also have parallels for the first pair of tubes (cf. n.1554). But that would leave the middle pair of tubes situated somewhere in the vagina between these two pairs of tubes, and we do not seem to have any parallels in other texts for this. However, since the middle pair is said by the text to be only slightly within the vagina,[1561] it is possible that this and the first pair together refer to what others might have considered to be a single pair of ducts. On the other hand one should also consider whether what underlies the description of the NĀG passage may not be a misunderstood theory postulating a pair of tubes at or near the vaginal opening, a pair near the cervix and a pair within the uterus, the last pair not directly susceptible to stimulation. Such a theory of course would be in keeping with some of the various theories we have already discussed, but in the case of the NĀG passage we would have to assume some significant misunderstandings of this theory by the author. Whatever the exact meaning of the whole passage may be, however, one particular is quite clear, namely that only two tubes seem to be connected with the sex of the child, i.e. actual procreation; there is no mention of a larger number of tubes or something similar.

13.13. But even though the NĀG passage cannot serve as evidence in this regard, we have nevertheless seen that there could indeed have been theories postulating the opening of a plural (i.e. not dual) number of tubular structures (☞ *sirā-*s "vessels" or ☞ *srotas-*s "channels") into the uterus.[1562] Now since three is also a plural number, one could try simply equating "plural" with "three". This, however, seems too facile an explanation; the plural

---

[1559] It should also be noted that ☞ *yonimukha-* might also refer not only to the uterus, but also to its "mouth" leading to the vagina, i.e. to its opening to the latter, and probably also to what encloses this opening, viz. the cervix.

[1560] Cf. e.g. §§12.37f., especially n.1485.

[1561] *kiñcid abhyantare caiva randhre.*

[1562] See §§6.17ff., §6.25, §§10.22-27 (with further references).

mentioned in the passages in question seems an indefinite number, which could mean that the number referred to is quite large, and moreover the passages mentioned seem to carry procreatory-menstrual fluid, whereas in the case of the theory of three passages (tubes) only two (i.e. a dual number) could be taken to carry this.

13.14. In connection with a plural number of passages attention may be drawn to the first verse of the passage of Bhoja quoted in §10.6. In the version of ĀY 4,147-150 (cf. n.923) it says (4,147):

> *garbho ruṇaddhi*[1563] *srotāṃsi rajaḥśukravahāni vai*[1564]
> *raktāj jarāyur bhavati nāḍī caiva rasātmikā*
> "The embryo/foetus verily obstructs the channels (☞ *srotas-*; plural) carrying *rajas-* and *śukra-*. From [menstrual] blood the ☞ *jarāyu-* is formed, and the tube (☞ *nāḍī-*) (scil. umbilical cord)[1565] indeed, whose essence/characteristic is ☞ *rasa-* (*rasātmaka-*, of which *rasātmikā-* is the feminine)."

The first part of the statement would seem to be in accord with the theory of (three) openings (one for semen, two for procreatory-menstrual fluid) just referred to and last discussed in §13.6; "blood" would then refer to the procreatory-menstrual fluid, from which the ☞ *jarāyu-* and the tube (umbilical cord) are formed (cf. e.g. §6.17 and Excursus III, also §9.6). The latter in the above lines seems to carry ☞ *rasa-*, which would indeed be in accord with the general theory on the nourishment of the embryo through *rasa-* coming through the umbilical cord (cf. §§10.5f.). Thus these verses seem to be in accord with what we have seen so far. One may, however, also hold that the ☞ *srotas-* (it would, if in accord with what we have seen so far, have to be one only) of semen (in contrast to the *srotas*-s of the procreatory-menstrual fluid) is not an internal structure, but the opening leading to the exterior (cf. §6.17), i.e. the vagina, whose "obstruction" would then probably refer to the opening of the uterus into this. In other words, what would be meant is that no semen can enter the uterus from outside (which leads to a problem similar to that discussed in §§8.20f.).[1566] Whether this explanation is viable is

---

[1563] Printed *ruṇāddhi*.

[1564] This whole line is not to be found in the quotation of NYC in the commentary on SU,Śā 4,24 (see n.923).

[1565] Cf. also Excursus III at the end of chapter 6.

[1566] One could even hold that this very same opening leading to the vagina is also meant with respect to *rajas-* (i.e. this is not allowed to leave the uterus), but against this we have the plural of *srotas-*, so that in the case of the procreatory-menstrual fluid at least this explanation does not seem feasible.

difficult to say; it should be pointed out that the fact that *srotas-* would refer to different sorts of structures in the case of the procreatory-menstrual fluid and semen might be taken to speak against it. Such a difficulty would not be there if we assume an internal discharge of semen of the woman into the uterus (cf. §§13.5ff.). It is ultimately also immaterial for these considerations whether the number of channels carrying semen and/or procreatory-menstrual fluid is fully in accord with the theories so far considered, as their number would not alter the deduction as to an internal discharge. Should, though, the actual numbers not tally, then we have no means of determining the actual number and distribution of channels presupposed by the ĀY passage, except that the sum would have to give a plural number.

13.15. However, the first line of the verse just quoted has *rasaraktavahāni* instead of *rajaḥśukravahāni* in the version Ḍalhaṇa gives in his commentary on SU,Śā 3,31f. (see n.923):

garbho ruṇaddhi srotāṃsi rasaraktavahāni vai.

What could *rasaraktavaha-* mean? Does it refer to channels "carrying *rasa-* and blood", to channels "carrying blood from *rasa-*" or to channels "carrying blood which is *rasa-*", on which latter attention may be drawn to the theory of blood and *rasa-* being the same referred to in n.1086? Before embarking on the discussion of this problem, it would perhaps not be out of place to direct our attention to several passages which we have hitherto neglected.

13.16. According to SU,Śā 9,9f., the "ducts" (☞ *dhamanī-*s) going transversely through the body subdivide into innumerable branches which go through the whole body lattice-like, ending at the pores of the skin (*romakūpa-* "hair-pit"). These branches carry sweat, but also augment/refresh (*abhitarpayanti*) or augment the *rasa-* (*raso yair upacīyate*) internally and externally,[1567] and carry the essences of the substances used in inunction, moistening, bathing and anointing into the body;[1568] this they obviously do

[1567] The commentators take this to refer to the skin: NYC (fol.70r):

rasaś cānnapānarasaḥ. annaṃ bahiś ca santarpaya[n]ti. bahis tvacam eva.

Ḍalhaṇa:

bahiś ca santarpayanti tvacam iti śeṣaḥ.

This of course brings us back to the problem of skin (☞ *tvac-*) and ☞ *rasa-* discussed in §§10.7ff.

[1568] The SU passage states:

tiryaggānāṃ tu catasṛṇāṃ dhamanīnām ekaikā śatadhā sahasradhā cottarottaraṃ vibhajyante. tās tv asaṅkhyeyāḥ. tābhir idaṃ śarīraṃ gavākṣitaṃ vibaddham ātataṃ ca. tāsāṃ mukhāni romakūpapratibaddhāni yaiḥ svedam abhivahanti rasaṃ cābhitarpayanty (var: *cāpi santarpayanty*) *antar bahiś ca. tair eva cābhyaṅgapariṣekāvagāhālepanavīryāṇy antaḥśarīram abhipratipadyante*

via the navel, whence they, like all *dhamanī*-s in the body, ultimately originate.[1569] Śā 7,3 mentions "vessels" (☞ *sirā*-s) originating in the navel and travelling transversely through the body (see also 7,23); Ni 5,3 (also AS,Ni 14,p.70a and AH,Ni 14,2f.) further mentions transverse *sirā*-s opening to the exterior of the body. Gayadāsa, commenting on SU,Ni 5,3, explains these *sirā*-s as subsuming, according to some, the *dhamanī*-s mentioned above, due to their similarity.[1570] Cakrapāṇidatta's commentary on CA,Śā 7,14 says that thick *sirā*-s open out to the exterior of the body and have hairs at their ends.[1571] Both *sirā*-s and *dhamanī*-s are mentioned in connection with the pores of the skin (*romakūpa*-) in SU,Ci 24,33 (see also 24,52), whereas according to AS,Śā 5,p.325b the hairs of the body are situated at the mouths of the *sirā*-s and *dhamanī*-s, which nourish the hairs, let out sweat, and also carry substances put on the skin into the body.

13.17. Now SU,Śā 3,31 tells us that, before the umbilical cord is formed and the various organs and limbs of the foetus are perceivable, this is nourished by the *upasneha*- "moistening/becoming wet" (or "moisture"?) of the transverse ☞ *dhamanī*-s carrying ☞ *rasa*- and passing through all parts of the body.[1572] But "*upasneha*- of the *dhamanī*-s" (*dhamanīnām upasnehaḥ*)

---

tvaci vipakvāni ... (9)
yathā svabhāvataḥ khāni mṛṇāleṣu biseṣu ca
dhamanīnāṃ tathā khāni raso yair upacīyate (10).

V.10 is quoted by Kāśirāma in his commentary on ŚĀR 1,5,38cd-39ab (in the edition 38 is wrongly printed as 37) with *bhavanti hi* in the place of *biseṣu ca*, and *antataś caret* in the place of *upacīyate*. NYC (fol.70r) reads *upanīyeta* (a mistake for *upanīyate*?) in the place of *upacīyate*. The openings (*khāni*) of the ☞ *dhamanī*-s referred to in 9,10 are obviously the same as the mouths (*mukhāni*) in 9,9, i.e. the pores of the skin; nevertheless, according to RAY/GUPTA/ROY, p.27 minute pores in the side-walls of the *dhamanī*-s themselves are meant. As regards *bisa*- and *mṛṇāla*- in 9,10 (which says that the *dhamanī*-s have openings like the openings in these), these terms are problematic and unclear; see on them MEULENBELD 1974, pp.482f.

1569 SU,Śā 9,4 states:
tāsāṃ tu khalu nābhiprabhavāṇāṃ dhamanīnām ūrdhvagā daśa. daśa cādhogāminyaḥ. catasras tiryaggāḥ.

1570 *tiryaggāḥ sirā iti. nābhiprabhavās trividhā hi sirāḥ sambhavanty ūrdhvagā adhogās tiryaggāś ca. anye tu sirāśabdena dhamanīr api sādṛśyād āhuḥ. uktaṃ ca tiryaggāṇāṃ catasṛṇām ekaikā* (obviously SU,Śā 9,9) *ityādi*.

1571 *etac ca keśādisaṅkhyānaṃ sthūlasirāmukhagatakeśādivibhāgena.*

1572 *asañjātāṅgapratyaṅgapravibhāgam āniṣekāt* (var: *asañjātāṅgapratyaṅgavibhāgaṃ tu garbhaṃ niṣekāt*) *prabhṛti sarvaśarīrāvayavānusāriṇīnāṃ rasavahānāṃ tiryaggatānāṃ dhamanīnām upasneho jīvayati.*

Ḍalhaṇa on *āniṣekāt prabhṛti*:
ā samantato bhāvena niṣecanam āniṣeko yonau śukrasya niṣecanam.

is ambiguous, as it could refer to *upasneha-* originating from the latter,. or else to *upasneha-* for the latter. If the former, then this could mean that the *dhamanī-*s are those of the mother, and open into the womb. This would fit the plural number of tubular structures maybe opening into the womb discussed in §§13.13ff., but on the other hand presents difficulties in that it is not easily reconciled with the descriptions in §13.16, which speak of structures branching out to the *outer* skin from the navel. Now one could hold against this that Śā 3,31 and 9,9f. need not necessarily fit together. But one must also ask how exactly the foetus is supposed to take up its nourishment in the form of *upasneha-*, emptied into the womb, if we follow this explanation. In this context Ḍalhaṇa says that the nourishment is taken up in the same manner as by plants situated on the bank of a body of water, i.e. obviously via the roots.[1573] How exactly Ḍalhaṇa interprets the SU passage and thus how exactly this statment is to be understood is unclear, but we might take it as a hint to consider whether the *dhamanī-*s mentioned might not be those of the foetus/embryo, which would thus be said to take in the nourishment in the womb through its *dhamanī-*s opening out via the pores of its skin. In other words, we would have *upasneha-* for (and not from) the *dhamanī-*s, namely those of the child, and not the mother. That structures within the foetus/embryo are meant would, alternatively, also follow if we took *upasneha-* in the meaning both Ḍalhaṇa and Gayadāsa give to it in their commentaries on SU,Ni 7,6 (in quite another context), namely as the exudation of liquid in the same manner as the liquid within a new earthen vessel exudes out through the walls of the receptacle.[1574] In these explanations *upasneha-* seems to refer to the substance exuded (which in English "exudation" can also mean), though in the case of one variant in Ḍalhaṇa's explanation (*upasnehe*) it seems rather to denote the action of exudation. Whatever the actual meaning of *upasneha-* presupposed be, if the simile be allowed to have relevance in the case of the nourishment of the foetus/embryo, then we would have an exudation from the walls of the *dhamanī-*s, meaning that these *dhamanī-*s would have to be within the child, as, if they opened into the womb, we would rather have a pouring out through their mouths instead.

[1573] *yathā pūrṇasaraḥsalilopasnehas tīrajātatarukadambakaṃ jīvayati tadvat prāṇadhāraṇaṃ karoti* (on Śā 3,31).

[1574] Ḍalhaṇa says:

> *upasnehavad iti. navaghaṭādyupasneho* (var: °*snehe*) *yathāṇutamasrotobhir bahiḥ sravad* (var: *bahiḥsrāvo*) *dṛśyate tadvat.*

Gayadāsa remarks:

> *upasnehavat. nūtanaghaṭād upasneho yathāṇubhir bahiḥsrotobhir bahiḥ sravad dṛśyate tadvat.*

Thus here "*upasneha*- of the *dhamanī*-s" could be taken to refer to *upasneha*- originating from (and not for) the latter, which are nevertheless part of the child's, and not the mother's, body. So, though ultimately it is still not clear what *upasneha*- actually means here, there is on the whole a strong possibility that, regardless of its actual meaning, the *dhamanī*-s with which it is connected are part of the child.

13.18. Against this one could hold that such ☞ *dhamanī*-s cannot be taken to be present in the embryo or foetus if its various limbs and organs are not yet visible. But according to SU,Śā 3,32 the *whole* body is present in the embryo from the very beginning with *all* its parts, but in so minute and fine a state that one cannot perceive these different parts.[1575] From this it could follow that such *dhamanī*-s were, at least by some authorities, taken to be present even before the limbs and organs were visible. And, in truth, CA,Śā 6,23 states first that the embryo is nourished by *upasneha*- and *upasveda*- ("exudation"?),[1576] then that it is nourished by *upasneha*- both by

[1575] *sarvāṇy aṅgapratyaṅgāni yugapat sambhavantīty āha dhanvantariḥ. garbhasya sūkṣmatvān nopalabhyante vaṃśāṅkuravac cūtaphalavac ca. ... evaṃ garbhasya tāruṇye sarveṣv aṅgapratyaṅgeṣu satsv api saukṣmyād anupalabdhiḥ. tāny eva kālaprakarṣāt pravyaktāni bhavanti.*

A verse version of this is to be found in BHĀ 1,3,314f. Though the medical texts several times refer to the simultaneous *appearance* of the various body parts (see for the relevant passages TIVĀRĪ 1,pp.149f.;158ff.; add ĀY 4,166, and note that on p.158 SU,Śā 3,32 is referred to as 3,18, on p.160,n.1 BH,Śā 4,30 as 4,31; see also KIRFEL 1954, p.352 and — on KĀ,Śā,Asa, p.46 — KĀ, Upodghāta,p.31), I could find only one other passage also referring to their being present in a minute, invisible state, namely BH,Śā 8,2 (on which see also TIVĀRĪ 1,pp.105;107), unless CA,Śā 3,17, which refers to the *bīja*- and *bījabhāga*- in the womb (see on this Excursus I at the end of chapter 2), too presupposes such a theory. Note, though, that Cakrapāṇidatta in his commentary on SU,Sū 15,19 says that all the ☞ *dhātu*-s are present in *bīja*--form in the procreatory fluids of male and female, it being unclear whether this refers to the fluids singly or to their combined form (see p.26). See on this subject also RASA, Upodghāta,p.53. Cf. also the statements on procreatory fluids being present in the body from the very beginning in §5.16, and also the interesting parallel in the form of a homunculus from early modern Europe given by KIRFEL 1954, p.354. I may here also draw attention to NICKEL, pp.67ff. on the controversy amongst Greek medical authorities, especially in the writings of Galen, with regard to the competing notions of preformation and epigenesis.

[1576] On what evidence TIVĀRĪ 1,pp.144;146 explains these two terms as two different sorts of osmosis is unknown to me, but it could be passages such as those referring to exudation from a new earthen vessel discussed in §13.17. She is, however, not consistent, as she uses other translations such as "unctuousness", "moistening", "moisture" or the like too (pp.144-146). CA,Śā 4,27 too says that the embryo/foetus is nourished through *upasneha*- and *upasveda*- from the mother; Cakrapāṇidatta explains this as follows:

upasneho dhātuniṣyandasambandhaḥ. upasvedaḥ śarīroṣmaṇā paraṃ gar-

means of the umbilical cord and via its pores.[1577] AS,Śā 2,p.291 too similarly refers to nourishment by means of both the umbilical cord and the pores of the skin.[1578] (These modes of nourishment have been discussed, with further references, by COMBA 1981, p.207,n.40, and TIVĀRĪ1,pp.141-146, so that I can refrain from going into further details here.) In the light of all this, the problem caused by the fact that SU,Śā 3,31 (and also AS,Śā 2,p.291) probably refers to nourishment from the very beginning, i.e. right from the

---

> *bhasya svedanam. upasvedaś ca garbhavṛddhikaro bhavaty eva. yathāṇḍajānāṃ pakṣater upasvedanaṃ vṛddhikaraṃ dṛṣṭam*
> "*upasneha-* is the (or: is that which has) relationship/connection of/with the *dhātu*--flow. *upasveda-* is causing the foetus to perspire (or: is perspiration of the foetus) in a high degree through the heat of the body. And *upasveda-* is indeed productive of growth in the embryo. As [for instance] *upasveda-* of the plumage of the egg-born (i.e. birds) is seen to be productive of growth."

(Could this last sentence, incidentally, tell us something about poultry-farming in Cakrapāṇidatta's time and area?) Indu in his commentary on the AS passage quoted in n.1578 has a different explanation:

> *upasnehaḥ snigdhatvam. upasveda utkledaḥ. tau cāhārayātau*
> "*upasneha-* is unctuousness/viscosity. *upasveda-* is becoming moist (i.e. discharge?) (or: wetness?). And both are become food."

[1577] The CA passage says:

> *vyapagatapipāsābubhukṣas tu khalu garbhaḥ paratantravṛttir mātaram āśritya vartayaty upasnehopasvedābhyāṃ garbhāśaye sadasadbhūtāṅgāvayavaḥ. tadanantaraṃ hy asya kaścil lomakūpāyanais upasnehaḥ kaścin nābhināḍyayanaiḥ.*

The punctuation above follows the commentary of Cakrapāṇidatta (*sadasadbhūtāvayava iti cchedaḥ*), according to which the passage seems to say that the nourishment by means of *upasneha-* and *upasveda-* takes place when the various body parts are both visible and not visible, and after that (*tadanantaram*) by means of the pores and the umbilical cord; *tadanantaram* is explained by Cakrapāṇidatta as referring to the state of all body parts being visible (*tadanantaram ity aṅgapratyaṅgavyaktau satyām*), though whether this explanation is correct is unclear. According to other editions (e.g. CA1,Śā 6,15, CA3, p.540) we should, however, read:

> ... *vartayaty upasnehopasvedābhyām. garbhas tu sadasadbhūtāṅgāvayavas tadanantaraṃ* ...,

i.e. the nourishment is at first by means of *upasneha-* and *upasveda-*, after which, when the body parts are not visible and visible, the other mode follows. Whichever manner of interpretation we may choose, it seems to follow that the first mode of nourishment is probably one in which the nourishment enters through the surface, the subsequent mode then combining both this and the nourishment via the now present umbilical cord.

[1578] *niṣekāt prabhṛti garbhāśayopasnehopasvedau vartanam. tato vyaktībhavadaṅgapratyaṅgasyāsya nābhyāṃ pratibaddhā nāḍī. nāḍyām aparā. tasyāṃ mātṛhṛdayam. tato mātṛhṛdayād āhāraraso dhamanībhiḥ syandamāno 'parām upaiti. tataḥ kramān nābhiṃ.* (What follows is given in n.1582.)

See also the version of this passage in Aruṇadatta's commentary on AH,Śā 1,56.

act of ejaculation,[1579] is of no great import, as nourishment is needed only when the embryo or foetus is actually there, be it in ever so minute a form.[1580] But if SU,Śā 3,31 thus does indeed refer to the *dhamanī*-s in the *child*, it cannot serve as an explanation for *srotas*-s which are *rasaraktavaha*- in the passage of Bhoja Ḍalhaṇa quotes (see §13.15), since this refers to structures blocked by the child.[1581]

13.19. On the other hand, the fact that Ḍalhaṇa quotes Bhoja in the context of his comments on SU,Śā 3,31f. could be taken to show that according to Ḍalhaṇa the SU passage is to be understood as referring to the ☞ *dhamanī*-s of the mother, i.e. to tubular structures which open into the womb. Ultimately, the problem of what is referred to in the SU passage seems not to be solvable with the means at our disposal, and the best that we can probably do at present seems to be simply to present the arguments for each of the two possible interpretations (see also §13.30). However, though we thus are unable to state with certainty whether or not SU,Śā 3,31 can be used as further proof for the theory of a plural number of tubular structures opening into the womb, we can at least use the aforegoing discussion as a point of departure for an investigation into a related matter. For whether the *dhamanī*-s be those of the mother or the child, the fact that ultimately they have a part in the process of nourishing the latter shows that the fluid mentioned in connection with them must enter into the uterus, whether it be discharged by or taken in through them. And this fluid could, in some way or the other, be linked with the procreatory-menstrual fluid.

13.20. But SU,Śā 3,31 uses *rasavaha*- "carrying ☞ *rasa*-" to qualify ☞ *dhamanī*- (see the text given in n.1572). If *dhamanī*-s are in any case and irrespective of in whose body they are, connected, as we have just seen, with something in the uterus, then from this it must follow that this *rasa*- is also found in the uterus. AS,Śā 2,p.291b mentions both *upasneha*- "moistening/becoming wet" (or "moisture"?) (cf. §§13.17f.) which enters the foetus through the pores of its skin, and *rasa*-, but it is unclear whether this *rasa*- too enters through the pores of the skin, and also what its relation to *upasneha*- (which may denote an action, but might also be a substance) is.[1582] But CA,Śā

[1579] If this is what *(ā)niṣekāt prabhṛti* (see n.1572 and n.1578) refers to, which seems most probable; I for one at any rate cannot think of anything else it could mean.

[1580] The objection is of course not valid only provided all parts of the body are actually present from the very beginning.

[1581] Note that the AS passage in n.1578 too mentions ☞ *dhamanī*-s, but these join the heart of the mother (see §13.20) and the ☞ *aparā*- and are a part of nourishment via the umbilical cord; they thus are of no consequence for the problem of the SU passage.

[1582] The relevant lines are (right after the text quoted in n.1578):

6,23 too, a few lines after the line quoted in n.1577, mentions nourishment through *rasa-*. The passage in n.1577, as we have seen in §13.18, speaks of nourishment both through the umbilical cord and the skin. Right after this the umbilical cord (*nāḍī-*) is mentioned (in the passage given in n.1641) as being linked with the heart of the mother (see on this below), but instead of learning what this umbilical cord conveys, we are told in the very next sentence:

> *sa tasya raso balavarṇakaraḥ sampadyate*
> "Its (obviously: the embryo's) *rasa-* becomes (is?) strength and colour creating."

This does not tell us anything about where this ☞ *rasa-* comes from; the order of the sentences might point to a special link, but no express link is made, thus the sentences need not be closely connected, but might just be individ-

---

> *tataś ca sa* (i.e. the ☞ *āhārarasa-* from the mother's heart — see below — entering the foetus through the umbilical cord) *punar garbhasya pakvāśaye svakāyāgninā pacyamānaḥ prasādabāhulyād dhātvādipuṣṭikaraḥ sampadyate. tathā romakūpair upasneho rasa eva ca payobhūtaḥ.*

The first line speaks of the *āhārarasa-* mentioned being "cooked". The next line is ambiguous:

> "Thus through the pores of the skin *upasneha-*, and ☞ *rasa-* indeed, which has become/is milk",

or:

> "Thus through the pores of the skin *upasneha-*. And *rasa-* indeed is that which has become/is milk",

or:

> "And thus through the pores of the skin *upasneha-*, *rasa-* indeed, which has become/is milk."

This could mean that *rasa-* enters through the skin, but it could also mean that the statement regarding *rasa-* merely refers to its transformation. Also, does "thus" mean that that which enters through the pores of the skin is also "cooked"? Or does it only refer to *dhātvādipuṣṭikaraḥ sampadyate*, which would mean that what enters through the pores nourishes the ☞ *dhātu-*s etc. too? Moreover, the relation between *upasneha-* and *rasa-* is not clear; are they the same, are they different substances, or is *upasneha-* not a substance? And where finally does the transformation of *rasa-* to milk occur? If *rasa-* is in the womb, then does it occur there, or after it has entered the body through the skin? Indu says:

> *tathā garbhāśayasyāpi sneho garbhasya romakūpaiḥ praviṣṭo dhātupuṣṭikaraḥ sampadyate. tathā garbhasya jātasyāpi sa eva mātur āhārarasaḥ payobhūto dhātupuṣṭikaraḥ sampadyate*,

i.e. that the *sneha-* "oleaginous substance" (this obviously explains *upasneha-*, showing that to Indu *upasneha-* is a substance; cf. also Indu's commentary in n.1576) of the womb enters through the pores of the skin of the embryo (*garbhasya romakūpaiḥ*, which is explicit) and nourishes the *dhātu-*s, and that the ☞ *āhārarasa-* of the mother, having become/in the form of milk, does the same — even when the foetus has been born (*garbhasya jātasyāpi*), which must refer to a transformation in the body of the mother, and not in that of the child.

ual statements placed one behind the other. After this, however, we are told that this is the food consisting wholly of *rasa-* (*sa ca sarvarasavān āhāraḥ*), the sentence after this saying that the *rasa-* of a pregnant woman is used for nourishing her own body, for her milk, and for the growth of the child.[1583] This does not tell us how this nourishment functions, but at any rate it is clear that *rasa-* is involved. Now one could also take food which is *sarvarasavat-* to be "food characterised by all the tastes (*rasa-*)", but since the statement on food being *sarvarasavat-* is linked through *ca* "and" with the preceding statement regarding the function of *rasa-* in the child, and since there *rasa-* is in the singular and also hardly anything but the ☞ *dhātu-* of this name, it seems reasonable to assume that *rasa-* in *sarvarasavat-* too refers to this same *rasa-*, i.e. that the food is indeed characterised as being made up fully of *rasa-*, which would also fit its nourishing function mentioned in the next sentence. One could now consider linking the two sentences mentioned with the preceding statement on the umbilical cord, which would mean that *this* mode of nourishment (through the umbilical cord) consists solely of *rasa-*. But we have seen that this link does not automatically follow from the order of the sentences. Even if it does, however, we still have the statement further above in the same passage that even after all the limbs have developed the foetus is nurtured by *both* modes of nourishment (cf. §13.18):

> *tadanantaraṃ hy asya kaścil lomakūpāyanair upasnehaḥ kaścin nābhināḍyayanaiḥ.*

Since the nourishment is described as consisting fully of *rasa-*, this should then pertain to both modes, from which it would follow that the nourishment taken in through the pores of the skin is *rasa-* too, and this latter must then be in the uterus. But even if this should not be the case according to the CA passage, we still have the statement of SU,Śā 3,31, which makes it very probable that *rasa-* is within the uterus. What are we to make of this? If we proceed further in the text of SU, we come to Śā 3,33, which lists the different parts of the foetus formed of the different factors which combine to create it: father (i.e. his semen), mother (i.e. her procreatory-menstrual fluid), ☞ *rasa-*, *ātman-*, *sattva-* and *sātmya-*. That these factors combine to form an embryo, each responsible for a different part of this, is a standard concept in Indian medical (and at least some classes of non-medical) literature. The first two are of course obvious factors, whereas the last three are metaphysical factors which are irrelevant in the context of our study.[1584] But what

---

[1583] These lines on the tripartition of *rasa-* in the pregnant woman have already been referred to in §10.6.

[1584] On these six factors in the formation of the embryo see n.6.

about *rasa*-? CA,Śā 3,4,1 and 3,4,5 refer to *rasa*- as something that can be taken in as a part of the food one ingests.[1585] What it actually is, is, however, unclear.[1586] CA,Śā 6,23, just quoted above, seems to make it probable that *rasa*- is the first element of the chain of ☞ *dhātu*-s. The passage also speaks of *upasneha*- (and *upasveda*-) and *rasa*- both in connection with nourishment through the skin and through the umbilical cord. It is a well-known fact that according to Indian medical theory on the nourishment of the embryo/foetus[1587] the umbilical cord connects the navel of the foetus with the heart (ĀY 4,167: chest-region) of the mother, carrying what is variously called ☞ *rasa*-,[1588] ☞ *āhārarasa*-/☞ *annarasa*-,[1589] *āhārarasavīrya*-,[1590] *āhārarasasāra*-[1591] and the like. Is the *rasa*- in the womb thus this *rasa*- which is the first *dhātu*- from food?

13.21. Some other passages from medical texts make this assumption problematic. Thus SU,Śā 3,31, directly before the passage quoted in n.1572, says that a tube (☞ *nāḍī*-) leading from the heart of the mother is joined with the umbilical cord; this tube is called a ☞ *rasa*--carrying tube, but carries *āhārarasavīrya*-.[1592] The latter could be taken to mean "strength (i.e. essence) which is *āhārarasa*- (cf. ĀY 4,149 in n.1590), but could also mean "strength (i.e. essence) of ☞ *āhārarasa*-", which is not exactly the same as *(āhāra)rasa*-. Further, CA,Śā 4,15 speaks of the embryo/foetus being connected with the mother's heart by structures through which *rasa*- flows (*rasavāhinībhiḥ*[1593] *saṃvāhinībhiḥ*).[1594] Similar structures through which *rasa*-

---

1585 3,4,5 also lists substances either containing a lot of *rasa*-, or whose ingestion leads to an increase of *rasa*-: the meat of goat, camel, deer and peacock, cow's milk, yoghurt, ghee, honey, oil, rock-salt, sugar-cane juice, *samudga*- and *śāli*--rice.

1586 Cf. in this context also n.410 and Excursus X at the end of chapter 7.

1587 On the nourishment of the embryo in general see n.921.

1588 E.g. CA,Śā 6,23, BHĀ 1,3,321, ĀY 4,168. Cf. also BH,Śā 5,8, KĀ,Sū,Lehādhyāya, p.2.

1589 E.g. BH,Śā 4,31, AS,Śā 2,p.291a. Cf. also AH,Śā 1,2 (= BYT 2,10), Cakrapāṇidatta on CA,Śā 4,21.

1590 E.g. SU,Śā 3,31, also ĀY 4,149 (*annād rasībhūtaṃ vīryaṃ* "strength, i.e. essence, from food which has become/is *rasa*-").

1591 E.g. Aruṇadatta quoting AS,Śā 2,p.291a in his commentary on AH,Śā 1,56.

1592 *mātus tu khalu rasavahāyāṃ nāḍyāṃ garbhanābhināḍī pratibaddhā. sāsya mātur āhārarasavīryam abhivahati. tenopasnehenāsyābhivṛddhir bhavati.*

1593 Var: *rasahāriṇībhiḥ*.

1594 It may be noted that here not the navel, but the heart of the embryo/foetus is said to

flows are also mentioned at other places;[1595] but what they actually carry is, however, stated by the texts to be *not rasa-*, but ☞ *ojas-* (in the eighth month of pregnancy).[1596] Could we from this postulate that the *rasa-* nourishing the embryo or foetus is actually *ojas-* too, which would be in accordance with *ojas-* and *rasa-* being synonyms?[1597] CAILLAT, pp.49f.;52-54 has actually collected material from Jaina Prakrit texts which say this very thing, the nourishment moreover taking place both via the surface of the body of the embryo/foetus and via the umbilical cord, as in §13.18.[1598] But is *rasa-* as a synonym of *ojas-* really the same as the first ☞ *dhātu-* of the alimentary chain?[1599] That some passages in medical texts seem to give an affirmative answer does not mean that this is also what it was originally, as these passages might contain later re-interpretations. In this regard it is also worth examining what the *rasa-* as part of the food one ingests referred to in §13.20 is, though of course even in this case we cannot be sure that *rasa-* as the nourishment of the child within the womb is the same as this *rasa-*. But *ojas-* might also be contained in *rasa-*. If this is so, *āhārarasavīrya-* and *āhārarasasāra-* mentioned in §13.20 might perhaps refer to *ojas-* in ☞ *āhārarasa-*.[1600] This explanation is, however, much too facile. Thus CA,Sū 30,10f. say that *ojas-* is the essence at the beginning (*sāram ādau*) of the embryo/foetus, is the *rasa-* from the *rasa-* of the embryo/foetus (*garbharasād rasaḥ*), is the first to enter the heart of the child being formed, and also the *sneha-* "unguent(?);

---

be connected with that of the mother. This is part of the standard description here called *dvaihṛdayya-*, usually, however, *dau(r)hṛda-* or *dohada-* (on which cf. LÜDERS, pp.44-47), i.e. the cravings of a pregnant woman. On this see TIVĀRĪ 1,pp.209ff., COMBA 1981, p.200,n.25 (on *dohada-* in general also n.482). The problem caused by the heart, and not the navel of the embryo/foetus being mentioned here may perhaps be solved by assuming that the heart is ultimately linked with the navel (cf. §13.16; also SU,Śā 9,5 referring to the ☞ *dhamanī-*s going upwards from the navel: *tās tu hṛdayam abhiprapannās tridhā jāyante*, and the parallel AS,Śā 6,p.330a); as this problem is, however, not really relevant to our study, I shall not pursue it further.

1595 E.g. CA,Śā 4,24 (*rasahāriṇībhiḥ* — var: *rasavāhinībhiḥ* — *saṃvāhinībhir*), AS,Śā 2,p.290a (*rasahāriṇībhir vāhinībhir*), KĀ,Śā,Asa, p.47 (*rasavahā-*).

1596 For references to this theory on *ojas-* see n.921.

1597 Cf. the discussions on ☞ *ojas-* and ☞ *rasa-* in Appendix II.

1598 On this mode of nourishment through *ojas-* (Prakrit *oyā-*) cf. also COMBA 1981, pp.204,n.34;207,n.40.

1599 This question is also raised in the discussion on ☞ *rasa-* in Appendix II.

1600 The expressions would then, of course, have to be understood as "*vīrya-/sāra-* of *āhārarasa-*".

fluid(?)" of or as the *rasa*- of the body. A part of the CA passage is also found in AS,Sū 19,p.145a; its relevant portion too states the same, except that here *ojas*- is clearly not the *sneha*- of the *rasa*- of the body, but *sneha*- identical with the latter[1601].[1602] Cakrapāṇidatta in his commentary on the CA passage takes this to mean that *ojas*- is present in the embryo from the very beginning, first as the essence when semen and the procreatory-menstrual fluid have just combined, then as the essence of the *rasa*- which signifies the combined state of these two fluids having formed the ☞ *kalala*-,[1603] and then within the heart when this is just being formed.[1604] Indu

---

[1601] As to this characterisation, since in mss *ḥ* before sibilants is often missing (cf. also n.1462), it is unclear whether the correct reading is that as printed in CA (*śarīrarasasnehaḥ*) or in AS (*śarīrarasaḥ snehaḥ*).

[1602] CA,Sū 30,9-12ab say:
yenaujasā vartayanti prīṇitāḥ sarvadehinaḥ (var: *sarvajantavaḥ*)
yad ṛte sarvabhūtānām jīvitaṃ nāvatiṣṭhate (9)
yat sāram ādau garbhasya yat tadgarbharasād (var: *yo 'sau gar°*) *rasaḥ*
saṃvartamānaṃ (var: *saṃvardhamānaṃ*) *hṛdayaṃ samāviśati yat purā* (10)
yasya nāśat tu (var: *yasyānāśān na*) *nāśo 'sti dhāri yad dhṛdayāśritam*
yac charīrarasasnehaḥ prāṇā yatra pratiṣṭhitāḥ (11)
tatphalā bahudhā (var: *vividhā*) *vā tāḥ phalantīva* (var: *phalantīti*) *mahāphalāḥ* (12ab).
AS,Sū 19,p.145a runs:
yat sāram ādau garbhasya yac ca garbharasād rasaḥ
saṃvartamānaṃ hṛdayaṃ samāśrayati yat purā
yac charīrarasaḥ snehaḥ prāṇo yatra pratiṣṭhitaḥ
yasyānāśān na nāśo 'sti prīṇitā yena dehinaḥ.
AS1,Sū 19,p.104 has *sāra* for *sāram*, *yat tad* for *yac ca*, and *yac charīrarasasnehaḥ prāṇā yatra pratiṣṭhitāḥ* for *yac charīrarasaḥ* ... . Hemādri in his commentary on AH,Sū 11,37-39 also quotes the first three lines, with different mss having both *yac ca* and *yat tad°*, with *samāviśati yat purā* (var: *°śati dehinaḥ*) for *samāśrayati yat purā*, and with the third line reading as in AS1, with the variants *yaḥ śarīrarasaḥ snehaḥ* ..., *yaḥ śarīre rasaḥ snehaḥ* ..., and *yaḥ śarīrarasasnehaḥ*

[1603] WEZLER 1993B, p.284,n.13 refers to a footnote on ☞ *kalala*- situated right here (i.e. this footnote, n.1603) in the pre-publication copy of my study; the matter of this footnote has been shifted to Appendix II.

[1604] *yat sāram ādau garbhasyeti śukraśoṇitasaṃyoge jīvādhiṣṭhitamātre yat sārabhūtaṃ tatrāpi tiṣṭhati. yat tadgarbharasād rasa iti garbharasāc chukraśoṇitasaṃyogapariṇāmena kalalarūpād rasa iti sārabhūtam. saṃvartamānaṃ hṛdayaṃ samāviśati yat pureti yadā hṛdayaṃ niṣpadyamānaṃ tadaiva vyaktalakṣaṇaṃ sad dhṛdayam adhitiṣṭhati yad ity arthaḥ. etena garbhāvasthātraye 'pi tad ojas tiṣṭhatīty ucyate. paraṃ garbhādau śukraśoṇitasārarūpatayā kalalāvasthāyāṃ tu rasasārarūpatayāvayavaniṣpattau tu svalakṣaṇayuktam eva bhavaty oja ity ojasaḥ sarvāvasthāvyāpakatvena mahattvam ucyate.*
See also Cakrapāṇidatta's commentary on SU,Sū 15,19 (15,18 in SU3).

explains the AS passage as meaning in the first case that it is the essence without which no life can enter into the combination of semen and procreatory-menstrual fluid, in the second case as being the essence of the first ☞ *dhātu-* of the embryo, and in the third case as being that which enters the heart before the [first?] *dhātu-* (see on this also Hemādri's commentary on AH,Sū 11,37-39 in §13.26 below).[1605] As regards the remark on *sneha-*, Cakrapāṇidatta takes *rasa-* (in *śarīrarasasnehaḥ*) to mean *sāra-* "essence", and both *rasa-* and *sneha-* to refer to *sāra-*; the *sneha-* is thus the *sāra-* of the *rasa-*s (plural!) of the body, which are equated with the *dhātu-*s. Indu says (provided the text is given correctly, which is not certain) that the *ojas-*, which is the *rasa-* of the body of it (seemingly: the embryo/foetus), is where the life-force (*prāṇa-*) together with the *sneha-* is rooted.[1606] No clear picture emerges from this, and matters are not made better by the ancient theory mentioned in §7.22, n.463 and p.557 below deriving all beings from *rasa-*, which brings us into the midst of the discussion on the relation between *rasa-* and water (and other substances) in §§7.11ff. and §§10.7ff. But the derivation of living beings from *rasa-* (which is not the same as saying that they are nourished by *rasa-*) also brings to mind BHĀ 1,3,175 already discussed in §10.6; this passage refers to a *rasa-* which seems to function like ☞ *āhāra-rasa-*, and enters the heart to mingle with another *rasa-* which is called *śarīrārambhaka-*, which means "causing the body to begin" or "the beginning/origin of the body". Now in §10.6 this latter *rasa-* was seen as basically the same as that called *sthāyin-* or *poṣya-* by the commentators, and as a result *śarīrārambhaka-* was considered in different meanings, but not the literal one. But what if the expression is meant quite literally, especially since §10.6 also mentions other BHĀ passages in which semen is called *śarīrārambhaka-*? BHĀ

1605 *yad garbhasyādau sāraṃ sāram iva sāraṃ. na hi tena vinā śuklaśoṇite jīvānupraveśaḥ. yac ca garbharasād rasaḥ garbhasya yo 'sau rasaḥ prathamo dhātus tasmād api sāraḥ prasādabhūtam. saṃvartamānaṃ yac ca purā dhātor dehe pravartamānaṃ prathamaṃ hṛdayam āśrayati paścāt kramād dehavyāptiṃ ca karoti.*

1606 Cakrapāṇidatta says:
śarīrarasasneha iti śarīrasārasāram. rasaśabdaḥ snehaśabdaś ca sāravacanaḥ. tena śarīrarasānāṃ dhātūnām api sāra ity arthaḥ.
Indu remarks:
yac ca śarīrarasas tasya sasneho yatra prāṇo vyāpako 'pi pratiṣṭhito baddhaḥ.
One could consider reading this as *yac ca śarīrarasas tasya sa sneho yatra* ..., which would then refer to the *ojas-* as the *sneha-* of the *rasa-* of the body and thus presuppose the reading *yac charīrarasasnehaḥ* in the AS passage in n.1602; the *prāṇa-* would be rooted in the *sneha-*. As regards the relation between *sneha-* and *ojas-* (and semen), see also the discussion in §10.18.

1,3,175 would then refer to a *rasa-* which would be in keeping with the theories that all beings originate from *rasa-*, but which would on the other hand cause difficulties with regard to the theory of the alimentary *dhātu-*s. Matters become even more intriguing when one considers SLAJE 1993, pp.269ff. On p.556 in Appendix II below attention is drawn to the fact that in the text discussed loc.cit. (Bhāskarakaṇṭha's *Cittānubodhaśāstra*) ☞ *dhātu-* and ☞ *bhūta-* seem to have been used synonymously, without it being clear in each particular case whether the elements *rasa-* etc. or the gross elements are meant. In the same text semen (*śukra-*) in its role of forming the embryo is also called the essence (*sāra-*) of these *dhātu-*s/*bhūta-*s, in one instance even of the *bhūta-*s which are *dehārambhaka-*, *deha-* of course being another term for "body" and as such a synonym of *śarīra-*.[1607] On the one hand this shows links with the theory of the chain of *dhātu-*s in medical texts, but on the other we also have points of contact with the notion of the *bīja-* "seed" found e.g. in the theory of *ālayavijñāna-* (cf. n.49). This latter also serves to draw attention to the role of *vijñāna-* during conception referred to in n.6, which is relevant here because in the text discussed by SLAJE 1993 it is stated that the *dhātu-*s or *bhūta-*s, whose essence semen is, are but forms (*rūpa-*) of or contain (*bhūteṣu vidyate*) *cit* "intellect, thought" (SLAJE 1993, p.270,n.98). Is therefore the *rasa-* of the BHĀ passage which is *śarīrārambhaka-* and situated within the heart actually an essence of other, unnamed substances, containing the intellect or faculty of cognition? This would clearly lead us to spheres different from those our study is primarily concerned with, and would also mean that *rasa-* here is used in a sense different from the ones we have been concerned with so far. I must of course point out that what I have written here represents no more than musings on what *śarīrārambhaka- rasa-* could mean in the context mentioned, as, given the potential points of contact with spheres not purely medical shown up, the whole clearly is in need of a detailed study, for which this is not the right place. But it is clear that all of this once again goes to prove the validity of what was written in DAS,R.P. 1992A, p.391,n.22: "As regards *rasa-* in various meanings not having to do with feelings generally ..., we lack a good, comprehensive study, a possible point of departure being its use in medical, tantric and alchemical works".[1608]

13.22. DAS,R.P. 1992A, p.391,n.22 mentions *ras* (the New Indo-Aryan equivalent of ☞ *rasa-*) referring to several body fluids too, amongst these also semen and the menstrual fluid. In this connection attention may be drawn to HĀ,Śā 1,p.415, last considered in §13.10, in the course of discussing which in

---

[1607] SLAJE 1993, p.270,n.99: *dehārambhakabhūtānāṃ sārabhāgo 'tra śukratām | yāti.*

[1608] Cf. also CA,Sū 26,8f., already mentioned in §7.15.

§§11.19f. the possibility was mentioned that *rasa-* in *rasendriya-* might, irrespective of what *indriya-* actually means, perhaps refer to the procreatory-menstrual fluid, so that the statement linking insufficient/weaker *rasendriya-* with the production of males, and insufficient/weaker *śukrendriya-* with that of females, would be in keeping with the general theory on the sex of the child, with maybe some sort of mixing up of theories, leading to different theories being placed side by side in the HĀ verse. The evidence from this HĀ verse alone is very weak though, as most of what can be said on this is basically not clearly provable. But we can see from p.534 in Appendix II below that there actually are theories in Jaina texts equating ☞ *ojas-* (Prakrit *oyā̆-*) with the procreatory-menstrual fluid, this *ojas-* also serving to nourish the child thus conceived within the womb (see §13.21). This evidence is much stronger, pointing not only to *ojas-* and *rasa-*, but also to the procreatory-menstrual fluid having the same functions and probably being identical according to some theories. Against this background, the statements on KĀ,Kh 9,17 and ĀY 4,151-153, which we discussed in §10.24 and §§10.27ff. (cf. also §10.31), assume a new importance. In these passages we found the procreatory-menstrual fluid in exactly the same function as *rasa-* in the passages from other texts referred to in §§10.5f.,[1609] according to which the *rasa-* of a pregnant woman divides into three parts nourishing the embryo or foetus, her milk (or breasts) and her body (or *rasa-*) respectively.[1610] All of which seems to show that in the works we are studying there seem to be, behind the theories generally connected with these texts, theories seeing *ojas-*, *rasa-* and the procreatory-menstrual fluid all both as nourishing and as reproductory fluids, seemingly largely synonymous, or at any rate connected in ways which make them seem to be so, and which are not clear to us.

13.23. The above also makes us consider whether several discrepancies we have noted in the course of this study may not stem from this relation between the fluids mentioned. Thus one may ask whether the theory on two different *rasa-*s (☞ *rasa-* and ☞ *āhārarasa-*) in connection with the chain of metabolic transformation of food (cf. §§7.37ff.) may not have had its origin in or been influenced by the concepts of *rasa-* as ☞ *ojas-* and *rasa-* as the first *dhātu-*, in which connection special attention may be drawn to BHĀ

1609 Even though in the case of the KĀ passage full surety seems unobtainable, the probability is great that the procreatory-menstrual fluid indeed is meant here. Should this not be the case, then the fluid meant would be blood in general (cf. §10.31).

1610 As regards the last function, in the KĀ passage it seems to nourish a chain of substances beginning with blood (and not *rasa-*). On it seemingly even turning to this blood see §§10.29f., which explain the relevant statement in a different manner.

1,3,175 in §13.21. Again, *ojas-* is usually said to be "somic" (☞ *saumya-*), but at times one also finds it described as fiery (☞ *āgneya-*).[1611] Now we know that *rasa-* is also characterised as "somic".[1612] But *ojas-* is also said to be a name of phlegm,[1613] which is "somic" too (see §7.15, §7.20). We have, however, seen in §10.18 and §13.21 that both *ojas-* and semen were taken to be *sneha-* by some, and thus might have been connected in a manner similar to that in which *ojas-*, *rasa-* and the procreatory-menstrual fluid are connected. But semen too is a "somic" substance.[1614] As regards "fiery", this is of course a characteristic of the procreatory-menstrual fluid (see §7.11). If, however, the characterisation of *ojas-* as "fiery" be taken to be somehow linked to the characterisation of another substance of the body as "fiery", then, given the male-oriented nature of Indian medicine (and ancient Indian thought in general), one may ask whether a substance only the female has is the right basis for such considerations. On the other hand we have seen in §13.22 that there do indeed seem to be links between the two; this should make us at least consider the possibility of a generalisation originating from the female (and not the male, as usual). But blood too is characterised as *ojas-*,[1615] and the discussion on ☞ *āgneya-* in Appendix II shows us that, problematic though the classification of blood may seem to be, some authorities at least held it to be "fiery", whilst according to others it was both "somic" and "fiery". Similar considerations apply in the case of the colour of *ojas-*, which is said to be slightly reddish-yellow, but also white or clear.[1616] The former colour reminds one of both blood and the procreatory-menstrual fluid. As regards the latter, the white colour of course reminds one of semen, but also of *rasa-*, which too is called clear at times (see §7.23). In the light of all this, we may ask whether the theory of there being two types of *ojas-*[1617] may not be connected with such conflicting notions.

13.24. What holds good for *ojas-* holds good for the procreatory-menstrual fluid too. We have already seen several times that some authorities seem to have presupposed the existence of two sorts of *ārtava-*; see on this

---

[1611] See p.530 and p.534 in Appendix II.

[1612] Cf. the discussion on ☞ *āgneya-* in Appendix II.

[1613] See the discussion on ☞ *ojas-* in Appendix II.

[1614] See ☞ *āgneya-* in Appendix II.

[1615] See on this the discussion on ☞ *ojas-* in Appendix II. Cf. also n.420 on the process of colouring of *rasa-* to give blood.

[1616] See Appendix II on ☞ *ojas-*.

[1617] This theory too is detailed in Appendix II.

§13.7 (with further references).[1618] In this connection the notion of the fluid having two different colours, red and clear, seems to play a role. We have so far as a rule discussed this matter in the light of the problem of female semen, different modes of discharge/ejaculation, and procreatory power. The discussion above lends a new angle to the whole, namely the possible connection of the procreatory-menstrual fluid with ☞ *ojas*- and ☞ *rasa*-. In this connection it may also not be out of place to return to the subject of the fluid contents of the womb during pregnancy. There are, as we have already seen, theories according to which the procreatory-menstrual fluid enters into the womb, collects there, and then, not being able to flow out of the body, gradually collects further up in the body, leading among other things to the breasts' filling with milk.[1619] Now one may make a case for assuming that the womb is filled with a fluid which is red in colour (see on this §6.17). On the other hand, red fluid is usually associated with the discharge during, but especially after the delivery of the child. In contrast, the fluid issuing before delivery, i.e. the amniotic fluid (which may also issue prematurely during the course of pregnancy if the amniotic sac is ruptured), may be reddish in colour, but is usually pale.[1620] The texts usually call it *garbhodaka-*, *garbhāmbhas-* or the like, i.e. "womb-water" or "water connected with the foetus".[1621] These two sorts of discharges might, apart from the problem of female semen, also have contributed to the notions on two sorts or states of the procreatory-menstrual fluid. And it may also be pointed out that the pale fluid of course would tally with notions on *rasa-* (and *ojas-* in a similar form) being present within the uterus.

13.25. There is, however, more that follows from the above. AS,Śā 2,p.291b tells us that the embryo/foetus is fed by *rasa-* which has turned to or is milk (*payobhūta-*), which means that it turns to milk either in the mother or within the child. It is unclear whether *rasa-* would in the first case be

---

[1618] The matter has also been referred to in §10.30.

[1619] See on this, both as regards a pregnancy and as regards a *gulma-*, §§6.17ff., §9.6 and §§10.24ff., and §5.9, §6.22, §9.4 and §10.2 respectively.

[1620] Cf. in this regard KĀ,Śā,Jā, p.56:
tantrīvarṇo 'lpaśaḥ srāvaḥ picchilaḥ putrajanmani
kiṃśukodakasaṅkāśaḥ putrikājanma śaṃsati
"If the woman is going to deliver a male child, she will get scanty, unctuous show resembling the colour of *guḍūci* (very light yellowish pink), however, if she is to deliver a female child, her show would resemble the colour of water obtained by washing the flowers of *palāśa* (dark red)" (translation of TIVĀRĪ 1,p.459).

[1621] For references see e.g. TIVĀRĪ 1,p.454. Add AH,Ut 1,10, AS,Ut 1,p.3b.

within the womb or not (see the discussion in n.1582), but in any case this is an intriguing statement, since milk either in the womb or within the unborn child does not seem to be a common notion. It is thus understandable that Indu explains the reference to milk as applying to the child after it has been born (and, obviously, is being breast-fed) (see n.1582), but there is nothing in the AS passage to expressly justify this explanation. Now we have seen above that milk is closely linked with ☞ *(āhāra)rasa-*,[1622] but not in the context of the contents of the womb. On the other hand, we have also seen that the procreatory-menstrual fluid is closely linked with milk, often seeming even to be identical with it (see §§10.5f., as well as the references just cited), while at the same time itself being closely linked with *(āhāra)rasa-* (see §10.19, §§11.8f., §11.23), its special link with blood (also as an ☞ *upadhātu-*; see §11.6) seeming to be a late or at any rate a variant theoretical development (cf. §11.34). The procreatory-menstrual fluid and *rasa-* were already seen (§13.22) to have a similar function within the womb; given the close links of milk with both these substances, its function in the AS passage just quoted seems to be another facet of the complex of ideas at work here. We also have passages referring to milk flowing due to the opening of ducts (☞ *dhamanī-*)[1623] or vessels (☞ *sirā-*)[1624] originating in the heart; the passages mentioned in n.299 also state that the ducts carrying milk reach the heart (from the navel), each then dividing into three parts. All this must of course remind us of *ojas-* and *rasa-*, since these are held to be present in the

---

[1622] See §§11.5ff., especially n.1091, with further references, also §11.33, §13.20 and TIVĀRĪ 1,pp.673-676.

Note also HĀ 1,8,pp.44f. (for variants see HĀ1, pp.86;243f.), which is a most interesting passage. It says first that *rasa-* from ingested matter (according to some mss: together with ingested water) mingles with bile in the vessels (*śirā-* = ☞ *sirā-*) of milk and is turned to milk by the digestive fire. This milk is red, but turns white when "cooked" with bile (TIVĀRĪ 1,pp.674f. obviously has not understood this last statement, and the same seems to be the case as regards Alix Raison in HĀ1, p.87). Virgins (*kumārī-*) and barren women (*vandhyā-*) do not have milk because of insufficient strength of their element(s) (*alpadhātubala-*); moreover, barren women do not have milk as their procreatory-menstrual fluid (*ārtava-*) is in excess (*adhika-*). We also learn that the tubes (☞ *nāḍī-*). of barren women are filled with wind (var: and of virgins with water); due to the exertion of birth the tubes are cleared and made free for milk to flow. From this it would seem to follow that milk is in principle always present and capable of flowing in a woman, but that it does not always do so because the conditions are not always favourable; cf. Excursus XII at the end of chapter 10.

[1623] SU,Śā 10,14, BHĀ 1,4,8.

[1624] AH,Ut 1,11 = AS,Ut 1,p.3b.

heart.[1625] Evidence which seems to speak for the notion that milk was always present in the body of a woman[1626] would seem to indicate that it was, maybe not only, but also, seen as more than a transient part of the body. In this connection not only the close links, but even maybe, according to some theories, the oneness of milk with another important vital substance of the body, namely semen,[1627] may also be recalled to mind (see §7.40, §8.23).[1628]

13.26. Thus we see that, apart from the different theories we have discussed so far in the course of this study, we will also have to consider another notion having to do with body fluids. It seems that the medical texts contain traces of a theory of the body containing some sort of vital fluid, which has been identified with several of the fluid substances we have considered in the course of our investigations (i.e. *ojas*-, *rasa*-, blood, procreatory-menstrual fluid, phlegm, semen, milk),[1629] and seems to have played a part in the formulation of or else influenced several of the various theories pertaining on the nature and function of these substances (cf. also n.1546). Highly interesting too is the fact that several of the fluids mentioned are connected with the heart, which has in the discussion above also been mentioned as the place of origin (in the mother) of the nourishment of the embryo or foetus through the umbilical cord. In this connection it is very intriguing that in the illustration mentioned in §6.14 the mysterious receptacle of semen (*śukrasthāna*-) is actually in the cardiac region; given the parallels between semen (of the male) and procreatory-menstrual fluid (and also semen of females), with regard to their place of collection,[1630] one does feel inclined to ask whether these, in their function as vital fluids (or aspects of

---

1625 See the discussion on ☛ *ojas*- and ☛ *rasa*- in Appendix II, also §§13.20f.

1626 See Excursus XII at the end of chapter 10, and n.1622; cf. too n.1098.

1627 The ducts associated with this are also connected with the heart in the passages mentioned in n.299 and just referred to above in connection with milk.

1628 Cf. in this regard also DAS,R.P. 1984A, pp.237ff. on the problem of milk, semen (on which see especially p.243,n.30) and *rasa*-, where non-medical (including Vedic) material too has been utilised. The remarks loc.cit. are, in comparison with what we have seen in this study, rather superficial, but nevertheless serve to give a basic idea of the problems involved, especially with regard to material not utilised here.

1629 This does not mean that all the substances mentioned have also been identified or linked with each other, at least in the material we have sifted. Thus I have for instance found no evidence for a closer link between milk and blood, especially not of the sort theoretically surmised in n.149, though cf. also n.1622.

1630 Cf. on this last §13.11, especially n.1552 (with further references).

one vital fluid) might not have been collected here too. No definite answer to this question seems, however, to be available, though AS,Śā 1,p.276a, discussed in §9.3, might be taken to refer to the heart as the place of storage of semen of the male and procreatory-menstrual fluid of the female; since, however, this passage is not unambiguous, its evidence is not above all doubt, though it would of course fit some of our deliberations nicely. Another passage of great interest is SU,Śā 5,11, which states that the ☞ *kaṇḍarā*-s joined to nape and heart, and going downwards, have as their extension (literally: "sprout at the tip") the penis.[1631] Unfortunately, we cannot garner more information from this than the fact that in this theory at least penis and heart were considered to be joined.[1632] But even apart from the problem of the place of collection, it is extremely difficult, and probably impossible, to determine the exact influence or influences due to the notion of a vital fluid of the sort described above within the body, as it is hidden beneath theories seeing the fluids mentioned as individual fluids in their own right (though they may be interconnected in various ways), with natures and functions of differing sorts and based on considerations not necessarily in keeping with the notion of a vital fluid. Thus what we actually have are tantalising hints and bits of information, many questions, and hazy ideas regarding the *possible* interaction and development of different theories. But at any rate we can point out that the notion of a vital fluid within the body, manifesting itself in various forms, leads into wider realms not only of Indian, but also of pan-Asian thought, into spheres vaster than the relatively narrow ones we are here primarily concerned with. This notion is on the one hand ancient, and has many parallels in Vedic, especially Upaniṣadic thought (in which context I may draw attention, too, to its connection with the notion of water as the basis of all life, which has also led to very lengthy and intricate discussions on the nature of water and ☞ *rasa*- especially in §§7.11ff. and §§10.7ff.); on the other hand it is also very wide-spread, playing a vital role especially in what is called Tantrism, but also having other Indian and extra-Indian ramifications. Given the connections between ☞ *ojas*- and ☞ *tejas*- "fiery energy" on

[1631] *grīvāhṛdayanibandhinīnām* (var: *grīvāhṛdayasambandhanīnām*) *adhobhāgagatānāṃ* (var. *adhomukhānāṃ*) *meḍhram [agraprarohaḥ]*. Āḍhamalla in his commentary on ŚĀR 1,5,45-47 too draws attention to this passage. BHĀ 1,3,268 too has this statement, but intriguingly leaves out ☞ *hṛdaya*- "heart", i.e. only the nape is mentioned.

[1632] The mention of the nape in this context is highly interesting too, given Tantric theories of the cranial location of semen (theories which are not in keeping with those generally found in Indian medicine, at least as far as the 'classical' texts are concerned). That the heart too is connected with both nape and penis also draws attention to a problem I discussed in DAS,R.P. 1992A, pp.411f.

the one hand and the heart and luminosity or some luminous energy on the other already in very ancient texts,[1633] we may also have to take into account matters going far beyond the subject of fluids only. But all I can do here is to draw attention to all this, as the matter cannot be taken up in the context of this study. I may, though, be permitted to point out that some relevant material is to be found in DAS,R.P. 1992A, which also contains many bibliographic references[1634] and thus might serve as a starting-point for further studies in this direction (cf. also the statement from DAS,R.P. 1992A in §13.21).[1635] It may also not be out of place to draw attention here to Hemādri's remarkable discussion of *ojas-* in its various forms and meanings in his commentary[1636] on AH,Sū 11,37-39. The whole is too lengthy to reproduce or discuss here in full, but I would like to mention one particular here, namely that according to Hemādri *ojas-* which is the ☞ *mala-* of male semen[1637] enters the heart of the embryo/foetus, being eight drops in quantity and of a slightly reddish-yellow colour due to its being mixed with the procreatory-menstrual fluid; this *ojas-* is then called blood (*jīvaśoṇita-*).[1638] (Cf. on this also §13.21.) All of which shows that a systematic study of the problem of (a) vital fluid(s) is bound to reveal a great deal of very interesting information.

13.27. But let us return to the root of all these remarks, namely *rasa-raktavaha-* as a characterisation of the ☞ *srotas*-s being blocked according to Ḍalhaṇa's version (in his commentary on SU,Śā 3,31f.) of Bhoja's verse in §13.15. In the light of what we have just seen, all the alternative translations considered for this expression make sense, i.e. we could have channels "carrying *rasa-* and blood", channels "carrying blood from *rasa-*", or channels "carrying blood which is *rasa-*", "blood" in each case seemingly referring (like it

---

1633 Thus e.g. in *Ṛgvedasaṃhitā* 6,9,6: *idáṃ jyótir hṛ́daya ā́hitaṃ yát* "this light which is deposited in the heart".

1634 Also to extra-Indian material.

1635 Of course what we also need are comprehensive studies, encompassing, *layer by layer*, the whole of Indian literature, on each of the fluids which seem to be able to function as a vital fluid, but it is probably utopian to expect such studies to be available any time soon. An attempt to study *ojas-* in this manner is GONDA, but this study is clearly not the last word on the subject of *ojas-* (much of the material is quite sketchy; this holds true especially for the medical material on pp.44-46).

1636 Referred to e.g. in the discussion on ☞ *ojas-* in Appendix II, and in n.812.

1637 Cf. on this the discussion on *ojas-* in Appendix II.

1638 *yad eva hi śukramalarūpam ojaḥ tad eva garbhahṛdayam anupraviṣṭam aṣṭabindupramāṇam ārtavānuviddhatvād īṣadraktapītaṃ jīvānuviddhaṃ jīvaśoṇitam ucyate.*

does, as we have by now seen, so often) to the procreatory-menstrual fluid, though in the light of the discussion on the vital fluid above one might nevertheless also be tempted to take "blood" as referring to blood in general, which would then, however, be the remnant of an ancient or at least not strictly 'medical' theory; especially the last option ("carrying blood which is ☞ *rasa*-") would, in case blood should indeed refer to blood in general, remind us of the notions referred to in n.1086 (cf. also §13.15).

13.28. There is, however, another possibility which might be considered. The readings *rajaḥśukravahāni vai* in ĀY 4,147 (see §13.14) and *rasaraktavahāni vai* in Ḍalhaṇa's commentary on SU,Śā 3,31f. (see §13.15) could both be corruptions of an original *rajoraktavahāni vai*, *rajoraktavaha*- meaning "carrying blood (*rakta*-) which is *rajas*-", in which case the reference would be to procreatory-menstrual fluid carrying channels only, and the verse would present absolutely no problems. Unfortunately, this possibility is only speculation, at least for now.

13.29. It is also worthwhile to return once again to the mode of nourishment of the embryo/foetus (see §§13.14ff.), especially the mode of transportation of nourishment from the mother's heart to its navel via the umbilical cord. The descriptions given in the texts present us with some difficulties. Not all the texts tell us exactly how the mother's heart and the foetus' navel are joined by the umbilical cord.[1639] Of those that do, SU,Śā 3,31 says only that the umbilical cord is joined to the tube (☞ *nāḍī*-) originating from the mother's heart.[1640] In the place of this latter single tube, CA,Śā 6,23 (cf. §13.18 and n.1588) mentions several vessels (☞ *sirā*-) joining the mother's heart and the *aparā*-, to which latter the umbilical cord is joined;[1641] AS,Śā 2,p.291a (cf. §13.18 and n.1589) similarly speaks of ducts (☞ *dhamanī*-).[1642] We thus seem to have two different descriptions of how the mother's heart is joined to the umbilical cord, once through only one tubular structure, once through several. TIVĀRĪ has obviously seen this difficulty too, for in the picture in 1,p.143 we find one tubular structure leading from the heart of the mother, but then giving rise to many branches, which then are joined to the *aparā*-. I have no idea whether the texts indeed presuppose this,

[1639] Seemingly via the ☞ *aparā*-, though this is not always expressly mentioned.

[1640] See n.1592; BHĀ 1,3,321 is similar.

[1641] *nābhyāṃ hy asya nāḍī prasaktā nāḍyāṃ cāparā. aparā cāsya mātuḥ prasaktā hṛdaye. mātṛhṛdayaṃ hy asya tām aparām abhisamplavate sirābhiḥ syandamānābhiḥ* (var: *saṃspandamānābhiḥ*).

[1642] *tato mātṛhṛdayād āhāraraso dhamanībhiḥ syandamāno 'parām upaiti. tataḥ kramān nābhim.*

or whether there were two different theories in vogue, but there is no denying that we do have mentions of a plural number of tubular structures. However, these passages seem parts of a closed system not leading directly into the uterus, and as such do not seem capable of serving as evidence for the existence of notions on a plural number of passages opening straight into the uterus. On the other hand, the fact that the embryo is nourished by substances entering the uterus even before the umbilical cord is formed (cf. §§13.17ff.) could be taken to mean that these substances enter the uterus through the same tubular structures that later are joined to the umbilical cord. Unfortunately, we have no description in the texts mentioning a plural number of tubular structures on how the substances that enter the uterus in the phase before the umbilical cord is formed actually do so, so that this is no more than just a possibility. Nevertheless, the possibility cannot be rejected, since one must ask what happens to the tubular structures that lead from the heart to the *aparā-* when the embryo is not nourished through them, for unless they simply vanish or become inactive/blocked, one would expect them to continue carrying something or other from the heart to the uterus, and this would be the case too if not a plural number, but only a single tubular structure should lead from heart to uterus.

13.30. Complicating matters is the only statement on the mode of entry of nourishment into the uterus before the umbilical cord is formed that we do find. SU,Śā 3,31 says, right after the statement just quoted in §13.29 on a *single* tubular structure leading from the heart of the mother to be ultimately joined to the umbilical cord, that a *plural* number of ☞ *dhamanī*-s is mentioned as carrying ☞ *rasa-*.[1643] There is, however, a great chance that these latter structures may be in the body of the embryo/foetus, not of the mother (see §§13.17f.); however, even if they be structures of the mother, they cannot be linked with the nourishment from the heart of the mother, since they are said to be transverse and passing through all parts of the body. Though this passage thus could eventually (though the probability is probably not very large) be further evidence for a plural number of tubular structures opening into the womb, the nature of these would be different from those we have considered in §13.29. It should also be noted that none of the passages (cf. §13.13) which led us to deduce that there might have been notions on a plural number of tubular structures opening into the uterus even hints at the heart having anything to do with these; thus if in the case of the passages discussed in §13.29 we may indeed suppose that there was a theory postulating a plural number of tubular structures leading into the uterus and connected

[1643] See the passage given in n.1572.

with the heart of a woman (see on the heart also §13.26), then there does not, at least on the basis of the present evidence, seem to be any link between this particular theory and other theories regarding the mode of ingress of substances into the uterus from the body of the woman and not mentioning the heart, i.e. we would seem to have two theories here (provided, of course, that the structures leading from the heart indeed do carry something into the uterus of a woman not pregnant).

13.31. If the evidence we have sifted does not allow any definite conclusion in this case too, we may at least console ourselves by pointing to some other material discussed which seems to corroborate some of our earlier postulates. In §6.25 and §§10.26ff.[1644] we discussed the possibility that there was a theory current according to which procreatory-menstrual fluid enters the womb until the foetus has grown so much that it blocks the ingress of the fluid, the fluid hindered from entering the womb then forming the ☞ *aparā-* and the umbilical cord (sort of pushing them into the womb from outside) and, still being pent up, moving ever upwards and giving rise also to milk. The passages discussed in this connection seem to postulate a plural number of tubular structures opening into the uterus, but actually the number of such structures is basically irrelevant in this context, as the same would be the case if only one or two tubular structures should have been postulated. Now the form of nourishment of the embryo/foetus that our texts seem to postulate is that this is first nourished through its skin by fluid within the womb, later, however, by means of the umbilical cord (see §§13.17ff.), and this would mesh very well with the theory of growth of the foetus blocking the ingress of fluid into the womb. But our silver linings always seem to have a cloud, and this case too is no exception. Thus, of the passages quoted in §§13.17ff., only SU,Śā 3,31 quite clearly refers to one mode of nourishment following upon the other. AS,Śā 2,p.291 also describes both modes one after the other, but ends the discussion by referring to the metabolic transformation of what has been taken up both through the umbilical cord and the pores of the skin (see n.1582); however, this is probably just an amplification on both modes of nourishment and does not mean that they continue simultaneously. CA,Śā 6,23, however, expressly says that both modes of nourishment are simultaneous. What makes all this all the more interesting is that no CA passage plays a role in the discussion on the foetus blocking the ingress of fluid, whereas SU,Śā 4,24 and AS,Śā 2,p.288a do (see §6.17). In other words, the description of the mode of nourishment in SU and AS meshes with the postulated theory of blockage of ingress of fluid in these texts,

[1644] With further references; see also n.277.

whereas the mode of nourishment described by CA would not, and CA does in fact have no material which seems of relevance for the postulated blockage of ingress of fluid. This could show that we have two different traditions at work here.

13.32. Let us now summarise the discussion in this chapter, in which we have not only considered new matter, but also tied up several loose ends. We began by discussing a passage from PA which describes three tubular structures opening into the uterus, two of which are linked with the male and female sex of the conceived child respectively; we deduced that they discharge procreatory fluid, though how this influences the sex of the child was unclear. The fluids discharged by the tubes might be different in quantity or in quality (see §13.3); since the predominance of one or the other tube has an effect on the reaction of the woman during intercourse, the time of orgasm might also play a role (which would fit theories already discussed before), but the heftiness of orgasm or the degree of excitability were considered too (§13.3). The theory of the right side being linked with males, the left side with females, was also mentioned in this context (loc.cit.). In this connection, attention was also drawn to the theory linking the sex of male and female children with intercourse on even and odd days of the *ṛtu-* respectively, and to explanations connecting different quantities of discharges with different days too, as well as to notions of the character of the child being linked with the time of intercourse, and to variations on these notions (§13.4). As to the third tubular passage opening into the uterus (and seemingly linked with the clitoris; cf. §§13.5f.), it was debated whether it too was taken to discharge something, and whether this discharge may be regarded as female "semen" (its internal discharge causing problems; cf. §13.11). It was remarked that an internal discharge made it most probable that the fluids discharged (procreatory-menstrual fluid and "semen") were previously collected and stored in some other part of the body (§13.11; cf. too §13.26). Attention was also drawn to previously discussed notions of three passages opening into the uterus, two carrying procreatory-menstrual fluid and one "semen" (§13.6); this led not only to the problem of the procreatory-menstrual fluid maybe having two different forms, but also to that of the time of discharge of the tubes carrying procreatory fluid (§13.7). It was debated whether the theory speaking only of two passages opening into the uterus is different from that speaking of three passages, or whether the third passage is also presupposed in the former, but simply not mentioned (§13.8f.). It was also asked whether a mixing up of different theories could be held responsible for the confusing statements of HĀ,Śā 1,p.414f. (§§13.9f.); no definite answer was found. A similar question was asked of a NĀG passage mentioning six different tubes (§13.12), of which, however, only two seem to be clearly connected with procreation and thus to open into the womb (it was debated too whether these two passages might have been held to be situated in the cervical region and thus be capable of stimulation through the penis, but opening into the interior of the womb; the discussion was inconclusive). After this statements according to which a plural number of tubular structures open into the uterus were taken up again (§13.13ff.), leading to a discussion which also touched upon the mode of nourishment of the embryo and came to the conclusion that there seem to be traces in the medical texts of an old theory of some vital fluid in the body, variously identified as *ojas*-, *rasa*-, blood, procreatory-menstrual fluid, phlegm, semen and milk (see §13.26), with which the notion of there being two forms of procreatory-menstrual fluid might eventually also be connected (§13.24).[1645] This vital fluid on the one hand seems to be connected specially with the heart, on the other to play a major role in the nourishment of the embryo/foetus, first by entering the uterus (seemingly steadily) and being absorbed through the skin of the embryo/foetus, then by entering the latter via the umbilical cord. This cord is, however, connected with the ☞ *aparā*-, which is connected with the heart of the mother, according to some theories through one, according to others through a plural number of tubular structures (see §13.30). This led to the deduction that the plural number of passages mentioned in some texts as leading into the uterus might perhaps be these same passages (§13.29). Lastly, the problem of blockage of passages through the foetus and the formation of the *aparā*- was taken up again, and it was deduced that, in the light of all the evidence sifted so far, the most plausible conclusion is that what is blocked is the ingress of fluid, leading to the *aparā*- being formed outside and sort of pushed into the uterus; attention was also drawn to there possibly having been variant traditions in this regard (§13.31)[1646].

[1645] Cf. also the remarks on possible causes for the theories on two sorts of ☞ *ojas*-, and on ☞ *rasa*- and ☞ *āhārarasa*- (see especially §13.23).

[1646] Cf. also n.301 and n.1062 on the possibility that the blockage was actually taken to be due to a membrane formed on the inner uterine wall.

CHAPTER 14
Conclusions

14.1. So we arrive at the end of a long tortuous road, and if the journey has been difficult and at times frustrating, it has, all in all, been worth its while, for it has helped us to come by information which is usually not so readily available, but which nevertheless tells us much about ancient Indian notions on the human body, the origin and development of human life and the relationship between the sexes. Above all, even though our sources, particularly the medical ones, are male-oriented,[1647] and even though we have not been able to tackle *all* related problems in detail,[1648] we did manage to gather a lot of information on the role the woman plays in conception, and on related subjects. Let us now summarise this.[1649]

14.2. According to the texts examined, an embryo is produced due to the mingling of the procreatory fluid of the female, which is the same as the menstrual fluid, with the semen of the male. Both these fluids are collected from all over the body and discharged (due to the action of agitated wind) by means of an orgasm, in the case of the female the orgasm being responsible for an ejaculation of procreatory-menstrual fluid into the uterus.[1650] After

---

1647 The latter sources as a rule presupposing a rich male as their ideal referential basis; cf. n.167.

1648 Examples are the problem of the exact meaning(s) of *bīja-* (cf. §§2.5f.), and of the egress of the embryo/foetus from the uterus into the general visceral cavity (cf. §6.20, also §9.4), in which regard we also discussed possible variant theories holding that the embryo is situated in the place of digestion together with food (§6.20).

1649 When references to paragraphs are given below, the discussion is either confined to the paragraphs mentioned, or else — and more often — the subject in question has been discussed in them last; references to other relevant parts of this study will then be found in these paragraphs. The same applies to notes referred to.

1650 An ejaculation of some sort (not necessarily the same in all respects) into the uterus is quite obviously also presupposed by the theory of conception in Siddha medicine (i.e. the Tamil system of medicine, which — as also its basic terminology would make us assume — is possibly but a local variant of the pan-Indian amorphous and heterogeneous conglomerate of ideas on medical matters, albeit in the local language and with local developments; on this matter see e.g. FILLIOZAT,J. 1974 pp.37f. — who cites the relationship between Greek and Yūnānī medicine by way of comparison — and also DAS,R.P. 1993B, pp.60-62;68f.). Cf. e.g. IYER, p.21: "The semen (male) and the sronitha (female fluid) thus secreted in the body enter the uterus in the form of the minutest visible atom (or particle) and jump with a force like that which is created when we beat forcibly a big drum ...". Cf. also the following extract from

collection from all over the body and prior to their discharge, both the fluids could have been considered first to be in a place of collection (which in the case of the female would probably not be the uterus). There seems to be evidence that this place of collection might have been taken to be the heart by some, but this evidence is inconclusive; there may, moreover, have been various competing theories on the place(s) of collection (§13.32). As to the sex of the embryo thus produced, there seem to have been several concurrent theories regarding the mode and circumstances due to which the mingling of fluids determines this. Probably the commonest theory is that it is the absolute quantitative relation of the two fluids to each other that determines the sex. Another theory, however, holds the day of intercourse during the fertile period of the woman to be responsible for the child's sex. Both theories are linked in the notion that different amounts of fluid are discharged on different days, but it is unclear whether this shows that they are but two facets of one and the same notion, or whether two different theories have thus been subsequently linked (§13.32), since there seem to have been other theories on the periodic waxing and waning of vital fluids of the body current too (cf. n.1546). According to yet another theory, the relative strength and potency of the fluids determines the child's sex (§11.34); in this case too it is unclear whether this is but another facet of one central notion, or a by origin independent theory.[1651] Apart from these there are notions pertaining to

---

a modern text of the Tibetan Vajrayāna Buddhist tradition: "When people engage in sexual intercourse this downward-voiding wind is also caused to move upwards by the union of the male and female sex organs. However, this movement does not take place within the central channel and, as a result, the ordinary inner fire ignites and blazes only momentarily. Because of this ignition and blazing the drops in the lower part of the body melt and flow downwards. A brief experience of bliss is felt until the drops leave the sex organ or, in the case of the woman, collect in the uterus" (GYATSO, p.57).

[1651] Cf. in this regard also LAQUEUR 1990, p.39: "Hippocrates argues for pangenesis, the view that each part of the body of each parent renders up some aspect of itself [cf. on this the theory of ĀYS in n.1086]; that the representatives of the various parts form a reproductive fluid or seed; and that conception consists of a blending, in various proportions and strengths, of these germinal substances. Hippocrates abandons any effort to attribute strong or weak seed respectively to actual males or females. Although males must originate from stronger sperm, 'the male being stronger than the female,' both are capable of producing more or less strong seed. What each emits is the result not of any essential characteristic of male or female, but of an internal battle between each sort of seed: 'what the woman emits is sometimes stronger, sometimes weaker; and this applies also to what the man emits.' Hippocrates insists on this point by repeating the claim and generalizing it to animals: 'The same man does not invariably emit the strong variety of sperm, nor the weak invariably, but sometimes the one and sometimes the other; the same is true in the woman's case.' This explains why any given couple produces both male and female offspring as well as stronger and weaker versions of

the time of intercourse determining the character of the child, linked especially, seemingly in later times, with the notion that the time of intercourse determines the child's sex, and giving rise to several variations (§13.32). Again, the sex of the child conceived is also taken to depend on which of the partners engaged in intercourse has an orgasm first, though there seem to be different versions and variations of this theory (§11.34).[1652] Whether the heftiness of orgasm or the degree of excitability too are considered to be factors is a matter of conjecture (§13.32), as is also whether the time and/or (maybe) heftiness or the like of orgasm is in any way taken to be linked with the quantity or quality of the procreatory fluid discharged (§13.3). As regards the latter, we have also discussed passages which might be saying that the procreatory fluid of male and female is each composed of different substances, also debating whether this might imply that this means that the fluids are composed of procreatory and non-procreatory parts; the evidence sifted seems inconclusive (§§11.15ff. and §§13.9f.)[1653]

14.3. The procreatory-menstrual fluid may look like blood,[1654] but, except for a very problematic statement which might perhaps be saying that both fluids are the same (§11.34), and some passages which might be taken as the basis for interpretative speculation in this direction (n.169, n.402 and n.1122; see also §7.10, §12.7 and n.884), all the evidence leads to the conclusion that blood and the procreatory-menstrual fluid are regarded as different

---

each; likewise for the beasts. If both partners produce strong sperm, a male results; if both produce weak sperm, a female is born; and if in one partner the battle has gone to the weak and in the other to the strong, then the sex of the offspring is determined by the quantity of sperm produced. A greater quantity of weak sperm, whether produced by the male or the female, can overwhelm a lesser quantity of strong sperm, of whatever origin, in the second round when the two meet in front of the uterus for renewed combat. Hippocrates is at pains to emphasize the fluidity of the situation and the interpenetration of male and female."

[1652] See also §12.5 on a problem connected with this notion.

[1653] This problem of different theories relating to the cause of a child's sex has parallels in mediaeval Western thought; for a short overview and references to relevant literature see BYNUM, p.187;212,n.100.

[1654] The redness of the procreatory fluid is a belief found even in modern-day South Asia. A particularly telling example is found in an esoteric Bengali song (*Māṭir piñjirār mājhe*) by the famous Hāsan Rājā (see e.g. SIṂHA, p.127):

bāpe māẏe bandi kailā khuśir mājhāre
lāle dhalāẏ bandi hailām piñjirār mājhāre
"Father and mother set me captive in the midst of [their] pleasure. In/Through red and white I became captive within the cage."

fluids,[1655] which is also substantiated by the fact that the former is a regular part of the chain of alimentary transformation of the ☞ *dhātu*-s, whereas the latter is not a part of the chain proper, but rather a product of a separate development from a member of this chain. Though the theory today generally regarded as valid derives the procreatory-menstrual fluid metabolically from blood,[1656] the 'classical' texts derive it not from blood, but from ☞ *(āhāra)rasa*- (§11.34); sporadic variant theories on the derivation of the procreatory-menstrual fluid might also have been in circulation (§8.14, §11.9, §11.22; cf. also n.1086). It is also either the same as or very closely linked to milk (which might have been taken to be always present in the body) (§13.25), though in the theory generally regarded as valid today the two are distinctly differentiated (§§11.6-8). There seem, however, also to be traces in the medical texts of an old theory of some sort of vital fluid in the body, which is variously identified as ☞ *ojas*-, ☞ *rasa*-, blood, procreatory-menstrual fluid, phlegm, semen and milk;[1657] attention may in this regard be drawn too to the discussion of and problems regarding *rasa*- and *āhārarasa*- (§13.23) and *rasa*- and water (§13.26) (and also ☞ *tvac*- "skin"; §10.7ff.), as well as to the oneness of *rasa*- and blood seemingly postulated by some not strictly 'medical' traditions (see n.1086 and §13.27). The varying identifications of the vital fluid seem to have led to conflicting statements on the different substances mentioned as being identified with it, as well as to various similarities and connections being postulated. The vital fluid is on the one hand intimately connected with the heart (☞ *hṛdaya*-),[1658] on the other seems to play an important role in the nourishment of the embryo/foetus, both by en-

---

[1655] Though their properties are at times mixed up; cf. e.g. §§7.5ff.

[1656] Cf. in this connection also CHÖPEL, pp.58;195f., where it is maintained (seemingly with respect to both male and female) that "the essence of blood is seminal [p.58: regenerative] fluid, but at first it is dissolved in blood [like butter in milk]. However, if it is churned by the action of male and female, the power of passion raises warmth in the blood, and the seminal [p.58: regenerative] fluid, like butter, comes out."

[1657] On the close connection between the latter two cf. §13.25. As regards milk, attention may also be drawn to WEZLER 1992A, pp.295ff., where a chain of successive transformation within (or, perhaps better: to) different entities in the context of the *sarvasarvātmakatvavāda*-begins with milk. In the examples adduced the milk is obviously that of animals (most probably cows), and the context too is quite different from that we have discussed. Nevertheless, here too milk seems to be seen as some sort of vital fluid, and it should be interesting to examine whether ultimately the different notions on milk as a vital fluid might not somehow be related.

[1658] On the extra-medical and even extra-Indian connotations the notion of the vital fluid and of its connection with the heart may have, cf. also §13.26.

tering the uterus and by means of the umbilical cord, which is ultimately connected with the heart of the mother (§13.32). Though no mention of the procreatory-menstrual fluid flowing through the umbilical cord is found, there are statements according to which it enters the uterus during pregnancy; the growth of the foetus then hinders its flow, leading to the formation of the ☞ *aparā-* and the umbilical cord, and, finally, due to its collecting ever further up, of milk, and this seems to imply that the ingress of the fluid into the uterus is stopped due to the growing foetus, and that the structures mentioned are sort of pushed into the womb from its wall due to the pressure of the pent up fluid outside (§13.32).

14.4. The discharge of the procreatory-menstrual fluid into the uterus seems most commonly to have been held to take place via two tubular structures opening out into this (note too the remarks on the place of collection in §14.2), though there also seem to have been notions on a plural (i.e. not dual) number of such structures (§13.32); some theories seem to link the sex of the child conceived with specific tubular structures or their discharges (§13.32).[1659] Though there seem to be no statements linking the discharge with the single tubular structure connected with the heart of the mother, and mentioned in connection with the nourishment of the embryo/foetus as leading to the *aparā-* (cf. §14.3), one may speculate that such a link might have been taken to exist in older times, if not with the single tubular structure, then at least with the heart, from which according to a variant theory a plural number of tubular structures leads to the *aparā-* (§13.32). As regards menstruation, it is a notion commonplace the world over that this is a process purifying the woman concerned; in our case we can say more concretely that it is a process in which old (i.e. most probably: no more fecund) procreatory-menstrual fluid is exchanged for new,[1660] and that the process of entering into the uterus is more or less the same as in an orgasm, i.e. the tubular structures open up, emitting the fluid (§5.15, §8.16, §8.23, §10.2). What

[1659] Whether any of these tubular structures may refer to the Fallopian tubes or not seems to me to be a matter of pure speculation; I have found no evidence that would support, but then again nothing that would speak against this. The evidence does not seem to allow us to say anything on the subject, which may thus be best left to apologists of the sort described in §1.7 to wax eloquent upon. See also Excursus II at the end of chapter 6.

[1660] Interestingly, ĀY 4,53-55, quoting the *Videhasaṃhitā* (see on Videha MEULENBELD 1974, p.426), tell us that this cleansing and the fact that they do not exercise, and thus do not aggravate their wind and bile, prevent baldness from appearing in women. See also §8.16 for another theory, attributed by ĀY to the work of Vṛddhātreya, which seemingly did not gain wide currency, on the cleansing of the procreatory-menstrual fluid through semen from the male, or maybe even through the embryo.

causes the discharge into the uterus is unclear, but it could be surmised that it is the pressure of the accumulated fluid that forces the openings into the uterus (which are, at least according to some theories, guarded by muscles; §13.6) to discharge (n.789), and one may also enquire whether similar pressures, due to various factors, within the body might not be responsible for (abnormal) discharges to the exterior at other times too. On the other hand, the procreatory-menstrual fluid seems to enter the uterus steadily during pregnancy (cf. §13.32), and there do not seem to be any statements on a periodicity of this, so that it seems to be in conflict with the theory of periodic entry of fluid into the uterus. Whether we indeed have conflicting theories here, or whether there is some other explanation which does away with this difficulty (such as, e.g., fluid pressure on the mouth of the uterus leading to the vagina forcing it to open periodically), I do not know.[1661] Attention may in this connection be drawn to §8.16, where we discussed a passage of Vṛddhātreya which might perhaps refer to a steady trickling of fluid into the uterus all the time (and not only during pregnancy); it bears examining whether this notion, if it should indeed exist, might be connected with the notions of non-periodic ingress of fluid during pregnancy, and whether the theory of steady ingress might be older than the theory of periodic ingress.[1662] As to the non-visibility of fluid during pregnancy, even though it continues to flow, this seems to be due to the closure of the uterus (n.277).

14.5. Apart from the procreatory-menstrual fluid, women also emit a substance called "semen" (mostly *śukra-*, but also *retas-* etc.). The nature of this seems to be one of the most problematic subjects of ancient Indian medical theory, and the statements on it made by different authorities are varied and more often than not contradictory.[1663] Thus there are controversies on

---

[1661] On the problem of internal circulation within the body see §6.19. On the problem of possible entry into the visceral cavity see §6.20.

[1662] One may also enquire about the relation of the theory of steady entry of fluid into the uterus in general (i.e. not only during pregnancy), should it indeed have existed, to that of orgasmic ejaculation, and also whether different sexologists might not have harboured different opinions, thus giving rise to much of the confusion evident in the discussion of §§12.2ff. But all this is speculation, and unless more evidence turns up we can do no more than merely draw attention to the matter and the questions connected with it.

[1663] The subject is even more problematic once one looks beyond the confines of medical and sexological literature. As a case in point, I may quote a half-verse found in *Viṣṇupurāṇa* 6,5,14ab = *Brahmapurāṇa* 234,14ab, according to which a child is *jāyamānaḥ purīṣāsṛṅmūtraśukrāvilānanaḥ* "one whose face is polluted by faeces, blood, urine and semen while being born". Clearly, "semen" (*śukra-*) here must be something the mother discharges at the time of birth; what could this be? (We may also ask what *asṛj-* "blood" refers to here: blood in gener-

whether this semen has the power of fecundity or not, whether it is the same as the procreatory-menstrual fluid or not, or whether the procreatory-menstrual fluid is actually present in two forms, one red and the other clear (which latter would then most probably correspond to "semen", though this is not certain).[1664] Added to this is the problem of its relation to semen held to be present in the body of both males and females, and which is the last "element" (☞ *dhātu-*) of the chain of alimentary transformation beginning with ☞ *rasa-*; its physiological origins too are unclear.[1665] As it is, female "semen" seems usually to be regarded as a fluid ejaculated by the woman due to an orgasm and reaching the exterior of the body (see on this ejaculate especially §§12.15f.), maybe after intermediate storage in some receptacle within the body (§13.32). On the other hand, the same terms (among these also "semen") seem to be in use, especially in sexological works, both for this fluid ejaculated as a result of orgasm, and non-orgasmic discharges due to sexual excitement (e.g. lubricating fluid) (§13.6; on "semen" being used to refer to the procreatory-menstrual fluid and vice versa see §12.3 and §12.6 respectively). As regards fluids of this latter sort, though there are sporadic mentions

---

al, or some other fluid only called so?) I may add, too, that there seems to be a lot of relevant material on the "semen" of women particularly (but not only) in Tantric texts. On material on the semen of women in modern South Asian thought cf. e.g. KAISER, p.104, BOTTÉRO, p.310. It has also to my knowledge not yet been examined whether and in what manner beliefs such as that expressed by CHÖPEL, p.204 might have played or still play a role in this context: "From the vagina of the women of Draviḍa a white fluid is secreted before copulation."

1664 One is reminded of these problems when reading in the work of WECK on Balinese medicine (pp.107f.) that a child is believed to be formed of a combination of male seed or semen ("Samen") and female seed or semen ("Samen"), the latter being called "kama ratih" (obviously the Sanskrit *kāmarati-*). Both fluids are the extract of food, but in colour white and red respectively. Elsewhere, however, we learn of the fluid of the female being blood (p.111). And p.112 speaks of a "fruit" (obviously, in the given context, the female procreatory fluid, whatever this be) falling from the upper part of the "tree" (the trunk of the body) to the "earth" (the genitals) and on the way being caught by the man (scil. his semen). When it is not caught, it falls to "earth", and there is menstruation. (On *kāmarati-* to describe fluids of women in the terminology of a religious group in modern Bengal heavily influenced by Tantrism see DAS, R.P. 1992A, pp.402ff.; cf. also p.410 on the body as a tree.)

1665 All these matters have been discussed throughout this study, especially in the part beginning with chapter 8 (cf. on this chapter the summary in §8.23), so that it does not seem feasible to give references to individual paragraphs or notes here. Note, however, that some difficulties might be linked with the possibly older theory mentioned in §14.3 of a vital fluid of the body variously identified with different fluids. As regards the difficulties occasioned by "semen" of females, it is interesting to observe how TIVĀRĪ 1,pp.45ff. attempts apologetically (in keeping with the remarks made in §1.7) to find explanations in accord with modern Western medical notions to account for the various statements on this substance.

of them, usually in sexological, rarely in medical works (some statements might incidentally be taken to allude to Bartholin's glands or something akin to them; n.1554), the fact that they are rarely mentioned makes it rather seem as if on the whole not much thought was given to them.

14.6. With regard to the mode of discharge of the fluid (not the procreatory-menstrual fluid) ejaculated due to an orgasm of the woman, there seem to be different theories on this. One theory seems to hold that there are three tubular structures opening into the uterus (and, at least according to some authorities, guarded by muscles), of which two discharge procreatory-menstrual fluid (cf. §14.4), one "semen" (§13.32). Since the latter fluid is as a rule taken to be visible externally, whereas the former leads to conception within the uterus, this naturally creates difficulties (also regarding ejaculation during pregnancy; cf. §12.38); on the other hand, it opens up intriguing perspectives as regards possible links with Tantric notions (n.1549). Another, less problematic, theory could hold that the fluid emitted on orgasm is ejaculated not into the uterus, but from the region of the portio vaginalis of the uterine cervix, and thus reaches the exterior without any difficulty (§12.38). We learn further that several parts of the female genitals are particularly susceptible to stimulation and thus linked with ejaculation due to orgasm; there seem to be two different theories in this regard, one mentioning the clitoris and the portio vaginalis,[1666] the other these two organs as well as the Gräfenberg spot (urethral ejaculation, though, does not seem to play a role with regard to the latter) (§12.37, §13.32).

14.7. Finally, mention must be made of several examined passages, from medical as well as sexological works, which had theories on various fluids and/or tubular structures linked with procreation and orgasm, and which were not clear in all respects; it was suspected that different notions might have been mixed up during their formulation (§13.32).

14.8. Stated in this compressed form, our findings may seem to some not as exciting as §14.1 rated them, but on the contrary rather banal, and maybe also a bit frustrating as the end-product of such a lengthy and intricate discussion. But anyone who has gone through the whole study above will appreciate how difficult it has been to put them together from bits and pieces of information buried in a vast corpus of textual matter. Moreover, in the course of extraction we have also had to tackle many difficult and fundamen-

---

[1666] On the uterine cervix cf. also LAQUEUR 1990, p.44: "Galen, for example, notes that 'the male penis ... as well as the neck of the uterus and the other parts of the pudendum' are richly endowed with nerves because they need sensation during sexual intercourse and that the testes, scrotum, *and* uterus are poorly endowed because they do not."

tal problems of Indian medicine, to discuss matters hardly ever dealt systematically with before; quite apart from the fact that we have thus come to learn much about how the body and its functions were viewed by medical specialists in ancient India,[1667] the matters discussed are, more specifically, of fundamental bearing for an understanding of the premises underlying the notions having to do with the main subject under consideration, even though they are themselves not primarily concerned with this subject. Though all this has made the whole process extremely laborious, such matters have not been mentioned in the précis given above.

14.9. I must here, however, again draw attention to the fact that the findings given above are based only on an analysis of a circumscribed body of texts which, different though they may be when compared with each other, nevertheless share common basic characteristics and are not too dissimilar. In general, I venture no statements as to theories in texts of a different nature, though I have sometimes made a few stray remarks; such texts might, for all I know, presuppose quite different theories with regard to some of the subjects discussed above. Nor do I in general venture to remark on the prehistory or later developments of the notions set forth, apart, once again, from the occasional stray remark.[1668] It should also be pointed out that much of what has been said above is based not on absolute certainty, but on a weighing of different probabilities, and thus leaves room for doubt and disagreement. We must also remember that the information was won from different works which, though similar in many respects, are neither in all respects similar to each other, nor even necessarily always homogeneous in themselves. Thus I might, in spite of having tried to avoid this pitfall, inadvertently have committed the same blunders of method I remarked on in §§1.7f.

14.10. Of the many deductions the above study has led to, one deserves special highlighting, namely that the evolution of a human being is here linked with sexual pleasure not only of the male, but of the female as well, since conception is regarded as being possible only if both the man and the woman have an orgasm. Now begetting progeny (especially, of course, male progeny) is one of the most fundamental requirements of *dharma*-, and

---

[1667] This obviously has implications for religious studies on Indian religion in general too; cf. in this context also SULLIVAN.

[1668] I may, however, be permitted to point to SLAJE 1995, a study which examines Vedic notions on procreation and comes to in many respects very similar conclusions as regards the fluid of the female involved. The conclusions of this study also show, incidentally, that MÜLLER 1960C, p.720 is wrong in maintaining that the notion of a procreatory substance of the female is relatively late ("In solchen Auffassungen zeigt sich, dass ein weiblicher Fortpflanzungsstoff erst nachträglich oder anhangsweise beachtet wurde").

we know that the texts dealing with *dharma-* practically require the man to try to obtain progeny as often as this is possible and allowable. The notion described above, however, enjoins upon such a man intent on obtaining progeny the duty of seeing to it that the physical and emotive needs of his wife are satisfied to such a degree that she obtains an orgasm during intercourse,[1669] and it would be highly interesting to ruminate upon what sort of a relationship between the sexes results from this, at least in theory.[1670] I am of course aware that we do not know how wide-spread the theories of our texts were in actual fact,[1671] as also to what degree they were actually regarded in practice, given the fact that the sordid realities of life were most probably often not at all in keeping with them,[1672] and I also realise that we as yet have no clear idea about whether and in what eventual form such theories might survive in modern India; but even so it does seem to me that here we have an aspect of Indian culture which we cannot afford to neglect, especially in the face of the distortions of many a fashionable, radical feministic study.

14.11. There are also some interesting questions which the notions discussed above lead to. Thus one may ask whether unwanted pregnancies were considered to be avoidable if the woman (and not necessarily the man) had no orgasm, and, if this notion indeed prevailed, what social significance it had. Even more important is the matter of pregnancy as the result of rape, for if conception was considered to be possible only as a result of an orgasm of the woman, a pregnancy as a result of rape would imply that she was a willing party to the act.[1673] Unfortunately, I could find nothing in Dharmaśāstra texts which could serve as an indication that this problem was con-

---

[1669] One could even go so far as to say that according to this *dharma-* is impossible without *kāma-* (cf. too n.1261). Though this is probably a bit too strongly put, one must admit that it is not entirely without justification.

[1670] It may be interesting to note that NĀG 12,2 warns males that a woman not satisfied during sexual intercourse might even kill her partner, as is said to have happened in Karṇāṭa:

nārī nīcaratodvignā svāminaṃ nāśayaty api
śrūyate caiva karṇāṭe kāntayā nihataḥ patiḥ.

[1671] In this connection it is intriguing that in the discussion on the relative positions of *dharma-*, *artha-* and *kāma-* with regard to each other in MBH 12,161, the arguments in favour of the absolute superiority of *kāma-* over the other two do not list the need to obtain progeny at all (12,161,27ff.; cf. too WILHELM 1966, p.301).

[1672] On this cf., for modern South Asia, e.g. KAKAR, pp.21ff.;68ff.

[1673] Cf. in this connection DEAN-JONES 1992, p.90,n.57, who draws attention to the contention of the Roman physician Soranus that "if a woman becomes pregnant from a rape she must have unconsciously enjoyed it", as well as DEAN-JONES 1994, pp.158-160.

sidered by them, but then I must also admit that my knowlege of this genre is rather meagre.[1674]

14.12. Finally, I may be permitted to remark that the ancient Indian notions unearthed and discussed above are not as strange in comparison with Western views on the subject as it may seem to us today. Especially LAQUEUR 1990[1675] shows us that many of these ideas, including that of the necessity of female orgasm for conception, were prevalent in the West until the eighteenth century, when the notion of the inherent inability of women to feel pleasure during sexual intercourse gained the upper hand.[1676]

1674 Another subject I can say nothing much about is the question of whether a child can be born without the agency of a male. In §8.6 attention was drawn to statements saying that this is possible, whereas according to §9.8 it is not possible. This is, however, all I can say on the subject. Attention may, though, be drawn to WINDISCH, p.24, who quotes a passage from the *Milindapañha* to the effect that excitement of a woman alone, without any intercourse, can bring about conception. Whether the later notions on the Buddha having been born without the agency of his father (cf. on these e.g. WINDISCH, chapter 9, i.e. pp.153-173, also pp.144;194) are in any way connected with such a theory is beyond my knowledge. (I may, however, point out that op.cit. does not comment upon the Buddha's birth through his mother's side being neither only Buddhistic — Indra too is born in this way — nor only Indian; cf. e.g. OLDENBERG, p.132,n.3.) Of course the mode of conception without the agency of a father mentioned above is different from the mode of conception by ingesting semen through the mouth (cf. §6.20, also n.1648), for, though the father does not indulge in sexual intercourse in this case, his fluid is nevertheless involved in the process. (On "supposed cases of conception without sexual connection mentioned in western writers" cf. RHYS DAVIDS, p.306.)

1675 See too LAQUEUR 1989.

1676 Unforgettable in this context is the advice given, in the middle of the eighteenth century, to the Austrian empress Maria Theresia by her royal physician when he was consulted on her seeming inability to have children (cited e.g. by LAQUEUR 1990, p.150): "Ceterum censeo vulvam Sanctissimae Maiestatis ante coitum esse titillandam" (LAQUEUR 1990: "titillandum"). Another version of the physician's advice, kindly conveyed to me by Klaus Mylius, runs: "Praeterea censeo vulvam Sacratissimae Maiestatis ante coitum diutius esse titillandam." Maria Theresia eventually had sixteen children.

APPENDIX I

Parallels in Greek and Yūnānī Medicine

The theory or theories to be found in Indian medicine and sexology discussed above seem to have many parallels in Greek medicine, ancient as well as in the form transmitted in Arabic.[1677] A comprehensive study of the Ancient Greek theories on reproduction has been made by LESKY, to whose monograph I draw attention; additional material based on more recent research is to be found in WEISSER 1983 (see also NICKEL's recent study of Galen's embryology), and, most recently, in DEAN-JONES 1994, which latter,[1678] though primarily concerned with theories on the female body and not reproduction as such, nevertheless has a wealth of very relevant material. LAQUEUR 1990 follows such theories right down to modern Europe, but unfortunately seems to have had no knowledge of LESKY and WEISSER 1983 (which has, inter alia, led to his paying hardly any attention to the Arabic tradition and its role in transmitting Greek theories to mediaeval Europe). LESKY and WEISSER have, however, been utilised by STIFTER; this study is concerned with the history of the knowledge on female ejaculation, and as such touches the subject of reproduction only quite peripherally (as is also the case as regards ROUSSELLE).

There were, as in India, several concurrent theories in Greek medicine, especially in its more ancient layers. According to one, semen (one gains the impression that, as a rule, only male semen is considered) originates from the brain and the spinal cord. This is an ancient theory also found all over the Middle East, in China — and also in India, especially in Tantric works, though generally[1679] *not* in medical literature (cf. also DAS,R.P. 1992A, pp.390f.). According to other Greek theories, semen is produced and collected from all over the body, either from bones, flesh and muscles, or, in another theory, from the four body fluids.[1680] It is obvious that in such a theory the testicles are more or less a transit station for semen (cf. §§6.12ff.), which, it may be mentioned, is ejaculated due to the heat of friction during

---

[1677] Cf. in this regard also n.5.

[1678] Which is incidentally an iconoclastic work taking a decided stance against many pet feminist 'readings' of ancient Greek authors.

[1679] I use the word "generally" because the possibility of this theory having nevertheless influenced some medical texts cannot be totally discounted; cf. e.g. n.1632.

[1680] On these two theories cf. also BOS, p.230.

intercourse. There is also the theory that both males and females have "semen",[1681] and that the combination of these produces an embryo, different sorts of combinations giving rise to males, females and hermaphrodites. Both sorts of "semen" were regarded by some as being basically of the same sort, but then again by others as being dissimilar. They are also, according to some theories, heterogeneous in composition, consisting of different components of varying effectivity as regards reproductive capability. The mingling of two fluids to produce a child is of course a parallel to Indian medical theory. But the most important point in this context is that the "semen" of females, i.e. the fluid mingling with male semen to produce a child, does not seem to have been brought into relationship with the catamenial or menstrual discharge as in India. The notions on the relationship between pleasure felt during sexual intercourse and the internal discharge of the woman are also problematic; cf. e.g. DEAN-JONES 1992, pp.82f., DEAN-JONES 1994, pp.78ff.;153ff.

In contrast to all these theories, we find Aristotle setting forth notions that, though they build in some respects upon the notions of his predecessors, essentially represent a break with them. According to Aristotle (on whose notions see also the recent detailed study by ALTHOFF, even though this unfortunately is only peripherally concerned with the topics of particular interest to us), food is, as in Indian medicine, "cooked", and from this is obtained *períttōma* "surplus", which reminds us at once of the alimentary ☞ *dhātu*-s of Indian medical theory,[1682] for here, through ever more "cooking" (*pakti*- or other forms of √*pac*) the food is refined ever more, at each stage giving a new "pure portion" (☞ *prasāda*-) corresponding to the "surplus" just mentioned. At the end we obtain, in the Aristotelian theory, blood (in which context the heart plays a major role, for which we also have Indian parallels; see n.668), and from this, as ultimately in Indian theory, semen as the ultimate product. (Whether semen is according to Aristotle actually produced in the testes — as the last step in "cooking" — is a matter of debate; cf. on these organs e.g. ALTHOFF, p.175, DEAN-JONES 1994, pp.61;187, and LAQUEUR 1990, p.32, with differing explanations on their actual function according to Aristotle).[1683] It must, however, be stressed that this theory,

---

1681 DEAN-JONES 1994, p.73,n.104 draws attention to a statement from which it seems to follow that the "semen" of women was thought to be situated all over the body, and particularly in the brain.

1682 See on both terms also ☞ *dhātu*- in Appendix II.

1683 On the great similarity between semen and brain according to Aristotle see e.g. ALTHOFF, pp.216f.

though similar in many points to the Indian theory of the chain of alimentary products, is by no means systemised on the *same* lines as this latter is. Aristotle also held that an embryo is produced by the union of male semen and female catamenial blood[1684] (not necessarily freshly secreted, according to LAQUEUR 1990, p.47, though according to ALTHOFF, p.204 sexual intercourse pulls it down into the uterus[1685]). This latter fluid, which enters the uterus through many small passages (cf. §14.4),[1686] is basically blood not fully "cooked", as the female, due to the heat in her body being less than that of the man,[1687] does not have the power of "cooking" blood to such an extent as to produce semen, which latter, as the "effective principle", basically determines the nature of the embryo (see on this also LAQUEUR 1990, pp.30;41f., NICKEL, pp.29ff., ALTHOFF, pp.189ff.)[1688], whereas female catamenial blood serves merely as the "matter" to build this up;[1689] the fluids of both male and female are not homogeneous by composition (cf. e.g. ALTHOFF, p.175). Most Greek theories held menstruation to be a cleansing of the body

---

[1684] Note also DEAN-JONES 1994, p.53: "He argues that no animal can produce two seminal residues, and since menses obviously parallel semen, they have to be a woman's seminal residue."

[1685] Loc.cit.: "Es heißt dort, daß bei Frauen der Geschlechtsverkehr die Katamenien herabziehe". However, statements on the connection between intercourse and menstruation also lend themselves to different interpretations, on which cf. e.g. DEAN-JONES 1994, pp.47ff.

[1686] On several openings into the uterus see ALTHOFF, p.201; on these openings especially in connection with theories of Galen see NICKEL, pp.56ff.

[1687] On the warmth necessary for the embryo within the mother's body to develop see ALTHOFF, pp.273f.;277.

[1688] In these passages the role of the pneuma within the seminal fluid is referred to. This pneuma is according to LAQUEUR 1990 breath, according to ALTHOFF (p.285) air ("Luft"). On the role of pneuma in general according to Aristotle and the Stoics see especially ALTHOFF, pp.283ff.; on pneuma in the theories of Galen see NICKEL, pp.34f. (also p.113). See also BOS, p.225, DEAN-JONES 1994, pp.60ff.;87;184ff.;216.

[1689] According to KEULS, written with the avowed aim of exposing the "phallocracy" of Athenian and also Western civilisation, "Aristotle was one of the fiercest misogynists of all times, obsessed with the need to prove that women play no genetic part in reproduction" (p.405). Several pre-Aristotelian authorities too held that women did not contribute seed or the like towards the formation of a child; cf. DEAN-JONES 1994, pp.149ff. This matter has also been expressed in the well-known simile of the field and the seed, for which we also have plentiful evidence from ancient India (cf. e.g. DESHPANDE, pp.225ff.), though not, to my knowledge, from medical texts, which on the contrary presuppose that both father and mother contribute something to the formation of the child, and that certain organs of the child are formed from the contribution of the mother, others from that of the father (see e.g. §5.11).

more or less in the form of a natural blood-letting;[1690] however, the Aristotelian notion of menstrual blood uniting with semen in conception is of course problematic in this context,[1691] and it is thus no wonder that his theory on the origin of menstrual blood is different.[1692] It also seems as if Aristotle held menstruation to be due to a steady accumulation of fluid in the uterus, whereas older theories held that the fluid was discharged into the uterus from the surrounding tissue only once a month and all at once (cf. e.g. DEAN-JONES 1994, pp.62ff.); here we have a parallel to the problems in Indian medicine referred to in §14.4. Noteworthy is also LESKY, p.24(1248), where Aristotle is quoted as saying that the "semen" of women (obviously a different fluid) is not capable of fertilisation, as it is simply a secretion of the uterus which some women have, while others do not (cf. also LAQUEUR 1990, p.255,n.36). This is a very striking parallel to some of the theories we have seen above in our study of Indian medical and sexological texts. I must, however, also point out that according to ALTHOFF, pp.183f.;203, Aristotle refers only to a vaginal secretion, i.e. (seemingly) a lubricating fluid (cf. on this also NICKEL, p.30).

The theories of Aristotle were influential in later times, but they never fully superseded the other theories, and were also not accepted by all. It may be noted that Greek medical theories were transmitted to mediaeval Europe (to a large extent via the medium of Arabic), and that medical scholars in Europe debated their correctness or incorrectness and applicability or non-applicability in practice as late as the eighteenth century A.D. (cf. also §14.12). It is also often held that Indian medicine too has been heavily influenced by Greek medicine. Whether or not this assumption is valid in other medical fields I leave open to debate, but the brief outline given above shows that there can hardly have been any really close contact between the systems as regards their notions in conception, though there clearly are similarities.[1693] One could, however, perhaps postulate an exception in the case

---

[1690] There were also various theories to explain why women had an excess of blood (blood in general) in their body and why it was necessary for them to get rid of this periodically.

[1691] On Galen's attempts to explain this process more satisfactorily see NICKEL, pp.32ff. (particularly 37ff.). Galen incidentally also held that women have "semen" too, and that this also plays a role in conception (on Galen's notions on female "semen" see especially NICKEL, pp.40ff.).

[1692] On yet another theory, in which the menstrual fluid is described as the "foam" of the humours in the body, see DEAN-JONES 1994, p.59.

[1693] Note in this respect also the musings on the different components of the procreatory fluids maybe being meant in an Indian theory (see §14.2, with further references), or on the

of the Aristotelian notions, for which one could consider building a case resting on closer contact with Indian medical theory. In this context one could of course argue that Aristotle's theories were developed newly by him and then transmitted to India and there developed further, but this does seem improbable, for the Indian system as we have analysed it is so well developed that it must have taken a long time to reach the stage we find in the 'classical' medical works, and this would seem to make it more likely that the borrowing, if there was indeed any, was the other way, i.e. that Aristotle, who after all breaks with many aspects of the older Greek theories, has borrowed from Indian sources, albeit quite sketchily or else from a theory or theories which is or are either (a) variant(s) or else (a) not yet fully developed forerunner(s) of the theory we have deduced above. If we leave aside the problem of postulating the existence of any sort of borrowing in the first place,[1694] then, if borrowing there really was, one could build a better case for the latter sort of borrowing. On the other hand, individual aspects of the Aristotelian notions could be used to build a case to prove the exact opposite, for instance with regard to the derivation of the catamenial fluid from blood, which one could take to have ultimately influenced Indian theory so that the procreatory-menstrual fluid is in later times derived from blood and no more from ☞ *(āhāra)rasa-* (cf. §14.3). But it must be admitted that the similarities between Aristotelian and Indian notions, though they are undoubtedly present, are on the whole so vague as to make borrowing only a very distant possibility. As it is, do similarities necessarily point to borrowing of one sort or the other, the postulation of which seems to be rather a fashion in some scholarly circles? For this leaves out of consideration all other possible developments, e.g. coincidence, or borrowing from a third, common source, which latter would then in most probability be Ancient Middle Eastern,[1695] especially since it now seems that not only part of Greek, but maybe also of

---

function of the testicles, which in Indian medical theory too seem to serve as mere transit stations for semen.

[1694] My lack of knowledge of Greek medicine does not allow me to make definite statements. Cf. in this connection, however, also the views of COMBA 1985, p.924: "... there is therefore some parallelism between the old Greek and Indian physiological doctrines; disputes arise about the same questions, similar answers are given to the problems. But the epistemological formulation of Aristotle gives to his statements a very different trend as regards the Hippocratic doctrinale *corpus* and as regards the Indian medicine. The Aristotelian system ended by devaluing the medicine as a τέχνη, subordinating it to the science of nature and excluding it from the encyclopaedia of sciences."

[1695] Cf. also VOGEL 1956, p.780 and 1959, p.33 on this subject and the problems connected with it.

Indian scientific knowledge was taken over from the Ancient Middle East and then developed further.[1696] Moreover, many doctrines of both Greek and Indian medicine seem to go back to a common Indo-European stock, so that the question of borrowing does not arise in such cases; this assumes even more importance in connection with the fact that we know that there was a large floating stock of medical lore in ancient Greece which was not incorporated in the pre-Aristotelian works, lore of which some might go back to Indo-European times and as such be a common heritage of both Greece and India.[1697] Then too, various notions common to both Greece and India seem in actual fact to be pan-Asian (or even spread further), and it seems a hopeless endeavour trying to determine where they originated. In any case, we are here concerned only with the theory or theories on reproduction and not other parts of medical theory. For example, ZYSK 1986 argues strongly that the use of dissection for the study of anatomy was introduced into India by the Greeks (see especially ZYSK 1986, pp.695f., though the remarks made there are tempered subsequently on pp.696f.),[1698] and in other fields of medicine too there may have been different sorts of cross-connections.[1699] This whole subject is, however, much too complex to deal with in a short, cursory manner, as is unfortunately often done, especially by many modern Indian scholars postulating Indian influence on Greek, though without having more than an extremely superficial knowledge of Greek ideas and theories, and usually no knowledge at all of the Greek language and its

---

1696 See especially FILLIOZAT,J. 1974, pp.320f. Cf. also BLACK, pp.96f., and see too KARTTUNEN, pp.112;118. I may also draw attention to FIŠER/FIŠEROVA, p.311.

1697 VOGEL 1995, p.85 holds, however, that the similarities between ancient Indian and ancient Greek notions are so great that there can be no other deduction than that the Indian notions go back to Greek notions which came to India, together with notions on astronomy and astrology, at and after the time of Alexander the Great, this being shown especially by the fact that all the relevant Indian texts are, at least in their present form, younger than the beginning of the Christian era.

1698 It is unfortunate that FIŠER/FIŠEROVA has not been utilised by ZYSK 1986, as it contains much that supplements the latter, and in some instances is probably also more correct, as e.g. in the case of the translation of *samyakprakuthita-* (SU,Śā 5,49) as "properly decomposed" (p.320) or "decomposed in the right way" (p.315) (ZYSK 1986, p.692: "completely putrid"; similarly DAS,R.P. 1983-1984 on the parallel passage AS,Sū 34,p.248a: "fully putrefied"). (On dissection in ancient India see also WEZLER 1992A, p.304,n.38 and CHATTOPADHYAYA 1991, pp. 209-214.)

1699 Cf. on this now ZIMMERMANN 1989A, pp.177ff., see also pp.70f. Attention may also be drawn to Ludwig Edelstein's remarks in ZIMMER,HEN.R., pp.xlviiiff., though on the subject of blood as a morbific entity see n.179 above.

literature.[1700] Moreover, as VOGEL 1995, pp.86f. has rightly pointed out, the similarities with Indian medicine are, on the Greek side, not so much with Hippocratic medicine as with the Hellenic philosophers in their statements on medicine.

Later developments of the Greek theories mentioned, mostly in the works of Galen and in Arabic works,[1701] are discussed in WEISSER 1983. In these later texts, especially those written in Arabic, we find detailed and intricate discussions on the nature of female "semen" as mentioned in older Greek works. It seems that this "semen" was often thought to be an ejaculate that females discharge during intercourse and which can be perceived.[1702] The problem of this is compounded by the post-Aristotelian discovery of the ovaries and the Fallopian tubes by Herophilos, from which time onward most medical authorities assumed that these organs were the equivalents in the female of the male testes and seminal ducts respectively,[1703] especially since Herophilos might have held the ovaries to be connected with the bladder (see below), which means ultimately that the secretion of the ovaries could not be held to play a role in the formation of the embryo. Nevertheless, the notion that females have "semen" was hardly doubted after this, but the notions on the presence of this fluid in the body are not of the same sort as in the Indian theory, if we regard the theory of the breaking down of food to ultimately produce semen in its applicability to both males and females. The discussion encompassed such problems as the colour and the nature of female "semen" (also in comparison with male semen), whether it played a part in procreation or not, whether, if it did, it did so — on the female side — alone or in conjunction with the menstrual fluid, or whether only the latter

[1700] On the subject of such postulated "borrowings" from India by Greeks see now KARTTUNEN, pp.103ff.; on Aristotle and Indian influence see p.117. See, however, also RAU 1990, who points out that we must, in our deliberations, also take into account material in Greek literature of the classical age not expressly marked as "Indian". (I may add that NARAIN levels quite some criticism at KARTTUNEN, but this has to do with different matters and is thus hardly of relevance in the given context.)

[1701] Which influenced European medicine up to the eighteenth century.

[1702] According to ROUSSELLE, p.27, the Hippocratic collection holds that: "Women were sure that they ejaculated and they told men so."

[1703] In 1667 Niels Stensen discovered the follicles in the ovaries of oviparous animals, postulating that the alleged testicles of women are organs fully analogous to the ovaries of such animals. This incorrect equation of follicle and ovum also characterised the first detailed description of the follicles of mammals by Regnier de Graaf in 1672. It was only in 1827 that this mistake was corrected through the research of Karl Ernst von Baers. (Data taken from NICKEL, p.49.)

fluid played a role in procreation. The nature of the role played, whether as a factor in conception or whether as a means for the growth of the embryo, was also a subject of debate.[1704] Many parts of this discussion have, as we saw above, parallels in India.

An interesting point of contention, for which I have, however, found no Indian parallel, was also whether the female procreatory fluid was discharged into the *vulva* and subsequently drawn from this into the uterus (again), or not.[1705] The nature of female "semen" with respect to various discharges of the female, especially during coitus, was also discussed; this also included the problem of whether it and the procreatory fluid of the female are one or not. In this connection it is interesting to note that Herophilos may have assumed that the Fallopian tubes open out into the urinary tract (see WEISSER 1983, p.93, and LAQUEUR 1990, p.5, the latter having no doubts on this score),[1706] which brings to mind our discussion on the discharge due to stimulation of the Gräfenberg spot; unfortunately, WEISSER 1983, maybe unaware of this aspect of the matter, has not discussed the problem in this light. Some manuscripts of Ibn Sīnā's *Ḥayawān* actually hold that a woman discharges her "semen" to the exterior through the urinary opening,[1707] which would be in keeping with what we know of the Gräfenberg spot.[1708] Other manuscripts, though, speak of a discharge through an opening beneath the urinary opening,[1709] i.e. through the vagina. Though this latter reading does look like an attempt to smoothe over a not well understood *lectio difficilior*, that does not automatically prove the former reading original; we thus cannot be sure that a urethral ejaculation is indeed referred to here.

There was also a lot of discussion on the factors responsible for the sex of the embryo. However, something approaching the Indian theory of the

---

1704 A very good discussion of the problems connected with the notion of female "semen" is found, with regard to Galen, in NICKEL, pp.40ff. (p.42 discusses the problem of female "pollution", i.e. the parallel to the "wet dreams" of males).

1705 This problem would of course crop up particularly if the female should be taken to ejaculate her fluid from somewhere not within the uterus proper (e.g. the portio vaginalis of the cervix; cf. §14.6). I lack the specialised knowledge needed to remark upon this with regard to Greek (and/or Yūnānī) medicine; all I can do is draw attention to the remarks in DEAN-JONES 1994, pp.156-160. See also n.1432 above.

1706 Cf. on this also the statements of Galen quoted by GRENSEMANN, pp.41f.

1707 *taqḏifu zarᶜa-hā ilā ḫāriǧi ᶜinda ṯaqbi 'l-baul* (p.389).

1708 WEISSER 1983, p.149 has not understood this passage.

1709 *... ilā ḫāriǧi ᶜinda ṯaqbin taḥta ṯaqbi 'l-baul.*

relative quantities of the procreatory fluids combining does not seem to have been considered, maybe because of the notion of an inherent dissimilarity of powers of the male and female procreatory fluids (cf., though, n.1651). There are some stray references to the effect of the quantity of a fluid, but only in very specialised contexts (see e.g. WEISSER 1983, pp.271f.;292-298; cf. also DEAN-JONES 1994, p.168,n.67); a similarity with Indian notions is as a rule not evident, except maybe in one case (WEISSER 1983, p.273). Neither does the sex of the child seem to have been connected with the days on which intercourse occurs, except in an obscure theory Ibn Sīnā refers to once (WEISSER 1983, p.289), which, however, is probably of a sort different from the Indian theory of a male being conceived on even, a female on uneven days. This latter theory is actually referred to by the physician ar-Rāzī, who, however, expressly labels it "Indian" (see WEISSER 1983, p.290, where this information is unfortunately not followed up to its Indian origin). There was, however, a current and ancient theory linking the sex of the child with one side of the body: right for boys, left for girls.[1710] This applied also to the testes and their female counterparts, i.e. the ovaries as suppliers of "semen". Note in this connection the theory in India pertaining to the sex of the child depending on which of the ducts opening into the uterus discharges. The connection of left with females, right with males is of course known in many different contexts also in India, especially from mantic literature (and also from mantic passages in medical literature, which I have not deemed it necessary to mention, as they do not fall within the purview of our study); cf. too §13.32. I may also draw attention to the theory linking the sex of the child to the right or left chamber of the uterus; a similar theory can of course not be expected in an Indian context, since there is no evidence of a theory in Indian medical texts of the bipartition of the uterus, the theory in Greek medicine maybe being due to the transference to humans of observations made while dissecting certain animals (cf. WEISSER 1983, pp.95f.), though other explanations have also been offered (cf. e.g. LAQUEUR 1990, p.253,n.20).[1711] Similar transferences due to observations of animals could have taken place in India also, but if they have, then probably (at least on the basis of the evidence at our disposal) in contexts not having to do with the reproductive system; cf. e.g. BODEWITZ 1992, pp.28f. on the dual of *kukṣí-*.[1712] We actu-

---

[1710] See also n.1531. For an interesting exception see WEISSER 1983, p.278.

[1711] The very ancient Greek notion that the uterus has a sort of life of its own and can wander within the female body (this theory was soon discarded; on it see also LAQUEUR 1990, p.251,n.3) also does not seem to have any parallels in India, even though it has parallels in other parts of the world, including Germany (cf. e.g. PLOSS/BARTELS/BARTELS, pp.424ff.).

[1712] This sort of transference is of a sort different from that found, for instance, in

ally do have what could be taken to be evidence for such transference, but this pertains to modern South Asia, and, though we might surmise that it represents an old (though probably non-medical) tradition, we cannot be sure on this count.[1713] Nevertheless, the fact that BODEWITZ 1992 presents evidence, which cannot simply be dismissed out of hand, that could be taken to show that in ancient India too certain peculiarities of animal anatomy were falsely taken to be valid for humans also, does bring to mind the remarks of NICKEL, pp.18ff.: here it is argued that for the ancient Greeks it often made no difference which animal (scil. mammal) was taken as the basis for examinations, since it was assumed that the (internal) anatomy of all mammals (including humans) was the same.[1714] One may also speculate that observations such as the rare uterus duplex in humans might have facilitated such deductions. In all such considerations, we must not, however, forget one important factor: in Greece there seems to have been a gradual evolution in the process of dissection. It seems to have been more or less absent, or at least not systematic, among the Hippocratic physicians,[1715] practised extensively with regard to animals by Aristotle, and systematised, also with regard to humans, in the Hellenistic period among the Alexandrians. This should caution us against speaking simply of "Greek medicine" in comparisons with Indian medicine, since neglecting such evolutionary factors in comparisons is bound to give us skewed results.

In the works discussed by WEISSER 1983 we also find discussions on whether female orgasm is a prerequisite for conception; Aristotle opined that it was not, though it could be helpful. The problem is of course linked with that of female "semen" and its ejaculation, whether this is capable of produc-

---

jaghana- "hind part/end, buttock" being used to denote the genitals of the human female, due to the fact that "the hind end (jaghana) of an animal contains its generative organs" (JAMISON 1996, p.41), as this is no case of observations on animal organs made through dissection being taken to be also representative of human organs not similarly scrutinised or scrutinisable.

1713 CARSTAIRS, p.83f.: "Hari Lal said: 'There is a fire in your stomach, whose heat warms the body, and everything you eat goes into two stomachs, solid food into one and liquids into another, and there they are cooked until they yield *ras*, and this in turn gives blood, and blood in time forms semen.' Rajmal described four stomachs, in series, in which food is progressively digested until blood, and then semen, is formed."

1714 The author is also of the opinion that much anatomical knowledge, especially knowledge on embryology, was obtained by the Greeks by vivisection, particularly of pregnant animals. According to DEAN-JONES 1994, p.22, the Alexandrians might even have vivisected humans. In any case, part of such dissection (and vivisection) seems to have been public, cf. on this e.g. STADEN. See also FIŠER/FIŠEROVA, pp.310ff.

1715 On this cf. also DEAN-JONES 1994, p.8.

ing a child or not, and its possible links with or differences from the procreatory fluid. Many authorities seem to have seen female orgasm primarily in its role of helping to draw male semen (and possibly also female semen; see above) into the uterus. Others seem to have held, however, that conception does require female orgasm, and some even held that conception takes place only if both the male and the female have their orgasm simultaneously.[1716] On the other hand, there was also the theory that the fluid of the female previously present in the uterus (for instance as the result of a previous orgasm) could also combine afterwards with semen subsequently entering the uterus. I do not know if parallels to this idea exist in India; the problem is interesting especially in connection with cases in which a woman conceived in spite of denying having felt any pleasure (e.g. in the case of rape; cf. §14.11). It may also be mentioned that in Yūnānī medical theory (just as in some Indian theories; cf.§5.7, also §11.28) conception was thought to be possible during the period of menstruation too; one theory even took this to be a factor in the development of the sex of the child (see e.g. WEISSER 1983, p.290). On the other hand, intercourse during menstruation was as a rule frowned upon, not only, but also due to religious strictures.

WEISSER 1983, pp.96f. mentions a most interesting theory of aṭ-Ṭabarī, according to whom the uterus has three openings for the ingress of semen. These openings are identified loc.cit. as those of the Fallopian tubes (for female "semen") and the orifice between uterus and vagina (for male semen). However, the Arabic text cited says that the latter opening is in "its (i.e. the uterus's) end/rear part" (*fī muʾaḫḫari-hā*); if this latter is in fact the end opposite to the vagina, then we would have a parallel to the Indian theory of three ducts. On the other hand, the end on which the vagina is situated may also be meant; moreover, *muʾaḫḫar* also means "tail", and the form of the uterus might possibly lead one to compare the part opening into the vagina with a tail.

The matter is, however, much more difficult. What WEISSER 1983 refers to is part of a passage of which parts are also discussed on p.574 below and shown to be quite problematic. The portion of relevance here says (FH, p.31,line 24 — p.32,line 2):

[1716] Similar discussions are not unknown in our times, especially in certain sub-cultures, but also (though then of course with a different focus) among serious researchers. However, such discussions remain controversial. It is nevertheless interesting to note that GHĀṆEKAR 1986, pp.64f. has drawn express attention to such discussions in his attempts to show that ancient Indian medical authorities already knew everything modern Western medicine does.

> *wa lahā ^{c}aṣabatāni nābitatāni min dāẖili*[1717] *tus(am)mayā-ni*;[1718] *qarnay-i 'r-raḥim*[1719]. *wa bihimā yağḏibu 'r-raḥimu 'z-zar^{c}a ilā dāẖil. wa fīhā ṯalāṯatu*[1720] *aw^{c}iyah, wāḥidun ^{c}alā ya-mīni-hā wa āẖaru ^{c}an yasāri-hā wa wāḥidun fī mu$^{\supset}$aẖẖari-hā. fa-$^{\supset}$iḏā waqa^{c}a 'z-zar^{c}u fī-hā inḍammat ^{c}alayhi. ṯumma yataġayya-ru 'z-zar^{c}. ...*
>
> "And it (the uterus) has two sinews/ligaments growing from within; they are called the horns of the uterus. And through them the uterus draws the seed inside. And in it (the uterus) there are three vessels, one to its right side, and another on its left side, and one at its end/rear part. Then when the seed falls into it, it (the uterus) encompasses it. Thereupon the seed is altered. ..."

SIGGEL 1941-1942, p.229 translates *wi^{c}ā$^{\supset}$* (plural: *aw^{c}iya*) as "Ausbuchtung", i.e. "bulge",[1721] but I doubt very much whether it can have this meaning. On the other hand, "vessel" is a very common meaning. Now we could also think of "receptacle" or the like, but then we would also have to ask what this could signify here, the more so since the word is in fact used to denote structures such as blood vessels or the like. So, though we cannot be fully sure on this count, it does seem as if three vessels of a similar sort were mentioned here, and the deduction from this would be not only that these vessels must have openings into the uterus, but that there must also be substances flowing through them. This would be in keeping with the statements of WEISSER 1983, in which context attention may also be drawn to NICKEL, p.45 (see also pp.41f.), where we are told of the theory of Galen (who builds on older theories) that the "testicles" of the female (i.e. her ovaries; see above) are connected via seminal vessels with the two horns of the uterus, into which they ejaculate semen during coitus. But it should be noted that the Arabic text does not tell us what flows through the vessels; instead, it says that the uterus draws "seed" into it through its two "horns".[1722] This means that not only

---

[1717] Printed *dāẖilin*.

[1718] Printed *yus°*.

[1719] I am using the transcription *raḥim* throughout, though of course one could also transcribe *riḥm*.

[1720] *ā* is missing in the printed text.

[1721] See also p.229,n.7 on his deductions from this.

[1722] The identification of the "horns" is no easy matter, since in Greek medicine different structures, actual or supposed (including the compartments of the uterus referred to above on

are we left in the dark as to what flows through the vessels, but we also run into difficulties in relating this to the procreatory fluid drawn in through the "horns". Moreover, we are not told what the "seed" drawn in is, i.e. whether it comes from the male or from the female. All of which goes to show that, quite apart from the problem of *muʾaḫḫar*, we have several other difficulties to cope with, so that it does not seem possible to compare this passage with Indian notions without further study.

But be that as it may, in any case the material from Greek and Arabic sources does not seem to indicate that the menstrual fluid was somehow connected with the ducts serving for the ingress of female "semen".

Finally, I may add that milk and the menstrual fluid were widely regarded as being identical or very closely connected, which is a remarkable parallel to Indian notions,[1723] even as regards the theory that breasts and uterus were somehow connected, so that during pregnancy part of the menstrual fluid could be transported to the breasts.[1724] A connection between uterus and breasts, uterine blood and milk is also postulated by WEISSER 1980, pp.149f.;153 (see also p.232).

---

p.506), might have been referred to by the term (cf. e.g. DEAN-JONES 1994, pp.66f.;69).

[1723] There were also various theories on the nourishment of the embryo or foetus through the menstrual fluid within the womb; see on these especially DEAN-JONES 1994, pp.200ff.

[1724] See especially DEAN-JONES 1994, pp.215ff., WEISSER 1983, pp.256-259, also p.275. On p.450,n.23 of the latter attention is drawn to ancient Egyptian parallels. Cf. too LAQUEUR 1990, pp.35f.

APPENDIX II

Discussion of Selected Technical Terms

Only terms whose detailed discussion in the main body of the text would have caused complications are included here. Other terms may be looked up with the aid of the indexes.

agni-

Fire. In medical texts often used technically to refer to the digestive fire(s). For relevant passages and further relevant literature on such "fires" (and the role ☞ *pitta-* "bile" plays with regard to them) see MEULENBELD 1974, p.437 (add BHĀ 1,3,168-173, ĀY 4,29-35) and MEULENBELD 1992, pp.2f.; see also EIM, pp.19-23, SHARMA,P.V. 1982, p.14, ŚĀSTRĪ, CONCEPT, pp.1ff.; 15ff., SHARMA,S.N., pp.1-22, DASH 1971 (especially pp.23ff.) (particularly the latter three works striving to bring the concepts of the texts in line with modern notions[1725]). The fact that several digestive fires are mentioned in the medical texts gives rise to problems regarding the actual mode of digestion; see on this the discussion in §7.18, with further references. As to the relationship between the digestive "fire" and bile (☞ *pitta-*), SU,Sū 21,9 expressly discusses the problem of whether or not they are identical. Cf. too AH,Śā 3,90 (on which see also n.473): *pittaṃ vahnir vahnijaṃ vā*. Ḍalhaṇa in his commentary on SU,Sū 35,27 seems to take bile and "fire" to be separate,[1726] AS,Sū 20,p.147b, however, not.[1727] Commenting on AH,Sū 12,10f., Aruṇadatta too equates bile and the digestive fire;[1728] see also the commentaries of Aruṇadatta, Śrīdāsa and Indu on AH,Sū 11,34f. Cakrapāṇidatta in his commentary on CA,Ci 15,217-235 calls "fire" the heat of bile (and not bile itself);[1729] cf. too his commentary on CA,Ni 1,20). According to the explanation of the text itself on ĀĀS 740f. (p.125), bile has the qualities (☞ *guṇa-*) of fire,[1730] which is not the same as saying that the two are identical. AS,Śā 6,p.332b says that some hold the form of bile called *pācaka-* to be identical with fire, while according to others fire is the internal heat of the ☞ *doṣa*-s,

---

[1725] On DASH 1971 see also p.585 below.

[1726] *agniḥ pittāśritaḥ.*

[1727] *labdhāgniśabdaṃ pittam annaṃ pacati sārakiṭṭau vibhajati.* This is at variance with the deduction of MEULENBELD 1992, pp.2f. regarding the separateness of bile and the digestive "fire" in another part of AS.

[1728] *agniśabdavācyaṃ yat pittaṃ tat pācakam* (var: *pācakam ity*) *ucyate. tac cānnaṃ pacati.*

[1729] *pittoṣmarūpasyāpi ca vahneḥ* ... .

[1730] *pittam agniguṇayuktam.*

☞ *dhātu*-s and ☞ *mala*-s together.[1731] See the discussion in BHĀ 1,3,124 too, also DASH 1971, pp.24ff. The problem is complicated by the fact that bile and fire may be equated within another referential system, in which "fire" (*agni-*, *vahni-* or other synonyms) has a different meaning (see n.473). DASH 1993, though it sets out to discuss this whole complex, has only very selective data and is tendentious in its argumentation, thus not offering a complete picture of the matter. That the whole problem is much more complicated and goes back to very ancient times is made clear by AV 18,3,5, (*ágne pittám apā́m asi* "Agni, you are the *pitta-* of the waters"), to which MÜLLER 1959/60, p.218,n.71 draws attention. See too the discussion on ☞ *grahaṇī-* below.

On the very ancient opposition, also found in medical texts, between *agni-* and *soma-*, see ☞ *āgneya-* below.

aṇḍa-

Literally "egg", from which a variety of meanings has developed, one of the most common being "testicle". Neologistically today also used for "ovum" and then projected backward into old medical texts; see on this §6.14.

The edition MĀDH has, in an Appendix (*Pariśiṣṭam*), a chapter called *Aṇḍādhāraroganidāna* "Pathology of the disease(s) of the *aṇḍa*--receptacle(s)" (pp.498f.); a footnote explains this as referring to "Disuses [sic] of fallopian tube". The colophon of the Appendix (p.508) and some laconic remarks on p.(2) of the Foreword (*Nivedan*) to the volume containing it inform us that the whole Appendix is actually a modern work by Brahmaśaṅkaramiśra. The verses on the *aṇḍādhāra*--disease seem to be verbatim borrowings from the popular BHAI (chapter 105, p.848), whose author lived probably in the latter half of the eighteenth century. This work, printed first in the late nineteenth century, has been edited and printed often, with varying contents. Intriguingly, some editions, including BHAI1, another much-used edition to which I had access, do *not* contain anything on the *aṇḍādhāra*--disease. In fact, many hold chapters 76-106 of BHAI to be not only nineteenth century additions, but also to stem from none other than Brahmaśaṅkaramiśra, the same person who wrote the appendix to MĀDH, and who redacted the text of edition BHAI!

An *aṇḍādhāra*--disease is, however, also described in ĀYV 79 (pp.549-551); the *nidāna-* portion has a wording different from that of BHAI, though the contents are similar, whereas the *cikitsā-* portion is more similar in wording, though not identical. ĀYV is a modern work with clear Western influence. It professes to be a compilation, but its sources are mostly unknown, as

---

[1731] *tatra pittaviśeṣaḥ pācakasaṃjño 'ntaragnir iti prāg abhihitaḥ. anye punar ācakṣate doṣadhātumalasannipātajanito 'ntarūṣmā yathānirdiṣṭādhiṣṭhānakarmāgnir iti.*

also the extent to which the ostensible compiler may actually have been not the compiler, but the author of the contents. Note in this context the *jarāyu*--disease mentioned in the immediate vicinity of the *aṇḍādhāra*--disease, where ☞ *jarāyu-* is clearly "uterus, womb". This meaning is to my knowledge *not* found in older medical texts, at least we have no clear evidence for this, for the only two cases where it could be taken to have this meaning are of doubtful nature;[1732] it is on the whole, in any case, quite uncommon, except in very late Sanskrit and in New Indo-Aryan, for instance in Bengali, where "uterus, womb" is the usual meaning, the other, older meanings being as a rule known only to the learned. The ostensible compiler of ĀYV was, however, a nineteenth century Bengali!

It must also be pointed out that neither BHAI nor ĀYV tell us what *aṇḍādhāra-* (also *ḍimbādhāra-*; *ḍimba-* = *aṇḍa-*) is actually supposed to mean, though according to both (the former in the *nidāna-* and *cikitsā-* sections, the latter only in the *cikitsā-* section) it is a disorder peculiar to women. The disease of the *aṇḍādhāra-* (seemingly as a disease of women) is also mentioned in RASAT 16,61; the same text (16,59;61) also refers to pain in the left *ḍimbāśaya-* "*ḍimba*--receptacle". The Sanskrit commentary explains these terms as "Ovary", and also quotes passages identical to those of ĀYV, with some minor deviations. It should be noted that RASAT was composed at about the beginning of the twentieth century.

In this connection the literal meaning of *aṇḍādhāra-*/*ḍimbādhāra-* (also *ḍimbāśaya-*) is intriguing: "egg-receptacle". Now *aṇḍa-* is, like German "Ei" and Arabic *bayḍa*,[1733] an extremely common word for "testicle",[1734] and the expression "egg-receptacle" itself seems also to be used in the meaning of "testicle", the common Sanskrit term being *aṇḍakośa-*/*aṇḍakoṣa-*, which is clearly a literal synonym of the other compounds mentioned above; GHATAGE, p.891a, even refers to its *dual* usage in the *Hastyāyurveda*.[1735] Thus this raises the very pertinent question of whether the *aṇḍādhāraroga-* mentioned above, if the notion of a disease with this or a similar name existing should be pre-modern, may not actually be a modern re-interpretation of a disease

---

[1732] *jarāyudoṣa-* in ŚĀR 1,7,182 is not explained by the text, and Āḍhamalla's commentary (cited on p.517) holds it to be a membrane covering the foetus. As regards *jarāyu-* in *jarāyupaṭala-* "*jarāyu*--membrane" = "afterbirth" in GAD,Kaumāratantra 4,37 (vol.3,p.487), this could mean either "membrane of the *jarāyu-*" or "*jarāyu-* which is a membrane".

[1733] Cf. on this term also the discussion on ☞ *yoni-* below.

[1734] Cf. similarly Lithuanian *paũtas* "egg; testicle".

[1735] BODEWITZ 1992, p.31,n.20 draws attention to the dual of uncompounded *kośa-* too; on this see below.

of the *testicles*. But whatever the "receptacle(s) of the *aṇḍa-/ḍimba-*" actually be (note in this context also the meaning "embryo" of *aṇḍa-/ḍimba-* mentioned below), and whatever the history and possible relationship between the statements of BHAI, ĀYV and RASAT may be, it should be clear from the above that *aṇḍādhāra-*, *ḍimbādhāra-* and *ḍimbāśaya-* are very modern terms (not listed, as far as I could ascertain, in any of the common dictionaries, including GHATAGE), and as such of no use to us with regard to the question of whether or not the ovaries or Fallopian tubes were known in ancient (or even mediaeval) India. But of course this will not deter those anxious to project modern notions onto ancient Indian medicine from making use of them in their argumentation.

On the other hand, SLAJE 1993, p.272,n.108 and FALK 1994, p.16 draw attention not only to the Vedic usage of *āṇḍá-*, but also to the later use of *aṇḍa-* (the latter word seemingly being a later, post-Vedic form of the former) to denote the embryo;[1736] SUNESON, p.119 also lists an *aṇḍa-* meaning "embryo" in the *Bhāgavatapurāṇa* (note also the similar meaning of the synonym *ḍimba-* listed by our dictionaries). SLAJE 1993, loc.cit. also mentions Bhāskarakaṇṭha's use of *aṇḍarasa-* in his *Cittānubodhaśāstra*, probably to denote the fluid which the embryo is at first (and not, as we might spontaneously assume from our modern point of view, to denote the fluid of the testicles, i.e. semen). These usages of *aṇḍa-* make one wonder whether *aṇḍādhāraroga-* and its synonyms might not actually denote a disorder of the uterus (or, maybe, the genital tract of the female as a whole), in which case too (as in the case of the possibility considered in the last paragraph) there is no need to refer to any meaning "ovary".[1737] But whether this explanation or the one in the last paragraph be correct or not, the fact remains that the expressions *aṇḍādhāraroga-* and the like arouse suspicions as to their authentic rooting in traditional Indian medicine.

aṇḍa- is also a synonym of ☞ *phala-* (literally "fruit"). The expression *phalakośa-/phalakoṣa-* is in use as a synonym of *aṇḍakośa-/aṇḍakoṣa-* "testicle" too (see above); *phalakośa-* even occurs in the dual in AH,Ni 11,22 = AS,Ni 11,p.57a, where *muṣka-* (dual) too is mentioned.[1738] Now with re-

---

[1736] According to FALK 1994, p.16, even more fitting than "embryo" would be "Brut", i.e. "brood".

[1737] Interestingly VŚS, p.476b lists ☞ *garbhāśaya-* as a synonym of *ḍimba-*. Other meanings (☞ *phupphusa-*, *plīhan-* etc.) given by dictionaries are irrelevant for the present discussion.

[1738] *muṣkau vaṅkṣaṇataḥ prāpya phalakośābhivāhinīḥ*
prapīḍya dhamanīr vṛddhiṃ karoti phalakośayoḥ.

Aruṇadatta and Śrīdāsa explain *phalakośa-* as *vṛṣaṇa-* "testicle", Indu (in his commentaries

gard to *aṇḍakośa-/aṇḍakoṣa-* and *phalakośa-/phalakoṣa-*, it is striking that both *aṇḍa-/phala-* and "*aṇḍa-/phala-*-receptacle" are seemingly used in the same meaning. I can think of three possible reasons for this: (a) *aṇḍa-/phala-* in the latter expression is used in the sense of "embryo", "germ" or something of the sort, and not as "testicle",[1739] or (b) the expression is used to collectively denote both testicles (*aṇḍa-/phala-*) together (which might include their enclosing scrotum), just as *muṣka-* and *vṛṣaṇa-* (both meaning "testicle") can be used, in the singular, or (c) we have in the expression a case of synonyms (both meaning "testicle") combining,[1740] since *kośa-/koṣa-* can also refer to the testicles. Explaining MĀDH 37,1cd-2ab (= AH,Ni 11,22 = AS,Ni 11,p.57a) (with *phalakoṣa-*), the *Madhukośa* does not give any synonym of *phalakoṣa-*, but says that *phalakoṣa-* (dual) might be a copulative compound composed of *phala-* and *koṣa-*; it is unclear what *phala-* and *koṣa-* would refer to then. Or, the commentary goes on, the compound may denote the *koṣa*-s (dual) of/in the *phala*-s (dual); in this case too it is unclear what each of the terms refers to. The text goes on to say that the dual of *phalakoṣa-* is an *upalakṣaṇa-*, i.e. implies something that has not been expressed,[1741] and therefore one *koṣa-* too could be meant.[1742] This explanation of the *Madhukośa* may imply that it takes *phalakoṣa-* to mean "testicle", but even that is nowhere expressly stated. See also the discussion on ☞ *phala-* below.

But whatever the correct explanation of the terms in general actually be, *aṇḍakośa-/aṇḍakoṣa-* and *phalakośa-/phalakoṣa-* seem as a rule not to refer to the scrotum only, i.e. the pouch enclosing the testicles, as the synonyms given in the native lexicons, namely *muṣka-*, *vṛṣaṇa-* and the like, too make clear. On the other hand, Gayadāsa's commentary on SU,Ni 12,6, referred to also on p.564, could be taken to be an example for *phalakośa-* (and

---

on both texts) adds to this the synonym *muṣka-*.

1739 This might also explain the equivalent *śiśna-* "penis" of *phalakośa-* given by VŚS, p.717b.

1740 I.e., "*aṇḍa-/phala-* which is a *kośa-/koṣa-*".

1741 I hesitate to use the term "synecdoche", as I am not sure whether this would fit here.

1742 *phalakoṣābhivāhinīr iti phalaṃ ca koṣaś ca phalakoṣau. athavā phalayoḥ koṣau. tadvahā dhamanīś ca prapīḍya sandūṣya vṛddhiṃ karoti. phalakoṣayor iti dvivacanam upalakṣaṇam. tenaikakoṣe 'pi dṛśyamānā vṛddhiḥ saṅgatā.* This explanation has been referred to in the explanation (given in the text itself) of ĀĀS 3449f. too. (On the masculine of *vṛddhi-* in the verse, namely *sa vṛddhiḥ saptadhā gadaḥ* in 2d, the *Madhukośa* says: *sa vṛddhir iti ktic-ktau ca saṃjñāyām ity anena ktijanto vṛddhiśabdaḥ puṃliṅgaḥ ktinnantas tu strīliṅgaḥ. vṛddhiḥ kuraṇḍo 'bhidhīyate.*)

muṣkakośa-) referring to the scrotum, in which connection it could be debated whether *kośa-/koṣa-* might not be used in the sense of "cover(ing)" here.[1743] Indeed, the meaning "scrotum" is given by many dictionaries, in spite of the dual forms which such dictionaries often expressly mention.[1744] The problem seems to be that English (like German) has no common term for collectively describing the whole aggregate of appendages hanging beneath the penis; probably "scrotum" (like "Hodensack") is but an attempt to give an equivalent, even though the term actually refers only to that which encloses the testicles. This terminology is applied to *muṣka-* and *vṛṣaṇa-* too when used to denote this aggregate; in these cases taking "scrotum" (or "Hodensack") in the literal sense of the word would clearly be absurd. Unfortunately, the problem is hardly perceived,[1745] so that we find a promiscuous use of "scrotum", for the expressions mentioned, in many modern works, seemingly with "scrotum" used *literally* to mean only the pouch enclosing the testes, though note that TIVĀRĪ 1,p.15 speaks of *muṣka-* as meaning "scrotum or testes", clearly taking these English terms to be synonyms, and that GHATAGE, p.893b refers to "the two scrota or testicles", which is mirrored by the explanation of *aṇḍa-* itself as "scrotum" in several modern Indian dictionaries. BODEWITZ 1992, p.31,n.20 sees in the fact that both singular and dual of *kośa-* are used in the same or similar meanings a parallel to the usage of *kukṣi-* to denote the belly. Should this be correct, then, following BODEWITZ 1992, p.29,[1746] we could hold that this refers primarily to the pouch containing the testicles, the scrotum, this pouch being originally regarded as twofold, which would then mean that the meaning "testicle" is probably not primary. This is, however, hardly the place to enter into a detailed discussion of this problem; what is relevant in our context is that the expressions mentioned are all associated with organs of the *male*.

[1743] Cf. on *koṣa-* "covering" WEZLER 1992B, pp.394;409,n.18.

[1744] Note that while MONIER-WILLIAMS (which has influenced most modern Indian dictionaries in English and Indian languages) gives only the meaning "scrotum", the German Sanskrit dictionaries have either the equivalent "Hodensack" for *aṇḍakośa-/aṇḍakoṣa-*, or only "Hoden" ("testicles"), which latter is ambiguous, as it could subsume the scrotum, but need not.

[1745] Note in this regard a similar problem with respect to the Lithuanian *paũtas* (see n.1734), whose plural *paũtai* can, according to NIEDERMANN/SENN/SALYS, p.683, mean "Hodensack".

[1746] Which holds that "there are a few other duals denoting a cavity which for one of other reason is regarded as twofold".

annarasa-

Literally "☞ *rasa-* of food (*anna-*)"; in medical texts often, though not necessarily always, a synonym of ☞ *āhārarasa-*. Cf. also §7.37 and §7.39.

aparā-

In CA,Śā 8,41f., SU,Śā 10,21, SU,Ci 15,17, AH,Śā 2,39 and AS,Śā 3, p.302a *aparā-* obviously refers to the afterbirth, most of which consists, besides the placenta, of the foetal envelopes, which explains why, e.g., Indu in his commentary on AS,Śā 3,p.302a and AH5,Śā 1,89cd = AH,Śā 1,90ab explains *aparā-* as *garbhaśayyā-* "place in which the embryo/foetus lies" (clearly not "womb" here!). In SU,Śā 4,24 (and the corresponding passage AS,Śā 2,p.288a, quoted in n.272)[1747] *aparā-* refers probably to this same conglomerate of placenta and foetal envelopes within the womb (Indu on AS,Śā 2,p.288a glosses it with *garbhaśayyā-* too). It may also be mentioned here that the major part of the afterbirth is formed by the chorion (☞ *jarāyu-*), this also forming a much larger part of the placenta than the decidua. This could explain why *jarāyu-*, if this indeed should refer to the chorion, is also often used in the sense of "afterbirth"[1748] and "foetal envelope (undifferentiated)". *jarāyu-* and *aparā-* are said to be identical by Gayadāsa in NYC, though he may at the same time also be differentiating between them (see Excursus III at the end of chapter 6); *jarāyu-* is here the foetal envelope, which is similar to the opinion of Āḍhamalla in his commentary on ŚĀR 1,7,182[1749] and of Ḍalhaṇa in his commentary on SU,Śā 4,7 (quoted in n.1878). Cakrapāṇidatta, commenting on CA,Śā 3,6, says expressly:

> *jarāyur amarā. yena veṣṭitā manuṣyādayaḥ prajāyante. jarāyuṇā veṣṭitā jāyanta iti jarāyujā manuṣyādayaḥ*
>
> "The *jarāyu-* is the *amarā-* (= *aparā-*, see below),[1750] enclosed with which humans etc. are born. Since (*iti*) they are born enclosed with the *jarāyu-*, humans etc. are [called] *jarāyu-*-born."

Aruṇadatta in his commentary on AH,Śā 1,84;1,90;2,39 calls *jarāyu-* a synonym of *aparā-* (similarly Śrīdāsa in his commentary on AH1, loc.cit.), whereas BHĀ 1,3,206 says that the *aparā-* is called *śrīvara-* in common language, a va-

---

[1747] On the *romarāji-* mentioned in this passage see DAS,R.P. 1991A, pp.34-40.

[1748] This was already postulated by BÖHTLINGK/ROTH 3,cols.52f.

[1749] Āḍhamalla: *jarāyur garbhasyāveṣṭanarūpaś carmaviśeṣaḥ*.

[1750] SHARMA,P.V. 1985, p.361 holds this postulated synonymity to be wrong.

riant, however, mentioning the *jarāyu-*.[1751]

AS,Śā 2,p.288a, a parallel to SU,Śā 4,24 discussed in §6.17, says:

*tatas tad adhaḥ pratihatam aparam aparaṃ copacīyamānam a-
parety āhuḥ. jarāyur ity anye.*

The first part of this statement refers to the blocking of *rajas-/ārtava-* due to pregnancy, this then flowing upwards, gradually accumulating in the body and forming the *aparā-* (and other things). The discussion in §6.17 makes it probable that *jarāyur ity anye*, i.e. "'*jarāyu-*', [say] some", means that the *aparā-* mentioned here is called *jarāyu-* by others, not that blocked *rajas-* forms the *aparā-* according to some, the ☞ *jarāyu-* according to others, as would seem to follow from Indu's commentary:

tad eva cārtavaṃ ... garbhasyāśaya evāparā garbhaśayyākhyā sampadyata iti kecid āhuḥ. anye punar ācāryās tad ārtavaṃ jarāyubhāvena pariṇamatīty āhuḥ

"'And that very *ārtava-* ... becomes the *aparā-*, called *garbhaśayyā-* (place in which the embryo/foetus lies) in the very abode of the embryo/foetus,' say some. Other authorities, however, say: 'That *ārtava-* turns into the *jarāyu-*.'"

Ḍalhaṇa, commenting on SU,Śā 4,24, says that women call the *aparā-* *amalā-* (var: *amarā-*), and Cakrapāṇidatta in his commentary on CA,Śā 6,23 says that commonly (*loke*) the *aparā-* is called *amarā-*. NYC too says that *aparā-* and *amarā-* (as well as *jarāyu-*; see above) are synonyms. In his commentary on SU,Ci, 15,17 Ḍalhaṇa says that the *aparā-*, the covering of the foetus called *jarāyu-*, is commonly called *āvala-*;[1752] note in this context also *āvaraṇā-* in the BHĀ variant quoted in n.1751, though it is unclear whether this is a synonym of *aparā-* or a qualification of *jarāyu-* mentioned there.

In his commentary on SU,Ci 37,125, Ḍalhaṇa equates *aparā-* and ☞ *ulba-/ulva-* as a "certain envelope of the foetus".[1753]

abīja-

A problematic term, not only because of the various meanings *bīja-* may have (see §§2.5f.), but also because the exact analysis of the compound too is not clear. For instance, in AS,Śā 1,p.278a semen is called *abīja-*, which

---

[1751] *aparā tu śrīvara* (var: *tv āvaraṇā jarāyūr* [sic]) *iti loke.*

[1752] *aparāṃ garbhāvaraṇaṃ jarāyusaṃjñaṃ yaṃ lokā āvalam ity ācakṣate.*

[1753] *aparā garbhāvaraṇaviśeṣaḥ. yasyā ulba iti loke prasiddhiḥ.*

obviously stands for "sterile".[1754] But what does *abīja-* here mean exactly? Does it mean "without *bīja-*", in which case it could refer to something in the semen, whatever this something may be? Or does it mean, rather, "through which the *bīja-* is not, i.e. is not formed", in which case *bīja-* would obviously denote the embryo, or, rather, its preliminary stage? Or does it mean "which is no *bīja-*", i.e. cannot serve as the starting point for a later development? Or is it simply a generalised term for "infertile; sterile"? Note also similar uses of *abīja-* referring to *ārtava-* in AS,Śā 1,p.278b, SU,Śā 2,5, as well as in the late *Nāḍījñānadarpaṇa* 249.

Semen (*retas-*) and female "blood" (*asra-*) "incapable with regard to the *bīja-*" (*bījāsamartha-*) are described in AH,Śā 1,10; Aruṇadatta in his commentary has the gloss *abīja-*.[1755] A variant reading, which Indu also follows, has *prajāsamartha-* "incapable with regard to progeny" for *bījāsamartha-*. It does indeed seem as if *bīja-* in *bījāsamartha-* referred to something usually produced due to the mingling of the procreatory fluids (which in this case is not produced), and Aruṇadatta himself says so in introducing this verse, which according to him describes the non-formation of the embryo (*garbha-*) even though the procreatory fluids mingle.[1756] Thus *abīja-* in Aruṇadatta's gloss must mean "through which the *bīja-* is not, i.e. is not formed". Of course it does not follow from this that this is always the meaning of *abīja-*, but at the very least we see that this meaning is very much to be reckoned with.

In SU,Sū 15,5,1 semen (*śukra-*) is said to be "that whose aim is the *bīja-*" or "for the sake of the *bīja-*" (*bījārtha-*);[1757] what this seems to mean is that semen serves to create the *bīja-*. Cakrapāṇidatta in his commentary says: "*bījārtha-* is to be known as: bringing forth the embryo",[1758] i.e. *bīja-* (in "that whose aim is the *bīja-*" or "for the sake of the *bīja-*") is here explained by "embryo" (*garbha-*) (in "bringing forth the embryo"). Ḍalhaṇa in his commentary says:

> *bījārthaṃ ceti bījaprayojanam. yathartukṣetrādisampattau bījam aṅkurajananaṃ syād evaṃ śukram api garbhajanakam ity arthaḥ*
> "'*bījārthaṃ ca*': *bījaprayojana-*. The meaning is: as the seed (*bīja-*) would be seedling-producing in the case of concord of

[1754] *vātapittaśleṣmakuṇapagandhigranthipūyakṣīṇamūtrapuriṣaretāṃsi tv abījāni bhavanti.*

[1755] *ārtavam api śukravad doṣadūṣitam abījam eva.*

[1756] *dampatyoḥ śūkrārtavasaṃyoge 'pi sati kvacid garbhasambhūtir na dṛśyate.*

[1757] *śukraṃ ... bījārthaṃ ca.*

[1758] SU3, p.260: *bījārthaṃ garbhajananaṃ jñeyam*, SU4, p.113: *bījārtho garbhajanano jñeyaḥ.*

season, field etc., so semen too is a producer of the embryo."
Since in the example Ḍalhaṇa gives the *bīja-* of the seedling is clearly itself something that produces something else (namely the seedling), it is difficult, if we take his example to be a parallel to the statement of SU he explains, to see his *bījaprayojana-* as referring to something itself bringing forth the human *bīja-*, and to analyse it as "cause of/means of attaining of the *bīja-*" or "whose motive/end is the *bīja-*", even though these translations would be in keeping with Cakrapāṇidatta's explanation. We must therefore ask whether Ḍalhaṇa presupposes a meaning "the *bīja-* which is the cause/means of attaining". It could, however, also be that the use of *bīja-* "seed" here referring to the seed of a plant is actually not in exact parallelity to the statement of SU, this *bīja-* of the plant in Ḍalhaṇa's commentary being parallel to the "semen" of SU, the *bīja-* of SU's *bījārtha-*, however, parallel to Ḍalhaṇa's "seedling" (*aṅkura-*), and thus referring to the embryo. The problem is complicated by Ḍalhaṇa's explanation of SU,Śā 2,33,[1759] where on the one hand season, field, water and seed (*bīja-*) combine to form a sprout, whilst on the other the time of *rajas-*, womb (☞ *garbhāśaya-*), ☞ *rasa-* and the *bīja-* which consists of the fluids of male and female (i.e. *ārtava-* and semen) parallelly combine to give a child.[1760]

GHĀṆEKAR 1986, p.21b has problems of a different sort with the term *abīja-* in SU,Śā 2,5 (loc.cit.: 2,4), because to him, as to most modern Indian scholars (cf. §1.7), the female procreatory fluid is "ovum". But since this itself is to him *bīja-* (p.43a), *abīja-* cannot of course mean that the fluid itself has no *bīja-*, but only that the womb (!) (*garbhāśaya-*) is not capable of accepting/obtaining the *bīja-* (*bīj'grahaṇ ke lie ayogya*). Commenting on this grammatically impossible explanation,[1761] he fabricates an elaborate theory according to which the text refers to two sorts of *ārtava-*, one, external, cleansing the womb, serving to lubricate the vagina and make it easier for the womb to accept semen and retain the embryo and called "Menstruation *yā* Menstrual blood" (though it obviously is a mixture of menstrual fluid, lubricating fluid and what not), the other part, internal, serving to create the em-

---

[1759] Also referred to in n.521.

[1760] *ṛtur aṅganāyā rajaḥsamayaḥ. kṣetraṃ garbhāśayaḥ. ambu punar āhārapākajo vyāpī rasadhātuḥ. bījaṃ strīpuṃsayor ārtavaśukre.*

[1761] The SU passage says: *ārtavam api tribhir doṣaiḥ śoṇitacaturthaiḥ pṛthag dvandvaiḥ samastaiś copasṛṣṭam abījaṃ bhavati*, i.e. the *ārtava-* itself is clearly *abīja-* here, as the Hindi translation too supposes (p.21a: *ārtav ... prajotpādan meṃ ayogya hotā hai*). (Note incidentally the way in which *śoṇita-* "blood" is mentioned here; though this need not necessarily itself be a ☞ *doṣa-*, it does seem at least to be closely allied to these.)

bryo and called "Ovum"![1762]

āgneya- and *saumya-*

The opposition of ☞ *agni-* and ☞ *soma-*, of *āgneya-* "fiery", i.e. "of the nature of/related to fire", and *saumya-* "somic", i.e. "of the nature of/related to *soma-*", is a theme which is found already in the literature of the oldest Vedic period, and has played an important role in all ages, as is impressively documented by e.g. the immense number of references to *agnīṣoma-* and its derivatives in GHATAGE, pp.424-432. The philosophical, mythological and ritual aspects of this opposition are very complex, and have to my knowledge not yet been studied in any detail (as an introduction see the works listed by MEULENBELD 1974, p.516, especially KIRFEL 1951;[1763] RUBEN 1961, p.82 seems incidentally not to have recognised a relevant reference). In AH,Ka 6,29½ this opposition is even linked with the Vindhyas and Himalayas respectively, which shows that it is also of ecological significance; this ecological significance of the opposition has been drawn attention to several times in ZIMMERMANN 1987, though the simple equation of *agni-* and *soma-* with "sun" and "moon" respectively[1764] seems to me to be rather apodictic, since it is clearly not the result of any in-depth study of the matter. Such a study will have to consider so many possible pairs in opposition,[1765] as well as hotly debated problems such as the evidence for the presence of *sóma-* "moon" in the Vedic Saṃhitās, that I for one dare not venture to speculate on the significance of *agni-* and *soma-* or *āgneya-* and *saumya-*.[1766] Now one can translate *agni-* and *āgneya-* as "fire" and "fiery" respectively without having to tackle such problems, simply leaving the problem of the interpretation of "fire" and "fiery" aside for future study, but one cannot do the same in the case of *soma-* and *saumya-*, as a similar translation of these terms must simultaneously also be an interpretation. Thus I have chosen to be not only

---

[1762] Cf. §2.3.

[1763] Though the latter's supposition that *soma-* refers to water still remains to be proved. On this supposition see also MÜLLER 1955, p.16 (continuation of n.58).

[1764] Thus for instance on pp.131;145;198;220; see also e.g. MÜLLER 1955, pp.15,n.58;21, MÜLLER 1963B, p.416, MÜLLER 1941-1942, p.480.

[1765] A sampling: sun and moon, Indra and Vṛtra/demons, sacrificial fire and drink, fire/lightning and water/rain (on the latter cf. also WITZEL, pp.788;826f.).

[1766] In studying these pairs one will probably also have to take into consideration such opposites as sun and moon in the human body in systems usually labelled as Tantric, and maybe also of fire and water in such systems as Balinese Tantric medicine (cf. on the latter e.g. WECK, pp.42f.;45ff.)

cautious, but also honest (since I am unable to offer a satisfactory interpretation), and not to translate *soma-* in this pair, either in general or with particular reference to the texts we are here concerned with.[1767] This is also why I am translating *saumya-* as "somic" and avoiding the translation "cool" which seems often to be favoured; "cool" seems implied too, but only as a part of what is ultimately meant, whatever that may be.

With regard to the special subject of study in this work, it may be noted that, even though it comes from "somic" ☞ *rasa-*, the procreatory-menstrual fluid is "fiery", whereas semen is "somic" and the embryo both "somic" and "fiery" (see n.423). Interestingly, *Aitareyāraṇyaka* 2,3,7 (on which see also GHURYE, p.268) has a different sort of opposition, for it says that the "blood" (*lohita-*) of women is the form of fire (*agni-*), the semen (*retas-*) of men the form of the sun (*āditya-*).[1768] Had this passage presupposed a link between semen and *soma-*, then it would most probably have connected semen with the moon, since one of the names of this is *soma-*, though that does not of course mean that *soma-* "moon" must be at the root of the original opposition of *agni-* and *soma-*, as is uncritically assumed by some scholars (see above). As regards blood in general, this is in Ḍalhaṇa's commentary on SU,Sū 14,7 (see n.451) and Cakrapāṇidatta's commentary on 14,4f. (SU3, p.232, SU4, p.101) called neither "hot" nor "cold" (*anuṣṇāśīta-*; see below), Cakrapāṇidatta further quoting Bhoja[1769], with a verse found in Āḍhamalla's commentary on ŚĀR 1,6,10f. too, to the effect that blood is normally neither "hot" nor "cold" (*anuṣṇaśīta-*);[1770] SU,Sū 21,17 states the same (*anuṣṇaśīta-*)[1771]. AH,Sū 27,1 (see n.376) also states that blood is neither "cold" nor "hot" (*aśītoṣṇa-*).[1772] ĀY 4,67, however, tells us that blood is both hot and cold (*uṣṇaṃ śītaṃ ca madhuraṃ ...*).[1773] On this subject see also §7.12.

[1767] ZIMMERMANN 1987, p.220 explains it not only by "moon", but also by "rain".

[1768] On which Sāyaṇa says that this is because of their red and white colour respectively.

[1769] On him cf. MEULENBELD 1974, pp.418f.

[1770] The verse as quoted by Cakrapāṇidatta runs:
anuṣṇaśītaṃ madhuraṃ snigdhaṃ raktaṃ ca varṇataḥ
śoṇitaṃ guru visraṃ syād vidagdhaṃ coṣṇam eva ca.
Instead of *vidagdhaṃ coṣṇam eva ca*, Āḍhamalla has *vidāhaś cāsya pittavat*, which tells us not that blood becomes hot when inflamed, but that its inflammation is like that of bile.

[1771] According to SU3, p.232 — but not p.372 or SU4, pp.101;162 — Cakrapāṇidatta presupposes the reading *anuṣṇāśīta-* "not hot and not cold" here.

[1772] Aruṇadatta, Indu, Candranandana, Śrīdāsa and Parameśvara in their commentaries take the negation to mean "slightly".

[1773] According to p.428,n.53 of the edition there is also a variant *anuṣṇaṃ* in one ms, but

Be all that as it may, what here is of primary interest to us are the terms "hot" and "cold"; it is unclear whether "hot" (*uṣṇa-*) and "cold" (*śīta-*) are to be taken to mean the same as "fiery" (*āgneya-*) and "somic" (*saumya-*) respectively, to be due to these latter, as Candranandana holds in his remarks on AH2,Sū 27,2,[1774] or to refer merely to "qualities" (☞ *guṇa-*)[1775] such as heavy and light, hard and soft, hot and cold, solid and flowing etc., which on the face of it seem basically to have nothing to do with the "somic" or "fiery" nature of a substance. In this connection cf. AS,Sū 36,p.251a (see n.377), where blood is said to be "neither cold nor hot" (*aśītoṣṇa-*), but at the same time "somic and fiery by nature" (*saumyāgneyaṃ prakṛtyā*), just as in Candranandana's commentary just quoted. The latter characteristic (*saumyāgneyatva-*) is mentioned also by Cakrapāṇidatta in his commentaries on SU,Sū 14,4f. (SU3, pp.232f., SU4, p.101: thrice, once in a quotation from Puṣkalāvata), Sū 15,9 (SU3, p.264, SU4,p.114) and Sū 21,17 (SU3, p.372, SU4, p.162) and Ḍalhaṇa in his commentary on SU,Sū 15,9 (*agnīṣomīyatva-*); in his commentary on SU,Sū 14,4f. Cakrapāṇidatta also says that blood is neither "hot" nor "cold" precisely because it is both "somic" and "fiery".[1776] In any case, most of the passages commented upon do not mention "somic" and "fiery" or have these in their vicinity, and thus need not necessarily presuppose the same as the commentators mentioned. As regards Ḍalhaṇa, the statement referred to above in his commentary on SU,Sū 15,9 is that of "some" (*kecid*), and thus not his own, and that holding blood to be only "fiery" in his commentary on SU,Sū 13,21 (*āgneyaraktadhātu-*) too is by "others" (*anye*). However, in SU,Sū 14,7 his reference to blood as "fiery" (see §7.12) seems to be his own opinion.[1777] But in §7.19 it is shown that he also construes a link between blood and fire in another system of reference (maybe that of the ☞ *mahābhūta-*s, on which possibility cf. below); in this regard Cakrapāṇidatta as quoted in §7.16 goes even further by postulating a link between blood and fire and water. Matters are not made easier by the statement of Gayadāsa that blood is "fiery" ("somic" not being mentioned) (see on this §7.12, also on the variants of SU,Sū 14,7 saying the same).

---

this is obviously a secondary "correction", and against the metre anyway.

1774 *tathāśītoṣṇam aśītaṃ nātyuṣṇaṃ sādhāraṇaṃ nṛsiṃharūpavad ubhayaguṇasvabhāvaṃ saumyāgneyaprakṛtyā.*

1775 Note that Candranandana also uses this expression.

1776 *ubhayātmakatvenaiva ca śoṇitam anuṣṇāśītaṃ bhavati.*

1777 In the same commentary he also refers to the characterisation of blood as "neither hot nor cold" (*anuṣṇaśīta-*) being the opinion of other authorities.

Cf. with regard to "hot" and "cold" and "fiery" and "somic" respectively, however, also SU,Sū 42,7,[1778] which states that the tastes (☞ *rasa-*) are in part *saumya-* "somic", in part *āgneya-* "fiery", and that the former are *śīta-* "cold", the latter *uṣṇa-* "hot". Does this mean that "somic" and "cold", "fiery" and "hot" respectively mean the same? But against this see AS,Sū 6,p.32a, where the heavenly water of the Ganges is said to be both "cold" *and* "somic";[1779] ☞ *ojas-* too is similarly characterised in SU,Sū 15,21. SU,Śā 6,16 tells us that coldness (*śaitya-*) is a characteristic of ☞ *soma-* (the term is not explained), with which *saumya-* is linked loc.cit., but this does not let us know what the exact relation between "cold" and "somic" is. Returning to the tastes (*rasa-*), I may mention their classification according to their potency (*vīrya-*), i.e. the potency ascribed to eight of the ☞ *guṇa-*s (heavy and light, hard and soft, hot and cold, etc.) mentioned above, more loosely these eight *guṇa-*s themselves.[1780] Three of the six tastes are associated with the hot, three with the cold potency; MEULENBELD 1987, pp.12f. links this with the theory of the tastes being composed of the five ☞ *mahābhūta-*s or gross elements, which theory is not unproblematic and probably of a late date (see e.g. MEULENBELD 1987, pp.5-7). Now in the classification of the different *guṇa-*s called *vīrya-*s (*guṇā vīryasaṃjñakāḥ*) in SU,Sū 41,11, the *guṇa-* "hot" (*uṣṇa-*) is called "fiery" (*āgneya-*); but here the classification of the *guṇa-*s is not on the basis of "somic" and "fiery", but of the five gross elements, "hot" here being "fiery" by belonging to the ☞ *mahābhūta-* or gross element fire (the *guṇa- śīta-* "cold" belongs to water). On the other hand, there is also an older theory mentioning just two *vīrya-*s instead of eight, namely only "hot" and "cold".[1781] SU,Sū 40,5, AH,Sū 9,17f., AS,Sū 17,p.126b, ŚĀR 1,2,17f., BHĀ 1,6,238 (see also the respective commentaries) and Cakrapāṇidatta in his commentary on CA,Sū 26,64f. tell us that the theory of "hot" and "cold" potencies is based upon the classification of substances as "fiery" or "somic" (ĀY 3,84 substitutes the terms "hot" and "cold" respectively), SU,Sū 40,5 (as well as SU,Sū 42,7[1782] and

---

[1778] Also quoted by ZIMMERMANN 1987, p.145, referred to on p.198, and said on p.240,n.37 to be "in opposition to" CA,Sū 26,53-56, without this statement being explained.

[1779] *śaityajīvanasaumyatvaiḥ pittaraktaviṣārtijit.*

[1780] See MEULENBELD 1987, pp.8;11-14, MEULENBELD 1974, pp.501f., SHARMA,P.V. 1982, pp.95ff. and 1985, pp.229ff., EIM, pp.157-159, COMBA 1990, p.31. (Note in this context also the statement of CA,Vi 8,35 that where other texts speak of *eight rasa-*s, CA speaks of the usual six; is there any connection between the eight *rasa-*s and the eight *vīrya-*s?)

[1781] See MEULENBELD 1987, p.12, MEULENBELD 1974, p.502, also SIDDH 1,26, ĀY 3,81-85, and Cakrapāṇidatta's commentary on SU,Sū 1,38 = SU3,Sū 1,27, SU4,Sū 1,36.

[1782] Cited above in connection with the tastes.

Śrīdāsa[1783]) expressly mentioning the "fiery" and "somic" nature of the world (*agnīṣomīyatvāj jagataḥ*).[1784] All this shows that we must reckon with two classifications, one based on "somic" and "fiery", the other on the five gross elements (☞ *mahābhūta-*), having been mixed up in medical theory and texts (see also §§7.14f.). That this is no mere speculation is borne out by SU,Sū 1,22, which expressly refers to these two different modes of classification:

> *loko hi dvividhaḥ sthāvaro jaṅgamaś ca. dvividhātmaka evāgneyaḥ saumyaś ca tadbhūyastvāt. pañcātmako vā*
>
> "The world is of two divisions, immovable and movable. [It is] verily of twofold nature, fiery and somic, due to their [respective?] preponderance [in substances].[1785] Or [it is] of the nature of five (obviously the gross elements)."

Cf. also the commentaries of Ḍalhaṇa and Cakrapāṇidatta, which try to harmonise both theories; in these commentaries "fiery" and "somic" are incidentally linked with "hot" and "cold" respectively.[1786] Interesting is also the classification of the seasons and the corresponding plants as "fiery" and "somic" in SU,Sū 36,5, right after which 36,6 classifies soils and plants grown on these on the basis of the five gross elements.

The notion of the "fiery" and "somic" nature of the world is not confined to medical texts; see e.g. GHATAGE, p.426a, MBH 8,24,84, 12,277,33 and 12,328,52, *Harivaṃśa* 30,45, 39,11 and App.1,37,53, *Bṛhajjābālopaniṣad* 2. Interesting in this context is also ŚB 11,1,6,19, where ☞ *soma-* and ☞ *agni-* are characterised as *anna-* "food" and *annāda-* "eater of food" respectively.

[1783] In his commentary on AH,Śā 3,62cd-63ab on p.117 of AHI.

[1784] On SU,Sū 40,5 see also DASH 1971, p.22, where, however, we have extremely naïve speculations based on the equation of ☞ *agni-* and ☞ *soma-* with *prāṇa-* and *anna-* respectively, probably due to such passages as *Bṛhadāraṇyakopaniṣad* 1,4,6 (cf. also the commentary), *Maitrāyaṇyupaniṣad* 6,10, VISHVA BANDHU, p.5, no.13 etc.; as regards the passage said on p.22,n.1 to be from the *Śatapathabrāhmaṇa*, I could not find this in ŚB. SHARMA,P.V. 1992, p.91 similarly writes that "The *Agni-soma* concept has been described well in the form of *Prāṇātman* and *Pākātman* gods", but the passage referred to on p.107,n.73, namely MBH 12,80*,1-4, merely states that *agni-* is called *pākātmạn-* because it digests food internally for upholding the *prāṇa-*s, from which what SHARMA,P.V. 1992 writes does not follow. On the other hand, ZIMMERMANN 1987, p.220 equates *agni-* and *soma-* here with "the sun and the moon, the fire and the rain"; this has already been remarked upon above in n.1767.

[1785] Cf. on this also BH,Śā 4,19f. mentioned below on p.526.

[1786] ZIMMERMANN 1987, p.198 also draws attention to this passage, but does not comment upon the fact that *vā* most probably shows that we have two contradictory divisions of the world here.

There are further difficulties linked with "fiery" and "somic". On a problematic theory in GARBH 2 linking *saumyaguṇa-* "somic quality (☞ *guṇa-*)" with both ☞ *rasa-* "taste" and the seven ☞ *dhātu-*s of the chain of metabolic change see n.959. See also the commentaries of Cakrapāṇidatta and Ḍalhaṇa on SU,Sū 6,15 and Ḍalhaṇa and Gayadāsa on SU,Ni 3,20 saying that the *prāṇa-*s "breaths" are *agnīṣomādi-* "*agni-*, *soma-* etc." (I could not find further information on this). Problematic is also *soma-* in CA,Sū 5,6:

> *laghūni hi dravyāṇi vāyvagniguṇabahulāni bhavanti. pr̥thvīsoma-guṇabahulānītarāṇi.*

Though the passage mentions only fire, earth and wind beside *soma-*, there being no fifth element,[1787] and neither any mention of the term ☞ *mahābhūta-* (or something similar) or the number five, the probability is very great (though the evidence does not allow us to clinch the issue) that *soma-* here is the gross element water, an explanation actually given by Cakrapāṇidatta's commentary and CART, where we find *toya-* obviously in explanation of *soma-*.[1788] SU,Sū 21,8 and CA,Sū 6,5, as well as Ḍalhaṇa in his commentary on SU,Sū 6,14, however, mention the triad of *soma-*, sun and wind, where *soma-* seems to mean the moon (cf. on this §7.20). Note in this context Gaṅgādhara's explanation of *bījadhātu-*[1789] in CA,Śā 4,7 (see p.556), referring to semen (which, as we have seen above, is "somic"), as *pitr̥yānena somalokagato diṣṭarūpaṃ bījaṃ tataḥ pratyāgataṃ punaḥ*, which refers to the well-known ancient belief that the dead return via the world of the moon, though it must also be noted that this belief, based on the theory of transmigration not mentioned in the oldest Vedic texts, is most probably younger than the oldest mentions of the opposition of *agni-* and *soma-*. In CA,Sū 12,12 we find the theory of Kāpya that it is *soma-* in the phlegm which is responsible for changes within the body;[1790] both Cakrapāṇidatta and CART remark that *soma-* here refers either to the "deity water" (*jaladevatā-*) or the moon. This brings to mind BH,Śā 4,18ff. Śā 4,18cd-19ab first state that the abdominal fire devours *rasa-*, blood, fat and flesh when no food is ingested; 19cd-20ab then say that all things are of the nature of *agni-* and *soma-*, this

---

1787 As in CA,Sū 25,40; see n.466.

1788 On the basis of the CA passage one could also deduce that we have here an older theory listing only four gross elements, though SHARMA,P.V. 1985, p.72 holds that "Caraka omits ākāśa here because it is not so favourable for stimulation of fire as such", whereas LARSON 1987, p.257,n.6 opines in a general manner that "The materialistic bias is undoubtedly also a reason for the absence of discussions of *ākāśa* in the medical literature".

1789 This is the reading of CA1.

1790 *soma eva śarīre śleṣmāntargataḥ kupitākupitaḥ śubhāśubhāni karoti.*

applying also to all corporeal living beings:

agnīṣomātmakaṃ sarvaṃ jagat sthāvarajaṅgamam
agnīṣomātmakāḥ sarve dehinas tu caturvidhāḥ.

(19cd ≈ *Harivaṃśa* App.1,37,53). This is followed by 20cd-21ab telling us that the fire devours all the intestines, both of the nature of sun and of moon (this is highly reminiscent of Tantric theories linking e.g. the tubular passages on the two sides of the spine with sun and moon respectively). 19cd-20ab look odd in the context of this passage, and one gets the feeling that they might have been inserted later to explain 20cd-21ab, but even if that should not be the case, it does very much look as if *agni-* and *soma-* were here identified with sun and moon.

For further remarks on "fiery" and "somic" see especially §§7.19f.

ānanda-

"Pleasure, joy". According to OLIVELLE, *ānanda-* is extremely often used "almost as a technical term for orgasmic rapture" (p.172), especially in the Vedic literature. In several of the contexts in which this term occurs in the material we have studied, *ānanda-* could very well have this meaning (cf. e.g. §§12.34f., n.1270 and n.1506). However, the matter requires study more detailed than has been possible here.

āsekya-

SU,Śā 2,38 describes the origin of a 'perverted' male called *āsekya-*, who attains an erection by drinking semen (cf. also n.20); the commentator Ḍalhaṇa says, probably correctly: by means of fellatio of someone else with his mouth, though Indu in his commentary on AS,Śā 2,p.293 says that he drinks his own or someone else's semen. See Āḍhamalla's comments on ŚĀR 1,7,171f. for another definition of the *āsekya-*. AS1,Śā 2,p.205 creates the impression that there are both male and female *āsekya-*s,[1791] but a comparison with AS,Śā 2,p.293 shows that in AS1 part of the text is missing;[1792] this portion has nothing to do with the *āsekya-*. The commentators of SU regard the *āsekya-* as a ☞ *ṣaṇḍ(h)a(ka)-*, Gayadāsa in NYC (fol.21r) also as a ☞ *napuṃsaka-*.[1793]

1791 *tad āsekyaṃ strīpuṃsau tadvidhau bhavataḥ.*

1792 The full text of AS runs: *tad āsekyaṃ nāma bhavati. tac chuklāsvādād dhvajocchrāyaṃ labhate. yadā punar ubhāv api bhavataḥ strīpuṃsau tadvidhau* (cf. the preceding *yadālpabījo 'lpabalaḥ pumān*) ... .

1793 See n.2007.

āhārarasa-

Besides the chain of seven elements (☞ *dhātu-*) obtained through the breaking down of food (usually *anna-* or *āhāra-*) by a process of successive transformation, we also find mention of a substance *āhārarasa-* (often also simply ☞ *rasa-*), the essential part of food (*āhāra-*). It is today usually taken to be an intermediate product between food and the first *dhātu-*, *rasa-*, but its actual identification in the old texts is not that easy. Also, *āhārarasa-* can at times simply mean "*rasa-* of food" = *rasa-*, in which cases no terminological differentiation between *rasa-* and *āhārarasa-* seems to be implied.

The problem of *āhārarasa-* is discussed in more detail in §§7.29ff.

uttaravasti-

According to MEULENBELD 1974, p.481 (with further bibliography) a fluid brought into the urethra by a clyster, but actually this can also refer to a fluid brought into the introitus of the vagina (as in §5.6). cf. e.g. SU,Ci 37,104f., AH,Sū 19,79, AS,Sū 28,p.223a (the urethra seems to be implied, however, in CA,Si 9,64f.).

upadhātu-

"Additional elements", on which see MEULENBELD 1974, pp.470f. The list today regarded as authoritative seems to appear for the first time in ŚĀR. The older history of this term, as well as what substances are called *upadhātu*-s, is a complicated matter, and has been discussed in some detail in §§11.5ff.

ulba-/ulva-

Often translated as "foetal envelope", but also as "amnion"; the latter could fit *Aitareyabrāhmaṇa* 1,3,16-18 stating that the embryo is covered by the *ulba-*, and then by the ☞ *jarāyu-* (often translated as "chorion").[1794] But

[1794] 16: *ulbaṃ vā etad dīkṣitasya yad vāsa ulbenaivainaṃ tat proṃuvanti*. 17: *kṛṣṇājinam uttaraṃ bhavaty*. 18: *uttaraṃ vā ulbāj jarāyu jarāyuṇaivainaṃ tat proṃuvanti*. I.e., the consecrated person (*dīkṣita-*) is covered by cloth and then the black antelope skin, called *ulba-* and *jarāyu-* respectively (the *dīkṣita-* is commonly regarded as an embryo reborn through the consecration). The text adds that as the baby is born together with the *ulba-*, but freed of the *jarāyu-*, so too does the consecrated person bathe covered with cloth, but freed of the black antelope skin (22: *unmucya kṛṣṇājinam avabhṛtham abhyavaiti tasmān muktā garbhā jarāyor jāyante*. 23: *sahaiva vāsasābhyavaiti tasmāt sahaivolbena kumāro jāyate*). FALK 1994, p.17 takes *jarāyu-* to mean *only* "placenta", since the human child is born covered by the foetal envelope, such a clear-cut usage of the word is not always found, maybe not even the rule, even in medical texts. Thus, commenting on *ulba-* in AH, Ut 1,1, the *Kairalī* commentary quotes the Sārakṛt (the author of the *Sāra*[1795]), to the effect that the *ulba-* is what the newborn is enveloped in, equating it with the ☞ *jarāyu-*;[1796] Indu in his commentary also equates the *ulba-* with the *jarāyu-* and says the same.[1797] Ḍalhaṇa in his commentary on SU,Śā 10,12 too equates *ulba-* and *jarāyu-*[1798], and also draws attention to Śā 2,54, which says that within the uterus the face of the embryo is covered by the *jarāyu-*. All this clearly creates problems when compared with *Aitareyabrāhmaṇa* 1,3,16-18. Śivadāsasena's commentary on AH,Ut 1,1 says something quite different, namely that *ulva-* refers to the slime in the face/mouth and throat of the newborn.[1799] TIVĀRĪ 1,p.650 explains the term in SU,Śā 10,12 and AS, Ut 1,p.1a (and AH, Ut 1,1) as meaning "vernix caseosa", which could fit these occurrences. According to ABEGG, p.93,n.7 *ulba-* in *Garuḍapurāṇasāroddhāra* 6,11 refers not to a covering membrane, but to the uterus (cf. §6.20 and BÖHTLINGK/ROTH 1,col.1006); this is reminiscent of the later usage of ☞ *jarāyu-*. See also COMBA 1981, p.208,n.41 on *ulba-/ulva-*. RASA, Upodghāta,pp.107f. explains *ulba-* as "water" in which the foetus/embryo is immersed and *jarāyu-* as that which envelops it.[1800]

---

but without the placenta ("der Mensch wird ohne *jarāyu-*, aber mit *ulba-* geboren, also ohne Mutterkuchen und mit Eihaut"). This is an impossible deduction, especially since we are told expressly that one *envelops/covers* the *dīkṣita-* (*proṃuvanti*) with the black antelope skin = *jarāyu-*. The deduction might stem in part from the laxness of many German scholars as to the terminology of the foetal envelopes, to which they refer by the singular "Eihaut" instead of the correct plural "Eihäute", there actually being three of these, viz. the amnion ("Schafhaut", "Wasserhaut"), chorion ("Lederhaut", "Zottenhaut") and decidua reflexa ("Siebhaut"). Thus BÖHTLINGK/ROTH 1,col.1006 and BÖHTLINGK 1991, 1,p.253 speak only of a singular "Eihaut", whereas BÖHTLINGK/ROTH 3,col.52 duly differentiates between an inner and an outer "Eihaut", "Amnion"/*ulba-* and "Chorion"/*jarāyu-* respectively (cf. also BÖHTLINGK 1991, 2,p.256).

[1795] I.e. the *Vyākhyāsāra*, whose author and Śrīdāsa were contemporaries and pupils of the same teacher, as Gerrit Jan Meulenbeld kindly informed me.

[1796] *yenopalipto jāyate garbhaḥ sa ulbo jarāyur iti sārakṛt*. On the child being enveloped by the *ulba-* see also n.1794.

[1797] *ulbo jarāyuḥ.´uktaṃ hi. veṣṭito jāyate yena garbhaḥ solbaḥ* [sic] *prakīrtitaḥ*.

[1798] *ulbaṃ jarāyuṃ*. Cf. also his remarks quoted in n.1878. Ḍalhaṇa's commentary on Ci 37,125 equates *ulba-* and ☞ *aparā-*; see n.1753.

[1799] *ulvo bālasya mukhakaṇṭhagataḥ kaphaḥ*. Similary, Śivadāsasena's commentary on *Cakradatta* [64],3 (p.668) says: *ulvaṃ bālakasya kaṇṭhagataśleṣmā*.

[1800] *ulbajarāyuśabdau cikitsāśāstre garbhanimajjanajale tadīyaveṣṭane ca sādhāraṇatayā prayuktau dṛśyete*.

ojas-

Vital fluid/force; see especially MEULENBELD 1974, p.451, also MAYRHOFER 1992- 1,p.278., as well as n.921 and n.1635. It is according to many ancient authorities the ultimate essence left over after the formation of semen, the last of the seven elements of the chain of alimentary transformation (☞ *dhātu-*), but is as a rule not termed an "element" in medical texts,[1801] though in non-medical texts it may, as e.g. in the commentary on AV 2,17,1 (see AV1) attributed to Sāyaṇa, which calls *ojas-* the *eighth dhātu-*.[1802] That such a notion was known in medical circles too is proved by Cakrapāṇidatta's opposition to it in his commentary on CA,Sū 30,7.[1803]

Apart from the references given under *dhātu-*, see too on *ojas-* from semen Narasiṃha's commentary (p.11) on RAS 1,6, and TA 21, TANT 21, ĀY 3,361-373;4,47-50 (see also EIM, pp.218-220, TIVĀRĪ, 1,p.48), though note that it appears *before* semen (*śukra-*) in the *dhātu*--lists in CA,Sū 26,43,1 and SU,Sū 42,10,1. At times *ojas-* is also called the ☞ *mala-* of semen, which has led Śrīdāsa in his commentary on AH1,Śā 3,63f. to state that *ojas-* is of two types, namely as the pure part (*prasāda-*) of semen, and as the *mala-* of this.[1804] BHĀ 1,3,183 states that *ojas-* is of two types depending on whether it is of the nature of ☞ *agni-* or ☞ *soma-*, i.e. "fiery" (☞ *āgneya-*) or "somic" (☞ *saumya-*).[1805] CART on CA,Sū 17,73-75 also tells us that *ojas-* is of two types, namely *para-* and *apara-*;[1806] see too Cakrapāṇidatta's commentary on the same and on CA,Śā 7,15 (see also the discussion in SHARMA,P.V. 1985, pp.160f.;260), as well as Aruṇadatta's commentary on AH,Sū 11,37-39 given below and Śrīdāsa's corresponding commentary. AS,Śā 8,p.345a mentions eight drops of *ojas-* characterised as *para-* in the heart, and *ojas-* of the

---

[1801] Cf. DAS,R.P. 1984A, pp.237f.;243,n.26 too.

[1802] *ojaḥ śarīrasthitikāraṇam aṣṭamo dhātuḥ. tad uktam ācāryaiḥ.*

> *X X X X X X X X ojo nāmāṣṭamī daśā*
> *kṣetrajñasya tad ojas tu kevalāśraya iṣyate*
> *yathā snehaḥ pradīpasya yathābhram aśanitviṣaḥ*

iti.

[1803] *etac caujа upadhāturūpaṃ kecid āhuḥ. dhātur hi dhāraṇapoṣaṇayogād bhavati. ojas tu dehadhārakaṃ sad api na dehapoṣakaṃ. tena nāṣṭamo dhātur ojaḥ.* In his commentary on SU,Sū 15,19 (15,18 in SU3) he says: *tac caujo dhātubhyo* (var in SU4: *dhātutvena*) *na bhinnam iti kṛtvā saptadhātukatvam.*

[1804] *dvividhaṃ hy ojaḥ. śuklasya prasādarūpam ekaṃ śuklasya malarūpaṃ dvitīyam.*

[1805] *agnīṣomātmakatvena* (var: *agnisomātmakatvena*) *dvirūpaṃ varṇitaṃ tu tat.* On this statement see also p.534 below.

[1806] *ojas tu dvividhaṃ param aparaṃ ca.*

nature of ☞ *rasa-* (*rasātmaka-*), one *prasṛti-* in quantity.[1807] Cakrapāṇidatta in his commentary on SU,Sū 15,19 (15,18 in SU3) tells us that the *para- ojas-*, having the measure of eight drops, is present in the heart, whereas the *apara-ojas-*, which is much more in quantity (*tato bahu*), refreshes (*prīṇayati*) the body through ☞ *dhamanī*-s emanating from the heart; this latter *ojas-* is half an *añjali-* in quantity. As proof of this what seems to be CA,Śā 7,15 is quoted, but not with the same words, for whereas CA,Śā 7,15 says that the quantity of *ojas-* of the nature of (scil. from?) phlegm (*ślaiṣmika-*) is half an *añjali-*,[1808] in Cakrapāṇidatta's quote this could apply to both phlegm (☞ *śleṣman-*) and *ojas-*, though the quotation could also be understood as referring to *ojas-* from *śleṣman-* (and not *śleṣman-* and *ojas-*).[1809] Cakrapāṇidatta's commentary on CA,Sū 30,7 says that the *para- ojas-* is little in quantity (*alpa-pramāṇa-*), and gives the quantity of the other *ojas-* as *one añjali-*, referring expressly to CA,Śā 7,15.[1810] Taken together, the statements in both of Cakrapāṇidatta's commentaries are thus clearly problematic.

In contrast to the theories above, Hemādri in his commentary on AH, Sū 11,37-39,[1811] seeing a problem in *ojas-* being called both the ☞ *mala-* and the essence (*sāra-*) of semen, tries to show that, despite appearances to the contrary, there is only one *ojas-*;[1812] in his long discussion he maintains, in a nutshell, that, though the *ojas-* which arises from the transformation (*pāka-* "cooking; ripening") of semen of a male combining with *ārtava-* in the womb of a woman is its *mala-*, the *sāra-* in this case being the embryo (*garbha-*), this *mala-* enters the embryo, and after several transformations finally emerges as the *sāra-* of the semen. In other words, it is a *mala-* with respect to the embryo and a *sāra-* with respect to ☞ *rasa-* etc.[1813]

The theory of two types of *ojas-* (*para-* and *apara-*) could have been caused by the characterisation of *ojas-* as *para-* in CA,Sū 30,7.[1814] As it is,

---

1807 *tadvad ojo 'pi paraṃ hṛdayastham aṣṭabindupramāṇam ādhāraḥ prāsṛtikasya rasātmakasyaujasaḥ kṣayavṛddhibhājaḥ.*

1808 *mastiṣkasyārdhāñjaliḥ. śukrasya tāvad eva pramāṇam. tāvad eva ślaiṣmikasyaujasa iti* (var: *tāvad eva caujasaḥ. strīṇām ārtavasya catvāro 'ñjalayaḥ*).

1809 *tāvac caiva śleṣmaṇa ojasaḥ pramāṇam.*

1810 Quoted as: *tāvad eva parimāṇaṃ ślaiṣmikasyaujasaḥ.*

1811 Cf. on this also n.812 and §13.26.

1812 *na caivam ojo'nekatvaṃ vācyam* (var: *śaṅkyam*).

1813 *malatvaṃ ca garbhāpekṣayā. rasādyapekṣayā tu sāratvam eveti.*

1814 *tat parasyaujasaḥ sthānaṃ tatra caitanyasaṅgrahaḥ*
hṛdayaṃ mahad arthaś ca tasmād uktaṃ cikitsakaiḥ (var: *cikitsite*).

para- seems to refer to the same as *pradhāna-* (characterising *ojas-*) in CA,Ci 24,35 (mentioned below on p.557), and it would thus have been interesting to see what Cakrapāṇidatta says on this latter characterisation; unfortunately, he does not comment upon it. Returning to *para-* as a characterisation of *ojas-* in CA, it may be pointed out that according to SU,Sū 15,19, *ojas-* is the *para-* ("supreme"?, "final"?) fiery energy (*tejas-*) of the ☞ *dhātu-*s,[1815] obviously as the purest part left over after semen has been refined. The statement is ambiguous in AH,Sū 11, 37,[1816] for *para-* here could be understood as a characterisation not of *tejas-*, but of *ojas-*.[1817] In AS,Sū 19,p.145a *para-* seems really to characterise *ojas-*, even though *tejas-* too is mentioned.[1818] Could this show a gradual development finally leading from a characterisation of *tejas-*, such as in SU, to a statement such as that of CA, in which no *tejas-* is mentioned? It bears looking into.[1819]

In his commentary on AH,Sū 11,37-39 already mentioned above, Hemādri has a quotation from an unidentified source[1820] which tells us that *ojas-* is used for

(a) *tejas-* "fiery energy" of the *dhātu-*s. See on this e.g. SU,Sū 15,19 and AH,Sū 11,37 and the discussion on them above.

(b) ☞ *rasa-*. See p.534.

(c) *jīvitaśoṇita-*. This probably refers to blood in general, as Āḍhamalla's commentary on ŚĀR 1,5,45-47 and Hemādri's commentaries (referring to *jīvaśoṇita-*) on AH,Ka 3,32 and Sū 18,26 too say; cf. also §7.12. Note too that AS,Sū 19,p.145a, AH,Sū 11,38, CA,Sū 17,74 and BHĀ 1,3,183 (wrongly ascribed to Caraka) tell us that *ojas-* is slightly reddish-yellow.[1821]

[1815] *dhātūnāṃ yat paraṃ tejas tat khalv ojas tad eva balam ity ucyate svaśāstrasiddhāntāt.*

[1816] Quoted also in BHĀ 1,3,184.

[1817] *ojas tu tejo dhātūnāṃ śukrāntānāṃ paraṃ smṛtam.* Aruṇadatta, Śrīdāsa, Candranandana, Parameśvara, Hemādri and Indu give different explanations of this line, reflecting its ambiguity; Indu comments upon it in the same manner as he comments upon AS,Sū 19,p.145a. The commentaries partly gloss *para-* with *pradhāna-*.

[1818] *tejo yat sarvadhātūnām ojas tat param ucyate.*

[1819] One could incidentally also make interesting deductions as to the age of CA,Sū 30,7 if it could be shown that this is the end point of a development which can be traced in the other texts quoted.

[1820] *dhātūnāṃ tejasi rase tathā jīvitaśoṇite*
śleṣmaṇi prākṛte vaidyair ojaḥśabdaḥ prakīrtitaḥ.

Āḍhamalla in his commentary on ŚĀR 1,5,45-47 has a similar explanation, except for the statement on *śleṣman-*.

[1821] Cf. also n.420 on the colouring of *rasa-*.

(d) ☞ *śleṣman-* "phlegm" in a balanced state. See also CA,Sū 17,117.

We have just seen above that the SU tradition as reflected in SU,Sū 15,19 holds *ojas-* to be the *para- tejas-* of the *dhātu*-s, as well as *bala-* "vigour" (see n.1815). Ḍalhaṇa in his commentary says that in other works/traditions (*tantrāntare*) *ojas-* is also used for *rasa-*, blood (*jīvaśoṇita-*) and heat (*ūṣman-*); cf. too Cakrapāṇidatta's commentary on SU,Sū 15,19 (15,18 in SU3). Explaining SU,Sū 15,21f. (15.20f. in SU3), Cakrapāṇidatta comments on *ojas-* denoting blood with diametrically opposed readings in SU3 and SU4, according to which Caraka says respectively that *ojas-* and blood are either not the same or the same.[1822] According to BH,Śā 5,1 *ojas-* and *tejas-* are different, both residing mainly in twelve parts of the body (to which the heart does *not* belong). Ḍalhaṇa in his commentary on SU,Sū 15,23, where we find both *ojas-* and *tejas-* mentioned, glosses *tejas-* with ☞ *pitta-* "bile", but in the BH passage one of the twelve parts of the body mentioned as receptacles of *ojas-* and *tejas-* is *pitta-* itself. Cakrapāṇidatta, commenting on CA,Ci 15,3, opines that *tejas-* may denote either the heat of the body (*dehoṣman-*), or else semen (*śukra-*) (here *tejas-* is not merely *derived* from semen), quoting an unidentified verse from a work on *śālākya-* in this context.[1823] Aruṇadatta in his commentary on AH,Sū 11,37-39 tells us that according to "others" (*anye*), *ojas-* is not the *tejas-* derived from semen, but phlegm (☞ *śleṣman-*), this explanation seeming to hinge on the meaning of *para-* in 11,37 being taken to mean not "final" or "supreme", but "different".[1824] SU,Sū 14,3 calls *rasa-* the *tejobhūta-* essence (*sāra-*) of digested food, which has caused Ḍalhaṇa difficulties (cf. his commentary).

Often *ojas-* is said to be in the heart (☞ *hṛdaya-*), where it is variously held by the commentators to be to the measure of six or eight drops.[1825] On *ojas-* in the heart see e.g. CA,Sū 17,74f. and Ci 24,35 with Cakrapāṇidatta's commentary (see also CART on the first CA passage), AS,Sū 19,p.145a and Śā 8,p.345a, ĀY 4,177f. (this referring to the foetus), AH,Sū 11,37 with the commentaries of Hemādri, Aruṇadatta, Śrīdāsa, Parameśvara and Candranandana (on AH2,Sū 11,40), BHĀ 1,3,184 and ŚĀR 1,5,47 with the commentaries of Āḍhamalla and Kāśirāma (as well as ŚRĪVĀSTAVYA, pp.185f.

[1822] SU3: *tasya* (scil. *jīvaraktasya*) *tv ojorūpatā carake noktā*, SU4: ... *carakeṇoktā*.

[1823] *dṛṣṭis tejomayī proktā śukraṃ tejaś ca kevalam*
tasmād dṛṣṭibalāpekṣī tejovṛddhiṃ samācaret.

[1824] *anye tv āhuḥ paraśabdenaitad dyotayaty anyad apy ojo 'sti na tad dhātūnāṃ śukrāntānāṃ tejaḥ. śleṣmākhyam iti*. Śrīdāsa quotes Aruṇadatta in the form ... *ojo 'sti na tad dhātūnāṃ tejaḥ śuklamalākhyam iti*, and does not mention *śleṣman-*.

[1825] See also p.530 above.

and EIM, p.105); see also n.420. I am mentioning these passages because often in Indian medical texts it is ☞ *rasa-* which is said to have the heart as its seat, and because they, in contradistinction to e.g. CA,Sū 30,5-12, CA,Ni 4,7, AH,Śā 3,19, AS,Sū 19,p.145a and Hemādri's commentary on AH,Sū 11,37-39 quoted above (cf. also EIM, p.218), do not tell us that *ojas-* is used in the sense of *rasa-*. Matters are complicated by AS,Śā 8,p.345a (already quoted on p.530), which differentiates betwees *ojas-* in the heart, and *ojas-* of the nature of *rasa-* (*rasātmaka-*). See on the problem of *ojas-* in the heart and its relation to *rasa-*, apart from the material given in the discussion on *rasa-* below, especially EIM, pp.218f.; see n.420 too. On the other hand, *ojas-* is *somātmaka-* "of the nature of ☞ *soma-*" (i.e. ☞ *saumya-* "somic") according e.g. to SU,Sū 15,21, AS,Sū 19,p.145a, AH,Sū 11,38 and ŚĀR 1,5,18,[1826] just as *rasa-* too is "somic"; however, *ojas-* is also characterised as "fiery" (☞ *āgneya-*) in BHĀ 1,3,183 (mentioned on p.530 above). Note also the different receptacles of *ojas-* and *rasa-* mentioned under ☞ *rasa-* below.

CART on CA,Sū 17,73-75 tells us that according to some *ojas-* is a variety of semen, an opinion the commentator does not favour.[1827] Cakrapāṇidatta's commentary on CA,Sū 30,7 is similar.[1828] We actually have evidence for *ojas-* referring to a procreatory fluid: a Jaina Prakrit text (TAND) calls not semen, but the procreatory fluid of women *oyā-* (i.e. *ojas-*) (e.g. p.5 = 147,l.1, p.6 = 148,ll.35f.;38); that the procreatory fluid is meant is clear not only from the context, but also e.g. from the synonyms *ratta-* (i.e. *rakta-*) on p.6 = 148,l.37 and *soṇia-* (i.e. *śoṇita-*) on p.16 = 158,l.7.[1829]

According to SU,Sū 15,21 *ojas-* is white (*śukla-*) in colour, which clashes with its reddish-yellow colour mentioned elsewhere (see p.532). As it is, SU,Sū 15,21f. is a problematic pair of verses, as the commentaries of Ḍalhaṇa and Cakrapāṇidatta (mentioning several variants) make clear; Cakrapāṇidatta's commentary (SU3, p.274 and SU4, p.119) also mentions a variant with *svaccha-* "clear" in the place of *śukla-*, and loc.cit. we find too a variant of 15,22cd which seems to equate *ojas-* and *rakta-* (this would be in accordance with one of the explanations of *ojas-* mentioned above), namely:[1830]

yad raktaṃ sarvabhūtānāṃ na kvacin nāvatiṣṭhate

"[The whole body is filled with *ojas-*,] which, the blood of all

---

[1826] In the latter passage, *ojas-* is called the ☞ *upadhātu-* of semen; cf. §11.5.

[1827] *anye tu śukraviśeṣam oja āhuḥ. tat tu na manoramam.*

[1828] *kecit tu śukraviśeṣam ojaḥ prāhuḥ. tac ca na manaḥ prīṇāti.*

[1829] See too CAILLAT, pp.49f., also §13.21 below.

[1830] Instead of *tadabhāvāc ca śīryante śarīrāṇi śarīriṇām.*

beings, is not not present anywhere [in the body]."

However, one could also analyse this as:

"[The whole body is filled with *ojas*-,] since the blood of all beings is not not present anywhere [in the body]."

This latter need not say that *ojas*- and blood are the same, but merely that they are closely linked. Yet another analysis is, however, possible; this would tell us only that the *ojas*- is coloured red:

"[The whole body is filled with *ojas*-,] which, red, is not not present anywhere [in the body] of all beings."

In any case, Cakrapāṇidatta's commentary on SU,Sū 15,19 (15,18 in SU3) says that *ojas*- is variously coloured.[1831]

See on the problem of *ojas*- also COMBA 1981, p.205,n.37 and GHĀṆEKAR 1977A, pp.94f.

kaṇḍarā-

It is difficult to say what exactly the structures denoted by this term are; cf. MEULENBELD 1974, p.451 (add GHĀṆEKAR 1986, p.153: "Tendons"). See too Cakrapāṇidatta's commentary on CA,Ci 15,17, as well as Gayadāsa's remarks on SU,Ni 1,25-29 and Śivadāsasena's commentary on DR,Āh 22. I have used the translation "large cords", as this seemed to be the most neutral term; ADRIAENSEN/BARKHUIS/RUIJTERS use "thick sinews" (p.306). According to Gayadāsa's comments on SU,Ni 1,25-29, *kaṇḍarā*-s and ☞ *snāyu*-s are basically the same, the former only being thick versions of the latter, though he also quotes the opinion of "Caraka and others" to the effect that *kaṇḍarā*-s are actually thick ☞ *sirā*-s.[1832]

kapha-

A synonym of ☞ *śleṣman*-.

kalala-

The very first stage of the embryo. The exact meaning of *kalala*- is unclear; COMBA 1981, p.197 translates it as "piccola massa", SCHMITHAUSEN, pp.37;215 as "proto-embryo(nic matter)", SUNESON, p.110,n.7 as "speck" and "indistinct fluidum", Reinhold F.G. Müller as "Flöckchen" (e.g. MÜLLER 1955,

---

[1831] *nānāvarṇatvam cāsya nānāvarṇasarvadhātusārarūpatvād eva.*

[1832] *snāyukaṇḍarayoś cābheda eva. yad āha sthūlā snāyuś ca kaṇḍareti. carakādayas tu sthūlā sirā kaṇḍarety āmananti.*

pp.24;25,n.98) and "Knöllchen" (e.g. MÜLLER 1955, pp.23f., MÜLLER 1963B, p.412,n.15), WINDISCH, p.87 as "geronnene Masse", DASGUPTA 2,p.314 as "jelly-like form", and the translators of ĀY 4,156 and 4,164 as "jelly like embryo" and "jelly like formation" respectively. For references to relevant primary and secondary literature see e.g. TIVĀRĪ 1,pp.147f.,[1833] COMBA 1981, p.197,n.17, SUNESON, pp.110ff., WINDISCH, pp.87-89, SCHMITHAUSEN, p.629 (under "kalala"; there is an addendum to p.37 on p.683 too); add RASAR,Rasendrakhaṇḍa,Yonivyādhi 80 (p.792). Ḍalhaṇa in his commentary on SU,Śā 2,49 says the *kalala-* resembles the mucus of the nose (*kalalaṃ siṅghāṇaprakhyam*). With regard to the expression *kalala-* in general, it may not be out of place to point out that the meaning being discussed here, though the one generally prevalent in Old Indo-Aryan, is not the only meaning of the word, and that in Middle Indo-Aryan (especially in Pali) the word also means "mud", which may indeed be its primary meaning and as such raises interesting questions as to its perhaps being a Middle Indo-Aryan loan in Old Indo-Aryan, and of its possible connections with Old Indo-Aryan *kardama-*. That in its turn leads to the question of whether these two words might not have been derived by the addition of different suffixes (*-la*/*-ma*) to a stem whose actual original form is yet to be ascertained.[1834] In this connection attention may also be drawn to *kalana-*, which according to our dictionaries is only lexicographically attested, but which is found also in SU4, p.117 (in Cakrapāṇidatta's commentary on SU,Sū 15,19),[1835] further in a variant to CA,Śā 4,9;[1836] SUNESON, pp.110,n.7;112 too notes an attestation from the *Vāyupurāṇa*. *kalana-* and *kalala-* could simply be scribal variants, but could also be true variants due to different suffixes (*-na* and *-la*).

kalā-

A unit of time, as well as (in medical texts usually) an anatomical term. On this latter (which poses several problems) see MEULENBELD 1974,

---

[1833] Who on p.148 attempts to equate *kalala-* with modern medical terms, on the — quite unproved — assumption that what it refers to *must* be something referred to by modern medicine too (cf. on this §1.7).

[1834] We would also have to take hyper-Sanskritisms into account.

[1835] *tad uktaṃ kalanopadarśane*. SU3, p.271 (Sū 15,19 is Sū 15,18 there), however, reads *tad uktaṃ kalalopadarśane*.

[1836] The accepted reading of CA is *sarvadhātukaluṣīkṛtaḥ kheṭabhūto bhavaty avyaktavigrahaḥ*, with the variants *sarvadhātukalanīkṛtaḥ* ... and *sarvadhātukalalīkṛtaḥ* ... . Cakrapāṇidatta cites this passage (SU3, p.271, SU4, p.117) as *sarvadhātukaluṣībhūte* (SU4: °*bhūto*) *bhavaty avyaktavigrahaḥ*.

pp.452f.[1837] (cf. also AS,Ut 40,p.314a, ĀY 4,243ff., ĀĀS 6863ff., BYT 2,46-50) and EIM, pp.110f. The best equivalent for the meaning in which it usually appears in medical texts seems to be "visceral part between the receptacles of the elements (☞ *dhātu-*)". It must, however, be pointed out that the meaning of *dhātu-* in this connection is problematic, as the *kalā-*s usually are said to hold flesh (☞ *māṃsa-*), blood, fat (☞ *medas-*), phlegm (☞ *śleṣman-*), faeces, bile (☞ *pitta-*) and semen, whereas the commentators as a rule explain *dhātu-* as referring to ☞ *rasa-*, blood, flesh etc. up to semen, i.e. as the usual chain of elements from the breakdown of food, though efforts are also made to accommodate the other substances mentioned.

As is clear from the foregoing, the number of *kalā-*s is as a rule given as seven, but according to BYT 2,50 (quoting from the *Vaśiṣṭhasaṃhitā*) = ĀY 4,257, women have three additional *kalā-*s.[1838]

MEULENBELD 1974, loc.cit. points out that *kalā-* may also refer to a layer of skin; cf. on this also §11.11.[1839]

For an interesting discussion of the term *kalā-* in various other meanings see Ḍalhaṇa's commentary on SU,Ut 66,7; cf. WEZLER 1993A, p.129,n.98 too. Note also the term ☞ *(candra)kalā-* in works on sexology.

kiṭṭa-

A synonym of ☞ *mala-*.

kṛmi-

Wormlike creatures, not necessarily true worms; see on this term especially MEULENBELD 1974, pp.622ff. Modern apologists often prefer to translate *kṛmi-* with words such as "micro-organism", "bacterium" or the like, obviously to bring in modern Western medical notions.

koṣṭha-

In anatomy, this as a rule refers to the internal cavity of the body. On problematic usages of the term see the discussion on ☞ *srotas-* below.

1837 Note that AS,Śā 5,p.318b and AH,Śā 3,9f., quoted by Meulenbeld, are both commented upon in nearly identical words by Indu.

1838 *strīṇāṃ daśa kalā jñeyā garbhasya stanayor api* (ĀY: *stanayos tathā*).

1839 Note that several modern Indian scholars prefer the translation "membrane" for the anatomical term *kalā-* in general.

klība-

As in the case of similar expressons referring to certain persons one would in many societies today tend to label 'abnormal', such as e.g. ☞ *dviretas-*, ☞ *napuṃsaka-*, ☞ *paṇḍaka-* and ☞ *ṣaṇḍha(ka)-*, it is very difficult to define *klība-* exactly; the term is often translated as "neuter" in the sense of "eunuch" or the like, in which case it might also mean "sterile", but it is also more often than not reproduced by "impotent" or "lacking manliness", which also would fit several occurrences very well. But the worth of such translations is debatable unless we have more in-depth studies of the term,[1840] as the data culled from medical texts makes clear.[1841]

SU,Ci 7,36 and AS,Ci 13,p.213a mention a structure or structures carrying semen *and srotas-*s coming from the testes, damage to the former causing "*klība*--hood" (*klaibya-*), but of the latter ☞ *dhvajabhaṅga-*, which is commonly taken to refer to lack of the faculty of erection (cf. on this also n.229 and n.308). From this it follows that, whatever *klaibya-* may mean, it is not the same as *dhvajabhaṅga-* in these passages. But *dhvajabhaṅga-* is one of the causes of *klaibya-* according to CA,Ci 30,154ff. (see especially 162 and 175). The causes of *dhvajabhaṅga-* are enumerated in 30,163-168ab,[1842] after which various symptoms (some looking very much like those of venereal diseases) are described, among them being also the destruction or decay of the glans penis, the penis and the testes;[1843] though it appears that these symptoms too relate to *dhvajabhaṅga-*, since the causes of this immediately precede them, we are told in 30,175cd that all this refers to the *klaibya-* which arises due to *dhvajabhaṅga-*. Right after this follows the statement (in 30,176ab) that this *dhvajabhaṅga-* is of five sorts according to some.[1844] Is *dhvajabhaṅga-* here the same as the *klaibya-* caused by *dhvajabhaṅga-* referred to in 30,175? Or did 30,176ab rather originally follow immediately upon 30,168ab, which would mean that 168cd-175 are later additions? Whatever

---

[1840] NAMOUCHI, pp.199f. is an attempt in this direction, but quite cursory and thus far from satisfactory, though we do see from it that the term seems to have several meanings. Cf. also DONIGER, p.58,n.150.

[1841] Cf. also n.1931.

[1842] These include an obstruction or keeping back of semen: *retasaś ca pratīghātād dhvajabhaṅgaḥ pravartate* (168ab).

[1843] 175ab: *viśīryate maṇiś cāsya meḍhraṃ muṣkāv athāpi ca.*

[1844] *etaṃ* (var: *evaṃ*) *pañcavidhaṃ kecid dhvajabhaṅgaṃ pracakṣate*. Çakrapāṇidatta says that this is according to whether it is caused by wind, bile (☞ *pitta-*), phlegm (☞ *kapha-*), blood, or the combination of all these (on the four substances mentioned cf. ☞ *doṣa-*).

the correct answer may be, it is clear that we have a problem here.

According to CA,Sū 19,4,5 *klaibya-* is caused by damage to the *bīja-* (semen?, the *bīja-* of the embryo?; cf. §§2.5f.), *dhvajabhaṅga-*, old age and *kṣaya-* (on this term cf. §§5.2f.) of semen.[1845] This is more or less echoed in CA,Ci 30,154f.[1846] (and explained in 155-191), though note that instead of *dhvajabhaṅga-* we have here damage to the penis (*dhvajopaghāta-*); Ci 30,162;175 obviously refer to the same cause by using *dhvajabhaṅga-*. BH,Sū 26,19 lists obstruction/retention of semen (*śukloparodha-*), weakness (*daurbalya-*), *dhvajabhaṅga-* and *kṣaya-* of semen (*śukrakṣaya-*). According to BHĀ 2,72,2ff.(pp.813f.) = YOG 2,Uttarakhaṇḍa,Vājīkaraṇādhikāra (pp.479f.), which seem to be an amplification of SU,Ci 26,9-15, there are seven sorts of *klaibya-*, amongst which *dhvajabhaṅga-* is also listed (this results from a *kṣaya-* of semen due to too much intercourse or non-usage of aphrodisiacs; thus also SU,Ci 26,11f.), but also non-erection of the penis[1847] due to mental and emotional factors (this is *mānasa-* "relating to the mind") as well as severing of the vessels (*śirā-* = ☞ *sirā-*) carrying *vīrya-* "virility; semen". SU,Ci 26,9f. too (similar in many respects to BH,Sū 26,19), listing different types of *klaibya-*, describe a *klaibya-* which is *mānasa-*. SU,Ci 26,11 also gives *saumyadhātukṣaya-* "*kṣaya-* of the somic (☞ *saumya-*) element" as one of the causes of *klaibya-*; this refers to *kṣaya-* of semen (cf. n.423). HĀ 3,47,p.359 says that there are five sorts of *klaibya-*, and that obstruction/retention (*nirodha-*), too much intercourse (*ativyavāya-*) and fatigue due to age (*vayaḥśrānte mānave*) cause loss (*hāni-*) of semen (*retas-*) and *klība*--hood (*klībatva-*). Three sorts of *klaibya-* are *mānasa-*, caused by *kṣaya-* of semen (*retas-*), and one of these has as its symptom *dhvajabhaṅga-*.[1848] According to SU,Ci 26,12 another sort of *klaibya-* is caused by damage to the relevant ☞ *marman-*,[1849] which Ḍalhaṇa explains as being due to severance of the tubes (☞ *nāḍī-*) carrying

---

[1845] *bījopaghātād dhvajabhaṅgāj jarāyāḥ śukrakṣayāc ca.*

[1846] *bījadhvajopaghātābhyāṃ jarayā śukrasaṅkṣayāt* (154cd)
klaibyaṃ sampadyate tasya śṛṇu sāmānyalakṣaṇam (155ab).

[1847] Literally: "the standard falls down" (*dhvajaḥ pataty ato nṛṇāṃ klaibyaṃ samupajāyate*).

[1848] The HĀ passage is also found in SC 4,2-6, with *sahajaṃ dhvajabhaṅgāc ca jāyate klībatā nare* in the place of *sahajaṃ śuṣkasaṃsvedāj jāyate klībatā nare*, *sparśane svedakampau ca* in the place of *sparśane svedakaṃ pañca*, and *vyavāyena manaḥkṣatiḥ* in the place of *vyavāyena manaḥkṣitiḥ*. Note that according to the first of the variant readings of SC mentioned, besides the *sahaja-* and the three *mānasa-* types there is a fifth sort of *klaibya-* caused by *dhvajabhaṅga-*.

[1849] *marmacchedena vā punaḥ | klaibyam etac caturthaṃ syān nṛṇāṃ puṃstvopaghātajam.*

semen.[1850] The corresponding BHĀ passage mentioned above has ☞ *sirā-* "vessel" in the place of *marman-*, and gives non-erection of the penis as a result.

CA,Ci 30,188 tells us that incurable *klaibya-* is according to "some" (*kecid*) said to arise not only from *dhvajabhaṅga-* and *kṣaya-*, but also due to amputation of the penis and removal of the testes (on this see n.229 too). Cf. in this regard also the *vātikaṣaṇḍhaka-* mentioned in CA,Śā 2,21 and discussed under ☞ *ṣaṇḍha-* below.

In CA,Ci 30,155-157 the general symptoms of all *klaibya*-s are given as lack of sexual desire (if *pravaṇa-* in *saṅkalpapravaṇa-* means "decayed, disappeared"), inability to have intercourse due to limpness of the penis (this is not termed *dhvajabhaṅga-*, but *liṅgaśaithilya-*), paucity of breath, fits of sweating, fruitless efforts at intercourse, withered (*mlāna-*) penis and lack of semen (*bīja-*, var: *vīrya-* "virility; semen").[1851] In their commentaries on SU,Sū 11,28 (SU3: 11,17, SU4: 11,24), Ḍalhaṇa and Cakrapāṇidatta explain *klība-* as *kṣīṇaśukra-* "whose semen is weakened/waning" and *alpaśukra-* "whose *śukra-* is scant", as opposed to other persons who are *aśukra-* "without semen". *klaibya-* is explained as lack of sexual desire in Indu's commentary on AS,Sū 19,p.140b,[1852] and maybe also in CA,Ci 30,155.[1853] MĀDH,Pariśiṣṭa,Klaibyanidāna 1ff. contains a detailed description of different types of *klaibya-*: both impotence (2: *aśaktaḥ surateḥ*) and sterility (4: *nirbījatā-*) are types of *klaibya-*, as is also *dhvajabhaṅga-* (9ff.); however, the text could be relatively modern,[1854] and as such its worth in the context of the 'classical' texts doubtful.

In AH,Śā 1,50 *klība-* is used more or less in the same context as ☞ *napuṃsaka-*. Here *klība-* refers to an embryo, obviously neither male nor female, and in this context it may not be out of place to point out that any study on *klība-* and *klaibya-* must also come to terms with such a usage of these words.[1855] Aruṇadatta explains *klība-* in AH,Śā 1,50 as denoting a

---

[1850] *marmacchedeneti śukravahanāḍīnāṃ chedanāt.*

[1851] Commenting on SU,Sū 15,5,1, which says that semen causes *dhairya-* "firmness; patience; fortitude", Ḍalhaṇa explains *dhairya-* as pertaining to heroism or valour, saying further that persons who are *klība-* are therefore *adhīra-*, i.e. without this characteristic (*dhairyaṃ śauryaṃ śūratvam. ata eva klībā adhīrāḥ*).

[1852] *klaibyaṃ ratyanabhimukhatā.*

[1853] *saṅkalpapravaṇa-*; see above.

[1854] See p.512 above.

[1855] This problem does not seem to have been seen by MÜLLER 1965, pp.37f., where *klība-*

☞ *ṣaṇḍha-*, neither male nor female.[1856] Indu in his commentary on AS,Śā 12,p.363b too says: *ṣaṇḍaḥ* (= *ṣaṇḍhaḥ*) *klībaḥ*, whereas in his commentary on AH5,Śā 1,49 = AH,Śā 1,50 he explains *klība-* as ☞ *napuṃsaka-*. In CA,Śā 2,19 both male and female *ṣaṇḍha-*s are described as being, among other things, *klība-*, impotent (*abala-*) and not arousable (*aharṣa-*); see on this passage the discussion on *ṣaṇḍha-* below. Cakrapāṇidatta in his commentary on CA,Śā 2,19 explains *klība-* as meaning "whose *bīja-* is corrupted" (*duṣṭabīja-*); in this connection cf. BYT 147,27f. (quoted in n.807), where *bījadoṣa-* "deficiency of the *bīja-*", "*ṣaṇḍha-*-hood" (*ṣāṇḍhya-*) and *klaibya-* of women are linked. Cakrapāṇidatta also says (though the text of CA itself does not) that the male and female *ṣaṇḍha-*s mentioned are sterile (☞ *abīja-*), pointing too to SU,Śā 2,44 (this passage is discussed in §8.7), which says that a *ṣaṇḍha-* has no semen (*śukra-*).

ĀĀS 6226 holds a *klība-* to be someone who is unable to have intercourse, and this state to be *klaibya-*, which is of seven sorts.[1857] From the following compilation of verses (pp.906ff.), among them some quoted above, we learn not only about various types of *klaibya-* already discussed above, but also that *dhvajabhaṅga-* is a sign of one form of *klaibya-*.

On *klība-* and *klaibya-* see also n.807 and n.1008. Matters are complicated by the fact that *klaibya-* also has a more general meaning (most probably "weakness" or "timidity" or the like), which is found in medical texts too, thus e.g. in CA,Śā 4,14, where *klaibya-* is one of the characteristics of women.

khagamukha-

In NĀG 19,4 *khagamukha-* seems to refer to the uterus or womb (see n.1556). But *khaga-* as a rule means "moving through the void", and is as such used mostly for birds, but also for other flying objects and animals, as well as for the wind and for heavenly bodies. Such a meaning of *khaga-* does not seem to fit here at all. On the other hand, *kha-* also refers to apertures as well as cavities of the body, and *ga-* can mean "leading to"; "leading to the aperture" could thus be a kenning for the female reproductive tract, i.e. the ☞ *yoni-*, in this case the vagina alone, leading to the *kha-*, the womb, from which it would follow that *khagamukha-* is the same as ☞ *yonimukha-* in the

---

in AH,Śā 1,5 and 1,50, referring to the embryo, is translated as "entmannt" (i.e. "emasculated"), though this must provoke one to ask how an embryo can be emasculated.

1856 *klībaḥ ṣaṇḍho na strī na pumān iti.*

1857 *klībaḥ syāt suratāsaktas tadbhāvaḥ klaibyam ucyate*
tac ca saptavidhaṃ p[r]oktaṃ bhiṣagvaryair vicakṣaṇaiḥ.

sense of "mouth, i.e. end, of the vagina". But *ga-* can also mean "situated in", and *khaga-* "situated in the [visceral] cavity" could refer to the whole genital tract, whose *mukha-* would also be the uterus. But both cases do leave one with a feeling of unease, and so the actual explanation may be quite different. In several parts of the world words for "bird" are used to refer to the male genitals (cf. e.g. SIMS-WILLIAMS/HAMILTON, pp.77f.). This is for instance also the case in Bengali (especially in East Bengal); here, however, the word *pakki/pokki*[1858] or even *pukki* (the Old Indo-Aryan equivalent of all is *pakṣin-* "bird") is used, as a rule not in cultured speech, both for the male and the female genitals (though probably not in both meanings in the same regions).[1859] Similarly, Vedic *śakuntikā́-* (feminine) and *śakuntaká-* (masculine), translated quite aptly as "birdikins" by JAMISON 1996, p.70, clearly have a sexual connotation, though the exact meaning is unclear.[1860] I know of no further Indian parallels for such a use of "bird" words, but I would submit that we at least consider the possibility of *khaga-* "bird" here being used in a similar way, i.e. as a synonym of *yoni-*, from which it would follow that *khagamukha-* is indeed a synonym of *yonimukha-*. Finally, there is also the possibility that *khagamukha-* is simply a mistake for *bhagamukha-*, which, since *bhaga-* is a known name for the female genitals, would obviously be a synonym of *yonimukha-*.

garbhāśaya-

Receptacle of the embryo or foetus. We have in our texts no clear and detailed description of this receptacle, though of course that has not stopped many modern Indian scholars from transposing current notions on the uterus into the ancient texts (cf. e.g. TIVĀRĪ 1,pp.9ff.). On the other hand, the texts clearly do presuppose some sort of internal receptacle opening to the exterior via the vagina, and even if we cannot be sure that the notions regarding this corresponded in all respects to what we today understand under "uterus" (cf. n.12; the matter deserves further investigation), it is methodically most unsound to simply deduce from the lack of clear descriptions, as Reinhold F.G. Müller does, that the texts do not know of any such organ,

---

1858 In the latter case the orthography follows the pronunciation, as *a* in the first syllable of a word followed by *i* or *u* is usually pronounced *o*.

1859 There are parallels for the same terms being used for male and female genitals, also in Indo-Aryan; cf., apart from the Vedic usage of the dual of *muṣka-* referred to on p.86, e.g. no.10917,8 in TURNER 1966 and TURNER 1985 (on similar uses of a family of related Slavic words to refer to both male and female genitals see REINHART).

1860 Cf. JAMISON 1996, p.70.

and that *garbhāśaya-* refers merely to some vague place where the embryo appears. Cf. MÜLLER 1942, p.43,n.2: "... örtlich allgemein als Ausdruck für den Bauch oder Unterleib. Das eigentliche Fortpflanzungsorgan war den altindischen Ärzten unbekannt", MÜLLER 1955, p.13,n.44: "Keimlage(r)", "wo der Keim (oder Embryo) sich einstellt", MÜLLER 1941-1942, p.162: "Keim-Einlagerung", "Stätte der Keimbildung im Innenleib", MÜLLER 1961B, p.94: "des Keimes (oder des Embryos) Einlagerung", MÜLLER 1963B, p.411: "Lagerung für den Keim", MÜLLER 1965, p.40: "Lager vom Embryo" and "Embryo-Lager"; see also MÜLLER 1965-1966A, pp.207;209,n.19, MÜLLER 1960C, p.720 and MÜLLER 1950, p.1442. According to MÜLLER 1963B, p.411,n.7, this does not seem to represent a receptacle ("Behälter"). MÜLLER 1965, p.40,n.154 is willing to concede that the Indian physicians may have known of the uterine cervix and may even have felt and seen it, but from this, he writes, one cannot deduce that they knew of the uterus — even in the case of (what he regards as later) descriptions of disorders which may be taken to be uterine prolapse.[1861] Now it is true that there might indeed have been ancient theories in India that did not show any knowledge of the existence of the uterus (see the discussion in §6.20), but of such theories, should our deductions as to their existence be sound, only traces seem to have survived, and in any case they can hardly simply be generalised in the way Müller has done.

See also the discussion on ☞ *yoni-* below.

gavínikā-

The meaning of this Vedic term is, like that of the obviously related and also dual *gavīní-* or *gavíní-*, unclear, but many take it to be "groin" (cf. also MAYRHOFER 1992-, p.480), and there is indeed much that speaks for this identification. But TEWARI, p.70, eager to bring the Fallopian tubes into the picture,[1862] says that these are meant by the term. This is stated apodictically, without a shred of evidence and no attempt at all at a philological study, and also in spite of the fact that taking such a meaning in a context such as e.g. AV 1,11,5[1863] is patently absurd, since this verse tells us that the *gavínike* are split apart to allow for easy delivery![1864]

[1861] "... ist bei den Schilderungen, welche als prolapsus uteri angesprochen werden können, eine indische Bewertung dieses Organes nicht zu sichern".

[1862] Cf. also Excursus II at the end of chapter 6 (pp.97-100).

[1863] Quoted in n.1370.

[1864] This, incidentally, also speaks against the equation of *gavīnyau* and ureters we find in SEN 1,p.75.

guṇa-

Quality (also translated as "property" and "attribute"). The term can be used in various senses, in medical texts, however, very often to refer to a series of qualities such as heavy and light, hard and soft, hot and cold, solid and flowing etc. See on these qualities, as well as on the various other meanings of *guṇa-*, MEULENBELD 1974, pp.460f. and 1987, p.8,n.19, and SHARMA,P.V. 1982, pp.87f. and 1985, p.32; cf. also HALBFASS, pp.113ff.;72f., KANNU, pp.97ff.;12ff.,FRAUWALLNER1,p.487("Eigenschaften");2,p.342("Eigenschaft"), BHADURI, pp.110ff., PREISENDANZ 2,pp.683f., NARAIN, pp.141ff., BRONKHORST 1994, LARSON 1987, p.249, CART on CA,Sū 1,49 and Excursus VI at the end of chapter 7, as well as n.522 and p.165 above. The discussion of the term ☞ *dravya-* below should also be referred to.

Some commentators of medical texts operate with explanations which ultimately amount to a non-differentiation between "quality" (*guṇa-*) and "that characterised by qualities" (*guṇin-*), i.e. "substance" (*dravya-*).[1865] In this connection see also the remarks on *guṇa-* "derived matter" in n.522. See also WEZLER 1985, pp.13f.

The gross elements (☞ *mahābhūta-*) may be referred to not only as *dravya-*s, but also as *guṇa-*s;[1866] cf. on this Excursus VI at the end of chapter 7 (also on the ☞ *dhātu-*s being called *guṇa-*s).

grahaṇī-

On the (imaginary?) organ *grahaṇī-* cf. §§7.27f., also MEULENBELD 1974, pp.619;219f, EIM, pp.20f.;96, DASH 1971, pp.35f., ŚĀSTRĪ, pp.xiii;77ff. It is usually taken to be the seat of the digestive fire (☞ *agni-*),[1867] and as a rule held to be situated just above the receptacle of digested food (*pakvāśaya-*) and below that of undigested food (*āmāśaya-*).[1868] Though mostly said to be part of the sixth ☞ *kalā-* (containing bile), it is by Cakrapāṇidatta in his commentary on CA,Ni 1,26 said to be situated in a part of the *āmāśaya-*, and seems as such to refer to a part within the *āmāśaya-*.[1869] One could maybe

---

[1865] Cf. e.g. n.514 and p.209 in Excursus VI at the end of chapter 7.

[1866] The gross elements themselves have *guṇa-*s; cf. n.456.

[1867] Thus also ĀĀS, p.123: *grahaṇīm agnyadhiṣṭhānam*, with reference to CA,Ci 15,56f.

[1868] Cf. in this regard also SU,Sū 21,10 in §7.27, but see also n.603 on the seat of the digestive fire.

[1869] *pittaṃ hy āmāśayapraviṣṭam eva bhavati. tac cāmāśayaikadeśasthaṃ grahaṇīrūpāmāśayaikadeśaṃ vahnisthānaṃ svasthānād gacchatīti yuktam* (commenting on *sa [śleṣmā] prakupitaḥ praviśyāmāśayam ... agniṃ nihatya paktisthānād ūṣmāṇaṃ bahir nirasya ...*) (cf. §7.34).

also hold that what is meant is that the *grahaṇī-* adheres externally to the *āmāśaya-*, but in his commentary on Ni 1,20 Cakrapāṇidatta says expressly that the *āmāśaya-* is the seat of the fire (*āmāśaya eva vahnisthānam*; *vahni-* = ☞ *agni-*), that, however, the *grahaṇī-* is the actual seat, situated in a part of the *āmāśaya-*, an entry of wind into the latter not automatically entailing a corruption of the former.[1870] From this it follows that according to Cakrapāṇidatta the former is indeed within the latter. On the fire within the *āmāśaya-* cf. n.603; see also n.570 and §7.34. I may also draw attention to the passages listed in §7.35 unequivocally stating that the abode of bile ☞ (*pitta-*) is the *āmāśaya-*, as well as the variant to CA,Ni 1,20 stating that the *āmāśaya-* is the seat of heat (*praviśyāmāśayam ūṣmaṇaḥ sthānam ūṣmaṇā saha*).[1871]

Gananath Sen (on whom see n.11) identified the *grahaṇī-* as the duodenum (SEN 2,p.208).

(candra)kalā-

According to the theory of the *(candra)kalā-*s, the susceptibility of different parts of women's bodies varies in accordance with the phases[1872] of the moon. See on this notion e.g. SCHMIDT 1922, pp.298ff., RAI, pp.89-93 (also the charts on pp.8f. of SMARAD);[1873] cf. also n.1558. The meaning of *(candra)kalā-* seems not to be the same in all texts, and might show the semantic development: "phase of the moon (connected with special susceptibility of a particular body part)" > "the special susceptibility of a particular body part (connected with a particular phase of the moon)" > "a particular body part having special susceptibility" (thus seemingly e.g. in ŚṚṄG 2,82ff.).[1874] A synonym of *(candra)kalā-* is *kāmakalā-*, in which connection attention may

---

[1870] *vahnisthānasyāmāśayaikadesatvenāmāśayapraveśe 'pi nāvaśyaṃ grahaṇīrūpavahnisthānaduṣṭir viśeṣeṇa labhyata iti.*

[1871] The difficulties occasioned by statements such as the above have of course not gone unnoticed; CONCEPT, p.7 (where the *grahaṇī-* is identified as the small intestine and the *āmāśaya-* as the stomach) tries to circumvent them by saying: "**Jāṭharāgni** is a concept comprehending many **Agnis** situated in different parts of the gastro-intestinal tract; i.e. **Āmāśaya, Grahaṇī,** etc."

[1872] How exactly "phase" is to be understood here is unclear. KAHLO, p.259,n.81 prefers to use "times/periods of the moon" ("Mondzeiten").

[1873] On similar notions found amongst the Hindu Balinese, cf. e.g. WECK, p.127.

[1874] Especially the latter meaning should make us reconsider whether the descriptions of women having a certain number of *kalā-*s in Jāyasī's *Padumāvatī* is really based on Arabian models, as KAPP, pp.203f.;206f.;209 assumes, though in this regard one must also reckon with the possibility of two different traditions having been mixed up.

be drawn to *manmathakalāśāstra-* in RAṬ 14,1 quoted in n.1264, which could be an indication that the knowledge of these *kalā-*s gradually came to be regarded as a separate branch of learning, though in the context loc.cit. the term does on the face of it seem rather to refer to works on sexology in general (*manmathakalā-* as "art(s) of Manmatha, i.e. Kāma"). Cf. also the name and subject matter of the work *Kāmakalāvilāsa* (cf. n.1331), which might point to 'Tantric' connections. A connection of some sort is indeed made very probable when one considers not only the many points of contact between texts belonging to the latter field and sexological works in general, but also that the subject is found even in quite recent works belonging to traditions called 'Tantric', thus e.g. in Bengali Sahajiyā Vaiṣṇava texts (see e.g. PARĀŚAR, pp.254f.; here the different locations of Kāma in both male and female in accordance with the different phases of the moon are described). Since such correspondences may, however, also point to the sexological tradition having been the giver, the contacts in this sphere between the two sorts of literature have to be examined in more detail. Nevertheless, such connections, also found in mediaeval Bengali Sufi literature[1875] (and thus probably much more widespread in South Asia as a whole), could also point to a connection with the common term for semen in esoteric (so-called Tantric) terminology, namely *candra-*, in this context then with reference to the man as well as the woman. See in this context also CHÖPEL, pp.41;214f. on the movement of the "physical essence" in the body of a woman.

jarāyu-

In very precise terminology the outer foetal envelope, perhaps the chorion; cf. on this the quotation from the *Aitareyabrāhmaṇa* in the discussion on ☞ *ulba-* above. But *jarāyu-* is often used in the sense of "afterbirth" and "foetal envelope (undifferentiated)" too. On this problem see the discussions on ☞ *aparā-* and ☞ *ulba-*; *jarāyu-* is also found explained as *ulba-/ulva-*.[1876] See too COMBA 1981, pp.191,n.7;196,n.13;208,n.41 and FILLIOZAT 1964, p.150 on various uses of *jarāyu-*. According to what KĀ,Śā,Gar 7 says, the *jarāyu-* could be the decidua vera, as the uterus (☞ *garbhāśaya-*) is over-

---

[1875] See on this e.g. CASHIN, p.106,n.156, p.163 (where the moon too is mentioned), p.173 and p.185. It may be noted that according to the data presented the notion of a vital fluid ☞ *rasa-* in the body also seems to be connected with this theory, in which connection attention may also be drawn to n.1546.

[1876] In the text quoted in n.1878, Ḍalhaṇa says that the *jarāyu-* has the form of the *ulba-* (*ulbākāra-*).

spread (*parivīta-*) by it;[1877] since the chorion is just beneath the decidua (the decidua reflexa), it may thus be that *jarāyu-* was (also?) used to denote both, the two not being differentiated. For the interesting explanation of *ulba-* and *jarāyu-* in RASA, Upodghāta,pp.107f. see p.529 above, and on the very late meaning "uterus, womb" of *jarāyu-* the discussion on ☞ *aṇḍa-* above. Note too that *jarāyu-* may also refer to a membraneous covering not necessarily connected with a foetus, also in medical texts, as in SU,Śā 4,7.[1878]

ḍimba-

A synonym of ☞ *aṇḍa-*. See also n.1737.

tejas-

Fiery energy. On its relationship to ☞ *ojas-* see the discussion on the latter above. Sometimes seemingly also used as a synonym of ☞ *agni-*.

tvac-

Skin. In medical texts, this often refers not to the skin as such, but to six skins which appear to be the several layers of the outer covering of the body (i.e. of the skin as such), though what exactly they imply has not yet been studied in detail. The number of skins is sometimes given not as six, but as seven. See on the skins e.g. CA,Śā 7,4, Ci 15,17, SU,Śā 4,4, Śā 5,6, AH,Śā 3,8f., AS,Śā 5,p.318a, BH,Śā 7,1, ŚĀR 1,5,19-22, BYT 2,85-89, ĀY 4,258-262, BHĀ 1,3,279-284, also GHĀṆEKAR 1986, pp.104ff., ĀĀS 6863;6880ff., YOGAC, pp.250f. There also seems to have been a controversy regarding the function of the first skin; see on this the discussion in §10.12. As regards the derivation of the skins from other substances, the data on this given by the medical texts are conflicting; cf. on this problem especially §11.8, and also n.1086.

The medical texts also refer to an undifferentiated entity *tvac-* which is a ☞ *dhātu-*, and which often seems to be, and in a few cases actually is expressly said to be, part of the chain of metabolic transformation of food, taking the place of ☞ *rasa-*. On the extremely complex problem of the relationship between *tvac-* and *rasa-* (and also water) see the lengthy discussion in §§10.7ff.

Commenting on AH,Sū 6,69-71, Hemādri differentiates between *tvac-*

1877 *jarāyuṇā parivītaṃ sa garbhāśaya ucyate.*

1878 *santatāṃś ca jarāyuṇā ... kalābhāgāṃs tu tān viduḥ.* Ḍalhaṇa: *jarāyur ulbākāro yena veṣṭitāḥ prāṇino jāyante kalāveṣṭako 'pi tadvad eva.*

and *carman-*: *na hi tvag eva carma. kiṃ tarhi. bāhyā tvak saṃhatiś carma*; the last line means: "the outer one is *tvac-*, the collection is *carman-*". This seems to imply that *tvac-* is only the outermost layer of the skin on the exterior of the body, whereas *carman-* is the totality of all the layers. However, one could also read this line as: *bāhyā tvaksaṃhatiś carma* (var: *bāhyatvaksaṃhatiś carma*) "the outer collection of the (var: the collection of the outer) *tvac*-s is *carman-*". Clearly, in all these cases one faces difficulties when also considering the theory of six (or seven) skins. But whatever Hemādri may actually be saying, one does indeed gain the impression that in Old Indo-Aryan in general *carman-* often (though not always) corresponds to "hide" rather than to "skin", not necessarily after it has been detached from the body (though in Vedic usage the latter might have been the case; see also JAMISON 1986, p.172).

dūṣya-

Usually refers to a ☞ *dhātu-* capable of being corrupted by a ☞ *doṣa-*, but other substances of the body capable of being corrupted can be called so too; on the latter see e.g. MEULENBELD 1991, p.104,n.5, as well as the commentaries of Gayadāsa and the *Madhukośa* respectively on SU,Ni 13,25cd = MĀDH 55,22ab.

doṣa-

Though in medical texts also used in the meaning "fault; deficiency; defect; disorder" or the like, *doṣa-* is as a technical term in general the commonest name for the three morbific entities wind, bile (☞ *pitta-*) and phlegm (☞ *śleṣman-*, *kapha-*). According to VOGEL 1995, p.77, the correct translation of the term should be "source of faults" ("Fehlerquelle"), since *doṣa-* is derived not from the simple verb √*duṣ*, but from the causative.

On *doṣa-* in general (also on when a *doṣa-* is called a ☞ *dhātu-*) see also VṚK, pp.276f.;519f., ROY, p.165, MEULENBELD 1974, pp.469f. and 1992, p.1, ZIMMERMANN 1989A, pp.129;134f., Cakrapāṇidatta on CA,Ci 28,248, and ŚĀR 1,5,23f. (24 = ĀĀS 6887) (whence we learn that a *doṣa-* can be not only a *dhātu-*, but also a ☞ *mala-*). On the term *doṣa-* subsuming not only the ☞ *dhātu*-s (as elements of the chain of transformation of food), but also the ☞ *mala*-s see MEULENBELD 1992, p.1; cf. also Gayadāsa's commentary on SU,Ni 2,4.[1879] ŚĀR 1,5,23 ≈ ĀĀS 6884 could be telling us that *doṣa-* refers to the morbific entities, the *dhātu*-s and the *mala*-s, though it could also be

[1879] *doṣaśabdo duṣṭeṣu dhātuṣv api vartate.*

understood as saying that the morbific entities are called *doṣa-*, *dhātu-* and *mala-*.[1880] To the various meanings of *doṣa-* mentioned by MEULENBELD 1992, loc.cit., add too the characterisation of *rajas-* and *tamas-* as *doṣa-*s relating to the *manas-* in CA,Sū 1,57, Vi 6,5 and Śā 4,34 (with *sattva-* in the place of *manas-*),[1881] AS,Sū 1,p.7b = AH,Sū 1,21;[1882] see also the discussion on *dhātu-* below, and n.681 — as well as FILLIOZAT,J. 1964, pp.170-172 and MÜLLER 1930, p.335,n.1, which criticise attempts to project the theory of the morbific entities into the oldest Vedic texts (cf. too MEULENBELD 1991, p.91).

There is evidence that blood too was a morbific entity, a *doṣa-*. On blood as a morbific entity cf. MEULENBELD 1990 and 1991, also VOGEL 1959, KIRFEL 1951, p.14, ZIMMER,HEI., p.201 (where each of the entities is connected with one of the four seasons), SU,Śā 7,9-15, Ci 35,6 (quoted also by Cakrapāṇidatta in his commentary on CA,Si 3,26) and Sū 21,10-18 (Cakrapāṇidatta on 21,16 expressly speaks of the *doṣatā-* of blood), Ḍalhaṇa on SU,Ut 12,38f., CART on CA,Sū 1,57, and SIDDH 1,34f., as well as n.179 and n.377.[1883] *Bālatantra* 1,3 too mentions blood on a par with wind, bile and phlegm.[1884] Śivadāsasena says on *sarvadoṣaharaṃ* in DR,Śākavarga [1],42e (= SU,Sū 46,214a): *sarvagrahaṇaṃ raktopasaṅgrahārtham*. The evidence seems to show that the concept of blood as a morbific entity is an archaic remnant, not a later development as Ludwig Edelstein presumes in ZIMMER,HEN.R., pp.xlixf. According to MÜLLER 1951, pp.48f., MÜLLER 1941-1942, pp.461;463, MÜLLER 1960C, p.718, MÜLLER 1961B, p.128 and MÜLLER 1965, p.19 (the latter two also referring to the Bower Manuscript) too, the notion of blood being a morbific entity is ancient, but belongs, according to MÜLLER 1958, p.75, to a variant tradition which could not displace the tradition of three *doṣa-*s; MÜLLER 1951, p.124,n.300 disagrees with a modern scholar who takes the theory of four morbific entities to be later than that of three morbific entities, and imported from Greece. As to the latter opinion, P.V. Sharma on p.xi of his introduction to SU too speaks of "Greek influence".

It is difficult to find an adequate English equivalent for *doṣa-*; "mor-

---

1880 *vātaṃ pittaṃ kaphaṃ doṣā dhātavaś ca malās tathā*
tatrāpi pañcadhā khyātāḥ pratyekaṃ prāṇadhāraṇāt.

1881 Cf. on these passages also ZIMMERMANN 1989B, p.11b.

1882 CART on CA,Sū 9,4 (see p.555 below) seems to speak for these particular substances being called *dhātu-*s too.

1883 See also the text of SU,Śā 2,5 quoted in n.1761, and the discussion on ☞ *raktapitta-* below.

1884 Note, however, that the term *doṣa-* itself as used in this verse probably means "defect" or something similar.

bific entity" is probably the best, as it is a neutral term. "Humour", which is common, creates difficulties inasmuch as it tempts one to transfer the concepts associated with the term in Greek medicine onto Indian medicine without further investigation. Translations such as "juice" or "sap" are happily hardly in use any more, though they do crop up now and then.[1885] Reinhold F.G. Müller in all his works (except for a few early ones) idiosyncratically translates *doṣa-* only as "Fehler" (i.e. "fault; defect"),[1886] with a lot of speculation on this to him only negative expression in medical contexts (see e.g. MÜLLER 1961B, p.105 and MÜLLER 1961A, with further references — also MÜLLER 1967, pp.101-103).

dravya-

Substance. On *dravya-* see Excursus VI at the end of chapter 7, n.522, n.959, §§10.10f. and also n.514 on "substances", as well as GANGOPADHYAYA, SHARMA,P.V. 1982, pp.86f. and 1985, pp.30f., ŚRĪVĀSTAVYA, pp.151ff., FRAUWALLNER 1,p.499("Substanz");2,p.345("Substanz"), BHADURI, pp.22-51;110ff. and eventually OETKE, p.156;[1887] on the problematic terms *dravya-* and ☞ *guṇa-* in other contexts see too WEZLER 1985, also HALBFASS, pp.51;78f., BRONKHORST 1994 and CA,Sū 1,51, as well as HOUBEN, pp.102f., where different specific implications of the terms and their relations in different philosophical systems are mentioned.[1888] Attention may also be drawn to what is said on the terms throughout overviews on Indian philosophy such as DASGUPTA,[1889] the latter (2,pp.360-373) also commenting especially on Indian

---

[1885] Thus e.g. BODE, p.153,n.14 and p.154,n.25 ("lichaamssap").

[1886] Sometimes in addition as "Krankheitsstoff" (i.e. "disease substance") or the like (cf. e.g. MÜLLER 1960C, p.718).

[1887] On Bhartṛhari's *Kriyāsamuddeśa* the latter says: "Es kann aber gar kein Zweifel darüber bestehen, daß *dravya* innerhalb dieses Textstückes so verwendet ist, daß zumindest unter anderem auch konkrete materielle Körper, die als Wirkfaktoren, *sādhana*, fungieren können, unter den durch diesen Ausdruck repräsentierten Begriff fallen."

[1888] Reinhold F.G. Müller in his works vehemently opposes translating *dravya-* as "substance", but what he has to say on the matter is rather confusing, and I cannot quite understand what he actually means to say; see e.g. MÜLLER 1965-1966B, MÜLLER 1965-66, p.63, MÜLLER 1961B, pp.105f., MÜLLER 1961C, pp.13;22,n.43, MÜLLER 1967, pp.107f. — on Müller see also §1.12 above. That "substance" might not in every case be an adequate translation for *dravya-* is also the opinion of FILLIOZAT,J. 1974, p.180.

[1889] This work is specially mentioned because it devotes quite some space to the philosophy of medical texts. There are of course various overviews of Indian philosophy, also such as have an encyclopaedic character, but there is no need to mention such works individually here,

medicine in this regard.

According to CA,Sū 1,48 the gross elements (☞ *mahābhūta-*), along with various other things, are *dravya*-s:[1890]

khādīny ātmā manaḥ kālo diśaś ca dravyasaṅgrahaḥ
sendriyaṃ cetanaṃ dravyaṃ nirindriyam acetanam.

This list of *dravya*-s corresponds to that of the Vaiśeṣika system, on which see e.g. HALBFASS, pp.89ff. KANNU, pp.57ff.;8ff. Cakrapāṇidatta in his commentary on CA,Sū 25,40 too equates the terms *dravya-* and *mahābhūta-* (cf. n.466). See also the material on *pañcendriyadravyāṇi* in Excursus VI at the end of chapter 7.

On *dravya-* in a sense approaching that of ☞ *dhātu-* cf. n.522, Excursus X at the end of chapter 7, and §§10.10f. On a more practical level, *dravya-* can, in medical texts, refer to the ingredients of a medicine.[1891] Moreover, *dravya-* can be used to refer to some particular substance or thing under discussion; an interesting example is found in the commentaries of Ḍalhaṇa and Cakrapāṇidatta on SU,Sū 40,18, where the various *dravya*-s of the verse[1892] are explained as the body, the substances of ingested food, food, the gross elements and the abdominal fire.[1893]

dviretas-

CA,Śā 2,18[1894] states that a *dviretas-* is characterised by the signs of both female and male (*strīpuṃsaliṅgin-*). But according to Cakrapāṇidatta's commentary, the "signs" (*liṅga-*) could refer only to those which men and women have in common (nose, eyes etc.), differentiating features, such as different sexual organs and secondary sexual characteristics, being absent; however, what could according to him alternatively be meant is a person either without or with only slightly developed secondary sexual characteris-

---

as they should be well known generally. See the indexes of all such works.

1890 CA,Sū 1,64 seems to presuppose the same; cf. n.459.

1891 At times to (originally) solid ones as opposed to liquid ones (*drava-*). Thus Ḍalhaṇa, commenting on SU,Sū 45,197: *dravyapradhānam ariṣṭaṃ dravapradhānam āsava ubhayapradhānaṃ madyaṃ*. (On *ariṣṭa-*, *āsava-* and *madya-* cf. MEULENBELD 1974, pp.441f.,445f.,486.)

1892 *dravye dravyāṇi yasmād dhi vipacyante na ṣaḍrasāḥ*
śreṣṭhaṃ dravyam ato jñeyaṃ śeṣā bhāvās tadāśrayāḥ.

1893 Ḍalhaṇa: *dravye pañcabhūtātmake dehe dravyāṇy āhāradravyāṇi.* Cakrapāṇidatta: *dravya āhārarūpe dravyāṇīty āhāragatāni pṛthivyādīni pacyante. ... yadā dravyaṃ dravyeṇa pacyata iti pāṭhas tadā dravyeṇeti jaṭharāgninā.*

1894 Quoted in full in n.230; see also n.103.

tics, the primary sexual characteristics being developed.[1895] In the cases in which the primary characteristics are developed, one could further hold that Cakrapāṇidatta's compound *upasthadhvaja-* refers to the characteristics of *both* sexes being present in the same individual; conversely, the copulative compound could also be taken to have an "or" between the individual members, not an "and" (i.e. the individual then has the primary characteristics of male or female). In any case, Cakrapāṇidatta seems unsure about the exact meaning of the term, and when we try to analyse it independently, we discover too that already the literal meaning of *dviretas-*, "one having two (i.e. both) *retas*-s", is unclear, for "having" could be taken in the sense of "having [at birth]", i.e. "arising from", referring to equal quantities of the procreatory fluids (each of these a *retas-*), or else it could mean "possessing", in which case *retas-* could even be used not necessarily in the old sense of "that which flows; flow; fluid" (as the first explanation would presuppose), but as "semen" (a common meaning, especially in later times), in which latter case we would then also be confronted with the problem of female semen is discussed in detail in our study. Gaṅgādhara (CA1, p.1902) seems indeed to take "having" in the sense of "possessing", for he says that males and females each have only one *retas-*, namely *śukra-* and *śoṇita-* respectively, whereas the *dviretas-* possesses both.[1896] However, though theoretically possible, this seems to be an extremely problematic explanation, for we would ultimately have a person able to generate a child in a woman and at the same time able to conceive or at least to menstruate; in any case, I do not know whether we can *a priori* assume this explanation of a modern scholar to in actual fact derive from some old tradition, though note Buddhaghoṣa's remarks on the ☞ *paṇḍaka-* (see p.562).

Further, Cakrapāṇidatta in the context of one of his explanations mentioned above (that no differentiating sexual characteristics at all are present), obviously taking into account the mention of *bīja-* in CA,Śā 1,18 (cf. n.57), not only analyses the term *dviretas-* as one whose *retas-* is *dvidhā* "twofold" (seemingly: during conception), thus giving rise to both male-hood and female-hood (which then apparently effectively cancel each other out), but

---

1895 *strīpuṃsaliṅgīti strīpuruṣasādhāraṇanāsikācakṣurādiliṅgayuktaḥ. yāni tu strīpuruṣayor asādhāraṇāny upasthadhvajastanaśmaśruprabhṛtīni tāni na sambhavanti. ... kiṃvā strīpuṃsaliṅgīti strīpuṃsayor yal liṅgam upasthadhvajarūpaṃ tadyukta eva strīpuruṣaliṅgī. uttarakālabhāvīni tv asya stanaśmaśruprabhṛtīni na bhavanti. ... kiṃvopasthadhvajau balavantau. yena tau garbhakāla eva bhavataḥ. stanādi tūttarakālabhāvitayā durbalam.*

1896 *dviretāḥ śoṇitaṃ ca śukraṃ ca dve retasī yasya sa tathā. pumāṃs tu śukraikaretāḥ strī tu śoṇitaikaretā iti kasmād dviretā ity asya uttaram.*

also equates *retas-* and *bīja-*,[1897] which leads additionally to the problem of *bīja-* discussed in §§2.5f. Cakrapāṇidatta's terminology is incidentally paralleled by GARBH 3, where the procreatory fluid of both father and mother is called not only *retas-*, but also *bīja-*.

In AS,Śā 2,p.292b *dviretas-* and ☞ *napuṃsaka-* seem to be synonyms referring to someone who is *dviprakṛti-*,[1898] which could mean "having both natures", but also "having two sources"; the latter might refer to the two procreatory fluids. The commentator Indu explains *dviprakṛti-* as referring to someone "having both characteristics" (*ubhayasvarūpa-*),[1899] who functions partly as a woman, partly as a man, and is not a male.[1900]

KOLHATKAR, p.150 draws attention to *dviretas-* in Vedic and holds that the term, referring to certain animals, "is used to explain the concept of cross-breed and the agents and the product of it". Whether this explanation is fully correct remains to be examined.

dhamanī-

Duct. Cf. MEULENBELD 1974, p.470 on this translation, as well as the discussion on ☞ *sirā-* below. On *dhamanī-* see also RASA, Upodghāta,pp.162f., GHĀṆEKAR 1986, pp.236f. and LIÉTARD/CORDIER, pp.197;241f.

dhātu-

In medical texts mostly the seven elements (enumerated in §2.4) which are formed in a process of successive transformation due to the breaking down of food.[1901] On this chain of seven elements see DAS,R.P. 1984A, pp.237ff.,[1902] EIM, pp.54-58 and MEULENBELD 1974, pp.470f. (to the latter's list of references add AH,Śā 3,61ff., AS,Śā 6,pp.333ff., SIDDH 1,12 and BH,Sū 11,1ff., also MÜLLER 1942, pp.48ff.). Often translated as "essential substance" or the like.

As WOLZ-GOTTWALD 1991[1903] rightly emphasises (p.39): "Ein Kör-

---

1897 *dvidhā strītvapuruṣatvopajanakatvena sthitaṃ reto bījaṃ yasya sa dviretāḥ.*

1898 *dviprakṛtir dviretā napuṃsakaṃ bhavati.*

1899 Cf. n.1231.

1900 *strīkāryam api stokaṃ sādhayati puṃskāryam api. na pumān ity arthaḥ.*

1901 According to Cakrapāṇidatta's commentary on CA,Sū 30,7, they are *dhātu-*s because they uphold (*dhāraṇa-*) and nourish (*poṣaṇa-*) the body; see n.1803.

1902 On n.26 there on the pure residue of semen cf. also CA,Ci 15,16 and ☞ *ojas-* above.

1903 An otherwise most problematic work; cf. DAS,R.P. 1992B.

per-Element geht in das andere über. Man stellt sich die dhātus nicht als abgeschlossene Elemente vor[,] sondern als Stationen eines Umwandlungsprozesses." It is probably a reason such as this which has made ZIMMERMANN 1987, pp.162ff. translate *dhātu-* with "tissue", seemingly in the Aristotelian sense (see on this n.12); "tissue" is also the translation ADRIAENSEN/BARKHUIS/RUIJTERS have used throughout. Similar considerations, coupled with linguistic speculation, seem to have led Reinhold F.G. Müller[1904] to rather idiosyncratic translations of *dhātu-* in medical texts, such as "Wurzelstoff" (e.g. in MÜLLER 1951, p.37 and *passim*, MÜLLER 1941, p.66,n.1), as well as "Umwandlung"/"Umsetzung" or "Umwandeln"/"Umsetzen" (see e.g. MÜLLER 1961C, MÜLLER 1961B, pp.106f., MÜLLER 1959/60, pp.206ff. and MÜLLER 1966, with further references), though he also makes what to him obviously seems a concession of sorts by stating that "für eine bequemere Verständlichkeit" one could translate *dhātu-* as "Schichten" or "Schichtung", though not as "Schicht" (MÜLLER 1961C, p.19); cf. too MÜLLER 1967, p.107.

Usually only the "elements" ☞ *rasa-* etc. are, with respect to the chain of transformation of food, technically called *dhātu*-s in Indian medical texts. But the term *dhātu-* basically means "constituent part"; as such it can be used in a variety of meanings both with regard to this chain and with regard to the body in other contexts.[1905] According to CA,Sū 28,4 and AS,Śā 6,p.333b, *dhātu-* subsumes both the elements (here: as the ☞ *prasāda*-s, i.e. pure parts) of the chain of transformation of food as well as the impurities (☞ *mala-*) of this chain[1906] (see also MEULENBELD 1974, p.470). Actually, this might be a remnant of a Vedic 'medical' usage of this term (see n.689); according to ALTHOFF, p.178, *períttōma* is used in a similar sense by Aristotle.[1907] Cakrapāṇidatta in his commentary on CA,Sū 9,4 says: "The *dhātu*-s are wind etc. (i.e. the three morbific entities or ☞ *doṣa*-s) and *rasa-* etc., and further *rajas*- etc."[1908] Since there seems to be no theory in medical texts calling *rajas*- a ☞ *mala-*,[1909] "*rajas-* etc." probably does not refer to the *mala*-s; is Cakra-

[1904] See on him §1.12.

[1905] On such different meanings see also NOBEL, pp.7-19 and EDHOLM, pp.103ff. Attention may be drawn too to CA,Vi 5,8 in n.2030 below and to Indu's statement quoted in n.764.

[1906] CA: *te sarva eva dhātavo malākhyāḥ prasādākhyāś ca* ..., AS: *tataś ca dhātvākhyāḥ prasādamalāḥ* ... . In the AS passage we also find ☞ *guṇa-* as a synonym of *dhātu-*; cf. Excursus VI at the end of chapter 7.

[1907] Cf. too pp.96,n.47;189,n.45 there, and Appendix I to our study.

[1908] *dhātavo vātādayo rasādayaś ca tathā rajaḥprabhṛtayaś ca.*

[1909] Though note that *mala-* is found in some non-medical texts to describe semen and the procreatory-menstrual fluid; cf. e.g. *Agastyasaṃhitā* 22,18 and the notes of the editor on this

pāṇidatta thinking of the problematic series of ☞ *upadhātu*-s, amongst which some theories also count *rajas*- (or *ārtava*- or "blood")? I must in this connection, though, also draw attention to what CART says on CA,Sū 9,4: "Through the word *dhātu*- [we have] the acceptation (i.e. meaning) of wind etc., further of *rajas*- and *tamas*-."[1910] I.e., the term *dhātu*- refers to the morbific entities as well as two of the three constituents of *prakṛti*-,[1911] one of these latter constituents being *rajas*-, which thus is a metaphysical entity and something quite different from the fluids we are considering in this study. Is Cakrapāṇidatta's commentary too to be understood in this sense? Since Śivadāsasena, the author of CART, is known to have been a follower of the school of Cakrapāṇidatta, this would seem very probable.[1912] See, however, also §5.3, as well as *raktadhātu*- in n.1086 and BHĀ 2,32,17-18(2), where *dhātu*- is a synonym of *ārtava*- (*dhāturūparakta*- and *ārtavarūparakta*- are there used in the same sense), as well as n.1108; cf. also §11.32 and §§12.2ff. on *dhātu*- describing fluids of women.

According to Gayadāsa's commentary on SU,Ni 2,4 the three morbific entities (☞ *doṣa*-), the *dhātu*-s (i.e. *rasa*- etc.) and the ☞ *mala*-s are all called *dhātu*-s.[1913] AH,Ni 16,41 = AS,Ni 16,p.82 refers to a list of substances, viz. the three morbific entities, the chain of elements *without rasa*- (i.e. only blood to semen), food, urine and faeces, preceding this by the term *dhātu*- (AH,Ni 16,31-40 = AS,Ni 16,pp.81b-82a). Similar problems arise in connection with the description of the ☞ *kalā*-s.

A remarkable definition is found in CA,Ni 5,3 and CA,Ci 21,15, where the three morbific entities and the four substances (themselves called *dhātu*-s; see also §10.7) skin (☞ *tvac*-), flesh (☞ *māṃsa*-), blood and serous fluid (☞ *lasīkā*-) are called the "seven *dhātu*-s". Seven *dhātu*-s are mentioned in AS,Śā 8,p.344b too, but here the sum of the three morbific entities (☞ *doṣa*-) singly (3), in pairs (3) and all together (1) is meant. In CA,Śā 4,7, semen of the male is called a *dhātu*-[1914] in the form of the *bīja*-;[1915] this could re-

---

verse (part 1, p.171,n.13).

[1910] *dhātuśabdena vātādīnāṃ tathā rajastamasor grahaṇam.*

[1911] On these see e.g. MEULENBELD 1974, p.513.

[1912] Note too the various passages from CA cited in the discussion on ☞ *doṣa*-, which are similar to CART's comment on CA,Sū 9,4 in that they also refer to the three morbific entities and *rajas*- and *tamas*-, but call these *doṣa*-s.

[1913] *dhātuvaiṣamyam iti doṣadhātumalavaiṣamyam ity arthaḥ.*

[1914] On semen as a *dhātu*- see also §12.4.

[1915] *bījarūpo dhātuḥ*. The variant *bījadhātuḥ*, however, may refer to *dhātu*- = *bīja*-, but al-

fer to semen as the *dhātu-* par excellence, since it is the ultimate product of the chain of *dhātu-*s,[1916] but *dhātu-* could also be seen rather in a more basic meaning "primitive matter", "basic constituent" or the like, since semen is one of the substances out of which the embryo and thus ultimately the whole human being is formed.[1917] Both *bījadhātu-* and *bīja-*, seemingly in the meaning of "semen", are found in KĀ,Śā,Asa, p.46 discussed in n.401. On *bīja-* and *dhātu-* cf. also n.1151, as well as, in another context, the synonymy of *dhātu-* and *bīja-* referred to by MUROJI, pp.137f., where *dhātu-* refers to the elements present in e.g., a piece of wood, which allow someone with supernormal mental powers to change this wood into those elements.

On other meanings of *dhātu-* (basically in a philosophical context) see CA,Śā 1,17. The gross elements (☞ *mahābhūta-*s) too may be called *dhātu-*s; this expression is common especially (but not only) in Buddhist terminology, though relatively rare in medical texts. In this regard I may also refer to SLAJE 1993, pp.269ff., where attention is drawn to a seemingly synonymous use of *dhātu-* and ☞ *bhūta-* (a common expression for the gross elements) in Bhāskarakaṇṭha's *Cittānubodhaśāstra*, without it being quite clear in each particular case whether the elements *rasa-* etc. or the gross elements are referred to, even in cases in which semen is called the essence (*sāra-*) of the *bhūta-*s or *dhātu-*s.[1918] See also n.59, n.1108, n.2030 and Excursus X at the end of chapter 7 on other meanings of *dhātu-*.

"Consciousness" (*cetanā-*) is also called a *dhātu-* in CA,Śā 4,8,[1919] which, equating it among other things with *brahman-*, *jīva-*,[1920] *bhūtātman-*, *indriyātman-* and *antarātman-*, states that in the newly formed embryo it takes up (*upādatte*) the gross elements (here not called ☞ *mahābhūta-*, but ☞ *guṇa-* and *dhātu-*),[1921] which would make it the base for the gross ele-

---

so to "*dhātu-* of the *bīja-* (i.e. the most elementary form of the embryo)".

1916 Cf. also *śarīradhātvātmā śukrabhūto* in Śā 4,7, as well as Cakrapāṇidatta's commentary on the passage.

1917 On *bīja-* see §§2.5f.

1918 It is obvious that this may lead to confusion, as pointed out e.g. in n.464. This is on top of the fact that the relationship between the gross elements and the elements of the chain of transformation of food is problematic anyway, as shown e.g. in Excursus VI at the end of chapter 7.

1919 See also Sū 25,9.

1920 Cf. on this n.6.

1921 On the problematic nomenclature of the gross elements in CA,Śā 4,8 see also Excursus VI at the end of chapter 7. According to the generally accepted opinion, *bhūtāni* of *yathā pralayātyaye sisṛkṣur bhūtāny akṣarabhūta ātmā sattvopādānaḥ pūrvataram ākāśaṃ sṛjati*

ments within the embryo; clearly, *cetanā-* is here a sixth *dhātu-* distinct from the other five *dhātu-*s (viz. the gross elements).[1922] According to CA,Śā 5,4f. and BH,Śā 5,11 the sixth *dhātu-* is *brahman-*, which is *avyakta-* "imperceptible/primary".[1923] The first five, i.e. the gross elements, are enumerated in CA,Śā 1,27 too and referred to as *khādi-* "emptiness (scil. ether) etc." in vv.16;24;63;66, Śā 1,16 mentioning *cetanā-* as the sixth *dhātu-*, and Śā 1,63 telling us that *khādi-*, *avyakta-*, *buddhi-* "intellect/perception" and *ahaṅkāra-* "conception of the individuality" are the eight *bhūtaprakṛti-*s, i.e. matrices of beings, which seem to be called *dhātu-*s in 1,17. Cakrapāṇidatta glosses *avyakta-*, which is discussed in detail in Śā 1,60ff. (and called, among other things, *ātman-* "principle of life" and *vibhu-* "omnipresent", but not *brahman-*) as *mūlaprakṛti-* "basic matrix"; see also his commentary on 1,17. I may also add that Cakrapāṇidatta in his commentary on CA,Ci 24,36, in which the *dhātu-*s present in the heart are mentioned, says that this refers to *sattva-*, *buddhi-*, *indriya-* etc. ("etc." probably referring to *ātman-*, mentioned along with the other three in Ci 24,35), even though Ci 24,34f. in this connection mention not only these, but also "☞ *rasa-*, wind etc." and "the chief (*pradhāna-*) ☞ *ojas-*" (what does *pradhāna-* signify in this context?; cf. *para-* on p.532 above). Refuting Vāryovida's theory that all beings are derived from *rasa-*,[1924] Hiraṇyākṣa tells us in CA,Sū 25,14f. that the *puruṣa-* (here obviously: corporeal being) is made up of six *dhātu-*s;[1925] he does not specify these six *dhātu-*s, though Cakrapāṇidatta in his commentary refers (WEZLER 1993A, p.120,n.67: "zu Recht") to Śā 1,16, which we have just mentioned, in his own explanation, however, using not the word *cetanā-*, but (just like CART) *ātman-* to characterise the sixth *dhātu-*.[1926] But Hiraṇyākṣa does seem to relate the six *dhātu-*s to the Sāṅkhya system;[1927] cf. in this context also SHARMA,P.V. 1985, pp.350;352 on Śā 1,16f.;24, and note that the commentaries cited use the term *dhātu-* in further philosophical contexts not

---

in the same passage also refers to the gross elements; whether this is indeed so has to my knowledge not yet been discussed in detail.

1922 See also the discussion of CA,Sū 11,12f. by FILLIOZAT,P.-S., pp.98f. On the role of the gross elements and *cetanā-* as *dhātu-*s in embryology see especially TIVĀRĪ 1,pp.107ff.

1923 Cf. also EDHOLM, pp.105f.

1924 See on this n.463 and §7.22.

1925 15ab: *ṣaḍdhātujas tu puruṣo rogāḥ ṣaḍdhātujās tathā.*

1926 Cakrapāṇidatta and CART, commenting on CA,Sū 25,9, equate *cetanādhātu-* and *ātman-*.

1927 15cd: *rāśiḥ ṣaḍdhātujo hy eṣa sāṅkhyair ādyaiḥ prakīrtitaḥ* (var: *parīkṣitaḥ*).

mentioned above to refer to the substances of which the body is made up, saying, however, that this is not relevant for medicine. In CA,Sū 25,16 Kuśika too speaks of six *dhātu*-s, though he does not specify them.[1928] CA,Vi 8,37 tells us that, though CA speaks of five *indriya*-s (sense-perceptions), other texts assume that there are six of these; that the *indriya*-s are in Indian philosophical speculation connected with the gross elements is well known. Cf. also ŚRĪVĀSTAVYA, pp.240f. On problems in Buddhist philosophical systems related to the equivalents of what is here called six *dhātu*-s see too "ṣaḍ°" in SCHMITHAUSEN, pp.652;650, also "skandha" on p.656 (also see HALBFASS, pp.53f., EDHOLM, pp.106f.).

Six *dhātu*-s are also mentioned in what is probably the oldest extant text with relevant medical material, namely in the *Suvarṇabhāsottamasūtra* (cf. §3.1). The text does not tell us what these six *dhātu*-s are, so it is unclear what exactly *dhātu*- refers to here; cf. on this problem NOBEL, p.8. I-tsing in the Chinese version of the text clearly takes *dhātu*- to refer to the elements of successive transformation due to the breaking down of food, referring, however, not to six, but to seven elements (NOBEL, p.4). His list has, moreover, the added peculiarity that, though the first six elements are those of the usual chain, the last one is not semen, but brain! All this has occasioned NOBEL, p.8 great problems; unfortunately, Nobel has not considered the obvious connections here with notions of the cranial location of semen not only in South Asia, especially in Tantrism, but also in other parts of Asia, including China (cf. on this e.g. DAS,R.P. 1992A, pp.390f.). An additional complication is pointed out by NOBEL, p.8, namely that the Tibetan translator of I-tsing drops the element fat in the list he presents us with, which thus consists of six elements. (Cf. on six *dhātu*-s also Excursus X at the end of chapter 7.)

dhvajabhaṅga-

Literally: "downfall/bending of the standard (scil. penis)", commonly taken to refer to impotence, i.e. lack of the faculty of erection. There are, however, many problems connected with the exact nature of this disorder; see on this the discussion on ☞ *klība*- above.

napuṃsaka-

According to WACKERNAGEL, pp.380f., *napuṃsaka*- (CA,Śā 2,22 has *napuṃs*-) refers to a being neither male nor female, and the general confu-

---

[1928] It is unclear whether Kuśika and Hiraṇyākṣa are identical; cf. WEZLER 1993A, p.121, n.69.

sion the medical texts show when attempting to describe or qualify the *napuṃsaka-* tends to substantiate this view, though an in-depth study of the matter is still wanting. In any case, it is difficult to find an exact correspondence in a Western language. Many scholars translate the term as "hermaphrodite", though on the difficulties of defining a hermaphrodite adequately see PERERA (cf. also KĀM 14(=2,9),pp.165ff. on the "third nature"). ANGOT, in what is probably the latest study on the subject,[1929] duly draws attention to the difficulties of adequately defining a *napuṃsaka-* and to the inadequacy of translations such as "eunuch", "hermaphrodite" or the like (and also deduces that the meaning of the term seems to have changed in course of time within the medical tradition),[1930] but adds to the confusion by referring several times to the "neutre" and to "neutralité", which brings the problem of the connotation of these words in French (and of their correspondences in other Western languages) into the discussion.[1931] On the nature of the *napuṃsaka-* see also MÜLLER 1963A and MÜLLER 1960B, according to which *napuṃsaka-* is not "neuter", but seen primarily in its role as "non-male" (cf. also MÜLLER 1965, pp.38f., KAHLO, p.259, and ZWILLING, cited on pp.560f. below). HĀ,Śā 1,p.416 says that a *napuṃsaka-* is a male of female character, neither male nor female.[1932] The commentaries have at times remarked in more detail on the nature of the *napuṃsaka-*; thus e.g. Cakrapāṇidatta in his commentary on CA,Śā 4,14, where this being is said to have merely an open-

---

[1929] Though this study cannot be called in-depth, especially as regards medical texts, and moreover does not take into account previous studies on the subject. Neither does it pay due attention to the problem of words such as ☞ *klība-*, ☞ *ṣaṇḍha-* and the like, which the texts often use as synonyms for *napuṃsaka-*.

[1930] According to ANGOT, anyone not clearly more or less male or more or less female is *napuṃsaka-*. He also stresses that the texts do not regard the *napuṃsaka-* as something 'abnormal'; I do not know whether a real in-depth study will allow us to assert this with such certainty or not. As it is, a disturbing tendency in ANGOT's study is the implicit assumption that in ancient India one held views similar to those several groups in the West have, namely that human sex is relative and a more or less social construct (p.19: "Le sexe est donc relatif; d'ailleurs l'humain quel que soit son sexe social, relève dans sa physiologie du masculin et du féminin simultanément ..."). One would have liked to have had proof that this view did actually prevail in ancient India.

[1931] According to ROSCOE, p.222, many civilisations seem not to have differentiated or differentiate between persons characterised by impotency, castration, homosexuality, asceticism and various sexual anomalies, rather putting all these into one and the same category, that of the "alternative gender", different from those of male and female capable of reproduction.

[1932] *samavīryarajastvena naraḥ strīprakṛtir bhavet*
napuṃsakam iti khyātaṃ na strī na puruṣo vadet.

This verse is also quoted by SC 4,33 and attributed to Hārīta.

ing not like the vulva/vagina.[1933] See also NYC's remark on p.581 below, according to which it seems to follow that a *napuṃsaka-* need not be sterile.

The term *napuṃsaka-* is often found equated or lumped together with ☞ *klība-*, ☞ *dviretas-*, ☞ *paṇḍaka-* and ☞ *ṣaṇḍ(h)a(ka)-* (cf. n.1931).

nāḍī-

MEULENBELD 1974, p.471 translates *nāḍī-* with "tubular vessel". I prefer "tube" to prevent any confusion with ☞ *sirā-*, which is translated as "vessel" by me. But it must be pointed out that *nāḍī-* seems also to be used in contexts in which one might consider using not "tube", but "tubule" (which denotes a small tube), in English. On the problem of finding an adequate translation for *nāḍī-* see also the discussion on ☞ *sirā-* below. On a probable equation of *nāḍī-* and ☞ *srotas-* see the discussion on the latter term below.

paṇḍaka-

Parallels from Buddhist texts to the descriptions of the 'abnormal' persons usually called ☞ *ṣaṇḍ(h)a(ka)*-s in medical works have been discussed by ZWILLING; the descriptions of the medical and the Buddhist texts, though parallels, agree fully only in some respects. It is interesting that the equivalent of *ṣaṇḍ(h)aka-* in the latter class of (Pali) texts is *paṇḍaka-*; *páṇḍaka-* is a term found already in the Vedic Saṃhitās, and the etymology of both this term and *ṣaṇḍ(h)a(ka)-* is problematic (cf. MAYRHOFER 1992- 2,p.70 and p. 680 respectively), as is also their relationship to the etymologically equally problematic *baṇḍa-* and related words (cf. MAYRHOFER 1992- 2,p.206), *baṇḍá-* being already Vedic too. According to ZWILLING, p.204 both *paṇḍaka-* and *ṣaṇḍ(h)a(ka)-*, as also ☞ *napuṃsaka-* and *tṛtīya- prakṛti-*,[1934] do *not* refer to a eunuch. P.205 says on the *paṇḍaka-* that "we are dealing with a variety of sexual dysfunctions and variations categorized under the general rubric '*paṇḍaka*,' and the reason for this is that they all share the common quality of being '*napuṃsaka*,' 'lacking maleness.' That is, for one reason or another they fail to meet the normative sex role expectations of the adult male." This is indeed the first impression that the discussion on ☞ *ṣaṇḍ(h)a(ka)-* leaves, but a final conclusion can only be reached when the use of the terms mentioned has been extensively examined, in detail and in a much larger mass of literature. It may nevertheless not be out of place here to point out that the descriptions of *ṣaṇḍ(h)a(ka)*-s in the medical texts, as well as his own obser-

---

[1933] *upasthaliṅgākāravirahitaṃ ca randhramātraṃ napuṃsakaliṅgaṃ bhavati.*

[1934] On which see, e.g., KĀM. 14(=2,9),pp.165f. Cf. too Excursus XIII at the end of chapter 12.

vations just cited, are clearly at variance with the statement of ZWILLING, p.209 that *paṇḍaka*-s were "a socially stigmatized class of passive, probably transvestite, homosexuals"; "stigmatized" they may have been (but do we have proof for such a decided statement, or are values of modern Western societies here being projected onto ancient India?), but as for the rest of the quotation, this does not at all do justice to the variety of 'abnormities' (both non-physical and physical) mentioned. ZWILLING, p.204 also explains *paṇḍaka-* as *apa-aṇḍa-ka-* (with elision of the initial *a* of *apa*, a not unknown feature). But this is problematic. Firstly, it is obviously based on a metaphoric usage of the *English* "has no balls", without any attempt to show that such a usage did actually apply in Old or Middle Indo-Aryan to denote "a weak or pusillanimous person" (loc.cit.), which is especially problematic in the light of the by no means clear role of the testicles in ancient Indian medical theory (cf. §§6.12ff., also p.498 in Appendix I). Secondly, the oldest mention of the term is Vedic (*páṇḍaka-*; OBERLIES, p.124,n.86 has overlooked this); since the etymology proposed gives **pāṇḍaka-* in Old Indo-Aryan, we would have to conclude that the word is actually Middle Indo-Aryan by origin and as such a loan already in Vedic, which is theoretically possible (cf. the Vedic word *páṇḍaga-*, which could be but a variant of *páṇḍaka-*), but requires solid arguments. Thirdly, we have the problem of *ṣaṇḍ(h)a(ka)-*; are this and *paṇḍaka-* etymologically related? (We also cannot neglect the possibility of confusion of *pa* and *ṣa* in particular cases, as the two are similar in some scripts.) Also, what is the relation of *ṣaṇḍ(h)a-* to the homonyms *ṣaṇḍa-* "group (of trees)" and *ṣaṇḍ(h)a-* "uncastrated bull" (on both see MAYRHOFER 1956-1980 3,pp.407f.)? Especially the last is problematic, though it might be an unrelated homonym (cf.WEZLER 1998, pp.265,268). Fourthly, we have the problem of *baṇḍa-* (see above) and related words (beginning with *v*); one could explain these as examples of the alternation of *p/v/b* common in Middle Indo-Aryan and also found already in Vedic (cf. OBERLIES, p.125,n.91 on *p/v*), but then one also must face the problem of the alternation in the cases discussed here being not in the middle of a word, but initial (on this cf. e.g. OBERLIES, p.125). In any case, all the problems enumerated above show that the proposed etymology of *paṇḍaka-* is problematic, though not *a priori* impossible.

WEZLER 1998, calling ZWILLING "voreingenommen" (p.263,n.18), regards ancient India as having known eunuchs, and derives *paṇḍaka-* from *apa-aṇḍa-ka-*, as an ancient Middle Indo-Aryan word. Should the term originally have referred to eunuchs and not been used metaphorically as ZWILLING assumes, then this derivation does become much more plausible. (The problem of *baṇḍa-* etc. is also discussed by WEZLER 1998, but the matter is quite complicated and requires various suppositions, so that there is probably scope for further discussion; the study itself speaks of a "Versuch" on p.275.)

ZWILLING, p.206 discusses Buddhaghoṣa's explanation of the *paṇḍaka-* according to which *paṇḍaka-*s can have *both* male and female sexual organs, though one of these may not always be visible.[1935] Buddhaghoṣa could hold that the *indriya-* "power" leading to the manifestation of various male or female characteristics is not necessarily linked with the actual presence of the male or female sexual organ(s) (though of course as a rule this link seems to be there). The matter clearly deserves detailed examination.

pitta-

Usually translated as "bile". But we must remember that, though the Sanskrit term by origin denotes the unmistakable secretion of the body that is called "bile" in English, this should not mislead us to conclude that *pitta-* and "bile" in the modern sense of the word are exact equivalents, since the functions ascribed to this secretion and its various forms are, as in ancient Greek medicine, quite different from our modern conceptions regarding bile. This difference has made many modern scholars of the sort described in §1.7, convinced that ancient Indian and modern Western concepts are in no way different, give most peculiar translations and explanations of *pitta-*. On the different kinds and functions of *pitta-* see MEULENBELD 1974, p.474, as well as ☞ *agni-* above (also n.420). On its various receptacles in the body see §7.35, on its connection with blood ☞ *raktapitta-* below. Bile is also connected with the organ called *tila-*; see on this Excursus VIII at the end of chapter 7.

peśī-

Usually translated as "muscle"; cf. MEULENBELD 1974, p.477. Though "muscle" seems indeed to fit most occurrences, and has therefore been retained in this study, MÜLLER 1965, p.30, vehemently opposes the notion of the muscle in the modern sense of the word being known to ancient Indian medicine, which according to him does not speak of contractile structures, but rather sees that which we today hold to be the effects of muscular activity as due to the working of wind; he prefers "Fleischstück" (piece/chunk of flesh), and points to the etymology of *peśī-* and its cognates. Indeed, MAYR-

---

[1935] Note also LAQUEUR 1989 and 1990, extensively discussing the notion long common in the West on men and women basically having the very same sexual organs, in the one case external, in the other internal (cf. also *ākṛter abhedād iti* in KĀM 6(=2,1),p.85). See too LAQUEUR 1990, pp.15;249,n.43 on the "uterus masculinus", the prostatic utricle, on which p.249,n. 43 says: "The uterus masculinus ... is the vestigal vagina, so named because it was once thought to represent the remains of a structure from which the uterus and upper vagina derive."

HOFER 1992- 2,pp.168f. would tend to support Müller's contention; this, however, creates problems when compared with the data of the medical texts, since *peśī*-s seem to have a function different from that of mere flesh. The matter is not made easier by the fact that usually the texts enumerate only the number of *peśī*-s. A definition of the term is hard to come by. Some works do define the *peśī*- (see e.g. PUROHIT, p.58 with footnote and p.165), but these definitions are not very helpful; in any case, they do not describe *peśī*-s as being contractile. The problem clearly has to be examined in detail, also with regard to the relationship between *peśī*- and ☞ *māṃsa*-.

prasāda-

Technical term for the pure part formed by the successive metabolic transformation of the ☞ *dhātu*-s of the chain of transformation of food. Each *dhātu*- (beginning with ☞ *rasa*-) is transformed into a pure part (*prasāda*-) and an impure part (☞ *mala*-, also *kiṭṭa*-), the pure part giving rise to the next *dhātu*- and so on. Further information on this matter is to be found in MEULENBELD 1974, pp.470f. and the other literature cited above in the discussion of the term *dhātu*-. In their commentaries on SU,Ni 10,18, both Gayadāsa and Ḍalhaṇa explain *prasāda*- as *dhāman*- "power"[1936] and *sāra*- "essence" (*sāra*- is indeed often used in the place of *prasāda*- in medical texts), Ḍalhaṇa further as ☞ *tejas*- "fiery energy; vital power". See also §7.19 and n.522.

phala-

Literally "fruit". According to Ḍalhaṇa's commentary on SU,Śā 5,41, *phala*- refers to the uterus (*garbhāśaya*-).[1937]

In SU,Sū 11,28 *phala*- appears as part of the compound *apavṛttodvṛttaphalayoni*-.[1938] Ḍalhaṇa explains *phala*- as ☞ *aṇḍa*-.[1939] Now one may hold that *aṇḍa*- need not necessarily refer to something only the male has (see the discussion on this word above), but should it nevertheless mean "testicle" here, then this would tally with the meaning of *phala*- found e.g. in RĀM 1,48,27 and 1,49,3ff. Cf. on *phala*- also Candranandana in his commen-

---

1936 "Abode" does not seem to fit here.

1937 *phalaṃ garbhāśayaḥ*.

1938 *naite kṣārakṛtyā durbalabālasthavirabhīru...tṛṣṇāmūrchāpadrutaklībāpavṛttodvṛttaphalayonayaḥ*.

1939 And ☞ *yoni*- as ☞ *garbhāśaya*- "uterus": *phalam aṇḍaṃ yonir garbhāśayaḥ*.

tary on *vr̥ṣaṇa-* in AH,Sū 6,71 (AH2,Sū 6,78),[1940] Indu in his commentary on *muṣka-* and *phalakośa-* in AS,Ni 11,p.57a and AH,Ni 11,22,[1941] and Aruṇadatta in his commentary on AH,Ni 11,22,[1942] where clearly the meaning "testicle" is presupposed. Cakrapāṇidatta in his commentary on the SU passage equates *phala-* and *aṇḍa-*.[1943] It is interesting to note that Cakrapāṇidatta uses the masculine *yasya* to refer to the persons having *phala-* and ☞ *yoni-* respectively. Now it is true that the masculine[1944] stands here because of *prāṇin-*, but one does wonder whether Cakrapāṇidatta would have used this mode of expression had he meant only females; in other words, it is very probable that according to Cakrapāṇidatta *phala-* (and ☞ *aṇḍa-*) refers to something males have. That *phala-* refers (or at least may refer) to something of the male is also shown by SU,Ni 12,6, according to which wind enters the *phalakośa-* (singular) and causes morbid swelling of the testicle(s?) (*muṣka-*).[1945] From this we may deduce either that *phalakośa-* refers to a testicle or the testicles collectively, or else to what ensheathes them, i.e. the scrotum, in which case *phala-* would mean "testicle". Gayadāsa seems to presuppose the latter, for he explains *phalakośa-* through *muṣkakośa-* in his commentary, also quoting a passage from Bhoja using *phalakoṣa-* (*phalakoṣaṃ prapadyate*). SU,Ci 19,21 too mentions *phalakoṣa-*, obviously as part of the urino-genital system, but gives no further information of use for identifying what the term refers to. The problem of *phalakośa-/phalakoṣa-* has been discussed above in connection with ☞ *aṇḍa-*. In any case, we see that in the examples quoted *phala-* either clearly refers or seems to refer to something of the male.

TIVĀRĪ, 1,pp.24f. unfortunately does not even consider such a meaning of *phala-* while discussing Ḍalhaṇa's commentary on SU,Sū 11,28 (quoted above). The only thing TIVĀRĪ, does tell us about Ḍalhaṇa's commentary is that according to Ḍalhaṇa *phala-* here stands for the uterus (p.24), which is actually wrong, since Ḍalhaṇa (echoing Cakrapāṇidatta), after his explanation just mentioned, says that others (*anye*) take *phalayoni-* (not *phala-*) to

---

1940 *vr̥ṣaṇau phalakośāparaparyāyau.*

1941 *muṣkau vr̥ṣaṇau phalakośau.*

1942 *phalakośayor vr̥ṣaṇayoḥ.*

1943 *udvr̥ttam ūrdhvagataṃ phalam aṇḍaṃ yonir vā yasya prāṇinaḥ. tathāpavr̥ttam adhogataṃ phalaṃ yonir vā yasya. nāsau kṣārakr̥tyaḥ.*

1944 The correct form when different persons of different genders, one of them in the masculine, are referred to (cf. e.g. SPEIJER, §28b).

1945 *kālāntareṇa phalakośaṃ praviśya muṣkaśopham āpādayati.*

refer to the uterus (☞ *garbhāśaya-*); these authorities (like the passage mentioned in n.56) thus seem to take *phala-* to refer to the embryo or foetus (i.e., *phalayoni-* would thus be "matrix of the embryo/foetus", though Ḍalhaṇa does not expressly say so). However, TIVĀRĪ 1,p.25 also writes, without referring to Ḍalhaṇa, that *phalayoni-* refers to the origin of the *phala-* "birth" (*prasav*), and as such to the uterus.[1946] The reason for Ḍalhaṇa's (and Cakrapāṇidatta's) explanation of *phala-* as ☞ *aṇḍa-* being ignored seems to be that TIVĀRĪ cannot use a male organ here. By contrast, TIVĀRĪ, 2,pp.320f. no more ignores Ḍalhaṇa's explanation of *phala-* as *aṇḍa-*, but chooses (just like that) to explain this as "ovary" (!; if at all, then it should logically have been "ovum").

In his commentary, Ḍalhaṇa also states that still others (*apare*) take *phala-* to mean *rajas-*, "because of its bringing about the fruition of (scil. in the form of?) the embryo".[1947] Cakrapāṇidatta in his commentary on SU,Sū 11,28 gives us only the first two explanations Ḍalhaṇa has, but not this last one.[1948] Aruṇadatta and Parameśvara, commenting on *prodvṛttaphalayoni-* in AH,Sū 30,6ab, also explain *phala-* as *rajas-*, giving no other explanation,[1949] whereas Hemādri explains *phalayoni-* only as ☞ *garbhāśaya-*; that all three do not consider *phala-* to be an organ of a male may be due to the fact that except for *bhīru-* all the other parts of the compound in which *phala-* appears refer to females,[1950] though the rest of the list with persons to whom one should not apply *kṣāra-* (AH,Sū 30,4cd-8ab) lists persons irrespective of sex as well as specifically male. Indu in his commentary on AH,Sū 30,6 and on the parallel AS,Sū 39,p.269b does not explain *phala-*, whereas Śrīdāsa in his commentary on the AH passage repeats Aruṇadatta's explanation of *phala-* as *rajas-* verbatim. Note that in all these cases the compound *prodvṛttaphalayoni-* (the meaning of whose individual members is not clear)[1951] is analysed either as a combination of *prodvṛtta-* and *phalayoni-*

[1946] *'phalayonayaḥ'* [thus in the plural, even though this stands only because *yoni-* is the last element of a compound referring to a plural number of persons] *ko 'phalasya (prasavasya) yonayaḥ' mān lene se 'phal' (prasav) kā sādhan'bhūt 'garbhāśay' arth sam'jhā jā sak'tā hai.*

[1947] *apare tu. phalaśabdaṃ rajasi vartayanti. kutaḥ. garbhasya phalasādhanatvāt.*

[1948] I may add that GHĀṆEKAR 1977A, p.64b faithfully reproduces all three of Ḍalhaṇa's explanations, expressly stating that *phala-* is taken to refer to *aṇḍa-* = the testes by some (*phal kā arth kuch log aṇḍ yā vṛṣaṇ karte haiṃ*).

[1949] *raja eva phalaṃ garbhākhyaphalasādhanopāyatvāt*, and *phalaṃ rajorūpaṃ*, respectively.

[1950] *bhīrugarbhiṇyṛtumatīprodvṛttaphalayoniṣu.*

[1951] On ☞ *yoni-* see below, s.v.

("whose *phalayoni-* is *prodvṛtta-*"), or *prodvṛttaphala-* and *yoni-* ("whose *yoni-* is *prodvṛttaphala-*"), or *prodvṛttaphala-* and *prodvṛttayoni-* ("whose *phala-* is *prodvṛtta-*" and "whose *yoni-* is *prodvṛtta-*"; *prodvṛtta-* applying to both *phala-* and *yoni-*).[1952] But Śrīdāsa also quotes a passage of Bhoja,[1953] from which it is clear (as Śrīdāsa too expressly remarks) that we also have to reckon with two different words, namely *prodvṛttayoni-* (Bhoia: *udvṛttayoni-*) and *phalayoni-*.[1954] Aruṇadatta, Śrīdāsa, Parameśvara and also Candranandana (who does not explain *phala-*) moreover directly or indirectly say that *prodvṛttaphalayoni-* (to the commentators seemingly the designation of the patient) refers to the disease *udāvṛttā-* (scil. ☞ *yoni-*) (var: *udāvartā-*) described in AH,Ut 33,33f.[1955] All this is not a little problematic, and there is no need to further and needlessly complicate it by bringing in unsubstantiated interpretations based on neologisms, for which, whether they refer to males or females, neither texts nor commentaries offer any indications.

According to VŚS, p.717b *phala-* may be used for the testicles[1956] and for the *rajas-* of women. On efforts by modern Indian scholars to show that *phala-* refers to the ovaries, see p.97 (in Excursus II at the end of chapter 6).

phupphusa-

This organ has not been identified properly, though many scholars hold it to refer to the lungs, a lung, or part of a lung, in keeping with the meaning of its relatives in New Indo-Aryan today (cf. TURNER 1966 and 1985, no.9090). On it see MEULENBELD 1974, p.458, PUROHIT, p.110, also ŚĀR 1,5,43 = ĀY 4,238 (attributed to Hariścandra).[1957] Some texts say (e.g. Āḍhamalla's commentary on ŚĀR 1,5,43, or Ḍalhaṇa's commentary on SU,Śā 4,25) that it is joined to the *hṛdayanāḍikā-*, i.e. the tubule[1958] of the heart

---

[1952] Ḍalhaṇa in his commentary on the SU passage similarly gives us these alternatives too, but with *udvṛtta-* and *apavṛtta-* in the place of *prodvṛtta-*.

[1953] On him cf. MEULENBELD 1974, pp.418f.

[1954] *udvṛttayonyāṃ phalayonyāṃ cety anye. uktaṃ ca bhojena. tatra sādhyeṣv apy upadrutam atibālam ... rajasvalāṃ kṣīyamāṇām udvṛttayoniṃ phalayoniṃ klībaṃ jvaradāhamūrchāpipāsāparītam api varjayet kṣāreṇeti.*

[1955] On this disease cf. e.g. TIVĀRĪ, 2,pp.34-36.

[1956] *vṛṣaṇagranthyau.* It is unclear whether this is taken from some traditional work (a lexicon?) or is a neologism (VŚS is a modern work). On *granthi-* cf. n.256.

[1957] See too n.2057 below.

[1958] A *nāḍikā-* is a small ☞ *nāḍī-* "tube".

(☞ *hṛdaya-*) or the like.

bhāva-

bhāva- "state of mind; sentiment; emotion", used often in a special technical sense in sexological works, in the discussion in KĀM, chapter 6(=2,1),pp.77ff. (see §§12.1f.), quite obviously describes a state of intense sexual pleasure, which seems to subsume an orgasm or, perhaps better, a feeling of satisfaction relying on this, the clearest proof being on p.80, which says[1959] that a woman has *bhāva-* right from the beginning of sexual intercourse, a man, however, only at the end.[1960] It is, however, unclear whether the term is used only in such a narrow sense. KĀM 6(=2,1),p.88 gives a list of terms used for *rati-*, namely *rasa-*,[1961] *rati-*, *prīti-*, *bhāva-*, *rāga-*, *vega-*,[1962] *samāpti-*, which all denote feelings or sensations connected with sexual intercourse, but are not necessarily true synonyms.[1963] Yaśodhara explains *bhāva-* here as:

> *kāmitākhyena bhāvena bhāvyamānatvād bhāvaḥ. kāmitākhyo 'pi bhāvyate phalarūpo 'neneti bhāvaḥ*
> "*bhāva-* because of the being produced (*bhāvyamānatva-*) through the emotion/sentiment (*bhāva-*) called *kāmitā̆-*[1964] ('longing'?). That called *kāmitā̆-* ('longed for'?) is also produced (*bhāvyate*) through it in the form of a result, thus *bhāva-*."

In his commentary on KĀM 2(=1,2),p.14 Yaśodhara also tells us that the *bhāva-* called *kāmitā̆-*, whose sign is the wish for coitus, is the reason for the [special] perception/impression of the faculty of the skin (viz. touch) (localised in the genitals), i.e. the cognition of the [desired] aim/object (scil. the orgasm).[1965] In other words, the *bhāva-* called *kāmitā̆-* is according to Yaśodhara the precondition for attaining *bhāva-*.

1959 As part of a theory ultimately refuted by Vātsyāyana.

1960 *sātatyād yuvatir ārambhāt prabhṛti bhāvam adhigacchati. puruṣaḥ punar anta eva.*

1961 Here clearly not in the same meaning ☞ *rasa-* mostly has in medical works.

1962 On this term see also n.1313.

1963 *rati-* itself clearly denotes an orgasm in Yaśodhara's commentary on 13(=2,8),p.160, but whether this is always the case remains to be examined.

1964 The length of the final vowel is not clear.

1965 Right after the passage quoted in n.1270 Yaśodhara's commentary says: *tasya vyañjanasya yo 'ntargataḥ* (var: *yo 'ntagatiḥ*) *sparśaviśeṣas tasmin viṣaye pratītir asāv arthapratītis tvagindriyabuddhiḥ. asyāḥ samprayogecchālakṣaṇaḥ kāmitākhyo bhāvaḥ kāraṇam.*

The symptoms of *bhāva-* are described in KĀM 13(=2,8),p.161 (see too KAN 2,8,19ff. and RATIR 7,36-40, the latter also quoted in SMAR, pp. 271f.).

bhūta-

See ☞ *mahābhūta-*.

majjan-

Marrow. It is discussed together with ☞ *medas-*.

marman-

Vital, vulnerable point of the body. On the *marman*-s see FEDOROVA,[1966] as well as MEULENBELD 1974, p.488, EIM, pp.118-120, and §7.20 above. However, there are also *marman*-s which are not listed in the usual *marman*--lists, as e.g. SU,Ci 7,37f. and CA,Si 9,4 make clear; on these see Excursus II at the end of chapter 6 (pp.97-100) and Excursus XI at the end of chapter 7 (pp.220-221). These latter *marman*-s are in part ☞ *srotas*-s. In this regard note that Tocharian A *marmañ* and Tocharian B *marmanma* mean "ducts (for blood)"; however, it is not at all clear whether these words are connected with Old Indo-Aryan *marman-*, though according to WINDEKENS, p.630 the possibility is not small. In any case it is intriguing to consider that *marman-* especially in the CA passage might have a meaning similar to the one in Tocharian, but this would have to be substantiated by more evidence.

mala-

Apart from referring to impurities in general, this word is also used technically for the impure part of the successive metabolic transformation of the ☞ *dhātu*-s of the chain of transformation of food (the pure part is called ☞ *prasāda-*). See on *mala-* in this sense (also called *kiṭṭa-* "waste matter") MEULENBELD 1974, pp.488-490 (on *mala-* designating an excited ☞ *doṣa-* or morbific entity — cf. p.490 — see also AH,Ni 2,3 and ĀĀS 6887 = ŚĀR 1,5,24), EIM, pp.116-118, ZIMMERMANN 1989A, pp.129ff., also n.522 and n.609 above. On the *mala-* of semen see the discussion on ☞ *ojas-* above; cf. also ĀY 4,46f.;49f. and Cakrapāṇidatta's commentaries on CA,Ci 15,16 and SU,Sū 14,10. See also n.1909 on *mala-* in what seems to be non-medical usage to refer to procreatory fluids.

1966 On this work cf. n.1968, though.

At times *mala-* seems to refer specifically to faeces (*purīṣa-* or synonyms).

mahābhūta-

A common expression for the five elements earth, water, fire, wind and ether; literally "great element" (sometimes also simply *bhūta-*). MEULENBELD 1974, p.490 prefers the translation "gross element" (followed by me too); see also loc.cit. for further references, to which add MITRA, pp.40-56, HALBFASS, pp.50f.[1967] Only four elements are presupposed by some Indian theories (probably older than those listing five elements), for instance in Vedic and Buddhist philosophy (cf. also RAU 1986, pp.9f.=219f., NOBEL, pp.12ff.). One may in this context also refer to the Greek theory of four elements, without any connection between the Greek and Indian theories (e.g. via a common ancestry) being necessarily implied. A well-known expression for the gross elements, especially — but not only — in Buddhist terminology, and as a rule rare in medical texts, is ☞ *dhātu-*. The terms ☞ *guṇa-* and ☞ *dravya-* too can refer to the gross elements.

See also n.60 and §§7.13f.

mahāsrotas-

Literally "large ☞ *srotas-*"; see the discussion on *srotas-* below.

māṃsa-

Flesh; the relatives of this word are still very much alive in New Indo-Aryan (cf. TURNER 1966 and 1985, no.9982). FEDOROVA translates it as "Muskel" ("muscle") throughout, without giving any reason.[1968] (Usually it is the word ☞ *peśī-* which is translated as "muscle".) Actually, this seems to be the continuation of a tradition going back at least to Palmyr Cordier, who translated the term as "tissu musculaire" (muscular tissue) (LIÉTARD/CORDIER,

---

[1967] Though strictly speaking not of direct relevance to our subject matter, attention may also be drawn to the arguments for the presence of all the gross elements in plants too as discussed in WEZLER 1987, pp.335ff.

[1968] As it is, one of the weak points of her otherwise much to be commended study is an unfortunate proclivity for apodictic statements. This is also very evident when she simply labels the five substances connected with the ☞ *marman-*s, namely flesh (☞ *māṃsa-*), vessels (☞ *sirā-*), cords (☞ *snāyu-*), bones (*asthi-*) and junctures (☞ *sandhi-*), as ☞ *dhātu-*s (for instance on pp.237,n.123;251f.;288ff.), obviously unaware of the problems connected with such terminology.

p.576,n.1), but without giving any reason either.

medas-

Commonly translated as "fat" (thus also by me), but this translation is not unproblematic. Gustave Liétard preferred the term "tissu cellulaire", i.e. cellular tissue, whereas Palmyr Cordier referred to it as "tissu adipeux", i.e. adipose tissue (LIÉTARD/CORDIER, pp.45;166;240 and p.576,n.1 respectively). On *medas-* and the difficulties its differentiation from ☞ *vasā-* causes see e.g. VṚK, p.281,n.1. According to SU,Śā 4,12f. (= ĀY 4,250f.), *medas-* is located in/on the abdomen (*udarastha-*) and within the small bones (in these it is said to be *sarakta-* "together with *rakta-*", which latter could refer to blood or to a red colour), whereas *majjan-* (marrow) is found within the large bones; *vasā-* is the oleaginous substance (*sneha-*) of pure (*śuddha-*) flesh (☞ *māṃsa-*).[1969] AS,Śā 5,p.319a says that *medas-* is located in/on the abdomen (*udara-*) and the small bones (where it is *sarakta-*)[1970] — and as the brain (*mastiṣka-*) within the skull (*kapāla-*[1971]). The sentence contain-

---

[1969] ... *medo hi sarvabhūtānām udarastham aṇvasthiṣu ca. mahatsu ca majjā bhavati.* (12)
sthūlāsthiṣu viśeṣeṇa majjā tv abhyantarāśritaḥ
athetareṣu (var: *tathetareṣu*) *sarveṣu saraktaṃ meda ucyate*
śuddhamāṃsasya yaḥ snehaḥ sā vasā parikīrtitā. (13)

[1970] *medo hi tasyām udare 'ṇvasthiṣu ca saraktaṃ bhavati.* Now *tasyām* refers to the third ☞ *kalā-*, which contains *medas-* (*tṛtīyā medodharā nāma*, immediately before the part just quoted), and one could be tempted to take *tasyām udare* to belong together and to mean "within it" (cf. the use of *udara-* in *asthnām udare* then, though with the genitive and not the locative as here), i.e. within the *kalā-* only, and not within/on the abdomen. However, Indu in his commentary separates *tasyām* and *udare*, i.e. *medas-* is within the *kalā-*, and this *medas-* is further characterised as being within/on the abdomen etc. (*tṛtīyāyāṃ medaḥ kartṛ saraktaṃ bhavaty udare 'ṇvasthiṣu*), and since this is also in keeping with the statement of SU, it does look more plausible. What seems to clinch the issue is the description of the *kalā-* immediately preceding: *dvitīyā raktadharā nāma māṃsasyābhyantaratas tasyāṃ śoṇitaṃ viśeṣataś ca sirāplīhayakṛtsu bhavati.* This parallel to the description of the *kalā-* containing *medas-* says that blood is within the respective *kalā-* (which is itself within the flesh), as well as particularly in some specified organs; it has only *tasyāṃ*, and that does make it extremely probable that *udare* is not to be construed with *tasyām* in the following passage.

[1971] On *kapāla-* see also DAS,R.P. 1991B, p.748,n.37; note that in line 22f. instead of:
cranial skeleton" (later; cf. "sconce,skull")?
one should read:
cranial skeleton (later; cf. 'sconce, skull')"?
What is meant is that "cranial skeleton" referring to all the bones of the head is later than "neurocranium" referring to the pan-shaped upper portion, a distinction which is relevant in the context there.

ing this last statement runs:

> *tad eva ca śirasi kapālapraticchannaṃ mastiṣkākhyaṃ mastuluṅgākhyaṃ ca sthūlāsthiṣu ca majjā.*

AS1, p.222 has the same reading (with *tu majjā* for *ca majjā*), i.e. without any period before *sthūlāsthiṣu*, which confirms the reading of AS and seems to mean that *medas-* is *majjan-* within the large bones, and that is exactly how Indu understands the passage.[1972] Placing a period before *sthūlāsthiṣu* would, however, mean that *sthūlāsthiṣu ca* (or *tu*) *majjā* is a new sentence saying that *majjan-* (and not *medas-* called *majjan-*) is within the large bones, which would be in accord with the SU passage. Aruṇadatta's commentary (and also Śrīdāsa's[1973]) on AH,Śā 3,9f. quotes a passage from an unknown source which says: *medo 'sthnām udare sthitam*; this could mean that the *medas-* of bones[1974] is situated in/on the abdomen, or that *medas-* is situated within the bones.[1975] The variants of this passage say that "*medas-* is situated within the abdomen (or: situated inside, within/on the abdomen)", and that "*medas-* and blood are situated within/on the abdomen";[1976] no bones are mentioned here. Further, *medas-* is within the small bones, whereas *majjan-* is within the large bones, and, according to the printed text, in the head between the brain (*mastuluṅga-*) and the skull (*kapāla-*), though one could consider whether this might not actually be a mistake for the statement that the brain is in the head within the skull.[1977] NYC on SU,Śā 4,12f. quotes a verse from an unidentified source (fol.36r) saying that both *medas-* and *majjan-* (are these equated here?) are within the bones, and not differentiating between large and small bones.[1978]

1972 *medo mastuluṅgatvaṃ yāti majjatvaṃ ca yāti.*

1973 Śrīdāsa also quotes and explains SU,Śā 4,12 as given in n.1969, but without *udarastham*. (Āḍhamalla in his commentary on ŚĀR 1,5,6f. quotes SU,Śā 4,12f.)

1974 In this context "of bones" would seem to mean "derived from bones".

1975 On the meaning of *udara-* then cf. n.1970.

1976 *medo 'ntar udare sthitam* and *medo'sram udare sthitam* respectively.

1977 The relevant part of what Aruṇadatta quotes runs:

> *... medo 'sthnām udare sthitam*
> *bhavaty* (var: *bhaved*) *aṇuṣu majjāntaḥ sthūlāsthiṣv atha mūrdhani*
> *mastuluṅgakapālāntaś ... .*

Should the actual reading of the last part be *mastuluṅgaḥ kapālāntaś* or *mastuluṅgaṃ kapālāntaś*, then we would get the alternative statement mentioned above. Śrīdāsa also reads *mastuluṅgakapālāntaś*, but gives as the reading of the line *bhavaty aṇuṣu* etc.: *aṇvasthiṣu ca majjā tu sthūlāsthiṣv eva mūrdhani*.

1978 *asthimadhyasthito majjā medo 'py atraiva saṃsthitam*

yoni(mukha)-

yoni- in medical texts very often refers to the whole female reproductive tract and encloses the uterus.[1979] Now *mukha-* may mean "upper part, top", and as such the *mukha-* "mouth" of the *yoni-* could actually be the uterus. This meaning would fit CA,Ni 3,14 well (see §5.9); I have in the course of this study also referred to other passages where it would fit. Passages such as SU,Śā 5,43f. could be responsible for this use of *mukha-* too, for in this passage the womb is not only described as being situated within the *yoni-*, but is also said to have the appearance of the mouth of the *rohita-* fish.[1980] *yoni-* may, however, refer to the uterus too (note that the general term for "uterine prolapse" is *yonibhraṃśa-*, which, however, might at times also refer, so it seems, to a vaginal prolapse[1981]), which is in keeping with the meaning "origin, matrix" of *yoni-*. SU,Śā 3,9 says that, when the *ṛtu-*[1982] is over, the *yoni-* closes (cf. §5.6); SCHMIDT 1922, p.286 opines that this refers to the "vulva", but the commentator Ḍalhaṇa explains it (like in his commentary on SU,Sū 11,28) as meaning ☞ *garbhāśaya-* (i.e. the womb), and the editor of AH in his note on the similar verse Śā 1,21cd-22ab (= AS,Śā 1,p.280b) has understood *yoni-* to mean the same. It must, however, also be mentioned that Indu in his commentaries on AH and AS says:

> *yonir bījaṃ na gṛhṇāti. garbhāśayaṃ na prāpayatīty arthaḥ*
> "The *yoni-* does not accept the *bīja-*[1983]. 'Does not make it reach the *garbhāśaya-*', is the meaning."

This is unclear, as *yoni-* and ☞ *garbhāśaya-* could be different here, or the same. But be that as it may, it is in truth rather hard to believe that a closure of the vulva might be meant, as this would allow neither intercourse[1984] nor urination, though as regards the latter, one could hold this to be possible if not the vulva, but the opening of the vagina into the vulva were meant,

---

tan majjagrahaṇenaiva meda[ḥ?] snehe pragṛhyate

[1979] Cf. also n.782 and TIVĀRĪ, 1,pp.6-9.

[1980] *yathā rohitamatsyasya mukhaṃ bhavati rūpataḥ*. See also TIVĀRĪ, 1,pp.9-12.

[1981] Cf. TIVĀRĪ 2,pp.59f.

[1982] See on this term §2.1.

[1983] Cf. on this term §§2.5f.

[1984] Medical authorities surely did not hold that, regardless of proscriptive injunctions in normative Dharmaśāstra texts, no one in actual life had intercourse outside the time of fecundity! In any case, forbidding anything shows that one has grounds to do so, i.e. that what one forbids is done by certain persons; we have also seen in §5.7 that the medical authorities even speak of children conceived through intercourse outside the *ṛtu-*. Cf. too n.133.

which would leave the urinary opening clear. But we must ask ourselves if the ancient authorities could indeed have meant this, since it is easily perceivable that the vagina can be entered at any time. Thus the "mouth" of the *yoni-* would most probably be its orifice leading to the vagina (eventually with/or the cervix surrounding this, the portio vaginalis). This is indeed how Cakrapāṇidatta understands *yonimukha-* in CA,Ni 3,14, for he says explicitly that it is the opening of the womb (*yonimukham iti garbhāśayadvāram*).[1985] We may also arrive at this meaning if we take *yoni-* to refer to the vagina, the "mouth" then lying at its upper end. In SU,Ci 37,115, where an ☞ *uttaravasti-* is to be administered in the direction of the *yonimukha-* (*yonimukhaṃ prati*) to cleanse the womb (37,116: *garbhāśayaviśuddhyartham*), either the uterus or its orifice could be meant.

On the other hand, the term *yonimukha-* may in many cases denote the genital opening leading to the exterior. This seems to be so in SU,Śā 10,21 too, where the *yonimukha-* is to be fumigated (*yonimukhaṃ dhūpayet*), or in 10,7, where urine and a slimy discharge from the *yonimukha-* are mentioned (*mūtraṃ prasicyate yonimukhāc chleṣmā cā*)? AS,Śā 5,p.325a refers to external muscles (☞ *peśī-*) at the "mouth" of the *yoni-*.[1986] SU,Śā 10,9 (*garbhe yonimukhaṃ prapanne*) and AS,Śā 3,p.301b (*garbhasya yonimukhapratipattau*) speak of the child attaining the *yonimukha-* during the last phase of delivery, when the woman delivering should press to her utmost. All this does seem to refer to the external genital opening. Similar seems to be the case in SU,Ni 8,4f., where different abnormal delivery positions of the foetus are mentioned and the term *yonimukha-* (also *yonidvāra-*) is used; but it is just possible that the orifice of the uterus could rather be meant, for an abnormal position may block this too. It is, however, not always clear whether *yonimukha-* in cases in which the external opening is meant refers to the "mouth" of the genitals in general, which could then refer to the vulva, or more precisely to the introitus of the vagina; cf. in this regard SU,Śā 10,7, which speaks of urine (*mūtra-*) too coming out of the *yonimukha-*. In any case, *yoni-* can refer only to the vagina not only in non-medical, but also in medical texts, e.g. in SU,Śā 5,45, where ☞ *garbhāśaya-* and *yoni-* are clearly differentiated, as also in CA,Ci 30,35 (see §5.10), where the "mouth" of the *yoni-* is mentioned side by side with the "mouth" of the uterus (*garbhāśayasya yonyāś ca mukham*).[1987] Note too CA,Ci 23,75, which mentions anointing

---

[1985] Cf. in this connection also SU,Ni 5,4 and its different explanations collected by TIVĀRĪ, 1,p.516.

[1986] The parallel passage SU,Śā 5,39 has *apatyapatha-*. Cf. also TIVĀRĪ, 1,pp.16ff.

[1987] Can this meaning of "mouth" of the *yoni-* then be taken for Ci 30,30;33 too?

the *yoni-*, which here thus probably refers to the vagina (or even only the vulva?). But the whole problem of the nomenclature has unfortunately not yet been studied in detail: cf. also n.137.

For curiosity's sake I may here also mention the opinion of Reinhold F.G. Müller, who deduced from what seemed to him to be a proven fact that the ancient Indians did not know of the uterus (☞ *garbhāśaya-*), that *yoni-* refers merely to the female genitals ("Scham"), at times also the vulva or the vagina only (e.g. MÜLLER 1955, p.13,n.43, also MÜLLER 1961B, p.127), *yonimukha-* referring to the labia ("Schamlippen") according to MÜLLER 1941-1942, p.466. Not only is all this simply apodictically stated as if it were self-evident, but even on the basis of Müller's own arguments it does not automatically follow that *yoni-* cannot describe the place where the embryo is situated, even if this should not be the uterus as we know it.

A nomenclature similar to *yonimukha-* might be found in the ninth century Arabic work of aṭ-Ṭabarī if SIGGEL 1941-1942, p.229 is correct. Translating *wa fī āḫiri fami-hā bayḍatāni* of FH, p.31,line 23, he writes: "An der äußersten Stelle ihres Mundes befinden sich zwei Eier", i.e. there are two eggs at the extreme spot of its (the uterus's) mouth. SIGGEL 1941-1942 explains "egg" as meaning ovary (p.229,n.6: "Keimdrüse") here, which would according to p.229,n.7 be the first known case of *bayḍa* "egg; testicle" being used to mean "ovary", as in later Arabic texts. From this it would logically follow that *fam* "mouth" refers to the internal cavity within the uterus, and not to its opening into the vagina. But that is problematic because of FH, p.31,line 22: *wa lahā famun ka-hai*ʾ*ati 'l-iḥlīl* "and it (the uterus, *raḥim*/*riḥm*) has a mouth by appearance like the urethra". That *fam* here refers to an orifice is also acknowledged by SIGGEL 1941-1942, p.229: "und hat einen Mund wie die Harnblase", and this is obviously also why "Mund" in connection with the two "eggs" is explained (p.229,n.5) as "die von ihrem Ausgang am weitesten entfernt liegende Stelle des weiblichen Genitaltraktus", i.e. the part within the female genital tract lying furthest from its opening, which would be the upper part of the uterus. But that would mean that we are here actually referring to something lying *opposite* the orifice, and I do not see how we could take *fī āḫiri fami-hā* to have this meaning. So either *fam* "mouth" does not refer to the orifice of the uterus (most probably: into the vagina), or else *bayḍa* "egg" does not signify the ovary. The former does seem unlikely in the light of the other mention of *fam*, from which it would then seem to follow that *bayḍa* is not used in the sense of "ovary"; this would also mean that, since *fam* cannot refer to the internal cavity of the uterus, we would have no parallel to the Indian nomenclature. But what could *bayḍa* "egg" refer to then? Is there any pair of organs near the opening of the uterus into the vagina that could be called so? Or does the extreme end of the "mouth" of the uterus actually refer to the opening of the genital tract to the exterior of the body, in which case the "eggs" might refer to Bartholin's glands? We might of course also consider reading *āḫari* instead of *āḫiri*, i.e. *wa fī āḫari fami-hā bayḍatāni*, which we would, because *āḫar* is not, due to its position, an adjectival qualification of *fam*, have to translate as "and at the other [end?, side?] of the mouth of it (the uterus) there are two eggs". This could be taken to refer to the uppermost end of the uterus, in which case "egg" might indeed mean "ovary" here, but I feel very uneasy about this translation. One could also consider whether some portion of the text might not have been lost after *fī āḫir*, and that *fami-hā bayḍatāni* is actually a mistake for *fa-humā bayḍatāni* "and so the two are eggs", referring to something mentioned in the lost portion of the text, but that does seem to be very problematic. As it is, there is quite obviously a problem here, so that the worth of this passage as a possible parallel to Indian terminology is highly uncertain.

yonivyāpad-

Disorder of the ☞ *yoni-*. For a list of such disorders and their terminology see TIVĀRĪ 2,pp.3ff. From the references given it is obvious that in several of the texts the names of the diseases mentioned are adjectives qualifying the feminine *yoni-*, also used, naturally in the feminine form, without the word *yoni-* (thus also in the case of ☞ *ṣaṇḍ(h)a-* and its feminine *ṣaṇḍ(h)ī-*, and of *vandhya-* and its feminine ☞ *vandhyā-*). This seems to have led to a usage, in the texts and/or their commentaries, on the one hand of these feminine forms as nouns, i.e. diseases names (the more or less standard usage today[1988]), on the other maybe as adjectives qualifying not *yoni-*, but the woman. Problems connected with such terminological confusion are discussed in this work in §5.14, §6.10, §8.7, §8.20 and §11.27, as well as n.1222.

raktapitta-

Literally "blood-bile"; on this disease see MEULENBELD 1974, pp.314-327;627f. (add: BYT 75,1ff.), also §7.5, §10.16 and n.428.

[1988] This usage is, however, not exclusive; cf., e.g., the explicit mention of the *yoni-* in the Hindi translation of SC 2,7: *jo strī maithun kar'neke samay atyaṃt ānaṃd pāve, viṣay'se tṛpt nahīṃ ho, us'kī yoni atyānaṃdā kah'lātī hai. aur jis'kī yonike bhītar caugird'se khaṅsī māṃs'kī gāṃṭh kaph aur rakt'ke up'drav'se ho jāve vah kaṃinī yoni kahī hai.* The verse translated (= SU,Ut 38,15, which, however, has *gacchati* for *vindati*) runs:

atyānandā na santoṣaṃ grāmyadharmeṇa vindati
kaṃinyāṃ kaṃikā yonau śleṣmāsṛgbhyāṃ prajāyate.

As a rule, *raktapitta-* is explained as a corruption of blood due to corrupted or excited bile (☞ *pitta-*). We might, however, also have evidence for a theory attributing it to an independent corruption or excitation of blood, this either corrupting the bile, or else combining with bile which is itself in an independently corrupted or excited state. Thus CA,Sū 24,5ff. tell us that corrupted blood gives rise to a host of diseases, among them *raktapitta-* (24,12); no mention of corrupt bile is made. Cf. also the AH, AS and the second of the HĀ passages cited and translated by MEULENBELD 1974, p.322, which speak of a simultaneous excitement or corruption of blood and bile and a subsequent commingling, from which it does not necessarily follow that bile corrupts blood. The first two (identical) passages (AH,Ni 3,3 and AS,Ni 3,p.23b) state:

> *pittaṃ raktasya vikṛteḥ* (var in AH: *vikṛtiḥ*) *saṃsargād dūṣaṇād api*
> *gandhavarṇānuvṛtteś ca raktena vyapadiśyate.*

MEULENBELD 1974's translation loc.cit. (neglecting the AH variant) makes bile corrupt blood:

> "The bile is (in this case), owing to the morbidly altered state of the blood (it has caused), its combination with and corruption of the blood and its conformity with it as to smell and colour, designated together with the blood."

In this he probably follows Aruṇadatta's commentary.[1989] The insertion "it has caused" is, however, not necessitated by the text, which we could also translate as:

> "The bile is, owing to [its] combination with the morbidly altered state of the blood, also owing to [its] corruption [by this][1990] and owing to its complying [with it] as to smell and colour, represented by (= designated as) blood."

Moreover, *vikṛti-* need not mean "morbidly altered state", but could, as Aruṇadatta expressly says in his commentary, be taken to refer to the "altered state" of blood in the sense that bile is generally held to be derived (as its ☞ *mala-*) from blood[1991] (it is also at times regarded as the homologue of

---

[1989] *yasmād raktaṃ pittenāśv eva dūṣyate tathā raktadūṣaṇaiḥ pittam api dūṣyate tātsthyāt.*

[1990] If it is the blood which is morbidly altered, then it follows that, if blood and bile mingle, it is the blood which must corrupt the bile and not vice versa.

[1991] *pittaṃ raktena vypadiśyate. yasmāt pittaṃ raktasya vikṛtir vikāro raktād utpannam ity arthaḥ.*

blood)[1992]. As such we could also translate AH,Ni 3,3 and AS,Ni 3,p.23b as:

> "The bile is, owing to [its being] an altered state of blood, owing to [its] combination [with this], also owing to [its] corruption [by this] ... ."

If we take the AH variant *vikṛtiḥ* for *vikṛteḥ*, the translation would be:

> "Bile is an altered state of blood. It is, owing to its combination [with this], also owing to [its] corruption [by this] ... ."

On the other hand SU,Sū 21,26, after 21,25 has told us that blood is excited due to the same factors as bile (which does not as such imply that it is itself excited due to bile), says that blood is never excited without the morbific entities, which is, however, ambiguous, as it could imply either that blood is corrupted only together with, or else only through the morbific entities (in which case it would, in this particular case, not be corrupted unless through bile).[1993] But even in the former case, the SU passage may not be in keeping with the alternative theory on the origin of *raktapitta-* due to blood corrupting bile, if such a theory should indeed exist.

In any case, the mention of blood in connection with the morbific entities makes it worth examining whether blood in *raktapitta-* may not itself function as a morbific entity (cf. the discussion on ☞ *doṣa-* above, with further references); as such, it could have corrupting powers and quite well corrupt another morbific entity (here bile).

raktasthāna-/raktāśaya-

Receptacle (*sthāna-*, *āśaya-*, also *āśraya-*)[1994] of blood within the body. Aruṇadatta in his commentary on AH,Śā 3,10f. says that the *āśaya-* of blood mentioned in the text refers to ☞ *rasa-*, i.e. not to a receptacle as such, while according to Śrīdāsa the throat is meant, which is supported by Indu in his commentary on AS,Śā 5,p.320a. Āḍhamalla in his commentary on ŚĀR 1,5,7-10 (= ĀY 4,240cd-243cd = ĀĀS 6870-6873 = BHĀ 1,3,215-218 in part) says that the spleen (*plīhan-*), to the left of the heart, is meant, and therefore the chest, but according to BYT 2,51 the receptacle of blood (*jīvāsrāśaya-*) is in the heart (☞ *hṛdaya-*). ŚĀR 1,5,9, ĀĀS 6872, CIK, p.5 say that the receptacle of blood (*jīvaraktāśaya-*) is the chest (*uras-*); so too does ĀY 4,242 (*jīva-*

[1992] Cf. MEULENBELD 1991, p.99.

[1993] *yasmād raktaṃ vinā doṣair na kadācit prakupyati*
tasmāt tasya yathādoṣaṃ kālaṃ vidyāt prakopaṇe.

[1994] On the various receptacles within the body see n.581.

raktāśraya-, var *jīvaraktāśaya-*). On the *āśaya-* of blood cf. also §6.14 and n.249 and the discussion on ☞ *hṛdaya-* below, on *jīvāsra-/jīvarakta-* §7.12. I may add that DASH 1971, p.102 is wrong in attributing to Cakrapāṇidatta's commentary on CA,Ci 24,36 the statement that the heart is the receptacle of blood; what Cakrapāṇidatta actually says is that the heart in particular is the place of the substances blood etc. (before that "*rasa-*, wind etc." are mentioned) that are present throughout the body.[1995] The significance of this is made clear by Cakrapāṇidatta's explaining this statement by citing CA,Sū 30,4 right after it; CA,Sū 30,4 says that the whole body with all its faculties (bodily or mental), as well as the *ātman-*, is dependent on/residing in the heart (*hṛdi saṃśritam*), which Cakrapāṇidatta in his commentary on 5f. explains as meaning that when the heart is in equilibrium, everything else is too, but when disorder strikes the heart, everything else is disordered too.[1996] This is clearly not the same as what DASH 1971 takes Cakrapāṇidatta to mean. In his commentary on SU,Sū 15,21f. (20f. in SU3) Cakrapāṇidatta characterises blood as being situated in the heart;[1997] this too is not the same as calling the heart the receptacle of blood (*raktasthāna-*, *raktāśaya-* or the like), though it does point to a special connection between heart and blood.

rasa-

As the first element (☞ *dhātu-*) of the chain of alimentary transformation, *rasa-* denotes "nutrient fluid". In this sense it is often translated as "chyle"; cf. also EIM, p.172.[1998]

This is of course not the only meaning of *rasa-* in medical texts; there are several others.[1999] A common one is "taste"; see on this especially e.g. MEULENBELD 1974, p.493, also SHARMA,P.V. 1982, pp.92-94, and most recently ZIMMERMANN 1987, *passim* (cf. also n.454). Other meanings, as well as the ones already mentioned, have been examined throughout this study. It is often extremely difficult to differentiate between various meanings of the term *rasa-*; see on this problem particularly §§7.13ff., §§10.7ff., §§11.19f.,[2000]

1995 *raktādīnāṃ tu sarvaśarīracarāṇām api viśeṣeṇa hṛdayaṃ sthānam uktam.*

1996 *hṛdayāśritatvaṃ hṛdaye prakṛtisthe ṣaḍaṅgādibhāvānāṃ prakṛtisthatvaṃ hṛdayopaghāte tadupaghāta ity arthaḥ.*

1997 *prāṇāyatanaṃ hi raktaṃ hṛdayasthaṃ jīvaraktam ucyate.*

1998 Attention may in this context also be drawn to the problem of ☞ *āhārarasa-*.

1999 On *rasa-* ("flavour, savour") in the meaning of *rati-* see the discussion on ☞ *bhāva-* above.

2000 These deal with the problem of whether *rasa-* may denote the female procreatory

§§13.21ff.[2001] and ☞ *āgneya-* above, particularly with regard to what appear to be traces of theories (some maybe quite ancient) on *rasa-* in the medical texts which do not tally smoothly with the well-known theories on *rasa-* as the first *dhātu-* (cf. e.g. the discussion on ☞ *tvac-*).

A special problem is the relationship between the *rasa-* and ☞ *ojas-*, inasmuch as both are at times held to be identical (see the discussion on *ojas-* above), and also, without necessarily being said to be identical, to be present in the heart (☞ *hr̥daya-*). On *rasa-* in the heart see n.420, n.668 and n.952, and especially the material on *hr̥daya-* below; on this theory in Vedic texts cf. also the material of ŚRĪVĀSTAV, pp.59-64.[2002] The Vedic material must, however, make us be wary in our identification of *rasa-*. Given the problems associated with this term, we cannot without further investigation simply assume that *rasa-* as a synonym of *ojas-* is the same as *rasa-* as the first of the chain of elements mentioned above, though in many cases it may appear to be so (cf. also Cakrapāṇidatta's commentary on CA,Sū 30,6f.). On the other hand, *rasa-* is usually taken to be "somic" (☞ *saumya-*), and *ojas-* too is at times characterised thus (see the discussion on *ojas-* for references). Conversely, a probably rather late diagram of medical nature seems to show different receptacles in the body for *rasa-* and *ojas-* respectively; see §6.14 (including n.249).

On the problem of where in the body *rasa-* is coloured to give blood, see the discussion in n.420.

lasīkā-

A fluid of the body (on which see particularly the discussion in §§10.10ff.); many modern Indian scholars arbitrarily translate this term by "lymph", without any serious discussion on the implications of applying such a special meaning here. I have preferred the ambiguous term "serous fluid" found in MEULENBELD 1974, p.489, in keeping with what seems to be the varying usage of *lasīkā-*, which, however, still requires an in-depth study.

---

fluid.

2001 With regard to the latter, note that DAS,R.P. 1992A draws attention to the use of *rasa-* (Bengali *ras*) in esoteric Bengali terminology to denote various vital fluids of the body. See on this also n.1875.

2002 However, the author's 'modernistic' explanations of this material seem despite some most interesting observations to be mostly irrelevant, since they appear to be based not on critical evaluation of textual evidence, but on apodictically presented speculation, often of apologetic nature, of the sort described in §1.7.

vandhyā-

"Barren woman". Note that "barren" may not necessarily refer to a woman incapable of becoming pregnant, but to one incapable of bearing a child (on which see also TIVĀRĪ 2,pp.281f.). On *vandhyā-* see also Śivadāsasena's remarks on AH,Ut 33,28 (29 in AH3) quoted above on p.28, as well as TIVĀRĪ 2,pp.56ff.;331,[2003] and KEL 9,1ff. Statements equating *vandhyā-* and *ṣaṇḍ(h)ī-* are problematic; cf. the discussion on ☞ *ṣaṇḍ(h)a-* below. Confusion may also arise due to *vandhya-* "barren" in its feminine form (*vandhyā*) being used not for a woman, but for her ☞ *yoni-*, as in SU,Ut 38,10 (on which cf. §6.10) and most probably SU,Ut 38,6.[2004] On this see also ☞ *yoni-vyāpad-*.

HĀ 3,48,p.361 (referred to in §5.3) mentions the term *kākavandhyā-* "crow-barren", a term usually taken to mean a woman capable of having only one child.

vasā-

MEULENBELD 1974, p.470 translates this as "muscle-fat". However, it seems to denote a substance present within or next to the fleshy part of the body (i.e. not necessarily only the muscles); this is probably what MÜLLER 1966, p.79 too had in mind when translating the term into German as "Speck". I propose the neutral translation "tissue-fat", but in this work also give Meulenbeld's translation, since a detailed examination of the term (which might even lead to another translation) remains pending. Further information on *vasā-* is given in the discussion on ☞ *medas-*; see also §11.8, especially n.1115.

śleṣman-

Synonym: *kapha-*. Usually translated as "phlegm". It should, though, be remembered that the English word is not used in the modern sense, where it as a rule describes only the thick, slimy matter secreted in the throat and nose. Rather, it is used as in ancient Greek medicine to describe a certain

---

2003 Cf. on op.cit., p.57 also n.214.

2004 *vandhyā* in 38,6 is most probably to be connected with *yoniḥ* in 38,8. No doubt at all arises if one follows the reading of SU,Ut 38,6cd-7ab as given in ĀĀS 5825:

udāvartā tathā vandhyā viplutā ca pariplutā
vātalā yonayaḥ proktā vātadoṣeṇa pañcadhā.

The SU edition (Ut 38,7) gives us a different reading for the last line, as well as for the two lines (in SU only one line) following.

slimy substance in the body which manifests itself in different ways, in which regard we must always remember that differentiations we today make were not necessarily made in ancient times, and vice versa. On the various kinds of *śleṣman-* or *kapha-* see MEULENBELD 1974, p.452, and on its receptacles in the body §§7.34f.

ṣaṇḍ(h)a(ka)-

Five different sorts of 'abnormal' persons are described in SU,Śā 2,38-45, only the fifth, a male without *śukra-* who acts and looks like a female, being called a *ṣaṇḍha(ka)-* (cf. §8.7). But Ḍalhaṇa in his commentary on Śā 2,38-41 calls the other four persons *ṣaṇḍha-*s too. Śā 2,44, which says that these four have semen, whereas the *ṣaṇḍha-* does not, is passed over by him with the laconic comment that the verse serves the purpose of easy memorising. In his commentary on 2,45 he, however, refers to the *ṣaṇḍha-* again. This verse tells us that in the 'abnormal' persons mentioned an opening of the vessels (☞ *sirā-*) carrying semen and an erection are obtained only by the 'abnormal' sexual methods mentioned.[2005] The problem is that Śā 2,44 has just told us that the *ṣaṇḍha-* does not have any *śukra-*; it thus seems as if 2,45 refers only to the other four persons. Ḍalhaṇa sees a problem too, but solves it in accordance with his assumption that all the 'abnormal' persons are *ṣaṇḍha-*s.[2006] Obviously Ḍalhaṇa means that the persons mentioned are *ṣaṇḍha-*s because even though they have semen, the vessels carrying this are usually blocked. Gayadāsa in NYC too calls all the 'abnormal' persons mentioned by SU *ṣaṇḍha-*s (fol.21r): *tatrāśekyādayaḥ ṣaṇḍhaviśeṣā eva jāyante*; ☞ *āsekya-* (NYC often has *āśekya-*) is the first of these persons. But he also refers to the first four as ☞ *napuṃsaka-*s, to the last one, however, as a semen-less *ṣaṇḍha-* (fol.21r).[2007] Commenting on 2,44f., he does explicitly state that the *ṣaṇḍha-* has no semen, but otherwise his commentary is like Ḍalhaṇa's (fol.22r).[2008] On fol.23r he speaks of the problem created by Nā-

---

2005 *anayā viprakṛtyā tu teṣāṃ śukravahāḥ sirāḥ*
harṣāt sphuṭatvam āyānti dhvajocchrāyas tato bhavet.

2006 *yadi teṣāṃ śukram asti tat kiṃ ṣaṇḍhatvam ucyate. ity āśaṅkyāhānayetyādi* "If they have semen, then why is the state of being a *ṣaṇḍha-* mentioned? Thus fearing [an objection], he (i.e. the author) says: 'Through this' etc. (i.e. 2,45)." (After this the meaning of *viprakṛtyā* is explained.)

2007 *yonidoṣādibhir bīje bījabhāge kiñcidduṣṭabījabhāgāvayave cātiduṣṭe 'pi napuṃsakāni. ṣaṇḍhas tu pañcamo bīje bījabhāge sakala eva duṣṭe. ayam eva śukrarahitaḥ. tasmin* (scil. obviously: *bīje*) *viśuddhe strī bhavati puruṣaś ca.* (On *bīja-* and *bījabhāga-* cf. Excursus I on pp.25-29.)

2008 *teṣāṃ caturṣu śukrasya bhāvaḥ. pañcame punar abhāvo viśeṣaḥ. ... aśukra[ḥ] ṣaṇ-*

rada mentioning fourteen types of *ṣaṇḍha-*, but Caraka only eight.[2009] This clearly refers to *Nāradasmṛti* 12,10ff. and CA,Śā 2,17ff. respectively. The *Nāradasmṛti* passage (to which SCHMIDT 1922, pp.471f. too draws attention) does indeed mention fourteen *ṣaṇḍha*-s, these being males who do not have the characteristics of a 'normal' male, i.e. (as the text tells us) have variant sexual practices, adulterate, are sterile, eunuchs, impotent etc.[2010] Some of these persons are found in the SU passage we are discussing, as well as in the CA passage mentioned. This latter too mentions some of the persons mentioned in the SU passage. However, it does *not* refer to eight *ṣaṇḍha*-s, but expressly calls only a few of the eight mentioned *ṣaṇḍa-* (i.e. *ṣaṇḍha-*), namely the male and the female *ṣaṇḍa-* (on this cf. below) and the *vātika-ṣaṇḍaka-* (on this too see below). According to BYT 139,3f., a *ṣaṇḍha-* is a male whose semen (*bīja-*, cf. §§2.5f.) is corrupted and/or incapable of fertilisation, the signs for this being premature ejaculation, and/or urine which is not exhilarating (or maybe: loud) (*hlādin-*) and frothy. Aruṇadatta in his commentary on AH,Śā 4,8 laconically equates *ṣaṇḍhatā-* "*ṣaṇḍha*--hood" with *nirapatyatā-* "childlessness", this is followed by Śrīdāsa. Indu in his commentary on the same passage says that *ṣaṇḍhatā-* refers to being unfit for women (*ayoṣidyogyatva-*) (scil. impotence?).

A *vātikaṣaṇḍhaka-* is mentioned in CA,Śā 2,21 as one whose testes have been destroyed.[2011] One would spontaneously take this to refer to a sterile male, but we cannot, given the uncertainty with regard to the presupposed function of the testes (see on this §6.12), say anything with surety in this regard; in any case, someone ejaculating only wind does not seem to be meant, since another person called *pavanendriya-* "wind-organ" (quite obviously with reference to the penis), is expressly described in 2,18 — as someone whose *śukrāśaya-* (see on this §§6.13ff.) the wind damages (*hatvā*) or (variant:) enters (*gatvā*).[2012]

All this shows that ascertaining the meaning of *ṣaṇḍ(h)a(ka)-* is problematic, matters not being made easier by Cakrapāṇidatta's laconic equation of *ṣaṇḍa-* and the equally difficult ☞ *napuṃsaka-* in his comments on CA,In

---

ḍhakaḥ. avidyamānaśukraḥ. yadi teṣāṃ śukram asti tat kiṃ ṣaṇḍhatvam ity āhānayā viprakṛtyeti. (This is followed by the explanation of *viprakṛtyā*.)

[2009] *nanu ca nāradena kanyādānavaraparīkṣāyāṃ caturdaśa ṣaṇḍhabhedāḥ paṭhitāḥ. carake punar aṣṭau. yad uktaṃ kasmād dviretā[ḥ] pavanendriyo vā ... .*

[2010] Cf. n.1931.

[2011] *vṛṣaṇau tu yasya nāśaṃ gatau.*

[2012] According to AS,Śā 2,p.293a a person called *vātendriya-* or "wind-organ" ejaculates only wind during intercourse.

12,16 (*ṣaṇḍā napuṃsakāḥ*), and of "*ṣaṇḍha*--hood" and "*napuṃsaka*--hood" in his commentary on CA,Sū 26,102 (*ṣāṇḍhyaṃ napuṃsakatā*), which Aruṇadatta echoes in his commentary on AH,Śā 6,2 (*ṣaṇḍhaṃ napuṃsakam*), while elsewhere (commenting on AH,Śā 1,50) equating *ṣaṇḍha*- and ☞ *klība*- (like Indu in his commentary on AS,Śā 12,p.363b). GHĀṆEKAR 1977A, p.64b seems unaware of these difficulties, for he says very assuredly that *klība*- designates someone with little or no *śukra*-, but not a *ṣaṇḍha*-, at the same time explaining *klība*- in SU,Sū 11,28 (loc.cit.: 24) as *napuṃsaka*-. See also §11.31 and n.807, as well as the material collected by SCHMIDT 1904, pp.277ff.

The medical texts also mention female *ṣaṇḍ(h)a*-s, namely *ṣaṇḍ(h)ī*-s. See on this §8.7, which refers too to the problem caused by the adjectival use of *ṣaṇḍ(h)a*- and *ṣaṇḍ(h)ī*- in the manner explained under ☞ *yonivyāpad*-.

Cakrapāṇidatta, commenting on CA,Ci 30,34f. (discussed in §8.7), where the *ṣaṇḍhī*- is described, quotes CA,Śā 4,30 which refers to a ☞ *vandhyā*- "barren woman", thus seemingly equating the two.[2013] Śivadāsasena's commentary on AH,Ut 33,39f. (40f. in AH3),[2014] which describes a female *ṣaṇḍa*-, also mentions the same part of CA,Śā 4,30 as Cakrapāṇidatta's commentary on CA,Ci 30,34f. Ḍalhaṇa in his commentary on SU,Ut 38,18[2015] says that the two terms basically mean the same, but that the former refers to a woman without, the latter with breasts. On *ṣaṇḍhī*- see also TIVĀRĪ 2,pp.51ff.;331. CA,Śā 2,19 too refers to male and female *ṣaṇḍha*-s,[2016] but describes these, if we follow Cakrapāṇidatta's explanation, as being slow (of passion? of procreatory fluid?), having only a little quantity of procreatory fluid (or: as having *bīja*- which is sluggish and of little quantity), impotent (or: without strength) (*abala*-), not arousable and ☞ *klība*-.[2017]

See also NAMOUCHI, pp.200-202. On p.201 attention is drawn to Kṣīrasvāmin's commentary on the *Amarakoṣa* equating *ṣaṇḍha*- and *varṣavara*- and explaining the latter as someone who holds back his semen.

A term related to *ṣaṇḍ(h)a(ka)*- could be ☞ *paṇḍaka*-. WEZLER 1998,

---

2013 Both CA passages as well as the commentary of Cakrapāṇidatta are also discussed in Excursus I following chapter 2.

2014 Quoted on p.28.

2015 Which refers to a breastless *ṣaṇḍī*- without *ārtava*-; cf. §6.1.

2016 These are not mentioned by name in the verse, but must be meant according to 2,17.

2017 *mandālpabījāv abalāv aharṣau klībau ca hetur vikṛtidvayasya* (*bīja*- would seem to be the respective procreatory fluid of male and female; see §2.6). However, it could be that the persons (in the dual) so characterised are not the *ṣaṇḍha*-s themselves, as supposed by Cakrapāṇidatta, but their parents; then parents of little passion etc. would be responsible for generating the *ṣaṇḍha*-s, who themselves would not be characterised by the text.

p.267,n.42 ruminates upon the possible etymology of *ṣaṇḍha-*.

sandhi-

Juncture. I follow MEULENBELD 1974, p.513 in using this neutral term.

sirā-

Vessel. I follow MEULENBELD 1974, p.514 (to whose references add GHĀṆEKAR 1986, pp.205ff.) in using this neutral term. Many modern Indian scholars use terms borrowed from modern Western medicine to explain *sirā-* (*śirā-*), ☞ *dhamanī-* and the like, without bothering to examine whether these fit,[2018] in this way always running the risk of drawing false parallels (cf. also ZYSK 1986, p.699,n.18);[2019] I have striven to use neutral English expressions which do not prejudicate the problem of finding out what such terms actually mean. On *sirā-*, ☞ *dhamanī-* and ☞ *nāḍī-* (obviously "nerve" according to ṢOḌ 2,8) see also the comments of Āḍhamalla on ŚĀR 1,5,3 (cf. too ŚĀR 1,5,38f. — printed as 37f. — and 1,5,47).[2020] Cf. also §13.16.

In his commentary on SU,Ni 1,25-29, Gayadāsa quotes the opinion of "Caraka and others" to the effect that ☞ *kaṇḍarā-*s are actually thick *sirā-*s (cf. n.1832).

soma-

The exact meaning of this word in medical texts is difficult to ascertain. Cf. on this problem ☞ *āgneya-* above and §§7.19f.

saumya-

"Somic", i.e. "of the nature of/related to ☞ *soma-*". On this term see the discussion under ☞ *āgneya-* above.

snāyu-

Cord. I follow MEULENBELD 1974, p.517 in using this neutral term.

[2018] Thus even ṢOḌ 2,9, on whose (modern) author Priyavrat Śarmā cf. §1.7.

[2019] The translations "vein" and "artery" for *sirā-* and *dhamanī-* respectively have been propagated particularly by Gananath Sen in various publications (cf. the remarks on SEN in n.11).

[2020] KAVAḌE, p.52 says expressly that it is impossible to define these terms accurately: *sirādhamanīnāḍyādayaḥ śārīrasaṃjñā api kenāpy ekenaiva viśiṣṭārthena prayuktā iti vaktuṃ na ko 'pi samutsaheta.*

ADRIAENSEN/BARKHUIS/RUIJTERS prefer "sinew" (p.306).[2021] See also the remarks on ☞ *kaṇḍarā-* above.

srotas-

In medical texts usually "aperture" or "channel", depending on the context.[2022] They are of two kinds, which I label "external" and "internal". The "external" ones are the apertures of the body leading to the exterior. The "internal" ones are, however, much more difficult to identify.

On the various meanings of *srotas-* see, apart from the discussion in §§6.17f., MEULENBELD 1974, pp.517f. (add: RASA, Upodghāta,pp.162f., GHĀṆEKAR 1986, p.239), PUROHIT, pp.68ff., EIM, pp.193-195 and DAS,R.P. 1984A, p.243,n.27f., as well as CA,Sū 30,12 and SU,Śā 9,3;13 (on these passages see below), CA,Vi 5,4 and BYT 2,137f. In ŚĀR 1,5,40-42 there is a description of the "openings" (*randhra-*) of the body (these include the pores of the skin), which the commentators Āḍhamalla and Kāśirāma unhesitatingly take to refer to *srotas-*s; the latter also draws attention to different numbers of (external) *srotas-*s being postulated by different authorities. See also ZYSK 1986, p.699,n.18. RANADE/PARAMJAPE, pp.56f. give us a peculiar explanation of *srotas-*. DASH 1971, pp.78f.;99ff. tries to show that *srotas-* means "capillary"; unfortunately, this very learned discussion is of the sort remarked on in §1.7, i.e. does not attempt to understand Indian medicine on its own terms, but to correlate it with modern Western medicine to show that notions of the latter are already presupposed by the former. In this regard it is interesting to note that the etymological connection made by the texts between *sravaṇa-* "flowing; streaming" and *srotas-* is on pp.78;99 reinterpreted to bring it in line with the theory of the latter meaning "capillary"; thus no reference at all is made to *sravaṇa-* "flowing", whose meaning is given on p.99 only as: "The term *sravaṇa* means 'to exude', 'to transude', 'to permeate through' or 'to filter through'" (p.78 is similar), irrespective of the fact that √*sru* primarily means "flow", and that *srotas-* too is basically "a flow" (since Vedic times very common e.g. to describe rivers, streams or the like).

Some *srotas-*s are also ☞ *marman-*s.

srotas- may, at any rate according to the commentaries on the 'classical'[2023] medical texts, at times subsume various hollow and tubular structures within the body, including both ☞ *sirā-*s and ☞ *dhamanī-*s. Thus Aruṇa-

---

[2021] Some scholars prefer translations such as "ligament" or "fibrous tissue".

[2022] ADRIAENSEN/BARKHUIS/RUIJTERS prefer the translation "duct" (e.g. on p.293).

[2023] See on this term §4.1.

datta in his commentary on AH,Śā 1,22f. (see n.264) glosses *dhamanī-* with *srotoviśeṣa-*, and Cakrapāṇidatta in his commentary on CA,In 5,41 seems to subsume *dhamanī-* under *srotas-*, whereas in his commentary on CA,Vi 5,4 he explains the line *sarvaṃ tad anyat tebhyaḥ*, in which *tebhyaḥ* obviously stands for *srotobhyaḥ*, with *tat sarvaṃ dhamanībhyo 'nyat*. See also Cakrapāṇidatta's commentary (quoted in n.198) on SU,Sū 15,12 equating *srotas-* and *dhamanī-*. NYC, fol.37v, in the commentary on SU,Śā 4,24, seems to equate ☞ *nāḍī-* and *srotas-* (cf. §6.25); here we also seem to find an identicalness postulated between *srotas-* and *dhamanī-* (see p.104). As regards the 'classical' texts themselves, SU,Śā 9,13 expressly says that *srotas-* does *not* subsume *sirā-* and *dhamanī-*,[2024] and SU,Śā 9,3 has the opposite theory of *sirā-* subsuming *srotas-* and *dhamanī-*, but attributed to others and refuted,[2025] whereas CA,Sū 30,12 is ambiguous.[2026] AS,Śā 6,p.332a on the other hand subsumes *sirā-* and *dhamanī-* under *srotas-*, but also says that this view is not shared by all medical authorities.[2027]

MEULENBELD 1974, p.517 states that CA,Vi 5,9, which is more or less reproduced in AS,Śā 6,p.332 a little after the AS passage just quoted, "uses the term *srotas* in a very broad sense, including *sirā*, *dhamanī*, *nāḍī*, etc.". This is indeed the traditional explanation of this passage, given also by Cakrapāṇidatta and Indu. The passage might, however, also be understood differently, namely that *srotas-*, ☞ *sirā-*, ☞ *dhamanī-*, ☞ *nāḍī-* etc. are the names of the various "places/spaces" (*avakāśa-*) of the ☞ *dhātu-*s in the body:[2028]

> (AS prefaces the passage with *anye punar āhuḥ*) *srotāṃsi sirā dhamanyo rasāyanyo* (this word is omitted by AS) *rasavāhinyo nāḍyaḥ panthāno* (AS: *panthāno 'yanāni*) *mārgāḥ śarīracchidrāṇi saṃvṛtāsaṃvṛtāni sthānāny āśayā* (AS: *āśayāḥ kṣayā*) *niketāś ceti śarīradhātvavakāśānāṃ* (AS: *śarīre dhātvavakāśānāṃ*) *lakṣyālakṣyāṇāṃ nāmāni bhavanti* (AS: *nāmānīti*).

Should *srotas-* indeed only be used in parallel to *sirā-*, *dhamanī-* etc., then it would not subsume them. However, CA goes on to say:

2024 *srotas tad iti vijñeyaṃ sirādhamanivarjitam.*

2025 *tatra kecid āhuḥ sirādhamanīsrotasām avibhāgaḥ. sirāvikārā eva hi dhamanyaḥ srotāṃsi ceti. tat tu na samyak. anyā eva hi dhamanyaḥ srotāṃsi ca sirābhyaḥ.*

2026 *dhmānād dhamanyaḥ sravaṇāt srotāṃsi saraṇāt sirāḥ* (referring to *mahāphala-* in 30,12ab).

2027 *tatra srotasām eva viśeṣāḥ sirādhamanyaḥ. anye tv anyatvam icchanti sirādhamanīsrotasām.*

2028 And not that only *srotas-* is another name for *avakāśa-*, the latter subsuming *sirā-*, *dhamanī-* etc.

> *teṣāṃ prakopāt sthānasthāś caiva mārgagāś ca śarīradhātavaḥ prakopam āpadyante. itareṣāṃ prakopād itarāṇi ca. srotāṃsi srotāṃsy eva dhātavaś ca dhātūn eva pradūṣayanti praduṣṭāḥ*
> "Through their excitement the ☞ *dhātu*-s of the body, those situated at their [respective] places as well as those going along their [respective] paths, attain excitement. And through the excitement of the ones the others. Corrupted, the *srotas*-s corrupt the *srotas*-s indeed, the *dhātu*-s the *dhātu*-s indeed."[2029]

Since only *srotas*-s and *dhātu*-s are mentioned here, *teṣāṃ* "their" at the beginning of this quotation seems to refer not only to the *avakāśa*-s, but also to the *srotas*-s, from which it follows that the traditional manner of explaining the passage given further above is most probably correct. i.e. this does tell us that *srotas*- subsumes all the *avakāśa*-s, enumerated as various hollow and tubular structures, within the body.[2030] I may add that AS,Śā 6,p.332b gives the passage just quoted in a different form:

> *te cāvakāśāḥ prakupitāḥ sthānasthān mārgasthāṃś ca dhātūn prakopayanti. te 'pi kupitās tāni srotāṃsi. srotāṃsi ca srotāṃsi dhātavaś ca dhātūn.*
> "And the excited *avakāśa*-s excite the *dhātu*-s of the body, those situated at their [respective] places and those situated on their [respective] paths. And the excited ones the *srotas*-s. And the *srotas*-s the *srotas*-s, and the *dhātu*-s the *dhātu*-s."

If *srotas*- is indeed only another term for all the *avakāśa*-s, then this would ultimately say the same as the CA correspondence, provided *teṣāṃ* there subsumes the *srotas*-s.[2031] One could of course hold that *avakāśa*- and *srotas*- are not the same here, but in that case it becomes very difficult to make sense of the whole. In this connection note also PUROHIT, p.133, who states that "ĀŚAYAS are receptacles and nothing but srotases"; "receptacles"[2032] are indeed subsumed under the term *srotas*- if our deductions above are correct. But PUROHIT obviously refers to "*āśayāḥ* [sic] *yāni srotāṃsi*" on p.6 of op.cit., quoted there as being a part of Indu's commentary on AS, without further details. I have managed to find the passage probably meant, which is

---

2029 See on this also SHARMA,P.V. 1985, p.328, MEULENBELD 1992, p.2.

2030 Note also that CA,Vi 5,8 lists not only the "elements" of the chain of alimentary transformation, but also other substances, viz. breath, water, food, urine, faeces, sweat, as being carried throughout the body by *srotas*-s; are all these meant here by the term *dhātu*-?

2031 Thus also Indu: *srotoduṣṭyā dhātudūṣaṇaṃ dhātuduṣṭyā srotodūṣaṇaṃ cāha.*

2032 On *āśaya*- "receptacle" see also n.581.

the commentary on AS,Śā 6,p.332b[2033]. Actually Indu writes:

sthānam āśayas tatra tiṣṭhantīti sthānasthāḥ. srotāṃsi ca srotāṃsīti yāni srotāṃsi duṣṭāni bhavanti tāny anyāny api dūṣayanti.

Since PUROHIT has given us only a garbled version of this, and then drawn conclusions from that, his remarks are worthless on methodological grounds.

Note also PUROHIT, p.172: "Srotas is the whole space which extends in the body from the origin, and which is carrying or producing something", which obviously is his translation of SU,Śā 9,13 (which he quotes on p.68); this verse, however, merely says that the "internal" *srotas*-s as described in Śā 9,12 (see §§6.18f.) permeate the whole body.[2034] On p.134 of op.cit. we find: "THE ALIMENTARY CANAL or 'Mahāstrotas' [sic] is a portion of the digestive tract and is called 'Antaḥ Koṣṭha'." This explanation of AH,Sū 12,46ab,[2035] quoted on p.7, is, however, correct only if what is meant is "the *koṣṭha*- within [the body]".[2036] Otherwise, it is wrong, for then what would be meant in the AH passage would be "the within" (*antar*) as a synonym of *koṣṭha*-, this being also called *mahāsrotas*- and holding the receptacles of undigested and digested food; and indeed, this latter explanation is in keeping with the parallel AS,Sū 22,p.166b.[2037] However, the term *koṣṭha*- used here is problematic, for usually this seems to refer to the whole internal cavity,[2038] but here it does look as if it referred only to the abdominal cavity.[2039] AS1,Sū 22,p.117 too describes the *koṣṭha*- = *mahāsrotas*- as *āmapakvāśayāśrayaḥ*; AS,Sū 22,p.166b, though, has *āmapakvāśayād āśrayaḥ*, most probably a mistake either for the reading of AS1 (in which case it would say the same as the AH passage), or else for *āmapakvāśayādyāśrayaḥ*, in which case the other visceral organs could be implied too, thus making *koṣṭha*- refer to the whole internal cavity. The parallel CA,Sū 11,48 is problematic too,

---

2033 *te cāvakāśāḥ prakupitāḥ* etc. quoted above.

2034 SU,Śā 9,13 runs:

mūlāt khād antaraṃ dehe prasṛtaṃ tv abhivāhi yat
srotas tad iti vijñeyaṃ sirādhamanivarjitam.

The singular is, as is common, used for the whole class of objects called *srotas*-. Ḍalhaṇa explains *mūlāt khād* of the verse as *hṛdayādicchidrāt*, i.e. "from the heart and other openings"; see on *mūla*- also the next paragraph below.

2035 *antaḥ koṣṭho* (not *antaḥkoṣṭho*) *mahāsrota āmapakvāśayāśrayaḥ*.

2036 I.e. *antaḥkoṣṭho* instead of *antaḥ koṣṭho*.

2037 *ābhyantaro mahāsrotaḥ ... koṣṭho 'ntar iti paryāyāḥ*.

2038 Cf. especially the secondary references given in n.581.

2039 Thus also HILGENBERG/KIRFEL, p.71: "Im Inneren ist der Unterleib das grosse Gefäss".

for here the *koṣṭha-* = *mahāsrotas-* is called *āmapakvāśayaḥ* "receptacle (masculine singular) of undigested and digested [food]";[2040] does *koṣṭha-* = *mahāsrotas-* really refer to only two(?)[2041] internal organs here (*āmāśaya-* and *pakvāśaya-*), or is it rather seen in its function of holding the contents of these organs, i.e. being, because it holds the organs themselves, thus ultimately an organ holding their contents too? The problem is compounded by the explanations of *koṣṭha-* in SU,Ni 7,6 by Ḍalhaṇa and Gayadāsa: both say that *koṣṭha-* here refers to the receptacle of the result of "cooking" (i.e. digestion) of food (*āhārapākādhāra-*);[2042] the text of SU itself does not explain what the term *koṣṭha-* signifies.

According to CA,Vi 5,8 and AS,Śā 6,p.331a the *mūla-* "base, basis" (on this see §6.18) of the "internal" *srotas-*s carrying breath is the heart and the *mahāsrotas-*; this would imply that the former cannot be a part of the latter, i.e. the latter would not subsume the thoracic cavity. But Indu in his commentary on AS,Śā 5,p.320 says (p.320b) that *all* the *āśaya-*s are within the *mahāsrotas-*,[2043] which would seem to imply that this term here refers to the whole internal cavity. We find the word *mahāsrotas-* in AS,Ni 11,p.58a too, but Indu does not comment upon it. But he does remark on *mahāsrotas-* in the parallel AH,Ni 11,37 (see AH5), calling it the *āmapakvāśayāśraya-* (cf. on this above); Aruṇadatta similarly calls it the *āmapakvāśayasthāna-*, whereas according to Candranandana and Toḍara[2044] it is the middle of the body, the abdomen.[2045] Cakrapāṇidatta and Ḍalhaṇa in their commentaries on CA,Ci 7,33 and SU,Ci 7,30-36 respectively quote a verse of Vaitaraṇa, according to which the bladder and uterus (☞ *garbhāśaya-*) of women are part of the *mahāsrotas-*:

bhagasyādhaḥ striyā bastir ūrdhvaṃ garbhāśayaḥ sthitaḥ
garbhāśayaś ca bastiś ca mahāsrotaḥsamāśritau.

Though it follows from this that the *mahāsrotas-* is something enclosing bladder and uterus, it does not tell us what it is exactly. Thus the relationship between *mahāsrotas-* and *koṣṭha-* remains unclear. In any case, it does not seem as if the equation of *koṣṭha-* and *srotas-* alone (i.e. not *mahāsrotas-*) in

---

2040 CA1, p.570, and CA3, p.99 have the same reading.

2041 Note the masculine singular. Cakrapāṇidatta: *āmasya pakvasya cāśaya āmapakvāśayaḥ*; CART: *āmapakvāśaya iti militā saṃjñā.*

2042 Gayadāsa has further remarks to make, on which see §7.27.

2043 *asṛgādyāśayāḥ kaṇṭhādayaḥ krameṇa mahāsrotasi vartante sapta.*

2044 See the note of the editor on AH,Ni 11,37.

2045 *mahāsrota udaraṃ śarīrasya madhyabhāgaḥ.*

RANADE/PARAMJAPE, pp.56f. (referred to above) is correct; at least I have come across no passages which substantiate this claim. Neither can we simply say, as DASH 1971, p.50 does, that the *mahāsrotas-* corresponds to the alimentary tract,[2046] or that it is "the major part of gastro-intestinal tract", as CONCEPT, pp.6f. holds. On a maybe peculiar use of *mahāsrotas-* by Indu to refer probably to the uterus see §9.4.[2047]

hṛdaya-

Heart. Not only ☞ *ojas-* and ☞ *rasa-*, but also several other substances (called ☞ *dhātu-*; see p.532) are said to be present in the heart. This creates problems especially with regard to *ojas-* and *rasa-*, as the discussion on these words above has shown. A further problem is caused by some statements that *rasa-* is coloured to give blood in the heart; see on this n.420. On a seemingly ancient theory connecting a vital fluid (variously identified with different fluids within the body) with the heart, see especially §13.26 (also §14.3). However, contrary to what we today would expect, the heart is only sporadically specially mentioned as a receptacle of blood (☞ *raktasthāna-/raktāśaya-*) in medical texts, though of course its connection with blood was known.

There seems to have been a theory on ingested food and drink first entering the *heart* and then the *āmāśaya-*. Aruṇadatta's commentary on AH,Śā 3,62ab says that the *sāra-* of food first goes to the heart as *prasāda-*.[2048] SU,Śā 6,25,[2049] AS,Śā 7,p.338b[2050] and AH,Śā 4,13[2051] tell us that the heart is the portal of the receptacle of undigested food (*āmāśaya-*), on which Aruṇadatta (who is followed by Śrīdāsa) says expressly that the heart is the "mouth" of the *āmāśaya-*, through which portal food and drink enter the latter.[2052] HILGENBERG/KIRFEL seem not to have understood this (p.190: "und ebenso tut es der Mageneingang und das Herz"). Food situated in the heart

---

2046 On the pages subsequent to loc.cit. this statement is quite obviously qualified, in which context *mahāsrotas-* and *koṣṭha-* are used throughout as synonyms without any awareness of the problems connected with such usage.

2047 Note also ĀĀS 4682, where the anus is called *mahatsrotas-*; the text too explains this word as *gudavivara-* "anal opening".

2048 More on this passage will be found in n.668.

2049 *āmāśayadvāraṃ sattvarajastamasām adhiṣṭhānaṃ hṛdayam.*

2050 *sattvarajastamodhiṣṭhānam āmāśayadvāraṃ hṛdayam.*

2051 *dvāram āmāśayasya ca sattvādidhāma hṛdayam.*

2052 *tac cāmāśayasya dvāraṃ mukham. tena hi dvāreṇānnapānam āmāśaye* (var: *āmāśayaṃ*) *praviśati.*

is also mentioned in a passage Ḍalhaṇa quotes as being an alternative reading of SU,Sū 46,122ff.[2053] See also GARBH 2, which says that the internal fire (*antarāgni-*) is in the place of fire (*agnisthāna-*) in the heart (*hṛdaya-*). One might be inclined to see a reference to this theory in AH,Sū 12,15 too, but *hṛdayasyānnavīryāc ca* there is ambiguous, as the fact that *annavīrya-* (probably = *rasa-*) is situated in the heart need not imply that the food (*anna-*) too is situated there. In any case, the notion that the heart is an organ into which food enters is common in many cultures of the world,[2054] so that its occurrence here should not surprise us; indeed, there probably is evidence for this belief from the oldest Vedic texts too,[2055] though I would not venture to remark definitively on this here. To those arguing apologetically as detailed in §1.7, however, such a theory must of course be extremely problematic, as for instance the statement of RANADE/PARAMJAPE in n.420 shows. NAGARATNAM/MADHAVI, pp.8ff. try to solve what is to them a great puzzle by identifying *hṛdaya-* "as the cardiac end of the oesophagus and Raktāśaya[2056] as heart"[2057] (p.18)! It would have been interesting to learn how they reconcile their identification of *hṛdaya-* as a part of the oesophagus with the numerous statements (see below) on it being the organ of emotions and the intellect, which incidentally is a commonplace in various cultures around the world too, occasioning modern researchers on these cultures more or less the same sort of problems as it occasions scholars studying ancient Indian medicine.[2058] One of the most ancient cultures from which we

---

[2053] The first line runs: *hṛdayasthānaniryāsavāhisrotomukhāni yat.*

[2054] Cf., just by way of an example, e.g. HDA, p.1795. KAPLONY, pp.39ff. lists several passages from ancient Egyptian literature which seem to show that the heart was held to be within the stomach. Unfortunately, it is not clear whether the translation "Bauch" ("stomach") is to be understood in a strict anatomical sense, or merely as referring to the body cavity. On the other hand, the differentiation between "stomach" and "body (cavity)" may be something important to us, but perhaps not to the ancient Egyptians.

[2055] Cf. e.g. THIEME 1994, p.367. On this belief in Indian texts cf. also DAS,R.P. 1993A, pp.38f.

[2056] On this see p.577 above.

[2057] And also "Śleśmāśaya [sic] as lungs, Cloma [sic] as pylorus and Phupphusa as pancreas". (On *śleṣmāśaya-* see §§7.33ff., on *kloman-* pp.213f., on ☞ *phupphusa-* above.)

[2058] Note in this regard e.g. the remarks of SCHIFFELER, p.Six,n.7: "The heart is regarded in traditional Chinese medicine as being both the seat of intelligence and emotional feeling, thus, xin can be translated as either 'heart' or 'mind.' To express this two-fold characteristic, I have translated it in a binal form as 'heart-mind.' N.B. the definitions of 'liver,' 'heart,' 'spleen,' 'lungs,' and 'kidneys' in Charles A.S. Williams' Encyclopedia of Chinese Symbolism and Art Motives, Repr. (New York: The Julian Press, 1960), pp. 253, 219, 361, 257, and 238."

have literary evidence for this belief is that of ancient Egypt, on which copious material has now been collected by KAPLONY, pp.31ff.

On the *hṛdaya-* as the organ of emotions and the intellect see e.g. ROŞU 1978, p.282 under "cœur", also DAS,R.P. 1992A, p.412,n.165, LÜDERS, p.46f., ZIMMER,HEN.R., pp.162-164.[2059] That this is an ancient notion in India is shown e.g. by *Ṛgvedasaṃhitā* 10,129,4.[2060] Indeed, VŚS, p.1194b gives us, apart from *vakṣas-* "chest", only *manas-* and *cetanāsthāna-* as synonyms of *hṛdaya-*.[2061] ALTHOFF (pp.83f.;98;131f.;135;258f.;264;274;277;284) may also be drawn attention to in this connection; here Aristotle's theory of sense perception is explained as referring to a process of wave-like movement (as heat impulses?) via the medium of blood to the heart, this latter also being (though maybe not according to Aristotle himself) the place where all the passages from the sense organs combine (pp.284f.). According to ALTHOFF, pp.70f.;138f.;143;280f., Aristotle also posits that food digested in stomach and intestines rises up in the form of a hot vapour until it reaches the brain, where it is cooled and condensed, this condensed portion, which ALTHOFF, p.70 calls "serum" (and which is highly reminiscent of ☞ *rasa-*), then going to the heart, where it is cooked further.[2062]

Āḍhamalla, in his commentary on ŚĀR 1,5,45-47 on the well-known

---

The same problem pertains with regard to the Persian *dil*, used as a loan-word in many Indian languages too; it is difficult to differentiate between "heart" and "mind" (this led me in DAS,R.P. 1992A, p.399,n.72 to use the makeshift translation "interior/core"). Cf. also Arabic *qalb* and *fuʾād* to describe both the organ "heart" and the seat of the psyche, in which regard attention may be drawn especially to SEIDENSTICKER; on p.191 of this work the author also says with regard to these words and *nafs* (usually translated as 'soul') that it "läßt sich für das Verhältnis von arabisch 'Herz' und *nafs* folgern, daß ersteres ursprünglich vorwiegend der Sitz der Emotionen war, letztere der Sitz des Begehrens. Das Hinzukommen der Sphäre 'Religiöses/Verstand' bei 'Herz' und dieser Sphäre und zusätzlich des intellektuellen Wollens bei *nafs* mag darauf hindeuten, daß der intellektuelle Bereich im Arabischen noch nicht fest besetzt war."

2059 See also Cakrapāṇidatta's commentary on CA,Sū 30,3-7.

2060 *sató bándhum ásati níravindan hṛdí pratī́ṣyā kaváyo manīṣā́*. Should Paul Thieme's interpretation of *bándhu-* "connection" as "umbilical cord" be correct (cf. FALK 1994, pp.13f.), then we might also have a second level of meaning here, namely the hint that the 'embryo' *sát-* "being" is connected with the 'mother' *ásat-* "non-being" through the umbilical cord leading to the latter's heart (cf. §13.21), since *hṛdí* "in the heart" could be understood as referring to the heart of the *kaví-*s as well as the heart of the *ásat-* (note also the locative of the latter word in connection with *sató bándhu-*). This second level, if indeed present, would accord with what seem to be references to intercourse in 10,129,5, and maybe also in the first part of 10,129,4.

2061 Otherwise, we find only descriptions of the organ *hṛdaya-* "heart" here.

2062 See on this also ALTHOFF, pp.169f.;172;274.

theory that the heart is the seat of consciousness (*cetanā-*), says that, though this is true, the heart is only a special seat of consciousness, whereas in general the whole body is the seat of consciousness.[2063]

In connection with the latter function of the heart, attention may also be drawn to *Yogavāsiṣṭha* 5,78,32ff., where two *hṛdaya-*s of the body are mentioned, the first, inferior one, being the organ one sees when the chest is opened, the second, superior one, being the consciousness/intellect (*saṃvid-*). It is obvious that this differentiation of two *hṛdaya-*s must have evolved from a notion in which both the organ and the consciousness/intellect were closely related, but which in some later time was no longer unquestioningly accepted and thus led to a differentiation. Nevertheless, even in modern times we find, in India, explanations seeking to uphold the traditional notions on the intellectual function of the heart and to bring these into line with modern notions, as e.g. ŚĀSTRĪ, pp.33ff.

As regards other functions of the heart, a discussion of these is beyond the scope of this study. It is clear, however, that the matter calls for detailed investigation, if only to prove or disprove the contention of Palmyr Cordier that the actual role of the heart was unknown to the ancient Indian physicians.[2064]

2063 *hṛdayaṃ ... cetanāsthānam iti. tad viśeṣeṇa caitanyapadaṃ kathitam. sāmānyena tu sakalaśarīram eva cetanāsthānam.* To prove the latter point, he quotes, incidentally, CA,Śā 1,136 in the form *cetanānām adhiṣṭhānaṃ mano dehaś ca sendriyaḥ ...* , i.e. *manas-* and the body together with the sense organs are the seat of the *cetanā-*s. It should be noted that the CA passage is generally found in the printed editions as *vedanānām adhiṣṭhānaṃ* etc., i.e., what is described is the seat of the *vedanā-*s (*vedana-* "perception, knowledge", *vedanā-* "feeling, sensation", also "pain"). Should the text presuppose *vedanā-* here, then the parallels to Buddhist terminology are unmistakable; cf. in this connection DAS,R.P. 1993A, p.21. In this connection attention may be drawn to SUGUNASIRI, who argues (albeit with some apologetic undertones) that the traditional Buddhist notion of the heart being the seat of consciousness is actually taken from the Vedic tradition, whereas the Buddha himself did not clearly commit himself on the question, though the evidence could point to his having seen the seat of consciousness as lying in what in Pali is called the *jīvitindriya-*, on which SUGUNASIRI, p.422 writes: "*jīvita* and *jīvitindriya* are, then, both coextensive with the whole body, and with each other."

2064 LIÉTARD/CORDIER, p.578,n.1: "Cependant la médecine indigène ignore la véritable circulation du sang, et le rôle du cœur lui est resté inconnu."

BIBLIOGRAPHY*

ABEGG, EMIL: Der Pretakalpa des Garuḍa-Purāṇa (Naunidhirāma's Sāroddhāra). Eine Darstellung des hinduistischen Totenkultes und Jenseitsglaubens. Aus dem Sanskrit übersetzt und mit Einleitung und Indices versehen. Berlin|Leipzig (Vereinigung wissenschaftlicher Verleger Walter de Gruyter & Co.) **1921**.

ADRADOS, FRANCISCO R.: "A Genre of Graeco-Indian Literature: The Erotic Tale." *Annals of the Bhandarkar Oriental Research Institute* (Poona) *74.**1993***, pp.125-139.

ADRIAENSEN, R.C.R. / BARKHUIS, R. / RUIJTERS, J.-L.: "An English Translation of Suśrutasaṃhitā, Nidānasthāna 1, 1-39, together with Gayadāsa's ⟨printed: Sayadāsa's⟩ Nyayacandrikā ⟨sic⟩." Proceedings of the International Workshop on Priorities in the Study of Indian Medicine held at the State University of Groningen 23-27 October 1983. Edited by G. Jan Meulenbeld. Groningen (Institute of Indian Studies, University of Groningen) **1984**. Rijksuniversiteit te Groningen, Publikaties van het Instituut voor Indische talen en culturen 4, pp.277-310.

Agastyasaṃhitā: Agastyasaṃhitā 1—23. Introduction, Critical Edition and Annotated Translation. Proefschrift ter verkrijging van het doctoraat in de Letteren aan de Rijksuniversiteit Groningen ... in het openbaar te verdedigen ... door Roelf Barkhuis ... . Groningen (Roelf Barkhuis) **1995**.

AGRAWALA, P.K.: The Unknown Kamasutras. Being an Anthology of the Ancientmost Love-Aphorisms by Seers of the Tradition and its Expositors. Varanasi (Books Asia) **1983**. Shilpa 1.

Aitareyabrāhmaṇa: Das Aitareya Brāhmaṇa. Mit Auszügen aus dem Commentare von Sāyaṇācārya und anderen Beilagen herausgegeben von Theodor Aufrecht. Bonn (Adolph Marcus) **1879**.

Aitareyāraṇyaka: Aitareyāraṇyakam. ŚrīmatSāyaṇācāryaviracitabhāṣyasametam. Etat pustakaṃ TaḷekaropāhvaNaraharaśāstribhiḥ saṃśodhitam. Puṇya (Ānandāśramamudraṇālaya) ^{3}**1959**. Ānandāśramasaṃskṛtagranthāvali 38.

Aitareyopaniṣad: Contained in RÖER, pp.143-247.

ALTHOFF, JOCHEN: Warm, kalt, flüssig und fest bei Aristoteles. Die Elementarqualitäten in den zoologischen Schriften. Stuttgart (Franz Steiner Verlag) **1992**. Hermes, Einzelschriften 57.

Altindische Grammatik: Jakob/Jacob Wackernagel. Altindische Grammatik. Göttingen (Vandenhoeck & ⟨III: und⟩ Ruprecht) **1957** ⟨II,2: **1954**; III **1930**⟩. ⟨The individual parts so far published are: Introduction générale. Nouvelle édition du texte paru en 1896, au tome I par Louis Renou, and, Band I. Lautlehre. Unveränderter Nachdruck der 1896 erschienenen ersten Auflage, and, Nachträge zu Band I von Albert Debrunner, and, Band II,1. Einleitung zur Wortlehre. Nominalkomposition. 2. unveränderte Auflage, and, Nachträge zu Band II1 ⟨sic⟩ von Albert Debrunner, and, Band II,2. Die Nominalsuffixe von Albert Debrunner, and, III. Band: Nominalflexion — Zahlwort — Pronomen von Albert Debrunner und Jacob Wackernagel.⟩

---

* When reproducing the names of editors and modern authors, titles, honorifics, degrees, residential information or the like given on the title pages have as a rule been left out.

Anaṅgaraṅga: Anaṅgaraṅgaḥ[.] MahākaviKalyāṇamallaviracitaḥ. Viṣṇuprasādabhaṇḍāriṇā saṃśodhitaḥ. Kāśī (Caukhambāsaṃskṛtapustakālaya) **1923**. HaridāsasaṃskṛtagranthamālāsamākhyaKāśīsaṃskṛtasīrizapustakamālā 9, Kāmaśāstravibhāga 1.

ANGOT, MICHEL: "La notion de *napuṃsaka* dans les textes médicaux, grammaticaux et rituels." *Bulletin d'Études Indiennes* (Paris) *11-12.* **1993-94**, pp.15-38.

Añjananidāna: ŚrīmadAgniveśamaharṣipraṇītam Añjananidānam. Sānvaya-'Vidyotinī'bhāṣāṭīkāvibhūṣitam. Ṭīkākāraḥ — Brahmaśaṅkaramiśraḥ. Banāras (Caukhambā Saṃskṛt Sīriz Āphis) **1952**. Haridāsa-saṃskṛta-granthamālā 227.

Annapānavidhi: Sushena's Ayurveda Mahodadhi — Annapanavidhi. Edited with Introduction by S. Venkatasubrahmanya Sastri. Tanjore (T.M.S. S.M. Library) **1950**. Tanjore Sarasvathi Mahal Series 20.

Ārogyacintāmaṇi: Ārogya Cintāmaṇi of Dāmodara Bhaṭṭācārya edited with introduction by S. Viswanatha Sarma. Madras (Government Oriental Manuscripts Library) **1951**. Madras Government Oriental Series 75.

Aṣṭāṅgahṛdaya: The Aṣṭāngahṛidaya ⟨sic⟩. A Compendium of the Ayurvedic System Composed by Vāgbhaṭa With the Commentaries (Sarvāngasundarā) of Aruṇadatta and (Āyurvedarasāyana) of Hemādri. Collated by Aṇṇā ⟨sic⟩ Moreśwar Kunte and Kṛiṣṇa Rāmchandra Śāstrī Navre. Edited by Hariśāstrī Parādkar Vaidya. Bombay (Nirṇaya-Sāgar Press) ^{6}**1939**.

——: The Ashtangahridaya composed by Vahatacharya with the commentary of Sreedasapandita. Part I. Edited by C.N. Narayanan Moose. Trivandrum (Government Press) **1940**. Sri Chithra Ayurveda Series 4, and, Aṣṭāṅgahṛdaya of Vāhaṭa with the commentary Hṛdayabodhikā of Śrīdāsapaṇḍita. Part II. (Sūtrasthāna Ch. XVI to XXX.) Edited by P.K. Narayana Pillai. Trivandrum (Government Press) **1950**. Trivandrum Sanskrit Series 155, and, Aṣṭāngahṛdaya ⟨sic⟩ of Vāhaṭa With the Commentary Hṛdayabodhikā of Śrīdāsapaṇḍita. Part III (Śārīrasthāna and Nidānasthāna). Edited By: K. Raghavan Pillai. Trivandrum (University of Kerala) **1962**. Trivandrum Sanskrit Series 201.

——: Aṣṭāṅgahṛdaya saṃhitāyā — uttaraṃ tantram. Śrīmat Śivadāsasena viracitayā Tattvabodha vyākhyāyopetam ⟨sic⟩. Śrīmaj Jyotiṣacandra Sarasvatī mahodayair nibaddhenopodghātena sambalitam. Kalakattā Greṣṭrīṭāvasathena-Sarasvatyupanāmakena Śrī-Jyotiṣacandra Devaśarmmaṇā sampāditam. Āgarā (Āgarā Viśvavidyālaya) **1942**. Śrīsvāmi-Lakṣmīrāma Nidhi Granthamālā 2.

——: ŚrīmadVāhaṭaviracitaṃ Aṣṭāṃgahṛdayaṃ—uttarasthānam. Kairalīvyākhyayā samullasitam. Aṣṭavaidyakulaprabhavena Vayaskarāgāranivāsinā śrīNārāyaṇaśarmātmajena Śaṃkaraśarmaṇā, Ceppāṭ Ke. Acyutavāryeṇa ca saṃśodhitam. Koṭṭaya (Vaidyasārathimudraṇālaya) **1942**. Vaidyasārathigranthāvali 1.

——: Ashtanga Hridaya Samhita with the Vakyapradeepika Commentary of Parameswara. Edited by Vayaskara N.S. Mooss. Kottayam (Vaidyasarathy Press) **1950-1963** (two volumes). Vaidyasarathy Series 9;15.

——: Aṣṭāṃg'hṛday (sūtr'sthān). Śrī Aruṇ'datt kṛt Sarvāṃg'sundarā[,] śrī Candr'nandan kṛt Padārth'candrikā[,] śrī Hemādri kṛt Āyurved Rasāyan ṭikā. Rām'prasād'jī Śarmā kṛt ṭippaṇī sahit. Bambaī (Śrīveṅkaṭeśvar Sṭīm Pres) **V.S. 2013**.

——: Aṣṭāṅga Hṛdaya Saṁhitā with The Śaśilēkha ⟨Part 3: Śaśilekha⟩ Commentary of Indu. Edited by Vayaskara N.S. Mooss. Kottayam (Vaidyasarathy Press) **1956-1978** (six volumes). Vaidya Sarathy ⟨Parts 3;5;6: Vaidyasarathy⟩ Series 13;17;S-3;18;19;22. ⟨Second revised edition of Parts 4-5: Kottayam (Vaidyasarathy Press) 1979-1981 (two volumes). Vaidyasarathy Series S-6;S-5.⟩

——: see also HILGENBERG/KIRFEL.

Aṣṭāṅgasaṅgraha: Aṣṭāṃgasaṃgrahaḥ śrīmadVāgbhaṭācāryanirmitaḥ. Ṣaḍbhiḥ sūtrādibhiḥ sthānair āyurvedaprakāśakaḥ. BhiṣagvaryaśiroratnaBālaśāstripadāṃbuje | miliṃdāyitacittena Janasthānanivāsinā. Tarṭe ity upanāmnā ca bhiṣakpadabhṛtā guroḥ | Sakhārāmatanūjena Gaṇeśena prayatnataḥ. Devadharakulotpannabhiṣajaḥ Kṛṣṇaśāstriṇaḥ | sāhyenāyaṃ yathābuddhi śodhitaś ca prakāśitaḥ. Muṃbāpura (Gaṇapatakṛṣṇājīmudraṇālaya) **Śaka 1810**.

——: Aṣṭāṅgasaṃgrahaḥ Induvyākhyāsahitaḥ. Ṭi. Rudrapāraśavena pariśodhya prakāśitaḥ. Trichur (Mangalodayam Press) **1913-1926** (three volumes).

Atharvavedasaṃhitā of the School of Śaunaka: Atharvavedasaṃhitā. Atharva Veda Sanhita herausgegeben von R. Roth und W.D. Whitney. Zweite verbesserte Auflage besorgt von Max Lindenau. Berlin (Ferd. Dümmlers Verlagsbuchhandlung) **1924**.

——: Atharvaveda (Śaunaka) With The Pada-pāṭha and Sāyaṇācārya's Commentary Edited And annotated with text-comparative data from original manuscripts and other Vedic works By Vishva Bandhu In collaboration with Bhīmdev, Vidyānidhi and Munīshvar Dev. Part I (Kāṇḍas I—V). Hoshiarpur (Vishveshvaranand Vedic Research Institute) [2]**1990**. Vishveshvaranad Indological Series 13. Vishveshvaranand Vedic Research Institute Publication 671.

Āyurvedābdhisāra: Ayurvedabdhisara. Part I. Edited by P. Sri Ramachandrudu. Part II. Edited by M. Gopal Reddy. Hyderabad (Sanskrit Academy, Osmania University) **1989**; **1991**. Sanskrit Academy Series 41;45.

Āyurvedamahodadhi: see *Annapānavidhi*.

Āyurvedaprakāśa: Āyurveda Prakāśa of Āchārya Śrī Mādhava edited with the Arthavidyotinī & Arthaprakāśinī Sanskrit & Hindi Commentaries by Gulrajsharma Mishra. Varanasi (Chowkhamba Vidya Bhawan) **1962**. The Vidyabhawan Ayurveda Granthamala 37.

Āyurvedasaukhya: Basic Principles of Āyurveda based on Āyurveda Saukhyaṁ of Ṭoḍarānanda by Bhagwan Dash, Lalitesh Kashyap. New Delhi (Concept Publishing Company) **1980**. Ṭoḍarānanda-Āyurveda Saukhyaṁ Series 2. ⟨I am quoting only from this volume; the other volumes of the text — published with varying titles — have no relevant information.⟩

Āyurvedasūtra: The Ayurvedasutram with the Commentary of Yoganandanatha edited by R. Shama Sastry. Mysore (University of Mysore) **1922**. Oriental Library Publications, Sanskrit Series 61.

Āyurvedavijñāna: Āyurvvedavijñānam. Uttarārddham. (Nidāna-cikitsitasthānam.) Vinodalāla Sena mahodayena saṃkalitam. Āśutoṣa Sena Pulinakṛṣṇa Sena kavirājābhyāṃ saṃskṛtaṃ prakāśitañ ca. Kalikātā (Āyurvvedayantra) [2]**1916**.

BAKKER, C.: "De geneeskunst der oude Arische volken." *Nederlandsch Tijdschrift voor Geneeskunde* (Haarlem) *76.**1932***, pp.564-581.

Bālatantra: Atha Bālatantram. (Nānāvidhauṣadhopacāraprayogasaṃdarbhitam.) Vidvadvara-Kalyāṇavaidyaviracitam. ... Varanandakumārarājavaidyaviracitayā hindīṭikayā sametam. Bambaī (Lakṣmīveṅkaṭeśvar Sṭīm-Pres) 1957, reprinted **1983**.

BANDYOPĀDHYĀY, SUREŚ CANDRA / DEBĪ, RAMALĀ: Bhāratīya samājer prāntabāsinī. Baidik yug theke ādhunik kāl paryanta bhāratīya samāj o sāhitye bārāṅganā. Kalikātā (Saṃskṛta Pustak Bhāṇḍār) **Beng. era 1392**.

BANERJI, SURES CHANDRA / BANERJI, RAMALA: The Castaway of Indian Society — history of prostitution in India since Vedic times, based on Sanskrit, Pāli, Prākrit and Bengali sources. Calcutta (Punthi Pustak) **1989**.

BHADURI, SADANANDA: Studies in Nyāya-Vaiśeṣika Metaphysics. Reprint Poona (Bhandarkar Oriental Research Institute) **1975**. Bhandarkar Oriental Series 5.

Bhaiṣajyaratnāvalī: ŚrīGovindadāsaviracitā — Brahmaśaṅkara Miśreṇa pariṣkṛtya parivarddhitā Bhaiṣajyaratnāvalī. 'Vidyotinī' bhāṣāṭīkā 'Vimarśa' ṭippaṇī sahitā. Ṭīkākāraḥ — Ambikādatta Śāstrī. Sampādakaḥ — Rājeśvaradatta Śāstrī. Banāras (Caukhambā-Saṃskṛt-Pustakālay) **1951**. Kāśī-Saṃskṛta-Granthamālā 152.

——: ŚrīGovindadāsaviracitā Bhaiṣajyaratnāvalī. Narendranāthamitrapādaiḥ saṃskṛtā parivarddhitā ca. Jayadeva-Vidyālaṅkāreṇa Candraprabhākhyayā vyākhyayā samalaṅkṛtā. Haridattaśāstribhiḥ saṃśodhitā viśeṣavacanādinā vivarddhitā ca. Lālacandrajī Vaidya ity etaiḥ sampāditā saṃvarddhitā ca. Dillī|Vārāṇasī|Paṭ'nā|Baṃg'laur|Madrās (Motīlāl Banār'sīdās) [8]1970, reprinted **1988**.

BHAT, M. RAMAKRISHNA: Varāhamihira's Bṛhat Saṁhitā with English Translation, Exhaustive Notes and Literary Comments. Part II. Delhi|Varanasi|Patna (Motilal Banarsidass) **1982**.

Bhāvaprakāśa: Bhāvaprakāśa of Śrī Bhāva Miśra Edited with the 'Vidyotinī' Hindī Commentary, Notes and Appendix By Brahmaśaṅkara Miśra and Rūpalālajī Vaiśya. Varanasi (Chaukhambha Sanskrit Sansthan) [5]**1969** (two volumes). The Kashi Sanskrit Series 130.

Bhelasaṃhitā: Bhela Saṁhitā. Editors: V.S. Venkatasubramania Sastri, C. Raja Rajeswara Sarma. Thanjavur (Literary Research Unit, T.M.S.S.M. Library)/New Delhi (Central Council for Research in Indian Medicine & Homoeopathy, Ministry of Health & Family Welfare[,] Govt. of India) **1977**. C.C.R.I.M.H. Pub[lication] 31.

Bheṣajakalpa: Bheṣaja Kalpaḥ. Editors: V.S. Venkatasubramania Sastri, C. Raja Rajeswara Sarma. Thanjavur (Literary Research Unit, T.M.S.S.M. Library)/New Delhi (Central Council for Research in Indian Medicine & Homoeopathy, Ministry of Health & Family Welfare, Govt. of India) **1977**. C.C.R.I.M.H. Pub[lication] 32.

BHISHAGRATNA, KUNJALAL: An English Translation of The Sushruta Samhita based on original Sanskrit Text with a full and comprehensive introduction, additional texts, different readings, notes, comparative views, index, glossary and plates (in three volumes). Translated and edited by Kunjalal Bhishagratna. Varanasi (The Chowkhamba Sanskrit Series Office) [2]**1963**. The Chowkhamba Sanskrit Series 30,1.

Bhojanakutūhala: Bhojanakutūhala of Raghunātha (Part I). Published by Śūranād Kunjan Pillai. Trivandrum (University of Travancore) **1956**. Trivandrum Sanskrit Series 178. ⟨To my knowledge incomplete.⟩

BLACK, J.A.: "The Babylonian Grammatical Tradition: The First Grammars of Sumerian." *Transactions of the Philological Society* (Oxford) *87.**1989***, pp.75-99.

BLOOMFIELD, MAURICE: A Vedic Concordance. Being an alphabetic index to every line of every stanza of the published Vedic literature and to the liturgical formulas thereof, that is an index to the Vedic mantras, together with an account of their variations in the different Vedic books. Delhi|Varanasi|Patna (Motilal Banarsidass) [2]**1964**. Harvard Oriental Series 10.

BODE, MAARTEN: "Empirie en dogma in een Ayurvedische praktijk in Kathmandu. Een pleidooi voor verlicht etnocentrisme in de medische antropologie." *Medische Antropologie* (Meppel) *7.**1995***, pp.140-157.

BODEWITZ, H.W.: "[Review of:] Klaus Mylius. Mallanāga Vātsyāyana: Das Kāmasūtra. Leipzig 1987." *Acta Orientalia* (Havnia) *50.**1989***, pp.256-258.

——: "Belly, Pelvis, Buttocks or Cheeks? Vedic *Kukṣī* (Dual)." *Indo-Iranian Journal* (Dordrecht | Boston | London) *35.1992*, pp.19-31.

BÖHTLINGK, OTTO (VON): Indische Sprüche. Sanskrit und Deutsch herausgegeben von Otto Böhtlingk. Zweite vermehrte und verbesserte Auflage. Neudruck der Ausgabe St. Petersburg 1870-1873. Osnabrück (Otto Zeller Verlagsbuchhandlung)/Wiesbaden (Antiquariat Otto Harrassowitz) **1966** (three volumes).

—: Sanskrit-Wörterbuch in kürzerer Fassung. Sieben Bände. First Indian Edition. Delhi (Motilal Banarsidass Publishers) **1991** (printed in 3 volumes).

BÖHTLINGK, OTTO / ROTH, RUDOLPH: Sanskrit-Wörterbuch herausgegeben von der Kaiserlichen Akademie der Wissenschaften, bearbeitet von Otto Böhtlingk und Rudolph Roth. Neudruck der St. Petersburg Ausgabe von 1855-1875. First Indian Edition. Delhi (Motilal Banarsidass Publishers) **1990** (seven volumes).

BOS, GERRIT: "Maimonides on the Preservation of Health." *Journal of the Royal Asiatic Society* (Cambridge) *Third Series 4.1994*, pp.213-235.

BOTTÉRO, ALAIN: "Consumption by Semen Loss in India and Elsewhere." *Culture, Medicine and Psychiatry* (Dordrecht) *15.1991*, pp.303-320.

Bower Manuscript: The Bower Manuscript. Facsimile leaves, Nagarī ⟨sic⟩ transcript, Romanised transliteration and English translation with notes. Edited by A.F. Rudolf Hoernle. Calcutta (Government Printing) **1893-1912**. Archæological Survey of India, New Imperial Series 22.

Brahmapurāṇa: Sanskrit Indices and Text of the Brahmapurāṇa by Peter Schreiner and Renate Söhnen. Wiesbaden (Otto Harrassowitz) **1987**. Purāṇa Research Publications, Tübingen 1.

Bṛhadāraṇyakopaniṣad: Bṛhadāraṇyakopaniṣat. RaṅgarāmānujaviracitaPrakāśikopetā. Ve.Śā.Saṃ.Rā. 'Śaṅkaraśāstrī Veṇegāvakara' ity etaiḥ saṃśodhitā. Puṇya (Ānandāśramamudraṇālaya) **1911**. Ānandāśramasaṃskṛtagranthāvali 64.

Bṛhadyogataraṅgiṇī: Trimallabhaṭṭaviracitā Bṛhadyogataraṅgiṇī. Etat pustakaṃ Karavīrasthaiḥ Ve.Śā.Rā.Rā. Hanumanta Pādhyeśāstrī ity etaiḥ saṃśodhitam. Puṇya (Ānandāśramamudraṇālaya) **1913-1914** (two volumes). Ānandāśramasaṃskṛtagranthāvali 71.

Bṛhajjābālopaniṣad: Contained in *Upaniṣatsaṅgraha* 1,pp.207-217.

Bṛhatsaṃhitā: The Bṛihat Saṃhitâ by Varâhamihira with the Commentary of Bhaṭṭotpala. Edited by Sudhákara Dvivedí. Part II. Benares (E.J. Lazarus & Co.) **1897**. The Vizianagram Sanskrit Series 12/Vol.10.

—: see also BHAT.

BRONKHORST, JOHANNES: "The Qualities of Sāṅkhya." *Wiener Zeitschrift für die Kunde Südasiens und Archiv für indische Philosophie* (Wien) *38.1994*, pp.309-322.

—: "Śvetaketu and the *Upanayana*." *Asiatische Studien/Études Asiatiques* (Bern | Berlin | Frankfurt am Main | New York | Paris | Wien) *50.1996*, pp.591-601.

BRUCKER, EGON: "Das Puṃsavana-Ritual aus der Sicht der heutigen Medizin." *Zeitschrift der Deutschen Morgenländischen Gesellschaft* (Stuttgart) *136.1986*, pp.428-433.

BUITENEN, J.A.B. VAN: "On the Archaism of the *Bhāgavata Purāṇa*." Krishna: Myths, Rites and Attitudes. Edited by Milton Singer. Honolulu (East-West Center Press) **1966**, pp.23-40;215-217.

BYNUM, CAROLINE WALKER: "The Female Body and Religious Practice in the Later Middle Ages." Fragments for a History of the Human Body. Part One. Edited by Michel Feher with Ramona Naddaff and Nadia Tazi. New York (ZONE) **1989**. ZONE 3, pp.160-219.

CAILLAT, COLETTE: "Sur les doctrines médicales dans le Tandulaveyāliya." *Indologica Tauri-*

nensia (Torino) *2.1974*, pp.45-55.

Cakradatta: Carakacaturānana śrīCakrapāṇi viracitaḥ Cikitsāsārasaṃgrahāparanāmā — Cakradattaḥ. Bhiṣakkulabhūṣaṇa śrīŚivadāsena kṛtayā Tattvacandrikāsamākhyayā ṭīkayā samalaṅkṛtaḥ. Jayadeva Vidyālaṅkāreṇa saṃśodhitaḥ. Lavapura (Mehara-candra-Lakṣmaṇadāsa) **1928**.

Caṇḍīmaṅgal: Kabikaṅkan mukunda-biracita caṇḍīmaṅgal. Sukumār Sen sampādita. Natun Dillī (Sāhitya Akādemi) [2]**1986**.

Carakasaṃhitā: Carakasaṃhitā sūtra-nidāna-vimāna-śārīrendriya-cikitsita[-] kalpa-siddhi-sthānātmakā. Āyurvedīyā bhagavatā Ātreyeṇa Punarvasunā upadiṣṭā Agniveśanā-madheyena tat śiṣyeṇa ⟨sic⟩ viracitā Carakābhidhena ṛṣiṇā pratisaṃskṛtā Jīvānanda Vidyāsāgara Bhaṭṭācāryyeṇa bahūni ādarśapustakāni samālocya saṃskṛtā prakāśitā ca. Calcutta (Sarasvatī Yantra) **1877**.

—: Caraka-saṃhitā. Mahāmuninā bhagavatĀgniveśena praṇītā maharṣi Carakeṇa pratisaṃskṛtā. CarakacaturānanaśrīmacCakrapāṇidattapraṇītayā Āyurvvedadīpikākhyaṭīkayā mahāmahopādhyāya-śrīGaṅgādharakaviratnakavirājaviracitayā Jalpakalpataru-samākhyayā ṭīkayā ca samalaṅkṛtā. Narendranātha Senaguptena ... Balāicandra Senaguptena ca sampāditā saṃśodhitā prakāśitā ca. Kalikātā (Dhanvantari-Ilekṭrik-Meśinayantra ⟨Volume 1: Dhanvantari-Ṣṭīma-Meśinayantra⟩) **Śaka 1849-1855** (three volumes).

—: Caraka-saṃhitā. (Mūl o baṅgānubād.) Satīś'candra Śarmmā Kabibhūṣaṇ karttṛk anūdita o prakāśita. Kalikātā (Bhaiṣajya-Ṣṭīm-Mesin-Yantra) **Beng. era 1311**.

—: The Charakasaṃhitā of Agniveśa Revised by Charaka and Dṛiḍhabala With the Āyurveda-Dīpikā Commentary of Chakrapāṇidatta. Edited by Jādavaji Trikamji Āchārya. Bombay (The Nirṇaya Sāgar Press) [3]**1941**.

—: MaharṣiPunarvasuśiṣyeṇa ṛṣivareṇa Agniveśena praṇītā mahāmuninā Carakeṇa Kāpilabalena Dṛḍhabalena ca pratisaṃskṛtā Carakasaṃhitā mahāmahopādhyāyaCara-kacaturānanaśrīCakrapāṇidattaviracitayā Āyurvedadīpikāvyākhyayā (tathā cikitsāsthānataḥ siddhisthānaṃ yāvat) śrī VāgbhaṭaśiṣyācāryavaraJajjaṭaviracitayā Nirantarapadavyākhyayā ca saṃvalitā. Haridattaśāstriṇā saṃśodhitā, pūritaJajjaṭaṭīkā-truṭitāṃśabhāgā ca. Dvitīyo bhāgaḥ. Lavapura (Motīlāla Banārasīdāsa) [2]**1941**.

—: The Caraka Saṃhitā, expounded by the Worshipful Ātreya Punarvasu, compiled by the great Sage Agniveśa and redacted by Caraka and Driḍhabala. Edited and published in six volumes with translations in Hindi, Gujarati and English by Shree Gulabkunverba Ayurvedic Society. Volume II. Jamnagar (Shree Gulabkunverba Ayurvedic Society) **1949**.

—: Caraka-Saṃhitā. Agniveśa's treatise refined and annotated by Caraka and redacted by Dṛḍhabala (Text with English Translation). Editor-Translator Priyavrat Sharma. Vol.II (Chikitsāsthānam to Siddhisthānam). Varanasi|Delhi (Chaukhambha Orientalia) **1983**.

Carakatattvapradīpikā: Carakatattvapradīpikā. ŚrīŚivadāsasenakṛtā Carakasaṃhitā-vyākhyā (sūtrasthānam). Sampādakau Priyavrataśarmā evaṃ Satyadeva Dube. ⟨Cover: Priyavrataśarmaṇā Satyadeva Dube ityākhyena ca sampāditā.⟩ Jay'pur (Śrīsvāmilakṣmīrām Ṭrasṭ) **1990**. Śrīsvāmilakṣmīrāmanidhigranthamālā 7.

CARSTAIRS, G. MORRIS: The Twice-Born. A Study of a Community of High-Caste Hindus. Second Impression. London (The Hogarth Press) **1961**.

CASHIN, DAVID: The Ocean of Love. Middle Bengali Sufi Literature and the Fakirs of Bengal. Stockholm (Association of Oriental Studies, Stockholm University) **1995**. Skrifter utgivna av Föreningen för Orientaliska Studier 27.

Catalogue of the India Office Library. Vol.II—Part I. Revised Edition. Sanskrit Books. By Pra-na Natha and Jatindra Bimala Chaudhuri. Revised and Edited by C.J. Napier. Section III (Kṛṣṇa-līlāmṛta—R). London (Her Majesty's Stationery Office) **1953**.

CHAKLADAR, H.C.: Social Life in Ancient India. A Study in Vatsyayana's Kamasutra. Calcutta (Susil Gupta) [2]**1954**.

Chāndogyopaniṣad: Chāndogyopaniṣat. ĀnandagirikṛtaṭīkāsaṃvalitaŚāṅkarabhāṣyasametā. Ratnāgirigrāmanivāsi...Bālaśāstrisūrisūnubhiḥ 'Āgāśe' ity upāhvaiḥ Ve.Śā. Saṃ.Rā.Rā. Kāśīnāthaśāstribhiḥ saṃśodhitā. Puṇya (Ānandāśramamudraṇālaya) [5]**1934**. Ānandāśramasaṃskṛtagranthāvali 14.

CHATTOPADHYAYA, DEBIPRASAD: "Tradition of Rationalist Medicine in Ancient India. Case for a Critical Analysis of the *Caraka-saṃhitā*." Sanskrit and World Culture. Proceedings of the Fourth World Sanskrit Conference of the International Association of Sanskrit Studies Weimar May 23-30, 1979. Edited by Wolfgang Morgenroth. Berlin (Akademie-Verlag) **1986**. Schriften zur Geschichte und Kultur des Alten Orients 18, pp.569-578.

——: History of Science and Technology In Ancient India. [Volume 2:] Formation of the Theoretical Fundamentals of Natural Science. ... (with invited contributions). Calcutta (Firma KLM) **1991**.

——: "Methodology: On Self-Criticism." Language, Logic and Science in India. Some Conceptual and Historical Perspectives. Contributors Debiprasad Chattopadhyaya, P.K. Mukhopadhyaya, Harshad R. Trivedi. D.P. Chattopadhyaya, Partha Ghose, Sibajiban Bhattacharyya. New Delhi (Project of History of Indian Science, Philosophy and Culture) **1995**. PHISPC Monograph Series on History of Philosophy, Science and Culture in India 4, pp.39-45.

CHÖPEL, GEDÜN: Tibetan Arts of Love. Introduced and Translated by Jeffrey Hopkins. With Dorje Yudon Yuthok. Ithaca, New York (Snow Lion Publications) **1992**.

Cikitsākalikā: Chikitsa Kalika of Tisatacharya. Edited by Vayaskara N.S. Mooss. Kottayam (Vaidyasarathy Press) **1950**. Vaidyasarathy Series 10.

Cikitsāsāramu: Vaidyaśikhāmaṇiyagu śrīmadGōpāladāsunicē raciyiṃpabaḍina Cikitsāsāramu. Āyurvēdādyanēka graṃthaṃbulanavalōkiṃci Puvvāḍa-Sūryanārāyaṇarāvugāricē āṃdhravacanarūpamuga raciyiṃpabaḍina Śārīraratnāvaḷī. Ī reṃḍu graṃthamulunu ... Nivṛtti-Vīrāsvāmiśāstrulavāricē lekhakapramādajanitadoṣaṃbulu lēkuṃḍunaṭula bariṣkariṃpabaḍi Puvvāḍa-Vēṃkaṭarāvupaṃtulavāricē Vartamānataraṃgiṇīmudrākṣaraśālayaṃdu reṃḍavatūri mudriṃpabaḍi prakaṭiṃpaṃbaḍiye. Madras (Vartamānataraṃgiṇīmudrākṣaraśāla) **1880**. ⟨Consulted at the India Office Library, London.⟩

Cikitsāsārasaṅgrahamu: Cikitsāsārasaṅgrahamu. Critically edited by S. Janakirama Sastry. Madras (Government Oriental Manuscripts Library) **1953**. Madras Government Oriental Series 105.

Cikitsātilaka: Cikitsā Tilakam of Srinivāsa ⟨sic⟩ edited with introduction by S. Venkatasubramanya Sastri. Madras (Government Oriental Manuscripts Library) **1953**.

COMBA, ANTONELLA: "Un capitolo della Śivagītā sulla medicina āyurvedica." Torino (Accademia delle Scienze) **1981**. Memorie dell'Accademia della Scienze di Torino, Serie V,5.1981, II — Classe di Scienze Morali, Storiche e Filologiche, pp.173-223.

——: "Some Priorities in Non-medical Texts." Proceedings of the International Workshop on Priorities in the Study of Indian Medicine held at the State University of Groningen 23-27 October 1983. Edited by G. Jan Meulenbeld. Groningen (Institute of Indian Studies, University of Groningen) **1984**. Rijksuniversiteit te Groningen, Publikaties van het Instituut voor Indische talen en culturen 4, pp.223-249.

——: "The Semen and its Origin in Old Indian and Greek Medicine." Proceedings of the Fifth World Sanskrit Conference. Varanasi, India: October 21—26,1981. Edited by R.N. Dandekar, P.D. Navathe. New Delhi (Rashtriya Sanskrit Sansthan) **1985**, pp.920-924.

——: "Carakasaṃhitā, Śārīrasthāna I and Vaiśeṣika Philosophy." Studies on Indian Medical History. Papers presented at the International Workshop on the Study of Indian Medicine held at the Wellcome Institute for the History of Medicine 2-4 September 1985. Edited by G. Jan Meulenbeld and Dominik Wujastyk. Groningen (Egbert Forsten) **1987**. Groningen Oriental Studies 2, pp.43-61.

——: "Universal (*sāmānya*) and Particular (*viśeṣa*) in Vaiśeṣika and in Āyurveda." *Journal of the European Āyurvedic Society* (Reinbek) *1.**1990***, pp.7-32.

——: La medicina indiana (Āyurveda). Torino (Promolibri) **1991**.

Concept of Jaṭharāgni ...: see The Concept of Jaṭharāgni ... .

CRAWFORD, PATRICIA: "Sexual Knowledge in England, 1500-1750." Sexual Knowledge, Sexual Science. The History of Attitudes to Sexuality. Edited by Roy Porter and Mikuláš Teich. Cambridge|New York|Melbourne (Cambridge University Press) **1994**, pp.82-106.

DANIÉLOU, ALAIN: Vâtsyâyana. Kâma Sûtra. Le bréviaire de l'amour. Commentaire Jayamangalâ en Sanskrit de Yashodhara. Extraits d'un commentaire en hindi de Devadatta Shâstrî. Traduits par Alain Daniélou. Monaco (Éditions du Rocher) **1992**. Collection Daniélou.

DAS, R.M.: Women in Manu and his Seven Commentators. Varanasi|Arrah (Kanchana Publications) **1962**.

DAS, RAHUL PETER: "More on the Dissection of Cadavers in Ancient India." *Ancient Science of Life* (Coimbatore) *3.**1983-1984***, p.48.

——: "[Review of:] G.U. Thite. Medicine. Its Magico-Religious Aspects according to the Vedic and Later Literature. Poona 1982." *Indo-Iranian Journal* (Dordrecht|Boston) *27.**1984*** ⟨= **1984A**⟩, pp.232-244.

——: "[Review of:] W.L. Smith. The One-Eyed Goddess. A Study of the Manasā Maṅgal. Stockholm 1980." *Indo-Iranian Journal* (Dordrecht|Boston) *27.**1984*** ⟨= **1984B**⟩, pp.319-341.

——: "Some Notes on Vrksayurveda." *Ancient Science of Life* (Coimbatore) *6.**1986-1987***, pp.6-9.

——: "Religionen und medizinische Ethik[:] 3. Hinduismus." Lexikon Medizin Ethik Recht. Herausgegeben von Albin Eser, Markus von Lutterotti, Paul Sporken unter Mitwirkung von Franz Josef Illhardt und Hans-Georg Koch. Freiburg|Basel|Wien (Herder) **1989**, cols.916-926;942.

——: "Miscellanea de Operibus Āyurvedicis." *Journal of the European Āyurvedic Society* (Reinbek) *1.**1990***, pp.47-68.

——: "The *romarāji-* in Indian Kāvya and Āyurvedic Literature." Panels of the VIIth World Sanskrit Conference. Kern Institute, Leiden: August 23-29, 1987. General Editor: Johannes Bronkhorst. Vol.VIII. Medical Literature from India, Sri Lanka and Tibet. Edited by G. Jan Meulenbeld. Leiden|New York|København|Köln (E.J. Brill) **1991** ⟨= **1991A**⟩, pp.1-65.

——: "On the Subtle Art of Interpreting." *Journal of the American Oriental Society* (Ann Arbor) *111.**1991*** ⟨= **1991B**⟩, pp.737-767.

——: "Problematic Aspects of the Sexual Rituals of the Bauls of Bengal." *Journal of the American Oriental Society* (Ann Arbor) *112.**1992*** ⟨= **1992A**⟩, pp.388-432.

——: "Indische Medizin und Spiritualität." *Journal of the European Āyurvedic Society* (Reinbek) *2.**1992*** ⟨= **1992B**⟩, pp.158-187.

——: "Heilsvorstellungen in der altindischen ('hinduistischen') Medizin." Religionen und medizinische Ethik. Mit Beiträgen von Rahul Peter Das, Mirko Frýba, Robert Jütte, Adel Theodor Khoury, Ina Rösing-Diederich. Ulm (Humboldt-Studienzentrum, Universität Ulm) **1993** ⟨= **1993A**⟩. Bausteine zur Philosophie 7, pp.11-40.

——: "On the Nature and Development of 'Traditional Indian Medicine'." *Journal of the European Āyurvedic Society* (Reinbek) *3.1993* ⟨= **1993B**⟩, pp.56-71.

——: "The *romarājī*- in Indian Kāvya and Āyurvedic Literature — Paralipomena." Festschrift Klaus Bruhn zur Vollendung des 65. Lebensjahres dargebracht von Schülern, Freunden und Kollegen herausgegeben von Nalini Balbir und Joachim K. Bautze. Reinbek (Dr. Inge Wezler) **1994**, pp.266-294.

——: "[Review of:] Kṛṣṇacaitanya Ṭhākur. Cikitsā bidhāne tantraśāstra. Kalikātā Beng. era 1394-1400." *Journal of the European Āyurvedic Society* (Reinbek) *4.1995*, pp.229-234.

——: "On the *Vṛkṣāyurveda* of Parāśara." *Journal of the European Āyurvedic Society* (Reinbek) *5.1997*, pp.196-215.

——: "Material aus medizinischen Texten zum Verzehr von Rind- und Büffelfleisch im Alten Indien." ⟨Forthcoming, most probably in a volume of essays on the subject of *ahiṃsā*-, edited by Albrecht Wezler.⟩ ⟨= **?**⟩

DASGUPTA, SURENDRANATH: A History of Indian Philosophy. Reprint Delhi | Varanasi | Patna | Bangalore | Madras (Motilal Banarsidass) **1988** (five volumes).

DASH, BHAGWAN: Concept of Agni in Āyurveda with Special Reference to Agnibala Parīkṣā. Varanasi (Chowkhamba Sanskrit Series Office) **1971**. The Chowkhamba Sanskrit Studies 81.

——: "*Pitta* Versus *Agni* — An Ayurvedic Perspective." Researches in Indian and Buddhist Philosophy. Essays in Honour of Professor Alex Wayman. Edited by Ram Karan Sharma. Delhi (Motilal Banarsidass Publishers) **1993**, pp.241-245.

DATTA, BHAKTI: Sexual Ethics in the Mahābhārata in the Light of Dharmaśāstra Rulings. London (Asia Publications) **1979**.

DATTA CAUDHURĪ, SUNIRMAL: Gaṅgā theke sur|mā. Cakrapāṇi baṃśer itihās. Kal|kātā (Nābhānā) **1988**.

Dattātreyopaniṣad: Contained in *Upaniṣatsaṅgraha* 1,pp.621-623.

DBANG 'DUS: Bod gangs can pa'i gso ba rig pa'i dpal ldan rgyud bzhi sogs kyi brda dang dka' gnad 'ga' zhig bkrol ba sngon byon mkhas pa'i gsung rgyun gYu thog dgongs rgyan. dBang 'dus kyis rtsom sgrig byas pa. Pe cin (Mi rigs dpe skrun khang) **1982**.

DE, SUSHIL KUMAR: Ancient Indian Erotics and Erotic Literature. Reprinted Calcutta (Firma K.L. Mukhopadhyay) **1969**.

DEAN-JONES, LESLEY: "The Politics of Pleasure: Female Sexual Appetite in the Hippocratic Corpus." *Helios* (Lubbock, Texas) *19.1992*, pp.72-91.

——: Women's Bodies in Classical Greek Science. Oxford (Clarendon Press) **1994**.

DESHPANDE, MADHAV M.: "Aryans, Non-Aryans, and Brāhmaṇas: Processes of Indigenization." *The Journal of Indo-European Studies* (McLean, Va.) *21.1993*, pp.215-236.

DEUSSEN, PAUL: Allgemeine Geschichte der Philosophie mit besonderer Berücksichtigung der Religionen. Erster Band, erste Abteilung: Allgemeine Einleitung und Philosophie des Veda bis auf die Upanishad's. Leipzig (F.A. Brockhaus) **1894**.

DONIGER, WENDY: The Laws of Manu. With an introduction and notes translated by Wendy Doniger with Brian K. Smith. Reprinted New Delhi | London | New York | Ringwood, Victoria | Toronto | Auckland (Penguin Books) **1992**. Penguin Classics.

Dravyaguṇa: Dravyaguṇaḥ. Mahāmahopādhyāya śrīCakrapāṇidattaviracitaḥ śrīŚivadāsasenakṛtaṭīkayā samalaṅkṛtaḥ. Jīvānandavidyāsāgarabhaṭṭācāryyeṇa saṃskṛtaḥ prakāśitaś ca. Kalikātā (Siddheśvarayantra) [2]**1897**.

EDHOLM, ERIK AF: "*Dhātuprasāda*." Sauhṛdyamaṅgalam. Studies in Honour of Siegfried Lienhard on his 70th Birthday. Edited by Mirja Juntunen, William L. Smith, Carl Suneson. Stockholm (The Association of Oriental Studies) **1995**, pp.91-116.

EMMERICK, RONALD E.: "Some Remarks on the History of Leprosy in India." *Indologica Taurinensia* (Torino) *12.**1984***, pp.93-105.

——: "Die Lepra in Indien." Aussatz • Lepra • Hansen-Krankheit. Ein Menschheitsproblem im Wandel. Teil II: Aufsätze. Herausgegeben von Jörn Henning Wolf. Würzburg (Deutsches Aussätzigen-Hilfswerk) **1986**, pp.185-199.

Encyclopaedia of Indian Medicine. Volume Two. Basic Concepts. Editor S.K. Ramachandra Rao with assistance of S.R. Sudarshan. Bombay (Popular Prakashan)/Bangalore (Dr. V. Parameshvara Charitable Trust) **1987**.S.147.

ESSER, A. ALBERT M.: "Modern-europäische und alt-indische Embryologie." *Münchener medizinische Wochenschrift* (München) *72.**1925***, pp.1643-1645.

FALK, HARRY: "Vedisch *upaniṣád*." *Zeitschrift der Deutschen Morgenländischen Gesellschaft* (Stuttgart) *136.**1986***, pp.80-97.

——: "Die Kosmogonie von RV X 72." *Wiener Zeitschrift für die Kunde Südasiens und Archiv für indische Philosophie* (Wien) *38.**1994***, pp.1-22.

FEDOROVA, MARIANA: Die Marmantheorie in der klassischen indischen Medizin. Inauguraldissertation zur Erlangung des Doktorgrades der Philosophie an der Ludwig-Maximilians-Universität zu München. München (Mariana Fedorova) **1990**.

FILLIOZAT, JEAN: The Classical Doctrine of Indian Medicine. Its origins and its Greek Parallels. Translated from the original French by Dev Raj Chanana. Delhi (Munshiram Manoharlal) **1964**.

——: Laghu-Prabandhāḥ. Choix d'articles d'indologie. Comité de publication Colette Caillat, Bruno Dagens, François Gros, Ludwik Sternbach. Leiden (E.J. Brill) **1974**.

——: La médecine indienne traditionnelle[,] son intérêt scientifique et pratique actuel. Table ronde sur L'apport des médecines asiatiques à la médecine universelle. Strasbourg, 21-23 Mai 1976. Strasbourg (Université Louis Pasteur) **1977**. Scientia Orientalis 5.

FILLIOZAT, PIERRE-SYLVAIN: "Caraka's Proof of Rebirth." *Journal of the European Āyurvedic Society* (Reinbek) *3.**1993***, pp.94-111.

Firdaus al-ḥikma fi 't-ṭibb: Firdausu'l-Ḥikmat or Paradise of Wisdom of 'Alí b. Rabban-al-Ṭabarí. Edited by M.Z. Siddiqi. Berlin (Subsidesed ⟨sic⟩ by the E.G.W. Gibb Memorial Trust and printed in the Buch- u. Kunstdruckerei "Sonne") **1928**.

——: see also SIGGEL 1941-1942.

FIŠER, IVO: Indian Erotics of the Oldest Period. Praha (Universita Karlova) **1966**. Acta Universitatis Carolinae, Philologica Monographia 14.

FIŠER, I. / FIŠEROVA, O.: "Dissection in Ancient India." History and Culture of Ancient India (For the XXVI International Congress of Orientalists). Editors: W. Ruben, V. Struve, G. Bongard-Levin. Moscow (Oriental Literature Publishing House) **1963**. ⟨Russian title: Istorija i kul'tura drevnej Indii (k XXVI Meždunarodnomu kongressu vostokovedov). Redakcionnaja kollegija: V. Ruben, V.V. Struve, G.M. Bongard-Levin. Moskva (Izdatel'stvo Vostočnoj Literatury) 1963⟩, pp.306-328.

FLEET, J.F.: "The Standard Height of an Indian Man." *The Journal of the Royal Asiatic Society of Great Britain and Ireland* (London) ***1911***, pp.208-209.

FRAUWALLNER, ERICH: Geschichte der indischen Philosophie. Salzburg (Otto Müller Verlag) **1953-1956** (two volumes). Reihe Wort und Antwort 6.

Gadanigraha: Gadanigraha of Śrī Vaidya Soḍhala With the 'Vidyotinī' Hindī Commentary By Indradeva Tripāṭhī. Edited by Gaṅgā Sahāya Pāṇḍeya. Varanasi (The Chowkhamba Sanskrit Series Office) **1968-1969** (three volumes). The Kashi Sanskrit Series 182.

GAEFFKE, PETER: "Karma in North Indian Bhakti Traditions." *Journal of the American Oriental Society* (New Haven) *105.1985*, pp.265-275.

GANGOPADHYAYA, MRINAL KANTI: "Vaisesika Categories in the *Caraka-Samhita*." CHATTOPADHYAYA 1991, pp.529-540.

Garbhopaniṣad: N.H. Keswani, N.K. Bhide. "Garbha Upanishad. A Brief Sanskrit Treatise on Ancient Indian Embryology." *Clio Medica* (Amsterdam) *1.1965-66*, pp.64-74.

—: Garbhopaniṣad. Publiée et traduite par Lakshmi Kapani. Paris (Librairie d'Amérique et d'Orient Adrien Maisonneuve) **1976**. Les Upanishad 21. ⟨Subsequently, but too late to incorporate it into my study, I came across an English translation by Asti Hustvedt of Kapani's French translation: "Upaniṣad of the Embryo." Fragments for a History of the Human Body. Part Three. Edited by Michel Feher with Ramona Naddaff and Nadia Tazi. New York (ZONE) **1989**. ZONE 5, pp.176-196.⟩

Garuḍapurāṇasāroddhāra: The Garuda Purâna (Sâroddhâra) with English Translation by Ernest Wood and S.V. Subrahmanyam and an Introduction from Sris Chandra Vasu. [Reprint] Varanasi|Delhi (Bharatiya Publishing House) **1979**. The Sacred Books of the Hindus.

GHĀṆEKAR, BHĀSKAR GOVIND: Suśrutasaṃhitā. Sūtrasthānam. 'Āyurvedarahasyadīpikākhyayā' hindīvyākhyayā samullasitā. Ṭīkākār Bhāskar Govind Ghāṇekar. [Reprint] Naī Dillī (Mehar'cand Lach'man'dās) **1977** ⟨= **1977A**⟩.

—: Suśrutasaṃhitā. Nidānasthānam. 'Āyurvedarahasyadīpikākhyayā' hindīvyākhyayā samullasitā. Ṭīkākār Bhāskar Govind Ghāṇekar. [Reprint] Naī Dillī (Mehar'cand Lach'man'dās) **1977** ⟨= **1977B**⟩.

—: Suśrutasaṃhitā. Śārīrasthānam. 'Āyurvedarahasyadīpikākhyayā' hindīvyākhyayā samullasitā. Ṭīkākār Bhāskar Govind Ghāṇekar. Naī Dillī (Mehar'cand Lach'man'dās Pab'likeśans) ^{9}**1986**.

GHATAGE, A.M.: An Encyclopaedic Dictionary of Sanskrit on Historical Principles. Volume One, and, Volume Two. General Editor A.M. Ghatage. Poona (Deccan College Postgraduate and Research Institute) **1976-1978**; **1981**.

GHOSH, MANOMOHAN: Glimpses of Sexual Life in Nanda-Maurya India. Translation of the Caturbhāṇī together with a critical edition of text. In two parts. Calcutta (Manisha Granthalaya) **1975**.

GHURYE, G.S.: Two Brahmanical Institutions. Gotra and Charana. Bombay (Popular Prakashan) **1972**.

GONDA, J.: Ancient-Indian *ojas*, Latin **augos* and the Indo-European nouns in *-es-/-os*. Utrecht (N.V. A. Oosthoek's Uitgevers Mij.) **1952**.

GRENSEMANN, HERMANN: Knidische Medizin. Teil I: Die Testimonien zur ältesten knidischen Lehre und Analysen knidischer Schriften im Corpus Hippocraticum. Berlin|New York (Walter de Gruyter) **1975**. Ars Medica, II. Abteilung 4,1.

GUPTA, PABITRAKUMAR: "Acharya Benoykumar Sarkar on Sanskritization." Ācārya binaẏ'kumār sar'kār. Pradyot Ghoṣ sampādita. Māl'daha (Praphesar Binaẏ'kumār Sar'kār Janmaśatabārṣik Kamiṭi) **1988**, pp.A-E.

GYATSO, GESHE KELSANG: Clear Light of Bliss. Mahamudra in Vajrayana Buddhism. Translated by Tenzin Norbu. Edited by Jonathan Landaw with Chris Kolb. London (Wisdom Publications) **1982**. Wisdom Advanced Books: Blue Series.

HAHN, MICHAEL: Candragomins Lokānandanāṭaka. Nach dem tibetischen Tanjur herausgegeben und übersetzt von Michael Hahn. Ein Beitrag zur klassischen indischen Schauspieldichtung. Wiesbaden (Otto Harrassowitz) **1974**. Asiatische Forschungen 39.

——: Joy For The World. A Buddhist Play by Candragomin. Translated with an Introduction and Notes by Michael Hahn. Berkeley (Dharma Publishing) **1987**. Tibetan Translation Series.

HĀLADĀRA, GURUPADAŚARMMAN: Vṛddhatrayī. Kalikātā (Bhārativikāśaśarmmahāladāra) **Beng. era 1362**.

HALBFASS, WILHELM: On Being and What There Is. Classical Vaiśeṣika and the History of Indian Ontology. Albany (State University of New York Press) **1992**.

HAMPIHOLI, VISHWANATH K.: Kāmashāstra in Classical Sanskrit Literature. Delhi (Ajanta Publications) **1988**.

Handwörterbuch des deutschen Aberglaubens. Herausgegeben von Hanns Bächtold-Stäubli unter Mitwirkung von Eduard Hoffmann-Krayer. Band 3. Freen — Hexenschuß. Reprinted Berlin|New York (Walter de Gruyter) **1987**.

HANEFELD, ERHARDT: Philosophische Haupttexte der älteren Upaniṣaden. Wiesbaden (Otto Harrassowitz) **1976**. Freiburger Beiträge zur Indologie 9.

HARA, MINORU: "A Note on the Buddha's Birth Story." Indianisme et Bouddhisme. Mélanges offerts à Mgr Étienne Lamotte. Louvain-La-Neuve (Université Catholique de Louvain, Institut Orientaliste) **1980**. Publications de l'Institut Orientaliste de Louvain 23, pp.143-157.

——: "The Holding of the Hair (*Keśa-grahaṇa*)." *Acta Orientalia* (Havnia) *47.* ***1986***, pp.67-92.

——: "Invigoration." Hinduismus und Buddhismus. Festschrift für Ulrich Schneider. Herausgegeben von Harry Falk. Freiburg (Hedwig Falk) **1987**, pp.134-151.

——: "*Deva-garbha* and *Tathāgata-garbha*." The Buddhist Forum. Volume III. 1991-1993. Papers in honour and appreciation of Professor David Seyfort Ruegg's contribution to Indological, Buddhist and Tibetan Studies. Edited by Tadeusz Skorupski & Ulrich Pagel. London (School of Oriental and African Studies, University of London) **1994**, pp.37-55.

Haramekhalā: Mādhukadhīrasahāyapaṇḍitaviracitā Haramekhalā saṭikā (prathamādicaturthaparicchedāntaḥ pūrvakhaṇḍaḥ). 'Bhaṭṭarāī' ity upāhva...Kṛṣṇaprasādaśarmaṇā sampāditā. Kāṣṭhamaṇḍapa (Nepālarāṣṭrīyābhilekhālaya) **V.S. 2026**. Purātattvaprakāśanamālā 64.

HARIPRAPANNAJĪ: Klomayāthātathyam. Mumbaī (Hariprapannajī) **1930**. ⟨Printed at the Nirṇay'sāgar Chāp'khānā.⟩

Hārītasaṃhitā: Harita Sanhita. A Complete System of Hindu Medicine by Maharshi Atraya. Edited and published by Binod Lal Sen. Calcutta (Kaviraj Binod Lal Sen) **n.d.**

——: La Hārītasaṃhitā. Texte médical sanskrit avec un index de nomenclature âyurvédique par Alix Raison. Pondichéry (Institut Français d'Indologie) **1974**. Publications de l'Institut Français d'Indologie 52.

Harivaṃśa: The Harivaṃśa being the Khila or Supplement to the Mahābhārata. For the first time critically edited by Parashuram Lakshman Vaidya. Poona (Bhandarkar Oriental Research Institute) **1969-1971** (two volumes).

Haṭhayogapradīpikā: The Hathayogapradīpikā ⟨sic; cover: Haṭhayogapradīpikā⟩ of Svātmārāma. With the commentary *Jyotsnā* of Brahmānanda and English translation. Reprinted Adyar, Madras (The Adyar Library and Research Centre) **1975**. The Adyar Library

General Series 4.

Ḥayawān: Ibn Sīnā. aš-Šifā'. aṭ-Ṭabī'iyyāt. 8 — al-Ḥayawān. Rāğa'a-hū wa qaddama lahū Ibrāhīm Madkūr. Bi-taḥqīq 'Abdalḥalīm Muntaṣir[,] Sa'īd Zāyid[,] 'Abdallāh Ismā'īl. Bi-munāsabat aḏ-ḏikrā 'l-alfiyyah liš-šaiḫ ar-ra'īs. al-Qāhirah (al-Hai'ah al-Miṣriyyah al-'Āmmah lit-Ta'līf wan-Našr) **1970**.

HEILIJGERS-SEELEN, DORY: The System of Five Cakras in *Kubjikāmatatantra* 14-16. Groningen (Egbert Forsten) **1994**. Groningen Oriental Studies 9.

HESSLER, FRANCISCUS/FRANZ: Suśrutas. Áyurvédas. Id est medicinae systema a venerabili D'hanvantare demonstratum a Suśruta discipulo compositum. Nunc primum ex Sanskrĭta in Latinum sermonem vertit, introductionem, annotationes et rerum indicem adjecit Franciscus Hessler. Erlanga (Ferdinandus Enke) **1844-1850** (three volumes).

—: "Allgemeine Übersicht der Heilkunde der alten Inder. Eine Studie." *Sitzungsberichte der mathematisch-physikalischen Classe der k.b. Akademie der Wissenschaften zu München* (München) *17.**1887***, pp.137-149.

HILGENBERG, LUISE: "Die Anschauungen von den Konstitutionstypen in der Medizin Altindiens und unserer Zeit." Studien zur Geschichte und Kultur des Nahen und Fernen Ostens. Paul Kahle zum 60. Geburtstag überreicht von Freunden und Schülern aus dem Kreise des Orientalischen Seminars der Universität Bonn. Herausgegeben von W. Heffening und W. Kirfel. Leiden (E.J. Brill) **1935**, pp.156-162.

HILGENBERG, LUISE / KIRFEL, WILLIBALD: Vāgbhaṭa's Aṣṭāṅgahṛdayasaṃhitā. Ein altindisches Lehrbuch der Heilkunde aus dem Sanskrit ins Deutsche übertragen mit Einleitung, Anmerkungen und Indices von Luise Hilgenberg und Willibald Kirfel. Leiden (E.J. Brill) **1941**.

HOERNLE, A.F.R.: The Suçruta-Saṁhitā or The Hindū System of Medicine According to Suçruta. Translated from the original Sanskrit by A.F.R. Hoernle. Fasciculus I. Calcutta (Asiatic Society) **1897**. Bibliotheca Indica [139] (New Series [Fascicle] 911).

—: "Studies in Ancient Indian Medicine." *The Journal of the Royal Asiatic Society of Great Britain and Ireland* (London) ***1907***, pp.1-18.

HOFFMANN, KARL: Der Injunktiv im Veda. Eine synchronische Funktionsuntersuchung. Heidelberg (Carl Winter Universitätsverlag) **1967**. Indogermanische Bibliothek. Dritte Reihe. Untersuchungen.

—: Aufsätze zur Indoiranistik herausgegeben von Johanna Narten. Wiesbaden (Dr. Ludwig Reichert Verlag) **1975-1976** (two volumes).

HOUBEN, JAN E.M.: The Saṃbandha-Samuddeśa (Chapter on Relation) and Bhartṛhari's Philosophy of Language. A Study of Bhartṛhari ⟨sic⟩ Saṃbandha-samuddeśa in the context of the Vākyapadīya with a translation of Helarāja's commentary Prakīrṇa-prakāśa. Groningen (Egbert Forsten) **1995**. Gonda Indological Studies 2.

ISHII, HIROSHI: "Seasons, Rituals and Society: The Culture and Society of Mithila, the Parbate Hindus and the Newars as seen through a Comparison of their Annual Rites." From Vedic Altar to Village Shrine. Towards an Interface between Indology and Anthropology. Edited by Yasuhiko Nagano, Yasuke Ikari. Osaka (National Museum of Ethnology) **1993**. Senri Ethnological Studies 36, pp.35-84.

IYER, T.G. RAMAMURTHI: The Hand Book of Indian Medicine. The Gems of Siddha System. Delhi (Sri Satguru Publications) ^{2}1981, reprinted **1986**.

JACOB, G.A.: Upaniṣadvākyakośaḥ. A Concordance to the Principal Upanishads and Bhagavadgîtâ. Bombay (Government Central Book Depôt) **1891**. Bombay Sanskrit Series 39.

JAINI, PADMANABH S.: Gender and Salvation. Jaina Debates on the Spiritual Liberation of Women. Berkeley|Los Angeles|Oxford (University of California Press) **1991**.

JAMISON, STEPHANIE W.: "Brāhmaṇa Syllable Counting, Vedic *Tvác* 'Skin', and the Sanskrit Expression for the Canonical Creature." *Indo-Iranian Journal* (Dordrecht|Boston|Lancaster|Tokyo) *29.**1986***, pp.161-181.

——: Sacrificed Wife/Sacrificer's Wife. Women, Ritual, and Hospitality in Ancient India. New York|Oxford (Oxford University Press) **1996**.

JOLLY, JULIUS: Medicin. Strassburg (Verlag von Karl J. Trübner) **1901**. Grundriss der Indo-Arischen Philologie und Altertumskunde 3,10.

Jvaranirṇaya: Jvaranirṇaya of Śrī. ⟨sic⟩ Nārāyaṇa Paṇḍita. Edited and Published by K. Raghavan Pillai. Trivandrum (Keraḷaviśvavidyālaya) **1963**. University of Kerala Sanskrit Series 202.

KAHLO, GERHARD: "Randbemerkungen zu Vātsyāyana und seinen Nachfolgern." *Annali dell'Istituto Universitario Orientale di Napoli* (Napoli) *Nuova Serie 16.**1966***, pp.251-267.

KAISER, RONALD: Die Professionalisierung der ayurvedischen Medizin und deren Rolle im indischen Medizinpluralismus. Bonn (Holos Verlag) **1992**. Kölner Ethnologische Arbeitspapiere 3.

KAKAR, SUDHIR: Intimate Relations. Exploring Indian Sexuality. New Delhi (Viking) 1989, reprinted **1990**.

Kalyāṇakāraka: The Kalyāṇa-Karākam ⟨sic⟩ of Ugrādityacharya ⟨sic⟩ Edited with introduction, translation, notes, indexes & dictionary by Vardhaman Parshwanath Shastri. Sholapur (Seth Govindji Raoji Doshi) **1940**. Sakharam Nemchand Granthamala 129.

Kāmakalāvilāsa: The Kāma Kalā Vilās of Punya Nanda with commentary. Edited with notes by Mukunda Rāma Shâstrî. Bombay|Śrīnagara (The Research Department, Jammu & Kashmir State) **1918**. Kashmir Series of Texts and Studies 12.

Kāmakuñjalatā: Kāmakunjalatā (A Collection of Old and Rare works on Kāma Śāstra) Edited by Ḍhuṇḍhirāja Śāstrī. Varanasi (The Chowkhamba Sanskrit Series Office) **1967**. The Chowkhamba Sanskrit Series 92.

Kāmaśāstra of Surūpa: Surūpa's Kāmaśāstra. An Erotic Treatise in the Tibetan Tanjur edited and translated by Claus Vogel. Helsinki (Societas Orientalis Fennica) **1965**. Studia Orientalia 30:3. ⟨This edition seems to be unknown to CHÖPEL, pp.38f.⟩

Kāmasūtra: ŚrīVātsyāyanapraṇītaṃ Kāmasūtram. Yaśodharaviracitayā Jayamaṅgalākhyayā ṭīkayā sametam. Vrajalālasūnoḥ Durgāprasādasya kṛte ... mudritam. Mumbayī (Nirṇayasāgarayantrālaya) **1891**.

——: see also DANIÉLOU and MYLIUS 1987.

Kandarpacūḍāmaṇi: Kandarpacūḍāmaṇiḥ. ŚrīVīrabhadradevanṛpati viracitaḥ Kuśalāparākhyena Rāmacandraśāstriṇā sampāditaḥ Sukhaprabhākhyayā vyākhyayā pariṣkṛtaś ca. Lāhaur (Banār'sīdās Sundar'lāl) **1926**.

KANNU, S. PEERU: The Critical Study of Praśastapādabhāṣya. Delhi (Kanishka Publishing House) **1992**.

KAPLONY, PETER: "Die Symbolik des Leibes und der Glieder im Alten Ägypten." Symbolik des menschlichen Leibes. Herausgegeben von Paul Michel. Bern|Berlin|Frankfurt/M.|New York|Paris|Wien (Peter Lang) **1995**. Schriften zur Symbolforschung 10, pp.21-48.

KAPP, DIETER B.: "Der Strībhedavarṇana-khaṇḍa in der Padumāvatī des Malik Muḥammad Jāyasī." *Indo-Iranian Journal* (The Hague) *16.**1975***, pp.183-221.

KARTTUNEN, KLAUS: India in Early Greek Literature. Helsinki (Societas Orientalis Fennica) **1989**. Studia Orientalia 65.

Kāśikā: Kāśikā. A Commentary on Pāṇini's Grammar. Part — I. (Adhyāyas 1-4). By Vāmana & Jayāditya. Editors Aryendra Sharma, Khanderao Deshpande, D.G. Padhye. Hyderabad (Sanskrit Academy, Osmania University) **1969**. Sanskrit Academy Series 17/A.14.

Kāśyapasaṃhitā: Kaśyapasaṁhitā (or Vṛiddhajīvakīya Tantra) by Vṛiddhajīvaka and revised by Vātsya. Edited by Jādavjī Trikamjī Āchārya and Somanāth Śarmā. Muṃbaī (Nirṇayasāgarayantrālaya) **1938**. Nepāl Sanskrit Series 1.

Kāṭhakasaṃhitā: Kâṭhakam. Die Saṃhitâ der Kaṭha-Çâkhâ herausgegeben von Leopold von Schroeder. Leipzig (F.A. Brockhaus) **1900-1910** (three volumes).

KATRE, SUMITRA MANGESH/S.M.: "On Some 'Laukika' Words Cited in the Commentaries of Cakrapāṇidatta (on Caraka) and Ḍalhaṇa (on Suśruta)." *Bulletin of the Deccan College Research Institute* (Poona) *18.1957 (Taraporewala Memorial Volume)*, pp.226-254.

——: Problems of Reconstruction in Indo-Aryan. Simla (Indian Institute of Advanced Study) **1968**.

KAVAḌE, VAIDYAPAÑCĀNANA KṚṢṆAŚĀSTRIN: Klomanirṇayaḥ. Puṇya (Vaidyapañcānana Kṛṣṇaśāstrin Kavaḍe) **Śaka 1851**. ⟨Printed at the Citraśālākhyamudraṇālaya.⟩

Kelikutūhala: Kelikutūhalam (bhāṣānuvādasahitam). Racayitā:— Mathurāprasāda Dīkṣitaḥ. Jhāsī (Ma.Ma.Paṃ. Mathurāprasād Dīkṣit) ^{2}**1954**.

KEULS, EVA C.: The Reign of the Phallus. Sexual Politics in Ancient Athens. Berkeley|Los Angeles|London (University of California Press) ^{2}**1993**.

KING, HELEN: "Sowing the Field: Greek and Roman Sexology." Sexual Knowledge, Sexual Science. The History of Attitudes to Sexuality. Edited by Roy Porter and Mikuláš Teich. Cambridge|New York|Melbourne (Cambridge University Press) **1994**, pp.29-46.

KIRFEL, WILLIBALD: Die fünf Elemente. Insbesondere Wasser und Feuer. Ihre Bedeutung für den Ursprung altindischer und altmediterraner Heilkunde. Eine medizingeschichtliche Studie. Walldorf—Hessen (Verlag für Orientkunde Dr. H. Vorndran) **1951**. Beiträge zur Sprach- und Kulturgeschichte des Orients 4.

——: "Ein medizinisches Kapitel des Garuḍapurāṇas." Asiatica. Festschrift Friedrich Weller zum 65. Geburtstag. [Herausgegeben von Johannes Schubert und Ulrich Schneider.] Leipzig (Otto Harrassowitz) **1954**, pp.333-356.

KLOPPENBURG, RIA: "Female Stereotypes in Early Buddhism: The Women of the Therīgāthā." Female Stereotypes in Religious Traditions. Edited by Ria Kloppenburg and Wouter J. Hanegraaff. Leiden|New York|Köln (E.J. Brill) **1995**. Studies in the History of Religions (*Numen* Book Series) 66, pp.151-169.

KOHL, JOSEF FRIEDRICH: "Ein Beitrag zum indischen Giftmädchenmotiv." *Zeitschrift der Deutschen Morgenländischen Gesellschaft* (Wiesbaden) *109.1959*, pp.324-339.

KOHLBRUGGE, H.: "Glück- und Unglückszeichen am menschlichen Körper." *Acta Orientalia* (Lugdunum) *20.1948*, pp.36-76.

KOLHATKAR, MADHAVI: "A Possible Explanation of a Marathi Proverb." *Bulletin of the Deccan College Post Graduate & Research Institute* (Pune) *51-52.1991-1992*, pp.149-151.

KULLMANN, WOLFGANG: "Aristoteles' Grundgedanken zu Aufbau und Funktion der Körpergewebe." *Sudhoffs Archiv* (Wiesbaden) *66.1982*, pp.209-238.

Kumāratantra: Étude de démonologie indienne. Le Kumāratantra de Râvaṇa et les textes parallèles indiens, tibétains, chinois, cambodgien et arabe par Jean Filliozat. Paris (Imprimerie Nationale) **1937**. Cahiers de la Société Asiatique, Première Série 4.

KURIYAMA, SHIGEHISA: "What is *nadipariksa*? On the identification of terms and the history of perceptual experience." Abstracts of Papers Presented at the 16th Annual Conference on South Asia[,] November 6, 7 & 8, 1987. Madison (South Asian Area Center, University of Wisconsin) **1987**, p.30.

Kūṭamudgara: Kūṭ'mudgar arthāt· cikitsākā bīj'rūp choṭāsā atikliṣṭ granth. Jis'ko Mur'-līdhar'śarmā sampādak ne sānvay'bhāṣāṭīkāse vibhūṣit kiyā. Bambaī (Śrīveṅkaṭeśvar Sṭīm-Mudraṇ'yantrālay) **V.S. 1966**.

LADAS, ALICE KAHN / WHIPPLE, BEVERLY / PERRY, JOHN D.: The G Spot And Other Recent Discoveries About Human Sexuality. New York (Holt, Rinehart and Winston) **1982**.

LAQUEUR, THOMAS W.: "'Amor Veneris, vel Dulcedo Appeletur'." Fragments for a History of the Human Body. Part Three. Edited by Michel Feher with Ramona Naddaff and Nadia Tazi. New York (ZONE) **1989**. ZONE 5, pp.90-131.

—: Making Sex. Body and Gender from the Greeks to Freud. Cambridge, Mass.|London (Harvard University Press) **1990**.

LARSON, GERALD JAMES: "Āyurveda and the Hindu philosophical systems." *Philosophy East and West* (Honolulu) *37.**1987***, pp.245-259.

—: "The Concept of Body in Ayurveda and the Hindu Philosophical Systems." Self as Body in Asian Theory and Practice. Edited by Thomas P. Kasulis with Roger T. Ames and Wimal Dissanayake. Albany (State University of New York Press) **1993**. SUNY Series, The Body in Culture, History and Religion, pp.103-121.

LEERTOUWER, LAMMERT: "Zur Definition des Objektes in der Religionshermeneutik." Beiträge zur Hermeneutik indischer und abendländischer Religionstraditionen. Arbeitsdokumentation eines Symposiums. Herausgegeben von Gerhard Oberhammer. Wien (Verlag der Österreichischen Akademie der Wissenschaften) **1991**. Österreichische Akademie der Wissenschaften, Philosophisch-historische Klasse, Sitzungsberichte 573. Beiträge zur Kultur- und Geistesgeschichte Asiens 6, pp.47-59.

LESKY, ERNA: Die Zeugungs- und Vererbungslehre der Antike und ihr Nachwirken. Wiesbaden (Franz Steiner Verlag) **1951**. Akademie der Wissenschaften und der Literatur in Mainz, Abhandlungen der Geistes- und Sozialwissenschaftlichen Klasse 1950,19.

LESLIE, CHARLES: "Interpretations of Illness: Syncretism in Modern Āyurveda." Paths to Asian Medical Knowledge. Edited by Charles Leslie and Allan Young. Berkeley|Los Angeles|Oxford (University of California Press) **1992**. Comparative Studies of Health Systems and Medical Care [32], pp.177-208.

LESLIE, I. JULIA: The Perfect Wife. The Orthodox Hindu Woman according to the *Strīdharmapaddhati* of Tryambakayajyan. Delhi (Oxford University Press) **1989**. Oxford University South Asian Studies Series.

—: "Some Traditional Indian Views on Menstruation and Female Sexuality." Sexual Knowledge, Sexual Science. The History of Attitudes to Sexuality. Edited by Roy Porter and Mikuláš Teich. Cambridge|New York|Melbourne (Cambridge University Press) **1994**, pp.63-81.

LIENHARD, SIEGFRIED: "Konstitution und Charakter nach den Lehren der altindischen Medizin." *Centaurus* (Copenhagen) *6.**1959***, pp.82-93.

—: Kokkoka. Ratirahasya. Geheimnisse der Liebeskunst. Aus dem Sanskrit übertragen von Siegfried Lienhard. Schmiden bei Stuttgart (Franz Decker Nachf.) **1960**.

—: "On the Meaning and Use of the Word *Indragopa*." *Indologica Taurinensia* (Torino) *6.**1978***, pp.177-188.

LIÉTARD, GUSTAVE / CORDIER, PALMYR: Un demi-siècle de recherches āyurvédiques. Gustave Liétard et Palmyr Cordier. Travaux sur l'histoire de la médecine indienne. Documents réunis et présentés par Arion Roşu. Paris (Collège de France, Institut de

Civilisation Indienne) **1989**. Publications de l'Institut de Civilisation Indienne 56.

Lohasarvasva: The Lohasarvaswam of Śrī Sureśwara With the 'Vidyotinī' Hindī Commentary By Śrī Pāvanī Prasād Sharmā. Edited by Brahmashankar Mishra. Varanasi (Chowkhamba Vidya Bhawan) **1965**. The Vidyabhawan Ayurveda Granthamala 46.

LÜDERS, HEINRICH: Philologica Indica. Ausgewählte kleine Schriften. Festgabe zum siebzigsten Geburtstage am 25. Juni 1939 dargebracht von Kollegen, Freunden und Schülern. Göttingen (Vandenhoeck & Ruprecht) **1940**.

Madanamahārṇava: Madanamahārṇava of Śrī Viśveśvara Bhaṭṭa edited by Embar Krishnamacharya and M.R. Nambiyar. Baroda (Oriental Institute) **1953**. Gaekwad's Oriental Series 117.

Mādhavanidāna: MahāmatiśrīMādhavakaraviracitaṃ Mādhavanidānam. Mahāmahopādhyāya-śrīVijayarakṣita-Śrīkaṇṭhadattābhyāṃ viracitayā 'Madhukośa'-vyākhyayā vibhūṣitam. Tathā Sudarśanaśāstrikṛtayā 'Vidyotinī' hindīṭīkayā navīnavaijñānika 'Vimarśena' ca samullasitam. Tac cedam Yadunandanopādhyāya cikitsakena susaṃskṛtya sampāditam. Vārāṇasī (Caukhambā Saṃskṛt Sirīj Āphis) [2]**1960-1961** (two volumes). Kāśī-Saṃskṛta-Granthamālā 158.

——: see also MEULENBELD 1974.

Mahābhārata: The Mahābhārata for the first time critically edited by Vishnu S. Sukthankar (Aug.1925-Jan.1943), S.K. Belvalkar (April 1943-March 1961), P.L. Vaidya (since April 1961). ... Poona (Bhandarkar Oriental Research Institute) **1933-1966** (nineteen volumes).

Maitrāyaṇyupaniṣad: Contained in *Upaniṣatsaṅgraha* 1,pp.176-194.

Mānavadharmaśāstra: Manusaṃhitā. Kullūkabhaṭṭakṛta ṭīkayā sahitā. Jīvānandavidyāsāgara Bhaṭṭācāryyeṇa saṃskṛtā. Kalikātānagara (Viḍan-yantra) **1874**.

——: see also DONIGER.

MAYRHOFER, MANFRED: Kurzgefaßtes etymologisches Wörterbuch des Altindischen. Heidelberg (Carl Winter Universitätsverlag) **1956-1980** (four volumes). Indogermanische Bibliothek, Reihe 2: Wörterbücher.

——: Etymologisches Wörterbuch des Altindoarischen. Heidelberg (Carl Winter Universitätsverlag) **1992-**. Indogermanische Bibliothek, 2. Reihe: Wörterbücher. ⟨Two volumes have appeared to date.⟩

MEULENBELD, GERRIT JAN: The Mādhavanidāna and its Chief Commentary, Chapters 1-10. Introduction, translation and notes. Leiden (E.J. Brill) **1974**. Orientalia Rheno-Traiectina 19.

——: "The Surveying of Sanskrit Medical Literature." Proceedings of the International Workshop on Priorities in the Study of Indian Medicine held at the State University of Groningen 23-27 October 1983. Edited by G. Jan Meulenbeld. Groningen (Institute of Indian Studies, University of Groningen) **1984** ⟨= **1984A**⟩. Rijksuniversiteit te Groningen, Publikaties van het Instituut voor Indische talen en culturen 4, pp.31-114.

——: "Comments on the Paper 'The Surveying of Sanskrit Medical Literature'." Proceedings of the International Workshop on Priorities in the Study of Indian Medicine held at the State University of Groningen 23-27 October 1983. Edited by G. Jan Meulenbeld. Groningen (Institute of Indian Studies, University of Groningen) **1984** ⟨= **1984B**⟩. Rijksuniversiteit te Groningen, Publikaties van het Instituut voor Indische talen en culturen 4, pp.115-120.

——: "Reflections on the basic concepts of Indian pharmacology." Studies on Indian Medical History. Papers presented at the International Workshop on the Study of Indian Medicine held at the Wellcome Institute for the History of Medicine 2-4 September 1985. Edited by G. Jan Meulenbeld and Dominik Wujastyk. Groningen (Egbert Forsten) **1987**. Groningen Oriental Studies 2, pp.1-17.

——: "Conformities and Divergences of Basic Āyurvedic Concepts in Veterinary Texts." *Journal of the European Āyurvedic Society* (Reinbek) *1.* ***1990***, pp.1-6.

——: "The constraints of theory in the evolution of nosological classifications: A study on the position of blood in Indian medicine (Āyurveda)." Panels of the VIIth World Sanskrit Conference. Kern Institute, Leiden: August 23-29, 1987. General Editor: Johannes Bronkhorst. Vol.VIII. Medical Literature from India, Sri Lanka and Tibet. Edited by G. Jan Meulenbeld. Leiden|New York|København|Köln (E.J. Brill) **1991**, pp.91-106.

——: "The Characteristics of a Doṣa." *Journal of the European Āyurvedic Society* (Reinbek) *2.* ***1992***, pp.1-5.

——: "The Many Faces of Āyurveda." *Journal of the European Āyurvedic Society* (Reinbek) *4.* ***1995***, pp.1-10.

MEYER, FERNAND: gSo-ba rig-pa. Le système médical tibétain. Reproduction par procédé photomécanique de l'édition corrigée de 1981. Paris (Éditions du Centre National de la Recherche Scientifique) **1983**. Centre régional de publications du C.N.R.S. Meudon-Bellevue. Greco Himalaya-Karakorum. Cahiers népalais.

——: "Théorie et pratique de l'examen des pouls dans un chapitre du *rGyud-bzhi*." Indo-Tibetan Studies. Papers in honour and appreciation of Professor David L. Snellgrove's contribution to Indo-Tibetan Studies. Edited by Tadeusz Skorupski. Tring (The Institute of Buddhist Studies) **1990**. Buddhica Britannica, Series Continua 2, pp.209-256. ⟨The bibliography for the whole volume is on pp.xiii-xxix.⟩

MEYER, JOHANN JAKOB: Trilogie altindischer Mächte und Feste der Vegetation. Ein Beitrag zur vergleichenden Religions- und Kulturgeschichte, Fest- und Volkskunde. Zürich (Max Niehans Verlag) **1937** (three parts).

——: Sexual Life in Ancient India. A Study in the Comparative History of Indian Culture. Reprinted London (Routledge & Kegan Paul) **1952**.

——: Dāmodaragupta's Kuṭṭanimatam (Lehren einer Kupplerin). Ins Deutsche übertragen von Johann Jakob Meyer. Leipzig (Lotus-Verlag) **n.d.** Altindische Schelmenbücher 2.

MITRA, JYOTIR: A Critical Appraisal of Āyurvedic Material in Buddhist Literature with Special Reference to Tripiṭaka. Varanasi (The Jyotiralok Prakashan) **1985**.

MITRA SHASTRY, AJAY: "Varāhamihira and Ayurveda." *Bulletin of the Indian Institute of History of Medicine* (Hyderabad) *22.* ***1992***, pp.83-92.

MONIER-WILLIAMS, MONIER: A Sanskrit-English Dictionary etymologically and philologically arranged with special reference to cognate Indo-European languages. New edition, greatly enlarged and improved with the collaboration of E. Leumann, C. Cappeller and other scholars. Reprinted Oxford (Clarendon Press) **1976**.

MÜLLER, REINHOLD F.G.: "Über Skelett-Darstellungen in Asia Major." *Asia Major* (Lipsia) *2.* ***1925***, pp.531-563.

——: "Die Medizin im Ṛg-Veda." *Asia Major* (Lipsia) *6.* ***1930***, pp.315-376.

——: "Zur anatomischen Systematik im *Yajus*." *Sudhoffs Archiv für Geschichte der Medizin und der Naturwissenschaften* (Leipzig) *27.* ***1934***, pp.20-31.

——: "Eine Konstitutionslehre der altindischen Medizin." *Zeitschrift für Rassenphysiologie* (München|Berlin) *12.* ***1941***, pp.59-72.

——: "Zu Vorstellungen altindischer Ärzte über Fortpflanzungs-Stoffe." *Quellen und Studien zur Geschichte der Naturwissenschaften und der Medizin* (Berlin) *8.* ***1941-1942***, pp.458-480.

——: Grundlagen altindischer Medizin. Halle (Saale) (Kaiserlich Leopoldinisch-Carolinisch

Deutsche Akademie der Naturforscher) **1942**. Nova Acta Leopoldina, Neue Folge Band 11, Nummer 74.

—: "Zum Blasensteinschnitt beim Weibe in der altindischen Medizin." *Zentralblatt für Gynäkologie* (Leipzig) *72.**1950***, pp.1441-1443.

—: Grundsätze altindischer Medizin. København (Ejnar Munksgaard) **1951**. Acta Historica Scientiarum Naturalium et Medicinalium 8.

—: Altindische Embryologie. Leipzig (Johann Ambrosius Barth) **1955**. Nova Acta Leopoldina, Neue Folge Band 17, Nummer 115.

—: Eigenwertungen in altindischer Medizin. Leipzig (Johann Ambrosius Barth) **1958**. Nova Acta Leopoldina, Neue Folge Band 20, Nummer 138.

—: "Über indische Vorstellungen von der Verdauung." *Mitteilungen des Instituts für Orientforschung* (Berlin) *7.**1959/60***, pp.198-223.

—: "Kritische Skizze zur Entwicklung der Medizin der Indoarier." *Wiener Zeitschrift für die Kunde Süd- und Ostasiens und Archiv für indische Philosophie* (Leiden|Köln|Wien) *4.* ***1960*** ⟨= **1960A**⟩, pp.14-35.

—: "Über ein Neutrum im Sanskrit." *Die Sprache* (Wien) *6.**1960*** ⟨= **1960B**⟩, pp.39-45.

—: "Altindische Embryologie." Atti del XIV° Congresso Internazionale di Storia della Medicina sotto l'Alto Patronato del Presidente della Repubblica. Volume II°. Roma — Salerno 13 — 20 Settembre 1954. Roma (Guerra & Belli) **1960** ⟨= **1960C**⟩, pp.717-722.

—: "*Doṣa* = Fehler." *Wissenschaftliche Zeitschrift der Martin-Luther-Universität Halle-Wittenberg* (Halle) *10.**1961*** ⟨= **1961A**⟩, p.1395.

—: "Wörterheft zu einigen Ausdrücken indischer Medizin." *Mitteilungen des Instituts für Orientforschung* (Berlin) *8.**1961*** ⟨= **1961B**⟩, pp.64-159.

—: "Dhātu (eine Ausdrucksbestimmung in der indischen Medizin)." *PHMA* ⟨= R(h)ēma⟩ (München) *7.**1961*** ⟨= **1961C**⟩, pp.3-23.

—: "Durst (und Hunger), bewertet durch indische Ärzte." *Communicationes ex Bibliotheca Historiae Medicae Hungarica* (Budapest) *21/22.**1961*** ⟨= **1961D**⟩, pp.41-53.

—: "Eine Wind-Lehre der Bheḷasaṃhitā." *Wiener Zeitschrift für die Kunde Süd- und Ostasiens und Archiv für indische Philosophie* (Leiden|Köln|Wien) *6.**1962***, pp.29-39.

—: "On the Indian Neuter." Jñānamuktāvalī. Commemoration Volume in Honour of Johannes Nobel. On the Occasion of his 70th Birthday offered by Pupils and Colleagues. Edited by Claus Vogel. [Reprinted] New Delhi (International Academy of Indian Culture) **1963** ⟨= **1963A**⟩. Sarasvati-Vihara Series 38, pp.153-158.

—: "Über das Herabsteigen des Keimes (zum Embriologie-Beginn ⟨sic⟩ im Śārirasthāna vom Aṣṭāṅgasaṃgraha des Vāgbhaṭa)." *Mitteilungen des Instituts für Orientforschung* (Berlin) *8.**1963*** ⟨= **1963B**⟩, pp.409-424.

—: "Beurteilung einer Siebengliederung von Krankheiten durch indische Wundärzte." *Communicationes ex Bibliotheca Historiae Medicae Hungarica* (Budapest) *32.**1964***, pp.19-29.

—: "Medizin der Inder in kritischer Übersicht." Indo-Asian Studies. Part 2 (Prof.Dr. Raghu Vira Memorial Volume) edited by Lokesh Chandra. New Delhi (International Academy of Indian Culture) **1965**. Śata-Piṭaka Series 37, pp.3-124.

—: "Einige Bemerkungen über Salz in Bewertung der Inder und ihrer Ärzte." *Clio Medica* (Amsterdam) *1.**1965-66***, pp.60-64.

—: "Die indische Bewertung des weiblichen Fortpflanzungsstoffes (*ārtava*)." *Centaurus* (Copenhagen) *11.**1965-1966*** ⟨= **1965-1966A**⟩, pp.205-209.

—: "*Indisch*: *Dravya* (medizingeschichtlich)." *Centaurus* (Copenhagen) *11.**1965-1966*** ⟨= **1965-1966B**⟩, pp.259-269.

——: "Eine indische Lehre über Zunahme und Abnahme bei körperlichen Umsetzungen." *Rivista degli Studi Orientali* (Roma) *41.1966*, pp.77-80.

——: "Über verschiedene Ergebnisse indischer Textuntersuchungen durch Sprachwissenschaftler oder einen Medizingeschichtler." *Rocznik Orientalistyczny* (Warszawa) *30,2.1967*, pp. 95-113.

MUROJI, YOSHIHITO G.: Vasubandhus Interpretation des *Pratītyasamutpāda*. Eine kritische Bearbeitung des *Pratītyasamutpādavyākhyā* (*Saṃskāra-* und *Vijñānavibhaṅga*). Stuttgart (Franz Steiner Verlag) **1993**. Alt- und Neu-Indische Studien 43.

MURTHY, K.R. SRIKANTHA: Vāgbhaṭa's Aṣṭāñga ⟨sic⟩ Hṛdayam (Text, English translation, Notes, Appendix and Indices). Translated by K.R. Srikantha Murthy. Varanasi (Krishnadas Academy) **1991-1995** (three volumes). Krishnadas Ayurveda Series 27.

MYLIUS, KLAUS: Geschichte der Literatur im alten Indien. Leipzig (Verlag Philipp Reclam jun.) **1983**.

——: Mallanāga Vātsyāyana. Das Kāmasūtra übersetzt von Klaus Mylius. Leipzig (Verlag Philipp Reclam jun.) **1987**.

Nāḍījñānadarpaṇa: ŚrīmadBhūdharabhaṭṭasaṃgrahīto Nāṛījñānadarpaṇaḥ, Nāṛījñāna-mala-mūtra-netra-jihvā-śabda-rūpaparīkṣādivividhaviṣayavibhūṣitaḥ. Jīvanacandra Kāvyatīrthakṛtayā 'Bhāṣāṭīkayā'ṭīkitaḥ. Rāmatejapāṇḍeyena sampāditaḥ. Kāśī (Paṇḍitpustakālay) **1949**.

Nāradasmṛti: Nārad'smṛti (ṭīkānugata bhābānubād-sambalita. Sampādak o anubādak — Nārāyaṇ'candra Smṛtitīrtha. Calcutta (Sanskrit College) **n.d.** ⟨Foreword dated Śaka 1873.⟩

NARAIN, HARSH: Evolution of the Nyāya-Vaiśeṣika Categoriology. Volume I. Early Nyāya-Vaiśeṣika Categoriology. Varanasi (Bharati Prakashan) **1976**.

Nāgarasarvasva: Kaviśekhar Padmaśrī viracit Nāgar'sarvāsv arthāt· sāṃsārik sukh'kā sādhan. Saṃskṛt mūl aur saral bhāṣāṭīkā sahit. Anuvādak Rāj'dhar Jhā. Kal'kattā (Śrīveṅkaṭeśvar Pustak Ejensī) **1929**.

NAGARATNAM, A. / MADHAVI, A.: "New Light on Ancient Indian Anatomy." *Bulletin of the Indian Institute of History of Medicine* (Hyderabad) *19.1989*, pp.1-20.

NAMOUCHI, NICOLE: Käufliche Liebe. Prostitution im alten Indien. Frankfurt am Main|Berlin|Bern|New York|Paris|Wien (Peter Lang) **1995**. Europäische Hochschulschriften 27,48.

NARAIN, A.K.: "[Review of:] Klaus Karttunen. India in Early Greek Literature. Helsinki 1989." *Journal of the American Oriental Society* (Ann Arbor) *112.1992*, pp.515-517.

Nāṭyaśāstra: Nāṭyaśāstra of Bharatamuni with the Commentary Abhinavabhārati By Abhinavaguptacarya ⟨sic⟩. Vol.III. Chapters. ⟨sic⟩ 19-27. General Editor D.R.S. Nagar. Edited by K.L. Joshi. Delhi|Ahmedabad (Parimal Publications) **1983**. Parimal Sanskrit Series 4.

NICKEL, DIETHARD: Untersuchungen zur Embryologie Galens. Berlin (Akademie-Verlag) **1989**. Schriften zur Geschichte und Kultur der Antike 27.

NIEDERMANN, MAX / SENN, ALFRED / SALYS, ANTON: Wörterbuch der litauischen Schriftsprache. II. Band: L—Pa. Heidelberg (Carl Winter Universitätsverlag) **1951**. Indogermanische Bibliothek, 2. Reihe: Wörterbücher.

NOBEL, JOHANNES: Ein alter medizinischer Sanskrit-Text und seine Deutung. Baltimore (American Oriental Society) **1951**. Supplement to the Journal of the American Oriental Society [71,3.1951] 11.

Nyāyacandrikā: Gayadāsa's hitherto unpublished commentary on the Śārīrasthāna of the Suśrutasaṃhitā; for details on this Nāgarī manuscript see: MEULENBELD 1974, p.398, and: Dominik Wujastyk. "Sanskrit Āyurvedic Manuscripts in the British Isles." *Journal of the European Āyurvedic Society* (Reinbek) *1.***1990**, pp.85-118 (see p.112).

OBERLIES, THOMAS: "Prakrit *thaṇavaṭṭa* — A propos Skt. *paṭṭa*." *Bulletin d'Études Indiennes* (Paris) *9.***1991**, pp.107-131.

OERTEL, HANNS: Euphemismen in der vedischen Prosa und euphemistische Varianten in den Mantras. München (Verlag der Bayerischen Akademie der Wissenschaften) **1942**. Sitzungsberichte der Bayerischen Akademie der Wissenschaften, Philosophisch-historische Abteilung 1942,8.

OETKE, CLAUS: "[Review of:] Giovanni Bandini. Die Erörterung der Wirksamkeit. Bhartṛharis Kriyāsamuddeśa und Helarājas Prakāśa zum ersten Male aus dem Sanskrit übersetzt. Wiesbaden 1980." *Zeitschrift der Deutschen Morgenländischen Gesellschaft* (Stuttgart) *136.* **1986**, pp.150-165.

O'FLAHERTY, WENDY DONIGER: Women, Androgynes and Other Mythical Beasts. Chicago|-London (The University of Chicago Press) **1980**.

OLDENBERG, HERMANN: Die Religion des Veda. Darmstadt (Wissenschaftliche Buchgesellschaft) ^{5}**1970**.

OLIVELLE, PATRICK: "Orgasmic Rapture and Divine Ecstasy: The Semantic History of *Ānanda*." *Journal of Indian Philosophy* (Dordrecht|Norwell, MA) *25.***1997**, pp.153-180.

Pādatāḍitaka: The Pādatāḍitaka of Śyāmilaka. Edited by G.H. Schokker. Part 1. The Hague|Paris (Mouton & Co.) **1961**. Indo-Iranian Monographs 9.

——: see also GHOSH and SCHOKKER/WORSLEY.

PAI, G.K.: "Puṁsavana in Bhāgavata Purāṇa." *Vishveshvaranand Indological Journal* (Hoshiarpur) *12.***1974**, pp.227-234.

Parahitasaṃhitā: Śrīnāthapaṇḍitaviracitā Parahitasaṃhitā. Śālākyatantram ⟨sic⟩ tathā śalyatantram. Parahita Samhita (A Medical Treatise of Parahita Family). Salakya Tantram and Salya Tantram. From 'Astangakanda' by Srinatha Pandita. Tirupati (Sri Venkateswara University) **1972**.

PARĀŚAR: Baiṣṇab'tantra ⟨cover: Baiṣṇab'tantram·⟩. Saṃkalan o anubād ŚrīParāśar. Kalikātā (Rājendra Lāibrerī) **n.d.** ⟨The picture of an almanac in an advertisement on the back cover bears the date Beng. era 1391.⟩

PENZER, N.M.: Poison-Damsels and other essays in Folklore and Anthropology. London (Chas. J. Sawyer) **1952**.

PERERA, L.P.N.: "Faculties of Sex and Related Phenomena in Buddhist Sexual Theory." Buddhist Philosophy and Culture. Essays in Honour of N.A. Jayawickrema. Editors David J. Kalupahana, W.G. Weeraratne. Colombo (N.A. Jayawickrema Felicitation Volume Committee) **1987**, pp.179-188.

PINGREE, DAVID: "Two *Karmavipāka* Texts on Curing Diseases and Other Misfortunes." *Journal of the European Āyurvedic Society* (Reinbek) *5.***1997**, pp.46-52.

PLOSS, HEINRICH / BARTELS, MAX / BARTELS, PAUL: Das Weib in der Natur- und Völkerkunde. Anthropologische Studien von weiland Dr. Heinrich Ploss und Dr. Max und Paul Bartels. Neu bearbeitet und herausgegeben von Ferd. Frhr. v. Reitzenstein. Elfte, stark vermehrte Ausgabe. Berlin (Neufeld & Henius) **1927** (three volumes).

POKARNA, K.L.: Social Beliefs, Cultural Practices in Health and Disease. Jaipur|New Delhi (Rawat Publications) **1994**. Indian Sociological Studies.

POLLOCK, SHELDON: "The Theory of Practice and the Practice of Theory in Indian Intellectual

History." *Journal of the American Oriental Society* (New Haven) *105.***1985**, pp.499-519.

——: "From Discourse of Ritual to Discourse of Power in Sanskrit Culture." *Journal of Ritual Studies* (Pittsburgh) *4.***1990**, pp.315-345.

PREISENDANZ, KARIN: Studien zu Nyāyasūtra III.1 mit dem Nyāyatattvāloka Vācaspati Miśras II. Stuttgart (Franz Steiner Verlag) **1994** (two volumes). Alt- und Neu-Indische Studien 46.

PRINTZ, WILHELM: "Buddha's Geburt." *Zeitschrift der Deutschen Morgenländischen Gesellschaft* (Leipzig) *79.***1925**, pp.119-132.

PUROHIT, G.V.: Āyurvedīyam Śārīram. Written to arrange the whole subject-matter on "Śārīra" in the ancient Ayurvedic books, with the short explanatory notes and illustrations. With English translation and Preface. Part I. Bombay (The Board of Research in Ayurveda) **1955**.

RAI, RAM KUMAR: Encyclopedia of Indian Erotics. Varanasi (Prachya Prakashan) **1983**. Indological Reference Series 3.

Rājamārtaṇḍa: The Rājamārtaṇḍa Of Śrī Bhojarāja with The 'Vidyotinī' Hindī Commentary By Śrī Pavanī Prasad ⟨sic⟩ Śarma ⟨sic⟩. Edited by Brahmaśankar Miśra. Varanasi (The Chowkhamba Vidyabhawan) **1966**. The Vidyabhawan Ayurved Granthamala 49.

Rāmāyaṇa: ŚrīmadVālmīkiRāmāyaṇam. [Colophon:] Idaṃ pustakaṃ Paṇaśīkaropāhva-Lakṣmaṇaśarmātmajena Vāsudevaśarmaṇā saṃskṛtaṃ .... Bombay (Nirnaya-sagar Press) **1921**.

RANADE, SUBHASH / PARAMJAPE, G.R.: Roga Vijñān and Vikṛti Vijñān. Edited by K.P. Sreekumari Amma. Trivandrum (Publication Division, Ayurveda College) **1984**. Kerala Government Ayurvedic Publication Series 8.

RAO, B. RAMA: "Some Views Based on the Survey of Medical Manuscripts and Practitioners." *Bulletin of the Indian Institute of History of Medicine* (Hyderabad) *22.***1992**, pp.11-17.

Rasahṛdayatantra: ŚrīmadGovindabhagavatpādaviracitaṃ Rasahṛdayatantram. ŚrīCaturbhujamiśraviracitayā Mugdhāvabodhinī samākhyayā vyākhyayā samullasitam. 'Kāle' ity upāhvaGurunāthātmajaTryambakena tathā Ācāryopāhvena Trivikramātmajena Yādavaśarmaṇā sampāditam. Jayadeva Vidyālaṅkāreṇa ṭippaṇyā samupaskṛtaṃ saṃśodhitañ ca. Lavapura (Motīlāla Banārasīdāsa) **1927**.

Rasaratnākara: SiddhaNityanāthapraṇītaḥ ⟨sic⟩ Rasaratnākaraḥ. (Samastarasagranthānāṃ śirobhūṣaṇam). MāthuravaiśyāyurvedoddhārakaŚālagrāmakṛtabhāṣāṭīkāvibhūṣitaḥ. Mumbayī (Śrīveṅkaṭeśvarastīm-mudraṇālaya) **V.S. 1966**.

Rasaratnasamuccaya: Rasaratna Samuchchaya by Śrī Vāgbhatāchārya ⟨sic⟩. Edited with the Suratnojjvalā Hindī Commentary and Notes, Introduction, Indices, Appendices etc; by Ambikādaṭṭa ⟨sic⟩ Śāstrī. Varanasi (Chowkhamba Sanskrit Series Office) [3]**1961**. The Haridas Sanskrit Series 91.

Rasārṇavakalpa: Rasārṇavakalpa (Manifold powers of the ocean of *rasa*). Text Edited and Translated into English by Mira Roy in collaboration with B.V. Subbarayappa. New Delhi (Indian National Science Academy) **1976**. Indian National Science Academy Monograph 5.

Rasataraṅgiṇī: Rasataraṅgiṇī. ŚrīkavirājaNarendranāthamitramahodayāntevāsinā prāṇācāryaśrīSadānandaśarmaṇā viracitā. ĀyurvedācāryaśrīHaridattaśāstriṇā kṛtayā Prasādanīvyākhyayā samudbhāsitā ca. Āyurvedācārya paṃ. Dharmānandaśāstriṇā 'Rasavijñāna' nāmakahindībhāṣayā vaktavyena ca vibhūṣitā. Sā ceyam ... Kāśīnāthaśāstriṇā sampāditā. Banāras (Motīlāl Banārasīdās) [6]**V.S. 2016**.

Rasavaiśeṣikasūtra: Bhadanta Nagaarjuna's ⟨sic⟩ Rasavaiseshikasutra with Naarasimha Bhaasya. Edited with an introduction by N.E. Muthuswami. Trivandrum (Government of Kerala) **1976**. Kerala Government Ayurvedic Publication Series 2.

Rasayogasāgara: The Rasayogasagara by Vaidya Pandit Hariprapannaji with Sanskrit & English Introduction & Notes. Bombay (The Author [i.e. Vaidya Pandit Hariprapannaji]) **1927-1930** (two volumes).

Ratirahasya: Ratirahasyam. Nāma kāmaśāstram. MahākaviśrīKokkokaviracitam. ŚrīKāñcīnāthakṛtayā Dīpikākhyayā ṭīkayā sanātham. ŚrīDevīdattaśarmaṇā ṭippaṇikayā viśadīkṛtya śodhitam. Kāśī (Phrīmaina Emḍa Kampanī) **1912**.

——: see also LIENHARD 1960 and UPADHYAYA.

Ratiśāstra: Rati-Sastram. Or the Hindu System of Sexual Science. (With text in Devanagri.) Translated from original text by Abinash Chandra Ghose With the aid of several Oriental Scholars. [Reprint of the 2nd Edition.] Delhi (Nag Publishers) **1977**. N.P. Series. ⟨On this work see also: *Kāmaśāstra* of Surūpa, p.37.⟩

Ratiśāstra compiled by Surendranāth Bhaṭṭācārya: Sacitra Ratiśāstra. Surendranāth Bhaṭṭācāryya praṇīta. Kalikātā (Akṣaỳ Lāibrerī) **n.d.**

Ratiśāstra of Nāgārjuna: Richard Schmidt. "Das Ratiśāstra des Nāgārjuna." *Wiener Zeitschrift für die Kunde des Morgenlandes/Vienna Oriental Journal* (Wien/Vienna) *23.***1909**, pp.180-190. ⟨On this work see also: *Kāmaśāstra* of Surūpa, p.37.⟩

RAU, WILHELM: Naturbeobachtung und Handwerkskunst im vorislamischen Indien. Stuttgart (Franz Steiner Verlag Wiesbaden) **1986**. Sitzungsberichte der Wissenschaftlichen Gesellschaft an der Johann Wolfgang Goethe-Universität Frankfurt am Main 22,4.

——: "[Review of:] Klaus Karttunen. India in Early Greek Literature. Helsinki 1989." *Orientalistische Literaturzeitung* (Berlin) *85.***1990**, col.715.

RAY, MANMATHA NATHA: An Index to the Proper Names occuring ⟨sic⟩ in Vālmīki's Rāmāyaṇa. Varanasi (Sampurnanand Sanskrit University) **1984**. The Princess of Wales Sarasvati Bhavana Studies (Reprint Series 5).

RAY, PRIYADARANJAN / GUPTA, HIRENDRANATH / ROY, MIRA: Suśruta Saṃhitā (A Scientific Synopsis). New Delhi (Indian National Science Academy) **1980**.

REINHART, JOHANNES: "Slav. *pyjь* 'membrum virile'." Iranian and Indo-European Studies. Memorial Volume of Otakar Klíma. Edited by Petr Vavroušek. Praha (Enigma Corporation) **1994**, pp.219-223.

RENOU, LOUIS: Grammaire sanscrite. Tomes I et II réunis. Phonétique — composition — dérivation — le nom — le verbe — la phrase. Deuxième édition revue, corrigée et augmentée. Paris (Librairie d'Amérique et d'Orient Adrien Maisonneuve) **1984**.

Ṛgvedasaṃhitā: Rig-Veda-Sanhita, the Sacred Hymns of the Brahmans; together with the Commentary of Sayanacharya. Edited by Max ⟨Volumes V-VI: F. Max⟩ Müller. London (WM. ⟨Volume I: W.; Volumes II-III: William⟩ H. Allen and Co.) **1849-1874** (six volumes).

RHYS DAVIDS, T.W.: The Questions of King Milinda. Translated from the Pâli by T.W. Rhys Davids. Reprinted Oxford (Oxford University Press)/ London (Humphrey Milford) **1925**. The Sacred Books of the East 35.

ROCHER, LUDO: "The Kāmasūtra: Vātsyāyana's Attitude Toward Dharma and Dharmaśāstra." *Journal of the American Oriental Society* (New Haven) *105.***1985**, pp.521-529.

RÖER, E.: The Taittaríya ⟨sic⟩ and Aittaréya ⟨sic⟩ Upanishads, with the Commentary of Sankara Áchárya, and the Gloss of Ánanda Giri, and the Swétáswatara Upanishad with the Commentary of Sankara Áchárya, Edited by E. Röer. Calcutta (Asiatic Society of

Bengal) **1850**. Bibliotheca Indica 7, Nos.22;33;34.

ROLLAND, PIERRE: "Un fragment médical 'védique': Le premier khaṇḍa du Vārāhapariśiṣṭa Bhūtotpatti." *Münchener Studien zur Sprachwissenschaft* (München) *30.1972*, pp. 129-138.

——: Le Vārāhagṛhyapuruṣa et autres extraits des Vārāhapariśiṣṭa à caractère domestique ⟨cover: Compleménts au rituel domestique védique. Le Vārāhagṛhyapuruṣa⟩. Aix-en-Provence (Éditions de l'Université de Provence) **1975**. Études indiennes 3.

ROSCOE, WILL: "Priests of the Goddess: Gender Transgression in Ancient Religion." *History of Religions* (Chicago) *35.1995-1996*, pp.195-230.

ROŞU, ARION: Les conceptions psychologiques dans les textes médicaux indiens. Paris (Institut de Civilisation Indienne) **1978**. Publications de l'Institut de Civilisation Indienne, série in-8°, 43.

——: "Le renouveau contemporain de l'Āyurveda." *Wiener Zeitschrift für die Kunde Südasiens und Archiv für indische Philosophie* (Leiden | Köln | Wien) *26.1982*, pp.59-82.

——: "Études āyurvédiques. II. Déontologie médicale et orthodoxie brahmanique." Sanskrit and World Culture. Proceedings of the Fourth World Sanskrit Conference of the International Association of Sanskrit Studies Weimar May 23-30,1979. Edited by Wolfgang Morgenroth. Berlin (Akademie-Verlag) **1986** ⟨= **1986A**⟩. Schriften zur Geschichte und Kultur des Alten Orients 18, pp.586-594.

——: "*Mantra* et *yantra* dans la médecine et l'alchimie indiennes." *Journal Asiatique* (Paris) *274.1986* ⟨= **1986B**⟩, pp.203-268.

ROUSSELLE, ALINE: Porneia. On Desire and the Body in Antiquity. Translated by Felicia Pheasant. New York (Barnes & Noble Books) **1996**.

ROY, MIRA: "Āyurveda." The Cultural Heritage of India. Volume VI. Science and Technology. Editors Priyadaranjan Ray and S.N. Sen. Calcutta (The Ramakrishna Mission Institute of Culture) **1986**, pp.152-176.

RUBEN, WALTER: Beginn der Philosophie in Indien. Aus den Veden. Berlin (Akademie-Verlag) [3]**1961**. Philosophische Studientexte. Texte der indischen Philosophie 1.

——: Die Entwicklung der Philosophie im alten Indien. Berlin (Akademie-Verlag) **1971**. Deutsche Akademie der Wissenschaften zu Berlin. Veröffentlichungen des Instituts für Orientforschung 67. Walter Ruben. Die gesellschaftliche Entwicklung im alten Indiēn IV.

SALOMON, CAROL: "The Cosmogonic Riddles of Lalan Fakir." Gender, Genre, and Power in South Asian Expressive Traditions. Edited by Arjun Appadurai, Frank J. Korom, and Margaret A. Mills. Philadelphia (University of Pennsylvania Press) **1991**. South Asia Seminar Series. Publications of the American Folklore Society, New Series, pp.267-304.

SALOMON, RICHARD: "Kṣemendra as Satirist: A New Look at the Deśopadeśa." *Acta Orientalia* (Havnia) *44.1983*, pp.11-32.

Saṃvarodayatantra: The Saṁvarodaya-Tantra. Selected Chapters. By Shinichi Tsuda. Tokyo (The Hokuseido Press) **1974**.

Sanskrithandschriften aus den Turfanfunden. Teil 1. Unter Mitarbeit von Walter Clawiter und Lore Holzmann herausgegeben und mit einer Einleitung versehen von Ernst Waldschmidt. Teil 3. Die Katalognummern 802-1014. Unter Mitarbeit von Walter Clawiter und Lore Sander-Holzmann herausgegeben von Ernst Waldschmidt. Wiesbaden (Franz Steiner Verlag) **1965**; **1971**. Verzeichnis der orientalischen Handschriften in Deutschland 10,1;10,3.

Śarīravijñāna: Śarīra-vijñānam. (Prathamo bhāgaḥ). Niśikānta-vaidyaśāstri-viracitam. Kāśī (Mahāmaṇḍala Yantra) **Beng. era 1328**. ⟨Consulted at the India Office Library, London. >

ŚARMĀ, PRIYAVRAT: Āyurved kā vaijñānik itihās. Vārāṇasī|Dillī (Caukhambhā Oriyanṭāliyā) [2]**1981**. Jay'kṛṣṇ'dās Āyurved Granth'mālā 1.

Śārṅgadharapaddhati: The Paddhati of Sarngadhara. A Sanskrit Anthology edited by Peter Peterson. Vol.I. The Text. Bombay (Government Central Book Depôt) **1888**. Bombay Sanskrit Series 37.

Śārṅgadharasaṃhitā: The Śārṅgadhara-Samhitā <sic> by Paṇḍit Sārṅgadharāchārya <sic> son of Paṇḍit Dāmodara. With the commentary Aḍhamalla's <sic> Dīpikā and Kāśirāma's Gūḍhārtha-Dīpikā. Edited with footnotes by Paraśurāma Śāstrī. Bombay (Nirṇaya-Sāgar Press) [2]**1931**.

SASTRI, HARAPRASAD: Magadhan Literature. Delhi (Sri Satguru Publications) [2]**1986**. Bibliotheca Indo-Buddhica 28.

ŚĀSTRĪ, HARIDATT: Dehadhātvagnivijñānam. Carak, suśrut, vāgbhaṭ ādi āyurvaidik saṃhitā granthoṃ meṃ sūcit mānav deh meṃ āhār'pācak paṃc'bhūtāgni tathā rasādidhātvagniyoṃ par viśeṣ anusandhān se prāpt niṣkarṣ ke ādhār par nirmit maulik granth. Dillī|Vārāṇasī|Paṭ'nā|Baṃg'laur|Madrās (Motīlāl Banār'sīdās) 1977, reprinted **1993**.

Śatapathabrāhmaṇa: The Çatapatha-Brâhmaṇa in the Mâdhyandina-çâkhâ with Extracts from the Commentaries of Sâyaṇa, Harisvâmin and Dvivedaganga edited by Albrecht Weber. Leipzig (Otto Harrassowitz) **1924**. Collectio Editionum Rariorum Orientalum, Noviter Impressarum 3.

Śataślokī: Śataślokī of Bopadeva. With the author's own Commentary Candrakala. Edited By Vayaskara N.S. Mooss. Kottayam (Vaidyasarathy Press) **1962**. Vaidyasarathy Series 14.

SCHARFE, HARTMUT: "Der Liebeskampf bei den indischen Erotikern." Kreative Aggression. Festschrift für George Bach. Hilarion Petzold, Hartmut Scharfe (Hrsg.). Paderborn (Junfermann-Verlag) **1985**. Integrative Therapie, Beiheft 9, pp.85-92.

SCHIFFELER, JOHN WM.: "Tenets of Traditional Chinese Medicine." *Asian Medicine. IASTAM Newsletter* (Atlanta, Georgia) *New Series 2.July **1992***, pp.Two-Six.

SCHMIDT, RICHARD: Liebe und Ehe im alten und modernen Indien (Vorder-, Hinter- und Niederländisch-Indien). Berlin (Verlag von H. Barsdorf) **1904**.

——: Beiträge zur indischen Erotik. Das Liebesleben des Sanskritvolkes nach den Quellen dargestellt. Berlin (Hermann Barsdorf Verlag) [3]**1922** <= [2]1911>.

——: "Die Benares-Ausgabe des Ratirahasyam." *Zeitschrift für Indologie und Iranistik* (Leipzig) *5.**1927***, pp.185-189.

——: Nachträge zum Sanskrit-Wörterbuch in kürzerer Fassung von Otto Böhtlingk. First Indian Edition. Delhi (Motilal Banarsidass Publishers) **1991**.

SCHMITHAUSEN, LAMBERT: Ālayavijñāna. On the Origin and the Early Development of a Central Concept of Yogācāra Philosophy. Tokyo (The International Institute for Buddhist Studies) **1987** (two volumes). Studia Philologica Buddhica, Monograph Series 4.

SCHOKKER, G.H. / WORSLEY, P.J: The Pādatāḍitaka of Śyāmilaka. Part 2. A Translation by G.H. Schokker and P.J. Worsley with a Complete Word-Index of the Four Ancient Sanskrit Bhanas by G.H. Schokker. Dordrecht|Boston (D. Reidel Publishing Company) **1976**.

SEIDENSTICKER, TILMAN: Altarabisch 'Herz' und sein Wortfeld. Wiesbaden (Otto Harrassowitz) **1992**.

SEN, GANANATH: The three volumes (of which I could procure only two) of Gananath Sen's work *Pratyakṣaśārīra* have seen various editions; the volumes and editions cited here are: Pratyaksha-Shariram. A Text-Book of Human Anatomy in Sanskrit Including Classical Nomenclature with A Short History of Ayurvedic Literature In English & Sanskrit Introductions. Part I. Calcutta (Kaviraj S.K. Sen) [4]**1940**, and, Pratyaksha-Shariram. A Text-Book of Human Anatomy in Sanskrit (Including History of Ayurveda, Classical Nomenclature & Elements of Physiology). Part II. New Edition. Calcutta (Kaviraj S.K. Sen) **1941**.

SHARMA, PRIYAVRAT/P.V.: Indian Medicine in the Classical Age. Varanasi (The Chowkhamba Sanskrit Series Office) **1972**. The Chowkhamba Sanskrit Studies 85.

——: Ḍalhaṇa and his Comments on Drugs. New Delhi (Munshiram Manoharlal) **1982**.

——: Caraka-Saṃhitā. Critical Notes (Incorporating the commentaries of Jejjaṭa, Cakrapāṇi, Gaṅgādhara and Yogīndranātha). Vol.III (Sūtrasthāna to Indriyasthāna). Varanasi | Delhi (Chaukhambha Orientalia) **1985**. Jaikrishnadas Ayurveda Series 36.

——: "Haramekhala-Tantra (The First Chapter on Medicine)." *Ancient Science of Life* (Coimbatore) *5.**1985-1986***, pp.143-150.

——: "Medical Data in Epics and Purāṇas." History of Medicine in India (From Antiquity to 1000 A.D.). New Delhi (Indian National Science Academy) **1992**, pp.87-116.

SHARMA, S.N.: Concept of *Jatharagni* in *Ayurveda* (a Patho-Physiological Study). Jaipur (Publication Scheme) **1992**.

SHARMA, SHIV: Ayurvedic Medicine Past and Present. Edited by Kailash Sharma. Varanasi (Krishnadas Academy) [3]**1987**. Krishnadas Ayurveda Series 18.

SHARPE, ELIZABETH: An Eight-hundred Year Old Book of Indian Medicine and Formulas. Translated from the original very old Hindi into Gujarati character and thence into English. Second A[sian]E[ducational]S[ervices] Reprint. New Delhi (Asian Educational Services) **1985**.

Siddhasāra: The Siddhasāra of Ravigupta. Volume 1: The Sanskrit Text edited by R.E. Emmerick. Wiesbaden (Franz Steiner Verlag) **1980**. Verzeichnis der orientalischen Handschriften in Deutschland, Supplementband 23,1.

Siddhayoga: ŚrīmadVṛndapraṇīto Vṛndamādhavāparanāmā Siddhayogaḥ. ŚrīkaṇṭhadattaviracitaVyākhyākusumāvalyākhyaṭīkāsametaḥ. Etat pustakaṃ Pādhye ity upanāmadhāribhiḥ Kṛṣṇaśāstritanūjaiḥ 'Haṇamantaśāstribhiḥ' saṃśodhitam. Puṇya (Ānandāśramamudraṇālaya) [2]**1943**. Ānandāśramasaṃskṛtagranthāvali 27.

SIGGEL, ALFRED: Alfred Siggel. "Gynäkologie, Embryologie und Frauenhygiene aus dem 'Paradies der Weisheit über die Medizin' des Abū Ḥasan ʿAlī b. Sahl Rabban aṭ-Ṭabarī nach der Ausgabe von Dr. M. Zubair aṣ-Ṣiddīqī, Berlin—Charlottenburg 1928, Buch- und Kunstverlag 'Die Sonne'. Übersetzt und erläutert." *Quellen und Studien zur Geschichte der Naturwissenschaften und der Medizin* (Berlin) *8.**1941-1942***, pp.216-272.

——: Die indischen Bücher aus dem Paradies der Weisheit über die Medizin des ʿAlī ibn Sahl Rabban aṭ-Ṭabarī. Übersetzt und erläutert. Wiesbaden (Franz Steiner Verlag) **1950**. Akademie der Wissenschaften und der Literatur in Mainz, Abhandlungen der Geistes- und Sozialwissenschaftlichen Klasse 1950,14.

SIṂHA, RAṆAJIT : Māṭir surer kh je. Kal'kātā (Pratikṣaṇ Pāb'likeśan) **1990**.

SIMS-WILLIAMS, NICHOLAS / HAMILTON, JAMES: Documents turco-sogdiens du IXe-Xe siècle de Touen-houang. London (Corpus Inscriptionum Iranicarum/School of Oriental and African Studies) **1990**. Corpus Inscriptionum Iranicarum 2,3.

SINGHAL, G.D. / GURU, L.V.: Anatomical & Obstetric Considerations in Ancient Indian Surgery. Based on Śārīra-Sthāna of Suśruta Saṁhitā. (An authentic interpretation of Śārīra-Sthāna of Suśruta-Saṁhitā with original Sanskrit for the modern doctors, anatomists, physiologists, surgeons, obstetricians and other specialists, Ayurvedic scholars, Historians of Medicine, Indologists and Orientalists). Allahabad (G.D. Singhal) **1973**. Ancient Indian Surgery Series 4.

SINGHAL, G.D. / PATTERSON, T.J.S.: Synopsis of Ayurveda based on a translation of the *Suśruta Saṁhitā* (The Treatise of Suśruta). Delhi (Oxford University Press) **1993**.

Śivatattvaratnākara: Śivatattva Ratnākara of Basavarāja of Keḷadi. Vol.I. Editor: S. Narayanaswamy Sastry. Mysore (Oriental Research Institute, University of Mysore) **1964**. Oriental Research Institute Publications Sanskrit Series 108.

SLAJE, WALTER: "Merkmale des Lebendigen: Zu einer naturphilosophisch begründeten Biologie in Bhāskarakaṇṭhas *Cittānubodhaśāstra*." *Journal of the European Āyurvedic Society* (Reinbek) *3.**1993***, pp.250-281.

——: "*Ṛtú-, Ṛtv(i)ya-, Ārtavá-*. Weibliche 'Fertilität' im Denken vedischer Inder." *Journal of the European Āyurvedic Society* (Reinbek) *4.**1995***, pp.109-148.

Smaratattvaprakāśikā: Richard Schmidt. "Revaṇārādhyas Smaratattvaprakāśikā." *Wiener Zeitschrift für die Kunde des Morgenlandes/Vienna Oriental Journal* (Wien/Vienna) *18.**1904***, pp.261-279.

Ṣoḍaśāṅgahṛdaya: Priyavrataśarman. Ṣoḍaśāṅgahṛdayam āryāvṛttabaddham āyurvedavivaraṇātmakam svopajñahindīvyākhyopetam. Vārāṇasī (Padmā Prakāśan) **V.S. 2044**.

SPEIJER, J.S.: Sanskrit Syntax. Reprinted Delhi (Motilal Banarsidass) **1980**.

SRINIVASAN, DORIS: "Vedic Rudra-Śiva." *Journal of the American Oriental Society* (New Haven) *103.**1983***, pp.543-556.

ŚRĪVĀSTAV, SUREŚ CANDR: Hṛdayopaniṣad. The Cardiorespiratory System as Described in Vedic Literature. (Vaidik ṛṣiyoṃ dvārā hṛd-phuphphus aṅg sambandhī sākṣātkār kiye gaye jñān kī samīkṣā). Gorakh|pur (Prācī Śodh Saṃsthān) **n.d.** ⟨Probably 1979⟩.

ŚRĪVĀSTAVYA, SANT|NĀRĀYAṆ: Carak|saṃhitākī dārś|nik pṛṣṭh|bhūmi. The Philosophical Background of Charaka-Samhita. Ilāhābād (Pīyūṣ Prakāśan) **1983**.

Śṛṅgāramañjarī: Srngaramanjari of Saint Akbar Shah. Based on old Sanskrit Manuscripts in Devanagari and Telugu Scripts. Edited with a critical study by V. Raghavan. Hyderabad (Archaeological Department, Hyderabad Government) **1951**. Hyderabad Archaeological Department Publication.

STADEN, HEINRICH VON: "Anatomy as Rhetoric: Galen on Dissection and Persuasion." *Journal of the History of Medicine and Allied Sciences* (New Haven, Connecticut) *50.**1995***, pp.47-66.

STERNBACH, LUDWIK: Gaṇikā-vṛtta-saṃgrahaḥ Or Texts on Courtezans in Classical Sanskrit. Compiled and Presented By Ludwik Sternbach. Hoshiarpur (Vishveshvaranand Vedic Research Institute) **1953**. Vishveshvaranand Indological Series 4.

——: Mahā-Subhāṣita-Saṁgraha. Being an extensive collection of wise sayings and entertaining verses in Sanskrit with introduction[,] English translation, critical notes and indices. Volume III. *Subhāṣita*-s — Nos.4209-6285. ā—i—ī (āṃ jñānaṃ — īhā dhanasya). Edited by S. Bhaskaran Nair. Hoshiarpur (Vishveshvaranand Vedic Research Institute) **1977**. Vishveshvaranand Indological Series 71. Vishveshvaranand Vedic Research Institute Publication 641.

STIFTER, KARL F.: Die dritte Dimension der Lust. Das Geheimnis der weiblichen Ejakulation. Frankfurt/Main|Berlin (Ullstein) **1988**.

Strīcikitsā: Atha Strīcikitsā. ... Vasatirāmasaṃgṛhītā, tatkṛtabhāṣānuvādasamalaṃkṛtā, ... Raghuvaṃśaśarmaṇā saṃśodhitā ca. Seyaṃ Haridāsātmajena ... Vrajavallabhaśarmaṇā ... mudrayitvā prākāśyam ānītā. Mumbāpurī (Śrīveṅkaṭeśvarasṭīm-mudraṇālaya) [4]**V.S. 1986**. ⟨Consulted at the India Office Library, London.⟩

Studies on Indian Medical History. Papers presented at the International Workshop on the Study of Indian Medicine held at the Wellcome Institute for the History of Medicine 2-4 September 1985. Edited by G. Jan Meulenbeld and Dominik Wujastyk. Groningen (Egbert Forsten) **1987**. Groningen Oriental Studies 2.

SUBHAKTHA, P.K.J.P.: "Cakrapāṇidatta." *Bulletin of the Indian Institute of History of Medicine* (Hyderabad) *22.***1992**, pp.53-59.

SUGUNASIRI, SUWANDA H.J.: "The Whole Body, not Heart, as 'Seat of Consciousness': The Buddha's View." *Philosophy East and West* (Honolulu) *45.***1995**, pp.409-430.

SULLIVAN, LAWRENCE E.: "Body Works: Knowledge of the Body in the Study of Religion." *History of Religions* (Chicago) *30.***1990-1991**, pp.86-99.

SUNESON, CARL: "Remarks on Some Interrelated Terms in the Ancient Indian Embryology." *Wiener Zeitschrift für die Kunde Südasiens und Archiv für indische Philosophie* (Wien) *35.* **1991**, pp.109-121.

Suśrutasaṃhitā: The Suśruta or System of Medicine taught by Dhanvantari, and composed by his Disciple Suśruta. Vol.I. Containing the three divisions called Sútra, Nidána, and Sháríra. Edited by Madhusúdana Gupta. Kalikātā (The Education Press) **1835**.

——: Suśruta-saṃhitā. Arthāt· Suśrut-ṛṣi-praṇīta prācīn śastrabidyā o śārīr'bijñāner pradhānatama baidyak grantha. Ḍallanācāryya kṛta Nibandhasaṃgraha, Cakrapāṇidatta kṛta Bhānumatī ṭīkā, baṅgānubād, ābaśyak sthale iṃreji pratiśabda, ḍāktār Oẏāij prabhṛti cikit'sak'barger mantabya, śārīr'saṃsthān ebaṃ śastra yantrādir pratikṛti samanbita. Bijaẏ'ratna Sen Kabirāj[,] Bhagabatīprasanna Sen Kabirāj o Niśikānta Sen Kabirāj karttṛk sampādita. Kalikātā (Maṇirām'yantra) **n.d.** ⟨Incomplete; see also p.22 of the introduction of SU. The date of the work given in the catalogue of the India Office Library in London, where I consulted this work, is Beng. era 1292.⟩

——: Sushrut-sañhitā (Sūtra Sthān) With Bhānumatī Commentary by Chakrapāṇi Datta. Edited by Jādavajī Trikamajī Achārya and Nandkishor Sharmā Bhishagāchārya. Agra (Agra University)[/Jaipur] (Swāmī Lakshmī Rām Trust) **1939**. Shrī Swāmī Lakshmī Rām Trust Series 1.

——: The Suśrutasaṃhitā of Suśruta with various readings, notes and appendix etc., edited with the co-operation of Jādavji Trikamji Āchārya by Nārāyaṇ Rām Āchārya. Bombay (Nirṇaya-Sāgar Press) **1945**.

——: Suśrutasaṃhitā of Suśruta With the Nibandhasaṅgraha Commentary of Śrī Ḍalhaṇāchārya and the Nyāyacandrikā Pañjikā of Śrī Gayadāsāchārya on Nidānasthāna. Edited from the Beginning to the 9th Adhyāya of Cikitsāsthāna by Jādavji Trikamji Āchārya and the rest by Nārāyaṇ Rām Āchārya. Reprinted Fourth Edition. Varanasi|Delhi (Chaukhambha Orientalia) **1980**. Jaikrishnadas Ayurveda Series 34.

——: see also ADRIAENSEN/BARKHUIS/RUIJTERS, BHISHAGRATNA, GHĀṆEKAR, HESSLER, HOERNLE 1897 and *Nyāyacandrikā*.

Suvarṇa(pra)bhāsottamasūtra: Suvarṇabhāsottamasūtra. Das Goldglanz-Sūtra. Ein Sanskrittext des Mahāyāna-Buddhismus. Nach den Handschriften und mit Hilfe der tibetischen und chinesischen Übertragungen herausgegeben von Johannes Nobel. Leipzig (Otto Harrassowitz) **1937**.

Śvetāśvataropaniṣad: Contained in RÖER, pp.248-375.

Taittirīyasaṃhitā: Die Taittirîya-Saṃhitâ. Herausgegeben von Albrecht Weber. Leipzig (F.A. Brockhaus) **1871-1872** (two volumes). Indische Studien 11;12.

—: The Taittirīya Saṃhitā of the Black Yajurveda With the Commentary of Bhaṭṭa Bhāskara Miśra. Edited by A. Mahadeva Sastri and K. Rangacharya. Vol.X (Kāṇḍa VII: Praśnas 4-5). Reprinted Delhi|Varanasi|Patna|Madras (Motilal Banarsidass) **1986**.

Taittirīyopaniṣad: Contained in RÖER, pp.1-141.

Tandulaveyāliya: Tandulaveyāliya. Ein Paiṇṇaya des Jaina-Siddhānta. Textausgabe, Analyse und Erklärung von Walther Schubring. Wiesbaden (Franz Steiner Verlag) **1969**. Akademie der Wissenschaften und der Literatur in Mainz, Abhandlungen der Geistes- und Sozialwissenschaftlichen Klasse 1969,6.

Tantrayukti: Tantra-Yukti. Edited by: Vayaskara N.S. Mooss. Kottayam (Vaidyasarathy Press) [2]**1979**. Vaidyasarathy Series S.7.

Tantrayuktivicāra: Vaidyanatha Neelamegha's Tantrayuktivichara. Edited by N.E. Muthuswami. With an exposition in English by K. Srikanta Moorthy. Trivandrum (Government of Kerala) [2]**1976**. Kerala Government Ayurvedic Publication Series 1.

TEWARI, P.V.: "Obstetrics and Gynecology in Vedic Literature." History of Medicine in India (From Antiquity to 1000 A.D.). New Delhi (Indian National Science Academy) **1992**, pp. 70-83.

The Concept of Jaṭharāgni and Dhātwagni in Indian Medicine (With special reference to malabsorption). New Delhi (Central Council for Research in Ayurveda and Siddha, Ministry of Health and Family Welfare, Govt. of India) **1987**.

THIEME, PAUL: Kleine Schriften [herausgegeben von Georg Buddruss]. Wiesbaden (Franz Steiner Verlag) **1971** (two volumes). Glasenapp-Stiftung 5.

—: "*barjaha-*." *Studien zur Indologie und Iranistik* (Reinbek) *19.* ***1994***, pp.367-375.

TIVĀRĪ, PREM'VATĪ: Āyurvedīya prasūti-tantr evaṃ strī-rog. Pratham-bhāg. Prasūti-tantr. Dvitīya-bhāg. Strī-rog. Vārāṇasī|Dillī (Caukhambhā Oriyanṭāliyā) **1986**; **1990**. Jay'-kṛṣṇ'dās Āyurved Granth'mālā 49. ⟨Also with English title: Premvati Tewari. Āyurvedīya Prasūti-tantra evaṃ Strī-roga. Part I. Prasūti-tantra (Obstetrics). Part II. Strī-roga (Gynecology). Varanasi|Delhi (Chaukhambha Orientalia) [1986]; [1990]. Jaikrishnadas Ayurveda Series 49.⟩

Triśatī: Triśatī. VaidyavaraśrīŚārṅgadharavinirmitā. ŚrīyutapaṇḍitaVaidyavallabhabhaṭṭaviracitasaṃskṛtaṭīkāsahitā. Seyaṃ Rāmaprasāda ity etasmāt samupalabhya, Kiśorīvallabhaviracitabhāṣāṭīkayā samalaṃkṛtya, Kṣemarājaśrīkṛṣṇadāsaśreṣṭhinā ... mudrayitvā prakāśitā. Mumbayī (Śrīveṅkaṭeśvara (Stīm) Mudraṇayantrālaya) **V.S. 1968**. ⟨Consulted at the India Office Library, London.⟩

TURNER, R.L.: A Comparative Dictionary of the Indo-Aryan Languages. London|New York|Toronto (Oxford University Press) **1966**.

—: A Comparative Dictionary of the Indo-Aryan Languages. Addenda and Corrigenda. Edited by J.C. Wright. London (School of Oriental and African Studies, University of London) **1985**.

UPADHYAYA, S.C.: The Hindu Secrets of Love. Kokashastra (Rati Rahasya) of Pandit Kokkoka. Translated from the Original Sanskrit by S.C. Upadhyaya. Foreword by V. Raghavan. [Reprinted] Bombay (D.B. Taraporevala Sons & Co.) **1981**.

Upaniṣatsaṅgraha: Upaniṣat-saṁgrahaḥ. Containing 188 Upaniṣads. Edited with Sanskrit Introduction by J.L. Shastri. Reprinted Delhi|Varanasi|Patna (Motilal Banarsidass) **1980**.

Vaidyakaśabdasindhu: Vaidyaka-Śabdasindhuḥ (Comprehensive Glossary of Āyurvedic Technical Terms with synonyms of Sanskrit, Latin, Hindi, Tamil, Telugu, Bengali etc. and having references of standard Āyurvedic texts). Compiler Umeśacandra Gupta. Revised &

Enlarged Second Edition By Nagendra Nātha Sena. Varanasi | Delhi (Chaukhambha Orientalia) [3]**1983**. Jaikrishnadas Ayurveda Series 56.

Vaidyamanorama: Vaidyamanorama enna cikitsākramaṃ. Pūrvvārddhaṃ. Kōṭṭayaṃ (Vaidyasārathi Prass) [Kollam era] 1126. Vaidyasārathi Granthāvali 3, and, Uttarārddhaṃ. Saṃśodhakan: Vayaskkara En. Ess. Müss. Kōṭṭayaṃ (Vaidyasārathi Prass) [2]**1978**. Vaidyasārathi Granthāvali M-1.

Vaidyamanoramā: Vaidya Manoramâ by Kâlîdâs and Dhârâkalpa. Edited by T. Nîlkanth Sharmâ, and Jâdavjî Tricumjî Âchârya. Bombay (Nirṇaya-Sāgara Press) **1913**. Âyurvedîya-Granthamâlâ 8-9. ⟨Consulted at the India Office Library, London.⟩

Vaidyāmṛta: Atha Vaidyāmṛt saṭīk. Saṃskṛt jis'kā śrī Bābā Jyotiḥ Svarūp jī ne bhāṣā meṃ tilak kiyā .... Benares (Benares Light Press) **1870**. ⟨Consulted at the India Office Library, London.⟩

Vaijayantī: The Vaijayantī of Yādavaprakāśa. For the first time critically edited by Gustav Oppert. Madras (Govt. Press) **1893**.

Vājasaneyisaṃhitā: Vājasaneyi-mādhyandina-śukla Yajurveda-saṃhitā Svādhyāyamaṇḍalasthānekapaṇḍitānāṃ sāhāyyena vividhaprācīnahastalikhitapustakapāṭhānusāreṇa Bhaṭṭācāryeṇa Śripādaśarmaṇā Dāmodarabhaṭṭasūnunā Sātavalekarakulajena saṃśodhitā. Aundha (Svādhyāyamaṇḍala) [2]**V.S. 2003**.

Vaṅgasenasaṃhitā: Baṅgasena-Saṃhitā (Cikitsāsāra Saṅgraha). Hindī vyākhyākār Rājīv Kumār Rāy. Sampādak Rām Kumār Rāy. Vārāṇasī (Prācya Prakāśan) **1983**. Dhanvantari Granth'mālā 1.

Vāsiṣṭhadharmaśāstra: ŚrīVāsiṣṭhadharmaśāstram. Aphorisms on the Sacred Law of the Âryas, as taught in the School of Vasishṭha. Edited, with critical notes, an anukramaṇikâ, indices of words and Vedic mantras, and an appendix of quotations as found in some Dharmanibhandas, by Alois Anton Führer. Bombay (Government Central Book Depôt) **1883**.

Vīrasiṃhāvaloka: Vīrasiṃhāvalokaḥ śrīmatTomaravaṃśāvataṃsena-Vīrasiṃhena saṃkalitaḥ (jyotiśśāstra-karmavipāka vaidyaśāstreṣūktaṃ sakalāmayānāṃ sanidānaṃ cikitsitam). Mumbaī (Lakṣmīveṅkaṭeśvaramudraṇālaya) **V.S. 1981**.

VISHVA BANDHU: Upaniṣadic Citations. Being a Concordance of basic extracts, useful for Researches in various branches of Vedic Sciences and Arts, drawn from the Upaniṣads and the Bhagavadgītā. In collaboration with Bhim Dev and Hans Raj. Hoshiarpur (Vishveshvaranand Institute, Panjab University) **1972**. Vishveshvaranand Indological Series 53. Vishveshvaranand Institute Publication 578.

Viṣṇupurāṇa: Kṣemarāja Śrīkṛṣṇadāsena sampāditasya mumbaī śrī veṃkaṭeśvarastīma mudraṇālayena prakāśitasya punarmudraṇam. ŚrīViṣṇumahāpurāṇam. The Viṣṇumahāpurāṇam. Rājendranāthaśarmaṇā bhūmikā pāṭhaśodhanābhyāṃ pariṣkṛtam .... Delhi (Nag Publishers) **1985**.

VOGEL, CLAUS: "Zur Entstehung der hippokratischen Viersäftelehre." *Hippokrates* (Stuttgart) *27.* ***1956***, pp.779-783.

——: "On the Ancient Indian and Greek Systems of Medicine." *Poona Orientalist* (Poona) *24.* ***1959***, pp.31-35.

——: "Die theoretischen Grundlagen der indischen Medizin." Heil und Heilung in den Religionen. Herausgegeben von Karl Hoheisel und Hans-Joachin Klimkeit. Wiesbaden (Harrassowitz Verlag) **1995**, pp.75-87.

Vṛkṣāyurveda of Surapāla: Das Wissen von der Lebensspanne der Bäume. Surapālas Vṛkṣāyurveda kritisch ediert, übersetzt und kommentiert von Rahul Peter Das. Mit einem

Nachtrag von G. Jan Meulenbeld zu seinem Verzeichnis 'Sanskrit Names of Plants and their Botanical Equivalents'. Stuttgart (Franz Steiner Verlag Wiesbaden) **1988**. Alt- und Neu-Indische Studien 34.

WACKERNAGEL, JACOB: Kleine Schriften. Herausgegeben von der Akademie der Wissenschaften zu Göttingen. Erster Band. Göttingen (Vandenhoeck & Ruprecht) **²1969**.

WADDELL, L.A.: "Ancient Indian Anatomical Drawings from Tibet." *The Journal of the Royal Asiatic Society of Great Britain and Ireland* (London) ***1911***, pp.207-208.

WALSH, E.H.C.: "The Tibetan Anatomical System." *The Journal of the Royal Asiatic Society of Great Britain and Ireland* (London) ***1910***, pp.1215-1244.

WATKINS, CALVERT: How to Kill a Dragon. Aspects of Indo-European Poetics. New York | Oxford (Oxford University Press) **1995**.

WAYMAN, ALEX: The Buddhist Tantras. Light on Indo-Tibetan Esotericism. New York (Samuel Weiser) **1973**.

WECK, WOLFGANG: Heilkunde und Volkstum auf Bali. Stuttgart (Ferdinand Enke Verlag) **1937**.

WEISBERG, MARTIN: "A doubting ob-gyn is convinced." *Sexual Medicine Today* (Washington) *6.**1982**,2*,p.15.

WEISSER, URSULA: Das "Buch über das Geheimnis der Schöpfung" von Pseudo-Apollonios von Tyana. Berlin | New York (Walter de Gruyter) **1980**. Ars Medica III,2.

——: Zeugung[,] Vererbung und pränatale Entwicklung in der Medizin des arabisch-islamischen Mittelalters. Erlangen (Verlagsbuchhandlung Hannelore Lüling) **1983**.

WEZLER, ALBRECHT: "Zum Verständnis des Kāmasūtra." *Zeitschrift der Deutschen Morgenländischen Gesellschaft* (Wiesbaden) *121.**1971***, pp.269-283.

——: "A Note on Mahābhāṣya II 366.26: *Guṇasaṃdrāvo Dravyam*. Studies on Mallavādin's Dvādaśāranayacakra II." Buddhism and Its Relation to Other Religions. Essays in Honour of Dr. Shozen Kumoi on His Seventieth Birthday. Kyoto (Heirakuji Shoten) **1985**, pp.1-33.

——: "Bemerkungen zu einigen von Naturbeobachtung zeugenden Textstellen und den Problemen ihrer Interpretation." *Studien zur Indologie und Iranistik* (Reinbek) *13/14.**1987***, pp. 321-346.

——: "The Warrior Taking to Flight in Fear. Some Remarks on Manu 7.94 and 95. (Beiträge zur Kenntnis der indischen Kultur- und Religionsgeschichte III)." *Indologica Taurinensia* (Torino) *14.**1987-88***, pp.391-432.

——: "On Two Medical Verses in the Yuktidīpikā." *Journal of the European Āyurvedic Society* (Reinbek) *1.**1990***, pp.127-148.

——: "Paralipomena zum Sarvasarvātmakatvavāda II: On the Sarvasarvātmakatvavāda and its Relation to the Vṛkṣāyurveda." *Studien zur Indologie und Iranistik* (Reinbek) *16/17.* ***1992*** ⟨= **1992A**⟩, pp.287-315.

——: "Sanskrit Prāṇabhṛt or What Supports What?" Ritual, State and History in South Asia. Essays in Honour of J.C. Heesterman. Edited by A.W. van den Hoek, D.H.A. Kolff, M.S. Oort. Leiden | New York | Köln (E.J. Brill) **1992** ⟨= **1992B**⟩. Memoirs of the Kern Institute 5, pp.393-413.

——: "Über Form und Charakter der sogenannten 'Polemiken im Staatslehrbuch des Kauṭalya'. (Untersuchungen zum *'Kauṭilīya' Arthaśāstra* II)." *Zeitschrift der Deutschen Morgenländischen Gesellschaft* (Stuttgart) *143.**1993*** ⟨= **1993A**⟩, pp.106-134.

——: "On a Prose Passage in the Yuktidīpikā of Some Significance for the History of Indian Medicine." *Journal of the European Āyurvedic Society* (Reinbek) *3.**1993*** ⟨= **1993B**⟩, pp.282-304.

——: "Der Tod als Mittel der Entsühnung (gemäss dem Dharmaśāstra)." Im Tod gewinnt der Mensch sein Selbst. Das Phänomen des Todes in asiatischer und abendländischer Religionstradition. Arbeitsdokumentation eines Symposions. Herausgegeben von Gerhard Oberhammer. Wien (Verlag der Österreichischen Akademie der Wissenschaften) **1995**. Österreichische Akademie der Wissenschaften, Philosophisch-historische Klasse, Sitzungsberichte 624. Beiträge zur Kultur- und Geistesgeschichte Asiens 14, pp.97-140.

——: "Sanskrit *paṇḍá-/páṇḍaka-*." *Zeitschrift der Deutschen Morgenländischen Gesellschaft* (Stuttgart) *148.**1998***, pp.261-276.

WHIPPLE, BEVERLY: "Research Concerning Sexual Response in Women." *The Health Psychologist* (Washington, D.C.) *17,3.Fall **1995***, pp.16-18.

WILHELM, FRIEDRICH: "Die Beziehungen zwischen Kāmasūtra und Arthaśāstra." *Zeitschrift der Deutschen Morgenländischen Gesellschaft* (Wiesbaden) *116.**1966***, pp.291-310.

——: "The Quotations in the Kāmasūtra of Vātsyāyana." *Indologica Taurinensia* (Torino) 7. ***1979***, pp.401-412.

WINDEKENS, A.J. VAN: Le tokharien confronté avec les autres langues indo-européennes. Volume I. La phonétique et le vocabulaire. Louvain (Centre International de Dialectologie Générale) **1976**. Travaux publiés par le Centre International de Dialectologie Générale de l'Université catholique néerlandaise de Louvain 11.

WINDISCH, ERNST: Buddha's Geburt und die Lehre von der Seelenwanderung. Leipzig (B.G. Teubner) **1908**. Abhandlungen der philologisch-historischen Klasse der Königlich Sächsischen Gesellschaft der Wissenschaften 26,2 = Abhandlungen der Königlich Sächsischen Gesellschaft der Wissenschaften 55,2.

WINTERNITZ, M.: Die Frau in den indischen Religionen. I. Teil: Die Frau im Brahmanismus. Sonderdruck aus dem Archiv für Frauenkunde und Eugenetik. Band II und III. Leipzig (Verlag von Curt Kabitzsch) **1920**.

WISE, T.A.: The Hindu System of Medicine. [Reprinted] Delhi (Mittal Publication) **1986**.

WITZEL, MICHAEL: "Meaningful Ritual. Vedic, Medieval, and Contemporary Concepts in the Nepalese Agnihotra Ritual." Ritual, State and History in South Asia. Essays in Honour of J.C. Heesterman. Edited by A.W. van den Hoek, D.H.A. Kolff, M.S. Oort. Leiden | New York | Köln (E.J. Brill) **1992**. Memoirs of the Kern Institute 5, pp.774-827.

WOLZ-GOTTWALD, ECKARD: "Zur Philosophie der traditionellen indischen Medizin Âyurveda. Ein phänomenologischer Interpretationsversuch über den Grundgedanken der Einheit von Mensch und Welt." *Salix* (Würzburg) *1.**1985***, pp.25-35.

——: Heilung aus der Ganzheit. Āyurveda als Philosophie in der Praxis. Gladenbach (Hinder + Deelmann) **1991**.

WOLZ-GOTTWALD, ECKARD / SCHMITT, JÜRGEN: "Ein Konzept von Gesundheit und Krankheit der traditionellen indischen Medizin Âyurveda. Ein Interpretationsversuch von Caraka-Saṃhitâ 4.1.86-156." *Sudhoffs Archiv* (Stuttgart) *70.**1986***, pp.18-26.

WULFF, DAVID M.: "Prolegomenon to a Psychology of the Goddess." The Divine Consort. Rādhā and the Goddesses of India. Edited by John Stratton Hawley and Donna Marie Wulff. Berkeley (Graduate Theological Union) **1982**. Berkeley Religious Studies Series, pp.283-297;363-364.

YAGI, TORU: "A Note on *bhojya-* and *bhankṣya-* ⟨sic; recte: *bhakṣya-*⟩." A Study of the Nīlamata — Aspects of Hinduism in Ancient Kashmir — edited by Yasuke Ikari. Kyoto (Institute for Research in Humanities, Kyoto University) **1994**. Report of the Research Project: Studies in Traditional Cultures in the Context of Ancient Indian and Indo-European Societies, pp.377-397

Yogaratnasamuccaya: The Yogaratnasamuchaya ⟨sic⟩ ⟨Pt.3: The Yogaratnasamuccaya of Anantakumāra⟩. Edited by C.N. Narayanan Moose ⟨Pt.3: V.A. Ramaswami Sastri⟩. Trivandrum (Government Press) **1940-1942**(Pt.1;2); (University of Travancore) **1947**(Pt.3). 1;2: Sri Chithra Ayurveda Series 5;6. 3: Trivandrum Sanskrit Series 152.

Yogacintāmaṇi: Yog꞉ciṃtāmaṇivaidyak꞉granth. Tākī bhāṣāṭīkā mār꞉vāḍīvac꞉nikāyukt śuddh꞉kar꞉kai Muṃbaī maiṃ paṇḍit Śrīdhar꞉śiv꞉lāl꞉kai prajñāyukt Jñān꞉sāgar chāp꞉-khānāmai chāp꞉kar prasiddh kiyā. Muṃbaī (Jñān꞉sāgar chāp꞉khānā) [3]**1887**.

Yogaratnākara: Yogaratnākara with 'Vidyotinī' Hindī Commentary by Lakṣmīpati Śāstrī. Edited by Brahmaśankar ⟨sic⟩ Śāstrī. Varanasi (The Chowkhamba Sanskrit Series Office) [2]**1973**. The Kashi Sanskrit Series 160.

Yogaśata(ka): Das Yogaśata. Ein Zeugnis altindischer Medizin in Sanskrit und Tibetisch. Herausgegeben und übersetzt mit Anmerkungen und Indices von Heinz Helmuth Michael Schmidt. Bonn (Heinz Helmuth Michael Schmidt) **1978**.

Yogavāsiṣṭha: The Yogavāsiṣṭha ⟨sic⟩ of Vālmīki. With the Commentary Vāsiṣṭhamahārāmāyaṇatātparyaprakāsha ⟨sic⟩. Part I. Containing Vairāgya, Mumukshu, Utpatti, Sthiti and Upashama. Edited by Wāsudeva Laxmaṇa Śāstrī Paṇsīkar. New Delhi (Munshiram Manoharlal) [3]**1981**.

ZIMMER, HEINRICH: Spiel um den Elefanten. Ein Buch von indischer Natur. Neuausgabe. Düsseldorf|Köln (Eugen Diederichs Verlag) **1976**.

ZIMMER, HENRY R.: Hindu Medicine. Edited with a Foreword and Preface by Ludwig Edelstein. Baltimore (The Johns Hopkins Press) **1948**. Publications of the Institute of the History of Medicine, the Johns Hopkins University. Third Series: The Hideyo Noguchi Lectures 6.

ZIMMERMANN, FRANCIS: The Jungle and the Aroma of Meats. An Ecological Theme in Hindu Medicine. Berkeley|Los Angeles|London (University of California Press) **1987**. Comparative Studies of Health Systems and Medical Care 20.

—: Le discours des remèdes au pays des épices. Enquête sur la médecine hindoue. Paris (Éditions Payot) **1989** ⟨= **1989A**⟩. Médecine et sociétés Payot.

—: "McKim Marriott. *Ethnosociology*." *Asian Medicine. IASTAM Newsletter* (Paris) *13.December* ***1989*** ⟨= **1989B**⟩, pp.10-11.

—: "[Review of:] Kenneth G. Zysk. Asceticism and Healing in Ancient India. Medicine in the Buddhist Monastery. New York|Oxford 1991." *Journal of the American Oriental Society* (Ann Arbor) *113.**1993***, pp.321-323.

ZWILLING, LEONARD: "Homosexuality As Seen in Indian Buddhist Texts." Buddhism, Sexuality, and Gender. Edited by José Ignacio Cabezón. Albany (University of New York Press) **1992**, pp.203-214.

ZYSK, KENNETH G.: "The Evolution of Anatomical Knowledge in Ancient India, with Special Reference to Cross-cultural Influences." *Journal of the American Oriental Society* (New Haven) *106.**1986***, pp.687-705.

—: "Buddhist Healing and Āyurveda: Some General Observations." Studies in Orientology. Essays in Memory of Prof. A.L. Basham. Editors S.K. Maity, Upendra Thakur, A.K. Narain. Agra (Y.K. Publishers) **1988**, pp.124-135.

—: "The Science of Respiration and the Doctrine of the Bodily Winds in Ancient India." *Journal of the American Oriental Society* (Ann Arbor) *113.**1993***, pp.198-213.

FURTHER RELEVANT LITERATURE WHICH COULD NOT BE UTILISED IN THIS STUDY*

ATCML: Approaches to Traditional Chinese Medical Literature. Proceedings of an International Symposium on Translation Methodologies and Terminologies. Edited by Paul Unschuld. Dordrecht | Boston | London (Kluwer Academic Publishers) **1989**.

BANAVĀRĪLĀLAGAUṚA: "Āyurvedadīpikāyāṃ śārīre suśruta[ḥ]." Researches in Ayurveda Past and Present (Felicitation Volume in Honour of Vaidya Pandit Gulraj Sharma Mishra). Editors Banwari Lal Gaur, Santosh Kumar Sharma "Khandal". Jaipur (Publication Scheme) **2000**, pp.101-106.

BEUSTERIEN, JOHN L.: "Jewish Male Menstruation in Seventeenth-Century Spain." *Bulletin of the History of Medicine* (Baltimore) *73.**1999***, pp.447-456.

COOPER, BRIAN: "A Fresh Look at the Origin of an Interesting Survival of the Dual in Russian." *Transactions of the Philological Society* (Oxford | Boston) *98.**2000***, pp.263-280.

DAS, RAHUL PETER: "Sanskrit *sardigṛdi-*." *Acta Orientalia Academiae Scientiarum Hungaricae* (Budapest) *51.**1998***, pp.303-307.

——: "Bemerkungen anläßlich einer neuen Arbeit zum *Pākadarpaṇa*." *Traditional South Asian Medicine* (Wiesbaden) *6.**2001***, pp.120-135.

——: "[Review of:] G. Jan Meulenbeld. A History of Indian Medical Literature. Volumes IA-IIB. Groningen 1999-2000." *Traditional South Asian Medicine* (Wiesbaden) *6.**2001***, pp.145-158.

DOSSI, BEATRICE: Samen, Seele, Blut. Die Zeugungstheorien des alten Indiens. München (Akademischer Verlag) **1998**. Ganesha 11.

FRANCO, ELI: "The Earliest Extant Vaiśeṣika Theory of *guṇas*." *Wiener Zeitschrift für die Kunde Südasiens und Archiv für indische Philosophie* (Wien) *44.* ***2000***, pp.157-163.

KALĀTATTVAKOŚA: Kalātattvakośa. A Lexicon of Fundamental Concepts of the Indian Arts. General Editor Kapila Vatsyayan. New Delhi (Indira Gandhi National Centre for the Arts)/Delhi (Motilal Banarsidass Publishers).
1: Eight Selected Terms. Editor Bettina Bäumer. **1988**.
2: Concepts of Space and Time ... . Editor Bettina Bäumer. **1992**.
3: Primal Elements — *Mahābhūta*. Editor Bettina Bäumer. **1996**.
4: Manifestation of Nature. *Sṛṣṭi Vistāra*. Editors Advaitavadini Kaul & Sukumar Chattopadhyay. **1999**.

KATZ, JOSHUA T.: "Testimonia Ritus Italici: Male Genitalia, Solemn Declarations, and a New Latin Sound Law." *Harvard Studies in Classical Philology* (Cambridge, Massachusetts | London) *98.**1998***, pp.183-217.

KOLLESCH, JUTTA: "The Corpus Medicorum Graecorum et Latinorum: Problems Related to the Philological Rendering of Medical Texts From Classic European Antiquity." ATCML, pp.159-166.

KOVACS, JÜRGEN: "Linguistic Reflections on the Translation of Chinese Medical Texts." ATCML, pp.85-96.

KULKARNI, NIRMALA: "On the Myth (TS. 2.5.1) Narrating the Origin of Menstruation." Research in Indology. A New Perspective. Edited by Rabindra Kumar Panda. Delhi (Bharatiya Kala Prakashan) **1998**, pp.125-137.

---

* See p.ix.

LA BARRE, WESTON: Muelos. A Stone Age Superstition About Sexuality. New York (Columbia University Press) **1984**.

LEIKOLA, ANTO: "Alcmaeon of Croton as the Founder of Western Biology." *Opuscula Instituti Romani Finlandiae* (Roma) *4.**1989***, pp.43-52.

MEULENBELD, G. JAN: A History of Indian Medical Literature. Groningen (Egbert Forsten) **1999**- ⟨two volumes with two parts each to date⟩. Groningen Oriental Studies 15.

MILSKY, CONSTANTIN: "In Search of a Term Translation Strategy for Chinese Medical Classics." ATCML, pp.77-81.

MITTWEDE, MARTIN: Der Āyurveda. Von den Wurzeln zur Medizin heute. Heidelberg (Karl F. Haug Verlag) **1998**.

MYLIUS, KLAUS: "Kokkokas *Ratirahasya* übersetzt und erläutert." *Journal of the European Āyurvedic Society* (Reinbek) *3.**1993***, pp.145-173, *4.**1995***, pp.163-193, *5.**1997***, pp.136-179.

—: "Kalyāṇamallas *Anaṅgaraṅga* übersetzt und erläutert (I)." *Traditional South Asian Medicine* (Wiesbaden) *6.**2001***, pp.97-119.

SANDER, L.: "Origin and Date of the Bower Manuscript, a New Approach." Investigating Indian Art. Proceedings of a Symposium on the Development of Early Buddhist and Hindu Iconography held at the Museum of Indian Art Berlin in May 1986. Berlin (Museum für indische Kunst, Staatliche Museen Preußischer Kulturbesitz) **1987**. Veröffentlichungen des Museums für indische Kunst 8, pp.313-323.

SCHARFE, HARTMUT: "The Doctrine of the Three Humors in Traditional Indian Medicine and the Alleged Antiquity of Tamil Siddha Medicine." *Journal of the American Oriental Society* (Ann Arbor) *119.**1999***, pp.609-629.

SELLMER, SVEN: "The Heart in the *Ṛg-veda*." On the Understanding of Other Cultures. Proceedings of the International Conference on Sanskrit and Related Studies to Commemorate the Centenary of the Birth of Stanisław Schayer (1899-1941), Warsaw University, Poland, October 7-10, 1999. Edited by Piotr Balcerowicz & Marek Mejor. Warsaw (Oriental Institute, Warsaw University) **2000**. Studia Indologiczne 7, pp.383-395.

SHAH, SHALINI: The Making of Womanhood. Gender Relations in the Mahābhārata. New Delhi (Manohar) **1995**.

SLAJE, WALTER: "Zur Erklärung der sog. 'Tobiasnächte' im vedischen Indien." *Studien zur Indologie und Iranistik* (Reinbek) *21.**1997***, pp.207-234.

—: "Bekanntes 'entdecken': Projektionen eigener Leitvorstellungen in fremde Denkmuster." Annäherung an das Fremde. XXVI. Deutscher Orientalistentag vom 25. bis 29. 9. 1995 in Leipzig. Im Auftrag der Deutschen Morgenländischen Gesellschaft herausgegeben von Holger Preissler und Heidi Stein. *(ZDMG-Suppl. 11)*. Stuttgart (Franz Steiner Verlag) **1998**, pp.17-24.

SYED, RENATE: "Zur Kenntnis der 'Gräfenberg-Zone' und der weiblichen Ejakulation in der altindischen Sexualwissenschaft." *Sudhoffs Archiv* (Stuttgart) *84.**1999***, pp.171-190.

UNSCHULD, PAUL U.: "Terminological Problems Encountered and Experiences Gained in the Process of Editing a Commentated *Nan-ching* Edition." ATCML, pp.97-107.

VANITA, RUTH / KIDWAI, SALEEM: Same-Sex Love in India. Readings from Literature and History. New York (St. Martin's Press) **2000**.

VISVANATHAN, SHIV: "Modern Medicine and its Non-Modern Critics: A study in discourse. Excerpts from a work in progress: A carnival for science." *Asian Medicine. IASTAM Newsletter* (Berkeley) *New Series 5.January **1995***, pp.10-11.

WANG-TOUTAIN, FRANÇOISE: "Bouddhisme et naissance." Dharmadūta. Mélanges offerts au Vénérable Thích Huyên-Vi à l'occasion de son soixante-dixième anniversaire dirigés par Bhikkhu Tampalawela Dhammaratana, Bhikkhu Pāsādika. Paris (Éditions You-Feng) **1997**, pp.271-287.

WHITE, DAVID GORDON: "Transformations in the Art of Love: Kāmakalā Practices in Hindu Tantric and Kaula Traditions." *History of Religions* (Chicago)*38.* ***1998-1999***, pp.172-198.

WINDISCH, ERNST: "Ueber den Sitz der denkenden Seele, besonders bei den Indern und Griechen, und eine Etymologie von gr. πραπίδες." Kleine Schriften herausgegeben von Karin Steiner und Jörg Gengnagel. Stuttgart (Franz Steiner Verlag) **2001**. Glasenapp-Stiftung 41, pp.403-451. ⟨Originally published in *Berichte über die Verhandlungen der Königlich-Sächsischen Gesellschaft der Wissenschaften, Philologisch-Historische Klasse* (Leipzig) *43.* ***1891***, pp.155-203.⟩

WISEMAN, NIGEL / ZMIEWSKI, PAUL: "Rectifying the Names: Suggestions for Standardizing Chinese Medical Terminology." ATCML, pp.55-66.

WUJASTYK, DOMINIK: "The Combinatorics of Tastes and Humours in Classical Indian Medicine and Mathematics." *Journal of Indian Philosophy* (Dordrecht | Hingham, MA) *28.* ***2000***, pp.479-495.

ZIMMERMANN, FRANCIS: "Terminological Problems in the Process of Editing and Translating Sanskrit Medical Texts." ATCML, pp.141-151.

ZWILLING, LEONARD / SWEET, MICHAEL J.: "The Evolution of Third-Sex Constructs in Ancient India: A Study in Ambiguity." Invented Identities. The Interplay of Gender, Religion and Politics in India. Edited by Julia Leslie, Mary McGee. New Delhi (Oxford University Press) **2000**. SOAS Studies on South Asia: Understandings and Perspectives, pp.99-132.

INDEX RERUM

(for cross-references also see the Index verborum; for individual substances also compare under generic headings*)

* For instance, for "fire", "water" etc. also look under "gross element", for "blood", "flesh", "bone", "marrow" etc. under "element", for "faeces", "urine" etc. under "impurity", etc.

INDEX VERBORUM

(for cross-references also see the Index rerum, for individual substances also compare under generic headings*)

---

* For instance, for *agni-*, *udaka-* etc. also look under *mahābhūta-*, for *rakta-*, *māṃsa-* etc. under *dhātu-*, for *purīṣa-*, *mūtra-* etc. under *mala-*, etc.

Persian

Prakrit

Sanskrit/Vedic

Tamil

Tibetan

Tocharian A

Tocharian B

INDEX LOCORUM

(see also the authors and works cited in the Index rerum; for text editions and abbreviations see pp.594ff. and pp.xiiiff. respectively)

The sign **C** *refers to a commentator*

Colophon

(modification of Vātsyāyana's famous verse near the end of his *Kāmasūtra*)

tad etat satyajijñāsoḥ pareṇaiva samādhinā
vihitaṃ lokayātrāyai na rāgārthe 'sya saṃvidhiḥ

SOME MORE MLBD PUBLICATIONS

ABUNDANCE THROUGH REIKI—*Paula Horan*

AWAKENING NATURE'S HEALING INTELLIGENCE (Expanding Ayurveda through the Maharishi Vedic Approach to Health)—*Hari Sharma*

AYURVEDA: A WAY OF LIFE—*Vinod Verma*

AYURVEDA: Secrets of Healing—*Maya Tiwari*

AYURVEDA: The Gentle Health System—*Hans H. Rhyner*

AYURVEDA: The Science of Self-Healing, a Practical Guide—*Vasant Lad*

AYURVEDA AND AROMATHERAPY—*Light Miller & Bryan Miller*

AYURVEDA AND LIFE IMPRESSIONS BODY WORK—*Donald Van Howten*

AYURVEDA AND PANCHAKARMA: The Science of Healing and Rejuvenation—*Sunil V. Joshi*

AYURVEDA AND THE MIND: The Healing of Consciousness—*David Frawley*

AYURVEDA FOR LIFE: Nutrition, Sexual Energy & Healing—*Vinod Verma*

AYURVEDA REVOLUTIONIZED: Integrating Ancient and Modern Ayurveda—*Edward F. Tarabilda*, Foreword *David Frawley*

AYURVEDIC BEAUTY CARE: Ageless Techniques to Invoke Natural Beauty—*Melanie Sachs*

THE AYURVEDIC COOKBOOK: A Personalized Guide to Good Nutrition and Health—*Amadea Morningstar & Urmila Desai*

AYURVEDIC HEALING: A Comprehensive Guide—*David Frawley*

AYURVEDIC HEALING FOR WOMEN: Herbal Gynecology—*Atreya*

AYURVEDIC REMEDIES FOR THE WHOLE FAMILY—*Light Miller*

BODY AND EMOTION: An Aesthetic of Illness and Healing in the Nepal Himalayas—*Robert R. Desjarlais*

CHINESE MEDICINE AND AYURVEDA—*Robert Svoboda & Arnie Lade*

THE COMPLETE REIKI HANDBOOK: Basic Introduction and Methods of Natural Application—*Walter Lubeck*

EMPOWERMENT THROUGH REIKI: The Path to Personal and Global Transformation—*Paula Horan*

ENERGETIC BODYWORK: Practical Techniques—*Rita J. McNamara*

ESSENTIALS OF AYURVEDA—*Priya Vrat Sharma*

THE HEALING POWER OF ESSENTIAL OILS: Fragrance Secrets for Everyday Use—*Rodolphe Balz*

HEALTH THROUGH BALANCE: An Introduction to Tibetan Medicine—*Yeshi Donden*, Ed. & Tr. *Jeffrey Hopkins*

HOMOEOPATHY: Heart and Soul—*Keith M. Souter*

INDIAN ALCHEMY OR RASAYANA: In the Light of Asceticism and Geriatrics—*S. Mahdihassan*

THE LOST SECRETS OF AYURVEDIC ACUPUNCTURE: An Ayurvedic Guide to Acupunture—*Frank Ros*

NATURAL HEALING THROUGH AYURVEDA—*Subhash Ranade*, Ed. *David Frawley*

THE NEPAL COOKBOOK—Illus. by *Padden Choedak Oshoe*

PATANJALI AND AYURVEDIC YOGA—*Vinod Verma*

THE PATH OF PRACTICE—*Maya Tiwari*

PLANETARY HERBOLOGY—*Michael Tierra*, Ed. *David Frawley*

POCKET MANUAL OF HOMOEOPATHIC MATERIA MEDICA: Comprising the Characteristic and Guiding Symptoms of All Remedies—*William Boericke*

PRAKRUTI: Your Ayurvedic Constitution—*Robert E. Svoboda*

RAINBOW REIKI: Expanding the Reiki System with Powerful Spiritual Abilities—*Walter Lubeck*

REIKI FIRE: New Information about the Origins of the Reiki Power—*Frank Arjava Petter*

REIKI: Way of the Heart—*Walter Lubeck*

REIKI WITH GEMSTONES—*Ursula Klinger Omenka*

SPIRITUAL BEAUTY CARE: Techniques and Practices to Enhance Your Inner and Outer Beauty—*Jacqueline Sinnige*, Tr. *Martine J. Koeman*

TIBETAN BUDDHIST MEDICINE AND PSYCHIATRY: The Diamond Healing—*Terry Clifford*, Foreword by *Dalai Lama*

YOGA AND AYURVEDA: Self-Healing and Self-Realization—*David Frawley*

THE YOGA OF HERBS: An Ayurvedic Guide to Herbal Medicine—*David Frawley & Vasant Lad*